Corinne Stockley, Chris Oxlade and Jane Wertheim

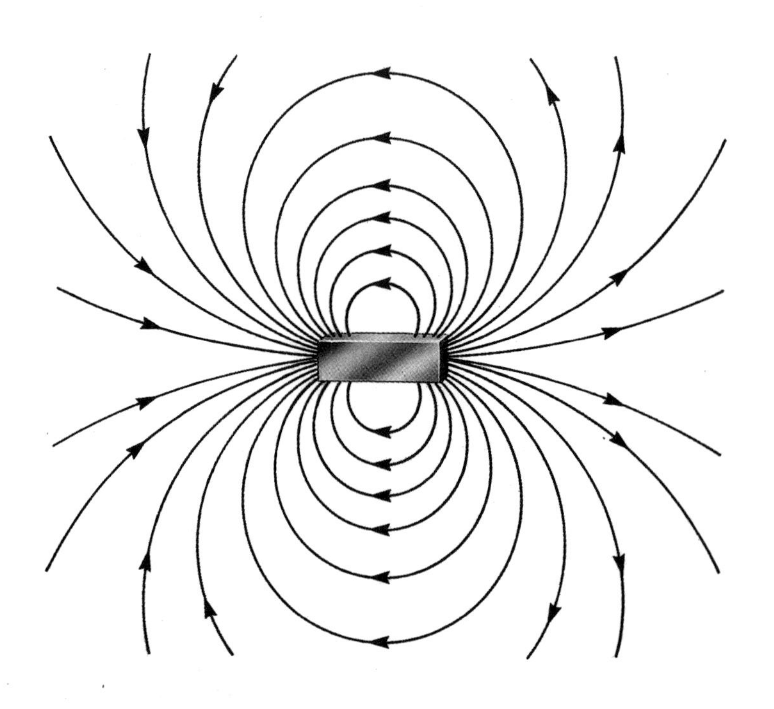

Part one

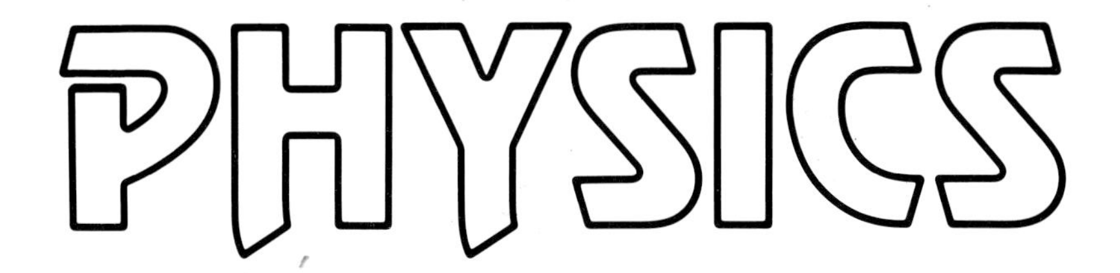

Designed by Chris Scollen, Stephen Wright and Roger Berry

Illustrated by Kuo Kang Chen, Guy Smith and Caroline Ewen

Scientific advisors: Dr. Tom Petersen, John Hawkins and Dr. John Durell

Additional designs by Iain Ashman, Anne Sharples and Camilla Luff

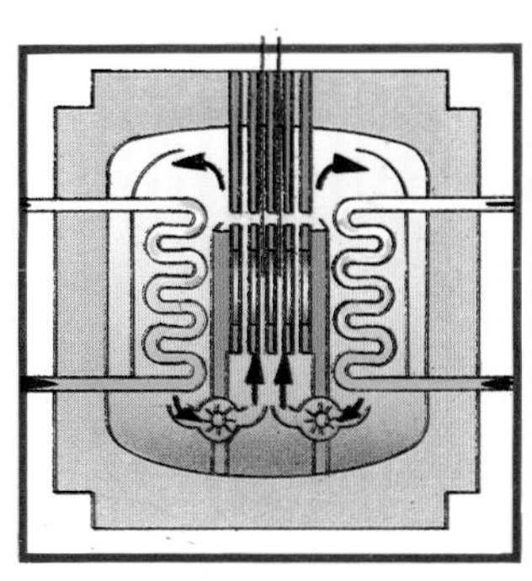

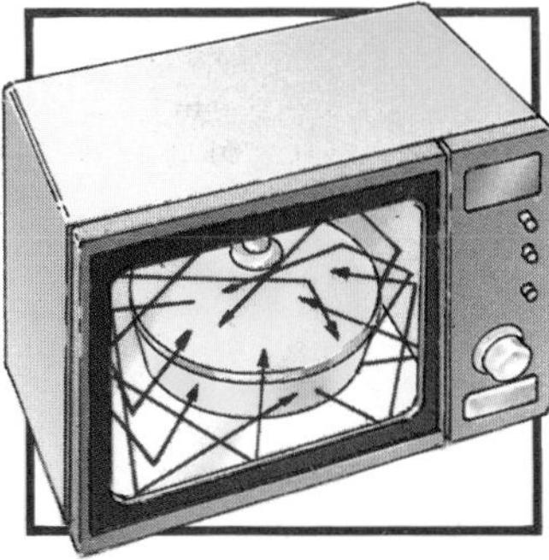

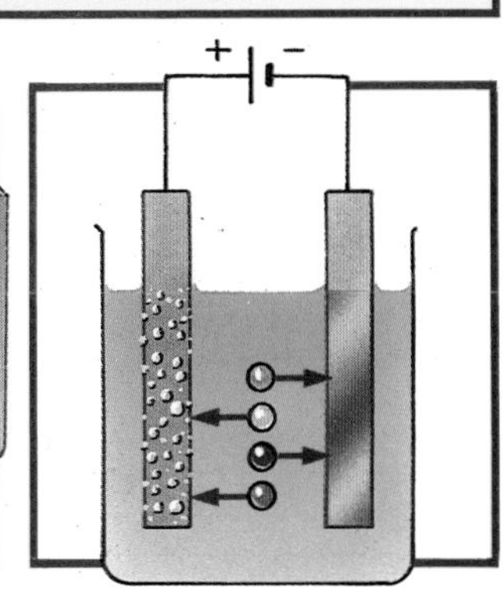

Contents

Mechanics and general physics

Heat

Waves

Electricity and magnetism

Atomic and nuclear physics

General

We are grateful to the following organizations for permission to use their illustrations.

British Standards Institution (p.88) NASA (p.93)
National Remote Sensing Centre (p.45)

Royal Marsden Hospital (p.40)
Toshiba (pgs. 1 and 45)
UK Atomic Energy Authority (pgs. 1, 94 and 95)

About physics

In this book, physics is divided into five main colour-coded sections, followed by a black and white section of general material relating to the whole subject.

Blue section
Mechanics and general physics

Yellow section
Heat

Red section
Waves

Green section
Electricity and magnetism

Pink section
Atomic and nuclear physics

Black and white section
General material – charts and tables, also information on the treatment of experimental results.

Physics is the study of the properties and nature of matter, the different forms of energy and the ways in which matter and energy interact in the world around us. The areas covered by the five different colour-coded sections are explained below.

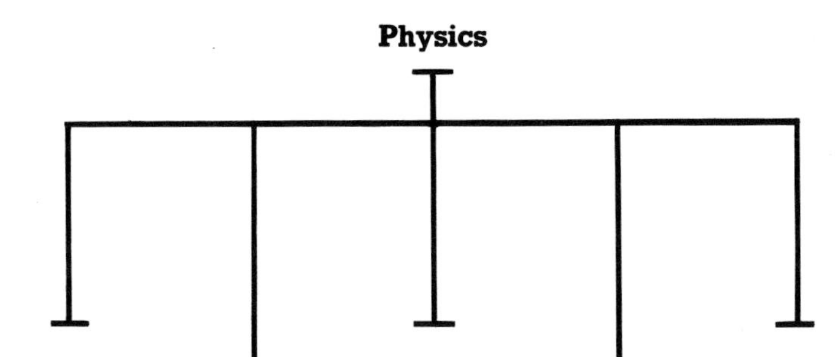

Mechanics and general physics section. Covers the main concepts of physics, e.g. forces, energy and the properties of matter.

Heat section. Explains heat energy in terms of its measurement and the effects of its presence and transference. Includes the gas laws.

Waves section. Look at the properties and effects of wave energy and examines sound, electromagnetic and light waves in detail.

Electricity and magnetism section. Explains the forms, uses and behaviour of the two linked phenomena.

Atomic and nuclear physics section. Examines atomic and nuclear structure and energy, radioactivity, fission and fusion.

Atoms and molecules

The Greeks believed that all matter was made up of tiny particles which they called **atoms**. This idea has since been expanded and theories such as the **kinetic theory** have been developed which can be used to explain the physical nature and behaviour of substances in much greater detail. Matter can exist in three different **physical states**. The state of a substance depends on the nature of the substance, its temperature and the pressure exerted on it. Changes between states are caused by changes in the pressure or temperature (for more about this, see **changes of state**, page 30).

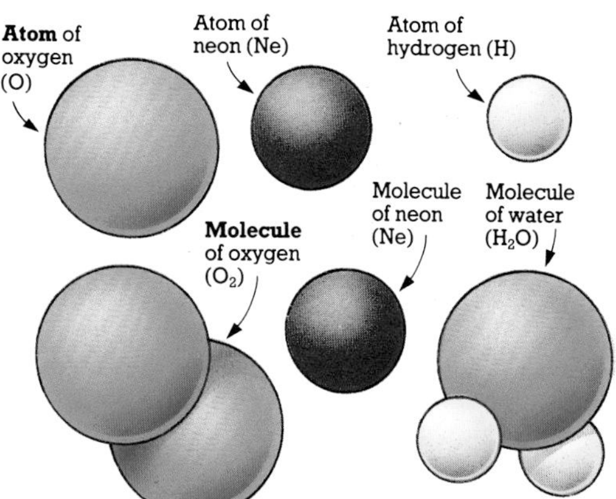

- **Atom**. The smallest part of a substance which can exist and still retain the properties of the substance. The internal structure of the atom is explained on pages 82-83. Atoms are extremely small, having radii of about 10^{-25} m and masses of about 10^{-10} kg. They can form **ions*** (electrically charged particles) by the loss or gain of **electrons*** (see **ionization**, page 88).

- **Molecule**. The smallest naturally-occurring particle of a substance. Molecules can consist of any number of **atoms**, from one (e.g. neon) to many thousands (e.g. proteins), all held together by **electromagnetic forces***. All the molecules of a pure sample of a substance contain the same atoms in the same arrangement.

- **Element**. A substance which cannot be split into simpler substances by a chemical reaction. All atoms of the same element have the same number of **protons*** in their **nuclei*** (see **atomic number**, page 82).

- **Compound**. A substance whose **molecules** contain the **atoms** of two or more **elements**, chemically bonded together, and which can thus be split into simpler substances. A **mixture** has no chemical bonding and is therefore not a compound.

Note that many substances do not have **molecules**:

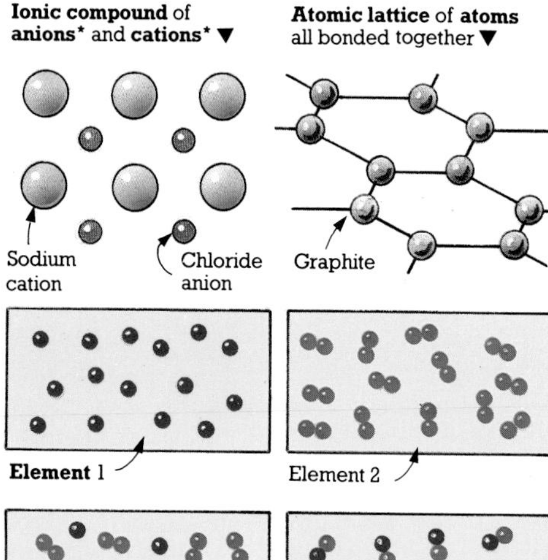

Mixture of 1 and 2 – no chemical bonding

Compound of 1 and 2 – elements bonded together

* **Anions, Cations**, 88 (**Ionization**); **Electromagnetic force**, 6; **Electrons**, 83; **Ions**, 88 (**Ionization**); **Nucleus, Protons**, 82.

- **Solid state**. A **state** in which a substance has a definite volume and shape and resists forces which try to change these.
- **Liquid state**. A **state** in which a substance flows and takes up the shape of its containing vessel. It is between the **solid** and **gaseous** states.
- **Gaseous state**. A **state** in which a substance expands to fill its containing vessel. Substances in this state have a relatively low density.
- **Gas**. A substance in the **gaseous state** which is above its **critical temperature** and so cannot be turned into a liquid just by increasing the pressure – the temperature must be lowered first, to create a **vapour**.
- **Vapour**. A substance in the **gaseous state** which is below its **critical temperature** (see **gas**) and so can be turned into a liquid by an increase in pressure alone.

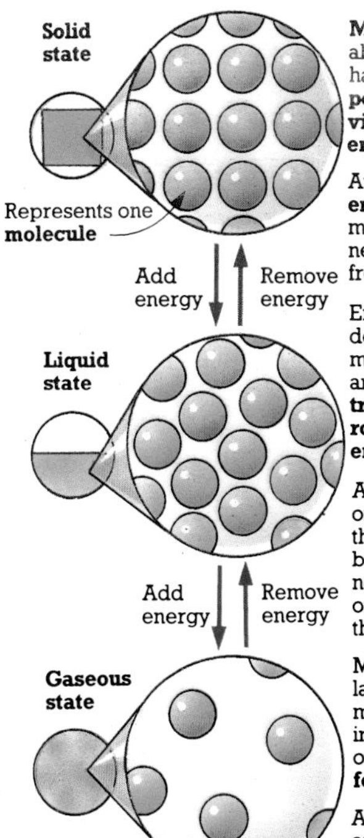

Molecules vibrate about mean positions, having **molecular potential energy*** and **vibrational kinetic energy***

Average **internal energy*** of molecule much less than that needed by it to break free from others.

Energy added breaks down regular pattern – molecules can move around and thus have **translational** and **rotational kinetic energy*** as well.

Average internal energy of molecule just about that needed for it to break free from neighbouring molecules, only to be captured by the next one.

Molecules have very large separation – they move virtually independently of each other – **intermolecular forces*** can be ignored.

Average internal energy of molecule much greater than a molecule needs to break free of others.

The kinetic theory

The **kinetic theory** explains the behaviour of the different physical states in terms of the motion of **molecules**. In brief, it states that the molecules of solids are closest together, have least energy and so move the least, those of liquids are further apart with more energy, and those of gases are furthest apart with most energy. See above right.

- **Brownian motion**. The observed random motion of small particles in water or air. It supports the kinetic theory, since it is clearly due to unseen impact with other moving particles (the water or air **molecules**).

Brownian motion of smoke particles as they are hit by **molecules** in the air.

- **Diffusion**. The mixing of two gases, vapours or liquids over a period of time. It supports the kinetic theory, since the particles must be moving to mix, and gases can be seen to diffuse faster than liquids.

Molecules of two **gases** diffuse together over time.

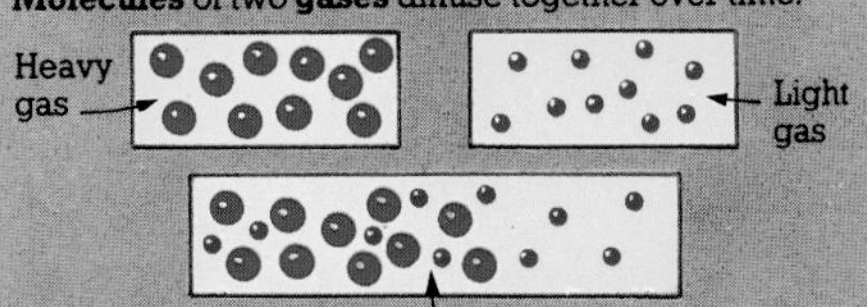

Light gas diffuses faster than heavy one

- **Graham's law of diffusion**. States that, at constant temperature and pressure, the rate of **diffusion** of a **gas** is inversely proportional to the square root of its density.

$$\text{Rate of diffusion} \propto \sqrt{\frac{1}{\text{density of gas}}}$$

* **Intermolecular force**, 7; **Internal energy**, 9; **Molecular potential energy**, 8; **Rotational**, **Translational** and **Vibrational kinetic energy**, 9 (**Kinetic energy**).

Forces

A **force** influences the shape and motion of an object. A single force will change its velocity (i.e. **accelerate*** it). Two equal and opposite forces will change its shape or size. It is a **vector quantity***, having both magnitude and direction, and is measured in **newtons**. The main types of force are **gravitational**, **magnetic**, **electric** and **nuclear**. See pages 104-107 for a comparison of the first three of these.

Forces are shown by arrowed lines (the length represents magnitude and the arrow direction).

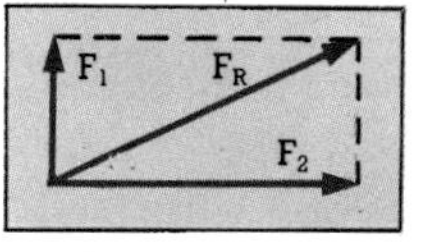

Effect of F_1 and F_2 is the same as F_R (the **resultant force**). F_1 and F_2 are the **components** of F_R.

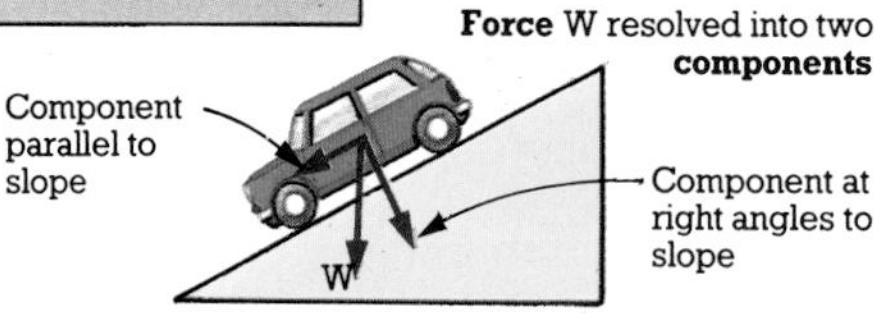

- **Newton (N)**. The **SI unit*** of force. One newton is the force needed to accelerate a mass of 1 kg by 1 m s^{-2}.

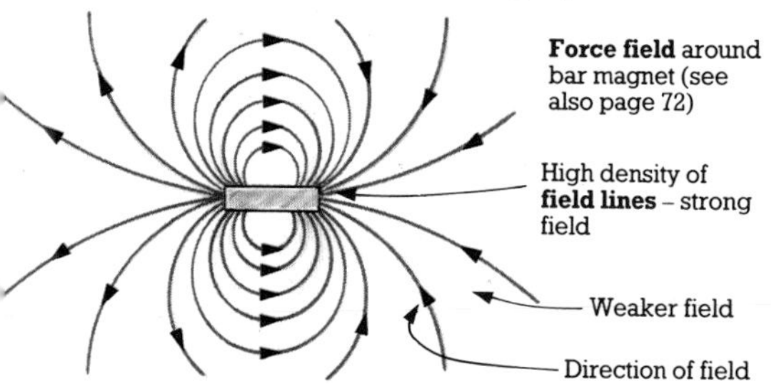

- **Force field**. The region in which a force has an effect. The maximum distance over which a force has an effect is the **range** of the force. Force fields are represented by lines with arrows to show their strength and direction (see also pages 58 and 72).

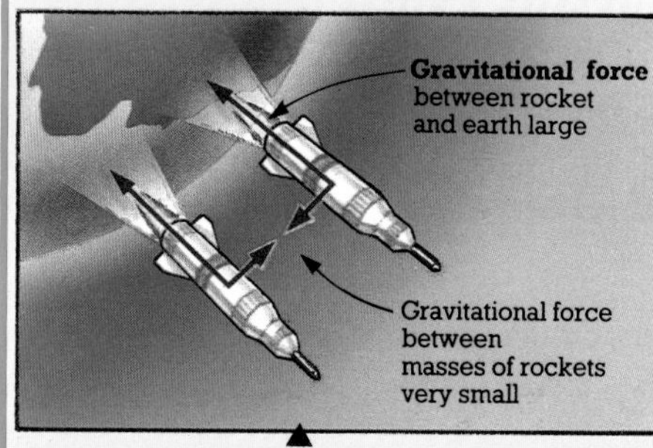

- **Gravitational force** or **gravity**. The force of attraction between any two objects which have mass (see also pages 18-19). It is very small unless one of the objects is very massive.

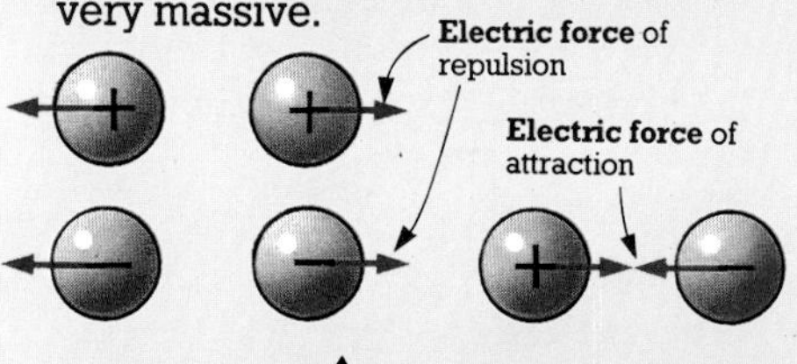

- **Electric** or **electrostatic force**. The force between two electrically-charged particles (see also page 56). It is repulsive if the charges are the same, but attractive if they are opposite.

- **Electromagnetic force**. A combination of the **electric** and **magnetic force**, which are closely related and difficult to separate.

- **Magnetic force**. A force between two moving charges. These moving charges can be electric **currents*** (see also page 70) or **electrons*** moving around in their **electron shells***.

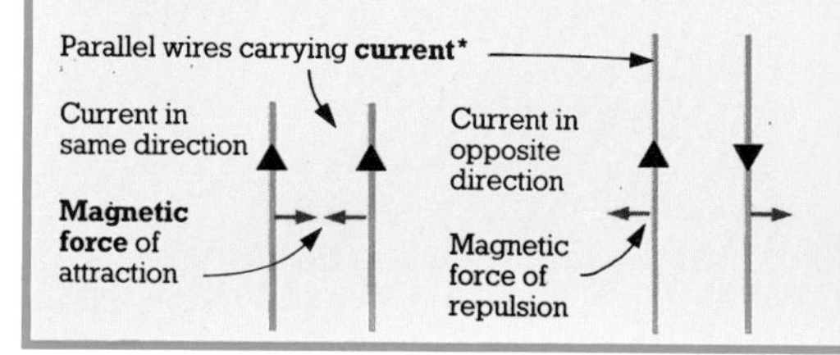

* **Acceleration**, 11; **Current**, 60; **Electrons**, **Electron shells**, 83; **SI units**, 96; **Vector quantity**, 108.

- **Intermolecular force**. The **electromagnetic force** between two molecules. The strength and direction of the force varies with the separation of the molecules (see picture, right).

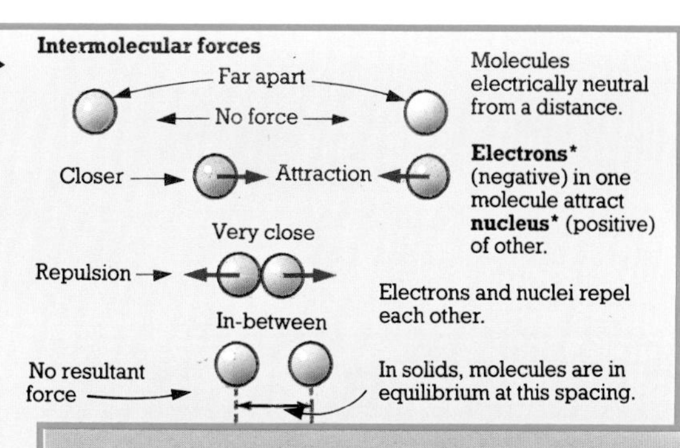

- **Tension force**. Equal and opposite forces which, when applied to the ends of an object, increase its length. They are resisted by the **intermolecular force**.

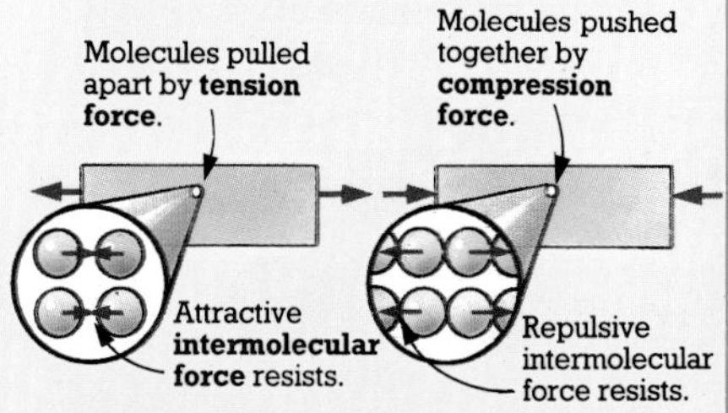

- **Compression force**. Equal and opposite forces which, when applied to the ends of an object, decrease its length. They are opposed by the **intermolecular force**.

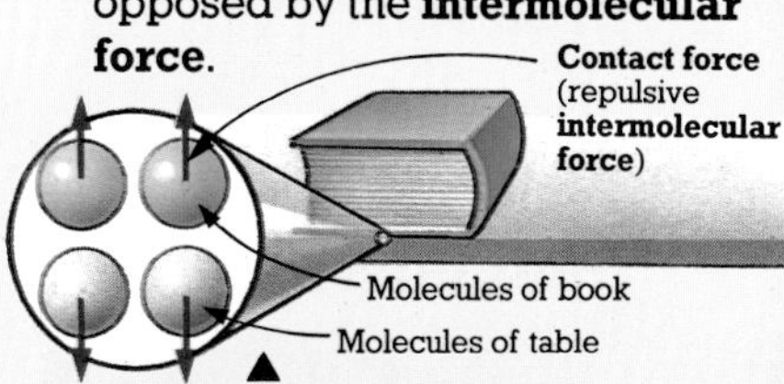

- **Contact force**. The **intermolecular force** of repulsion between the molecules of two objects when they touch.
- **Nuclear force**. The force of attraction between all the particles of an atomic nucleus (the **protons*** and **neutrons***). It prevents the **electric force** of repulsion between the protons from pushing the nucleus apart (see also page 84).

- **Frictional force** or **friction**. The force which acts to oppose the motion of two touching surfaces over each other, caused by the **intermolecular force** between the molecules of the two surfaces. There are two types, the **limiting frictional force** and the **dynamic frictional force**.

- **Limiting** or **static frictional force**. The maximum value of the **frictional force** between two surfaces. It occurs when the two surfaces are on the point of sliding over each other.

- **Dynamic frictional force** or **sliding frictional force**. The value of the **frictional force** when one surface is sliding over another at constant speed. It is slightly less than the **limiting frictional force**.

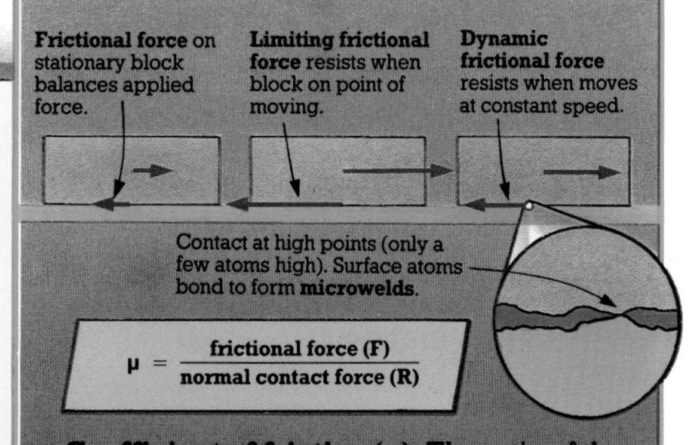

$$\mu = \frac{\text{frictional force (F)}}{\text{normal contact force (R)}}$$

- **Coefficient of friction** (μ). The ratio of the **frictional force** between two surfaces to that pushing them together (the **normal contact force**). There are two values, the **coefficient of limiting friction** and the **coefficient of dynamic friction**.

* **Electrons**, 83; **Neutrons**, **Nucleus**, **Protons**, 82.

Energy

Work is done when a force moves an object. **Energy** is the capacity to do work. When work is done on or by an object, it gains or loses energy respectively. Energy exists in many different forms and can change between them (energy **conversion** or **transformation**), but cannot be created or destroyed (**law of conservation of energy**). The **SI unit*** of energy and work is the **joule (J)**.

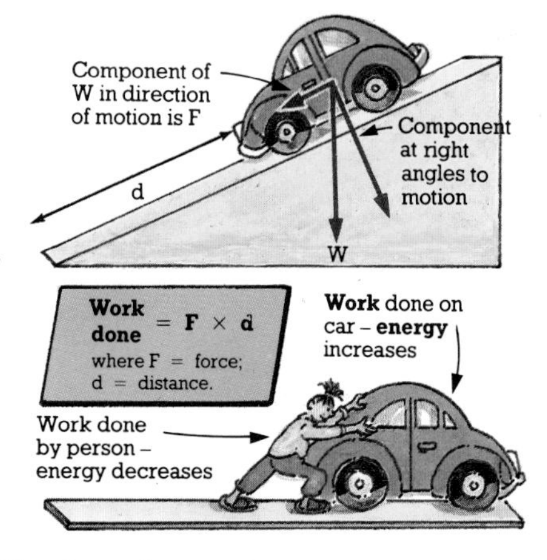

- **Potential energy (P.E.)**. The energy of an object due to its position in a **force field***, which it has because work has been done to put it in that position. The energy has been "stored up". The three forms of potential energy are **gravitational potential energy**, **electromagnetic potential energy** and **nuclear potential energy** (depending on the force involved).

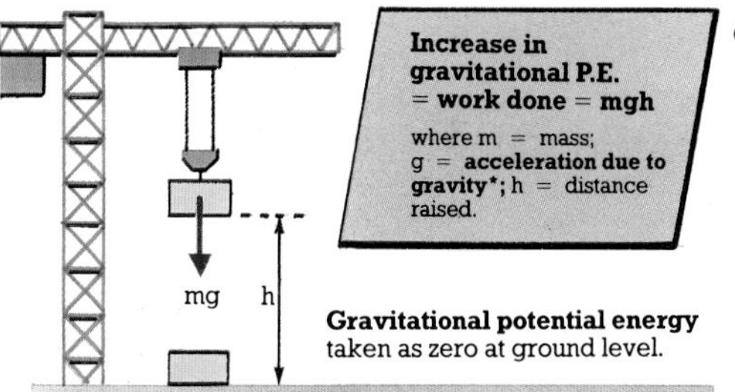

- **Gravitational potential energy**. The **potential energy** associated with the position of an object relative to a mass which exerts a **gravitational force*** on it. If the object is moved further from the mass (e.g. an object being lifted on the earth), work is done on the body and its gravitational potential energy is raised.

- **Electromagnetic potential energy**. The **potential energy** associated with the position of a body in a **force field*** created by an **electromagnetic force***.

- **Elastic potential energy** or **strain energy**. An example of the **molecular potential energy**, stored as a result of stretching or compressing an object. It is the work done against the **intermolecular force***.

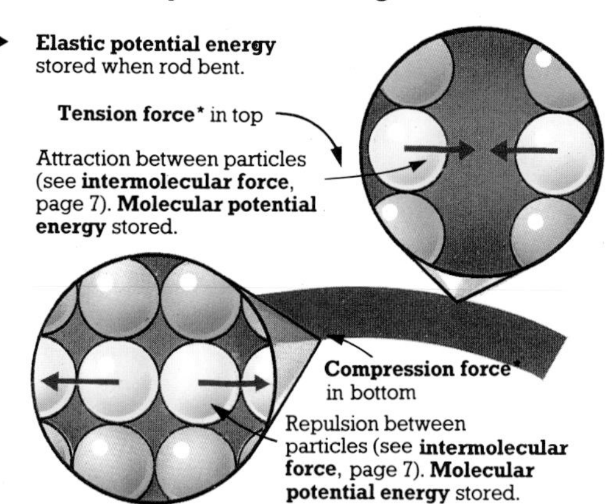

- **Molecular potential energy**. The **electromagnetic potential energy** associated with the position of molecules relative to one another. It is increased when work is done against the **intermolecular force***.

- **Chemical energy**. Energy stored in substances such as fuels, food, and chemicals in batteries. It is released during chemical reactions, e.g. as heat when a fuel burns, when the **electromagnetic potential energy** of the atoms and molecules changes.

* **Acceleration due to gravity**, 18; **Compression force**, 7; **Electromagnetic force**, **Force field**, **Gravitational force**, 6; **Intermolecular force**, 7; **SI units**, 96; **Tension force**, 7.

- **Nuclear potential energy**. The **potential energy** stored in an atomic **nucleus***. It is released during **radioactive decay***.
- **Kinetic energy** (**K.E.**). The energy associated with movement. It takes the form of **translational**, **rotational** and **vibrational energy**.

▶ **Kinetic energy** of two objects linked by spring

$$K.E. = \frac{1}{2}mv^2$$

where m = mass; v = velocity.

Translational

Vibrational

Rotational

Mechanical energy of pendulum constant (if resistive forces neglected).

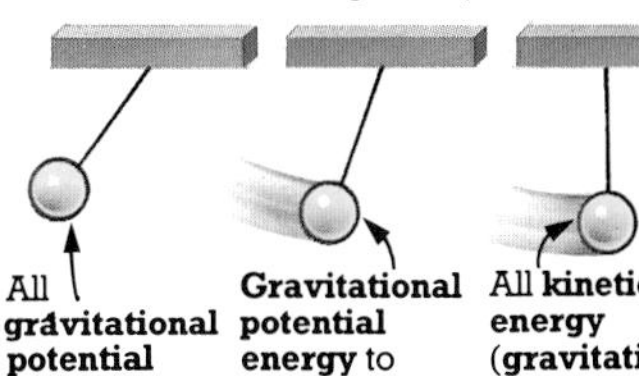

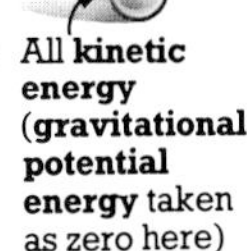

◀ • **Mechanical energy**. The sum of the **kinetic energy** and **gravitational potential energy** of an object.

- **Internal** or **thermal energy**. The sum of the **kinetic energy** and the **molecular potential energy** of each molecule in an object. If the temperature of an object increases, so does its internal energy.

▼

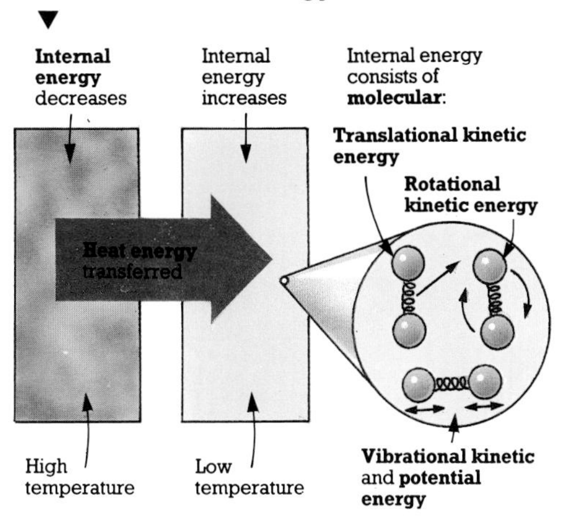

- **Heat energy** or **heat**. The energy which flows from one place to another because of a difference in temperature (see pages 28-33). When heat energy is absorbed by an object, its **internal energy** increases.
- **Wave energy**. The energy associated with wave action. For example, the energy of a water wave consists of the **gravitational potential energy** and **kinetic energy** of the water molecules.
- **Electric** and **magnetic energy**. The types of energy associated with electric charge and moving electric charge (current). They are collectively referred to as **electromagnetic energy**.
- **Radiation**. Any energy in the form of **electromagnetic waves*** or streams of particles. See also pages 28 and 86-87.
- **Power**. The rate of doing work or the rate of change of energy. The **SI unit*** of power is the **watt** (**W**), which is equal to 1 joule per second.

▼

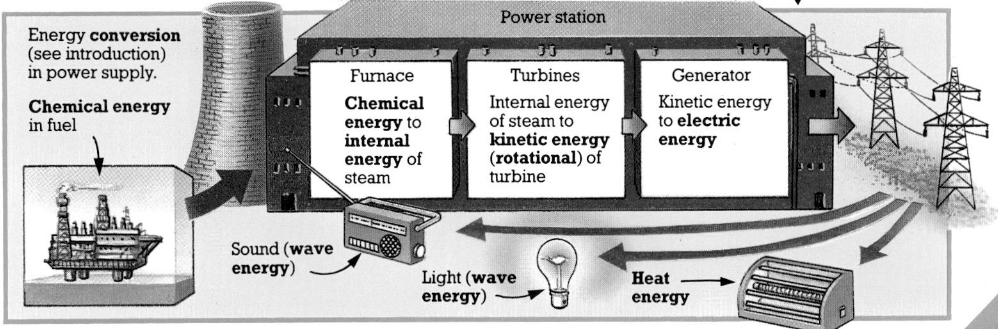

* **Electromagnetic waves**, 44; **Nucleus**, 82; **Radioactive decay**, 87; **SI units**, 96.

Motion

Satellite spinning in orbit

Rotational motion (motion around **centre of mass**)

Translational motion (motion of centre of mass)

Motion is the change in position and orientation of an object. The motion of a **rigid** object (one which does not change shape) is made up of **translational motion**, or **translation**, i.e. movement of the **centre of mass** from one place to another and **rotational motion**, or **rotation**, i.e. movement around its centre of mass. The study of the motion of points is called **kinematics**.

Linear motion

Linear or **rectilinear motion** is movement in a straight line and is the simplest form of **translational motion** (see introduction). The linear motion of any rigid object is described as the motion of its **centre of mass**.

- **Centre of mass**. The point which acts as though the total mass of the object were at that point. The centre of mass of a **rigid** object (see introduction) is in the same position as its **centre of gravity** (the point through which the earth's gravitational force acts on the object).

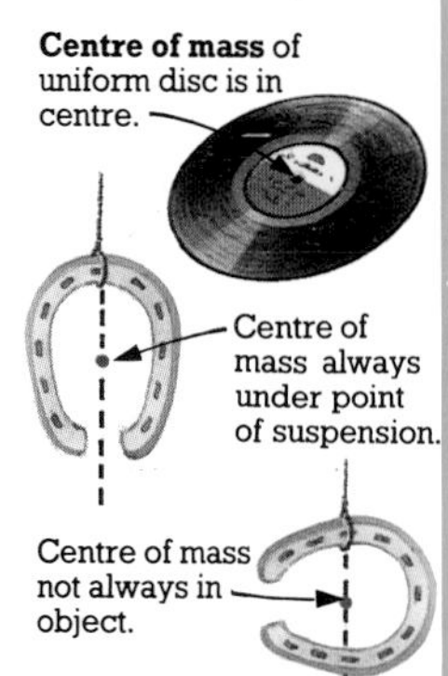

- **Displacement**. The distance and direction of an object from a fixed reference point. It is a **vector quantity***. The position of an object can be expressed by its displacement from a specified point.

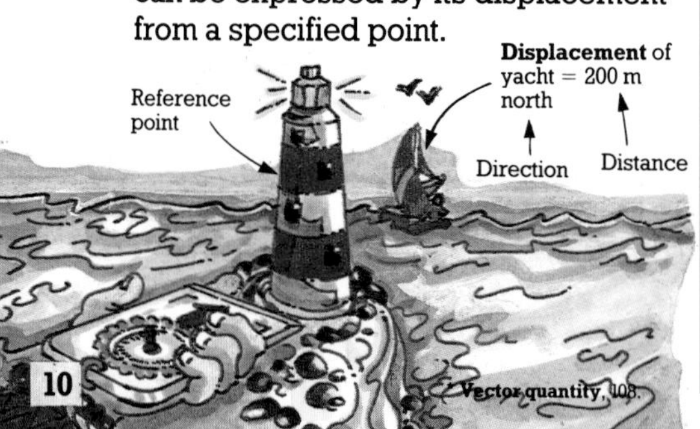

- **Speed**. The distance an object travels in a certain length of time. If the speed of an object is constant, it is said to be moving with **uniform speed**. The **average speed** of an object over a time interval is the distance travelled by the object divided by the time interval. The **instantaneous speed** is the speed at any moment.

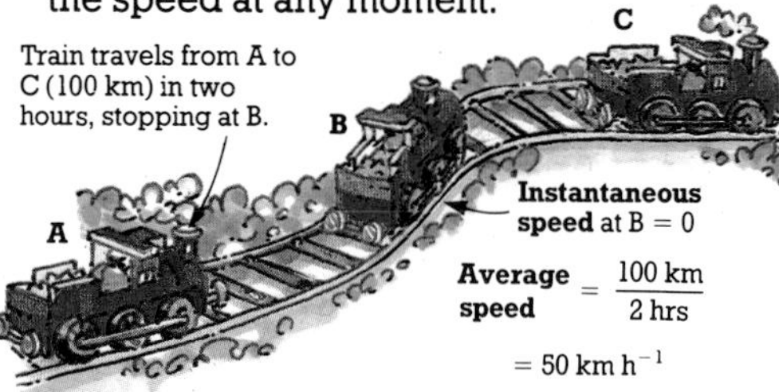

- **Velocity**. The **speed** and direction of an object (i.e. its **displacement** in a given time). It is a **vector quantity***. **Uniform velocity**, **average velocity** and **instantaneous velocity** are all defined in a similar way to **uniform speed** etc. (see speed).

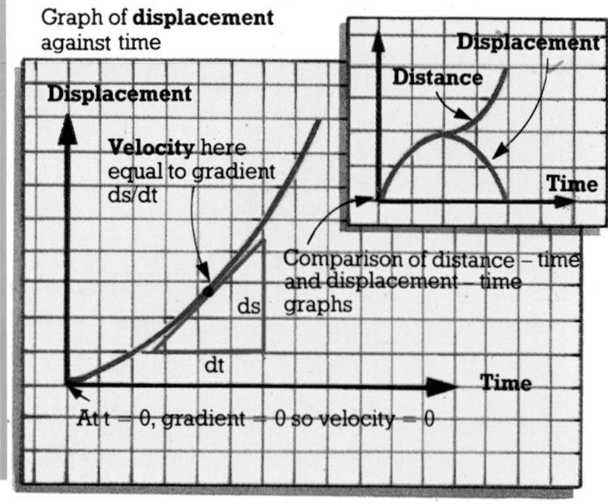

*Vector quantity, 108.

- **Relative velocity**. The **velocity** which an object appears to have when seen by an observer who may be moving. This is known as the velocity of the object relative to the observer.

40 m s^{-1}

Relative velocity of B (seen from A) = 70 m s^{-1} to left.

30 m s^{-1}

Relative velocity of A (seen from B) = 70 m s^{-1} to right.

- **Acceleration**. The change of **velocity** of an object in a certain time. It is a **vector quantity***. An object accelerates if its **speed** changes (the usual case in **linear motion**) or its direction of travel changes (the usual case in **circular motion***). **Deceleration** in one direction is acceleration in the opposite direction to the motion (negative acceleration). An object whose velocity is changing the same amount in equal amounts of time is moving with **uniform acceleration**.

Graphs of **velocity** against time

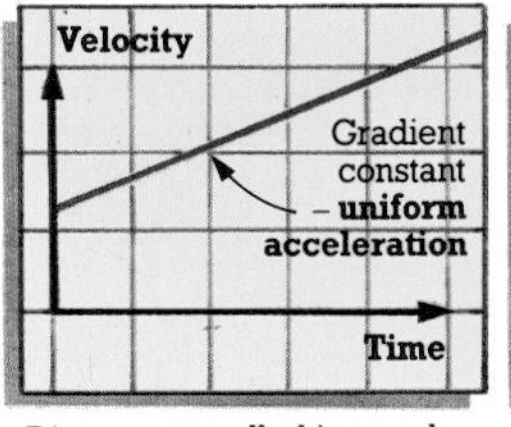

Distance travelled in equal time intervals increases.

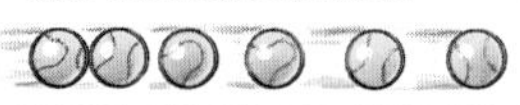

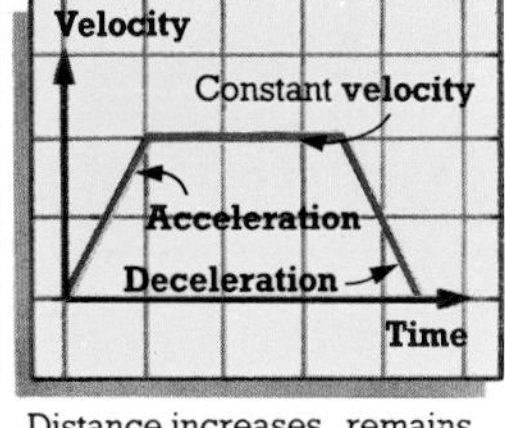

Distance increases, remains constant, then decreases.

$$\mathbf{v} = \mathbf{u} + \mathbf{a}t$$
$$\mathbf{s} = \tfrac{1}{2}(\mathbf{u} + \mathbf{v})t$$
$$\mathbf{s} = \mathbf{u}t + \tfrac{1}{2}\mathbf{a}t^2$$
$$\mathbf{v}^2 = \mathbf{u}^2 + 2\mathbf{as}$$

where t = time; u = initial **velocity** at time = 0; v = final velocity after t; s = **displacement** after t; a = **acceleration** (constant).

- **Equations of uniformly accelerated motion**. Equations which are used in calculations involving **linear motion** with **uniform acceleration**. A **sign convention** must be used. The equations use **displacement**, not distance, so changes of direction must be considered.

Sign convention Right chosen as positive

– +

Displacement 0

Negative displacement

Positive displacement

Object moving to left has negative **velocity**.

Object moving to right has positive velocity.

Velocity becoming more positive means positive **acceleration**.

Velocity becoming more negative means negative acceleration (**deceleration**).

- **Sign convention**. A method used to distinguish between motion in opposite directions. One direction is chosen as positive, and the other is then negative. The sign convention must be used when using the equations of motion (see above).

Rotational motion

Rotational motion is the movement of an object about its **centre of mass**. In rotational motion, each part of the object moves along a different path, so that the object cannot be considered as a whole in calculations. It must be split into small pieces and the **circular motion*** of each piece must be considered separately. From this, the overall motion of the object can be seen.

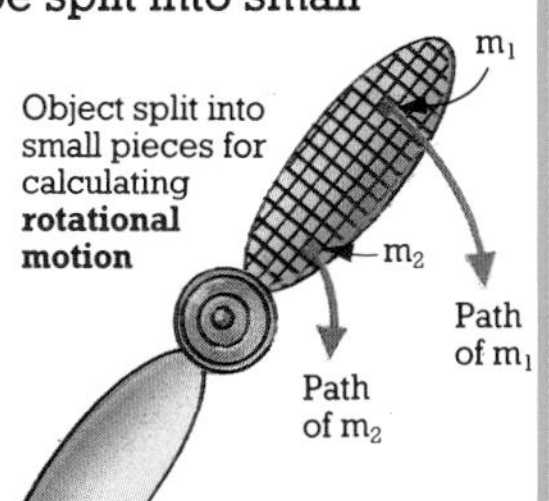

* **Circular motion**, 17;
Vector quantity, 108.

Dynamics

Dynamics is the study of the relationship between the motion of an object and the forces acting on it. A single force on an object causes it to change speed and/or direction (i.e. **accelerate***). If two or more forces act and there is no resultant force, the object does not accelerate, but changes shape.

Two equal but opposite forces. No resultant force – no acceleration, but rope stretches.

Forces not equal. Rope still stretches, but also accelerates to left due to resultant force.

- **Newton's laws of motion**. Three laws formulated by Newton in the late 1770's which relate force and motion.

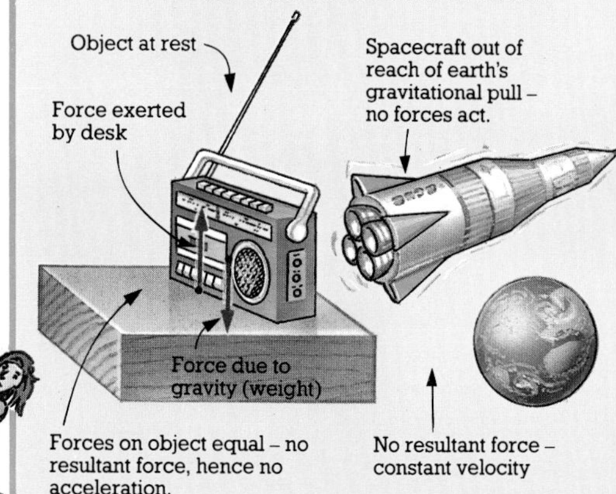

- **Newton's first law**. If an object is at rest, or if its speed and direction are constant, then the resultant force on it is zero.

- **Mass**. A measurement of the **inertia** of an object. The force needed to accelerate an object by a given amount depends on its mass – a larger mass needs a larger force.

- **Inertia**. The tendency of an object to resist a change of velocity (i.e. to resist a force trying to accelerate it). It is measured as **mass**.

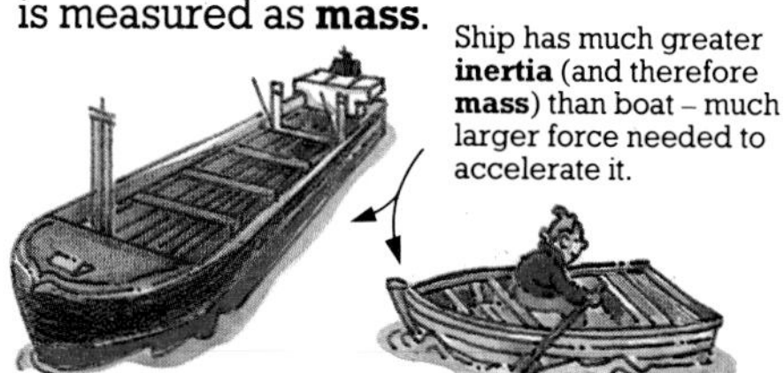

Ship has much greater **inertia** (and therefore **mass**) than boat – much larger force needed to accelerate it.

- **Momentum**. The **mass** of an object multiplied by its velocity. Since velocity is a **vector quantity***, so is momentum. See also **law of conservation of linear momentum**.

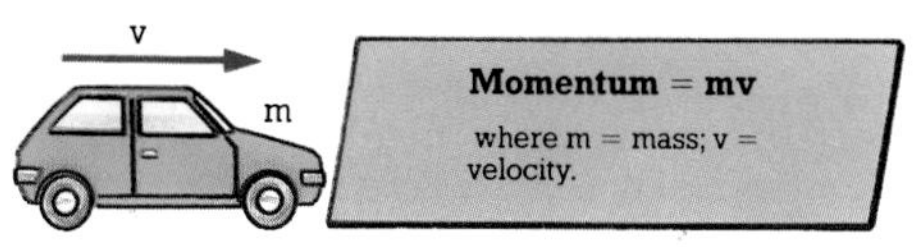

Momentum = mv

where m = mass; v = velocity.

- **Impulse**. The force acting on an object multiplied by the time for which the force acts. From **Newton's second law**, impulse is equal to the change in **momentum** of an object. An equal change in momentum can be achieved by a small force for a long time or a large force for a short time.

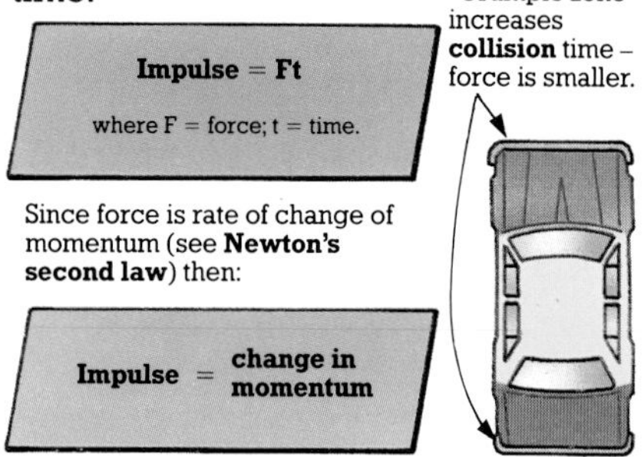

Impulse = Ft

where F = force; t = time.

Since force is rate of change of momentum (see **Newton's second law**) then:

Impulse = change in momentum

"Crumple zone" increases **collision** time – force is smaller.

- **Collision**. An occurrence which results in two or more objects exerting forces on each other. This is not the everyday idea of a collision, because the objects do not necessarily have to be in contact.

* **Acceleration**, 11; **Vector quantity**, 108.

- **Newton's second law**. If the **momentum** of an object changes, i.e. if it accelerates, then there must be a resultant force acting on it. Normally, the **mass** of the object is constant, and the force is thus proportional to the acceleration of the object. The direction of the acceleration is the same as the direction of the force.

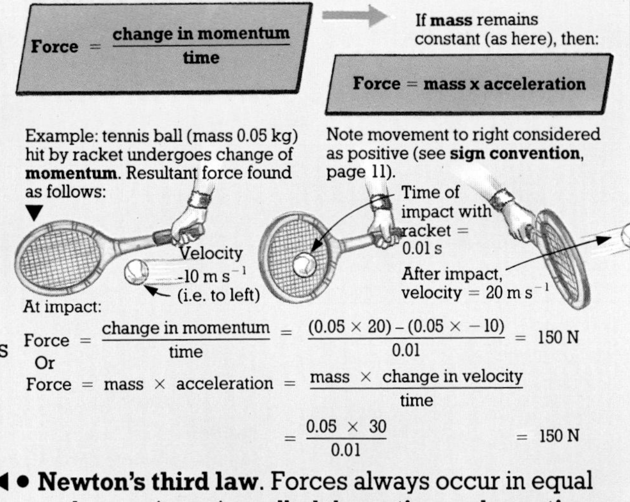

Racket exerts force on ball, accelerating it in opposite direction.

Ball exerts equal and opposite force on racket (felt as sudden slowing down of racket).

- **Newton's third law**. Forces always occur in equal and opposite pairs called the **action** and **reaction**. Thus if object A exerts a force on object B, object B exerts an equal but opposite force on A. These forces do not cancel each other out, as they act on different objects.

- **Law of conservation of linear momentum**. When two or more objects exert forces on each other (are in **collision**), their total **momentum** remains constant, provided no external forces act. If the time for the collision is very small and the system is considered just before and just after the collision, forces such as friction can be ignored.

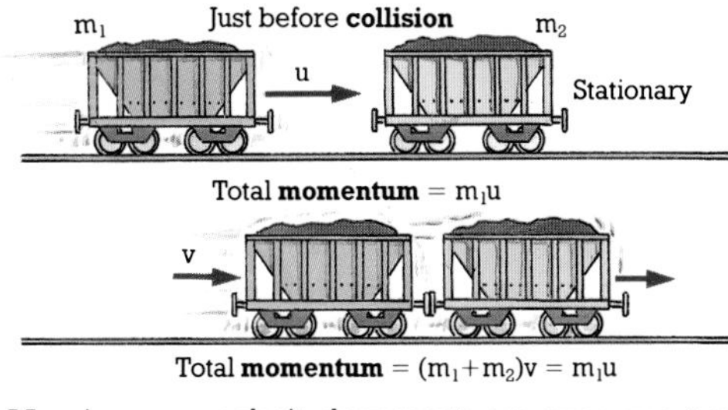

Mass increases – velocity decreases to conserve momentum.

Rocket engine

Oxygen

Fuel

Combustion chamber

Air taken in and compressed.

Jet engine

Stream of gas – **momentum** is conserved so engines gain same amount of momentum as gas, but in opposite direction.

Fuel burnt.

- **Rocket engine**. An engine which produces a high velocity stream of gas through a nozzle by burning fuel held on board. The **mass** of gas is small, but its high velocity means it has a high **momentum**. The rocket gains an equal amount of momentum in the opposite direction (see **law of conservation of linear momentum**). Rocket engines are used in space because other engines require air.

- **Jet engine**. An engine in which air is drawn in at the front to burn fuel, producing a high velocity jet of gas. The principle is the same as that for the **rocket engine**, except that the gas is produced differently and the engine cannot be used in space because it requires air.

Turning forces

A single force produces an **acceleration*** (see **dynamics**, page 12). In **linear motion***, it is a **linear acceleration**. In **rotational motion***, **angular acceleration*** (spinning faster or slower) is caused by a turning force or **moment** acting away from the axis of rotation (the **fulcrum**).

- **Moment** or **torque**. A measure of the ability of a force to rotate an object about an axis (the **fulcrum**). It is the size of the force multiplied by the perpendicular distance from the axis to the line along which the force acts (see diagram, right). The **SI unit*** of moment is the **Newton metre (N m)**.

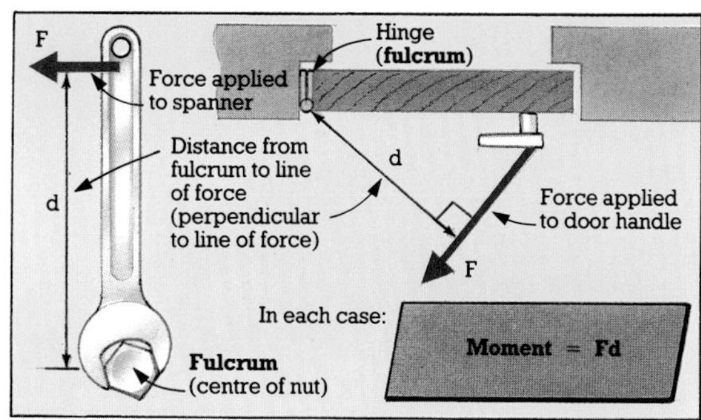

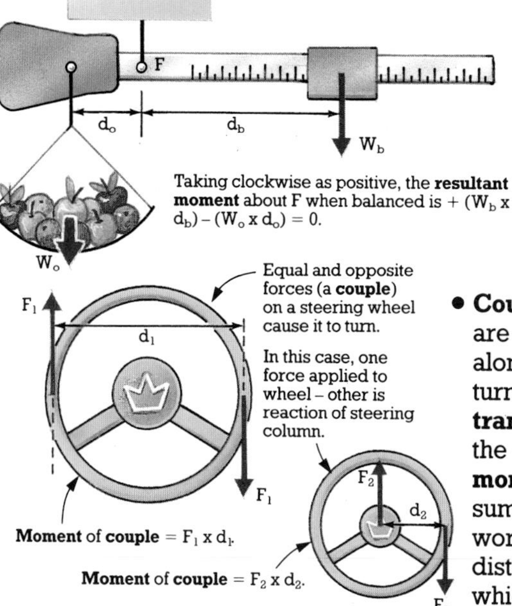

When considering moments, the axis about which they are taken must be stated and a **sign convention*** must be used to distinguish between clockwise and anti-clockwise moments. The **resultant moment** is the single moment which has the same effect as all the individual moments acting.

- **Couple**. Two parallel forces which are equal and opposite but do not act along the same line. They produce a turning effect only, with no resultant **translational motion*** (movement of the **centre of mass***). The **resultant moment** produced by a couple is the sum of the moments produced and works out to be the perpendicular distance between the lines along which the forces act, multiplied by the size of the forces.

Equilibrium

When an object is not accelerating, then the resultant force (the combined effect of all the forces acting on it) is zero and it is said to be in **equilibrium**. It can be in **linear equilibrium** (i.e. the **centre of mass*** is not accelerating) and/or **rotational equilibrium** (i.e. not accelerating about the centre of mass). In addition, both cases of equilibrium are either **static** (not moving) or **dynamic** (moving).

* **Acceleration**, 11; **Angular acceleration**, 17; **Centre of mass**, 10; **Linear motion**, 10; **Rotational motion**, **Sign convention**, 11; **SI units**, 96; **Translational motion**, 10.

- **Linear equilibrium**. The state of an object when there is no acceleration of its **centre of mass***, i.e. its speed and direction of motion do not change. The resultant force on the object when it is in linear equilibrium must be zero (see also **Newton's third law**, page 13).

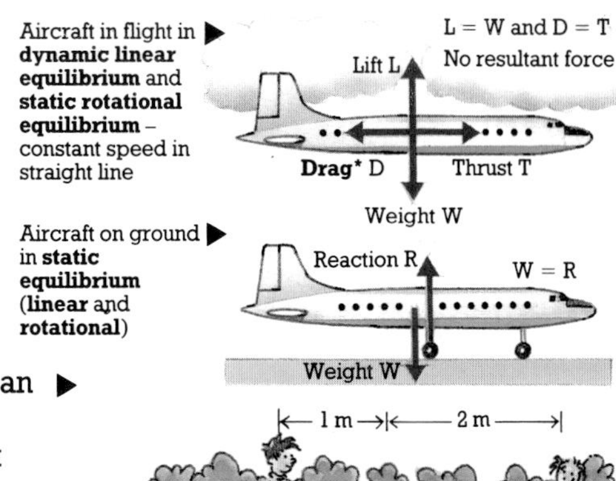

- **Rotational equilibrium**. The state of an object when there is no **angular acceleration***, i.e. it spins at constant **angular velocity***. If an object is in rotational equilibrium, the **resultant moment** (see **moment**) about any axis is zero. This is known as the **principle of moments**.

1 m 2 m

800 N 400 N

Beam in **static rotational equilibrium**, since 800 x 1 = 400 x 2

- **Toppling**. A condition which occurs if the vertical line through the **centre of mass*** of an object does not pass through the base of the object. If this occurs, a **couple** of the weight and reaction rotates the object further over (see diagram, left).

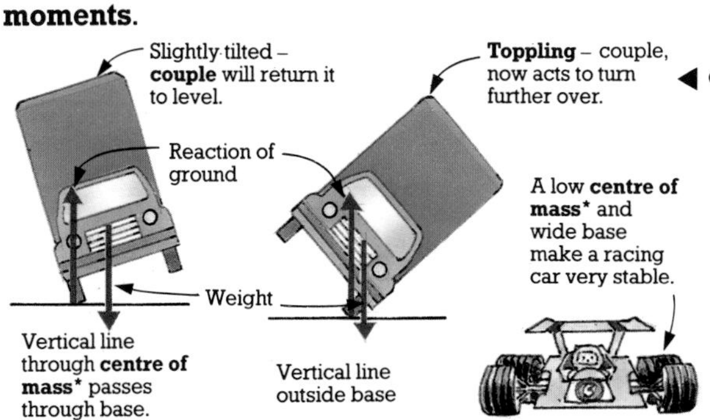

- **Stable equilibrium**. A state in which an object moved a small distance from its equilibrium position returns to that position. This happens if the **centre of mass*** is raised when the object is moved.

- **Unstable equilibrium**. A state in which an object moved a small distance from its equilibrium position moves further from that position. This happens if the **centre of mass*** is lowered when the object is moved.

- **Neutral equilibrium**. A state in which an object moved a small distance from its equilibrium position remains in the new position. This happens if the **centre of mass*** remains at the same height.

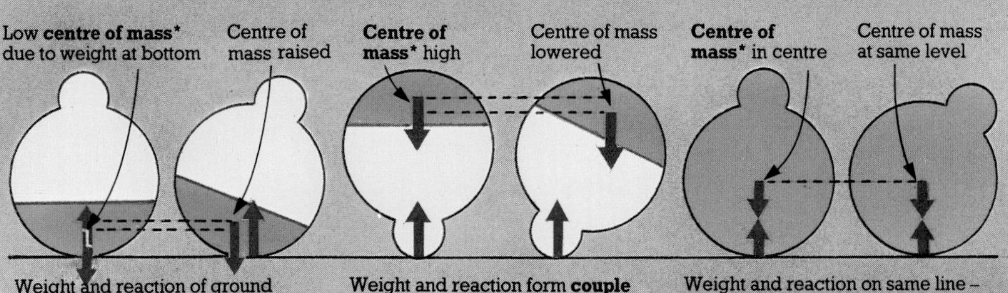

Weight and reaction of ground form **couple** to turn toy upright.

Weight and reaction form **couple** which turns toy further over.

Weight and reaction on same line – no **couple** toy stays in new position.

* **Angular acceleration**, **Angular velocity**, 17; **Centre of mass**, 10; **Drag**, 19 (**Terminal velocity**).

Periodic motion

Periodic motion is any motion which repeats itself exactly at regular intervals. Examples of periodic motion are objects moving in a circle (**circular motion**), the swing of a pendulum and the vibration of molecules. **Wave motion*** consists of the periodic motion of particles or fields.

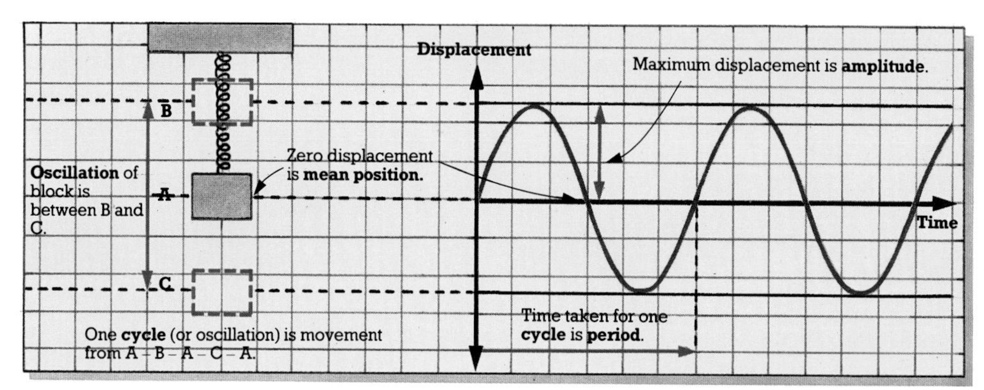

- **Cycle**. The movement between a point during a motion and the same point when the motion repeats. For example, one rotation of a spinning object.
- **Oscillation**. Periodic motion between two extremes, e.g. a mass moving up and down on the end of a spring. In an oscillating system, there is a continuous change between **kinetic energy*** and **potential energy***. The total energy of a system (sum of its kinetic and potential energy) remains constant if there is no **damping** .
- **Period (T)**. The time taken to complete one **cycle** of a motion, e.g. the period of rotation of the earth about its axis is 24 hours.
- **Frequency (f)**. The number of **cycles** of a particular motion in one second. The **SI unit*** of frequency is the **Hertz** (**Hz**), which is equal to one cycle per second.

$$f = \frac{1}{T}$$

where f = **frequency**;
T = **period**.

- **Mean position**. The position about which an object **oscillates**, and at which it comes to rest after oscillating, e.g. the mean position of a pendulum is when it is vertical. The position of zero displacement of an oscillating particle is usually taken as this point.
- **Amplitude**. The maximum displacement of an **oscillating** particle from its **mean position**.
- **Damping**. The process whereby **oscillations** die down due to a loss of energy, e.g. shock absorbers in cars cause oscillations to die down after a car has gone over a bump.

Damping in an **oscillating** system

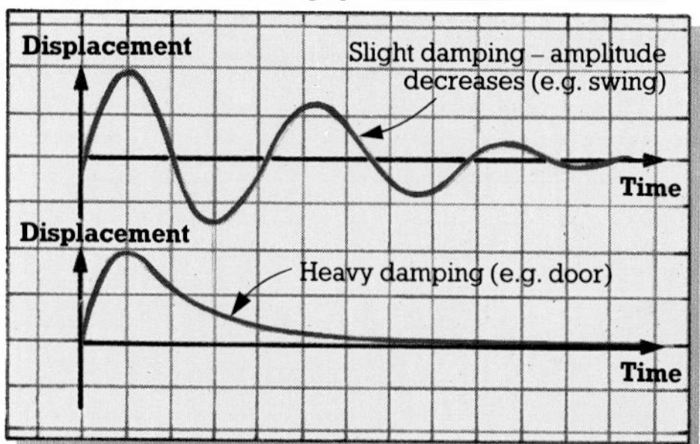

* **Kinetic energy**, 9; **Potential energy**, 8; **SI units**, 96; **Wave motion**, 34.

- **Natural** or **free oscillation**. The **oscillation** of a system when left after being given a start. The **period** and **frequency** of the system are called the **natural period** and **natural frequency** (these remain the same as long as the **damping** is not too great).

- **Forced oscillation**. The **oscillation** of a system when given a repeated driving force (a force applied to the system) at regular intervals. The system is made to oscillate at the **frequency** of the driving force, irrespective of its **natural frequency**.

- **Resonance**. The effect exhibited by a system in which the **frequency** of the driving force (a force applied to the system) is about the same as the **natural frequency** of the system. The system then has a large **amplitude**.

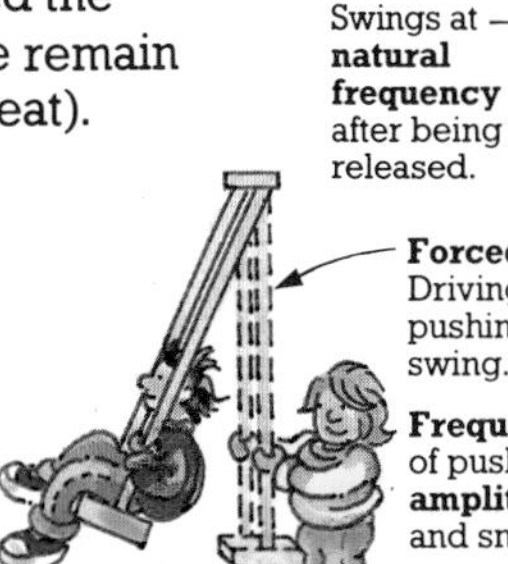

Circular motion

Uniform circular motion is the motion of an object in a circle at constant speed. Since the direction (and therefore the velocity) changes, the object is constantly accelerating towards the centre (**centripetal acceleration**), and so there is a force acting towards the centre. Circular motion can be considered in terms of **angular velocity**.

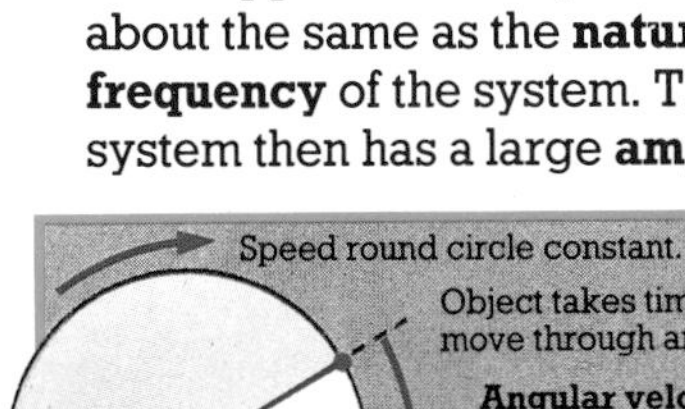

Angular velocity is a measure of the angle moved through per second. It is measured in radians per second.

Angular velocity $= \theta/t$ rad s^{-1}

Angular acceleration involves a change in angular velocity (i.e. speed round circle changes).

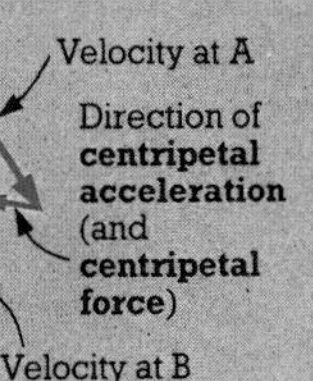

$$a = \frac{v^2}{r}$$

where a = **centripetal acceleration**; v = velocity around circle; r = radius of circle.

- **Centripetal acceleration** (**a**). The acceleration of an object in circular motion (see above) acting towards the centre of the circle.

- **Centripetal force**. The force which acts on an object towards the centre of a circle to produce **centripetal acceleration**, and so keeps the object moving in a circle.

- **Centrifugal force**. The equal and opposite reaction (see **Newton's second law**, page 13) to the **centripetal force**. Note that it does not act on the object moving in the circle and is not considered.

Gravitation

Gravitation is the effect of the **gravitational force*** of attraction (see also page 104) which acts between all objects in the universe. It is noticed with massive objects like the planets, which remain in orbit because of it. The gravitational force between an object and a planet, which pulls the object downwards, is called the **weight** of the object.

- **Newton's law of gravitation**. States that there is a gravitational force of attraction between any two objects with mass which depends on their masses and the distance between them. The **gravitational constant** (**G**) has a value of 6.7×10^{-11} N m^2 kg^{-2}, and its small value means that gravitational forces are negligible unless one of the masses is very large.

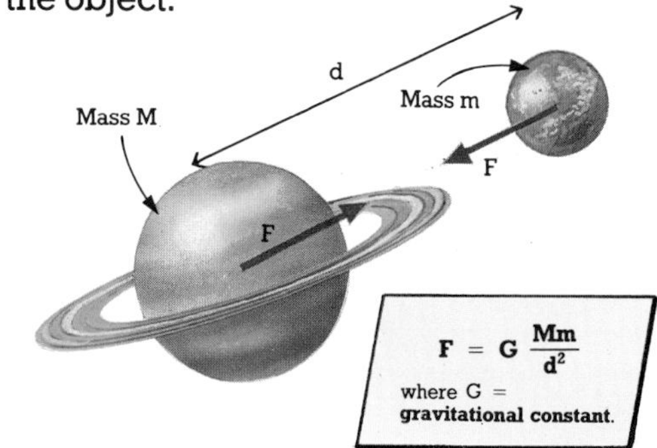

$$F = G\frac{Mm}{d^2}$$

where G = **gravitational constant**.

- **Weight**. The gravitational pull of a massive object (e.g. a planet) on another object. The weight of an object is not constant, but depends on the distance from, and mass of the planet. Hence, although the mass of an object is independent of its position, its weight is not.

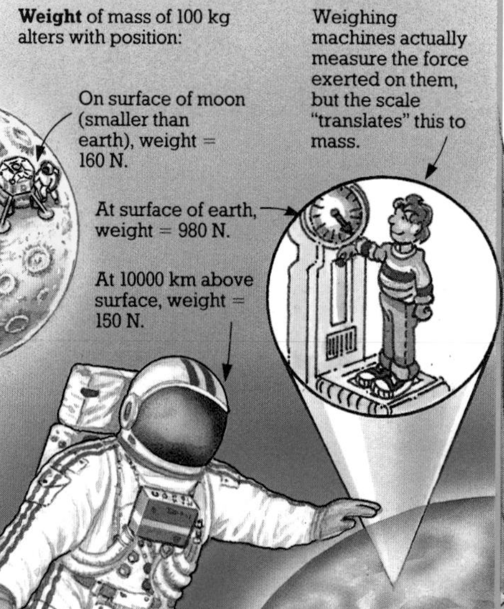

- **Acceleration due to gravity** (**g**). The **acceleration*** produced by the gravitational force of attraction. Its value is the same for any mass at a given place. It is about 9.8 m s^{-2} on the earth's surface, and decreases above the surface according to **Newton's law of gravitation**. The value of 9.8 m s^{-2} is used as a unit of acceleration (the **g-force**).

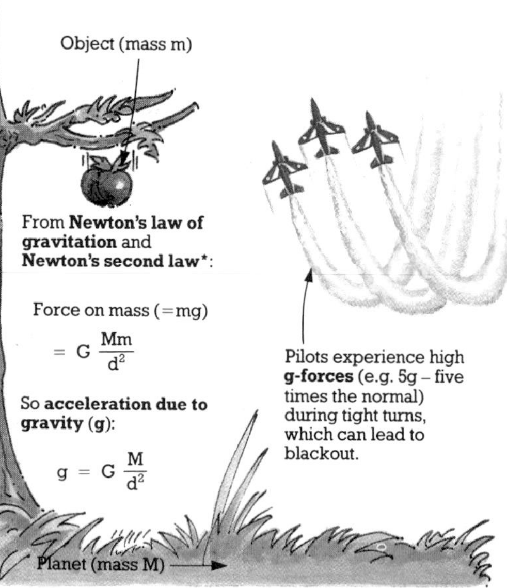

* **Acceleration**, 11; **Gravitational force**, 6; **Newton's second law**, 13.

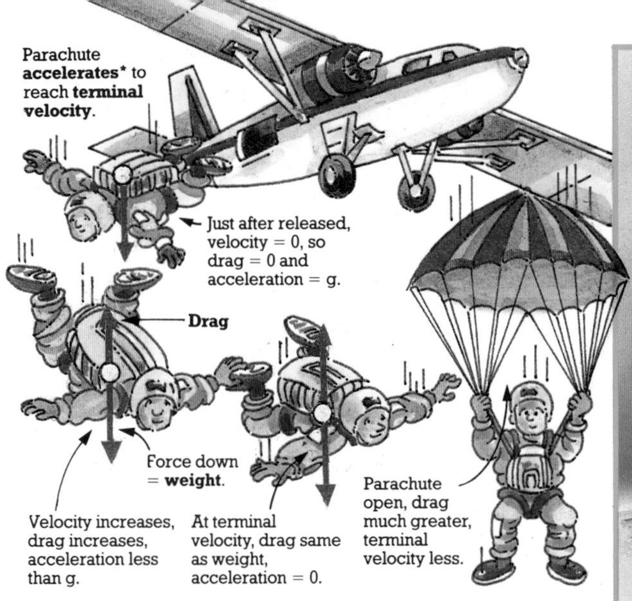

- **Terminal velocity**. The maximum, constant velocity reached by an object falling through a gas or liquid. As the velocity increases, the resistance due to the air or liquid (**drag**) increases. Eventually, the drag becomes equal to the **weight** of the object.

- **Escape velocity**. The minimum velocity at which an object must travel in order to escape the gravitational pull of a planet without further propulsion. It is about 40000 km h^{-1} on earth.

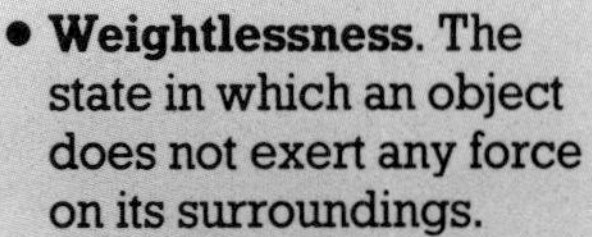

- **Weightlessness**. The state in which an object does not exert any force on its surroundings.

- **True weightlessness**. **Weightlessness** due to an object being in a gravity-free region.

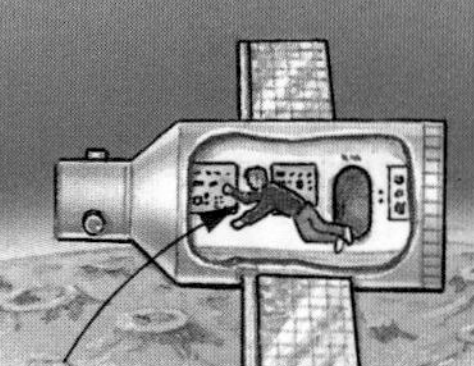

- **Apparent weightlessness**. The state of an object when it is as if there were no gravitational forces acting. This occurs if two objects **accelerate*** independently in the same way.

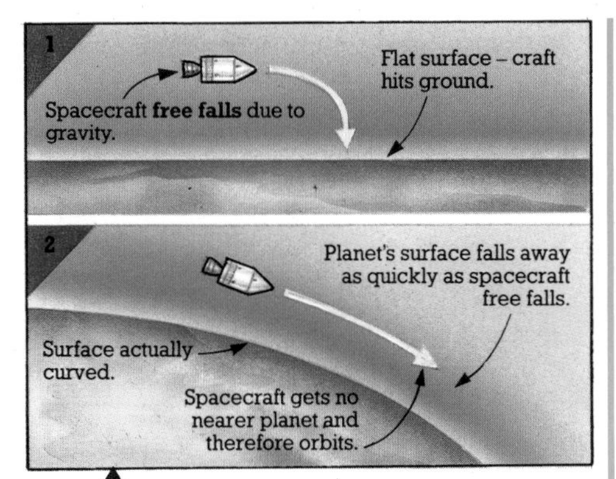

- **Free fall**. The unrestricted motion of an object when it is acted upon only by the gravitational force (i.e. when there are no resistive or other forces acting, e.g. air resistance).

- **Geo-stationary** or **parking orbit**. The path of a satellite which orbits the earth in the same direction as the rotation of the earth so that it stays above the same place on the surface all the time. The satellite is said to have a **period*** of 24 hours.

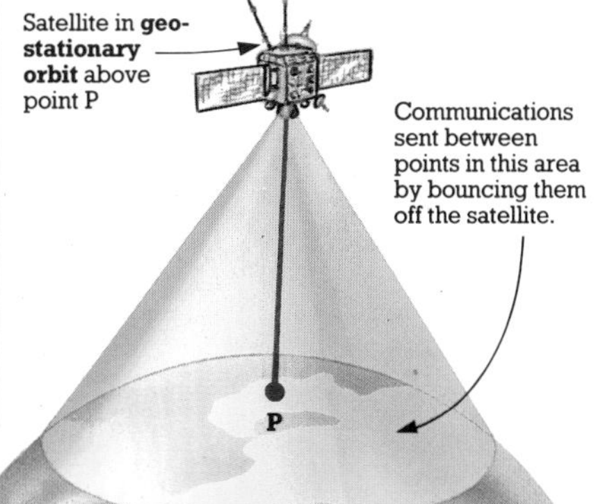

* **Acceleration**, 11; **Period**, 16.

Machines

A **machine** is a device which is used to overcome a force called the **load**. This force is applied at one point and the machine works by the application of another force called the **effort** at a different point. For example, a small effort exerted on the rope of a **pulley** overcomes the weight of the object being raised by the pulley.

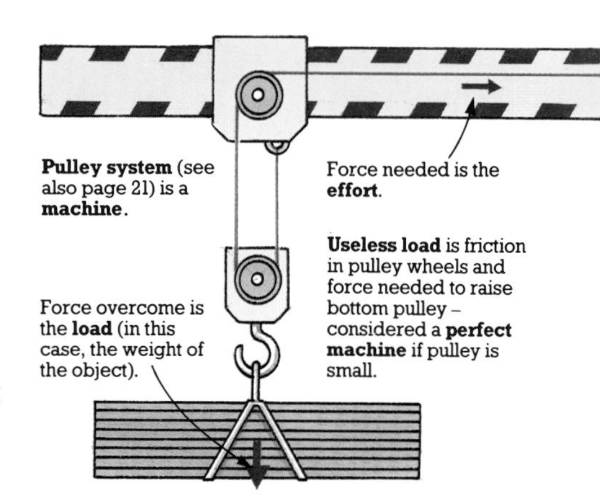

- **Perfect machine**. A theoretical machine, with a **useless load** of zero. Machines in which the useless load is negligible compared to the load can be considered as perfect machines.
- **Useless load**. The force needed to overcome the **frictional forces*** between the moving parts of a machine and to raise any of its moving parts.

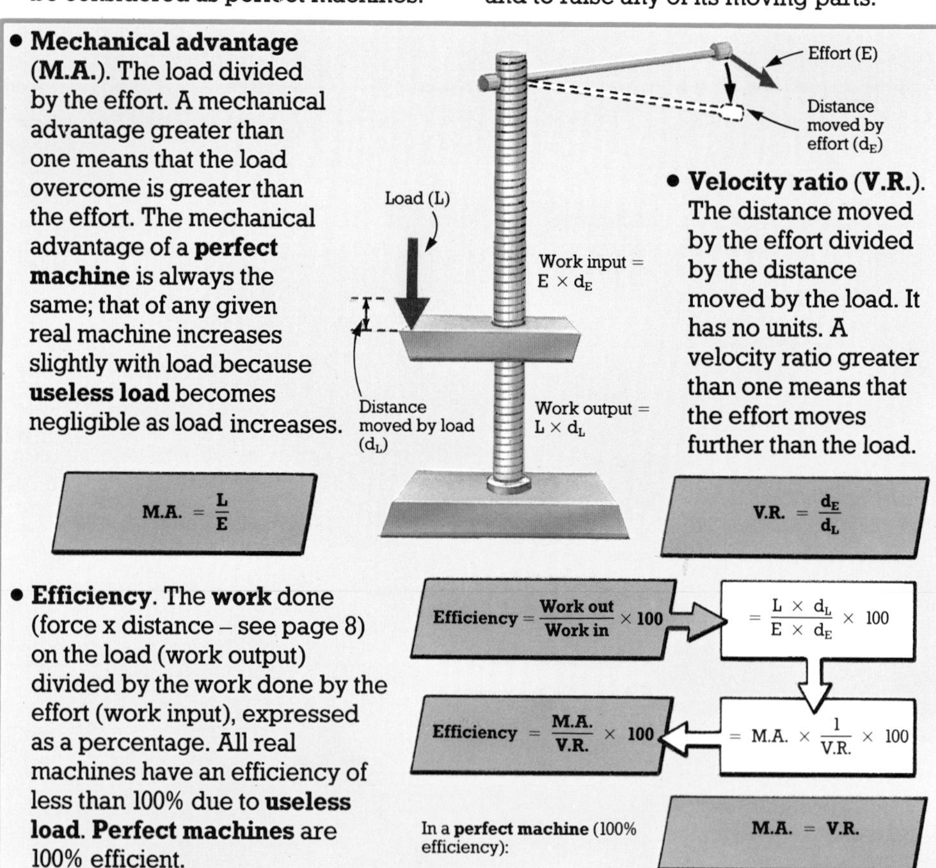

- **Mechanical advantage (M.A.)**. The load divided by the effort. A mechanical advantage greater than one means that the load overcome is greater than the effort. The mechanical advantage of a **perfect machine** is always the same; that of any given real machine increases slightly with load because **useless load** becomes negligible as load increases.
- **Velocity ratio (V.R.)**. The distance moved by the effort divided by the distance moved by the load. It has no units. A velocity ratio greater than one means that the effort moves further than the load.
- **Efficiency**. The **work** done (force x distance – see page 8) on the load (work output) divided by the work done by the effort (work input), expressed as a percentage. All real machines have an efficiency of less than 100% due to **useless load**. **Perfect machines** are 100% efficient.

* **Frictional force**, 7.

Examples of machines

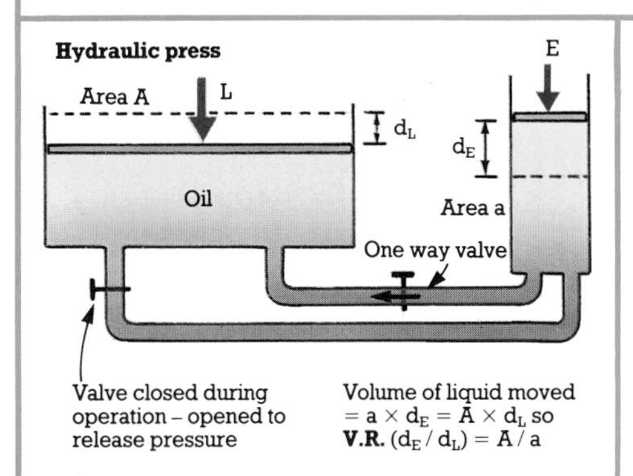

- **Hydraulic press**. A large and small cylinder connected by a pipe and filled with fluid, used to produce large forces.

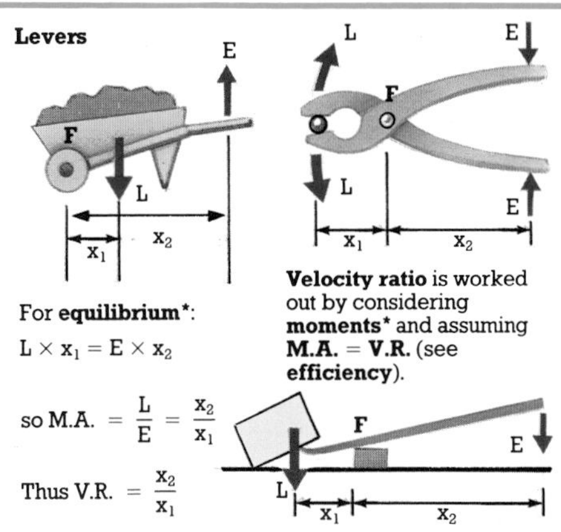

- **Lever**. Any rigid object which is pivoted about an axis called the **fulcrum** (**F**). The load and effort can be applied on either or the same side.

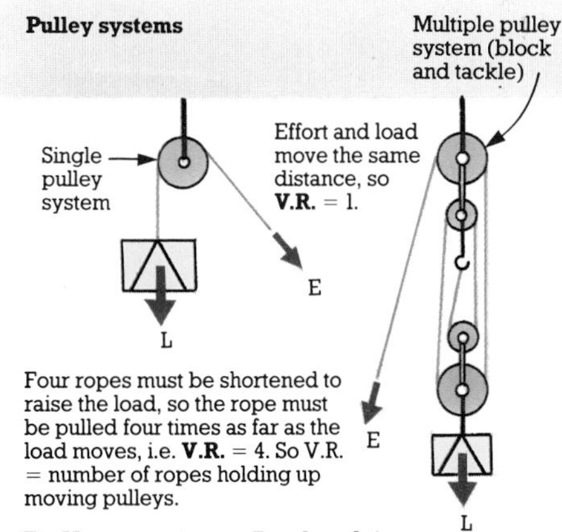

- **Pulley system**. A wheel (or combination of wheels) and a rope, belt or chain which transmits motion.

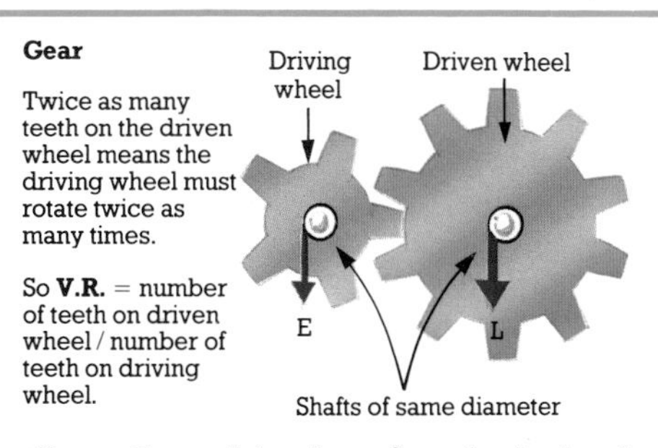

- **Gear**. A combination of toothed wheels used to transmit motion between rotating shafts.

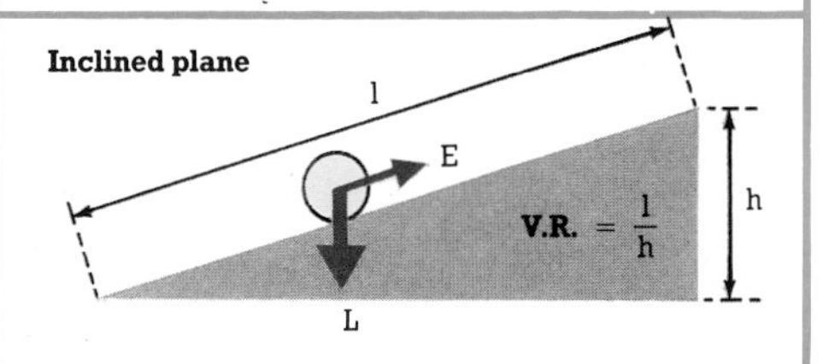

- **Inclined plane**. A plane surface at an angle to the horizontal. It is easier to move an object up an inclined plane than to move it vertically upwards.

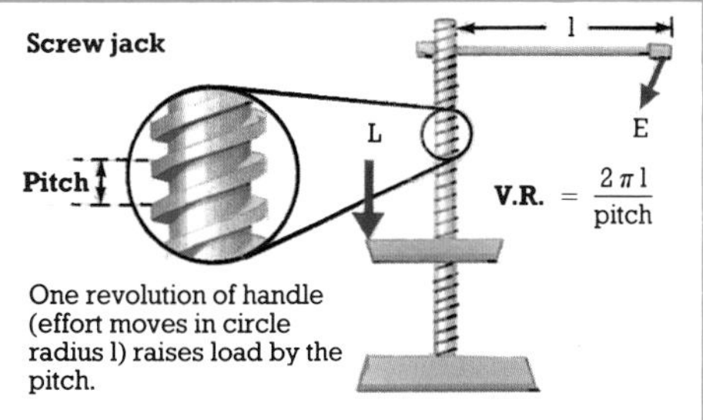

- **Screw jack**. A system in which a screw thread is turned to raise a load (e.g. a car jack). The **pitch** is the distance between each thread on the screw.

* **Equilibrium**, **Moment**, 14.

Molecular properties

There are a number of properties of matter which can be explained in terms of the behaviour of molecules, in particular their behaviour due to the action of the forces between them (**intermolecular forces***). Among these properties, and explained below, are **elasticity**, **surface tension** and **viscosity**. See also pages 4-5 and 24-25.

- **Elasticity**. The ability of a material to return to its original shape and size after distorting forces (i.e. **tension** or **compression forces***) have been removed. Materials which have this ability are **elastic**; those which do not are **plastic**. Elasticity is a result of **intermolecular forces*** – if an object is stretched or compressed, its molecules move further apart or closer together respectively. This results in a force of attraction (in the first case) or repulsion (in the second), so the molecules return to their average separation when the distorting force is removed. This always happens while the strength of the force is below a certain level (different for each material), but all elastic materials finally become plastic if it exceeds this level (see **elastic limit** and **yield point**).

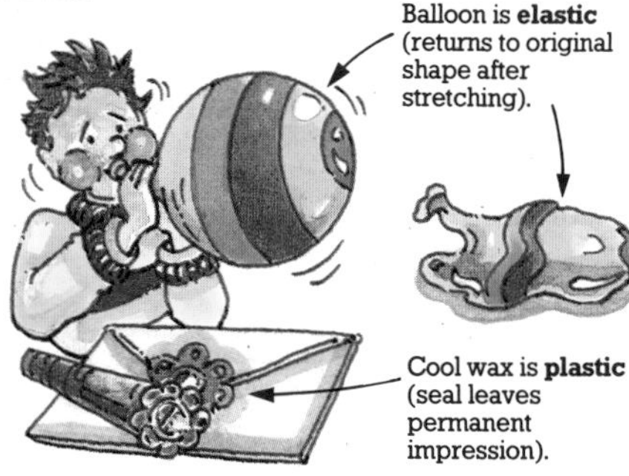

- **Hooke's law**. States that, when a distorting force is applied to an object, the **strain** is proportional to the **stress** (see diagram below). As the strength of the force increases, though, the **limit of proportionality** (or **proportional limit**) is reached, after which Hooke's law is no longer true (see graph, page 23).

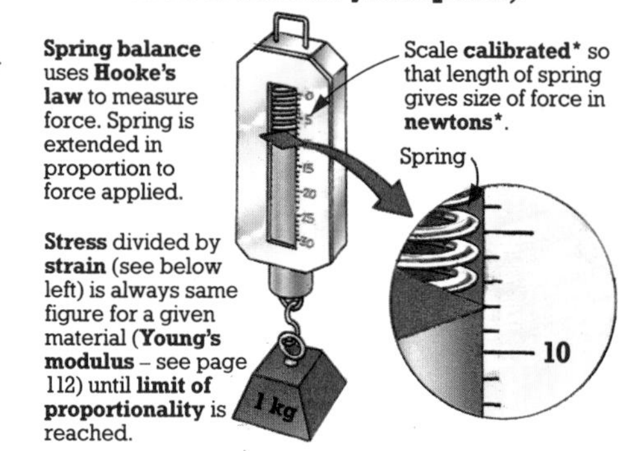

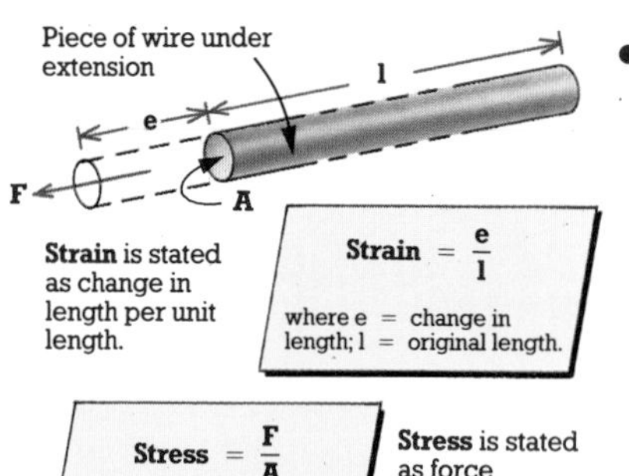

Strain is stated as change in length per unit length.

$$\text{Strain} = \frac{e}{l}$$

where e = change in length; l = original length.

$$\text{Stress} = \frac{F}{A}$$

where F = force applied; A = cross-sectional area.

Stress is stated as force applied per unit area.

- **Elastic limit**. The point, just after the **limit of proportionality** (see **Hooke's law**), beyond which an object ceases to be **elastic**, in the sense that it does not return to its original shape and size when the distorting force is removed. It does return to a similar shape and size, but has suffered a permanent strain (it will continue to return to this new form if forces are applied, i.e. it stays elastic in this sense). The **yield stress** of a material is the value of the **stress** at its elastic limit. See graph, page 23.

* **Calibration**, 344; **Compression force**, **Intermolecular force**, 7; **Newton**, 6; **Tension force**, 7.

- **Yield point**. The point, just after the **elastic limit**, at which a distorting force causes a major change in a material. In a **ductile*** material, the internal structure changes – bonds between molecular layers break and the layers flow over each other. This change is called **plastic deformation**.

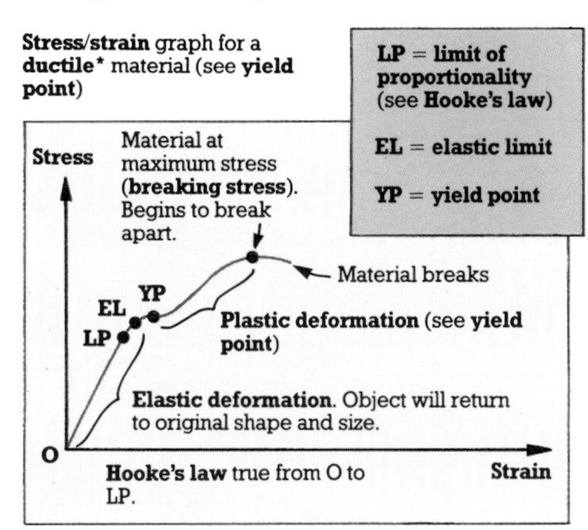

(the material becomes **plastic**). It continues as the force increases, and the material will eventually break. A **brittle** material, by contrast, will break at the yield point. The **yield value** of a material is the value of the **stress** at its yield point.

- **Viscosity**. The ease of flow of a fluid. It depends on the strength of the **frictional force*** between different layers of molecules as they slide over each other.

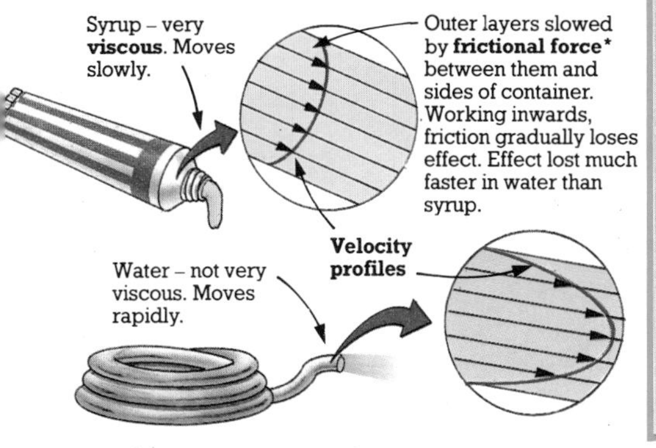

- **Surface tension**. The skin-like property of a liquid surface, resulting from **intermolecular forces*** which cause it to contract to the smallest possible area.

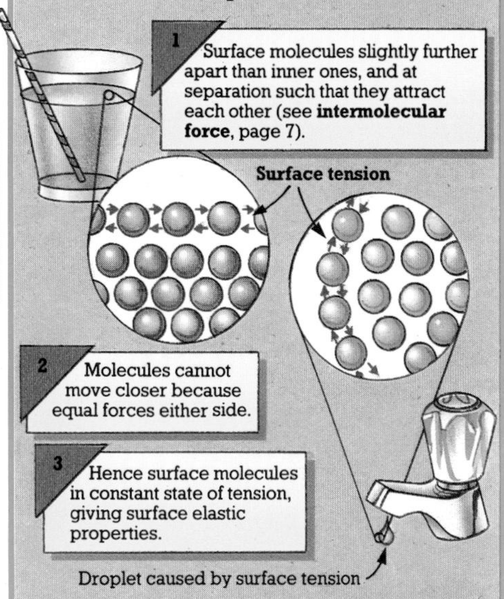

- **Adhesion**. An **intermolecular force*** of attraction between molecules of different substances.

Capillary action or **capillarity** is a result of **adhesion** or **cohesion**.

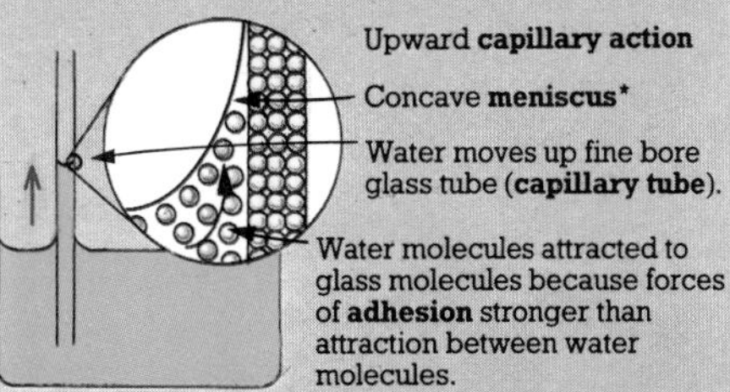

- **Cohesion**. An **intermolecular force*** of attraction between molecules of the same substance.

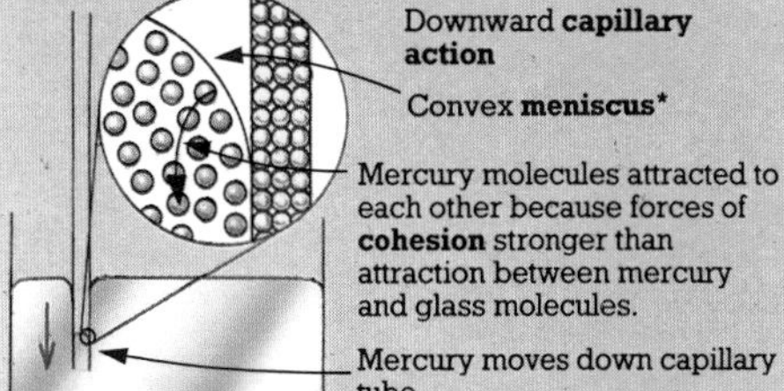

* **Ductile**, 344; **Frictional force**, **Intermolecular force**, 7; **Meniscus**, 345.

Density

The **density** (ρ) of an object depends on both the mass of its molecules and its volume (see formula, right). For example, if one substance has a higher density than another, then the same volumes of the substances have different masses (the first greater than the second). Similarly, the same masses have different volumes.

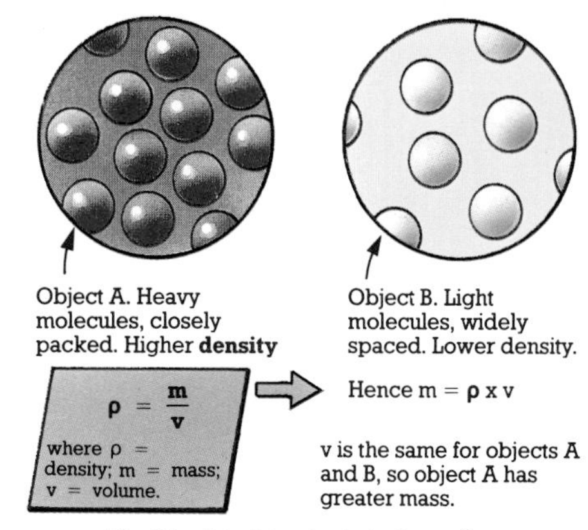

The **SI unit*** of density is the kg m^{-3}.

- **Relative density** or **specific gravity**. The density of a substance relative to the density of water (which is 1000 kg m^{-3}). It indicates how much more or less dense than water a substance is, so the figures need no units, e.g. 1.5 (one and a half times as dense). It is found by dividing the mass of any volume of a substance by the mass of an equal volume of water.

- **Density bottle**. A container which, when completely full, holds a precisely measured volume of liquid (at constant temperature). It is used to measure the density of liquids (by measuring the mass of the bottle and liquid, subtracting the mass of the bottle and dividing by the volume of liquid).

Density bottle

Fine bore tube (**capillary tube**) in glass stopper. Bottle filled, stopper inserted, excess liquid rises through tube and runs out – ensures same volume each time.

- **Eureka can**. A can used to measure the volume of a solid object with an irregular shape, in order to calculate its density. The volume of water displaced is equal to the volume of the object. The density of the object is its mass divided by this volume.

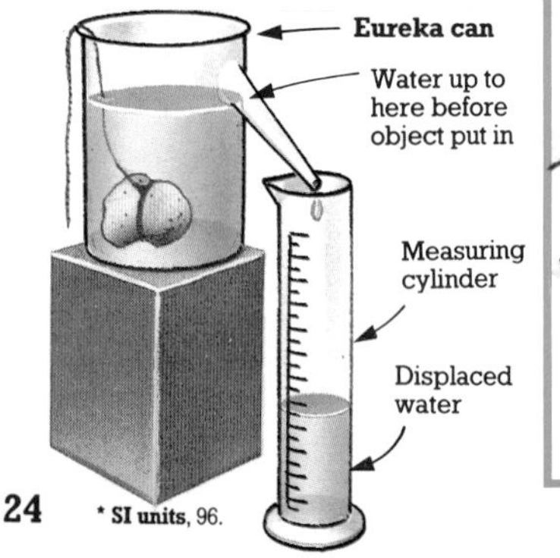

Pressure

Pressure is the force, acting at right angles, exerted by a solid, liquid or gas on unit area of a substance (solid, liquid or gas).

$$\text{Pressure} = \frac{\text{force}}{\text{area}}$$

The **SI unit*** of pressure is the **pascal** (**Pa**).

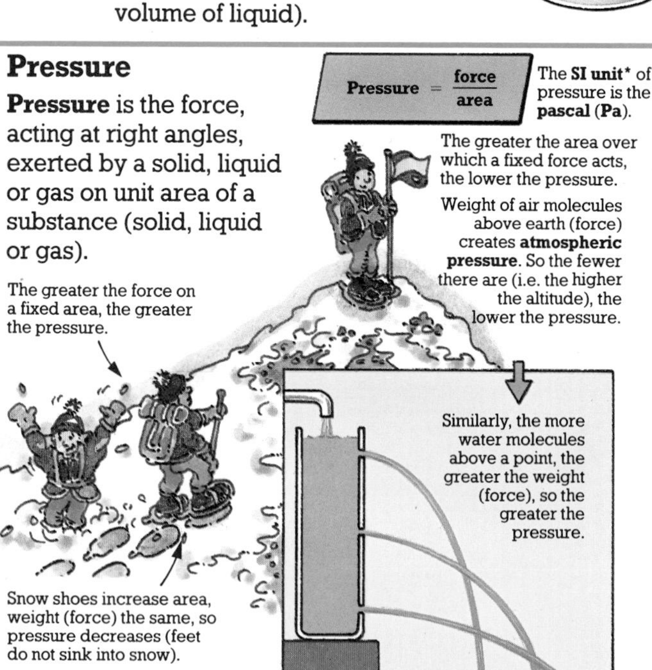

* **SI units**, 96.

Objects in fluids

An object in a fluid experiences an upward force called the **upthrust**. According to **Archimedes' principle**, this is equal to the weight of the fluid the object displaces. The **principle of flotation** further states that, if the object is floating, the weight of displaced fluid (upthrust) is equal to its own weight (floating here means stationary at any point in the fluid). It can be shown (see below) that whether an object sinks, rises or floats in a fluid depends entirely on density.

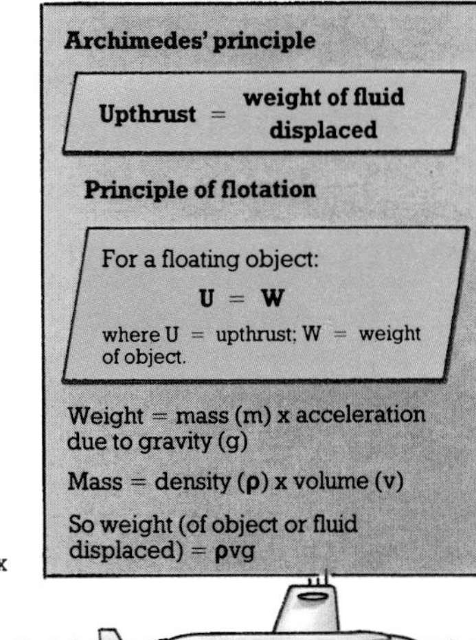

Submarines demonstrate two principles above right. Altering air/water mix in ballast tanks alters density (density of object made of more than one material is average density of different materials). See below.

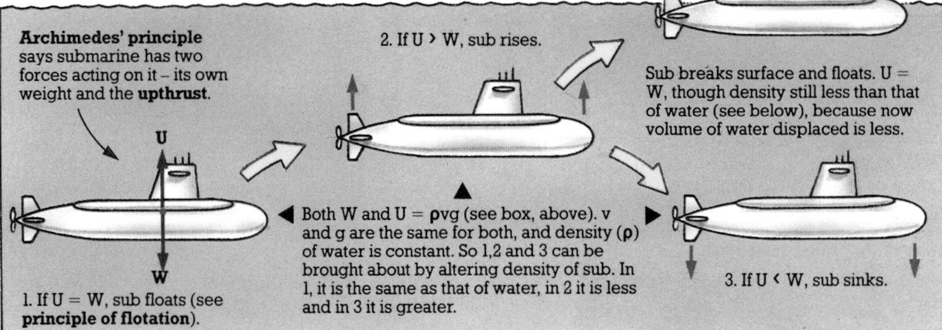

- **Barometer**. An instrument used to measure **atmospheric pressure** (see picture, left). There are several common types.

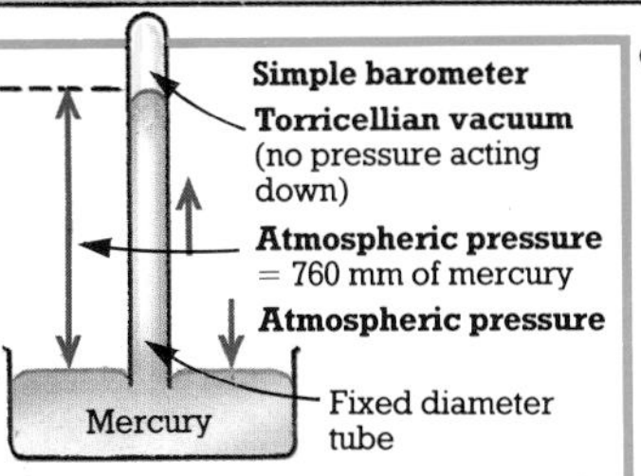

- **Hydrometer** or **aerometer**. An instrument which measures the density of a liquid by the level at which it floats in that liquid. If the liquid is very dense, the hydrometer floats near the surface, as only a small volume of liquid need be displaced to equal the weight of the hydrometer.

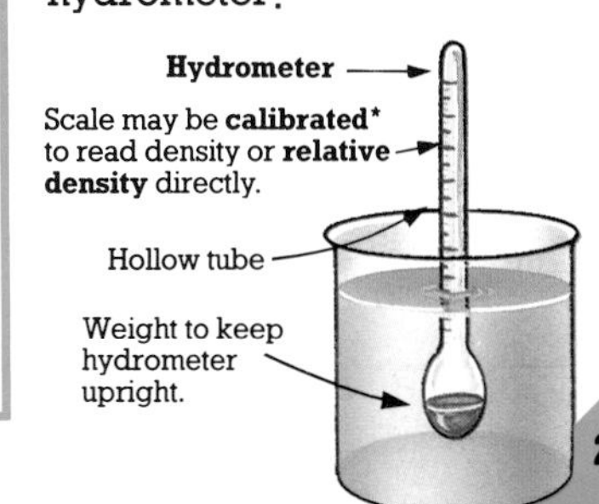

- **Manometer**. A U-shaped tube containing a liquid. It is used to measure difference in fluid pressures.

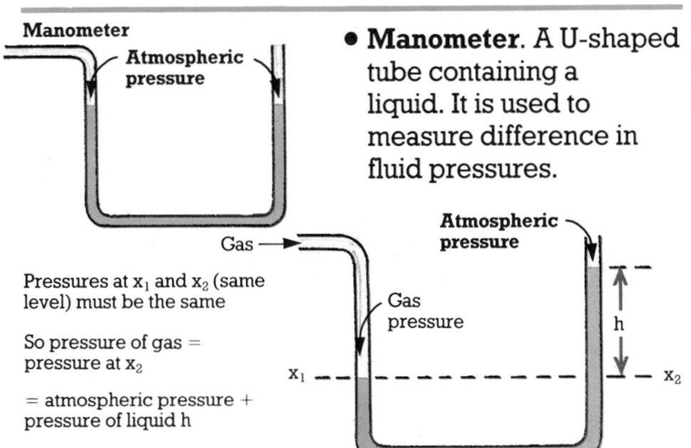

* **Calibration**, 344.

Temperature

The **temperature** of an object is a measurement of how hot the object is. It is measured using **thermometers** which can be **calibrated*** to show a number of different temperature scales. The internationally accepted scales are the **absolute temperature scale** and the **Celsius scale**.

- **Thermometer**. An instrument used to measure temperature. There are many different types and they all work by measuring a property which changes with temperature – a **thermometric property**. **Liquid-in-glass thermometers**, for example, measure the volume of a liquid (they are **calibrated*** so that increases in volume mark rises in temperature).

- **Liquid-in-glass thermometer**. A common type of **thermometer** which measures temperature by the expansion of a liquid in a fine bore glass tube (**capillary tube**). A glass bulb holds a reservoir of the liquid, which is usually either mercury or coloured alcohol. These are very responsive to temperature change – mercury is used for higher temperature ranges and alcohol for lower ones.

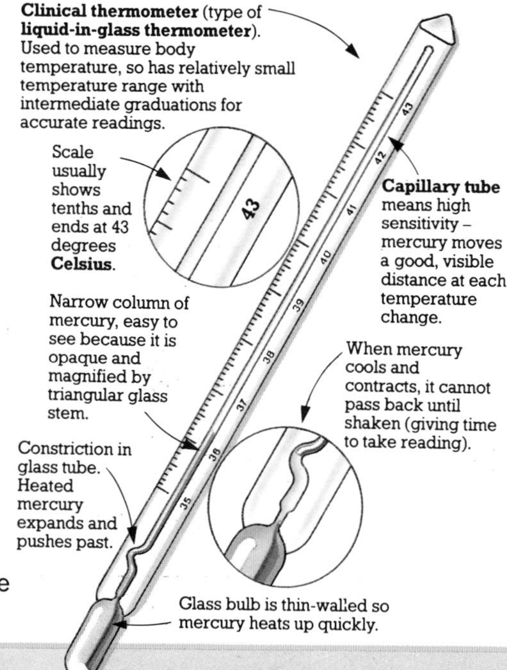

Using **fixed points** to **calibrate*** the **Celsius scale** on a **thermometer**.

Upper fixed point

Hypsometer (double-walled copper vessel)

Position of end of mercury thread marked as 100°C.

Manometer* – measures steam pressure (should be **atmospheric pressure***).

Steam out

Mercury bulb in steam

Steadily boiling water

Lower fixed point

Position of end of mercury thread marked as 0°C.

Funnel

Pure, melting ice

Beaker

Thermometer

Upper fixed point

Fundamental interval

Lower fixed point

- **Fixed point**. A temperature at which certain recognisable changes always take place (under given conditions), and which can thus be given a value against which all other temperatures can be measured. Examples are the **ice point** (the temperature at which pure ice melts) and the **steam point** (the temperature of steam above water boiling under **atmospheric pressure***). Two fixed points are used to **calibrate*** a thermometer – a **lower** and an **upper fixed point**. The distance between these points is the **fundamental interval**.

* **Atmospheric pressure**, 24; **Calibration**, 344; **Manometer**, 25.

- **Maximum** and **minimum thermometers**. Special **liquid-in-glass thermometers** which record the maximum or minimum temperature reached over a period of time. They contain a metal and glass **index** (see picture) which is pushed up or pulled down (respectively) by the liquid **meniscus***. The index stays at the maximum or minimum position it reaches during the time the thermometer is left. It is reset using a magnet.

Maximum thermometer
Mercury
Convex **meniscus***
Index at highest position so far reached by mercury.
Maximum temperature reading

Minimum thermometer
Coloured alcohol
Index at lowest position so far reached by alcohol.
Concave **meniscus***
Minimum temperature reading

Other types of thermometer

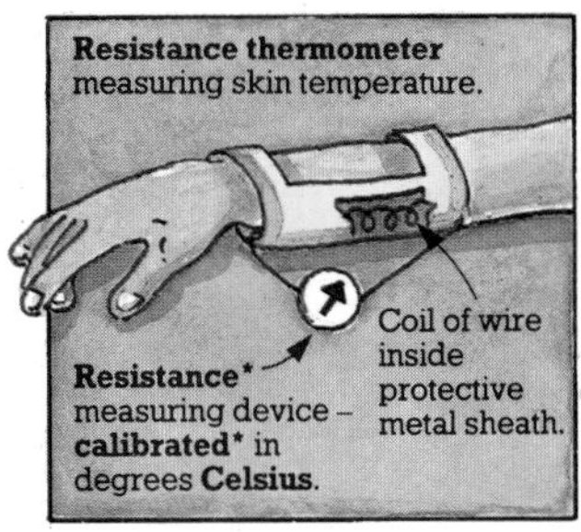

- **Resistance thermometer**. Measures temperature from change in **resistance*** it causes in a wire.

Similar devices, e.g. under aeroplane wings, use resistance change in **thermistors***.

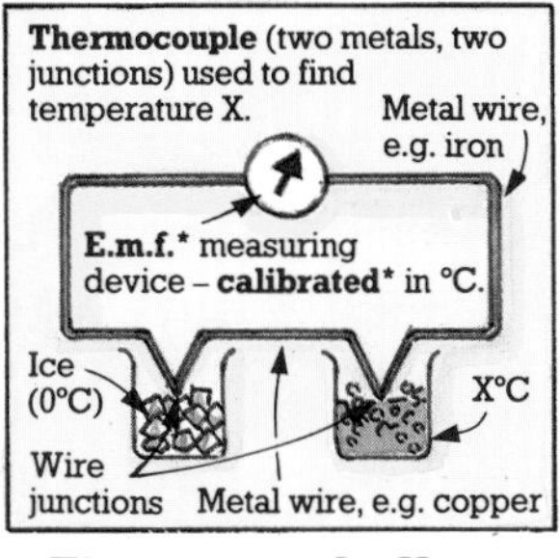

- **Thermocouple**. Uses the **e.m.f.*** produced across metal junctions to measure temperature difference.

- **Absolute** or **thermodynamic temperature scale**. A standard temperature scale, using units called **kelvins (K)**. The zero value is given to the lowest possible temperature theoretically achievable, called **absolute zero**. It is impossible to have a lower temperature, as this would require a negative volume (see graph) which cannot exist.

- **Celsius scale (°C)**. A standard temperature scale identical in graduations to the **absolute temperature scale**, but with the zero and one hundred degree values given to the **ice point** and **steam point** respectively (see **fixed point**).

- **Fahrenheit scale (°F)**. An old scale with the values 32°F and 212°F given to the **ice point** and **steam point** respectively (see **fixed point**). It is rarely used in scientific work.

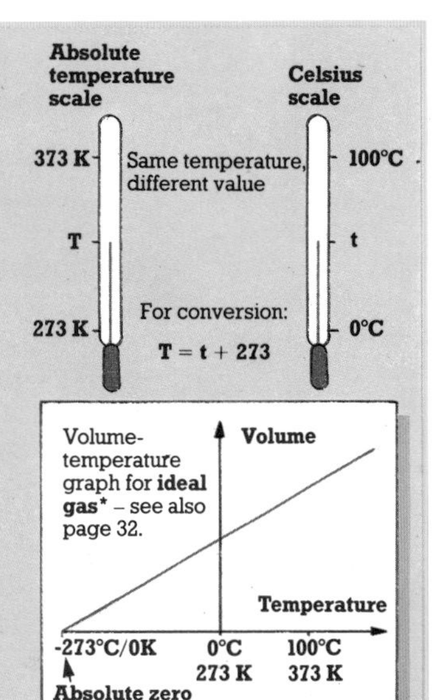

* **Calibration**, 344; **Electromotive force (e.m.f.)**, 60; **Ideal gas**, 33; **Meniscus**, 345; **Resistance**, 62; **Thermistor**, 65.

Transfer of heat

Whenever there is a temperature difference, **heat energy** (see page 9) is transferred by **conduction**, **convection** or **radiation** from the hotter to the cooler place. This increases the **internal energy*** of the cooler atoms, raising their temperature, and decreases the energy of the hotter atoms, lowering theirs. It continues until the temperature is the same across the region – a state called **thermal equilibrium**.

- **Conduction** or **thermal conduction**. The way in which heat energy is transferred in solids (and also, to a much lesser extent, in liquids and gases). In good **conductors** the energy transfer is rapid, occurring mainly by the movement of free **electrons*** (electrons which can move about), although also by the vibration of atoms – see **insulators** (bad conductors).

Heat energy moves up metal spoon (good **conductor**).

Heated **electrons*** gain **kinetic energy***. Move out fast in all directions.

Electrons collide with atoms, passing on heat energy.

Hot atoms vibrate, but only collide with neighbours.

- **Insulators**. Materials such as wood and most liquids and gases, in which the process of **conduction** is very slow (they are bad **conductors**). As they do not have free **electrons***, heat energy is only transferred by conduction by the vibration and collison of neighbouring atoms.

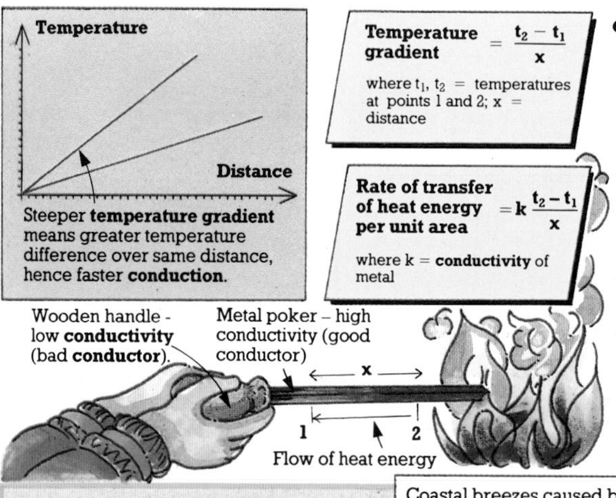

- **Conductivity** or **thermal conductivity**. A measure of how good a **conductor** a material is (see also page 112). The rate of heat energy transfer through an object depends on the conductivity of the material and the **temperature gradient**. This is the temperature change with distance along the material. The higher the conductivity and the steeper the gradient, the faster the energy transfer.

- **Convection**. A way in which heat energy is transferred in liquids and gases. If a liquid or gas is heated, it expands, becomes less dense and rises. Cooler, denser liquid or gas then sinks to take its place. Thus a **convection current** is set up.

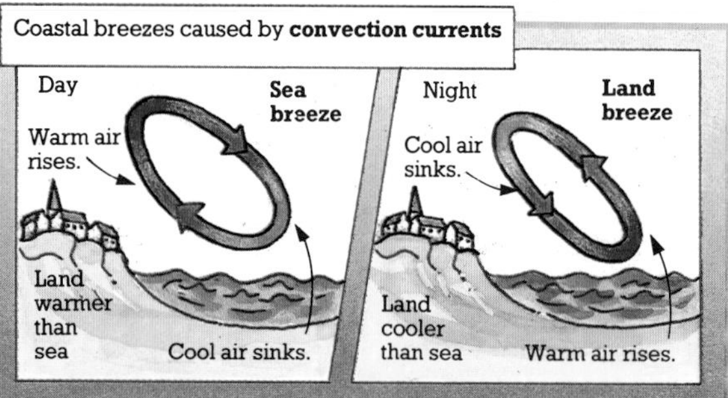

* **Electrons**, 83; **Internal energy**, **Kinetic energy**, 9.

- **Radiation**. A way in which heat energy is transferred from a hotter to a cooler place without the **medium*** taking any part in the process. This can occur through a vacuum, unlike **conduction** and **convection**. The term radiation is also often used to refer to the heat energy itself, otherwise known as **radiant heat energy**. This takes the form of **electromagnetic waves***, mainly **infra-red radiation***. When these waves fall on an object, some of their energy is absorbed, increasing the object's **internal energy*** and hence its temperature. See also **Leslie's cube**, below.

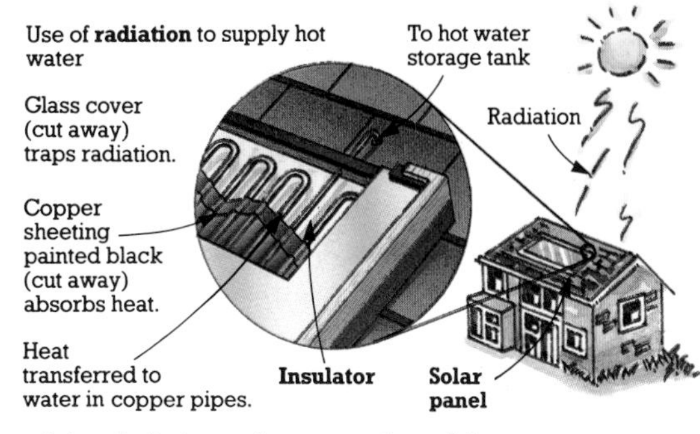

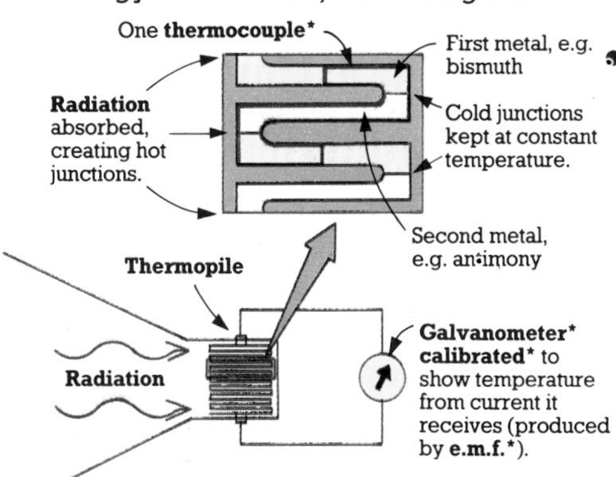

- **Thermopile**. A device for measuring **radiation** levels. It consists of two or more **thermocouples*** (normally over 50) joined end to end. Radiation falls on the metal junctions on one side and the temperature difference between these hot junctions and the cold ones on the other side produces an **e.m.f.*** across the thermopile, the size of which indicates how much radiation has been absorbed.

Leslie's cube used to compare powers of **radiation** – numbers show best (1) to worst (4) surface.

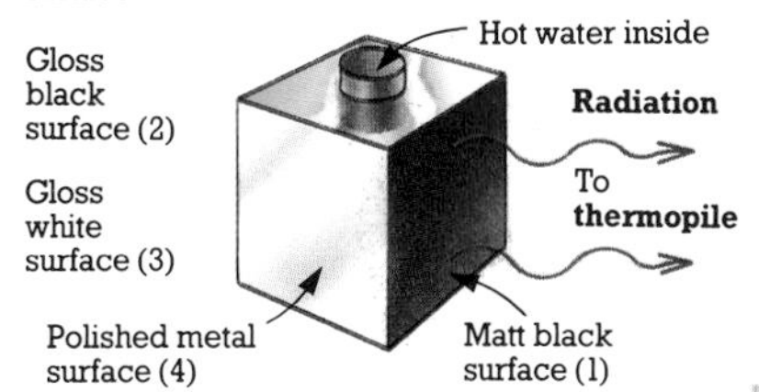

- **Leslie's cube**. A thin-walled, hollow cube (good **conductor**) with different outside surfaces. It is used to show that surfaces vary in their ability to **radiate** and absorb heat energy. Their powers of doing so are compared with an ideal called a **black body**, which absorbs all radiation that falls on it, and is also the best radiator.

- **Greenhouse effect**. The warming effect produced when **radiation** is trapped in a closed area, e.g. a greenhouse. The objects inside absorb the sun's radiation and re-emit lower energy radiation which cannot pass back through the glass. Carbon dioxide in the atmosphere forms a similar barrier, and its level is increasing, hence the air is slowly getting warmer.

- **Vacuum flask**. A flask which keeps its contents at constant temperature. It consists of two glass containers, one inside the other, with a vacuum between them (stopping heat energy transfer by **conduction** and **convection**) and shiny surfaces (minimising transfer by **radiation**).

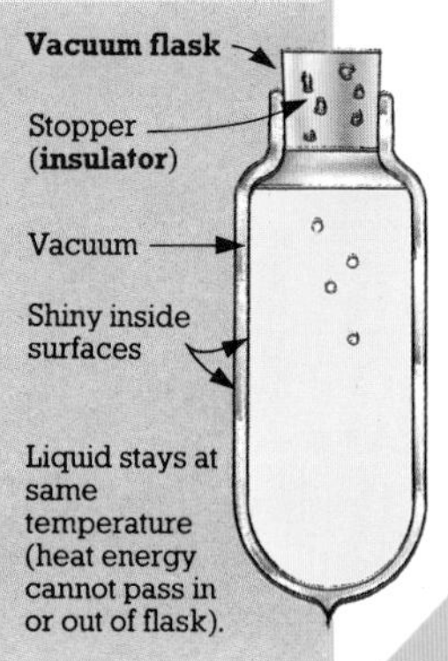

* **Calibration**, 344; **Electromagnetic waves**, 44; **Electromotive force (e.m.f.)**, 60; **Galvanometer**, 77; **Infra-red radiation**, 45; **Internal energy**, 9; **Medium**, 345; **Thermocouple**, 27.

Effects of heat transfer

When an object absorbs or loses **heat energy** (see pages 28-29), its **internal energy*** increases or decreases. This results in either a rise or fall in temperature (the amount of which depends on the **heat capacity** of the object) or a **change of state**.

- **Heat capacity (C).** The heat energy taken in or given out by an object when its temperature changes by 1 K. It is a property of the object and depends on both its mass and the material(s) of which it is made, hence its value is different for every object and has to be worked out individually.

$$Q = C(t_2 - t_1)$$

where Q = heat energy lost or gained; C = **heat capacity**; t_1 and t_2 = temperatures.

The **SI unit*** of **heat capacity** is the joule per kelvin ($J\ K^{-1}$).

- **Specific heat capacity (c).** The heat energy taken in or given out when 1 kg of a substance changes temperature by 1 K. It is a property of the substance alone, i.e. there is a set value for each substance. This value changes if a **change of state** occurs. See also page 112.

$$Q = mc(t_2 - t_1)$$

where m = mass; c = **specific heat capacity**; Q, t_1, t_2 as above.

The **SI unit*** of **specific heat capacity** is the joule per kilogram per kelvin ($J\ kg^{-1}\ K^{-1}$).

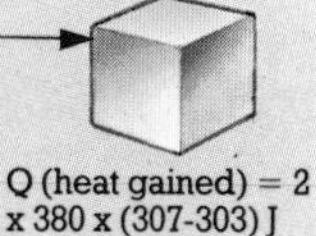

Mass (m) of 2 kg brass (**specific heat capacity** 380 J $kg^{-1}\ K^{-1}$) heated for a set time. Temperature rises from 303 K (t_1) to 307 K (t_2).

Q (heat gained) = 2 x 380 x (307-303) J

So Q = 3040 J

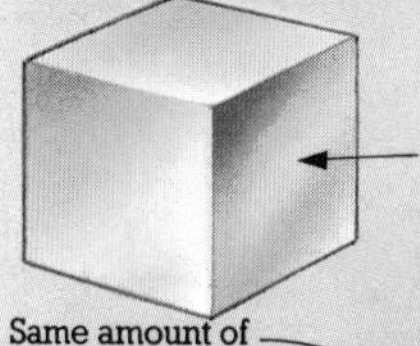

Thus same amount of heat energy taken in by 16 kg of brass would raise temperature by 0.5 K.

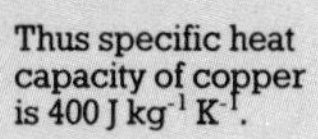

Same amount of heat energy given to mass of 2 kg of copper causes temperature rise of 3.8 K.

Thus specific heat capacity of copper is 400 J $kg^{-1}\ K^{-1}$.

Changes of state

A **change of state** is a change from one **physical state** (the solid, liquid or gaseous state) to another (for more about **physical states**, see page 4). While a change of state is happening, there is no change in temperature. Instead, all the energy taken in or given out is used to make or break molecular bonds. This is called **latent heat (L)** – see graphs, page 31. The **specific latent heat (l)** of a substance is a set value, i.e. the heat energy taken in or given out when 1 kg of the substance changes state.

- **Vaporization.** The change of state from liquid to gaseous at a temperature called the **boiling point** of the liquid (it is said to be **boiling**). The term is also used more generally for any change resulting in a gas or vapour, i.e. including also **evaporation** and **sublimation**.
- **Freezing.** The change of state from liquid to solid at the **freezing point** (the same temperature as the **melting point** of the solid).

- **Melting.** The change of state from solid to liquid at a temperature called the **melting point** of the solid.

- **Condensation.** The change of state from gas or vapour to liquid.

* **Internal energy**, 9; **SI units**, 96.

- **Evaporation**. The conversion of a liquid to a vapour by the escape of molecules from its surface. It takes place at all temperatures, the rate increasing with any one or a combination of the following: increase in temperature, increase in surface area or decrease in pressure. It is also increased if the vapour is immediately removed from above the liquid by a flow of air. The **latent heat** (see **changes of state**) needed for evaporation is taken from the liquid itself which cools and in turn cools its surroundings.

- **Sublimation**. The conversion of a substance from a solid directly to a gas, or vice-versa, without passing through the liquid state.

Dry ice (solid carbon dioxide) **sublimes**.

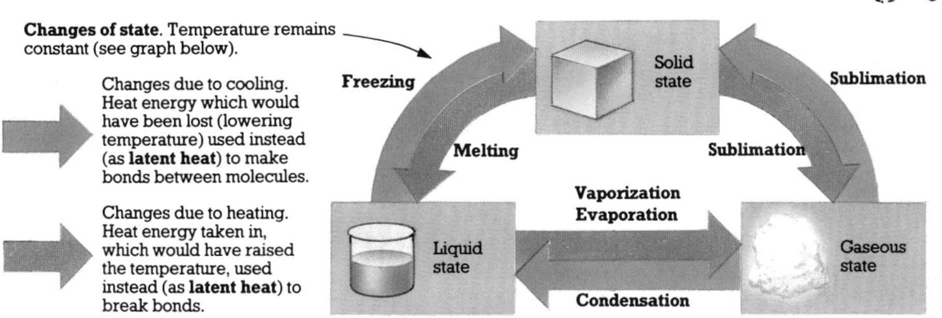

Changes of state. Temperature remains constant (see graph below).

Changes due to cooling. Heat energy which would have been lost (lowering temperature) used instead (as **latent heat**) to make bonds between molecules.

Changes due to heating. Heat energy taken in, which would have raised the temperature, used instead (as **latent heat**) to break bonds.

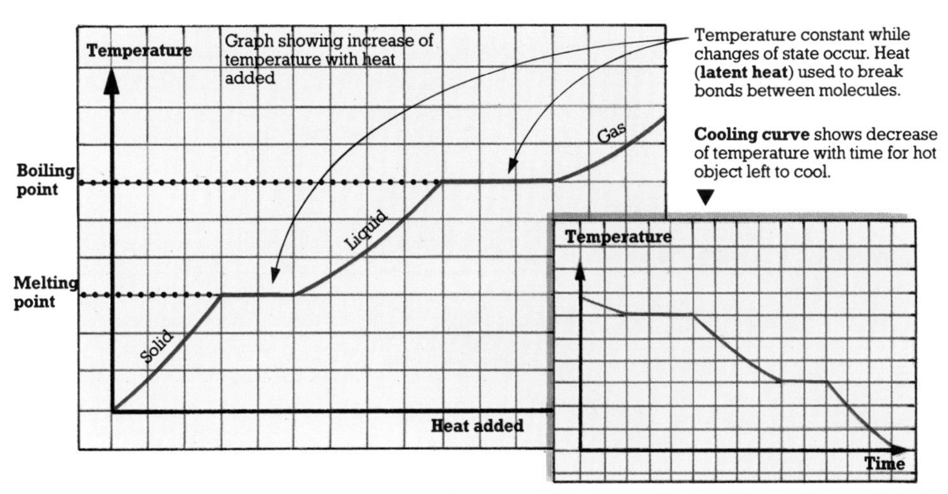

Temperature constant while changes of state occur. Heat (**latent heat**) used to break bonds between molecules.

Cooling curve shows decrease of temperature with time for hot object left to cool.

- **Specific latent heat of vaporization**. The heat energy taken in when 1 kg of a substance changes from a liquid to a gas at its **boiling point**. It is the same as the heat given out when the process is reversed.

- **Specific latent heat of fusion**. The heat energy taken in when 1 kg of a substance changes from a solid to a liquid at its **melting point**. It is the same as the heat given out when the process is reversed. See also page 112.

$$Q = ml$$

where Q = heat energy lost or gained by object; m = mass, l = **specific latent heat**.

The **SI unit*** of **specific latent heat** is the joule per kilogram ($J\ kg^{-1}$).

* **SI units**, 96.

Expansion on heating

Most substances expand when heated – their molecules move faster and further apart. The extent of this expansion (**expansivity**) depends on **intermolecular forces***. For the same amount of heat applied (at constant pressure), solids expand least, as their molecules are closest together and so have the strongest forces between them. Liquids expand more, and gases the most.

Expansion of solids on heating must be taken into account in building work.

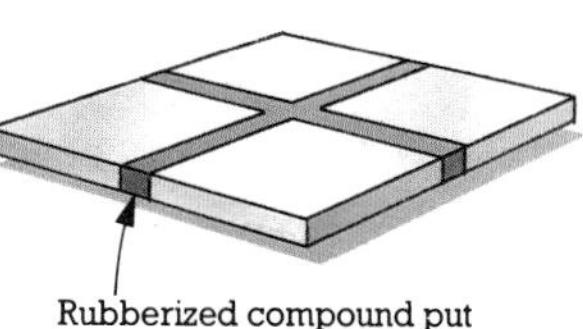

Rubberized compound put between paving stones

- **Bimetallic strip**. A device which shows the expansion of solids on heating. It is made up of two different strips of metal, joined along their (equal) length. When heated or cooled, both metals expand or contract (respectively), but at different rates, so the strip bends. Such strips are used in **thermostats**.

Thermostat (temperature regulator)

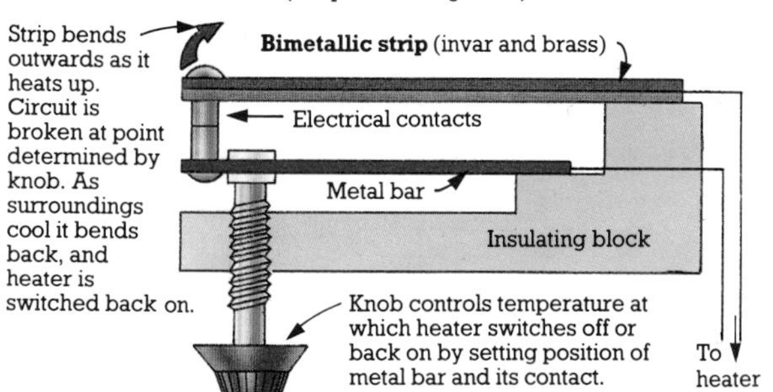

- **Linear expansivity** (α). A measurement of the fraction of its original length by which a solid expands for a temperature rise of 1K.

- **Superficial** or **areal expansivity** (β). A measurement of the fraction of its original area by which a solid expands for a temperature rise of 1K.

For solids or liquids:

$$\text{Expansivity (linear, superficial or cubic)} = \frac{\text{change in (length, area or volume)}}{\text{original (length, area or volume)} \times \text{temperature rise}}$$

Note the only relevant measurement for liquids is **cubic expansivity**. It is either **real** or **apparent** (see page 33), hence so is change in volume in formula.

For gases:

$$\text{Cubic expansivity} = \frac{\text{change in volume at constant pressure}}{\text{volume at 0°C (273K)} \times \text{temperature rise}}$$

- **Cubic** or **volume expansivity** (γ). A measurement of the fraction of its original volume by which a substance expands for a temperature rise of 1K. It is the same for all gases (at constant pressure) when they are assumed to behave as **ideal gases**. Since gases expand by very large amounts, the original volume is always taken at 0°C so that proper comparisons can be made (this is not felt to be necessary with solids or liquids as the changes are so small).

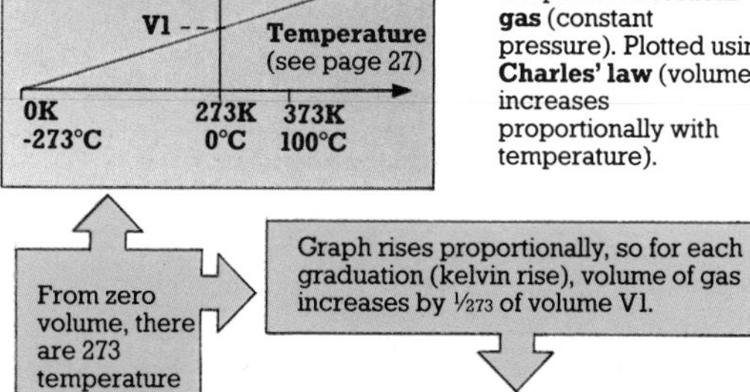

◀ Change in volume with temperature of **ideal gas** (constant pressure). Plotted using **Charles' law** (volume increases proportionally with temperature).

Thus, for an **ideal gas**:

$$\text{Cubic expansivity} = \frac{1}{273}\ K^{-1}$$

* **Intermolecular force**, 7.

- **Real** or **absolute cubic expansivity**. An accurate measurement of the fraction of its volume by which a liquid expands for a temperature rise of 1K.
- **Apparent cubic expansivity**. A measurement of the fraction of its volume by which a liquid apparently expands for a temperature rise of 1K. In fact, the heat applied also causes very slight expansion of the container, so its calibrated measurements are no longer valid.
- **Anomalous expansion**. The phenomenon whereby some liquids contract instead of expanding when the temperature rises within a certain range (e.g. water between 0°C and 4°C).

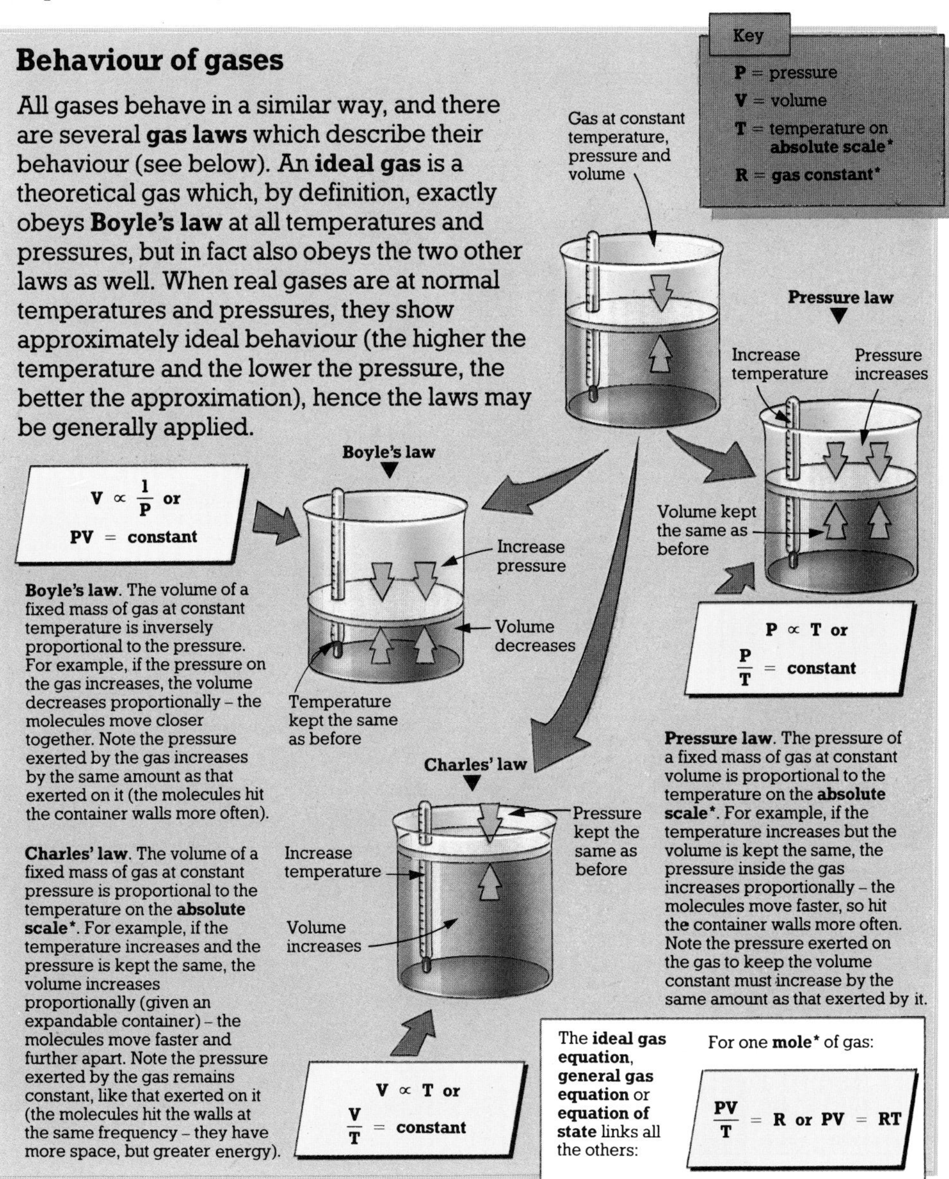

Behaviour of gases

All gases behave in a similar way, and there are several **gas laws** which describe their behaviour (see below). An **ideal gas** is a theoretical gas which, by definition, exactly obeys **Boyle's law** at all temperatures and pressures, but in fact also obeys the two other laws as well. When real gases are at normal temperatures and pressures, they show approximately ideal behaviour (the higher the temperature and the lower the pressure, the better the approximation), hence the laws may be generally applied.

$$V \propto \frac{1}{P} \text{ or } PV = \text{constant}$$

Boyle's law. The volume of a fixed mass of gas at constant temperature is inversely proportional to the pressure. For example, if the pressure on the gas increases, the volume decreases proportionally – the molecules move closer together. Note the pressure exerted by the gas increases by the same amount as that exerted on it (the molecules hit the container walls more often).

Charles' law. The volume of a fixed mass of gas at constant pressure is proportional to the temperature on the **absolute scale***. For example, if the temperature increases and the pressure is kept the same, the volume increases proportionally (given an expandable container) – the molecules move faster and further apart. Note the pressure exerted by the gas remains constant, like that exerted on it (the molecules hit the walls at the same frequency – they have more space, but greater energy).

$$V \propto T \text{ or } \frac{V}{T} = \text{constant}$$

$$P \propto T \text{ or } \frac{P}{T} = \text{constant}$$

Pressure law. The pressure of a fixed mass of gas at constant volume is proportional to the temperature on the **absolute scale***. For example, if the temperature increases but the volume is kept the same, the pressure inside the gas increases proportionally – the molecules move faster, so hit the container walls more often. Note the pressure exerted on the gas to keep the volume constant must increase by the same amount as that exerted by it.

The **ideal gas equation**, **general gas equation** or **equation of state** links all the others:

For one **mole*** of gas:

$$\frac{PV}{T} = R \text{ or } PV = RT$$

* **Absolute temperature scale**, 27; **Gas constant**, 113; **Mole**, 96.

Waves

All **waves** transport energy without permanently displacing the **medium*** through which they travel. They are also called **progressive** or **travelling waves**, as the energy travels from a source to surrounding points (but see also **stationary wave**, page 43). There are two main types – **mechanical waves**, such as sound waves, and **electromagnetic waves** (see page 44). In all cases, the **wave motion** is regular and repetitive (i.e. **periodic motion** – see page 16) in the form of **oscillations** – regular changes between two extremes. In mechanical waves it is particles (molecules) that oscillate and in electromagnetic waves it is electric and magnetic fields.

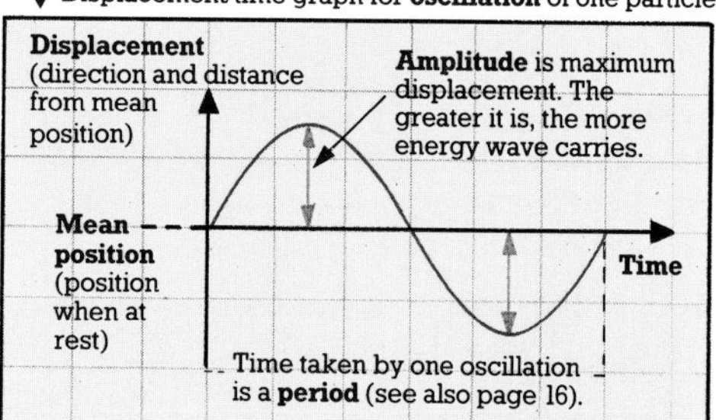

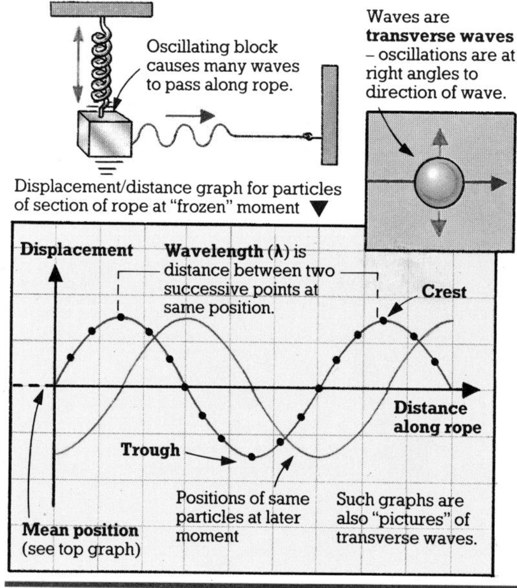

- **Transverse waves**. Waves in which the oscillations are at right angles to the direction of energy (wave) movement, e.g. water waves (oscillation of particles) and all **electromagnetic waves*** (oscillation of fields – see introduction).
- **Crests** or **peaks**. Points where a wave causes maximum positive displacement of the **medium***. The crests of some waves, e.g. water waves, can be seen as they travel.
- **Troughs**. Points where a wave causes maximum negative displacement of the **medium***. The troughs of some waves, e.g. water waves, can be seen as they travel.

- **Wavefront**. Any line or section taken through an advancing wave which joins all points which are in the same position in their oscillations. Wavefronts are usually at right angles to the direction of the waves and can have any shape, e.g. **circular** and **straight wavefronts**.

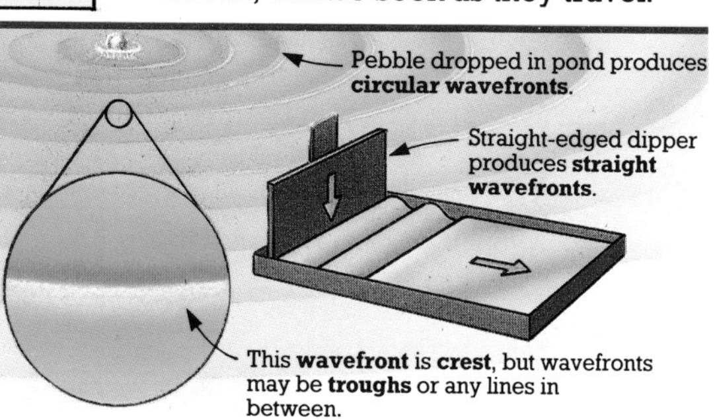

* **Electromagnetic waves**, 44; **Medium**, 345.

- **Longitudinal waves**. Waves in which the oscillations are along the line of the direction of wave movement, e.g. sound waves. They are all **mechanical waves** (see introduction), i.e. it is particles which oscillate.

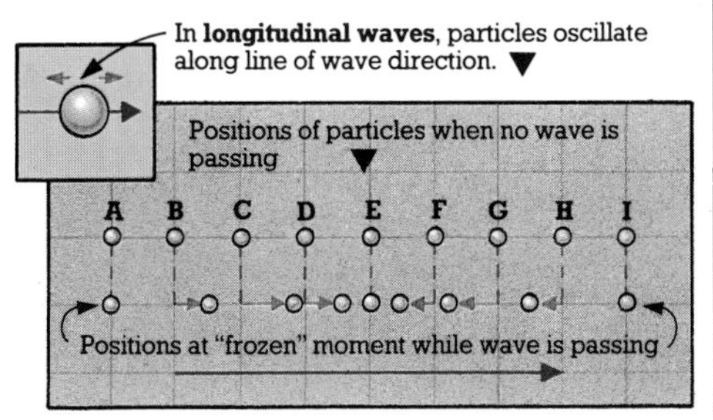

Graph of particles above at "frozen" moment. Graph not in this case "picture" of wave (see second graph, page 34).

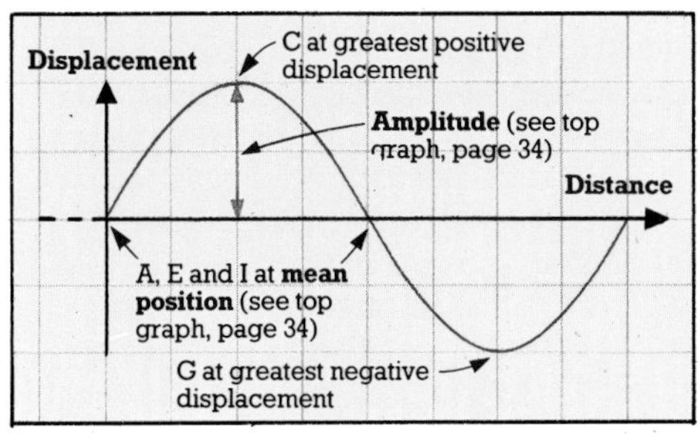

Graph of pressure or density against distance for **longitudinal wave** shows **compressions** and **rarefactions**.

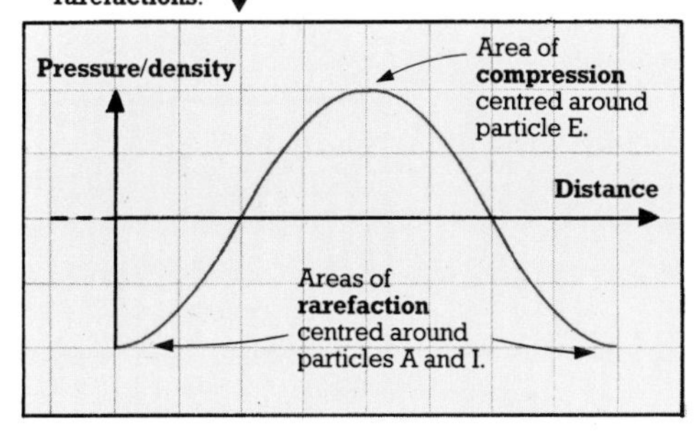

- **Compressions**. Regions along a **longitudinal wave** where the pressure and density of the molecules is higher than when no wave is passing.
- **Rarefactions**. Regions along a **longitudinal wave** where the pressure and density of the molecules is lower than when no wave is passing.

* **Damping**, 16; **Medium**, 345.

- **Wave velocity**. The distance moved by a wave in one second. It depends on the **medium*** through which the wave is travelling – the denser the medium, the slower the wave.

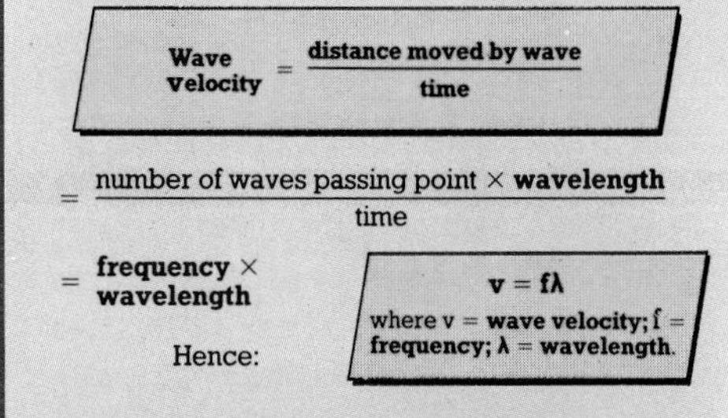

- **Frequency (f)**. The number of oscillations which occur in one second when waves pass a given point (see also page 16). It is equal to the number of **wavelengths** (see second graph, page 34) per second.
- **Attenuation**. The gradual decrease in **amplitude** (see top graph, page 34) of a wave as it passes through matter and loses energy. The amplitudes of oscillations occurring further from the source are less than those of oscillations nearer to it. This can be seen as an overall **damping***, although this term actually refers to the gradual decrease in amplitude of the oscillations of any single particle.

Graph showing **attenuated** wave

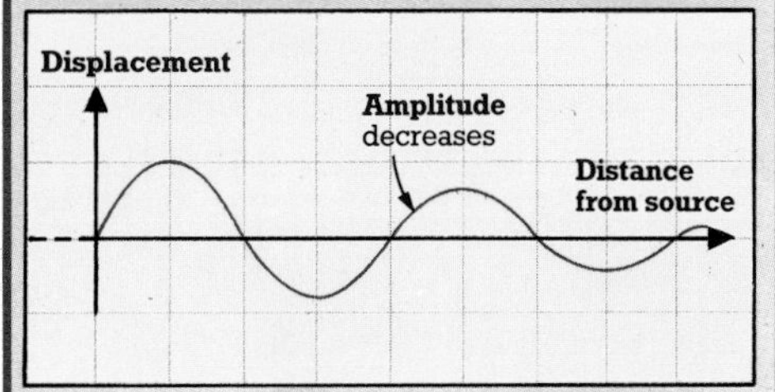

- **Wave intensity**. A measurement of the energy carried by a wave. It is worked out as the amount of energy falling on unit area per second. It depends both on the **frequency** and **amplitude** of the wave.

Reflection, refraction and diffraction

An obstacle or a change of environment causes a wave to undergo **reflection**, **refraction** or **diffraction**. These are different types of change in wave direction and often also result in changes in the shape of the **wavefronts***. For more about the reflection and refraction of light waves, see pages 46-53.

Light source

Ripple tank

Drop of water produces circular wavefronts (straight wavefronts produced by moving paddle with straight, flat surface).

Sponge beach – absorbs wave energy, stopping waves reflecting back off side of tank.

Shadows of ripples

Barriers and other devices are put into tank to produce changes in wave direction.

- **Ripple tank**. A tank of water used to demonstrate the properties of water waves.

- **Reflection**. The change in direction of a wave due to its bouncing off a boundary between two **media***. A wave that has undergone reflection is called a **reflected wave**. The shape of its wavefronts depends on the shape of the **incident wavefronts** and the shape of the boundary. For more about the reflection of light waves, see pages 46-49.

Examples of **reflected wave** shapes

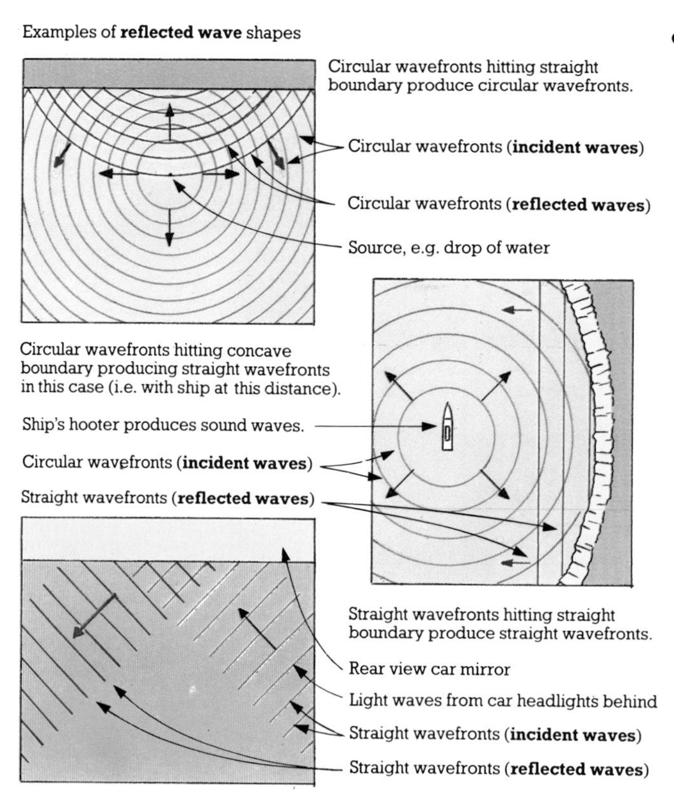

- **Incident wave**. A wave that is travelling towards a boundary between two **media***. Its wavefronts are called **incident wavefronts**.

- **Diffraction**. The bending effect which occurs when a wave meets an obstacle or passes through an aperture. The amount the wave bends depends on the size of the obstacle or aperture compared to the **wavelength*** of the wave. The smaller the obstacle or aperture by comparison, the more the wave bends.

* **Medium**, 345; **Wavefront**, **Wavelength**, 34.

- **Refraction**. The change in direction of a wave when it moves into a new **medium*** which causes it to travel at a different velocity. A wave which has undergone refraction is called a **refracted wave**. Its **wavelength*** increases or decreases with the change in velocity, but there is no change in **frequency***. For more about the refraction of light waves, see pages 50-53.

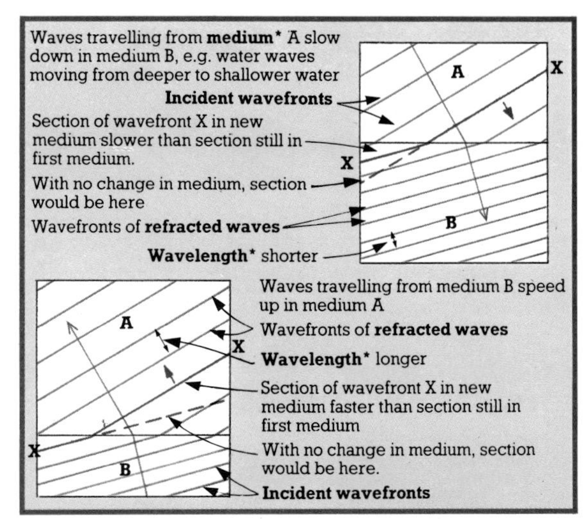

Other examples:

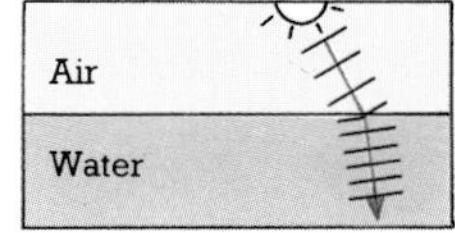

Waves slow down on entering a denser medium.

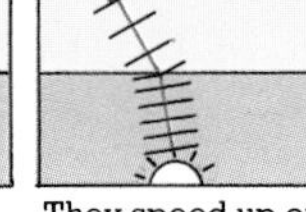

They speed up on entering a less dense medium.

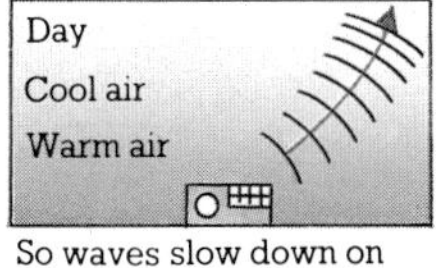

So waves slow down on entering a cooler medium ▶

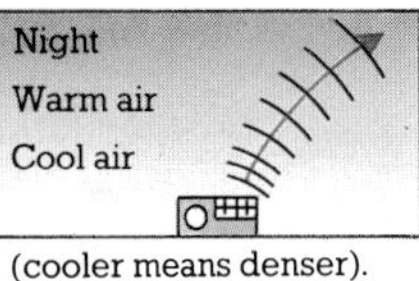

(cooler means denser). They speed up on entering a warmer medium.

- **Refractive index (n)**. A number which indicates the power of **refraction** of a given **medium*** relative to a previous medium. It is found by dividing the velocity of the **incident wave** in the first medium by the velocity of the **refracted wave** in the given medium (subscript numbers are used – see formula). The **absolute refractive index** of a medium is the velocity of light in that medium compared to the velocity of light in a vacuum (or, generally, in air). For more about refractive index and light, see page 50.

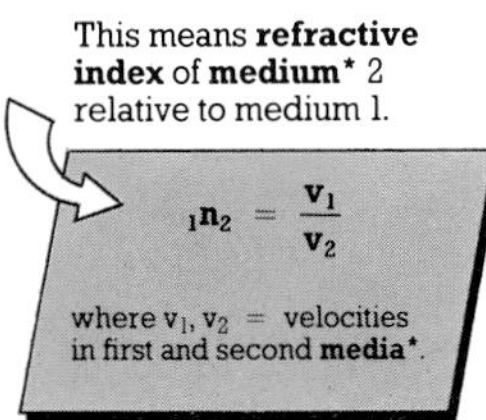

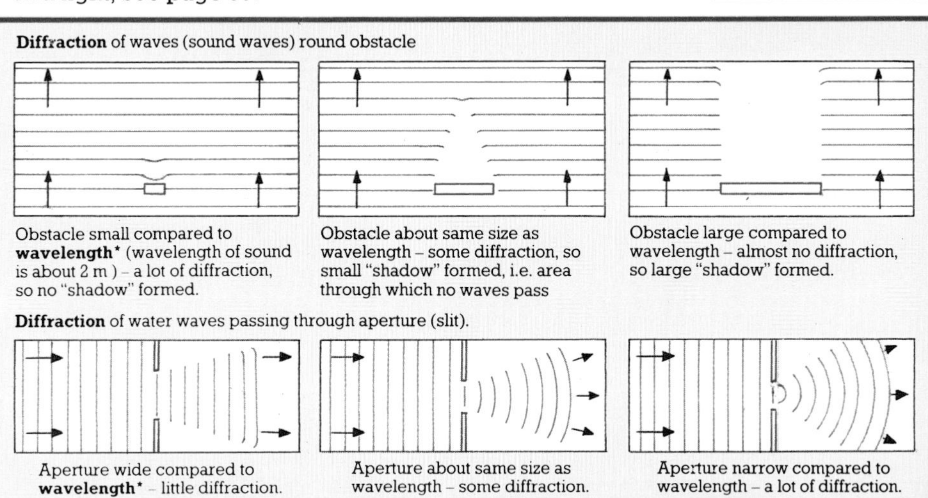

Diffraction of waves (sound waves) round obstacle

Obstacle small compared to **wavelength*** (wavelength of sound is about 2 m) – a lot of diffraction, so no "shadow" formed.

Obstacle about same size as wavelength – some diffraction, so small "shadow" formed, i.e. area through which no waves pass

Obstacle large compared to wavelength – almost no diffraction, so large "shadow" formed.

Diffraction of water waves passing through aperture (slit).

Aperture wide compared to **wavelength*** – little diffraction.

Aperture about same size as wavelength – some diffraction.

Aperture narrow compared to wavelength – a lot of diffraction.

* **Frequency**, 35; **Medium**, 345; **Wavelength**, 34.

Wave interference

When two or more waves travel in the same or different directions in a given space, variations in the size of the resulting disturbance occur at points where they meet (see **principle of superposition**, page 39). This effect is called **interference**. When interference is demonstrated, e.g. in a **ripple tank***, sources which produce **coherent waves** are always used, i.e. waves with the same wavelength and frequency, and either **in phase** or with a constant **phase difference** (see **phase**). This ensures that the interference produces a regular, identifiable **interference pattern** of disturbance (see picture, page 39). The use of non-coherent waves would result only in a constantly-changing confusion of waves.

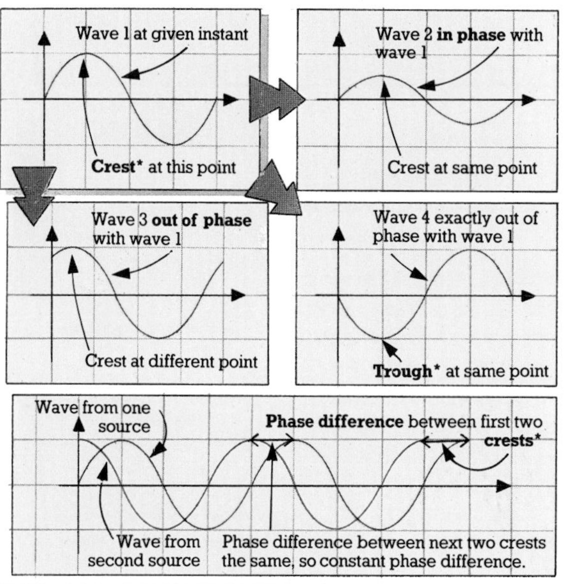

- **Phase**. Two waves are **in phase** if they are of the same frequency and corresponding points are at the same place in their oscillations (e.g. both at **crests***) at the same instant. They are **out of phase** if this is not the case, and exactly out of phase if their displacements are exactly opposite (e.g. a crest and a **trough***). The **phase difference** between two waves is the amount, measured as an angle, by which a point on one wave is ahead of or behind the corresponding point on the other. For waves exactly out of phase, the phase difference is 180°; for waves in phase, it is 0°.

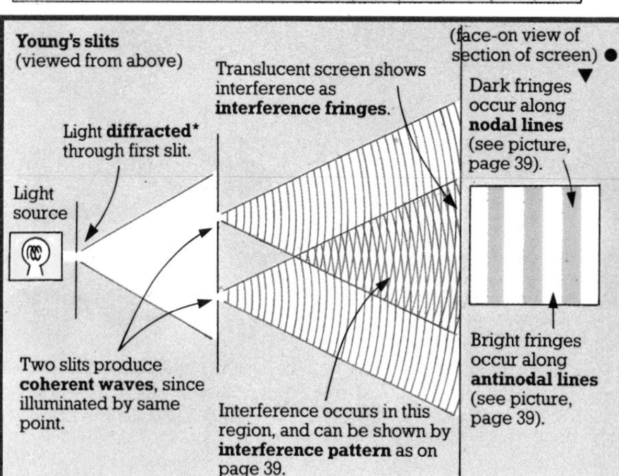

- **Young's slits**. An arrangement of narrow, parallel slits, used to create two sources of **coherent** light (see introduction). They are needed because coherent light waves cannot be produced (for studying interference) as easily as other coherent waves, as light wave emission is usually random. The interference of the light **diffracted*** through the slits is seen on a screen as light and dark bands called **interference fringes**.

* **Crests**, 34; **Diffraction**, **Ripple tank**, 36; **Troughs**, 34.

- **Principle of superposition**. States that, when the **superposition** of two or more waves occurs at a point (i.e. two or more waves come together), the resultant displacement is equal to the sum of the displacements (positive or negative) of the individual waves.

- **Constructive interference**. The increase in disturbance (reinforcement) which results from the **superposition** of two waves which are **in phase** (see **phase**).

- **Destructive interference**. The decrease in disturbance which results from the **superposition** of two waves which are exactly **out of phase**.

Constructive interference

First wave Coincides with

Amplitude* A1

B1

Second wave **in phase** with first.

A2 B2

Resultant wave has greater amplitude.

A3 B3

If two waves of amplitude A1 coincide in phase, resultant amplitude is double original amplitude.

According to **principle of superposition**, **A1** + **A2** = **A3**.
This is true of displacement at any other point, e.g. **B1** + **B2** = **B3**.

Destructive interference

First wave Coincides with

Amplitude* A1

B1

Second wave exactly **out of phase** with first.

A2 B2

Resultant wave has smaller amplitude.

A3 B3

If two waves of amplitude A1 coincide exactly out of phase, resultant amplitude is zero.

- **Nodes** or **nodal points**. Points at which **destructive interference** is continually occurring, and which are consequently regularly points of minimum disturbance, i.e. points where **crest*** meets **trough*** or **compression*** meets **rarefaction*** A **nodal line** is a line consisting entirely of nodes.

- **Antinodes** or **antinodal points**. Points at which **constructive interference** is continually occurring, and which are consequently regularly points of maximum disturbance, i.e. points where two **crests***, **troughs***, **compressions*** or **rarefactions*** meet. An **antinodal line** is a line consisting entirely of antinodes.

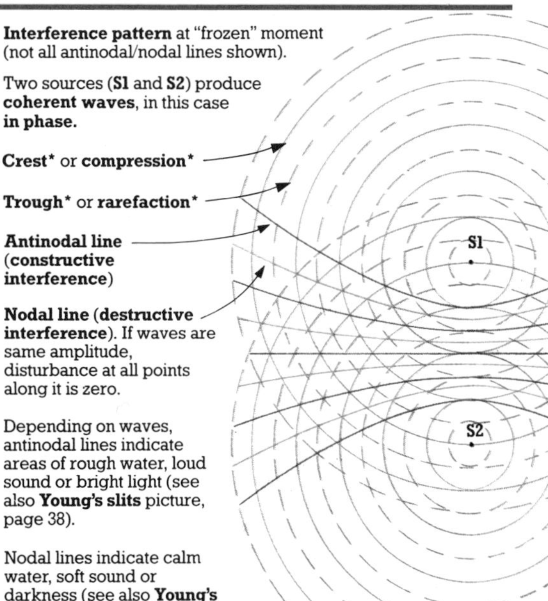

* **Amplitude**, 34; **Compressions**, 35; **Crests**, 34; **Rarefactions**, 35; **Troughs**, 34.

Sound waves

Sound waves, also called **acoustic waves**, are **longitudinal waves*** – waves which consist of particles oscillating along the same line as the waves travel, creating areas of high and low pressure (**compressions*** and **rarefactions***). They can travel through solids, liquids and gases and have a wide range of **frequencies***. Those with frequencies between about 20 and 20,000 **Hertz*** (the **sonic range**) can be detected by the human ear and are what is commonly referred to as sound (for more about perception of sound, see pages 42-43). Others, with higher and lower frequencies, are known as **ultrasound** and **infrasound**. The study of the behaviour of sound waves is called **acoustics**.

- **Ultrasound**. Sound composed of **ultrasonic waves** – waves with **frequencies*** above the range of the human ear, i.e. above 20,000 **Hertz***. They have a number of uses.

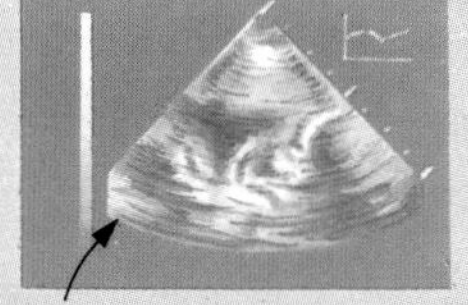

Scan of baby in mother's womb

Ultrasound is used in **ultrasound scanning** of the human body (uses **echoes** – see page 41).

Bone, fat and muscle all reflect **ultrasonic waves** differently. Reflected waves (**echoes**) converted into electrical pulses which form an image (scan) on a screen.

- **Infrasound**. Sound composed of **infrasonic waves** – waves with **frequencies*** below the range of the human ear, i.e. below 20 **Hertz***. At present they have few technical uses, as they can be distressing to humans.

Behaviour of sound waves

- **Speed of sound**. The speed at which sound waves move. It depends on the type and temperature of the **medium*** through which they travel. The speed of sound waves travelling through dry air at 0°C is 331 $m\ s^{-1}$, but this increases if the air temperature increases.
- **Subsonic speed**. A speed below the **speed of sound** in the same **medium*** and under the same conditions.
- **Supersonic speed**. A speed above the **speed of sound** in the same **medium*** and under the same conditions.

Ordinary commercial aeroplanes fly at **subsonic speeds**.

Concorde can fly at **supersonic speed**.

- **Sonic boom**. A loud bang heard when a **shock wave** (see below right) produced by an aeroplane moving at **supersonic speed** passes a listener.

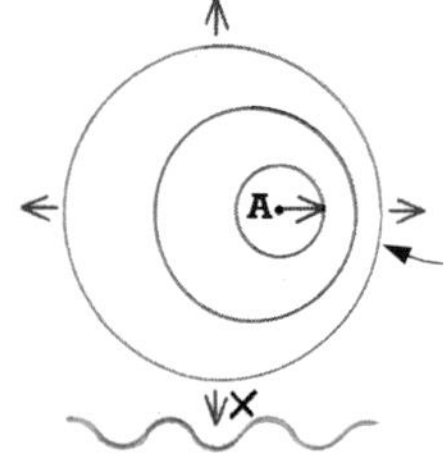

As aeroplane (A) travels forward, it creates **longitudinal waves*** in air, i.e. areas of high and low pressure (**compressions*** and **rarefactions***).

Wavefronts* can "get away" from aeroplane and begin to disperse.

Listener at X will hear waves as sound (a "whoosh" of air – as well as separate sound of engines).

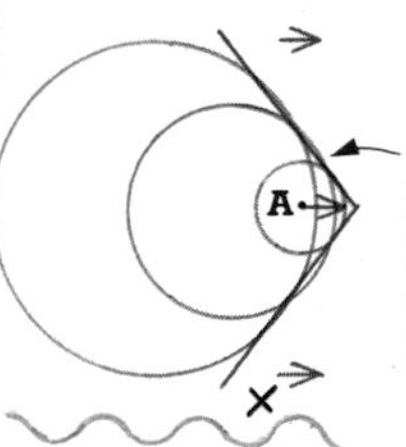

Supersonic aeroplane (A) overtakes its wavefronts while creating more, so they overlap.

Causes large build up of pressure (**shock wave**) pushed in front of aeroplane and unable to "get away", like bow wave of ship (if ship travelling faster than water waves it creates).

Listener at X will hear wave as sudden loud **sonic boom**.

* **Compressions**, **Frequency**, 35; **Hertz**, 16 (**Frequency**); **Longitudinal waves**, 35; **Medium**, 345; **Rarefactions**, 35; **Wavefront**, 34.

- **Echo**. A sound wave which has been reflected off a surface, and is heard after the original sound. Echoes, normally those of **ultrasonic waves**, are often used to locate objects and determine their exact position (by measuring the time the echo takes to return to the source). This technique has a number of names, each normally used in a slightly different context, though the distinctions between them are unclear.

Sonar (derived from **so**und **na**vigation and **r**anging)

Ultrasonic waves emitted by device below ship.

Waves reflected back by submarine (**echoes**).

Echoes picked up by sensing equipment below ship. Converted to electrical pulses which form an image of submarine on a screen in the ship.

Ultrasound scanning (see page 40) is one example. Others are **echo-sounding** and **sonar**, both of which have marine connotations (echo-sounding normally refers to using echoes to measure the depth of water below a ship, sonar to using them to detect objects under water). **Echolocation** usually describes the way animals use echoes to find prey or avoid obstacles in the dark.

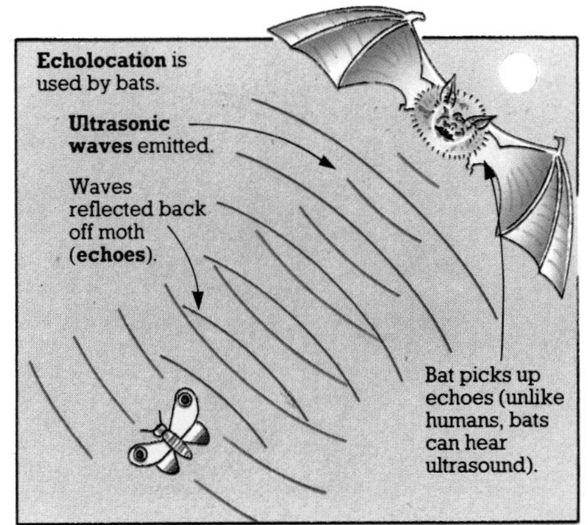

- **Reverberation**. The effect whereby a sound appears to persist for longer than it actually took to produce. It occurs when the time taken for the **echo** to return to the source is so short that the original and reflected wave cannot be distinguished. If the wave is reflected off many surfaces, then the sound is enhanced further.

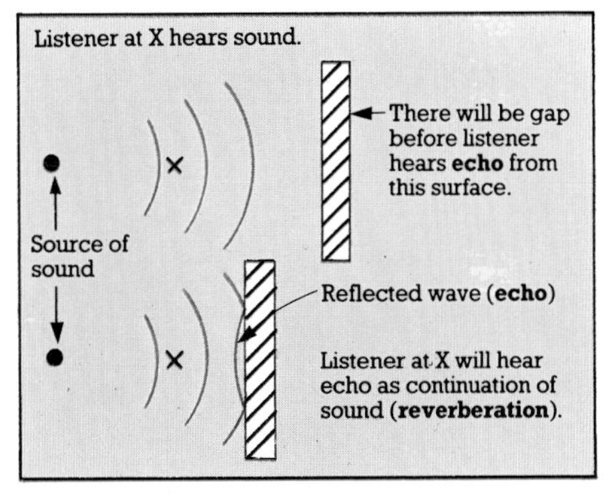

- **Doppler effect**. The change in **frequency*** of the sound heard when either the listener or the source moves relative to the other. If the distance between them is decreasing, a higher frequency sound is heard than that actually produced. If it is increasing, a lower frequency sound is heard.

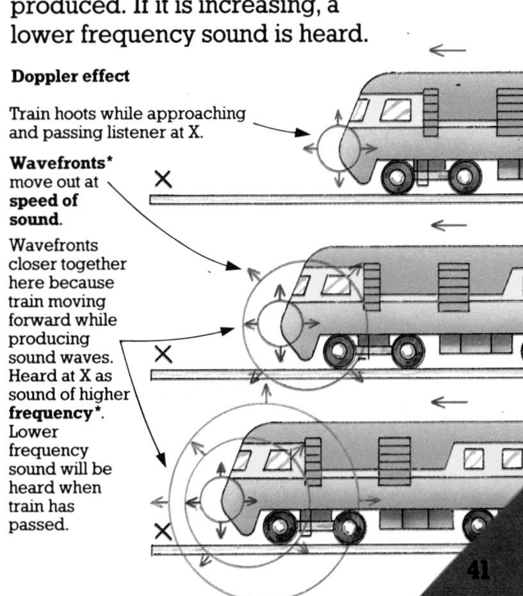

* **Frequency**, 35; **Wavefront**, 34.

Perception of sound

Sounds heard by the ear can be pleasant or unpleasant. When the waveform of a **sound wave** (see pages 40-41) repeats itself regularly, the sound is usually judged to be pleasant. However, when it is unrepeated and irregular, the sound is thought of as a **noise**. Every sound has a particular **loudness** and **pitch** and many, especially musical sounds, are produced by **stationary waves**.

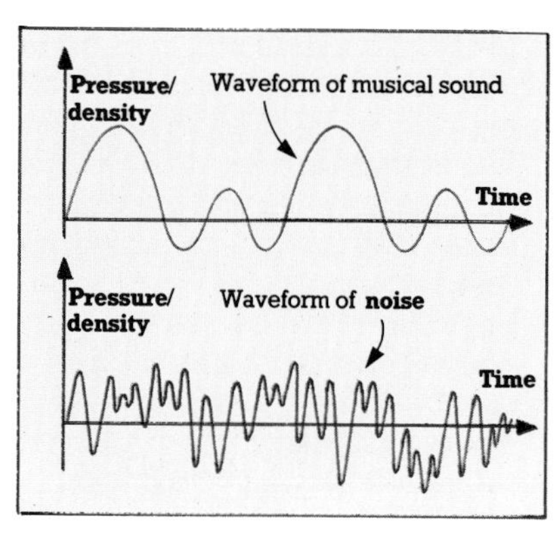

- **Loudness**. The size of the sensation produced when sound waves fall on the ear. It is subjective, depending on the sensitivity of the ear, but is directly related to the **wave intensity*** of the waves. It is most often measured in **decibels** (**dB**), but also, more accurately, in **phons** (these take into account the fact that the ear is not equally sensitive to sounds of all **frequencies***).

- **Pitch**. The perceived **frequency*** of a sound wave, i.e. the frequency heard as sound. A high pitched sound has a high frequency and a low pitched sound a low frequency.

- **Beats**. The regular variation in **loudness** with time which is heard when two sounds of slightly different **frequency*** are heard together. It is a result of **interference*** between the two waves. The **beat frequency** is equal to the difference in frequency between the two sounds (see picture). The closer together the frequency of the sounds, the slower the beats.

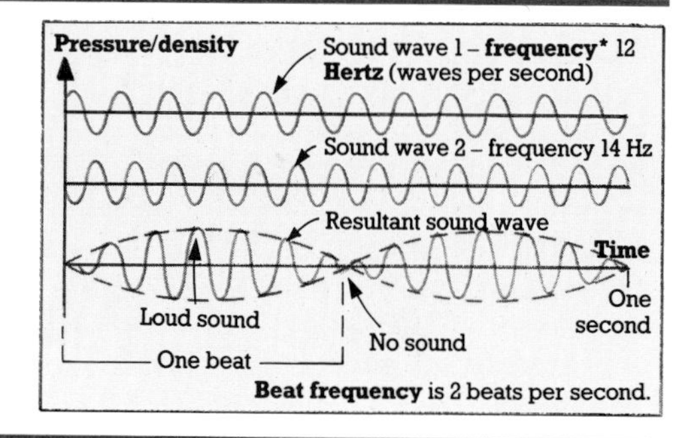

Beat frequency is 2 beats per second.

Musical sounds

All music is based on some kind of **musical scale**. This is a series of **notes** (sounds of specific **pitch**), arranged from low to high pitch with certain **intervals** between them (a musical interval is a spacing in **frequency***, rather than time). The notes are arranged so that the maximum number of pleasant sounds can be obtained. What is regarded as a pleasant sound depends on culture – Eastern music is based on a different scale to Western music.

Western **musical scale** is based on **diatonic scale** – consists of 8 **notes** (white notes on piano) ranging from lower to upper C.

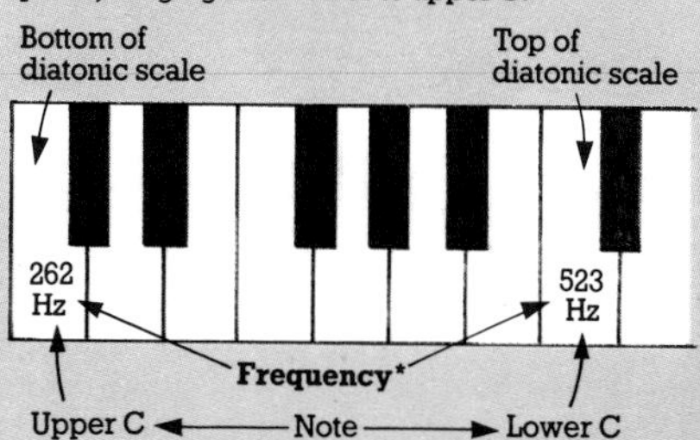

Black notes have **frequencies*** between those of notes on **diatonic scale**. Together with these, they form **chromatic scale**.

* **Frequency**, 35; **Interference**, 38; **Wave intensity**, 35.

- **Stationary** or **standing wave**. A wave that does not appear to move. It is not in fact a true wave, but is instead made up of two waves of the same velocity and **frequency*** continuously moving in opposite directions between two fixed points (most commonly the ends of a plucked string or wire). The repeated crossing of the waves results in **interference*** – when the waves are **in phase***, the resultant **amplitude*** is large, and when they are **out of phase***, it is small or zero. At certain points (the **nodes**), it is always zero. The amplitude and frequency of a stationary wave in a string or wire determines those of the sound waves it produces in the air – the length and tension of the string or wire determines the range of frequencies, and hence the pitch of the sound produced.

Formation of **stationary wave**

Displacement

A

B

Waves A and B (same **amplitude***) moving in opposite directions between two fixed points.

A + B

Resultant wave at time t = 0. Amplitude doubled.

A

B

Resultant wave at time t = 1. Amplitude zero.

A + B

At time t = 2, resultant wave has same amplitude as at t = 0, but is transposed, i.e. **crests*** where there were **troughs*** and vice versa.

Nodes

Antinodes (see below)

Resultant wave continuously moving rapidly between position held at t = 0 and that held at t = 2. Wave observed appears stationary.

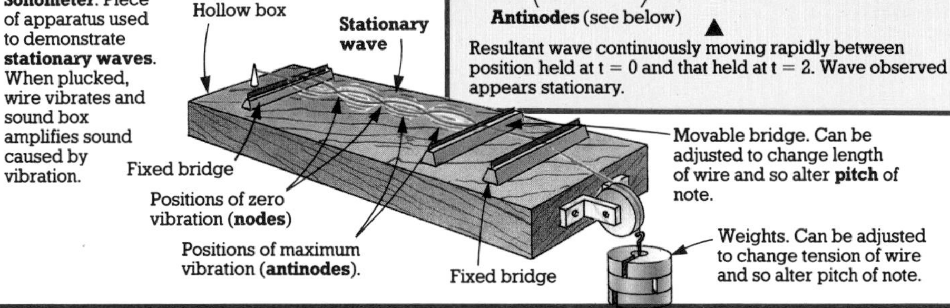

Sonometer. Piece of apparatus used to demonstrate **stationary waves**. When plucked, wire vibrates and sound box amplifies sound caused by vibration.

- **Modes of vibration**. The same note played on different instruments, although recognizable as the same note, has a quite distinct sound quality (**timbre**), characteristic to the instrument. This is because, although the main, strongest vibration is the same for each note whatever the instrument (its **frequency*** is the **fundamental frequency**), vibrations at other frequencies (**overtones**) are also produced at the same time. The set of vibrations specific to an instrument are its modes of vibration.

Lowest **mode of vibration** (**fundamental frequency**) of note on given instrument.

If frequencies of **overtones** are simple multiples of fundamental frequency, also called **harmonics**.

1st overtone (2nd harmonic, i.e. frequency doubled. Note fundamental is 1st harmonic)

2nd overtone (note this is 4th harmonic, i.e. in this case there is no 3rd harmonic)

Combined modes of vibration (i.e. all three put together). Characteristic waveform of note for this instrument.

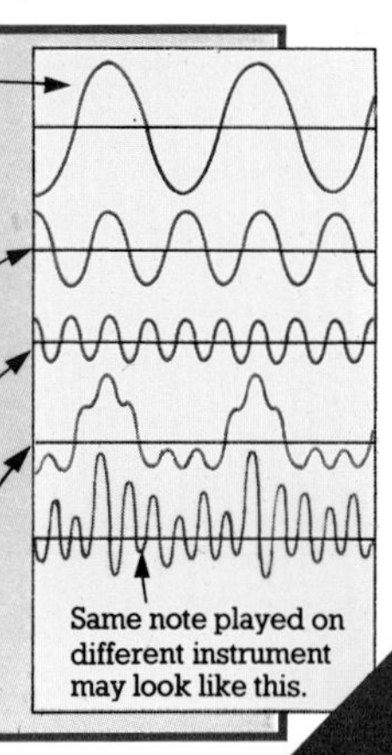

* **Amplitude**, **Crests**, 34; **Frequency**, 35; **In phase**, 38 (**Phase**), **Interference**, 38; **Out of phase**, 38 (**Phase**); **Troughs**, 34.

Electromagnetic waves

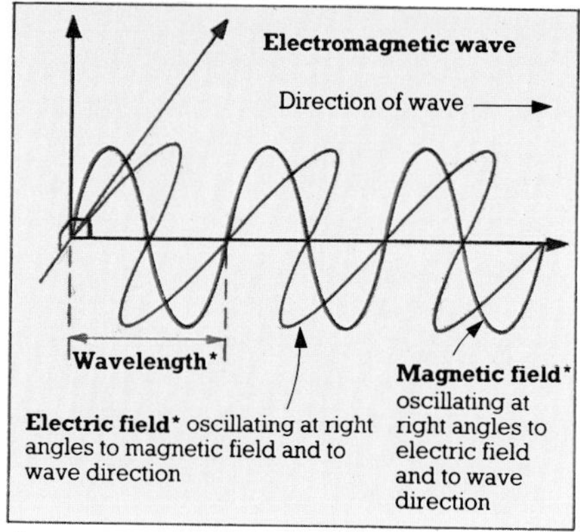

Electromagnetic waves are **transverse waves***, consisting of oscillating **electric** and **magnetic fields***. They have a wide range of **frequencies***, can travel through all **media***, including vacuums, and, when absorbed, cause a rise in temperature (see **infra-red radiation**). **Radio waves** and some **X-rays** are emitted when free **electrons*** are accelerated or decelerated, e.g. as a result of a collision. All other types occur when electrons move between shells (see page 85). All except radio waves occur as random pulses called **photons** (see **quantum theory**, page 85), rather than a continuous stream.

Electromagnetic spectrum (range of electromagnetic waves). Five main sections are **wavebands**, i.e. particular ranges of **frequencies* and wavelengths*** within which the waves all have the same characteristic properties.

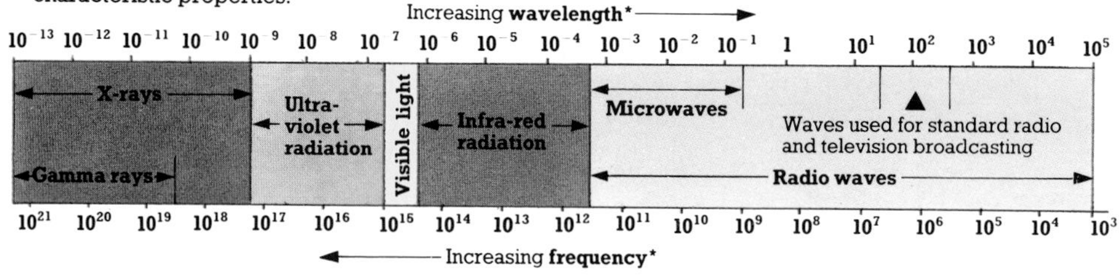

- **Gamma rays** (**γ-rays**). Electromagnetic waves emitted by **radioactive*** substances (see also page 86). They are in the same **waveband** and have the same properties as **X-rays**, but are produced in a different way and are at the top end of the band with regard to energy.

X-radiography produces pictures (**radiographs**) of inside of body. **X-rays** pass through tissue but absorbed by denser bones, so bones appear opaque on film.

Human elbow

- **X-rays** or **Röntgen rays**. Electromagnetic waves which **ionize*** gases they pass through, cause **phosphorescence** and bring about chemical changes on photographic plates. They are produced in **X-ray tubes*** and have many applications.

- **Ultra-violet radiation** (**UV radiation**). Electromagnetic waves produced, for example, when an electric current is passed through **ionized*** gas between two **electrodes***. They are also emitted by the sun, but only small quantities reach the earth's surface (most "lose" their energy by ionizing atoms in the atmosphere). These small quantities are vital to life, but larger amounts are dangerous. Ultra-violet radiation causes **fluorescence**, e.g. when produced in **fluorescent tubes***, and also a variety of chemical reactions, e.g. tanning of the skin.

* **Electric field**, 58; **Electrode**, 66; **Electrons**, 83; **Fluorescent tube**, 80 (**Discharge tube**); **Frequency**, 35; **Ionization**, 88; **Magnetic field**, 72; **Medium**, 345; **Radioactivity**, 86; **Transverse waves**, **Wavelength**, 34; **X-ray tube**, 81.

- **Phosphorescence**. A phenomenon shown by certain substances (**phosphors**). When they are hit by short **wavelength*** electromagnetic waves, e.g. **gamma rays** or **X-rays**, they absorb these and emit **visible light**, i.e. waves of longer wavelength. This emission may continue after the gamma or X-rays have stopped. If it only occurs briefly afterwards in rapid flashes, these are called **scintillations** (see also **scintillation counter**, page 90).

- **Infra-red radiation** (**IR radiation**). Electromagnetic waves produced by hot objects. They produce by far the largest temperature rise of any electromagnetic waves as they are most easily absorbed (see introduction and **radiation**, page 29). They can be used to form **thermal images** on special infra-red sensitive film, exposed by different degrees of heat, rather than light.

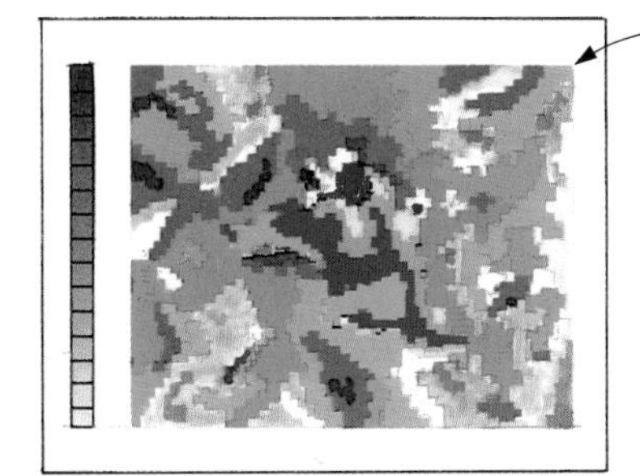

Thermal image of lake water used as coolant for **nuclear power station*** (taken from a satellite).

Warmest water brown, coolest blue. Each colour represents about one degree temperature difference.

- **Fluorescence**. A phenomenon shown by certain substances. When they are hit by **ultra-violet radiation**, they absorb this and emit **visible light**, i.e. light waves of a longer **wavelength***. This emission stops as soon as the ultra-violet radiation stops.

- **Visible light**. Electromagnetic waves which the eye can detect. They are produced by the sun, by **discharge tubes*** and by any substance heated until it glows (emission of light due to heating is called **incandescençe**). They cause chemical changes, e.g. on photographic film, and the different **wavelengths*** in the **waveband** are seen as different colours (see page 55).

- **Microwaves**. Very short **radio waves** used in **radar** (**ra**dio **d**etection **a**nd **r**anging) to determine the position of an object by the time it takes for a reflected wave to return to the source (see also **sonar**, page 41). **Microwave ovens** use microwaves to cook food rapidly.

Magnetron connected to normal electricity supply produces **microwaves**.

Microwaves pass through food container, but not metal lining of oven. Cause water, fat or sugar molecules in food to oscillate, creating heat energy and cooking the food.

Waves reflected off oven walls, so cook food evenly.

Microwave oven

- **Radio waves**. Electromagnetic waves produced when free **electrons*** in radio aerials are made to oscillate (and are hence accelerated) by an **electric field***. The fact that the frequency of the oscillations is imposed by the field means that the waves occur as a regular stream, rather than randomly. They are used to communicate over distances.

Radio waves with short **wavelengths*** can penetrate ionosphere, hence are used to communicate over long distances via satellites.

Radio waves with long **wavelengths*** reflected within ionosphere, hence are used to transmit information from place to place on same area of earth's surface.

Ionosphere (region of **ionized*** gas around the earth)

Earth

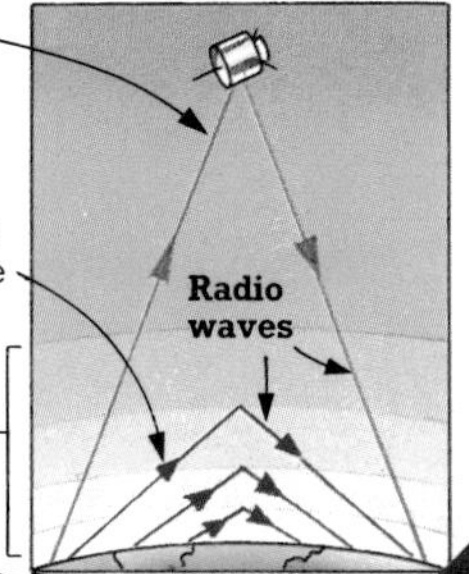

* **Discharge tube**, 80; **Electric field**, 58; **Electrons**, 83; **Ionization**, 88; **Nuclear power station**, 94; **Wavelength**, 34.

Light

Light consists of **electromagnetic waves*** of particular **frequencies*** and **wavelengths*** (see pages 44-45), but is commonly referred to and diagrammatically represented as taking the form of **rays**. Such a ray is actually a line (arrow) which indicates the path taken by the light waves, i.e. the direction in which the energy is being carried.

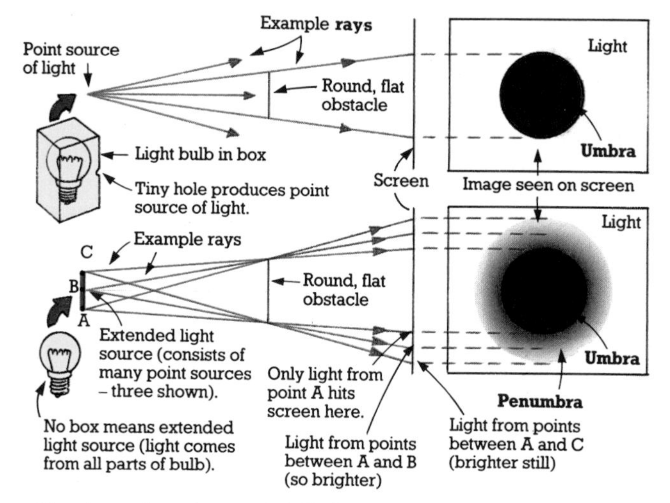

- **Shadow**. An area which light rays cannot reach due to an obstacle in their path. If the rays come from a point they are stopped by the obstacle, creating a complete shadow called an **umbra**. If they come from an extended source, a semi-shadow area called a **penumbra** is formed around the umbra.

Reflection of light

Reflection is the change in direction of a wave when it bounces off a boundary (see page 36). Mirrors are usually used to show the reflection of light (see right and also pages 48-49). It must be noted that when an object and its image are drawn in mirror (and **lens***) diagrams, the object is assumed to be producing light rays itself. In fact the rays come from a source, e.g. the sun, and have already been reflected once, i.e. off the object.

Incident ray. Ray of light before reflection (or **refraction***).

Angle of incidence (i). Angle between **incident ray** and **normal** at **point of incidence**.

Point of incidence. Point at which **incident ray** meets boundary and becomes **reflected ray** (or **refracted ray***).

Normal. Line at right angles to boundary through chosen point, e.g. **point of incidence**.

Angle of reflection (r). Angle between **reflected ray** and **normal** at **point of incidence**.

i

r

Reflected ray

Laws of reflection of light

1. The **reflected ray** lies in the same plane as the **incident ray** and the **normal** at the **point of incidence**.

2. The **angle of incidence** (i) = the **angle of reflection** (r)

- **Regular reflection**. The reflection of parallel **incident rays** (see above) off a flat surface such that all the **reflected rays** are also parallel. This occurs when surfaces are very smooth, e.g. highly polished surfaces such as mirrors.

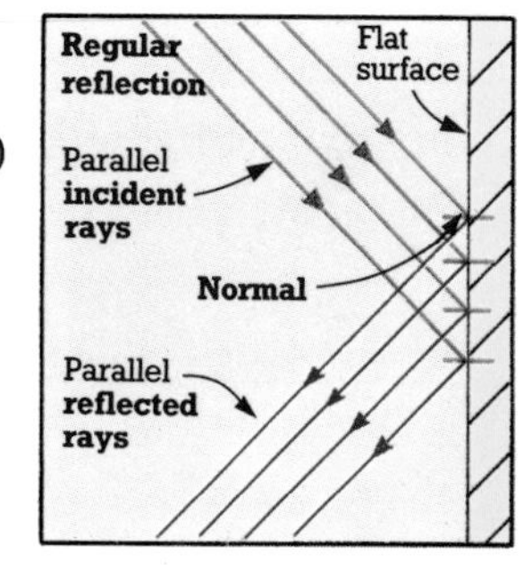

* **Electromagnetic waves**, 44; **Frequency**, 35; **Lenses**, 52; **Refracted ray**, **Refraction**, 50; **Wavelength**, 34.

- **Eclipse**. The total or partial "blocking off" of light from a source. This occurs when an object casts a **shadow** by passing between the source and an observer. A **solar eclipse** is seen from the earth when the moon passes between the earth and the sun, and a **lunar eclipse** is seen when the earth is between the sun and the moon.

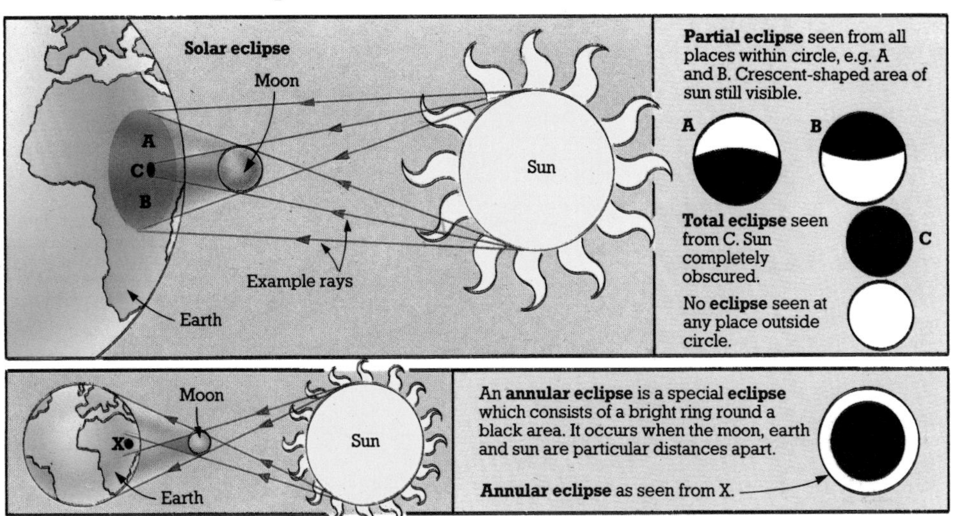

- **Diffuse reflection**. The reflection of parallel **incident rays** (see left) off a rough surface such that the **reflected rays** travel in different directions and the light is scattered. This is the most common type of reflection as most surfaces are irregular when considered on a scale comparable to that of the **wavelength*** of light (see page 44).

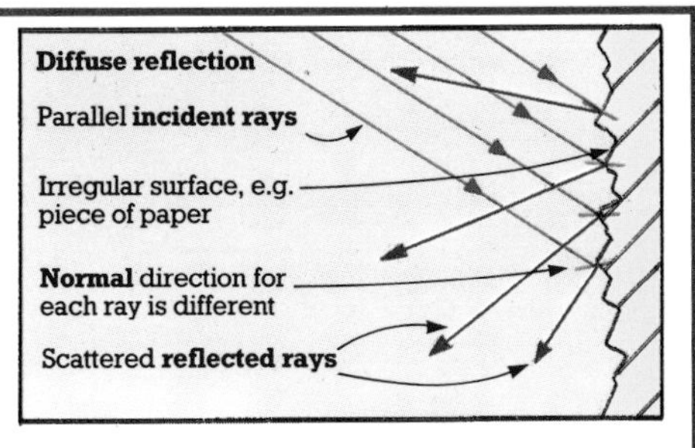

- **Plane mirror**. A mirror with a flat surface (see also **curved mirrors**, pages 48-49). The image it forms is the same size as the object, the same distance behind ("inside") the mirror as the object is in front, and **laterally inverted** (the left and right sides have swapped round).

Reflection in **plane mirror**

Image distance = object distance

Image seen where reflected rays appear to come from (**virtual image***).

Image distance

Object distance

Incident ray A

Reflected ray A seen here. Eye assumes it has come from point 1.

1

2

Rays hitting at right angles are reflected back along same line

Plane mirror

Reflected ray B seen here. Eye assumes it has come from point 2.

Incident ray B

Object (from which light rays are assumed to have originated – see reflection of light, page 46).

- **Parallax**. The apparent movement of two objects relative to one another, as seen by a moving observer. For example, two objects at different distances from a moving train appear to move past each other to someone looking at them from the train window – the further away of the two objects appearing to move faster and hence a greater distance.

* **Virtual image**, 49 (**Image**); **Wavelength**, 34.

Reflection of light (continued)

Light rays are reflected from curved surfaces, as from flat surfaces, according to the **laws of reflection of light** (see page 46). The images formed by reflection from **curved mirrors** are particularly easily observed. There are two types of curved mirror – **concave** and **convex mirrors**. For all diagrams showing reflection of light, the object is assumed to be the source of the light (see **reflection of light**, page 46) and certain points (see below), together with known facts about light rays passing through them, are used to construct the paths of the reflected rays.

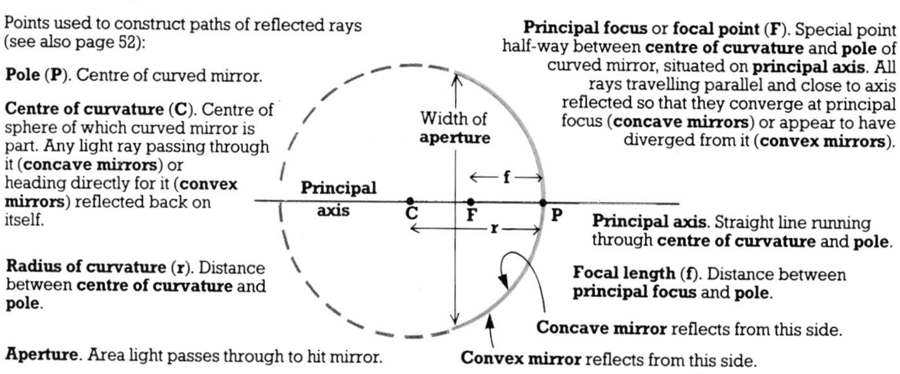

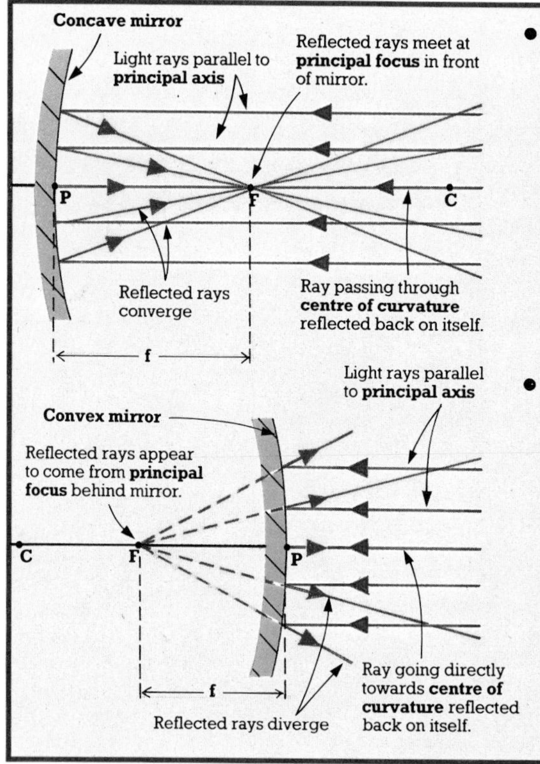

- **Concave** or **converging mirror.** A mirror with a reflecting surface which curves inward (part of the inside of a sphere). When light rays parallel to the **principal axis** fall on such a mirror, they are reflected so that they converge at the **principal focus** in front of the mirror. The size, position and type of **image** formed depends on how far the object is from the mirror.

- **Convex** or **diverging mirror**. A mirror with a reflecting surface which curves outwards (part of the outside of a sphere). When light rays parallel to the **principal axis** fall on such a mirror, they are reflected so that they appear to diverge from the **principal focus** behind ("inside") the mirror. The **images** formed are always upright and reduced, and **virtual images** (see **image**).

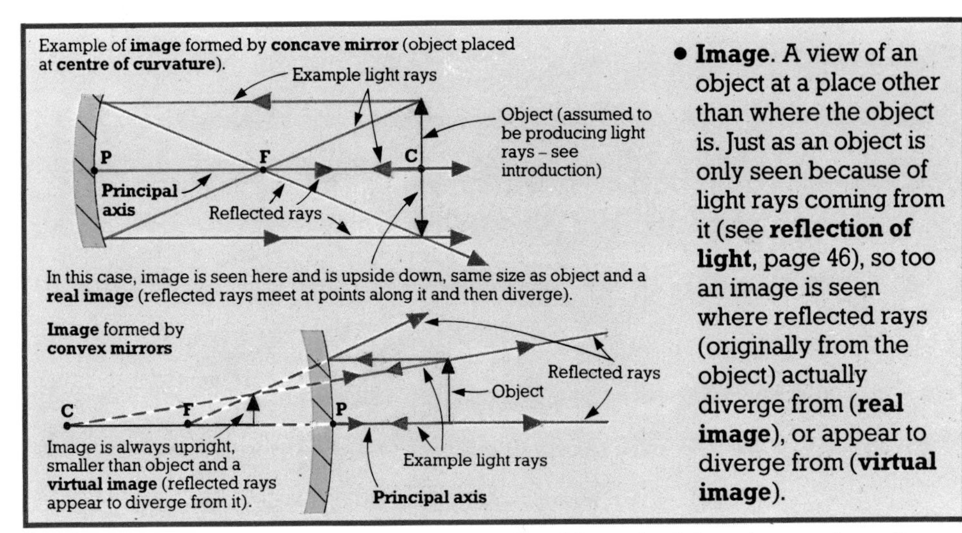

- **Image**. A view of an object at a place other than where the object is. Just as an object is only seen because of light rays coming from it (see **reflection of light**, page 46), so too an image is seen where reflected rays (originally from the object) actually diverge from (**real image**), or appear to diverge from (**virtual image**).

- **Mirror** or **lens formula**. Gives the relationship between the distance of an object from the centre of a curved mirror or **lens***, the distance of its **image** from the same point and the **focal length** of the mirror or lens. An image may be formed either side of a mirror or lens, so a **sign convention*** is used to give position.

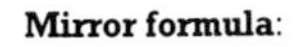

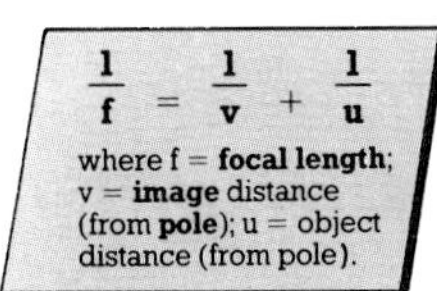

$$\frac{1}{f} = \frac{1}{v} + \frac{1}{u}$$

where f = **focal length**; v = **image** distance (from **pole**); u = object distance (from pole).

Real is positive sign convention

1. All distances are measured from the mirror as origin.
2. Distances of objects and **real images** are positive.
3. Distances of **virtual images** are negative.

- **Linear magnification**. The ratio of the height of the image formed by a mirror or **lens*** to the object height.

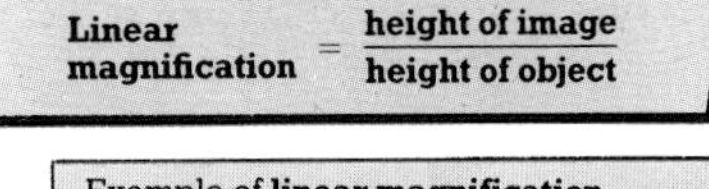

$$\textbf{Linear magnification} = \frac{\textbf{height of image}}{\textbf{height of object}}$$

- **Principle of reversibility of light**. States that, for a ray of light on a given path due to reflection, **refraction*** or **diffraction***, a ray of light in the opposite direction in the same conditions will follow the same path. Light rays parallel to the **principal axis**, for example, are reflected by a **concave mirror** to meet at the **principal focus**. If a point source of light is placed at the principal focus, the rays are reflected parallel to the axis.

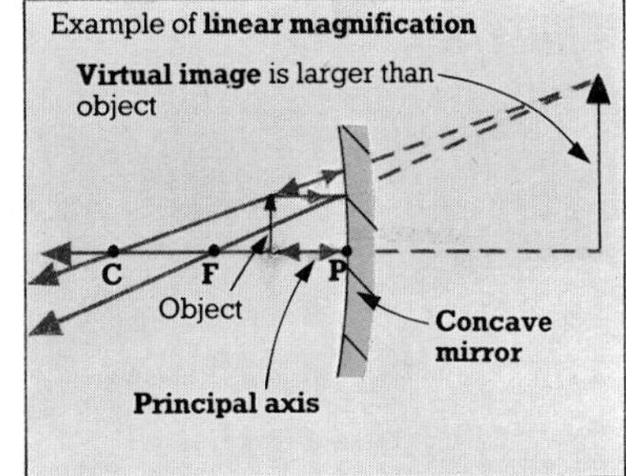

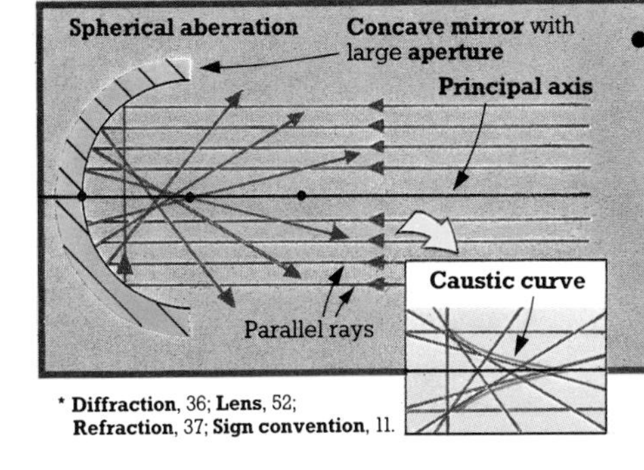

- **Spherical aberration**. An effect seen when rays parallel to the **principal axis** (and different distances from it), hit a curved mirror and are reflected so that they intersect at different points along the axis, forming a **caustic curve**. The larger the **aperture**, the more this is seen. It is also seen in **lenses*** with large apertures.

* **Diffraction**, 36; **Lens**, 52; **Refraction**, 37; **Sign convention**, 11.

Refraction of light

Refraction is the change in direction of any wave as a result of its velocity changing when it moves from one **medium*** into another (see also page 37). When light rays (see page 46) move into a new medium, they are refracted according to the **laws of refraction of light**. The direction in which they are refracted depends on whether they move into a denser or less dense medium and are consequently slowed down or speeded up (see diagram below).

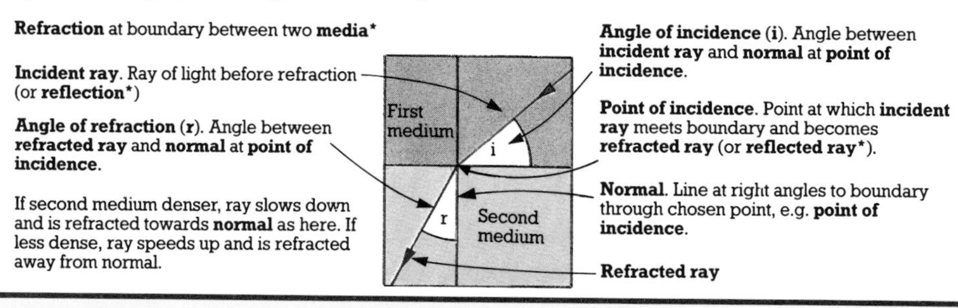

Laws of refraction of light

1. The **refracted ray** lies in the same plane as the **incident ray** and **normal** at the **point of incidence**.

2. (**Snell's law**). The ratio of the **sine*** of the **angle of incidence** to the sine of the **angle of refraction** is a constant for two given **media***. This constant is the **refractive index** (**n** – see page 37). When referring to light, this is also known as the **optical density** and, as with refractive index in other cases, can also be calculated by dividing the velocity of light in one medium by its velocity in the second medium. See also **apparent depth** picture.

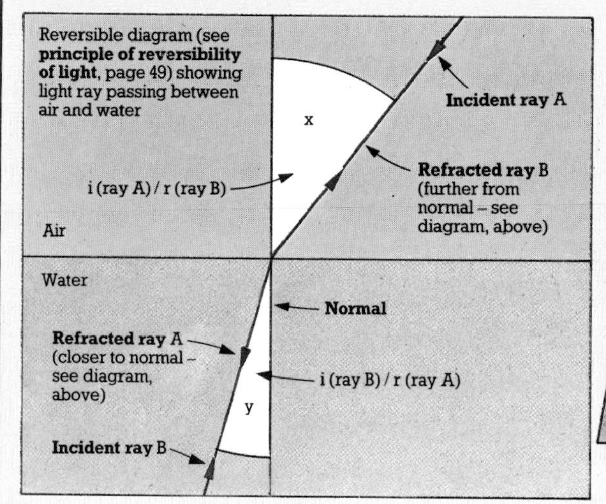

For either direction, **refractive index*** of second **medium*** relative to first is written $_1n_2$.

According to **Snell's law**:

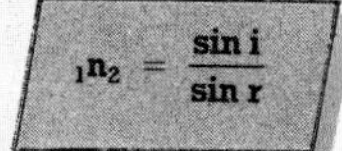

Media may also be specified by subscript letters – $_an_w$ means refractive index of water relative to air and $_wn_a$ means that of air relative to water.

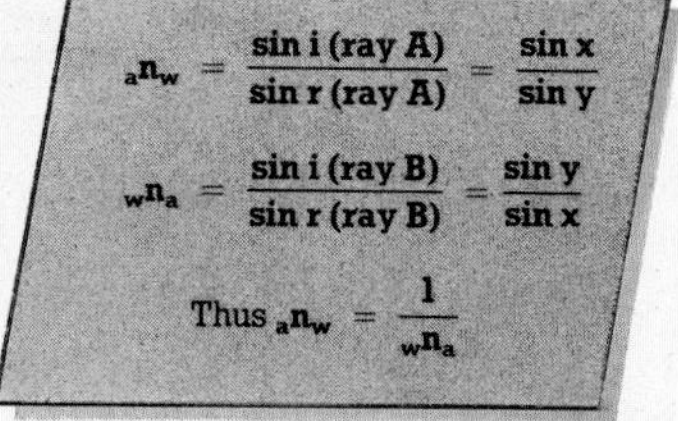

If no subscripts are given, the value is the **absolute refractive index***.

* **Absolute refractive index**, 37 (**Refractive index**); **Medium**, 345; **Reflected ray**, **Reflection**, 46; **Sine**, 345.

- **Apparent depth**. The position at which an object in one **medium*** appears to be when viewed from another medium. The brain assumes the light rays have travelled in a straight line, but in fact they have changed direction as a result of refraction. Hence the object is not actually where it appears to be.

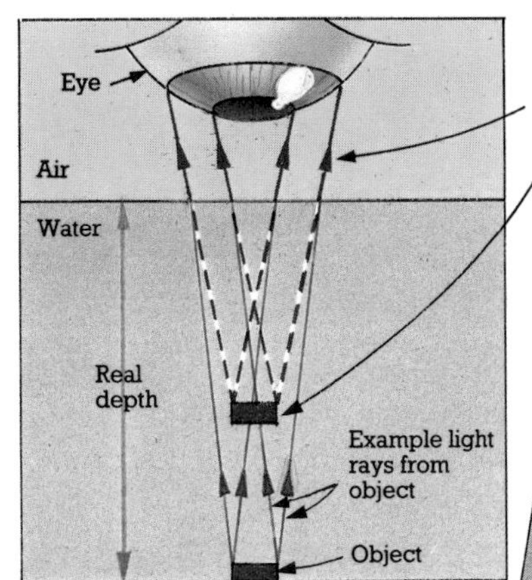

Apparent depth

Rays refracted when they leave water

Brain assumes light rays have taken straight paths (dotted lines), so object seen here.

Real depth and apparent depth can also be used to calculate **refractive index***:

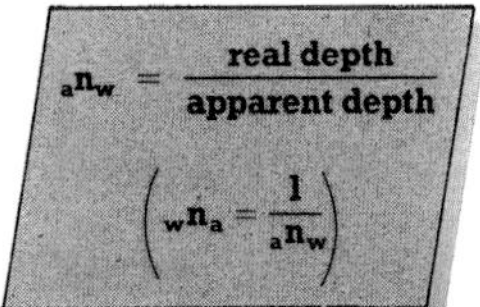

- **Critical angle (c)**. The particular **angle of incidence** of a ray hitting a less dense **medium*** which results in it being refracted at 90° to the **normal**. ▶ This means that the refracted ray (**critical ray**) travels along the boundary, and does not enter the second medium.

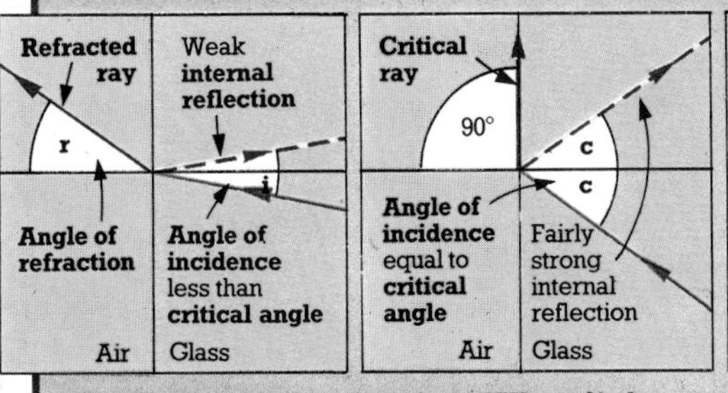

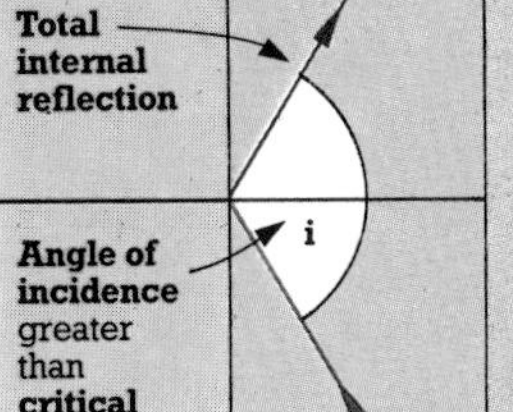

Critical angle can also be used to calculate **refractive index***:

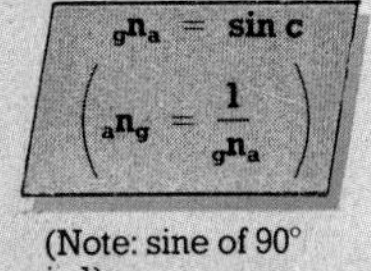

(Note: sine of 90° is 1)

- **Total internal reflection**. When light travelling from a dense to a less dense **medium*** hits the boundary between them, some degree of reflection back into the denser medium (**internal reflection**) always accompanies ▶ refraction. When the **angle of incidence** is greater than the **critical angle**, total internal reflection occurs, i.e. all the light is internally reflected.

Optical fibres transmit light by **total internal reflection**. Bundles of such fibres have a number of uses, e.g. in communications and in medicine (e.g. in **endoscopes**, used by doctors to see inside the body)

Bundle of fibres

Glass fibre

Light rays

Outer layer of less dense glass

Angle of incidence greater than **critical angle**, so **total internal reflection** occurs.

- **Prism**. A transparent solid which has two plane refracting surfaces at an angle to each other. Prisms are used to produce **dispersion*** and change the path of light by refraction and **total internal reflection**.

Prism refracting light ray

Angle of deviation, i.e. angle between ray entering prism and ray emerging from it.

Path of ray if it had not passed through prism.

Prism causing **total internal reflection**

Angle of incidence greater than **critical angle**

Angle of deviation = 90°

* **Dispersion**, 54 (**Colour**); **Medium**, 345; **Refractive index**, 37.

Refraction of light (continued)

Light rays are refracted at curved surfaces, e.g. **lenses**, as at flat surfaces, according to the **laws of refraction of light** (see page 50). Unlike with flat surfaces, though, images are formed. There are two basic types of lens, **concave** and **convex lenses**, which can act as **diverging** or **converging lenses** depending on their **refractive index*** relative to the surrounding **medium***. For all diagrams showing image production by refraction, the object is assumed to be the light source (see **reflection of light**, page 46), and certain points (see below), together with known facts about light rays passing through them, are used to construct the paths of the refracted rays. The positions of objects and images can be determined using the **mirror (lens) formula***.

Points used to construct paths of refracted rays (see also page 48). All lenses shown are considered thin lenses (i.e. thickness of lens small compared to **focal length**). Though rays bend both on entering and emerging, they are drawn as bending only once, at a vertical line running through **optical centre**.

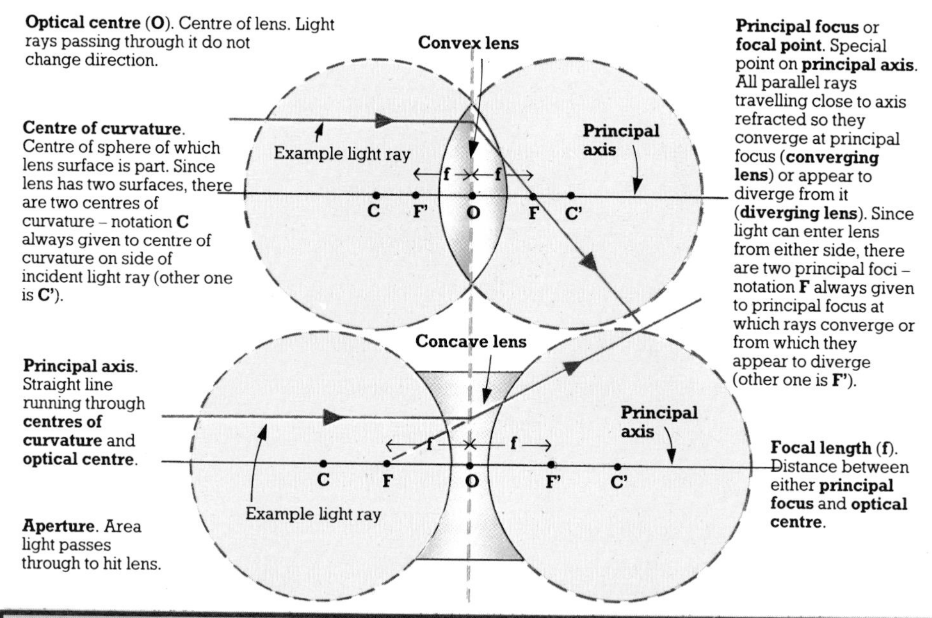

Glass **convex lens** in air acts as **converging lens**.

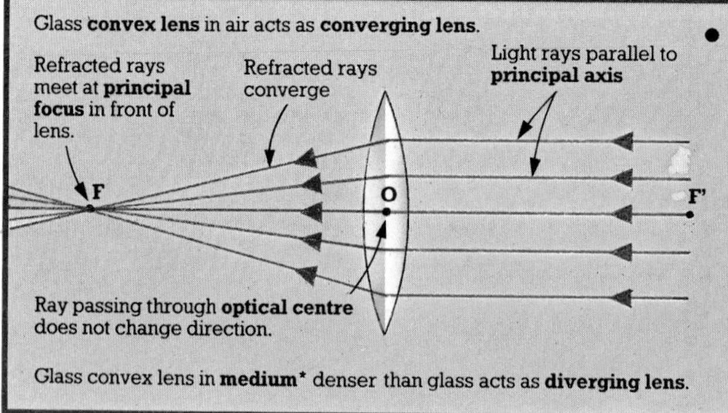

Glass convex lens in **medium*** denser than glass acts as **diverging lens**.

- **Converging lens.** A lens which causes parallel rays falling on it to converge on the **principal focus** on the other side of it. Both **concave** and **convex lenses** can act as converging lenses depending on the **refractive index*** of the lens relative to the surrounding **medium***.

* **Medium**, 345; **Mirror formula**, 49; **Refractive index**, 37.

• **Convex lens**. A lens with at least one surface curving outwards. A lens with one surface curving inwards and one outwards is convex if its middle is thicker than its outer edges (it is a **convex meniscus**). A glass convex lens in air acts as a **converging lens**. The size, position and type of image it forms (**real*** or **virtual***) depends on how far it is from the object.

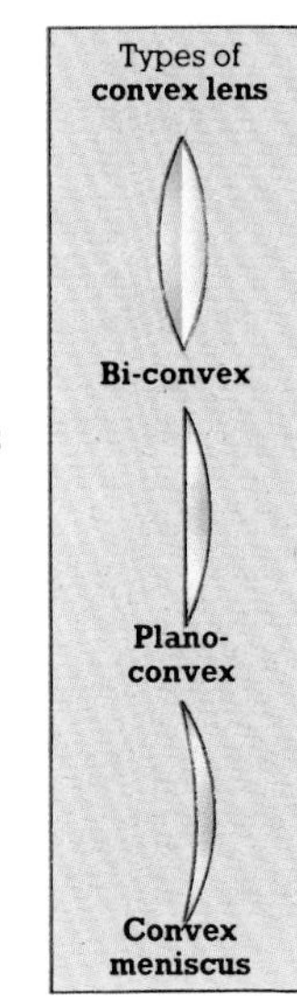

• **Concave lens**. A lens which has at least one surface curving inwards. A lens with one surface curving inwards and one outwards is concave if its middle is thinner than its outer edges (it is a **concave meniscus**). A glass concave lens in air acts as a **diverging lens**. The position of the object in relation to the lens may vary, but the image is always of the same type.

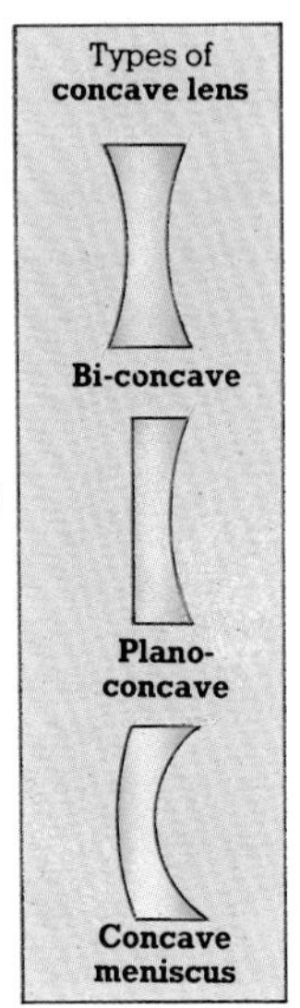

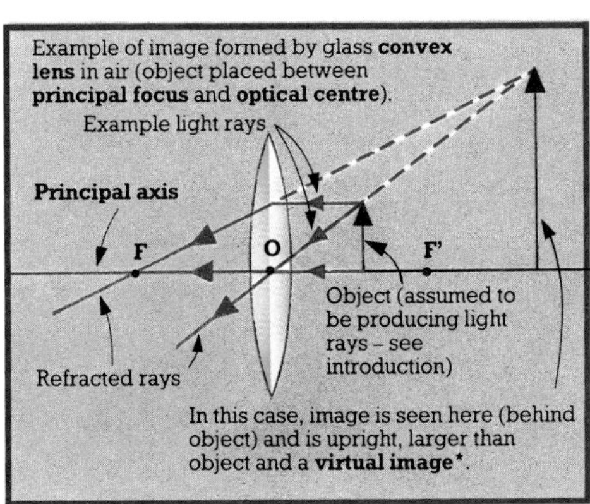

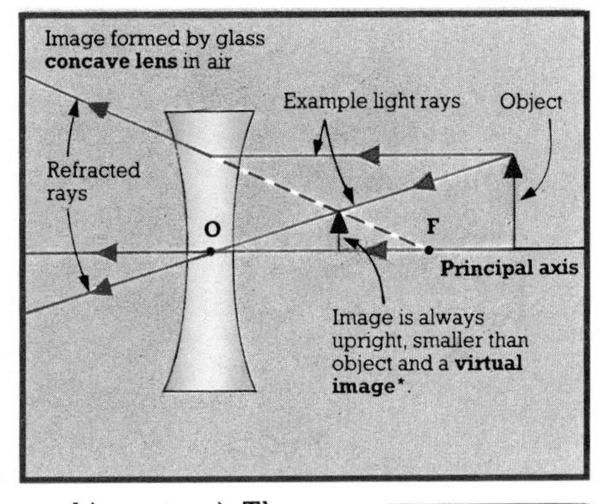

• **Power (P)**. A measure of the ability of a lens to converge or diverge light rays, given in **dioptres** (when **focal length** is measured in metres). The shorter the focal length, the more powerful the lens.

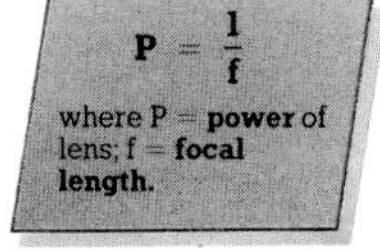

$$\mathbf{P} = \frac{1}{f}$$

where P = **power** of lens; f = **focal length.**

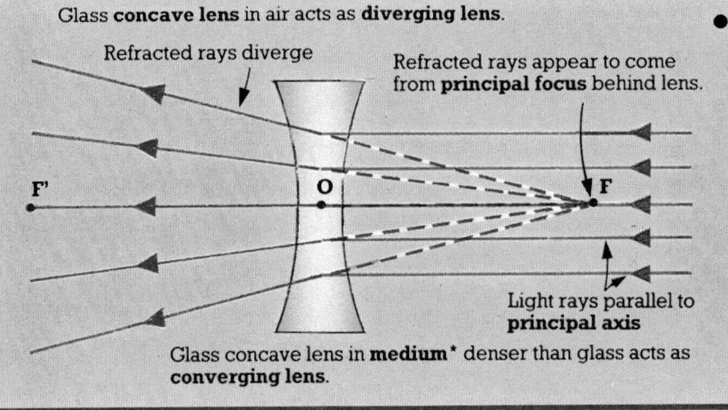

• **Diverging lens**. A lens which causes parallel rays falling on it to diverge so that they appear to have come from the **principal focus** on the same side as the rays enter. Both **concave** and **convex lenses** can act as diverging lenses, depending on the **refractive index*** of the lens relative to the surrounding **medium***.

* **Medium**, 345; **Real image**, 49 (**Image**); **Refractive index**, 37; **Virtual image**, 49 (**Image**).

Optical instruments

An **optical instrument** is one which acts on light, using one or more **lenses*** or **curved mirrors*** to produce a required type of image. Listed below are some of the more common optical instruments.

- **Camera**. An optical instrument that is used to form and record an image of an object on film. The image is inverted and a **real image***.

As before (see page 52), refraction by lenses shown as one change of direction only, this time on line through optical centre of whole lens assembly.

Camera (reflex) ▶

Prism directs light to eye.

Mirror directs light to prism and eye, so object can be seen. Flips up when picture taken.

Film. Areas hit by light undergo chemical reaction. Permanent image produced by developing film.

Shutter. Moves away as picture is taken to allow light onto film.

Diaphragm. Series of overlapping metal pieces. Adjusted to alter size of **aperture** (central hole) and hence amount of light allowed through.

Example light rays from top and bottom of object at distant point.

Lens assembly. Produces inverted image on film. Can be moved to focus on objects at different distances.

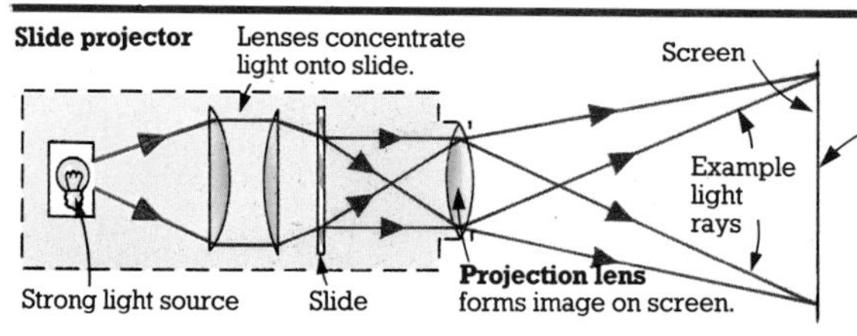

◀ • **Slide projector**. An optical instrument which produces a magnified image of a slide.

- **Microscope**. An optical instrument which magnifies very small objects. If it has only one **lens***, it is a **simple microscope** or **magnifying glass**. If it has more, it is a **compound microscope**.

Compound microscope ▼

Eyepiece lens. Produces final image seen by eye (see below). A **simple microscope** consists of this lens alone.

Image formed by objective lens (enlarged, inverted and a **real image***). Acts as object for eyepiece lens.

Objective lens

Object on transparent microscope slide

Image formed by eyepiece lens (enlarged, inverted and a **virtual image***).

Strong light source

Colour

When all the different **wavelengths*** of **visible light** (see page 45) fall on the eye at the same time, white light is seen. However, white light can also undergo **dispersion**, whereby it is split into the **visible light spectrum** (i.e. its different wavelengths) by **refraction***. This may occur accidentally (see **chromatic aberration**), or it may be produced on purpose, e.g. with a **spectrometer**.

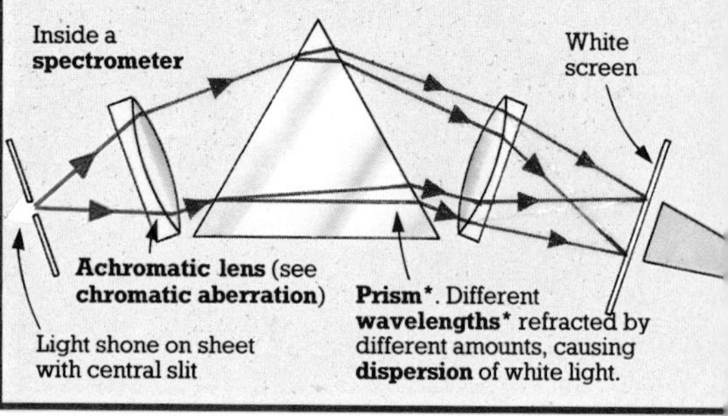

* **Curved mirrors**, 48; **Lenses**, 52; **Prism**, 51; **Real image**, 49 (**Image**); **Refraction**, 50; **Virtual image**, 49 (**Image**); **Wavelength**, 34.

- **Telescope**. An optical instrument used to make very distant (and therefore apparently very small) objects appear larger.

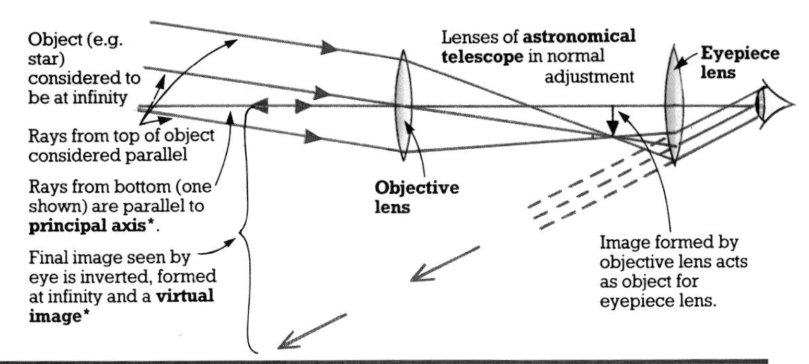

- **Visual angle**. The angle, at the eye, of the rays coming from the top and bottom of an object or its image. The greater it is, the larger the object or image appears. Optical instruments which produce magnification, e.g. **microscopes**, do so by creating an image whose visual angle is greater than that of the object seen by the unaided eye. The **angular magnification** or **magnifying power** (see below) of such an instrument is a measurement of the amount by which it does so.

$$\textbf{Angular magnification} = \frac{\textbf{visual angle of image}}{\textbf{visual angle of object}}$$

- **Chromatic aberration** or **chromatism**. The halo of colours (the **visible light spectrum** – see below, left) sometimes seen around images viewed through lenses. It results from **dispersion** (see **colour**). To avoid this, good quality optical instruments contain one or more **achromatic lenses** – each consisting of two lenses combined so that any dispersion produced by one is corrected by the other.

- **Visible light spectrum**. A display of the colours that make up a beam of white light. Each colour band represents a very small range of **wavelengths*** – see **visible light**, page 45.

Face-on view

Visible light spectrum ▼

Red
Orange
Yellow
Green
Blue
Indigo
Violet

Primary colours

△ = **secondary colours** (combinations of primary colours)

Complementary colours are any two that produce white light when mixed, e.g. red and cyan.

Red
Green
Blue
△ Yellow
△ Cyan
△ Magenta

- **Primary colours**. Red, blue and green light – colours that cannot be made by combining other coloured light. Mixed equally, they give white light. By mixing them in the right proportions, every colour in the **visible light spectrum** can be produced. Note these are the pure primary colours – those referred to in art (red, blue and yellow) only act as primary colours because the paints are impure.

- **Colour mixing**. If white light is shone onto a pure coloured filter, only light of the same colour (range of **wavelengths***) as the filter passes through (the other colours are absorbed). This is **subtractive mixing** or **colour mixing by subtraction**. If light of two different colours, filtered out in this way, is shone onto a white surface, a third colour (a mixture of the two) is seen by the eye. This is **additive mixing**, or **colour mixing by addition**.

* **Principal axis**, 52; **Virtual image**, 49 (**Image**); **Wavelength**, 34.

Static electricity

Electricity is the phenomenon caused by the presence or movement of electrically charged particles (**electrons*** or **ions***) which exert an **electric force***. A material is said to have a negative electric charge if it has a surplus of electrons, and a positive charge if it has a deficit of electrons. An electric **current** (see page 60) is the movement of electrons through materials and is therefore a transfer of charge – this can be contrasted with **static electricity**, which can be said to be electricity "held" by a material with electric charge.

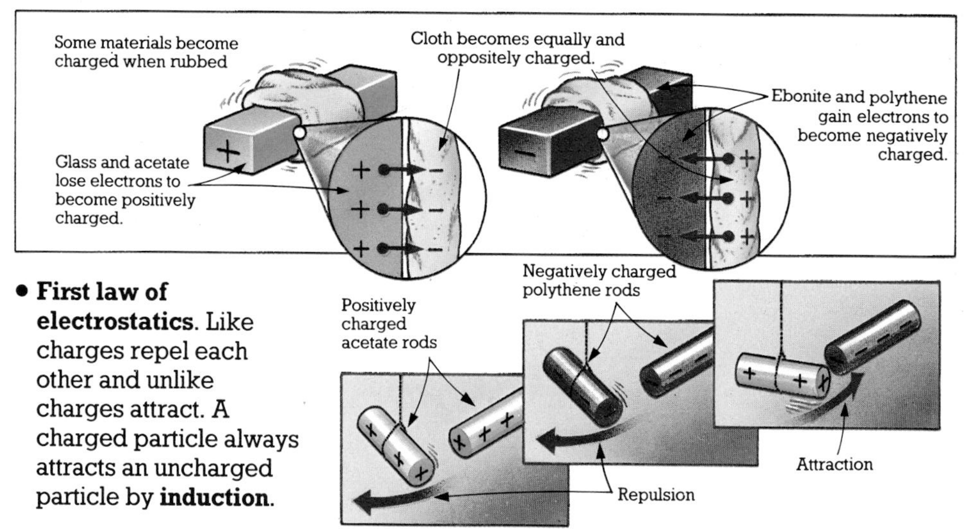

- **First law of electrostatics**. Like charges repel each other and unlike charges attract. A charged particle always attracts an uncharged particle by **induction**.

- **Conductor**. A material containing a large number of electrons which are free to move (see also **conductivity**, page 62). It can therefore **conduct** electricity (carry an electric **current** – see introduction). Good conductors are metals, e.g. copper, aluminium and gold.

- **Insulator**. A material with very few or no electrons free to move (i.e. a bad **conductor**). Some insulators become electrically charged when rubbed because electrons from the surface atoms are transferred from one substance to the next, but the charge remains on the surface.

- **Electroscope**. An instrument for detecting small amounts of electric charge. A **gold leaf electroscope** is the most common type. When the leaf and rod become charged, they repel and the leaf diverges from the rod. The greater the charges, the larger the divergence of the leaf. A **condensing electroscope** contains a **capacitor*** between the cap and the case which increases the sensitivity. ▶

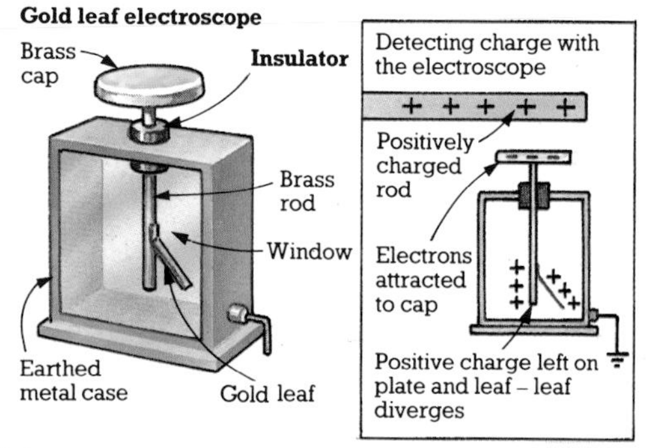

* **Capacitor**, 59; **Electric force**, 6; **Electrons**, 83; **Ions**, 88 (**Ionization**).

- **Induction** or **electrostatic induction**. A process by which a **conductor** becomes charged with the use of another charge but without contact. Generally charges are induced in different parts of an object because of repulsion and attraction. By removing one type of charge the object is left permanently charged.
- **Proof plane**. A small disc made of a **conductor** mounted on a handle made of an **insulator**. It is used to transfer charge between objects.
- **Surface density**. The amount of charge per unit area on the surface of an object. It is greater where the surface is more curved, which leads to charge being concentrated at sharp points (see **point action**). Only a sphere has constant surface density.

Charging a **conductor** by **induction**

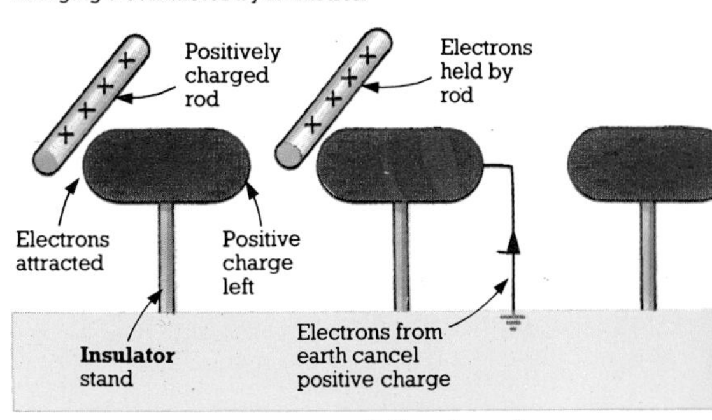

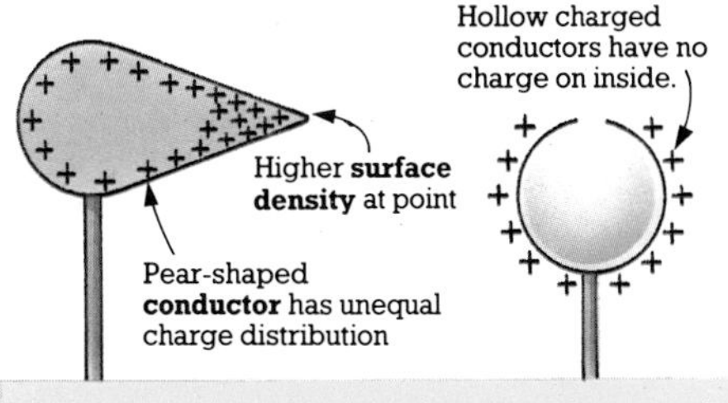

- **Point action**. The action which occurs around a sharp point on the surface of a positively-charged object. Positive ions in the air are repelled by the large charge at the point (see **surface density**). These collide with air molecules and knock off electrons to produce more positive ions which are also repelled. The result is an **electric wind** of air molecules.

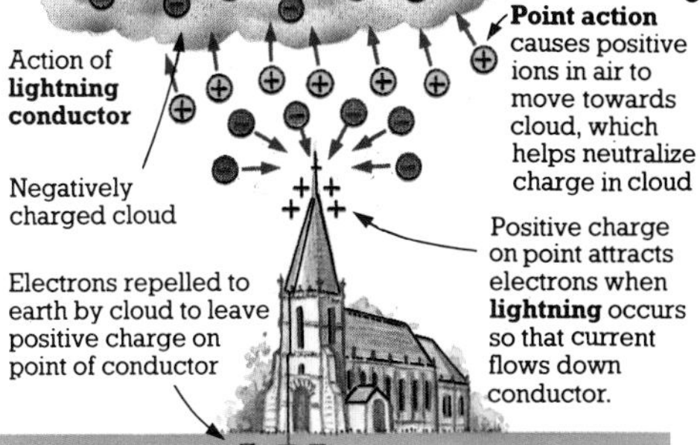

- **Lightning**. The sudden flow of electricity from a cloud which has become charged due to the rubbing together of water and air molecules. A **lightning conductor** is used to help cancel the charge on the cloud by **point action** and to conduct the flow of electricity down to earth so that it does not flow through buildings. The lightning stroke is similar to the effect in a **discharge tube***.

- **Van de Graaff generator**. A machine used to produce electric charge from **mechanical energy***. A moving band collects charge either by friction or from another electrical source and deposits it on an insulated conductor.

Van de Graaff generator

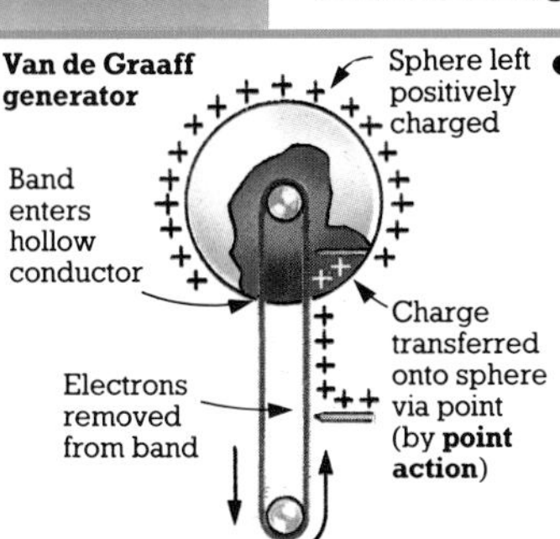

- **Electrophorus**. An instrument consisting of a negatively-charged **insulator** and a brass plate on an insulating handle. It is used to produce a number of positive charges from one negative charge.

* **Discharge tube**, 80; **Mechanical energy**, 9.

Potential and capacitance

A difference in charge between any points causes an **electric field**, i.e. a **force field*** in which charged particles experience an **electric force***. The intensity of an electric field at a point is the force on a unit positive charge at that point and the direction is that along which it would move (see also pages 104-107). Charged objects in an electric field have **potential energy*** (this is called **potential** when referring to electricity) because of their position.

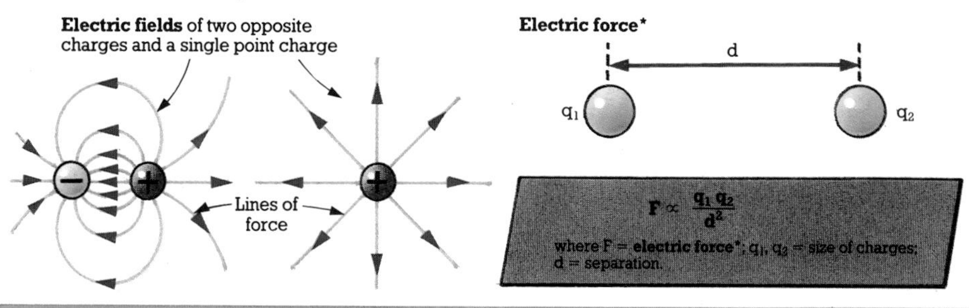

- **Potential**. The energy associated with a charge at a point in an electric field because of the force acting on it (similar to the **potential energy*** of a mass due to the **gravitational force***). The energy of a charge depends on its size and the potential of its position. A positive charge tends to move towards points of lower potential. This is moving down the **potential gradient**. Potential cannot be measured, but the **potential difference** between two points can. ▼

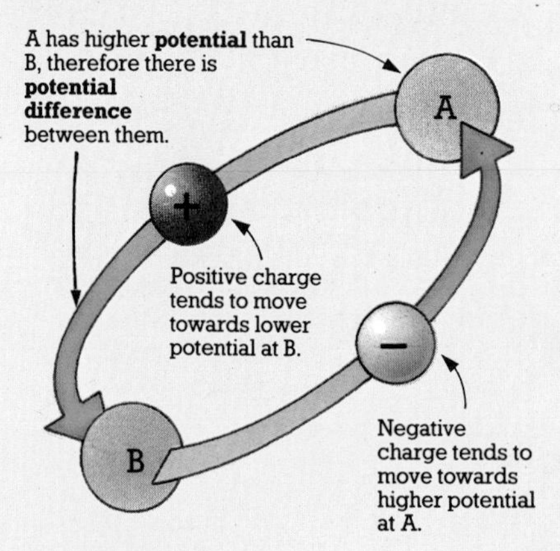

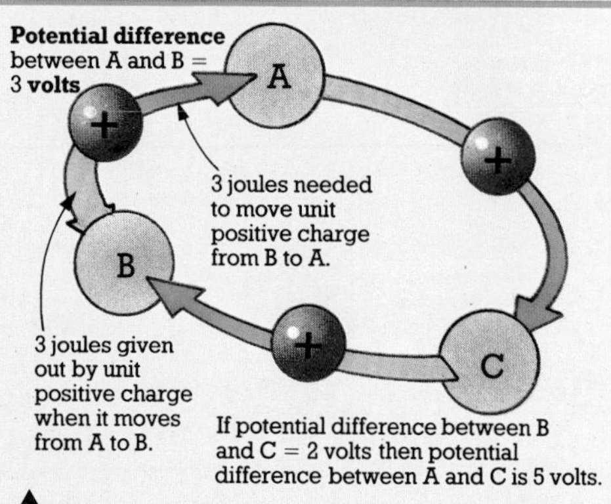

▲

- **Potential difference**. A difference in **potential** between two points, equal to the energy change when a unit positive charge moves from one place to another in an electric field. The unit of potential difference is the **volt** (potential difference is sometimes called **voltage**). There is an energy change of one joule if a charge of one **coulomb*** moves through one volt. A reference point (usually a connection to the earth) is chosen and given a potential of zero.
- **Equipotential**. A surface over which the **potential** is constant.

* **Coulomb**, 60; **Electric force, Force field, Gravitational force**, 6; **Potential energy**, 8.

Capacitance

When a **conductor*** is given a charge it undergoes a change in **potential**. **Capacitance** or **capacity** is the ratio of the charge gained by an object to its increase in potential. An object with a higher capacitance requires a larger charge to change its potential by the same amount as an object with a smaller capacitance.

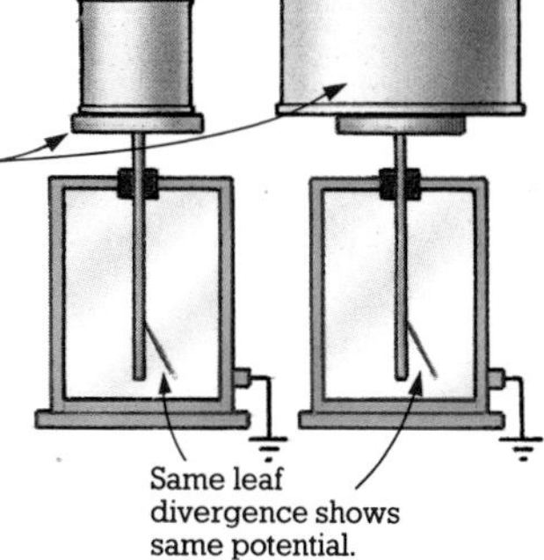

$$C = \frac{Q}{V}$$

where C = **capacitance**; Q = charge; V = **potential**.

- **Farad**. The unit of capacitance. It is the capacitance of an object whose **potential** is increased by one **volt** when given a charge of one **coulomb***.

- **Capacitor**. A device for storing electric charge, consisting of two parallel metal plates separated by an insulating material called a **dielectric**. The capacitance of a capacitor depends on the dielectric used, so one is chosen to suit the capacitance needed and the physical size required.

Capacitor

Metal plates – capacitance increases with size.

Dielectric – capacitance depends on material used.

Plate separation – capacitance increases as gap gets smaller.

- **Dielectric constant**. The ratio of the capacitance of a **capacitor** with a given **dielectric** to the capacitance of the same capacitor with a vacuum between the plates. The value is thus the factor by which the capacitance is increased by using the given dielectric instead of a vacuum (note that air is almost the same).

Variable capacitor

Vanes swivel to change area between them.

Variable capacitors are used in tuning circuits in radios.

- **Variable capacitor**. A **capacitor** consisting of two sets of interlocking vanes, often with an air **dieletric**. The size of the interlocking area is altered to change the capacitance.

- **Paper capacitor**. A **capacitor** made with two long foil plates separated by a thin waxed paper **dielectric**. **Polyester capacitors** are made in a similar way. ▶

Paper or **polyester capacitor**

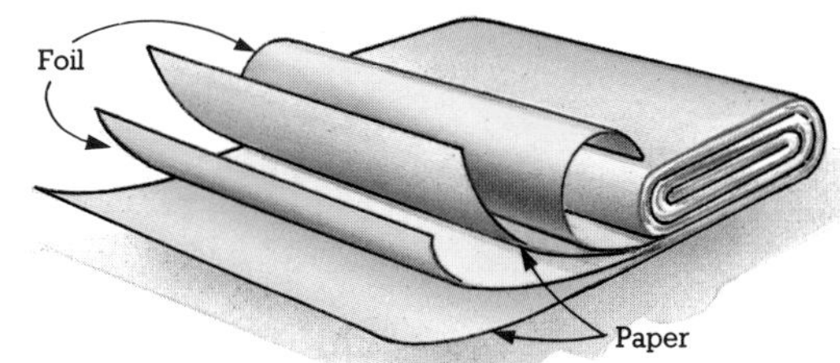

- **Electrolytic capacitor**. A **capacitor** with a paste or jelly **dieletric** which gives it a very high capacitance in a small volume. Due to the nature of the dielectric, it must be connected the correct way round to the electricity supply.

- **Leyden jar**. A **capacitor** consisting of a glass jar with foil linings inside and out. It was one of the first capacitors invented.

* **Conductor**, 56; **Coulomb**, 60.

Electric current

An electric **current** (**I**) is the flow of electrons (negatively charged particles – see page 83) through a material and thus a transfer of electric charge. Electrons only move if they are in an **electric field*** which creates a difference in **potential*** between two places. Therefore a **potential difference*** is needed to produce an electric current. A **circuit** is a closed loop, consisting of a source of current and one or more components, around which the current flows.

- **Electromotive force** ▶ (**e.m.f.**). The **potential difference*** produced by a **cell***, **battery*** or **generator***, which causes current to flow in a circuit. A source of e.m.f. has two **terminals** (where wires are connected), between which it maintains a potential difference. A **back e.m.f.** is an e.m.f. produced by a component in the circuit which opposes the main e.m.f.

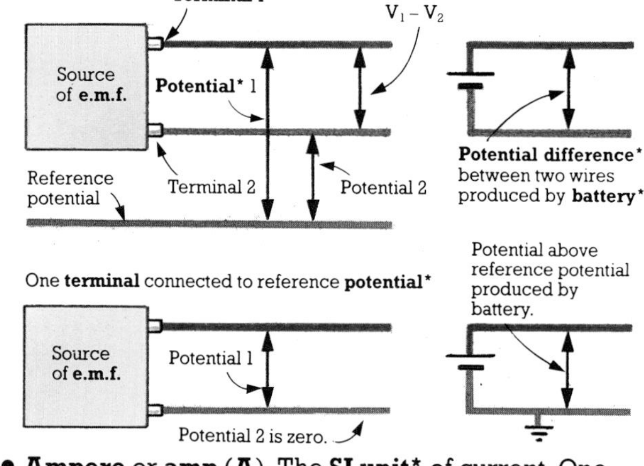

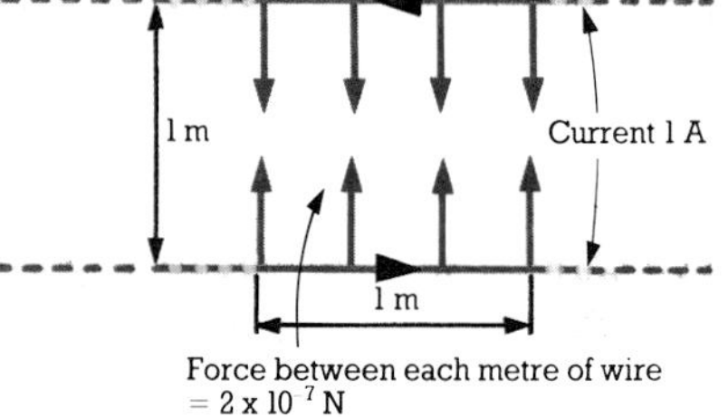

Force between each metre of wire = 2×10^{-7} N

- **Ampere** or **amp** (**A**). The **SI unit*** of current. One ampere is the current which, when flowing through two infinitely long wires one metre apart in a vacuum, produces a force of 2×10^{-7} newtons per metre of wire (see also page 96). Current is accurately measured by a **current balance**, which measures (by adapting the theory above) the force between two coils of wire through which current is flowing. **Ammeters*** are **calibrated*** using current balances.

- **Coulomb**. The **SI unit*** of electric ▶ charge. It is equal to the amount of charge which passes a point in a conductor if one **ampere** flows through the conductor for one second.

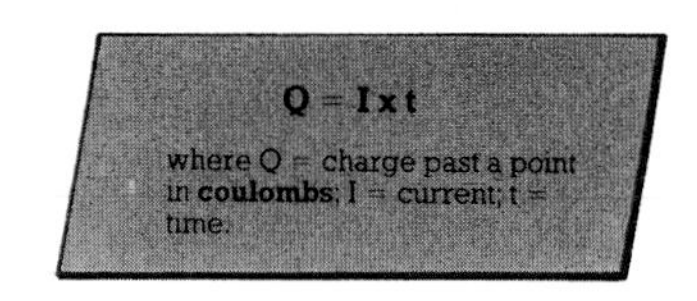

- **Direct current** (**d.c.**). Current which flows in one direction only. Originally current was assumed to flow from a point with higher **potential*** to a point with lower potential. Electrons actually flow the other way, but the convention has been kept.

Cell* or **battery*** provides **e.m.f.** and causes **direct current** to flow.

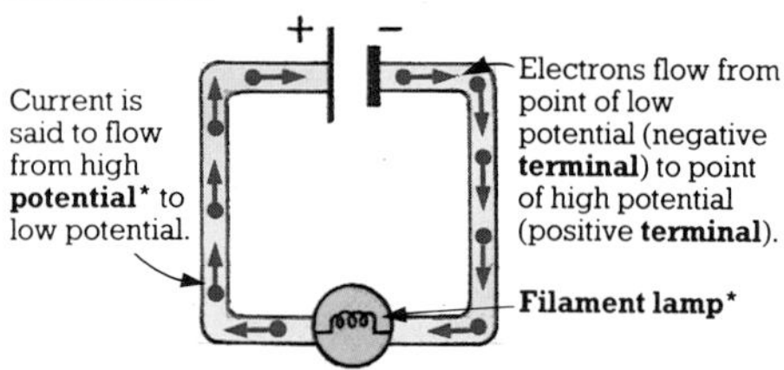

* **Ammeter**, 77; **Battery**, 68; **Calibration**, 344; **Cell**, 68; **Electric field**, 58; **Filament lamp**, 64; **Generator**, 78; **Potential**, **Potential difference**, 58; **SI units**, 96.

- **Alternating current (a.c.).** Current whose direction in a circuit changes at regular intervals. It is caused by an alternating **electromotive force**. Plotting a graph of current against time gives the waveform of the current. Alternating currents and electromotive forces are generally expressed as their **root mean square** values (see picture, right).

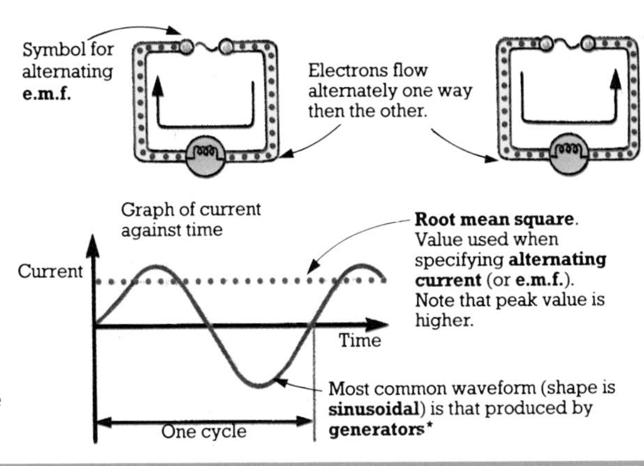

Electricity supply

Electricity for domestic and industrial use is produced at power stations by large **generators***. These produce **alternating current** at a frequency of 50 or 60 Hz.

Alternating current, unlike **direct current**, can be easily **transformed** (see **transformer**, page 79) to produce larger or smaller **potential differences***. This means that high voltages and thus low currents can be used for transmission, which reduces power losses in the transmission cables considerably.

Power station

Step-up transformer* at power station increases e.m.f. to between 100 and 400 kV.

Substation reduces e.m.f. to between 10 and 30 kV.

Factories usually have own transformers because they need higher e.m.f. than houses.

Turbines driven by steam turn generators to produce **alternating current** at 50 or 60 Hz with **e.m.f.** between 10 and 30 kV.

Transmission lines

Small substation reduces e.m.f. to 110 V or 240 V.

Two wires from substation to houses

All domestic electricity ▶ supplies consist of at least two wires from a substation along which **alternating current** flows. In some cases, one of the wires is connected to earth at the substation so that there is only one wire with **potential*** above earth. In some countries there is an additional wire connected to earth as a safety measure.

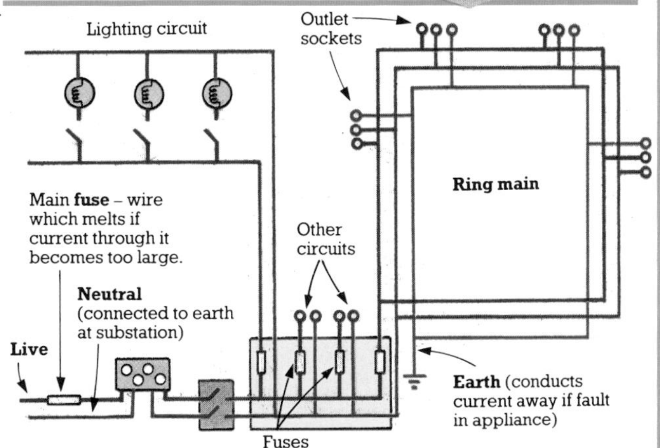

* **Generator**, 78; **Potential**, **Potential difference**, 58; **Step-up transformer**, 79.

Controlling current

The strength of a current flowing in a circuit depends on the nature of the components in the circuit as well as the **electromotive force***. The **resistance** of the components and the magnetic and electric fields they set up all affect the current in them.

- **Ohm's law**. The current in an object at constant temperature is proportional to the **potential difference*** across its ends. The ratio of the potential difference to the current is the **resistance** of the object. The object must be at constant temperature for the law to apply since a current will heat it up and this will change its resistance (see also **filament lamp**, page 64). Ohm's law does not apply to some materials, e.g. **semiconductors***.

- **Resistance (R)**. The ability of an object to resist the flow of current. The value depends on the **resistivity** of the substance from which the object is made and its shape. The unit of resistance is the **ohm (Ω)**. Electrons moving in the object hit atoms and give them energy, heating the object and using up energy from the source of **electromotive force***.

- **Resistivity (ρ)**. The ability of a substance to resist current. Good **conductors*** have a low resistivity and **insulators*** have a high resistivity. It is the inverse of the **conductivity** of the substance and increases with temperature. See also page 112.

- **Resistor**. A device with particular **resistance** value, used to produce a required **potential difference***. Resistors can have values from less than one **ohm** up to many millions of ohms. The most common type is the **carbon resistor**, made from compressed carbon of known **resistivity**.

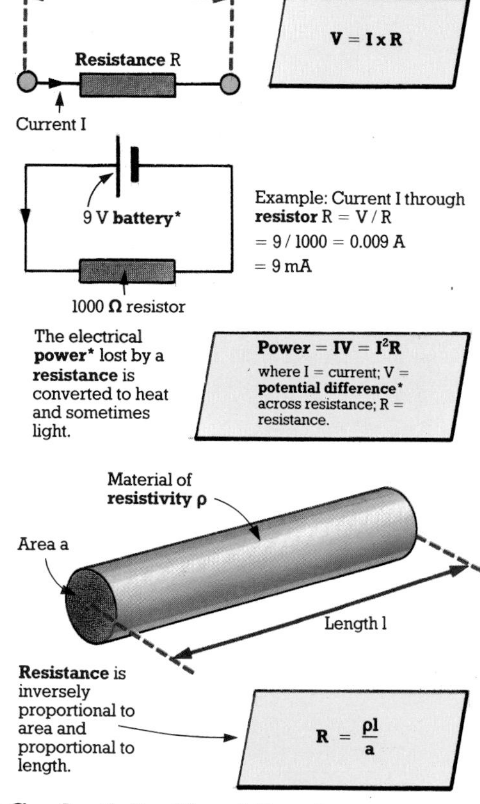

- **Conductivity**. The ability of a substance to allow the flow of current (see also **conductor** and **insulator**, page 56). It is the inverse of the **resistivity**.

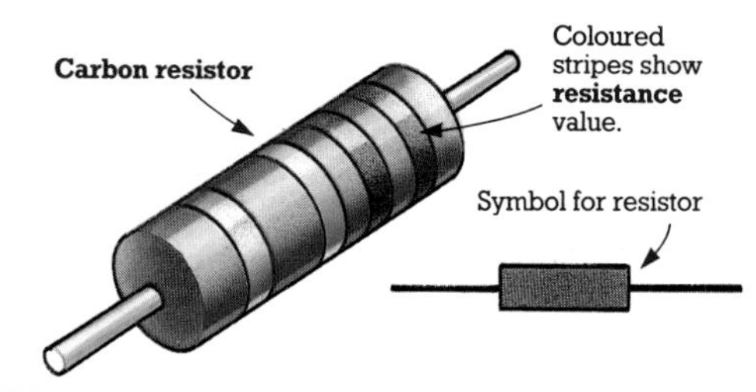

* **Battery**, 68; **Conductor**, 56; **Electromotive force**, 60; **Insulator**, 56; **Potential difference**, 58; **Power**, 9; **Semiconductors**, 65.

- **Internal resistance (r).** The ▶ **resistance** of a **cell*** or **battery*** to the current it causes. It is the resistance of the connections in the cell and some chemical effects (e.g. **polarization***). The current in a circuit may therefore be less than expected.

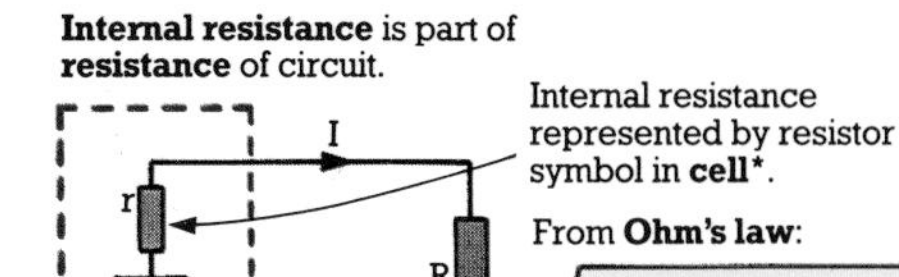

From **Ohm's law**:

$$V = I(R + r)$$

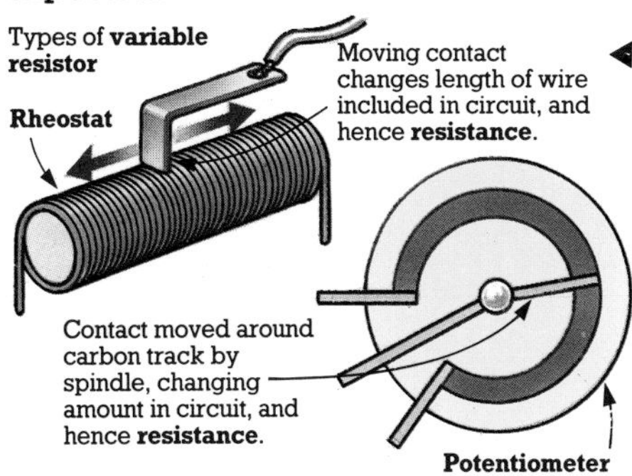

- **Variable resistor.** A device whose **resistance** can be changed mechanically. It is either a coil of wire of a particular **resistivity** around a drum along which a contact moves (for high currents) or a carbon track with a moving contact. A variable resistor can be used as a **potential divider** if an extra contact is added. It is then a **potentiometer**.

- **Potential divider** or **voltage divider.** ▶ A device used to produce a **potential difference*** from another, higher potential difference.

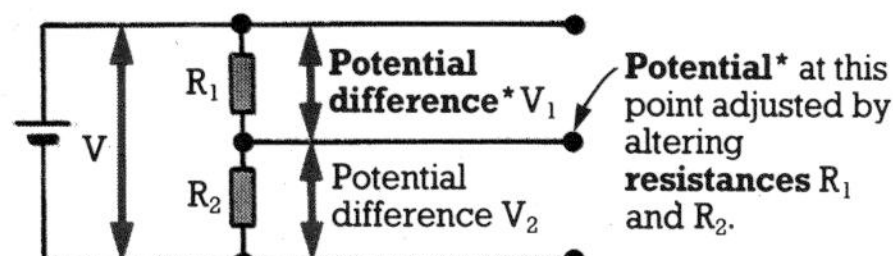

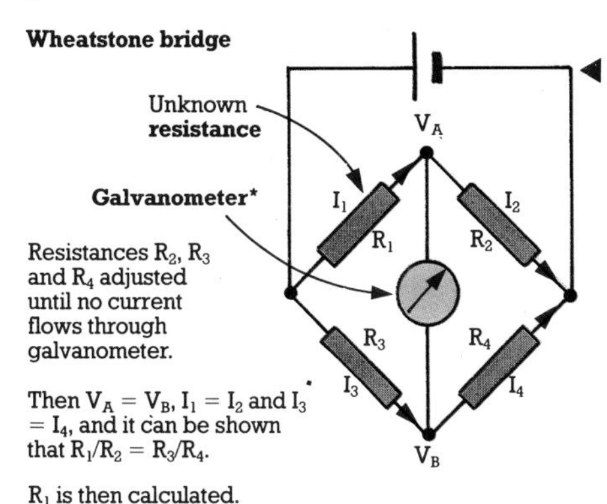

- **Wheatstone bridge.** A circuit used to measure an unknown **resistance** (see diagram). When the **galvanometer*** indicates no current, the unknown value of one **resistor** can be calculated from the other three. The **metre bridge** is a version of the wheatstone bridge in which two of the resistors are replaced by a metre of wire with a high resistance. The position of the contact from the galvanometer on the wire gives the ratio R_3/R_4 in the circuit shown.

- **Kirchhoff's laws.** Two laws which ▶ summarize conditions for the flow of current at an instant. The first states that the total current flowing towards a junction is equal to the total current flowing away from the junction. The second states that the sum of the products of **resistance** and current for each component in a circuit is equal to the **electromotive force*** applied to the circuit.

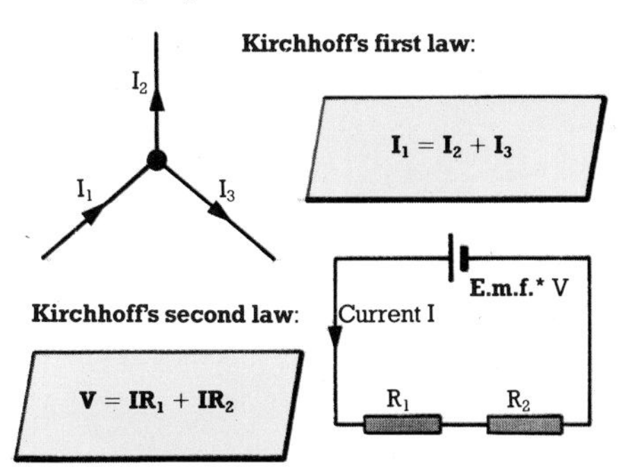

$$I_1 = I_2 + I_3$$

$$V = IR_1 + IR_2$$

* **Battery**, **Cell**, 68; **Electromotive force (e.m.f.)**, 60; **Galvanometer**, 77; **Polarization**, 68; **Potential**, **Potential difference**, 58.

Controlling current (continued)

- **Series**. An arrangement of components in which all of the current passes through them one after the other.

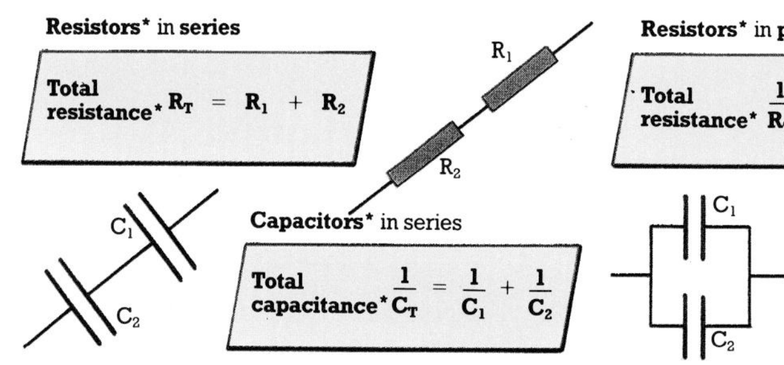

- **Parallel**. An arrangement of components in which current divides to pass through all at once.

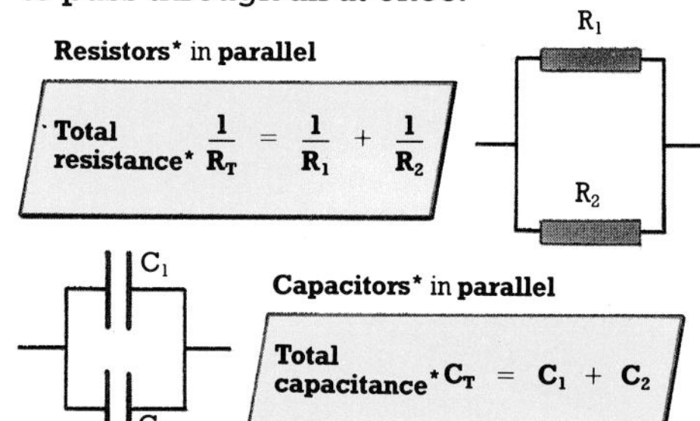

- **Impedance**. The ratio of the **potential difference*** applied to a circuit to the **alternating current*** which flows in it. It is due to two things, the **resistance*** of the circuit and the **reactance**. The effect of impedance is that the **e.m.f.*** and current can be out of phase.
- **Reactance**. The "active" part of **impedance** to **alternating current***. It is caused by **capacitance*** and **inductance** in a circuit which alter the **electromotive forces*** as the current changes.
- **Inductance**. The part of the **impedance** of a circuit due to changing magnetic fields affecting the current (see also **electromagnetic induction**, page 78). This happens in a device called an **inductor**.

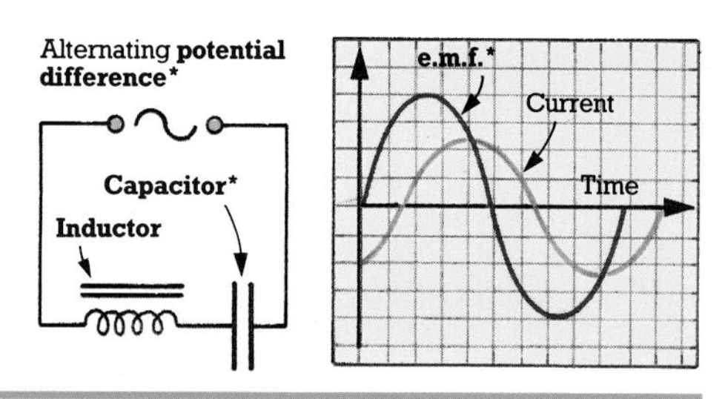

- **Filament lamp**. A lamp consisting of a coil of tungsten wire (the **filament**) inside a glass bulb containing argon or nitrogen gas at low pressure. When current flows through the coil, it heats up rapidly and gives out light. Tungsten is used because it has a very high melting point and the bulb is gas-filled to reduce evaporation of the tungsten.

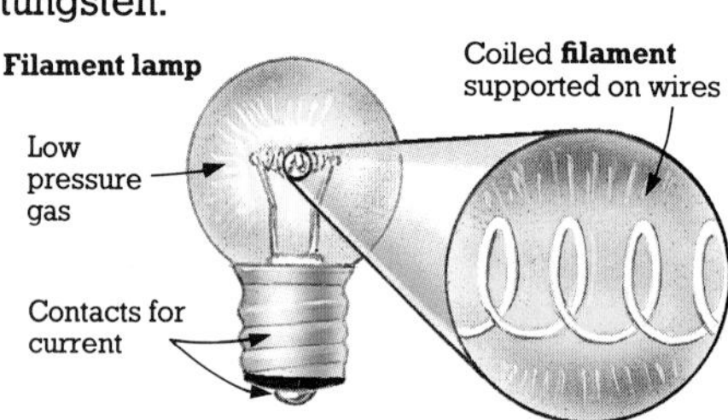

- **Switch**. A device, normally mechanical (but see also **transistor**), which is used to make or break a circuit. A **relay*** is used when a small current is required to switch a larger current on and off.

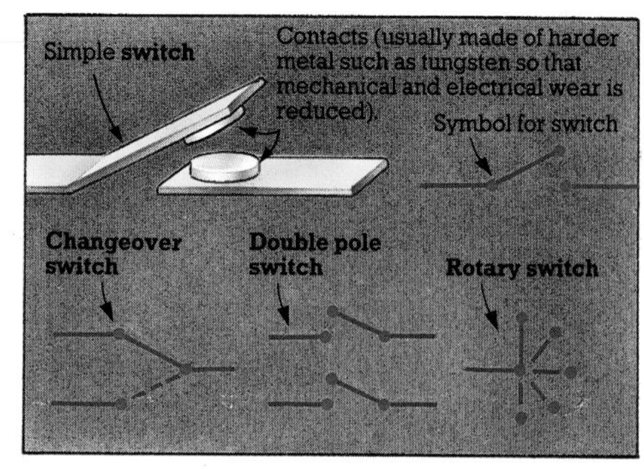

* **Alternating current**, 61; **Capacitance**, **Capacitor**, 59; **Electromotive force** (e.m.f.), 60; **Potential difference**, 58; **Relay**, 75; **Resistance**, **Resistor**, 62.

Semiconductors

Semiconductors are materials whose **resistivity*** is between that of a **conductor** and an **insulator** (see page 56) and decreases with increasing temperature or increasing amounts of impurities (see **doping**). They are widely used in electronic circuits (see also page 111).

- **Doping**. The introduction of a small amount of impurity into a semiconductor. Depending on the impurity used, the semiconductor is known as either a **p-type** or **n-type**. Combinations of the two types are used to make **diodes** and **transistors**.

Construction of **diode**
n-type semiconductor
p-type semiconductor
p-n junction
Symbol for diode

Forward biased – diode has low **resistance***.
Current flow

Reverse biased – diode has very high resistance.
Negligible current flow

- **Diode**. A device made from one piece of **p-type** semiconductor (see **doping**) and one piece of **n-type** semiconductor joined together. It has a very low **resistance*** in one direction (when it is said to be **forward biased**) and a very high resistance in the other direction (**reverse biased**).

- **Half-wave rectification**. The use of a **diode** to remove all the current flowing in one direction from **alternating current***. Current only flows one way around the circuit.

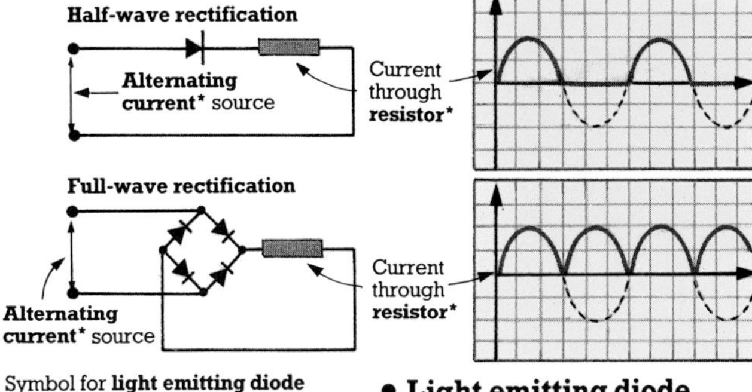

- **Full-wave rectification**. The conversion of **alternating current*** to **direct current***. It is used when direct current is required from mains electricity.

Symbol for **light emitting diode**
Numeric display of shaped LEDs

- **Light emitting diode (LED)**. A **diode** with a higher **resistance*** than normal, in which light is produced instead of heat.

- **Thermistor**. A semiconductor device whose **resistance*** varies with temperature, used to electronically detect temperature changes.

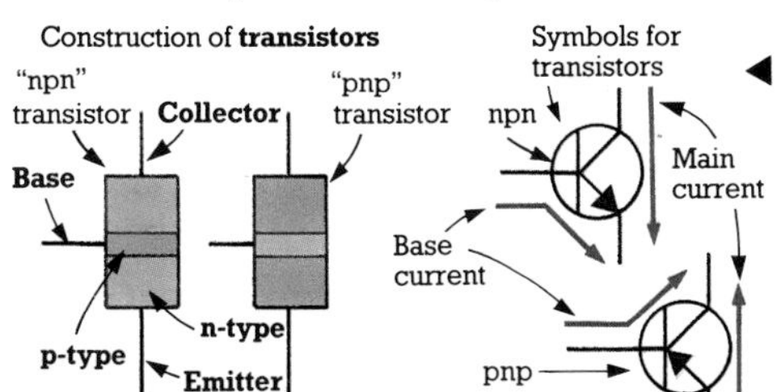

Symbols for transistors
npn
Main current
Base current
pnp

- **Transistor**. A semiconductor, normally made from a combination of the two types of semiconductor. There are three connections, the **base**, **collector** and **emitter** (see diagram). The **resistance*** between the collector and emitter changes from very high to very low when a small current flows into the base. This small base current can therefore be used to control a much larger collector to emitter current.

* **Alternating current**, 61; **Direct current**, 60; **Resistance**, **Resistivity**, **Resistor**, 62.

Electrolysis

Electrolysis is the process whereby electric current flows through a liquid containing **ions*** (atoms which have gained or lost an **electron*** to become charged) and the liquid is broken down as a result. The current is conducted by the movement of ions in the liquid and chemicals are deposited at the points where the current enters or leaves the liquid. There are a number of industrial applications.

- **Electrolyte**. A compound which conducts electricity when either molten or dissolved in water. All compounds made from ions or which split into ions when dissolved (**ionization***) are electrolytes. The concentration of ions in an electrolyte determines how well it conducts electricity.

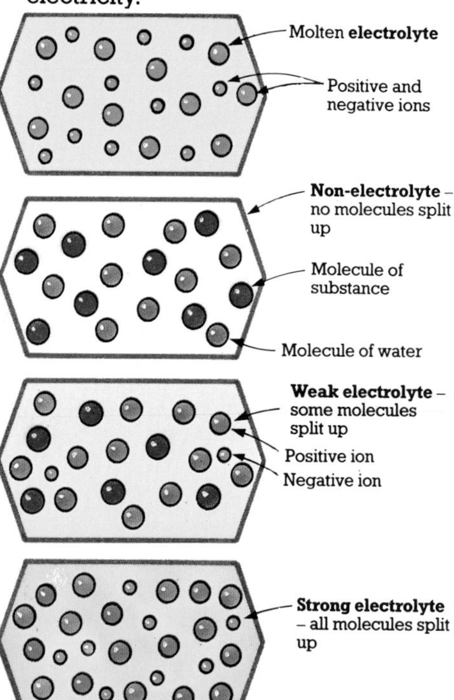

- **Electrode**. A piece of metal or carbon placed in an **electrolyte** through which electric current enters or leaves during electrolysis. Two are needed – the **anode** (positive electrode) and the **cathode** (negative electrode). An **active electrode** is one which is chemically changed by electrolysis; an **inert electrode** is one which is not changed.

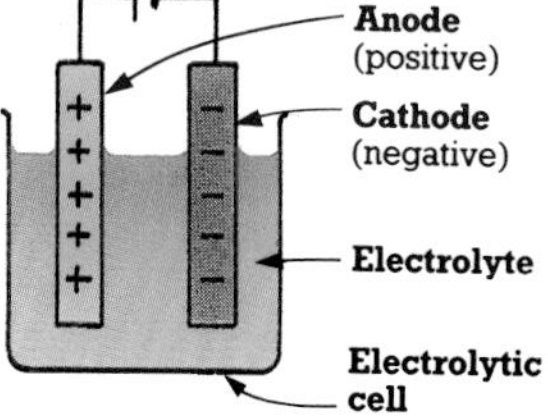

- **Electrolytic cell**. A vessel in which electrolysis takes place. It contains the **electrolyte** and the **electrodes**.

- **Ionic theory of electrolysis**. A theory which attempts to explain what happens in the **electrolyte** and at the **electrodes** during electrolysis. It states that the **cations** (positive ions) are attracted towards the **cathode** and the **anions** (negative ions) towards the **anode**. There they lose or gain electrons respectively to form atoms (they are said to be **discharged**). If there are two or more different anions, then one of them will be discharged in preference to the others. This is called **preferential discharge**. ▼

Electrolysis of copper sulphate solution

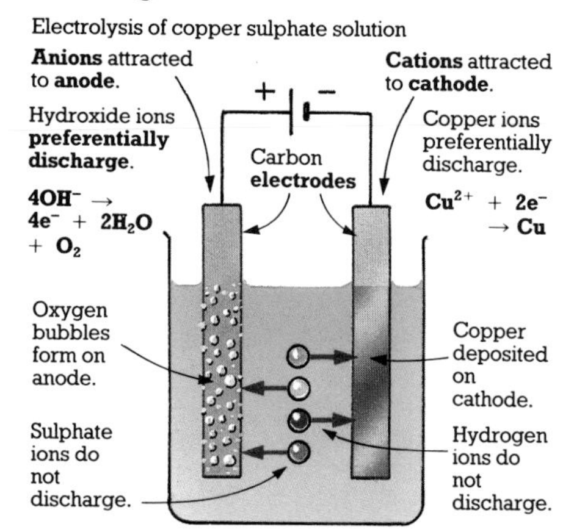

* **Electrons**, 83; **Ions**, 88 (**Ionization**).

• **Faraday's laws of electrolysis**. Two laws which relate the quantity of electricity which passes through an **electrolyte** to the amount of the substances formed. **Faraday's first law** states that the amount of the substances is proportional to the quantity of electricity (the **electrochemical equivalent** of a substance is the mass liberated by one ampere flowing for one second). **Faraday's second law** states that the amount of the substance deposited is inversely proportional to the size of the charge on its ion.

Electrolysis of copper sulphate solution with copper **electrodes** (**copper voltameter**)

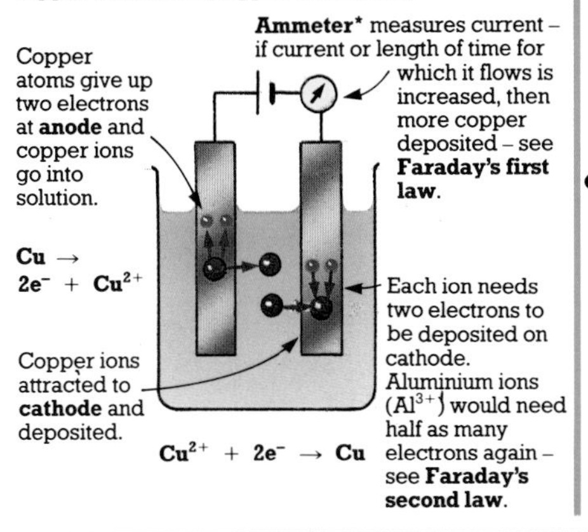

• **Voltameter** or **coulometer**. An **electrolytic cell** used for investigating the relationships between the amount of substance produced at the **electrodes** and the current which passes through the cell. For example, the **copper voltameter** (see below left) contains copper sulphate and copper electrodes.

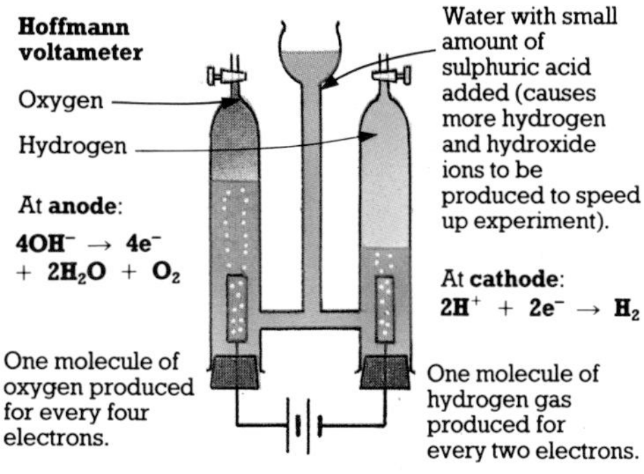

• **Hoffmann voltameter**. A type of ▲ **voltameter** used for collecting and measuring the volumes (and hence the masses) of gases liberated during electrolysis. For example, electrolysis of acidified water produces hydrogen and oxygen in a two to one ratio (note that this also indicates the chemical composition of water, i.e. H_2O).

Uses of electrolysis

• **Electroplating** or **electrodeposition**. The coating of a metal object with a thin layer of another metal by electrolysis. The object forms the **cathode** and ions of the coating metal are in the **electrolyte**.

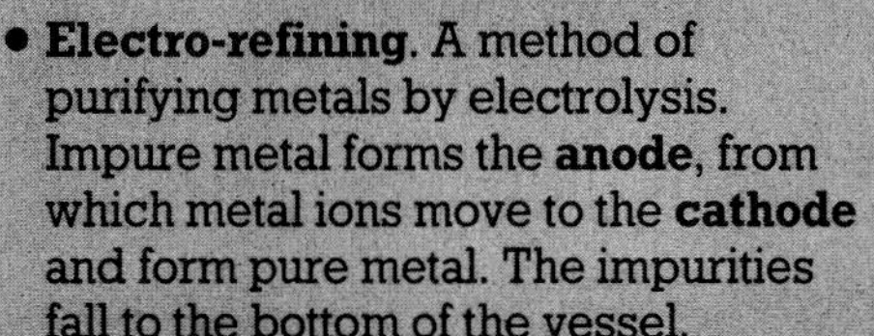

• **Electro-refining**. A method of purifying metals by electrolysis. Impure metal forms the **anode**, from which metal ions move to the **cathode** and form pure metal. The impurities fall to the bottom of the vessel.

• **Metal extraction**. A process which produces metals from their molten ores by electrolysis. Very reactive metals are obtained by this process, e.g. sodium and aluminium. ▼

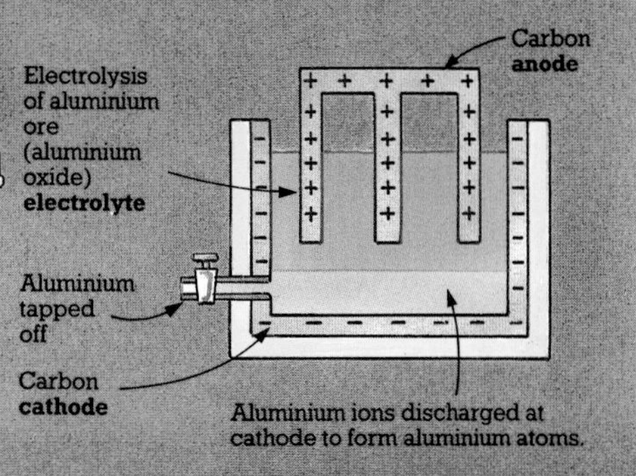

* **Ammeter**, 77.

Cells and batteries

The Italian scientist Volta first showed that a **potential difference*** exists between two different metals when they are placed in certain liquids (**electrolytes***) and therefore that a **direct current*** can be produced from chemical energy. This arrangement is called a **cell**, **electrochemical** or **voltaic cell**. The potential difference (caused by chemical changes in the cell) is called an **electromotive force*** and its size depends on the metals used. A **battery** is a number of cells linked together.

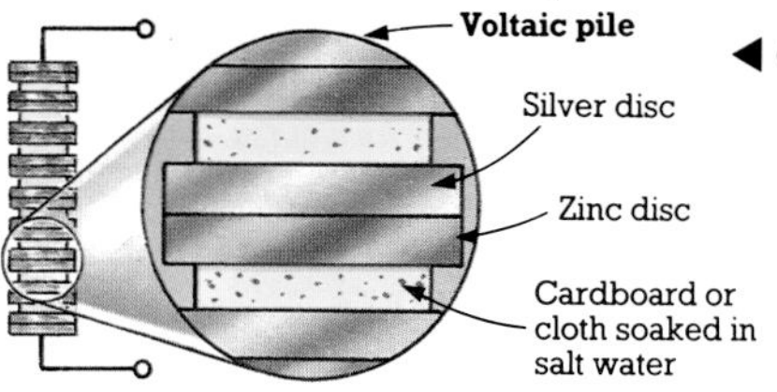

◀ • **Voltaic pile**. The first battery made, consisting of a pile of silver and zinc discs separated by cardboard or cloth soaked in salt water. This arrangement is the same as a number of **simple cells** linked together.

• **Simple cell**. Two plates of different metals separated by a salt or acid solution **electrolyte*** (normally copper and zinc plates and dilute sulphuric acid). The simple cell only produces an **electromotive force*** for a short time before **polarization** and **local action** have an effect. ▶

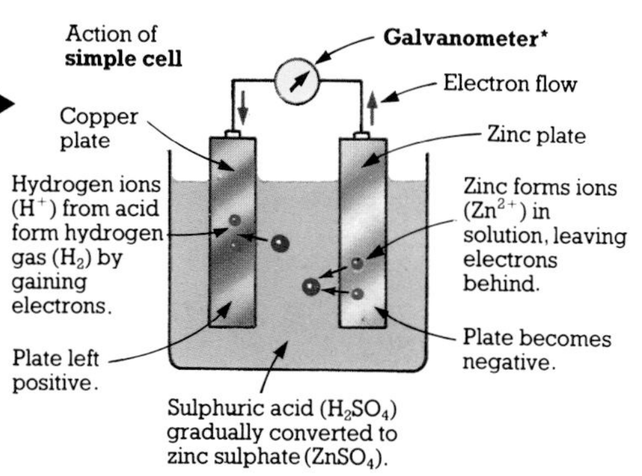

• **Polarization**. The formation of bubbles of hydrogen on the copper plate in a **simple cell**. This reduces the **electromotive force*** of the cell, both because the bubbles insulate the plate and also because a **back e.m.f.*** is set up. Polarization can be eliminated by adding a **depolarizing agent**, which reacts with the hydrogen to form water.

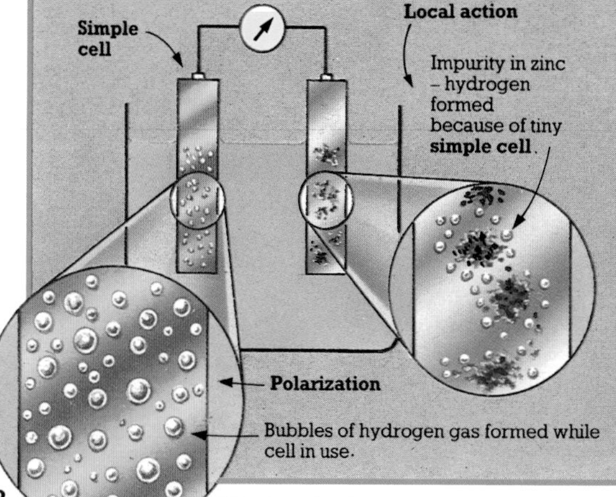

• **Local action**. The production of hydrogen at the zinc plate in a **simple cell**. Impurities (traces of other metals) in the zinc plate mean that tiny simple cells are formed which produce hydrogen due to **polarization**. Hydrogen is also produced as the zinc dissolves in the acid (even when the cell is not working). Local action can be prevented by coating the plate with an **amalgam***.

* **Amalgam**, 344; **Back e.m.f.**, 60 (**Electromotive force**); **Direct current**, 60; **Electrolyte**, 66; **Galvanometer**, 77; **Potential difference**, 58.

- **Capacity**. The ability of a cell to produce current over a period of time. It is measured in **ampere hours**. For example, a 10 ampere hour cell should produce one ampere for 10 hours.

- **Leclanché cell**. A cell in which **polarization** is overcome by manganese oxide (a **depolarizing agent**). This removes hydrogen more slowly than it is formed, but continues working to remove excess hydrogen when the cell is not in use. The cell provides an **electromotive force*** of 1.5 V.

Leclanché cell
Zinc rod
Carbon rod
Porous pot
Carbon and manganese oxide (**depolarizing agent**)
Ammonium chloride solution

- **Standard cell**. A cell which produces an accurately known and constant **electromotive force***. It is used in laboratories for experimental work.

- **Primary cell**. Any cell which has a limited life because the chemicals inside it are eventually used up and cannot be replaced easily.

- **Dry cell**. A version of the **Leclanché cell** in which the ammonium chloride solution is replaced by paste containing ammonium chloride, meaning that it is portable. The cell provides an **electromotive force*** of 1.5 V. Dry cells deteriorate slowly due to **local action**, but still have a life of many months.

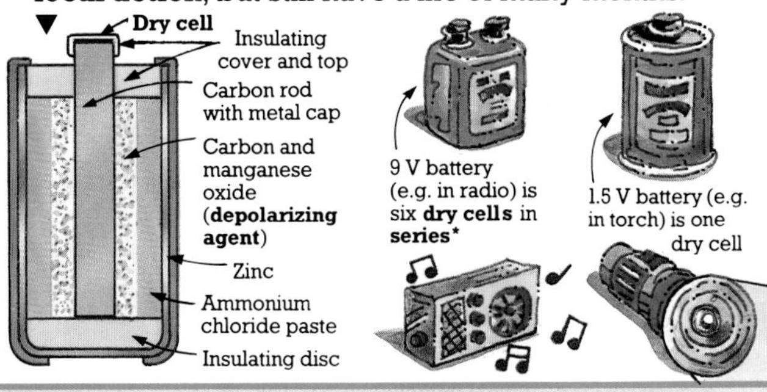

- **Secondary cell**. Also known as an **accumulator** or **storage cell**. A cell which can be recharged by connection to another source of electricity. The main types are the **lead-acid accumulator** and the **alkaline cell**.

- **Lead-acid accumulator**. A **secondary cell** containing a dilute sulphuric acid **electrolyte***, and plates made from lead and lead compounds.

- **Alkaline cell**. A **secondary cell** containing an **electrolyte*** of caustic potash solution. The plates are normally made of nickel and cadmium compounds (it is then called a **nickel-cadmium cell**).

The cell can give out a very large current because it has a low **internal resistance***, and is therefore mainly used in vehicles for starting and lighting.

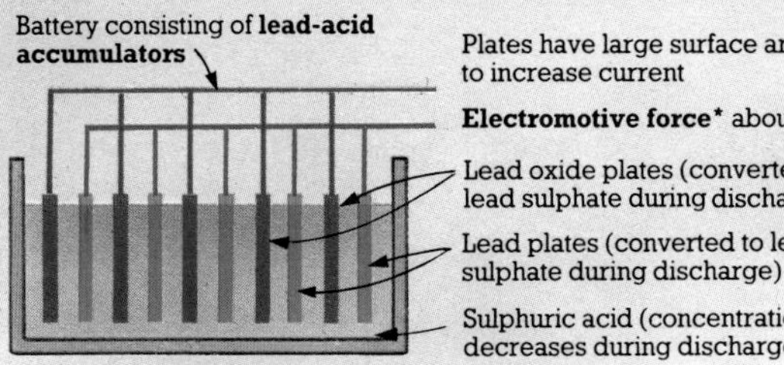

* **Electromotive force**, 60; **Electrolyte**, 66; **Internal resistance**, 63; **Series**, 64.

Magnets

All **magnets** have a **magnetic field*** around them, and a **magnetic force*** exists between two magnets due to the interaction of their fields. Any material which is capable of being **magnetized** (can become a magnet) is described as **magnetic** (see **ferromagnetic**) and becomes magnetized when placed in a magnetic field. The movement of charge (normally **electrons***) also causes a magnetic field (see **electromagnetism**, pages 74-76).

Suspended bar magnet

North pole points to **magnetic north***.

South pole points to **magnetic south***.

First law of magnetism

Unlike **poles** attract.

Like **poles** repel.

- **Pole**. A point in a magnet at which its **magnetic force*** appears to be concentrated. There are two types of pole – the **north** or **north seeking pole** and the **south** or **south seeking pole** (indentified by allowing the magnet to line up with the earth's **magnetic field***). All magnets have an equal number of each type of pole. The **first law of magnetism** states that unlike poles attract and like poles repel.

- **Magnetic axis**. A line joining the **north** and **south poles** of a magnet, about which its **magnetic field*** is symmetrical.

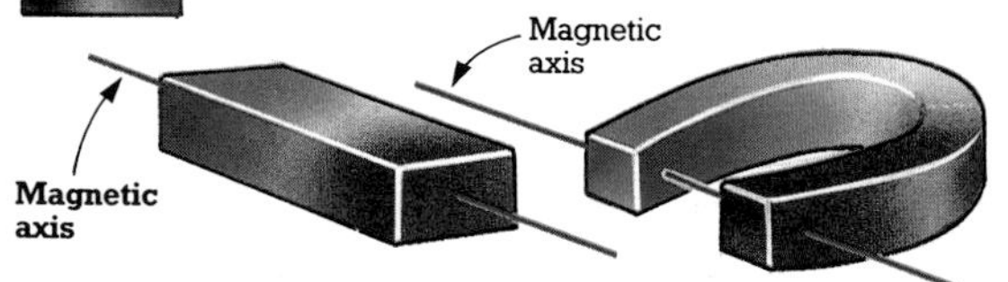

- **Ferromagnetic**. Describes a material which is strongly magnetic (i.e. is magnetized easily). Iron, nickel, cobalt and alloys of these are ferromagnetic, and are described as either **hard** or **soft**. **Sintered** materials (made by converting various mixtures of powders of the above metals into solids by heat and pressure) can be made magnetically very hard or soft by changing the metals used.

- **Hard**. Describes a **ferromagnetic** material which does not easily lose its magnetism after being magnetized, e.g. steel. Magnets made from these materials are called **permanent magnets**.

- **Soft**. Describes a **ferromagnetic** material which does not retain its magnetism after being magnetized, e.g. iron. Magnets made from these materials are called **temporary magnets**. **Residual magnetism** is the small amount of magnetism which can be left in magnetically soft materials.

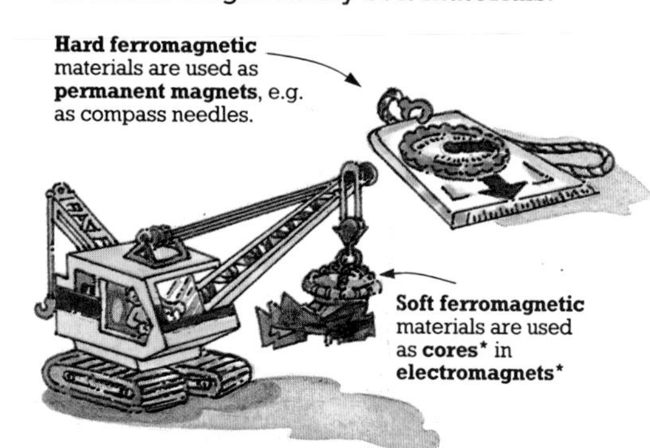

* Core, 74; Electromagnet, 75; Electrons, 83; Magnetic field, 72; Magnetic force, 6; Magnetic north, Magnetic south, 73.

- **Susceptibility**. A measurement of the ability of a substance to become magnetized. **Ferromagnetic** materials have a high susceptibility.

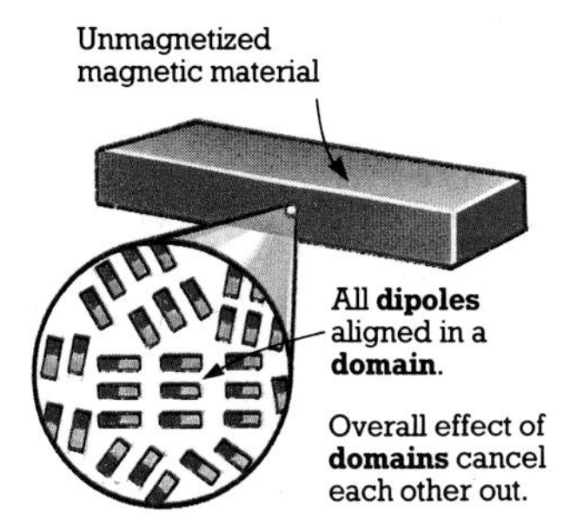

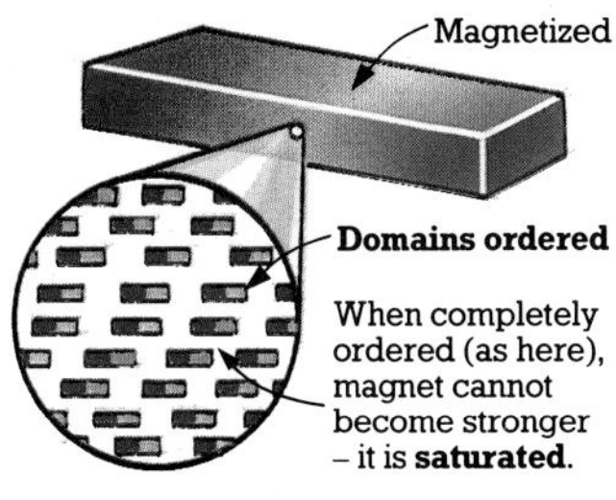

Domain theory explains why magnets retain their **poles** when broken.

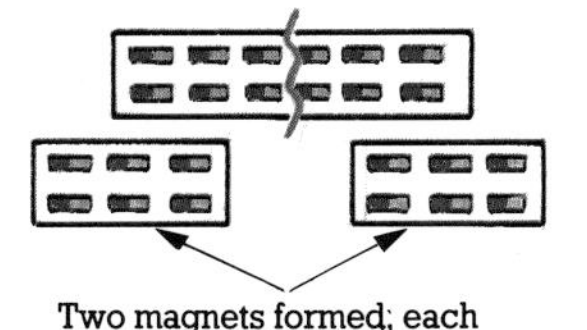

Two magnets formed; each with a **north** and **south pole**.

▲

- **Domain theory of magnetism**. States that **ferromagnetic** materials consist of **dipoles** or **molecular magnets**, which interact with each other. These are all arranged in areas called **domains**, in which they all point in the same direction. A ferromagnetic material becomes magnetized when the domains become **ordered** (i.e. aligned).

Magnetization

When an object is magnetized, all the **dipoles** become aligned (see **domain theory**). This only happens when the object is in a **magnetic field*** and is called **induced magnetism**.

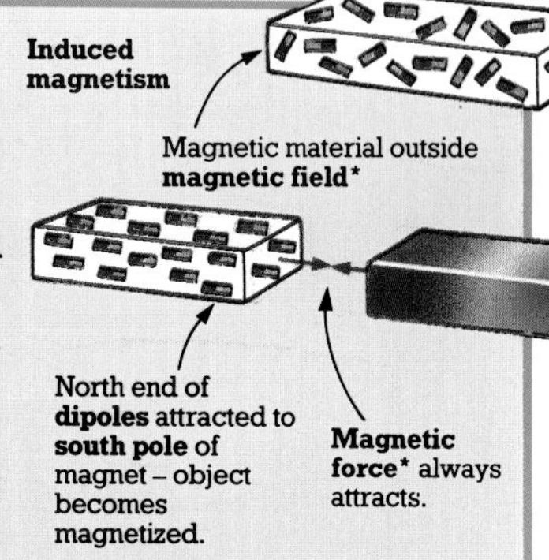

- **Single touch**. A method of magnetizing an object by stroking it repeatedly with the **pole** of a **permanent magnet** (see **hard**). Magnetism is induced in the object from the **magnetic field*** of the magnet.

Methods of magnetization (all involve **induced magnetism**)

Single touch **Divided touch**

Consequent poles are produced when like poles are used in **divided touch**.

- **Divided touch**. A method of magnetizing an object by stroking it repeatedly from the centre out with the opposite **poles** of two **permanent magnets** (see **hard**). Magnetism is induced in the object from the **magnetic field*** of the magnets.

- **Demagnetization**. The removal of magnetism from an object. This can be achieved by placing the object in a changing **magnetic field***, such as that created by a coil carrying **alternating current***. Alternatively, the **dipoles** (see **domain theory**) can be excited to point in random directions by hammering randomly or by heating above 700°C.

- **Self-demagnetization**. ▶ Loss of magnetism by a magnet because of the attraction of the **dipoles** (see **domain theory**) for the opposite **poles** of the magnet. It is reduced using pieces of soft iron (called **keepers**) arranged to form a closed loop of poles.

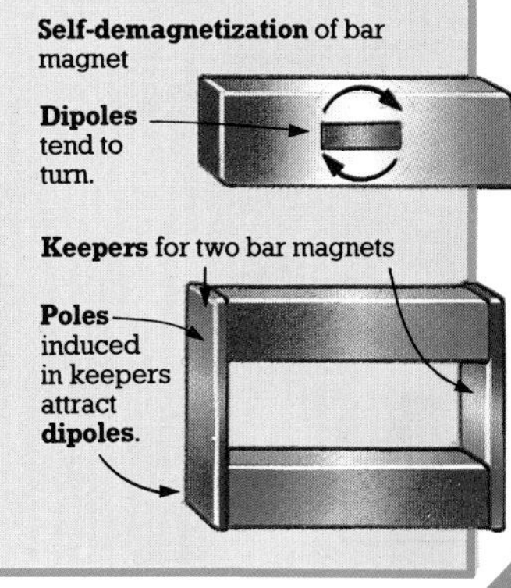

* **Alternating current**, 61; **Magnetic field**, 72; **Magnetic force**, 6.

Magnetic fields

A **magnetic field** is a region around a **magnet** (see page 70) in which objects are affected by the **magnetic force***. The strength and direction of the magnetic field are shown by **magnetic field lines**.

- **Magnetic field lines** or **flux lines**. Lines which indicate the direction of the magnetic field around a magnet. They also show the strength of the field (see **magnetic flux density**). The direction of the field is that in which a **north pole*** would move due to the field. Magnetic field lines are plotted by sprinkling iron filings around a magnet or by recording the direction of a **plotting compass** (a small compass with no directions on it) at various points.

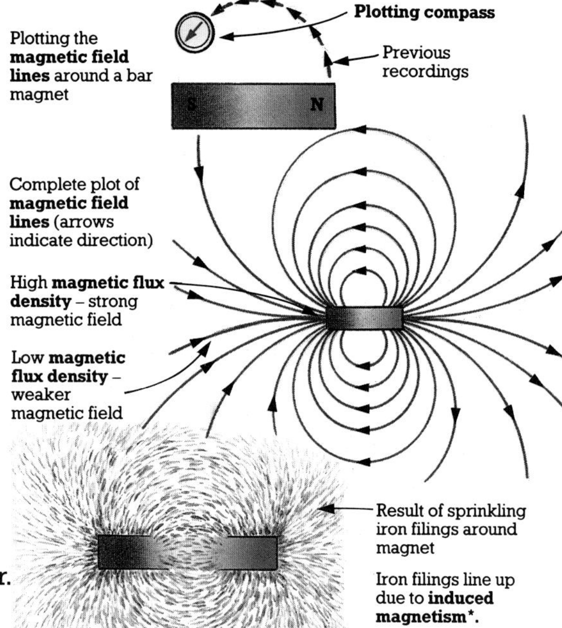

- **Magnetic flux density**. A ▶ measurement of the strength of a magnetic field at a point. This is shown by the closeness of the **magnetic field lines** to each other. Magnetic flux density is normally highest around the **poles***.

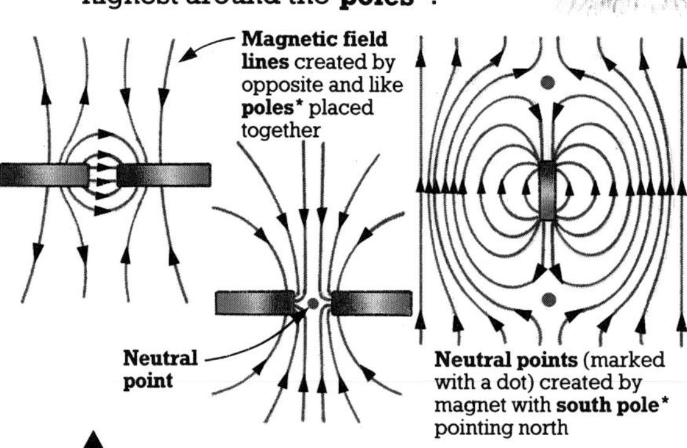

▲

- **Neutral point**. A point of zero magnetism (the **magnetic flux density** is zero). It occurs where two or more magnetic fields interact with an equal but opposite effect. A bar magnet suspended along the **magnetic meridian**, with the **south pole*** pointing to the north, has two neutral points in line with its **magnetic axis***.

- **Diamagnetism**. Magnetism displayed by some substances when placed in a strong magnetic field. A piece of diamagnetic material tends to spread **magnetic field lines** out and lines up with its long side perpendicular to them. This effect is caused by **electrons*** being disturbed.

- **Paramagnetism**. Magnetism displayed by some substances when placed in a strong magnetic field. A piece of paramagnetic material tends to concentrate **magnetic field lines** through it and lines up with its long side parallel to them. It is caused by **dipoles*** moving slightly towards alignment.

* **Dipole**, 71; **Electrons**, 83; **Induced magnetism**, 71; **Magnetic force**, 6; **Magnetic axis**, **Pole**, 70.

The earth's magnetism

The earth has a magnetic field which acts as though there were a giant bar magnet in its centre, lined up approximately between its geographic north and south poles, although the angle is constantly changing. The **north pole*** of a compass points towards a point called **magnetic north,** its south pole to **magnetic south.**

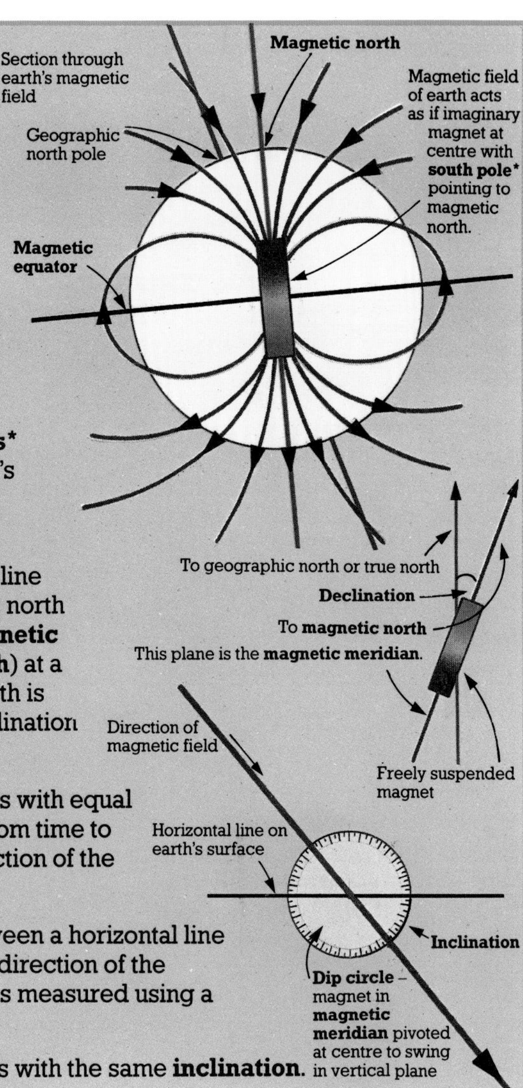

- **Magnetic meridian**. The vertical plane containing the **magnetic axis*** of a magnet suspended in the earth's magnetic field (i.e. with its **north pole*** pointing to **magnetic north**).
- **Declination**. The angle between a line taken to true north (the geographic north pole) and one taken along the **magnetic meridian** (towards **magnetic north**) at a point. The position of magnetic north is gradually changing and so the declination alters slowly with time.
- **Isogonal lines**. Lines joining places with equal **declination**. These are redrawn from time to time because of the changing direction of the earth's magnetic field.
- **Inclination** or **dip**. The angle between a horizontal line on the surface of the earth and the direction of the earth's magnetic field at a point. It is measured using a **dip circle** (see picture).
- **Isoclinal line**. A line linking places with the same **inclination**.

- **Permeability**. A measure of the ability of a substance to "conduct" a magnetic field. Soft iron is much more permeable than air, so the magnetic field tends to be concentrated through it.

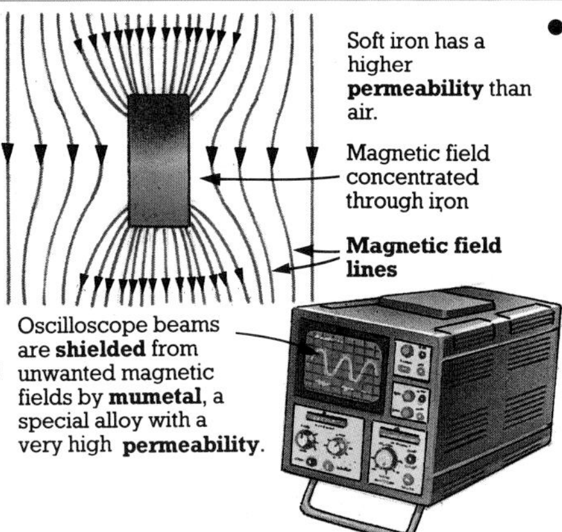

- **Shielding** or **screening**. The use of soft magnetic material to stop a magnetic field from reaching a point, effectively by "conducting" the field away. This is used in sensitive instruments, e.g. oscilloscopes.

* **Magnetic axis**, **Pole**, 70.

Electromagnetism

An electric current flowing through a wire produces a **magnetic field** (see pages 72-73) around the wire, the shape of which depends on the shape of the wire and the current flowing. These magnetic fields can be plotted in the same way as for **permanent magnets***. This effect, called **electromagnetism**, is used in very powerful magnets and also to produce motion from an electric current.

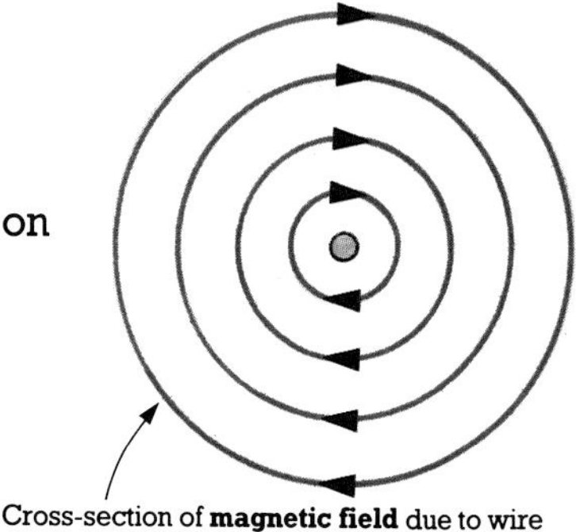

Cross-section of **magnetic field** due to wire carrying current directly into paper.

- **Ampere's swimming rule**. States that the north end of a compass needle placed near a current-carrying wire will be deflected towards the left hand of a person imagined to be swimming in the direction of the current and facing the wire.

Ampere's swimming rule

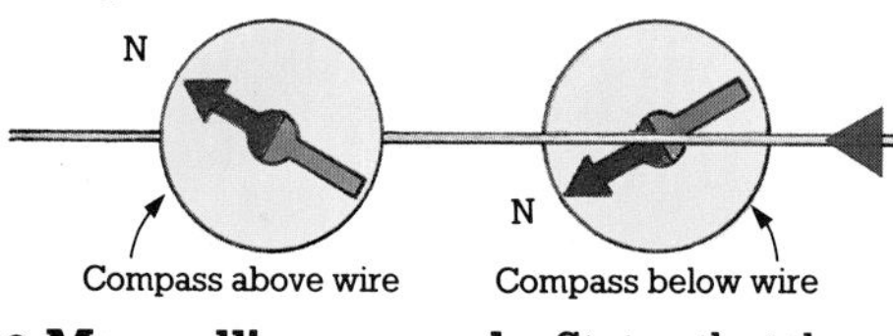

- **Maxwell's screw rule**. States that the direction of the magnetic field around a current-carrying wire is the way a screw turns when being screwed in the direction of the current.

Maxwell's screw rule

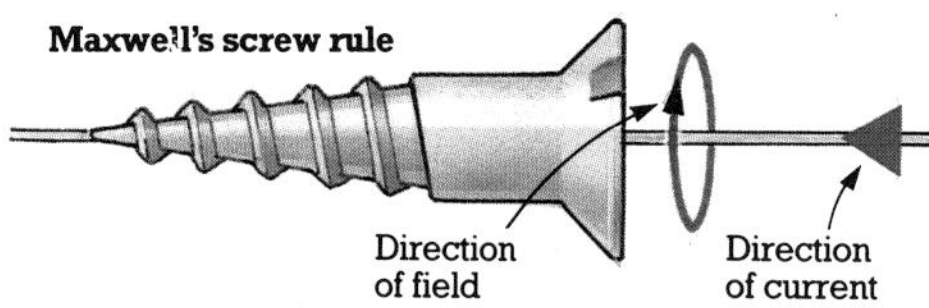

- **Right-hand grip rule**. States that the direction of a magnetic field around a wire is that from the base to the tips of the fingers if the wire is gripped by the right hand with the thumb pointing in the direction of the current.

Right hand grip rule

Direction of current

Direction of field

- **Coil**. A number of turns of current-carrying wire, produced by wrapping the wire around a shaped piece of material (a **former**). Examples are a **flat coil** and a **solenoid**.

- **Flat coil** or **plane coil**. A **coil** of wire whose length is small in comparison with its diameter.

- **Solenoid**. A **coil** whose length is large in comparison with its diameter. The magnetic field produced by a solenoid is similar to that produced by a bar magnet. The position of the **poles*** depends on the current direction (see diagram).

- **Core**. The material in the centre of a **coil** which dictates the strength of the field. Soft **ferromagnetic*** materials, most commonly soft iron, create the strongest magnetic field and are used in **electromagnets**.

Solenoid ▼

Clockwise current looking at end gives south **pole***.

Region inside is **core**. This solenoid is air-cored.

Anticlockwise current looking at end gives north pole.

Magnetic field of **solenoid** above (slice taken lengthways) ▼

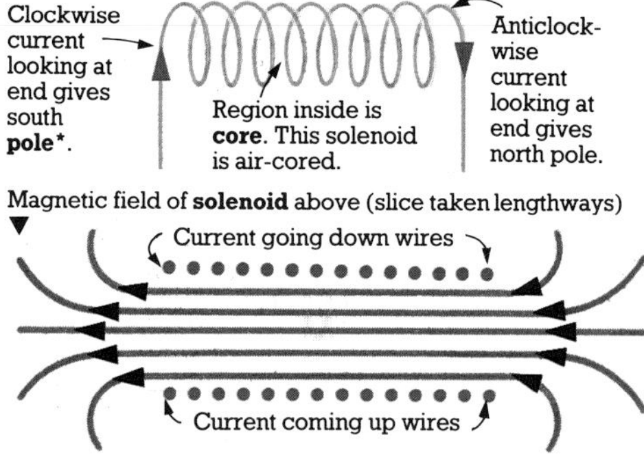

* **Ferromagnetic**, 70; **Permanent magnets**, 70 (**Hard**); **Pole**, 70.

- **Electromagnet**. A **solenoid** with a **core** of soft strongly **ferromagnetic*** material. This forms a magnet which can be switched on and off simply by turning the current on and off. Practical electromagnets are constructed so that two opposite **poles*** are close to each other, producing a strong magnetic field. The electromagnet has a number of applications, some of which are shown below.

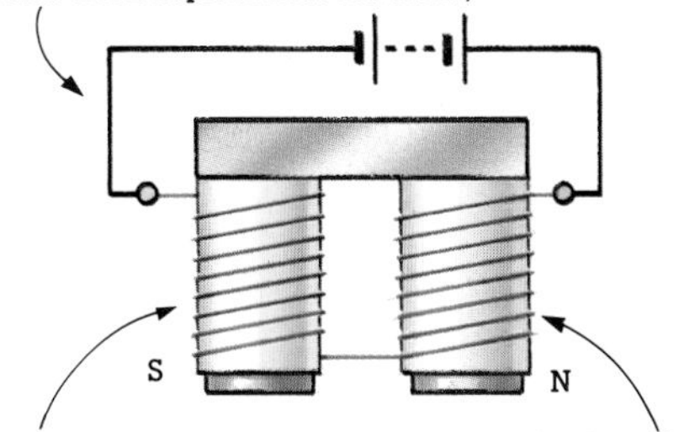

Applications of electromagnets

Electromagnets have a large number of applications, all of which use the fact that they attract metals when they are switched on and therefore convert **electric energy*** to **mechanical energy***. In two of the following examples, sound energy is produced from the mechanical energy.

- **Electric buzzer**. A device which ▶ produces a buzzing noise from **direct current***. A metal arm is attracted by an **electromagnet**, moves towards it, and in doing so breaks the circuit carrying current to the electromagnet. The arm is thus released and the process is repeated. The resulting vibration of the arm produces a buzzing noise. In the **electric bell**, a hammer attached to the arm repeatedly strikes a bell.

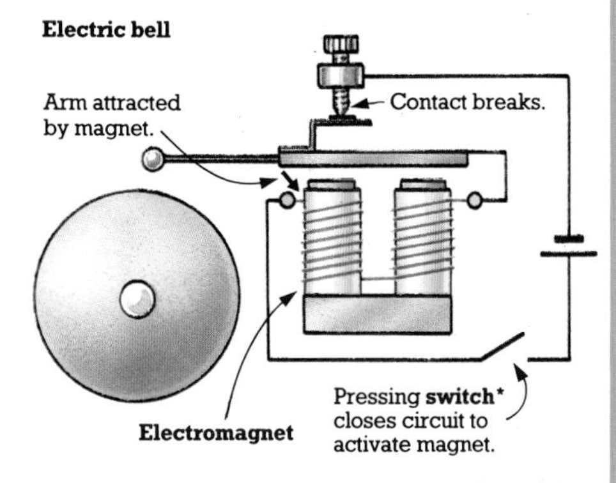

- **Earpiece**. A device used to transform electrical signals to sound waves. The **permanent magnet*** attracts the metal diaphragm, but the strength of this attraction is changed as changing current (the incoming signals) flows through the coils of the **electromagnet**. The diaphragm thus vibrates to produce sound waves.

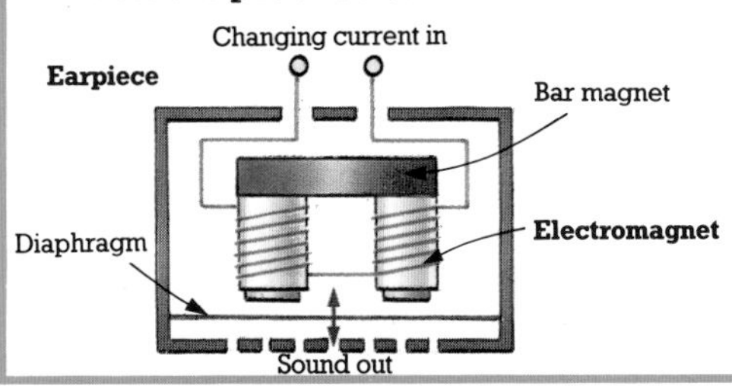

- **Relay**. A device in which a **switch*** is closed by the action of an **electromagnet**. A relatively small current in the **coil** of the electromagnet can be used to switch on a large current without the circuits being electrically linked.

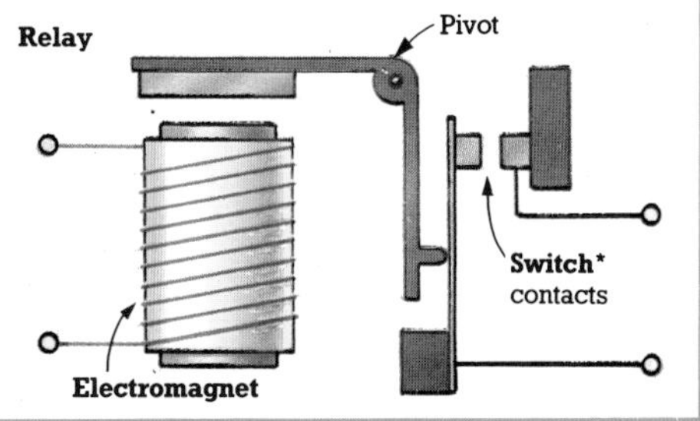

* **Direct current**, 60; **Electric energy**, 9; **Ferromagnetic**, 70; **Mechanical energy**, 9; **Permanent magnets**, 70 (**Hard**); **Pole**, 70; **Switch**, 64.

Electromagnets continued – the motor effect

The **motor effect** occurs when a current-carrying wire goes through a magnetic field. A force acts on the wire which can produce movement. This effect is used in **electric motors**, where **mechanical energy*** is produced from **electric energy***. The effect can also be used to measure current (see page 77), since the force depends on its magnitude.

- **Fleming's left-hand rule** or **motor rule**. The direction of the force on a wire carrying a current through a magnetic field can be worked out with the left hand (see diagram).

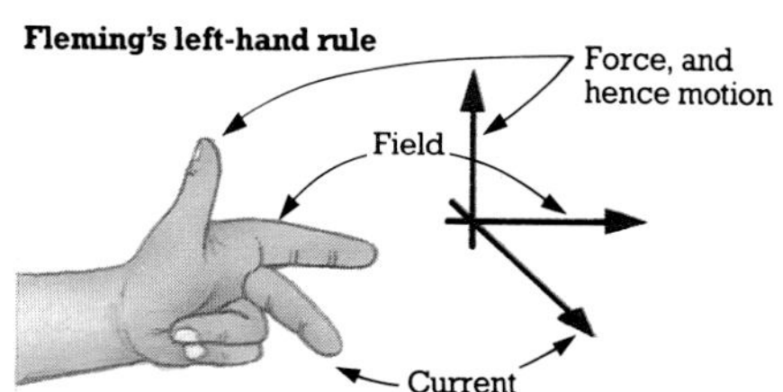

- **Electric motor**. A device which uses the **motor effect** to transform **electric energy*** to **mechanical energy***. The simplest motor consists of a current-carrying, square-shaped, flat **coil***, free to rotate in a magnetic field (see diagram). Motors produce a **back e.m.f.*** opposing the e.m.f. which drives them. This is produced because once the motor starts, it acts as a **generator*** (i.e. the movement of the coil in the field produces an opposing current).

Simple **electric motor**

Commutator. A ring split into two or more pieces, via which current enters and leaves the **coil*** of an **electric motor**. It ensures that the current enters the coil in the correct direction to make the motor rotate in one direction continuously.

Brushes. Contacts, normally made of carbon, through which current enters the **commutator** in an **electric motor**.

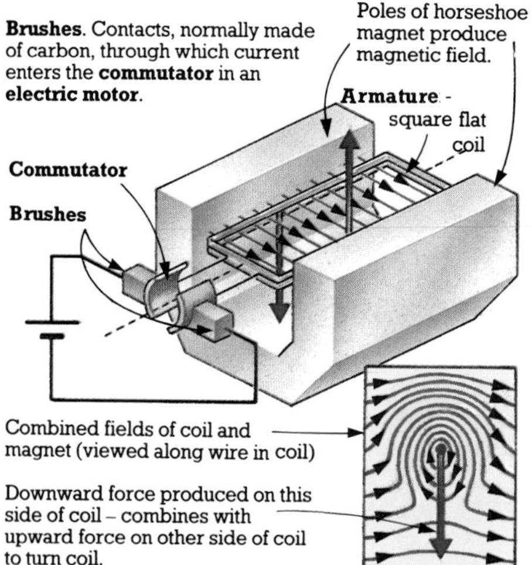

- **Barlow's wheel**. A spiked, brass wheel in a magnetic field. Current enters at its centre and leaves from a spike which dips into mercury. The motor effect makes the wheel turn as the spikes dip into the mercury in turn.

- **Field windings**. Sets of **coils*** around the outside of an **electric motor**, which take the place of a permanent magnet to produce a stronger magnetic field. This increases the power of the motor.

- **Loudspeaker**. A device which uses the **motor effect** to transform electrical signals into **sound waves***. It consists of a **coil*** in a radial magnetic field (the direction of the field at any point is towards the centre). As the current changes, the coil, which is attached to a paper cone, moves in and out of the field (see diagram). The paper cone vibrates the air, producing sound waves which depend on the strength and frequency of the current.

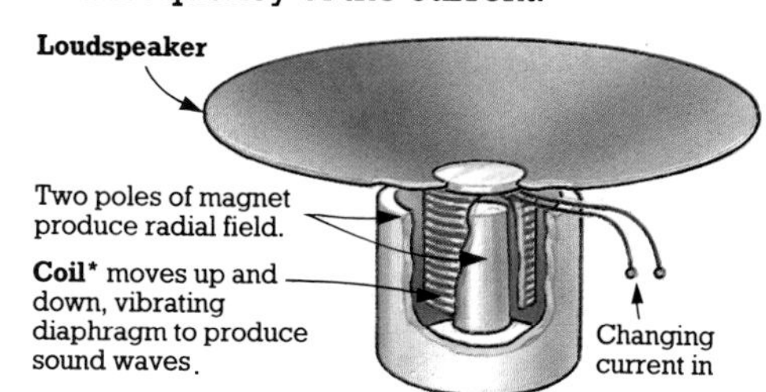

* **Back e.m.f.**, 60 (**Electromotive force**); **Coil**, 74; **Electric energy**, 9; **Generator**, 78; **Mechanical energy**, 9; **Sound waves**, 40.

Electric meters

Current can be detected by placing a suspended magnet near a wire and observing its deflection. This idea can be extended to produce a device (a **meter**) in which the deflection indicates on a scale the strength of the current. The current measuring device can then be adapted to measure **potential difference***.

- **Galvanometer**. Any device used to detect a **direct current*** by registering its magnetic effect. The simplest is a compass placed near a wire to simply show whether a current is present. The **moving coil galvanometer** uses the **motor effect** to show a deflection on a scale (see diagram).

Moving coil galvanometer

Coil* of wire carries current

Counterweight for pointer

Soft iron cylinder conducts magnetic field.

Return spring

Pointer

Scale

Horseshoe magnet

Radial magnetic field due to soft iron cylinder.

Force on coil (see **motor effect**, page 76) increases with current.

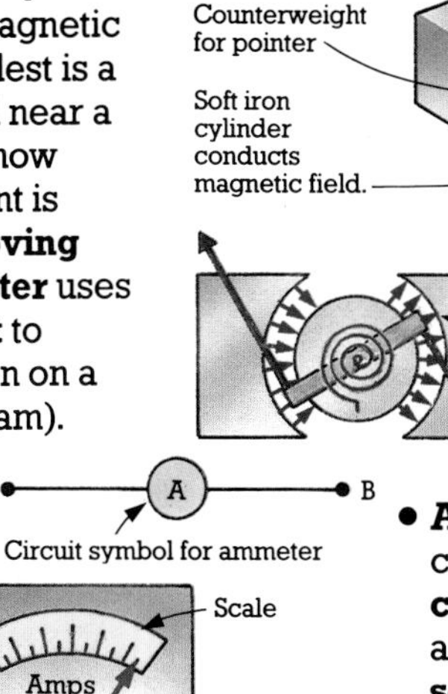

Shunt – value chosen so that when maximum current to be read flows, current through galvanometer gives **full scale deflection**.

- **Ammeter**. A device used to measure current. It is a version of the **moving coil galvanometer**, designed so that a certain current produces a **full scale deflection**, i.e. the pointer moves to its maximum position. To measure higher currents a **shunt** is added (see diagram). A larger current now produces full scale deflection on the new scale.

- **Voltmeter**. A device used to measure the **potential difference*** between two points. It is a **galvanometer** between the two points with a high **resistance*** in **series***. A certain potential difference produces the current for a **full scale deflection** (see **ammeter**). To measure higher potential differences, a **multiplier** is added (see diagram).

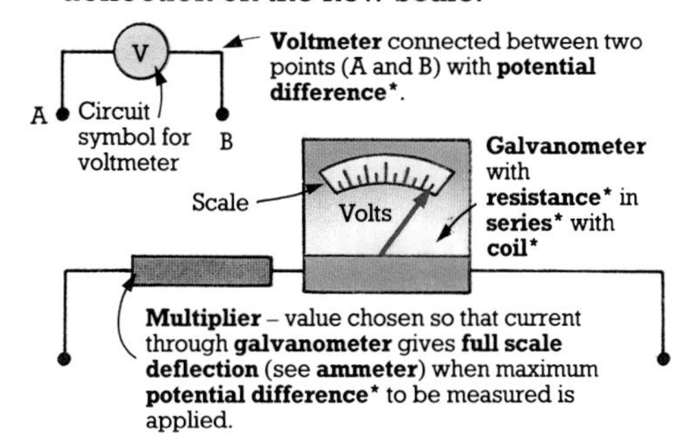

Multiplier – value chosen so that current through **galvanometer** gives **full scale deflection** (see **ammeter**) when maximum **potential difference*** to be measured is applied.

- **Multimeter**. A **galvanometer** combined with the **shunts** (see **ammeter**) and **multipliers** (see **voltmeter**) necessary to measure currents and **potential differences***.

- **Moving iron meter**. A **meter** in which the current to be measured induces magnetism in two pieces of iron which attract or repel each other to produce a deflection.

* **Coil**, 74; **Direct current, 60**; **Potential difference**, 58; **Resistance**, 62; **Series**, 64.

Electromagnetic induction

Michael Faraday found that, as well as a current passing through a magnetic field producing movement (see **motor effect**, page 76), movement of a **conductor*** in a magnetic field produces an **electromotive force*** in the conductor. This effect, called **electromagnetic induction**, happens whenever a conductor is placed in a changing magnetic field.

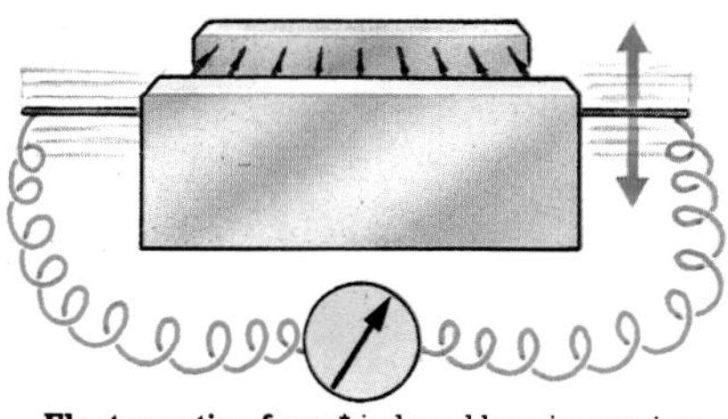

Electromotive force* induced by wire moving through magnetic field produces current, which then causes current through **galvanometer***.

- **Faraday's law of induction**. States that the size of an induced electromotive force in a **conductor*** is proportional to the rate at which the magnetic field changes.
- **Lenz's law**. States that an induced electromotive force always acts to oppose the cause of it, e.g. in an **electric motor***, the e.m.f. produced because it acts as a **generator** opposes the e.m.f. driving the motor.
- **Fleming's right-hand rule** or **dynamo rule**. The direction of an induced current can be worked out from the direction of the magnetic field and the movement by using the right hand (see diagram).

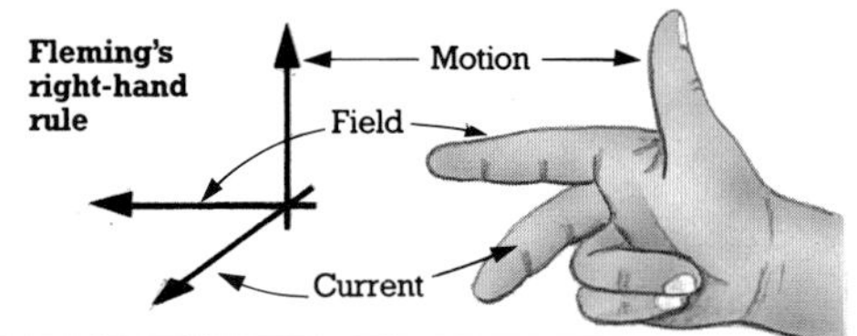

- **Generator** or **dynamo**. A device used to produce electric current from **mechanical energy***. In the simplest generator (see diagram), an alternating electromotive force is induced in a **coil*** as it rotates in a magnetic field. A generator for **direct current*** has a **commutator***, as on an **electric motor***, which means the current always flows in the same direction.

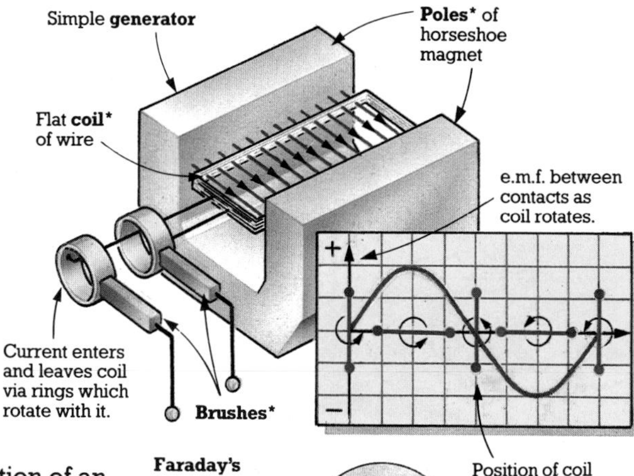

- **Mutual induction**. The induction of an electromotive force in a **coil*** of wire by changing the current in a different coil. The changing current produces a changing magnetic field which induces a current in any other coil in the field. This was first demonstrated with **Faraday's iron ring**.

Faraday's iron ring

Primary circuit

Secondary circuit

Soft iron ring – "conducts" fields between coils.

Closing or opening **switch*** causes change in magnetic field in ring which induces current in secondary circuit.

* **Brushes**, 76; **Coil**, 74; **Commutator**, 76; **Conductor**, 56; **Direct current**, 60; **Electric motor**, 76; **Electromotive force**, 60; **Galvanometer**, 77; **Mechanical energy**, 9; **Pole**, 70; **Switch**, 64.

- **Self-induction**. The induction of an electromotive force in a **coil*** of wire due to the current inside it changing. For example, if the current in a coil is switched off, the resulting change in the magnetic field produces an electromotive force across the coil, in some cases much higher than that of the original.

- **Eddy current**. A current set up in a piece of metal when a magnetic field around it changes, even though the metal may not be part of a circuit. Eddy currents can cause unwanted heat energy, e.g. in the iron core of a **transformer**.

Transformers

A **transformer** consists of two **coils*** of wire wound onto the same **core*** of soft **ferromagnetic*** material. It is used to change an alternating electromotive force in one of the coils to a different e.m.f. in the other coil, e.g. in electricity supply, see page 61. Hardly any energy is lost between the two circuits in a well designed transformer.

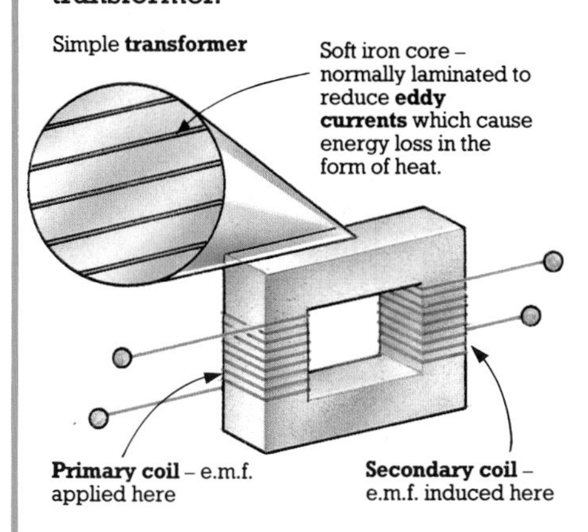

- **Primary coil**. The **coil*** in a **transformer** to which an alternating electromotive force is applied in order to produce an electromotive force in the **secondary coil**.

- **Secondary coil**. The **coil*** in a **transformer** in which an alternating electromotive force is induced by the electromotive force applied to the **primary coil**. Some transformers have two or more secondary coils.

- **Turns ratio**. The ratio of the number of turns in the **primary coil** in a **transformer** to the number of turns in the **secondary coil**. The turns ratio is also the ratio of the electromotive force in the primary coil to the electromotive force in the secondary coil.

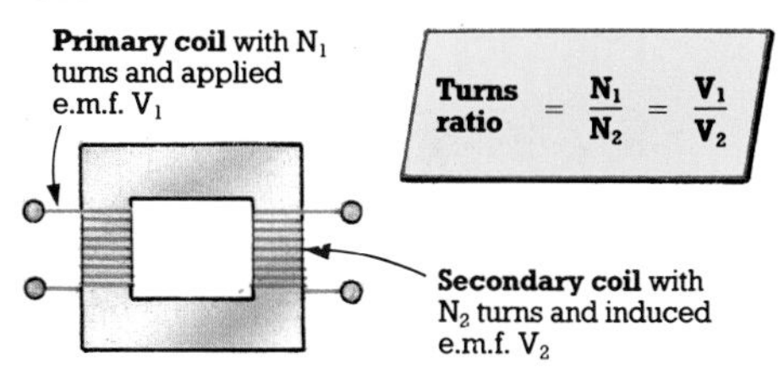

$$\text{Turns ratio} = \frac{N_1}{N_2} = \frac{V_1}{V_2}$$

- **Step-up transformer**. A **transformer** in which the electromotive force in the **secondary coil** is greater than that in the **primary coil**. The **turns ratio** is less than one.

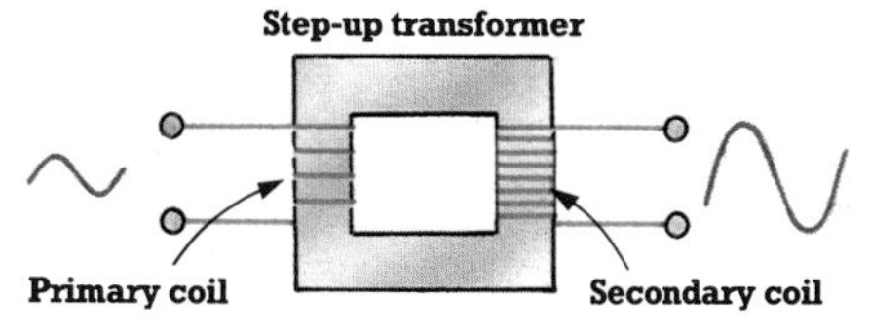

- **Step-down transformer**. A **transformer** in which the electromotive force in the **secondary coil** is less than that in the **primary coil**. The **turns ratio** is greater than one.

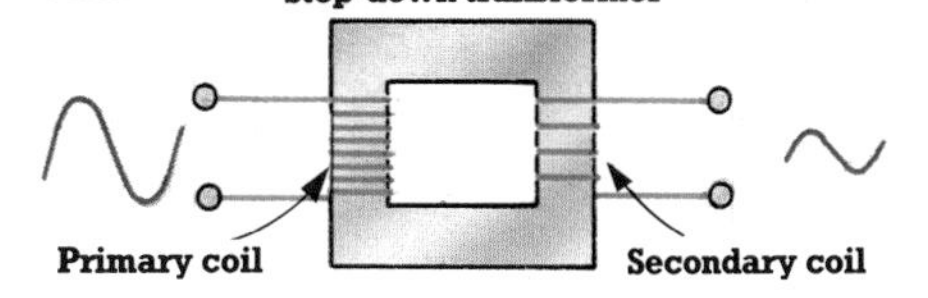

* **Coil**, **Core**, 74; **Ferromagnetic**, 70.

Cathode rays

A **cathode ray** is a continuous stream of **electrons** (negatively-charged particles – see page 83) travelling through a low pressure gas or a vacuum. It is produced when electrons are freed from a metal **cathode*** and attracted to an **anode***. Cathode rays have a number of applications, from the production of **X-rays*** to **television**, all of which involve the use of a shaped glass tube (called an **electron tube**) containing a low pressure gas or a vacuum for the rays to travel in. The rays are normally produced by an **electron gun**, which forms part of the tube.

- **Electron gun**. A device which produces a continuous stream of electrons (a cathode ray). It consists of a heated **cathode*** which gives off electrons (this is called **thermionic emission**) and an **anode*** which attracts them to form a stream.

- **Maltese cross tube**. An **electron tube** in which the cathode ray is interrupted by a cross which casts a "shadow" on a **fluorescent*** screen at the end of the tube. This shows the electrons are moving in straight lines.

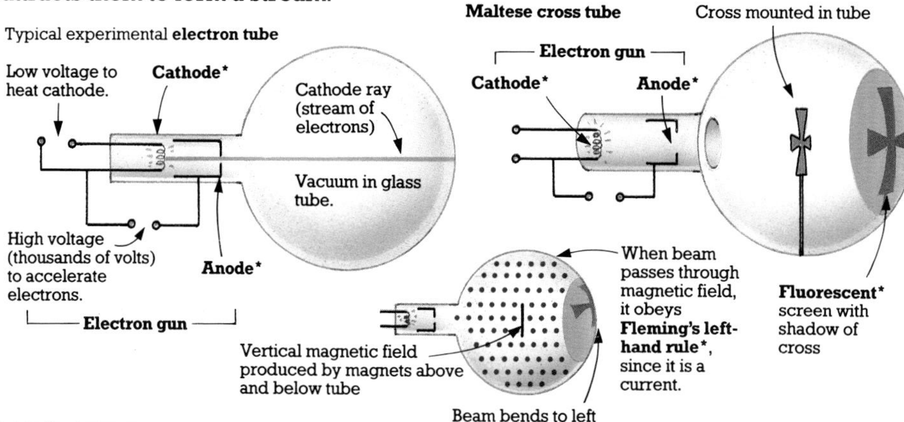

- **Discharge tube**. A gas-filled glass tube in which **ions*** and electrons are attracted by the **electrodes*** and move towards them at high speed. As they do so, they collide with gas atoms, causing these atoms to split into more ions and electrons, and emit light at the same time. The colour of the light depends on the gas used, e.g. neon produces orange light (used in advertising displays) and mercury vapour produces blue-green light (used for street lighting). Discharge tubes use up to five times less electricity than other lighting. A **fluorescent tube** is a discharge tube filled with mercury vapour, which emits **ultraviolet radiation***. This hits the inside of the tube, causing its coating of special powder to give out **visible light*** (see **fluorescence**, page 45).

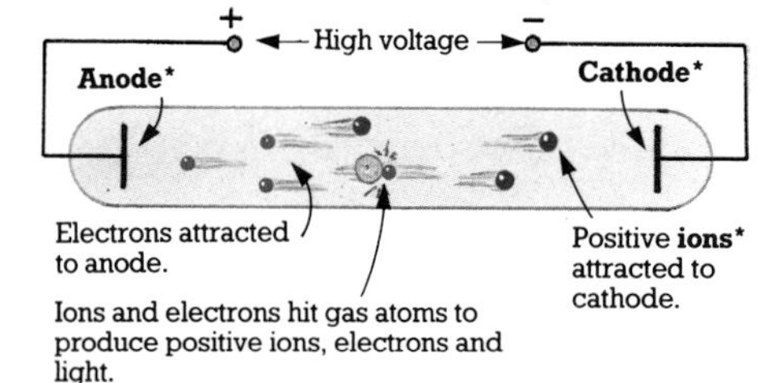

* **Anode, Cathode**, 66 (**Electrode**); **Fleming's left-hand rule** 76; **Fluorescence**, 45; **Ions**, 88 (**Ionization**); **Ultraviolet radiation**, 44; **Visible light**, 45; **X-rays**, 44.

- **X-ray tube**. A special **electron tube** used to produce a beam of **X-rays***. A cathode ray hits a tungsten target which stops the electrons suddenly. This causes X-rays to be produced in a beam.

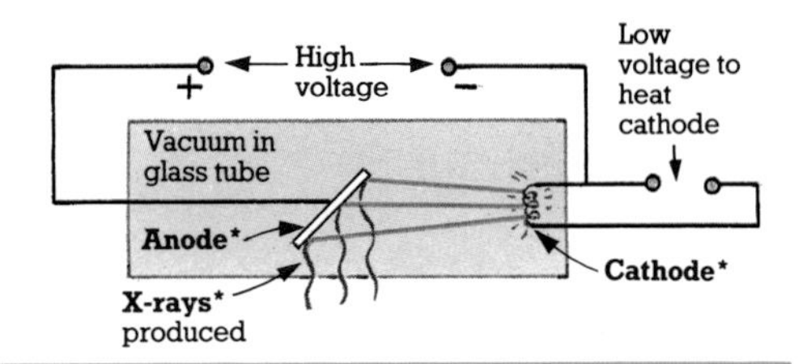

The cathode ray oscilloscope

The **cathode ray oscilloscope (CRO)** is an instrument used to study currents and **potential differences***. A cathode ray from an **electron gun** produces a spot on a **fluorescent*** screen. In normal use, the ray is repeatedly swept across the back of the screen at a selected speed and so produces a visible trace across the front. If a signal is fed into the oscilloscope, the vertical position of the beam will change according to the strength of the signal and the trace on the screen then shows this change over time.

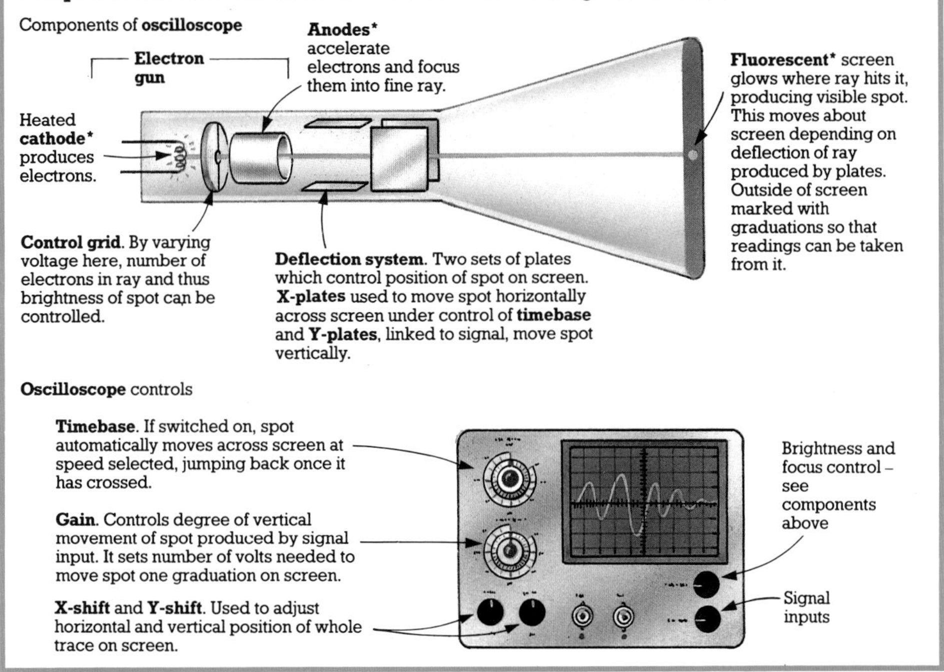

Television

Television pictures are reproduced by using an **electron tube** in which the cathode ray scans across the screen varying in strength according to the signal. Different levels of light, according to the strength of the ray, are given off from different parts of the screen to produce a picture.

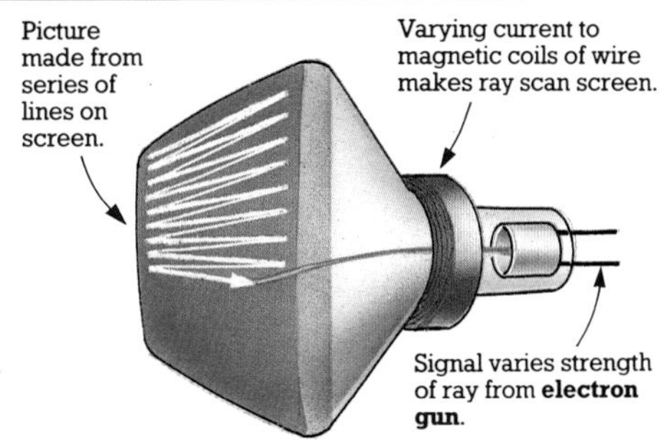

* **Anode**, **Cathode**, 66 (**Electrode**); **Fluorescence**, 45; **Potential difference**, 58; **X-rays**, 44.

Atomic structure

A great deal has been learnt about the physical nature of atoms (see also page 4) since Greek philosophers first proposed that all matter was made of basic indivisible "building blocks". It is now known that an atom is not indivisible, but has a complex internal structure, consisting of many different smaller particles (**subatomic particles**) and a lot of empty space.

- **Rutherford-Bohr atom**. A "solar system" representation of an atom, devised by Ernest Rutherford and Niels Bohr in 1911. It is now known to be incorrect (**electrons** have no regular "orbits" – see **electron shells**).

Rutherford-Bohr atom model

Positively-charged "sun" (heavy **nucleus** – thought at the time to be just protons; neutron not yet discovered)

Negatively-charged "planets" (**electrons**) kept in "orbits" by **electric force*** of attraction

Empty space

- **Nucleus** (pl. **nuclei**) or **atomic nucleus**. The central core of an atom, consisting of closely-packed **nucleons** (**protons** and **neutrons**).

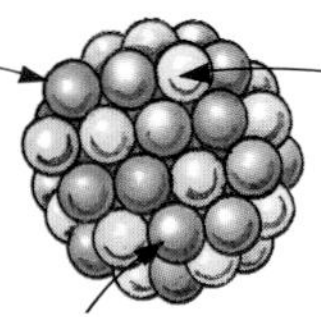

Nucleus – (almost all the mass of the atom, but very tiny – its radius is approx. 1/10000th that of the atom)

Proton (mass approx. 1836 times that of an **electron**)

Neutron (mass approx. 1840 times that of an **electron**)

- **Protons**. Positively charged particles in the **nucleus**. The number of protons (**atomic number**) identifies the element and equals the number of **electrons**, so atoms are electrically neutral.
- **Neutrons**. Electrically neutral particles in the **nucleus**. The number of neutrons in atoms of the same element can vary (see **isotope**).

- **Mass number (A)**. The number of **protons** and **neutrons** (**nucleons**) in a **nucleus**. It is the whole number nearest to the **relative atomic mass** of the atom, and is important in identifying **isotopes**.
- **Atomic number (Z)**. The number of **protons** in a **nucleus** (hence also the number of **electrons** around it). All atoms with the same atomic number are of the same element (see also **isotope**).
- **Neutron number (N)**. The number of **neutrons** in a **nucleus**. See also graph, page 87.

The **mass** and **atomic numbers** are often written with the symbol of an element:

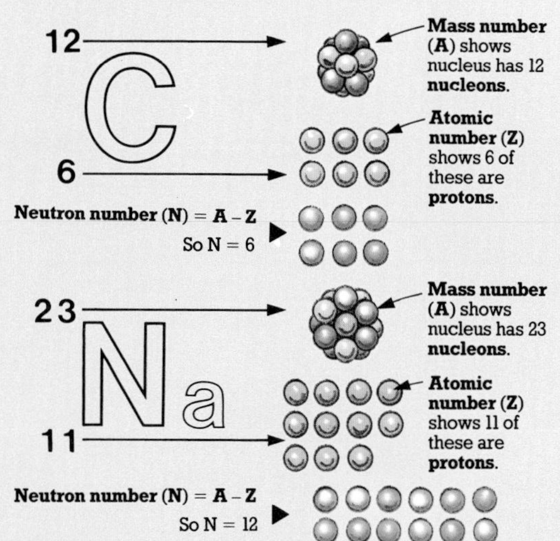

*Electric force, 6.

• **Electrons**. Particles with a negative charge and very small mass. They move around the **nucleus** in **electron shells**. See also **proton**.

• **Electron shells** or **energy levels**. Regions of space around a **nucleus** containing moving **electrons**. An atom can have up to seven (from the inside, called the **K**, **L**, **M**, **N**, **O**, **P** and **Q shells**). Each can hold up to a certain number of electrons (the first four, from the inside, can take up to 2, 8, 18 and 32 electrons respectively). The further away the shell is from the nucleus, the higher the energy of its electrons (hence the name energy level). The **outer shell** is the last shell with electrons in it. If this is full or has an **octet** (8 electrons), the atom is very stable (see page 84).

The positions of electrons in their shells cannot be exactly determined at any one time, but each shell consists of **orbitals**, or **probability clouds**. Each of these is a region in which one or two electrons are likely to be found at any time.

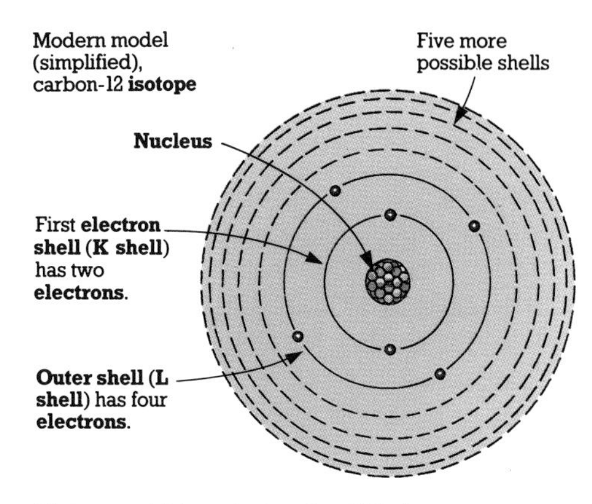

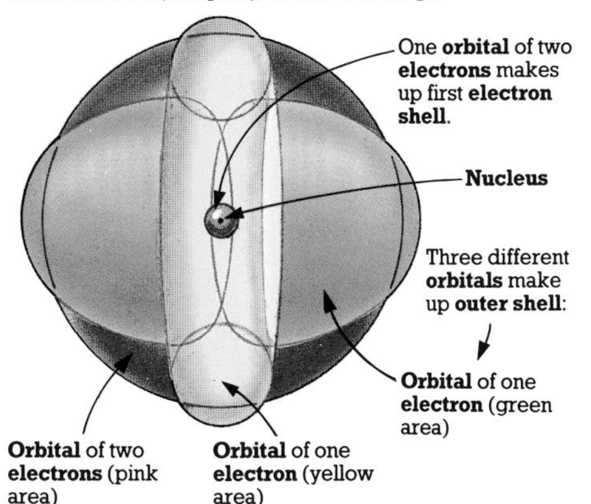

• **Isotopes**. Different forms of the same element, with the same atomic number, but different **neutron numbers** and hence different **mass numbers**. There are isotopes of every element, since even if only one natural form exists (i.e. the element is **monoisotopic**), others can be made artificially (see **radioisotope**, page 86).

Mass numbers are used with names or symbols when **isotopes** are being specified:

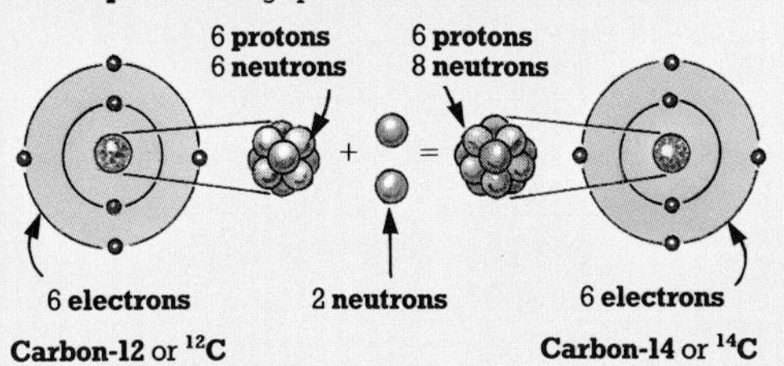

• **Relative atomic mass**. Also called **atomic mass** or **atomic weight**. The mass of an atom in **unified atomic mass units (u)**. Each of these is equal to 1/12 of the mass of a carbon-12 atom (**isotope**). The relative atomic mass of a carbon-12 atom is thus 12 u, but no other values are whole numbers, e.g. 26.9815 u (aluminium).

The relative atomic mass takes into account the various isotopes of the element, if these occur in a natural sample. Natural chlorine, for example, has three chlorine-35 atoms to every one of chlorine-37, and the relative atomic mass of chlorine (35.453 u) is a proportional average of the two different masses of these isotopes.

Atomic and nuclear energy

All things, whether large objects or minute particles, have a particular **energy state**, or level of **potential energy*** ("stored" energy). Moreover, they will always try to find their lowest possible energy state, called the **ground state**, which is the state of the greatest stability. In most cases, this involves recombining in some way, i.e. adding or losing constituents, and in all cases it results in the release of the "excess" energy – in large amounts if the particles are atoms, and vast amounts if they are nuclei. The greater the **binding energy** of an atom or nucleus, the greater its stability, i.e. the less likely it is to undergo any change.

- **Mass defect**. The mass of an atom or nucleus is less than the sum of the masses of its parts when these are apart. The difference is the mass defect. It is the mass of the **potential energy*** lost when the parts came together (see **binding energy** and formula below).

Einstein showed that energy has mass. Hence any loss of **potential energy*** also results in a loss of mass – the mass of the energy itself. **Einstein's mass-energy formula**:

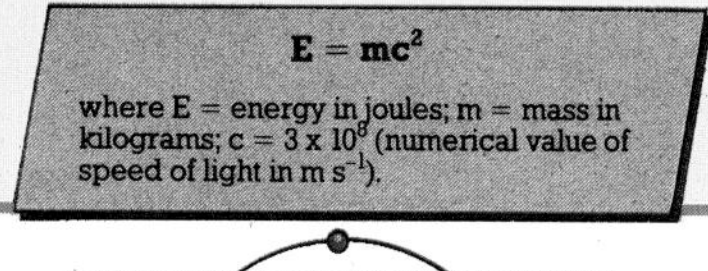

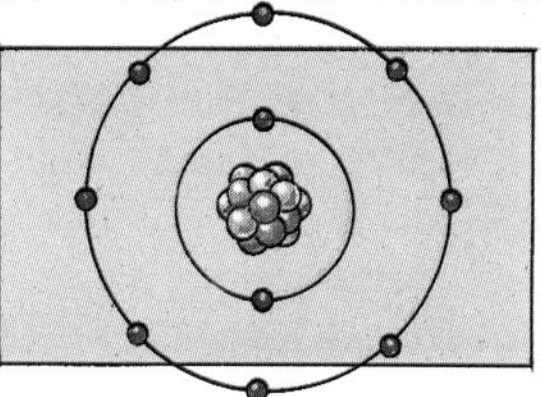

Atoms with full **outer shells***.

Greatest stability, highest **B. E.** or **B.E. per nucleon** (see below).

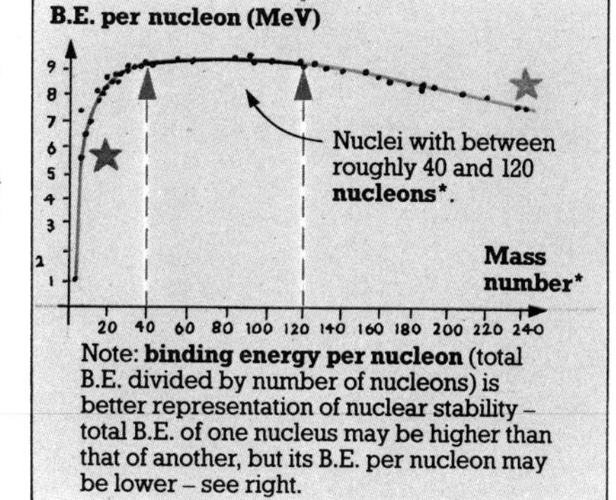

Note: **binding energy per nucleon** (total B.E. divided by number of nucleons) is better representation of nuclear stability – total B.E. of one nucleus may be higher than that of another, but its B.E. per nucleon may be lower – see right.

- **Binding energy (B.E.)**. The energy input needed to split a given atom or nucleus into its constituent parts (see pages 82-83). The **potential energy*** of an atom or nucleus is less than the total potential energy of its parts when these are apart. This is because, when they came together, the parts found a lower (collective) **energy state** (see introduction and **nuclear force**), and so lost energy. The binding energy is a measure of this difference in potential energy – it is the energy needed to "go back the other way" – so the greater it is, the lower the potential energy of an atom or nucleus and the greater its stability. Binding energy varies from atom to atom and nucleus to nucleus.

- **Nuclear force**. The strong attractive force which holds the parts of a nucleus (**nucleons***) together and overcomes the **electric force*** of repulsion between the **protons***. Its effect varies according to the size of the nucleus (see graph) as the force only acts between immediately neighbouring nucleons. The greater the attractive effect of the nuclear force, the higher the **binding energy** of the nucleus (i.e. the more energy was lost when the parts came together).

* **Electric force**, 6; **Mass number**, 82; **Nucleons**, 82 (**Nucleus**); **Outer shell**, 83 (**Electron shells**); **Potential energy**, 8; **Protons**, 82.

- **Quantum theory**. States that energy takes the form of minute, separate "chunks" called **quanta** (sing. **quantum**), rather than a steady stream. The theory was originally limited to energy emitted by bodies (i.e. **electromagnetic wave*** energy), though all other kinds of energy (see pages 8-9) are now generally included. Electromagnetic quanta are now specified as **photons**. The theory further states that the amount of energy carried by a photon is proportional to the **frequency*** of the emitted electromagnetic radiation (see pages 44-45).

Energy carried by **quantum** (**photon**):

$$E = hf$$

where E = energy in joules; h = **Planck's constant** (6.63×10^{-34} J s^{-1}); f = **frequency*** in Hertz.

Electron volt (**eV**). Unit of atomic energy, equal to **kinetic energy*** gained by one electron moved through **potential difference*** of 1V.

$$1 \text{ eV} = 1.6 \times 10^{-19} \text{ J}$$

Megaelectron volt (**MeV**). Unit of nuclear energy, equal to 1 million eV.

$$1 \text{ MeV} = 1.6 \times 10^{-13} \text{ J}$$

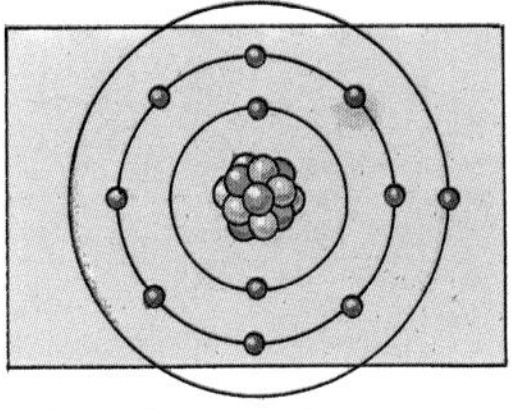

Few **electrons*** in outer shell, or nearly a full outer shell, i.e. just one or two electrons missing.

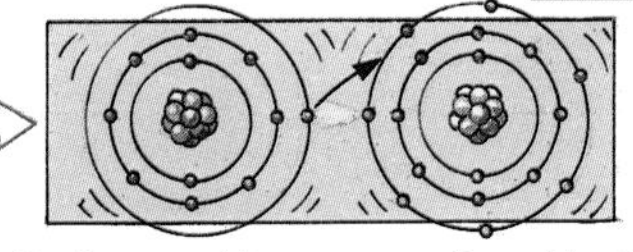

Putting unstable atoms together with other unstable atoms causes them to react – electrons transferred, i.e. recombinations take place.

Change in energy state of atom or nucleus is atomic or nuclear **transition**. Change resulting in change in chemical properties (i.e. different element) is **transformation** or **transmutation**.

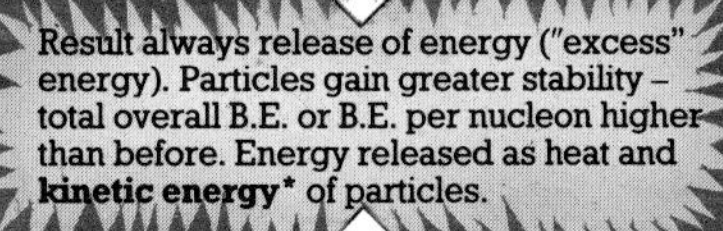

Result always release of energy ("excess" energy). Particles gain greater stability – total overall B.E. or B.E. per nucleon higher than before. Energy released as heat and **kinetic energy*** of particles.

Less stable, low B.E. or B.E. per nucleon.

As shown, finding higher B.E. (lower energy state) may occur with very little "help", e.g. putting unstable atoms together or in the case of spontaneous radioactivity. In other cases, energy may need to be added (heating atoms, nuclei or particles).

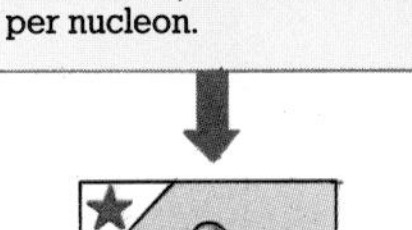

Few **nucleons*** and hence relatively large surface area. Only a few nucleons (relative to whole number) have "pulling" **nuclear force** acting on them from neighbours on all sides, so overall effect of nuclear force is less.

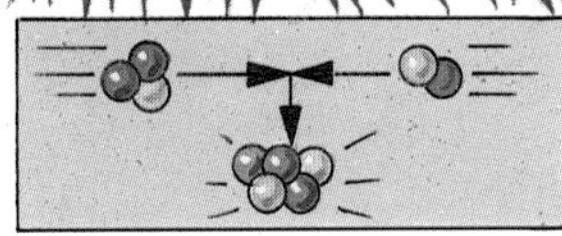

Heating nuclei gives them high **kinetic energy*** and means two will join when they collide (see **fusion**, page 93).

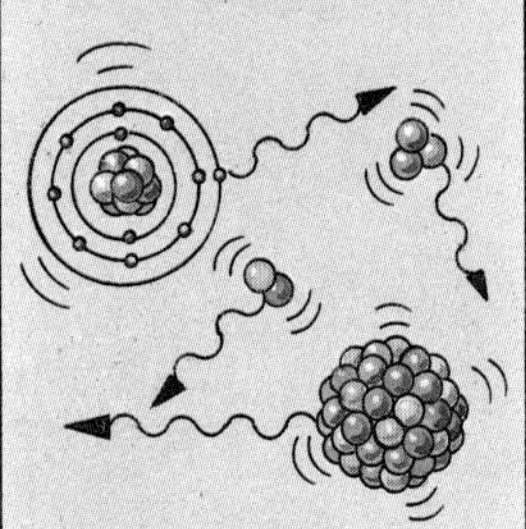

Or

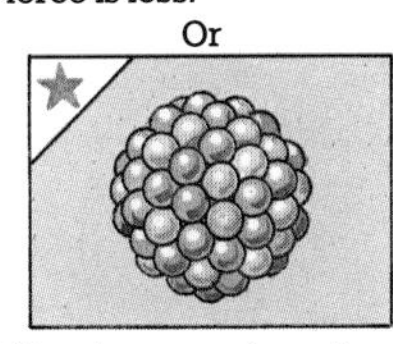

Very large number of nucleons. Means more **protons*** – **electric force*** of repulsion has more effect, overall effect of nuclear force is less.

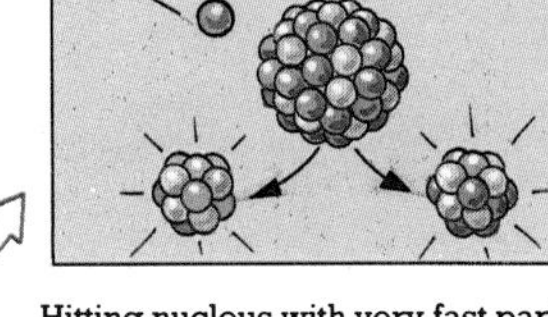

Hitting nucleus with very fast particle causes it to split (see **fission**, page 92).

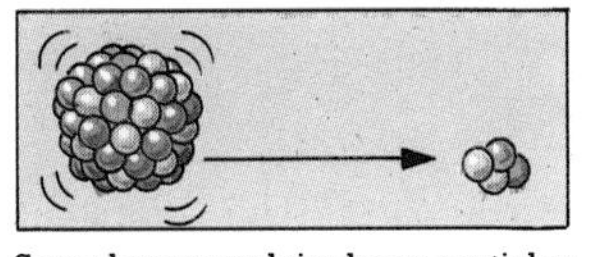

Some heavy nuclei release particles spontaneously (see **radioactivity**, page 86).

If not enough energy added, energy re-released as **photons** (see **quantum theory**). In atoms, electrons "fall" back and type (**frequency***) of photons emitted depends on shells between which they move. **X-rays*** (highest frequency, most energy) if innermost shells, **UV radiation*** if shells further out, and so on (see spectrum, page 44). With nuclei, always **γ-rays*** (more energy involved).

* **Electric force**, 6; **Electromagnetic waves**, 44; **Electrons**, 83; **Frequency**, 16, 35; **Kinetic energy**, 9; **Nucleons**, 82 (**Nucleus**); **Potential difference**, 58; **Protons**, 82; **Ultra-violet** (**UV**) **radiation, X-rays**, 44; **Gamma** (γ) **rays**, 44, 86.

Radioactivity

Radioactivity is a property of some unstable **nuclei** (see pages 82 and 84), whereby they break up spontaneously into nuclei of other elements and emit **radiation***, a process known as **radioactive decay**.

There are three types of radiation emitted by radioactive elements – streams of **alpha particles** (called **alpha rays**), streams of **beta particles** (**beta rays**) and **gamma rays**. For more about the detection and uses of radiation, see pages 88-91.

- **Radioisotope** or **radioactive isotope**. Any radioactive substance (all substances are effectively **isotopes** – see page 83). There are several naturally-occurring radioisotopes, most of which still exist because they have very long **half-lives** (e.g. uranium-238), though one, carbon-14, is continually produced by **cosmic rays** (see **background radiation**, page 88). Other radioisotopes are produced by **nuclear fission***, and more still are produced in research centres, where nuclei are hit by very fast particles (ranging from **protons*** to uranium nuclei). These are speeded up in **particle accelerators** of many different types, e.g. **cyclotrons**.

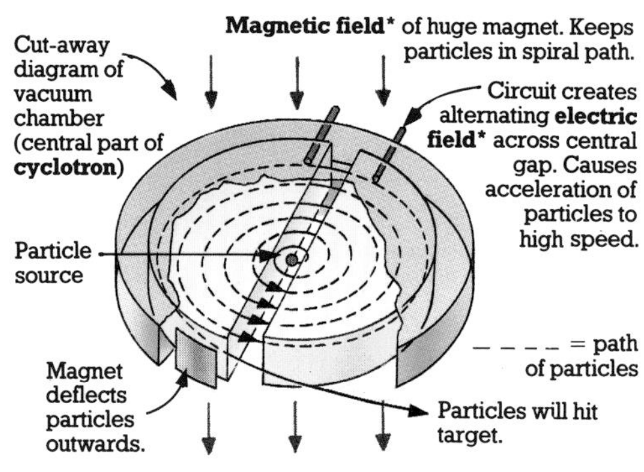

- **Alpha particles** (**α-particles**). Positively-charged particles ejected from some radioactive nuclei (see **alpha decay**). They are relatively heavy (two **protons*** and two **neutrons***), move relatively slowly and have a low penetrating power.

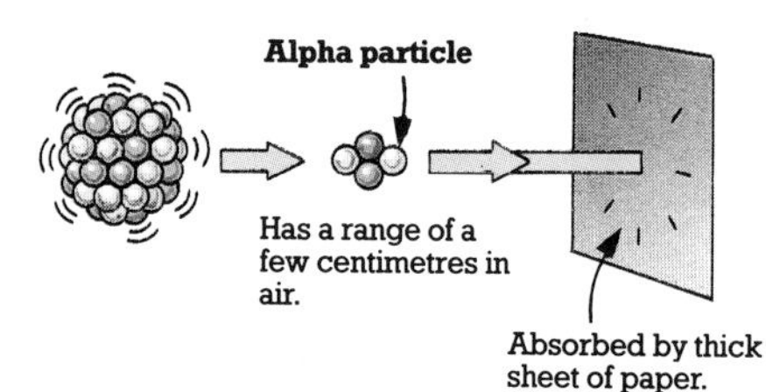

- **Beta particles** (**β-particles**). Particles ejected from some radioactive nuclei at about the speed of light. There are two types – **electrons*** and **positrons**, which have the same mass as electrons, but a positive charge. See **beta decay**.

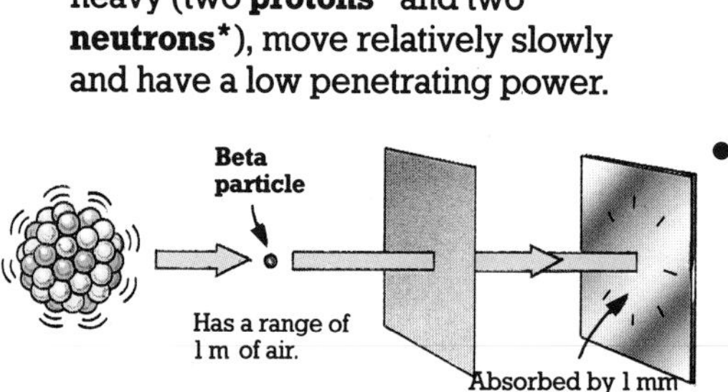

- **Gamma rays** (**γ-rays**). Invisible **electromagnetic waves** (see also page 44). They have the highest penetrating power and are generally, though not always, emitted from a radioactive nucleus after an **alpha** or **beta particle**.

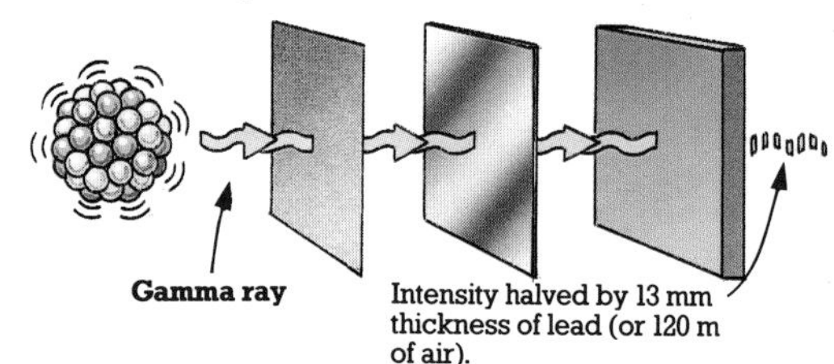

* **Electric field**, 58; **Electrons**, 83; **Magnetic field**, 72; **Neutrons**, 82; **Nuclear fission**, 92; **Protons**, 82; **Radiation**, 9.

- **Radioactive decay**. The spontaneous splitting up of a radioactive nucleus. It shoots out bits of itself as **alpha** or **beta particles**, normally followed by **gamma rays**. When a nucleus ejects such a particle, i.e. undergoes a nuclear **disintegration**, energy is released (see page 84), and a different nucleus (and atom) is formed. If this is also radioactive, the decay process continues until a stable (non-radioactive) atom is reached. Such a series of disintegrations is called a **decay series, radioactive series, decay chain** or **transformation series**.

- **Half-life (T½)**. The time it takes for half the atoms in a sample to undergo **radioactive decay**, and hence for the radiation emitted to be halved. This is all that can be accurately predicted – it is impossible to predict the decay of any single atom, since they decay individually and randomly. The range of half-lives is vast, e.g. strontium-90, 9 minutes; uranium-238, 4.5 x 109 years.

The rate of **radioactive decay** is measured in **becquerels (Bq)**. One becquerel equals one **disintegration** per second. An older unit, the **curie**, equals 3.7×10^{10} becquerels.

Decay series, showing **radioactive decay** of thorium-232 to stable lead-208 ▼

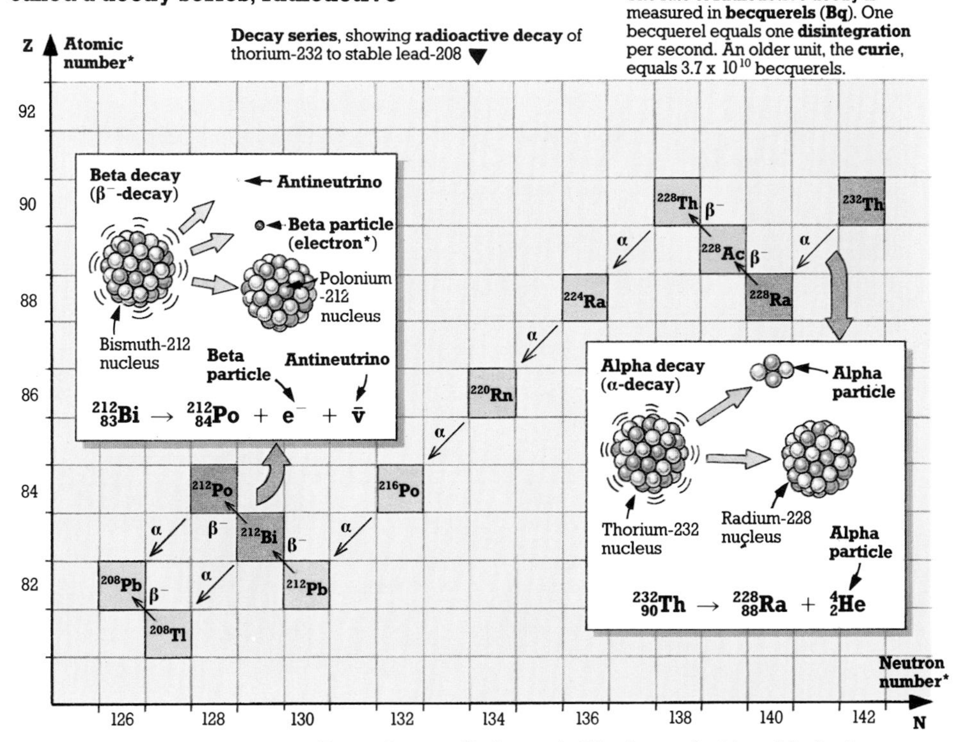

- **Alpha decay (α-decay)**. The loss of an **alpha particle** by a radioactive nucleus. This decreases the **atomic number*** by two and the **mass number*** by four, and so a new nucleus is formed.

- **Beta decay (β-decay)**. The loss of either kind of **beta particle** by a radioactive nucleus. The electron (β^- or e^-) is ejected (with another particle called an **antineutrino**) when a **neutron*** decays into a **proton***. The positron (β^+ or e^+) is ejected (with another particle called a **neutrino**) when a proton decays into a neutron. Beta decay thus increases or decreases the **atomic number*** by one (the **mass number*** stays the same).

* **Atomic number**, 82; **Electrons**, 83; **Mass number, Neutron number, Neutrons, Protons**, 82.

Detecting and measuring radioactivity

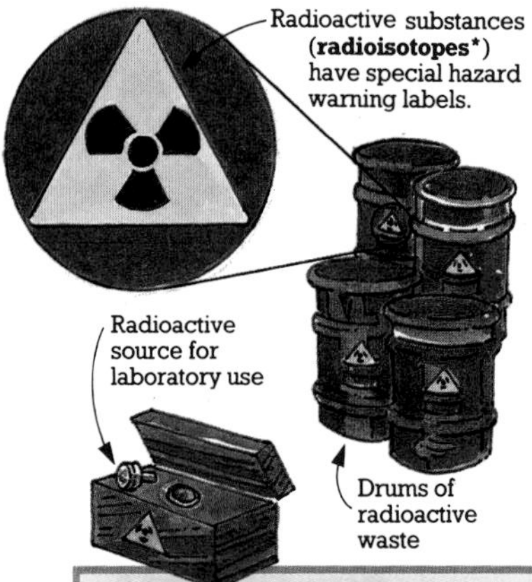

There are a number of devices which detect and measure the radiation emitted by radioactive substances (**radioisotopes***). Some are used mainly in laboratories (to study artificially produced radioisotopes); others have a wider range of uses (e.g. as monitoring devices for safety purposes) and can also be used to detect **background radiation**. Most of the devices detect and measure the radiation by monitoring the **ionization** it causes – see **Geiger counter** and **pulse electroscope**, right, and **cloud** and **bubble chambers**, page 90.

- **Background radiation**. Radiation present on earth (in relatively small amounts), originating both from natural and unnatural sources. One notable natural source is carbon-14, which is taken in by plants and animals. This is constantly being produced from stable nitrogen-14 due to bombardment by **cosmic rays** (**cosmic radiation**) entering the atmosphere from outer space. These are streams of particles of enormously high energy. The **background count** is a measure of the background radiation.

- **Ionization**. The creation of **ions** (electrically-charged particles), which occurs when atoms (electrically neutral) lose or gain **electrons***, creating **cations** (positive ions) or **anions** (negative ions)

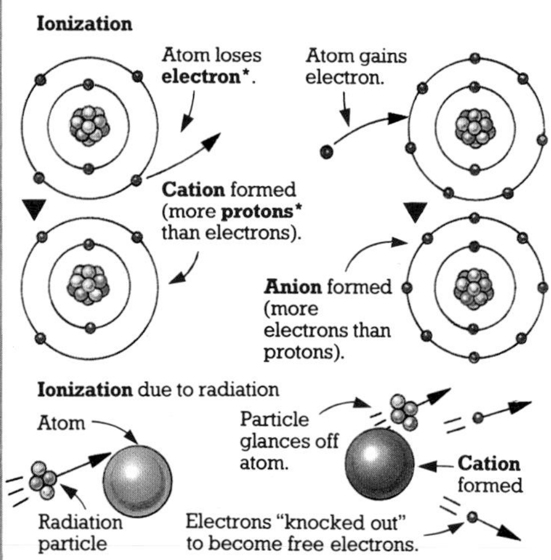

respectively. In the case of radiation, **alpha** and **beta particles*** ionize the atoms of substances they pass through, usually creating cations. This is because their energy is so high that they cause one or more electrons to be "knocked out". For more about ions and ionization, see pages 130-131.

* **Alpha particles**, **Beta particles**, 86; **Electrons**, 83; **Gamma rays**, 86; **Protons**, 82; **Radioisotope**, 86.

Detection devices

- **Geiger counter**. A piece of apparatus consisting of a **Geiger-Müller tube**, a **scaler** and/or **ratemeter** and often a loudspeaker. The tube is a gas-filled cylinder with two **electrodes*** – its walls act as the **cathode***, and it has a central wire **anode***. The whole apparatus indicates the presence of radiation by registering pulses of current between the electrodes. These pulses result from the **ionization** the radiation causes in the gas (normally low pressure argon, plus a trace of bromine). A scaler is an electronic counter which counts the pulses and a ratemeter measures the count rate – the average rate of pulses in counts per second.

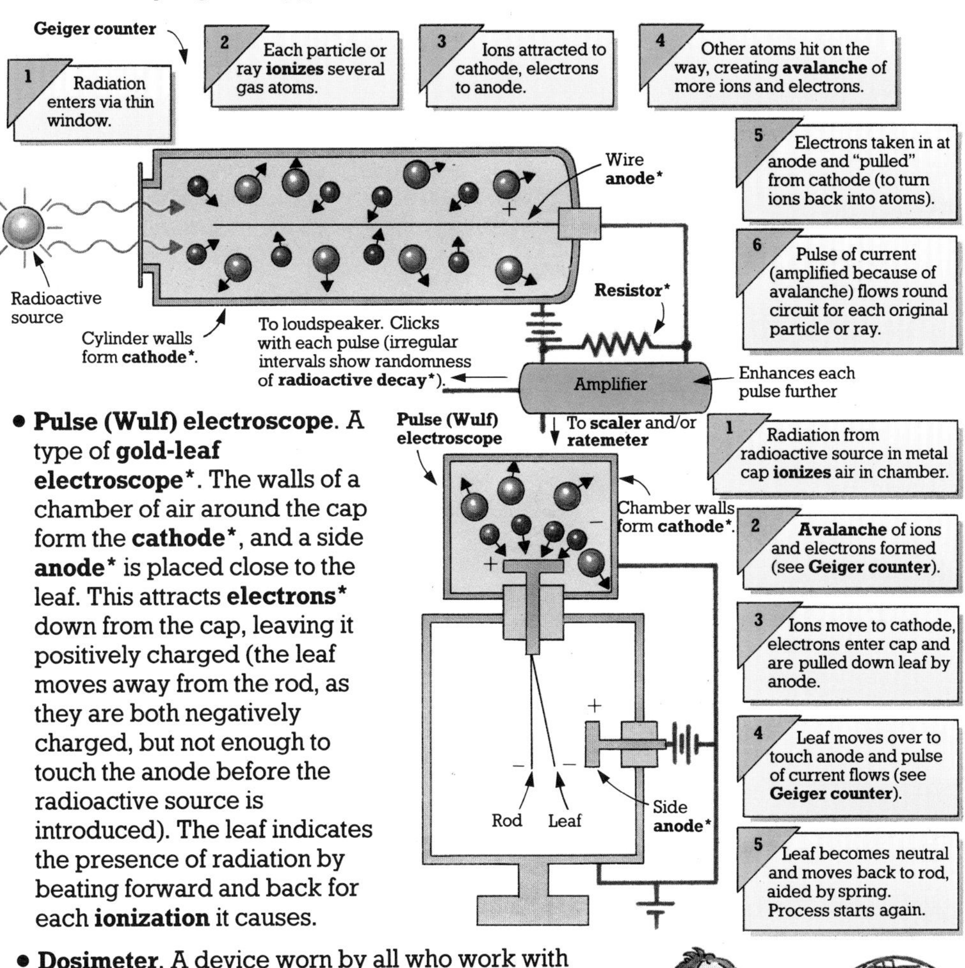

- **Pulse (Wulf) electroscope**. A type of **gold-leaf electroscope***. The walls of a chamber of air around the cap form the **cathode***, and a side **anode*** is placed close to the leaf. This attracts **electrons*** down from the cap, leaving it positively charged (the leaf moves away from the rod, as they are both negatively charged, but not enough to touch the anode before the radioactive source is introduced). The leaf indicates the presence of radiation by beating forward and back for each **ionization** it causes.

- **Dosimeter**. A device worn by all who work with radioactive material. It contains photographic film (which radiation will darken). This is developed regularly and the amount of darkening shows the **dose** of radiation the wearer has been exposed to.

Dosimeter

* **Anode**, **Cathode**, **Electrode**, 66; **Electrons**, 83;
Gold-leaf electroscope, 56 (**Electroscope**); **Radioactive decay**, 87; **Resistor**, 62.

Detection devices (continued)

- **Cloud chamber**. A device in which the paths taken by **alpha** and **beta particles*** show up as tracks. This happens when the vapour in the chamber (alcohol or water vapour) is turned into **supersaturated** vapour by cooling (in one of two different ways – see below). This is vapour below the temperature at which it should condense, but which does not condense because there are no dust or other particles for droplets to form around.

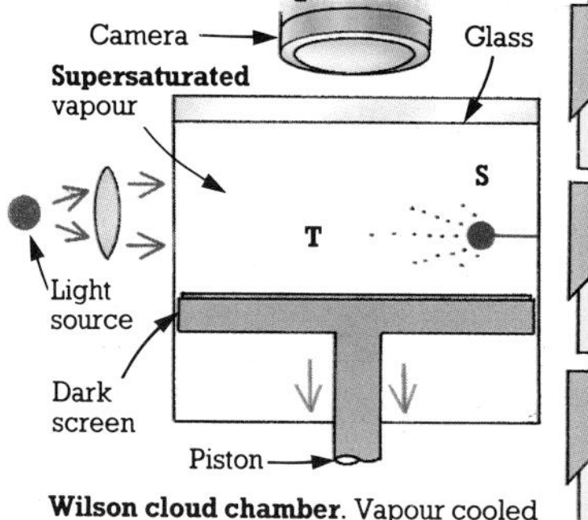

Wilson cloud chamber. Vapour cooled by sudden increase in volume (withdrawal of piston).

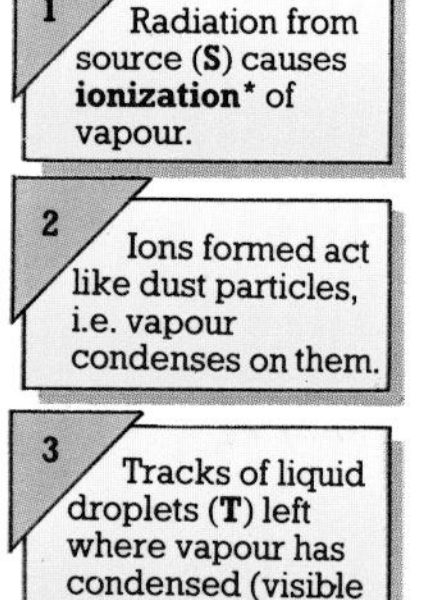

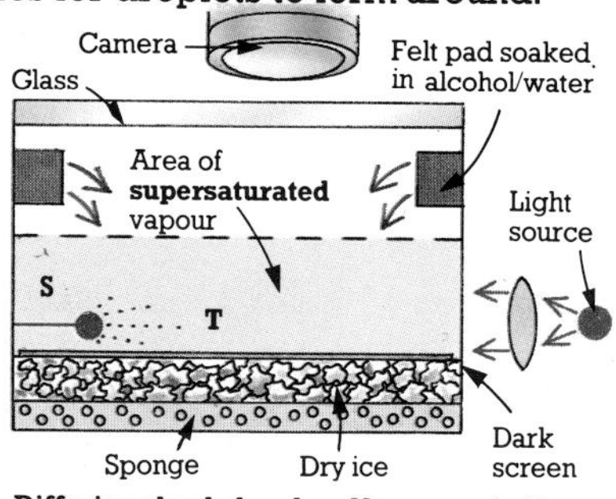

Diffusion cloud chamber. Vapour cooled by a base of dry ice (solid carbon dioxide). Vapour **diffuses*** downwards.

Cloud chamber tracks (produced at irregular intervals, showing random nature of **radioactive decay***).

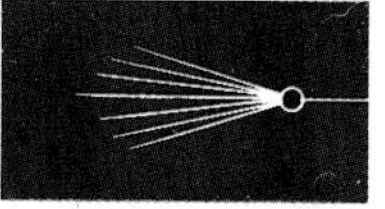

Tracks made by heavy **α-particles*** are short, straight and thick.

Tracks made by light **β-particles*** are long, straggly and thin.

Gamma rays* do not create tracks themselves, but can knock **electrons*** out of single atoms. These then speed away and create tracks like β-particle tracks (see left).

- **Bubble chamber**. A device which, like a cloud chamber, shows particle tracks. It contains **superheated** liquid (usually hydrogen or helium) – liquid heated to above its boiling point, but not actually boiling because it is under pressure. After the pressure is suddenly lowered, nuclear particles entering the chamber cause **ionization*** of the liquid atoms. Wherever this occurs, the energy released makes the liquid boil, producing tracks of bubbles.

Bubble chamber tracks. Generally curved, because magnetic field set up to deflect particles (leads to better identification).

- **Scintillation counter**. A device which detects **gamma rays***. It consists of a **scintillation crystal** and a **photomultiplier** tube. The crystal is made of a **phosphor*** (e.g. sodium iodide). Phosphors emit light flashes (**scintillations**) when hit by radiation.

Scintillation counter

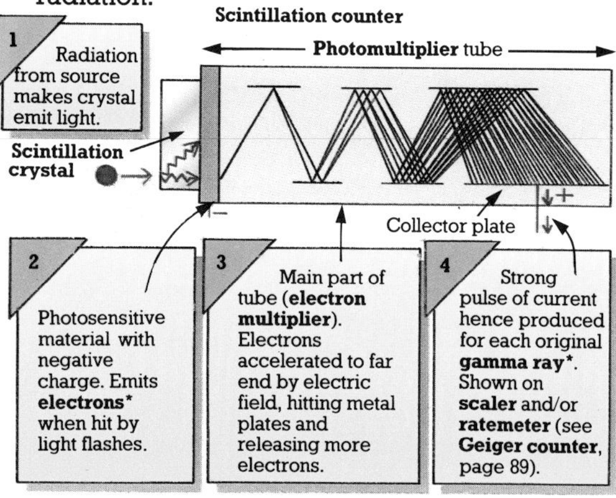

* **Alpha particles**, **Beta particles**, 86; **Diffusion**, 5; **Electrons**, 83; **Gamma rays**, 86; **Ionization**, 88; **Phosphor**, 45 (**Phosphorescence**); **Radioactive decay**, 87.

Uses of radioactivity

The radiation emitted by **radioisotopes*** (radioactive substances) can be put to a variety of uses, particularly in the fields of medicine, industry and archaeological research.

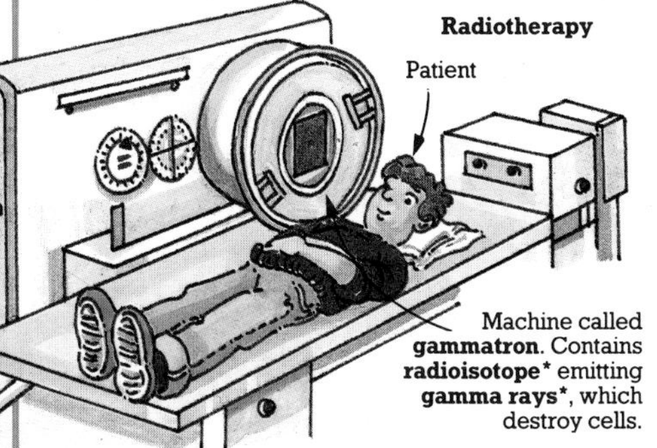

- **Radiology**. The study of radioactivity and **X-rays***, especially with regard to their use in medicine.

- **Radiotherapy**. The use of the radiation emitted by **radioisotopes*** to treat disease. All living cells are susceptible to radiation, so it is possible to destroy malignant (cancer) cells by using carefully controlled doses of radiation.

- **Radioactive tracing**. A method of following the path of a substance through an object, and detecting its concentration as it moves. This is done by introducing a **radioisotope*** into the substance and tracking the radiation it emits. The radioisotope used is called a **tracer**, and the substance is said to be **labelled**. In medical diagnosis, for example, high levels of the radioisotope in an organ may indicate the presence of malignant (cancer) cells. The radioisotopes used always have short **half-lives*** and decay into harmless substances. ▶

Patient has swallowed **labelled** substance.

Scanner (**gamma camera**) picks up **radioisotope*** in body.

Sample **scan** (doctor can locate diseased areas from different colours).

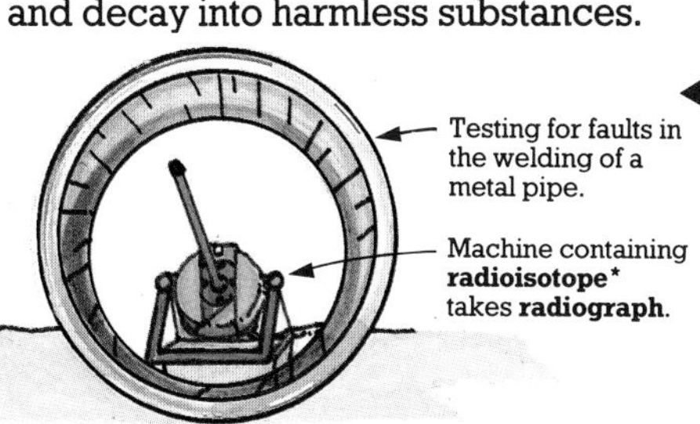

- ◀ **Gamma radiography** (γ-**radiography**). The production of a **radiograph** (similar to a photograph) by the use of **gamma rays*** (see also **X-radiography**, page 44). This has many uses, including quality control in industry.

- **Radiocarbon dating** or **carbon dating**. A way of calculating the time elapsed since living matter died. All living things contain a small amount of carbon-14 (a **radioisotope*** absorbed from the atmosphere), which continues to emit radiation after death. This emission gradually decreases (carbon-14 has a **half-life*** of 5,700 years), so the age of the remains can be calculated from its strength. ▶

Preserved organic material from an archaeological site can be dated by **radiocarbon dating**.

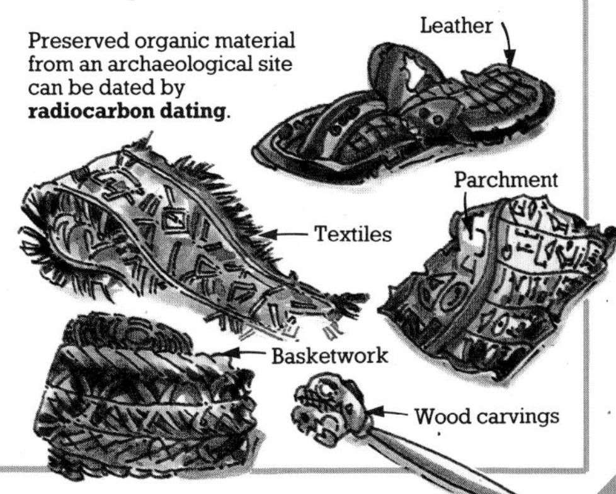

* **Gamma rays**, 86; **Half-life**, 87; **Radioisotope**, 86; **X-rays**, 44.

Nuclear fission and fusion

The central **nucleus** of an atom (see page 82) holds vast amounts of "stored" energy (see pages 84-85). **Nuclear fission** and **nuclear fusion** are both ways in which this energy can be released. They are both **nuclear reactions** (reactions which bring about a change in the nucleus).

- **Nuclear fission**. The process in which a heavy, unstable nucleus splits up into two (or more) lighter nuclei, roughly equal in size, with the release of two or three **neutrons*** (**fission neutrons**) and a large amount of energy (see also page 84). The two lighter nuclei are called **fission products** or **fission fragments** and many of them are **radioactive***. Fission is made to happen (see **induced fission**) in **fission reactors*** to produce heat energy. It does not often occur naturally (**spontaneous fission**).

Induced fission of uranium-235

Fast-moving **neutron*** collides with ^{235}U nucleus

Unstable ^{236}U nucleus formed

Fission products lanthanum-148 and bromine-85 formed (other pairs of nuclei of similar mass may be formed instead).

^{236}U nucleus undergoes fission

Three neutrons released

Energy released (see page 84)

Nuclear equation for reaction, above right (see **mass** and **atomic numbers**, page 82):

$$^{235}_{92}U + ^{1}_{0}n \rightarrow ^{236}_{92}U \rightarrow ^{148}_{57}La + ^{85}_{35}Br + 3^{1}_{0}n + \text{energy}$$

- **Spontaneous fission. Nuclear fission** which occurs naturally, i.e. without assistance from an outside agency. This may happen to a nucleus of a heavy element, e.g. the **isotope*** uranium-238, but the probability is very low compared to that of a simpler process like **alpha decay*** occurring instead.

Induced fission causing **chain reaction**

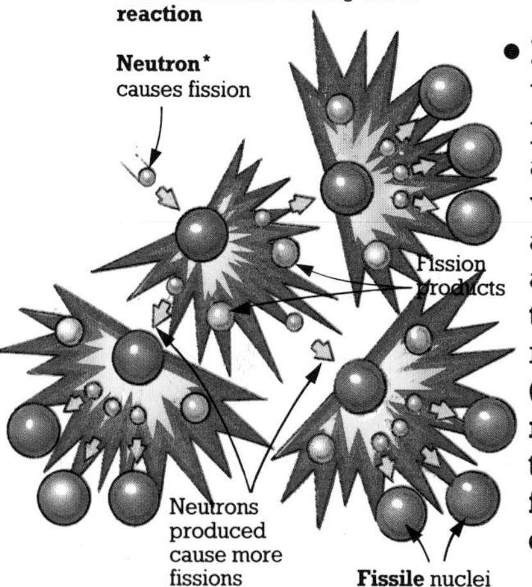

- **Induced fission. Nuclear fission** of a nucleus made unstable by artificial means, i.e. by being hit by a fast particle (often a **neutron***), which it then absorbs. Not all nuclei can be induced to fission in this way; those which can, e.g. those of the **isotopes*** uranium-235 and plutonium-239, are described as **fissile**. If there are lots of fissile nuclei in a substance (see also **thermal** and **fast reactor**, page 95), the neutrons released by induced fissions will cause more fissions (and neutrons), and so on. This is known as a **chain reaction**. A well controlled chain reaction is allowed to occur in a **fission reactor***, but that occurring in a **fission bomb** is uncontrolled and extremely explosive.

* **Alpha decay**, 87; **Fission reactor**, 94; **Isotopes**, 83; **Neutrons**, 82; **Radioactivity**, 86.

- **Critical mass**. The minimum mass of a **fissile** substance needed to sustain a **chain reaction** (see **induced fission**). In smaller **subcritical masses**, the surface area to volume ratio is too high, and too many of the **neutrons*** produced by the first fissions escape into the atmosphere. Nuclear fuel is kept in subcritical masses.

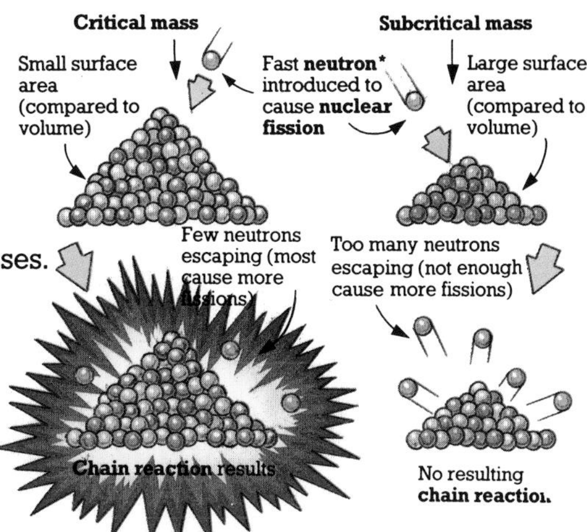

- **Fission bomb** or **atom bomb** (**A-bomb**). A bomb in which two **subcritical masses** (see above) are brought together by a trigger explosion. The resulting **chain reaction** (see **induced fission**) releases huge amounts of energy.

- **Nuclear fusion**. The collision and combination of two light nuclei to form a heavier, more stable nucleus, with the release of large amounts of energy (see also page 84). Unlike **nuclear fission**, it does not leave **radioactive*** products. Nuclear fusion requires temperatures of millions of degrees Celsius, to give the nuclei enough **kinetic energy*** for them to fuse when they collide (because of the high temperatures, fusion reactions are also called **thermonuclear reactions**). It therefore only occurs naturally in the sun (and stars like it), but research is being carried out with the aim of achieving controlled, induced fusion in **fusion reactors***.

- **Fusion bomb** or **hydrogen bomb** (**H-bomb**). A bomb in which uncontrolled **nuclear fusion** occurs in a mixture of tritium and deuterium (hydrogen **isotopes***). A trigger **fission bomb** creates the high temperature needed (fusion bombs are also called **fission-fusion bombs**). The energy released is about 30 times that from a fission bomb of the same size.

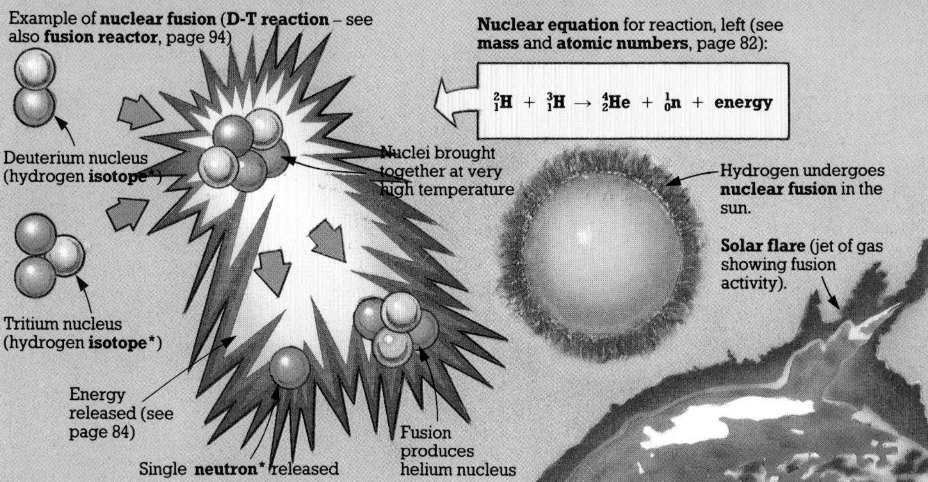

* **Fusion reactor**, 94; **Isotopes**, 83; **Kinetic energy**, 9; **Neutrons**, 82; **Radioactivity**, 86.

Power from nuclear reactions

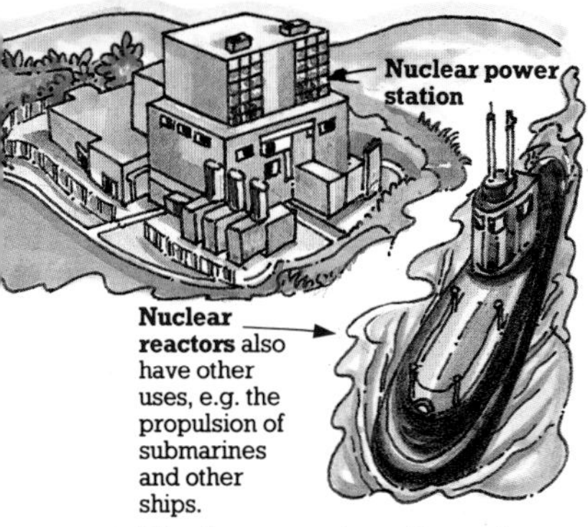

Nuclear reactors also have other uses, e.g. the propulsion of submarines and other ships.

A **nuclear reactor** is a structure inside which nuclear reactions produce vast amounts of heat. There are potentially two main types of reactor – **fission reactors** and **fusion reactors**, though the latter are still being researched. All present-day **nuclear power stations** are built around a central fission reactor and each generates, per unit mass of fuel, far larger amounts of power (electricity) than any other type of power station.

- **Fission reactor**. A **nuclear reactor** in which the heat is produced by **nuclear fission***. There are two main types in use in nuclear power stations – **thermal reactors** and **fast reactors** (see following page), both of which use uranium as their main fuel. This is held in long cylinders packed in the **core** (centre of the reactor). The rate of the **chain reaction*** (and hence power production) is closely controlled by **control rods**. See diagram below for the use of a fission reactor to generate power.

Schematic diagram of **fission reactor** and power station complex

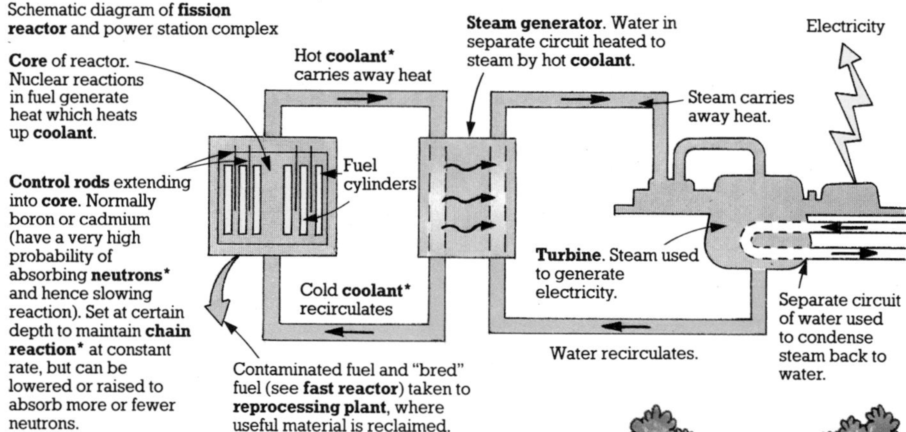

- **Fusion reactor**. A type of **nuclear reactor**, being researched but as yet undeveloped, in which the heat would be produced by **nuclear fusion***. This would probably be the fusion of the nuclei of the hydrogen **isotopes*** deuterium and tritium – known as the **D-T reaction** (see picture, page 93). There are several major problems to be overcome before a fusion reactor becomes a reality, but it would produce about four times as much energy per unit mass of fuel as a **fission reactor**. Also, hydrogen is abundant, whereas uranium is scarce, and dangerous and expensive to mine.

Dangerous **radioactive*** waste (spent fuel) from **fission reactors** must be buried. **Fusion reactors** would not produce such waste.

* **Chain reaction**, 92 (**Induced fission**); **Coolant**, 344; **Isotopes**, 83; **Neutrons**, 82; **Nuclear fission**, 92; **Nuclear fusion**, 93; **Radioactivity**, 86.

Types of fission reactor

- **Thermal reactor**. A **fission reactor** containing a **moderator** around the fuel cylinders. This is a substance with light nuclei, such as graphite or water. It is used to slow down the fast **neutrons*** produced by the first fissions in the natural uranium fuel – the neutrons bounce off the light nuclei (which themselves are unlikely to absorb neutrons) and eventually slow down to about 2200 m s^{-1}.

Slowing the neutrons improves their chances of causing further fissions (and continuing the **chain reaction***). Fast neutrons are likely to be "captured" by the most abundant nuclei – those of the **isotope*** uranium-238 (see **fast reactor**), whereas slow neutrons can travel on until they find uranium-235 nuclei. These will undergo fission when hit by neutrons of any speed, but make up a smaller percentage of the fuel (despite the fact that it is now often enriched with extra ^{235}U).

Types of **thermal reactor**

Pressurized water reactor (PWR) ▼

Control rods
Pressurizer
To turbine
Steam generator
From turbine
Steel vessel
Outer concrete **shield**
Inner steel **shield**
Fuel cylinders
Coolant* is pressurized water, also acting as **moderator**.

▼ **Advanced gas-cooled reactor (AGR)**

Concrete **shield**
Control rods
Fuel cylinders
To turbine
Steam generator
From turbine
Steel vessel
Graphite **moderator**
Pressurized carbon dioxide **coolant***

- **Fast reactor** or **fast breeder reactor (FBR)**. A **fission reactor**, inside which the **neutrons*** which cause the fission are allowed to remain as fast neutrons (travelling at about 2×10^7 m s^{-1}). The fuel used is always enriched with extra nuclei of uranium-235 (see **thermal reactor**) and plutonium-239. Both of these will fission easily when hit by fast neutrons, unlike uranium-238, which is far more likely to "capture" the neutrons (becoming ^{239}U) and undergo **radioactive decay***. The final product of this decay, however, is ^{239}Pu. Fast reactors are also called "breeders" because this decay process of ^{239}U to ^{239}Pu is allowed to happen in a blanket of ^{238}U around the main fuel. Hence more fuel is created and can be stored.

Fast reactors have a more compact core and run at higher temperatures than thermal reactors. They are also more efficient, using up a much greater proportion of their fuel before it becomes contaminated.

Fast reactor ▼

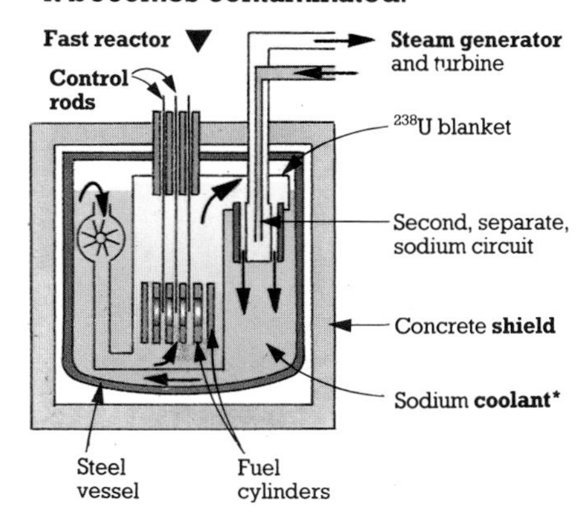

* **Chain reaction**, 92 (**Induced fission**); **Coolant**, 344; **Isotopes**, 83; **Neutrons**, 82; **Radioactive decay**, 87.

Quantities and units

Physical quantities are such things as **mass***, **force*** and **current***, which are used in the physical sciences. They all have to be measured in some way and each therefore has its own **unit**. These are chosen by international agreement and are called **International system** or **SI units** (abbreviated from the French Systeme International). All quantities are classified as either **basic quantities** or **derived quantities**.

- **Basic quantities**. A set of quantities from which all other quantities (see **derived quantities**) can be defined (see table, right). Each basic quantity has its **basic SI unit**, in terms of which any other SI unit can be defined.

Basic quantity	Symbol	Basic SI unit	Abbreviation
Mass	m	kilogram	kg
Length	l	metre	m
Time	t	second	s
Current	I	ampere	A
Temperature	T	kelvin	K
Quantity of substance	–	mole	mol
Luminous intensity	-	candela	cd

Basic SI units

- **Kilogram (kg)**. The SI unit of mass. It is equal to the mass of an international prototype metal cylinder kept at Sèvres, near Paris.
- **Metre (m)**. The SI unit of length. It is equal to the length of 1 650 763.73 **wavelengths*** of a certain type of radiation emitted by the krypton-86 atom.
- **Second (s)**. The SI unit of time. It is equal to the duration of 9 192 631 770 **periods*** of a certain type of radiation emitted by the caesium-133 atom.
- **Ampere (A)**. The SI unit of electric current (see also page 60). It is equal to the size of a current flowing through parallel, infinitely long, straight wires in a vacuum that produces a force between the wires of 2×10^{-7} N every metre.
- **Kelvin (K)**. The SI unit of temperature. It is equal to 1/273.16 of the temperature of the **triple point** of water (the point at which ice, water and steam can all exist at the same time) on the **absolute temperature scale***.
- **Mole (mol)**. The SI unit of the quantity of a substance (note that this is different from mass because it is the number of particles of a substance). It is equal to the amount of substance which contains 6.02×10^{23} (this is **Avagadro's number**) particles (e.g. atoms or molecules).
- **Candela (cd)**. The SI unit of intensity of light. It is equal to the strength of light from 1/600 000 square metres of a **black body*** at the temperature of freezing platinum and at a pressure of 101 325 N m^{-2}.

Prefixes

A given SI unit may sometimes be too large or small for convenience, e.g. the metre is too large for measuring the thickness of a piece of paper. Standard fractions and multiples of the SI units are therefore used and written by placing a prefix before the unit (see tables below and right). For example, the millimetre (mm) is equal to one thousandth of a metre.

Standard fractions and multiples (those involving powers of 10^3, e.g. 10^3, 10^6, 10^{-3}).

Fraction	Prefix	Symbol
10^{-3}	milli-	m
10^{-6}	micro-	μ
10^{-9}	nano-	n

Multiple	Prefix	Symbol
10^3	kilo-	k
10^6	mega-	M
10^9	giga-	G

Other fractions and multiples in use

Fraction or multiple	Prefix	Symbol
10^2	hecto-	h
10^1	deca-	dc
10^{-1}	deci-	d
10^{-2}	centi-	c

* **Absolute temperature scale**, 27; **Black body**, 29 (**Leslie's cube**); **Current**, 60; **Force**, 6; **Mass**, 12; **Period**, 16; **Wavelength**, 34.

- **Derived quantities**. Quantities other than **basic quantities** which are defined in terms of these or in terms of other derived quantities. The derived quantities have **derived SI units** which are defined in terms of the **basic SI units** or other derived units. They are worked out from the defining equation for the quantity and are sometimes given special names.

Derived quantity	Symbol	Defining equation	Derived SI unit	Name of unit	Abbreviation
Velocity	v	$v = \frac{\text{distance}}{\text{time}}$	$m\ s^{-1}$	-	-
Acceleration	a	$a = \frac{\text{velocity}}{\text{time}}$	$m\ s^{-2}$	-	-
Force	F	$F = \text{mass} \times \text{acceleration}$	$kg\ m\ s^{-2}$	Newton	N
Work	W	$W = \text{force} \times \text{distance}$	N m	Joule	J
Energy	E	Capacity to do work	Joule	-	-
Power	P	$P = \frac{\text{work done}}{\text{time}}$	$J\ s^{-1}$	Watt	W
Area	A	Depends on shape (see page 101)	m^2	-	-
Volume	V	Depends on shape (see page 101)	m^3	-	-
Density	ρ	$\rho = \frac{\text{mass}}{\text{volume}}$	$kg\ m^{-3}$	-	-
Pressure	P	$P = \frac{\text{force}}{\text{area}}$	$N\ m^{-2}$	Pascal	Pa
Period	T	Time for one cycle	s	-	-
Frequency	f	Number of cycles per second	s^{-1}	Hertz	Hz
Impulse	-	$\text{Impulse} = \text{force} \times \text{time}$	N s	-	-
Momentum	-	$\text{Momentum} = \text{mass} \times \text{velocity}$	$kg\ m\ s^{-1}$	-	-
Electric charge	Q	$Q = \text{current} \times \text{time}$	A s	Coulomb	C
Potential difference	V	$V = \frac{\text{energy transferred}}{\text{charge}}$	$J\ C^{-1}$	Volt	V
Capacitance	C	$C = \frac{\text{charge}}{\text{potential difference}}$	$C\ V^{-1}$	Farad	F
Resistance	R	$R = \frac{\text{potential difference}}{\text{current}}$	$V\ A^{-1}$	Ohm	Ω

Equations, symbols and graphs

All **physical quantities** (see pages 96-97) and their units can be represented by **symbols** and are normally dependent on other quantities in some way. There is therefore a relationship between them which can be expressed as an **equation** and shown on a **graph**.

Equations

An **equation** represents the relationship between two or more physical quantities. This relationship can be expressed as a **word equation** or as an equation relating **symbols** which represent the quantities. The second case is used when a number of quantities are involved, since it is then easier to manipulate. Note that the meaning of the symbols must be stated.

Word equation

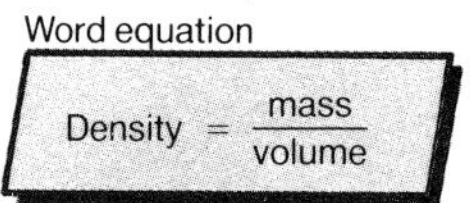

$$\text{Density} = \frac{\text{mass}}{\text{volume}}$$

Symbol equation

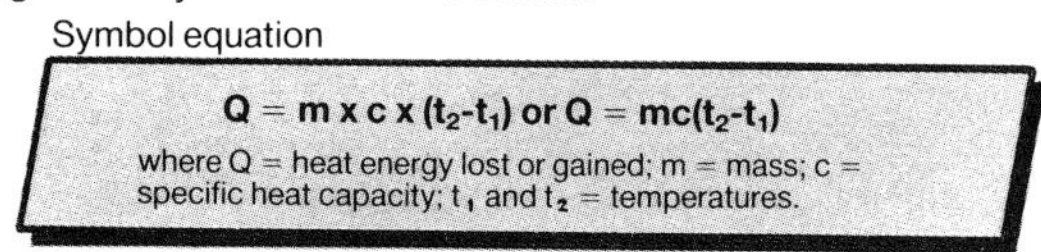

$$Q = m \times c \times (t_2 - t_1) \text{ or } Q = mc(t_2 - t_1)$$

where Q = heat energy lost or gained; m = mass; c = specific heat capacity; t_1 and t_2 = temperatures.

Graphs

A **graph** is a visual representation of the relationship between two quantities. It shows how one quantity depends on another. Points on a graph are plotted using the values for the quantities obtained during an experiment or by using the equation for the relationship if it is known. The two quantities plotted are called the **variables**.

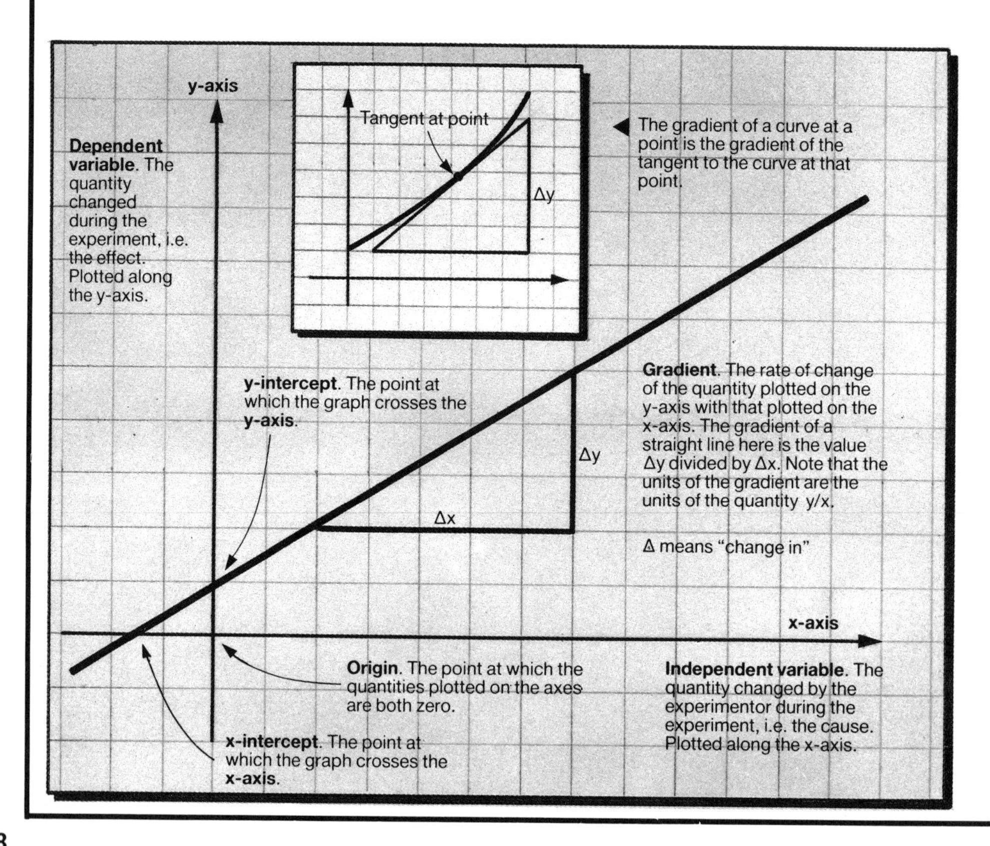

Symbols

Symbols are used to represent **physical quantities**. The value of a physical quantity consists of a numerical value and its unit. Therefore any symbol represents both a number and a unit.

Symbols represent number and unit

$m = 2.1$ kg $\qquad s = 400$ J kg^{-1} K^{-1}

"Current through resistor = I" (i.e. do not need to say I amps since the unit is included)

Note that a symbol divided by a unit is a pure number.

$m = 2.1$ kg means that m/kg = 2.1

This notation is used in tables and to label graph axes.

Any number in this column is a length in metres.

l/m	t^2/s^2
0.9	3.6
1.0	4.0
1.1	4.4
1.2	4.8

Any number in this column is a time squared measured in seconds squared.

Any number on scale is a force in newtons.

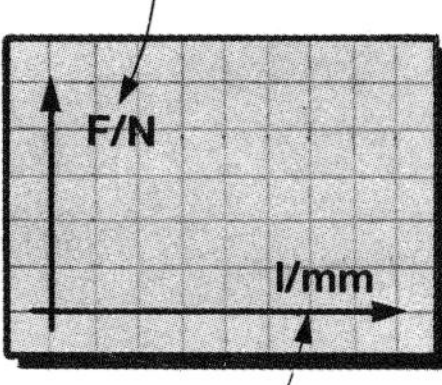

Any number on scale is a length in millimetres.

Plotting graphs

1. The quantity controlled during an experiment should be plotted along the x-axis and the quantity which changes as a result along the y-axis.

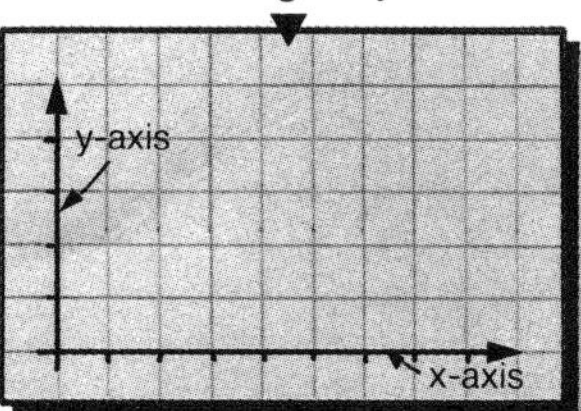

2. The scales on the axes should be chosen so that values are easy to find (squares on the paper representing multiples of three should be avoided).

3. The axes should be labelled by the symbol representing the quantity (or the name of the quantity) and its unit, e.g. length/mm.

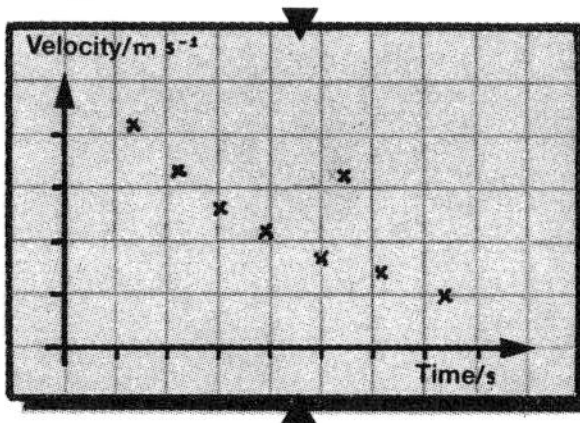

4. Points on the graph should be marked in pencil with a × or a ⊙.

5. A smooth curve or straight line should be drawn which fits the points best (this is because physical quantities normally are related in some

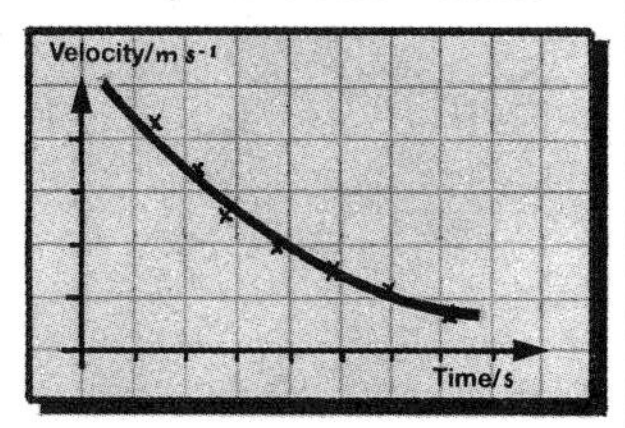

definite way). Note that joining the points up will not often produce a smooth curve. This is due to experimental errors.

Information from graphs

A straight line graph which passes through the origin shows that the quantities plotted on the axes are proportional to each other (i.e. if one is doubled then so is the other).

The amount of scatter of the points about the smooth curve gives an indication of the errors in the data due to inaccuracies in the procedure, the equipment and the measuring (this happens in any experiment).

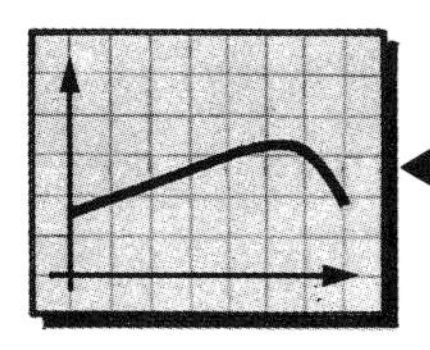

A graph shows the region in which the relationship between two quantities is linear (i.e. one always changes by the same amount for a fixed change in the other). It is that in which the graph is a straight line.

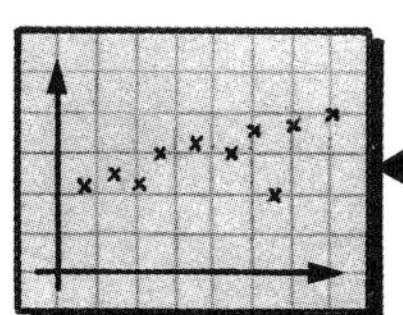

Individual points a long way from the curve are probably due to an error in measuring that piece of data in the experiment. However, the point should not be ignored – it should be checked and re-measured if possible.

Measurements

Measurement of length

The method used to measure a length depends on the magnitude of the length. A metre rule is used for lengths of 50 mm or more. The smallest division is normally 1 mm and so lengths can be estimated to the nearest 0.5 mm. For lengths less than 50 mm, the error involved would be unacceptable (see also **reading error**, page 103). A **vernier scale** is therefore used. For the measurement of very small lengths (to 0.01 mm) a **micrometer screw gauge** is used.

- **Vernier scale**. A short scale which slides along a fixed scale. The position on the fixed scale of the zero line of the vernier scale can be found accurately. It is used in measuring devices such as the **vernier slide callipers**.

Method of reading position of zero line on **vernier scale**:

1. Read the position of the zero line approximately – in this case 8.3 cm.

2. Find the position on the vernier scale where the marks coincide – in this case 2.

3. Add this to the previous figure – the accurate reading is 8.32 cm.

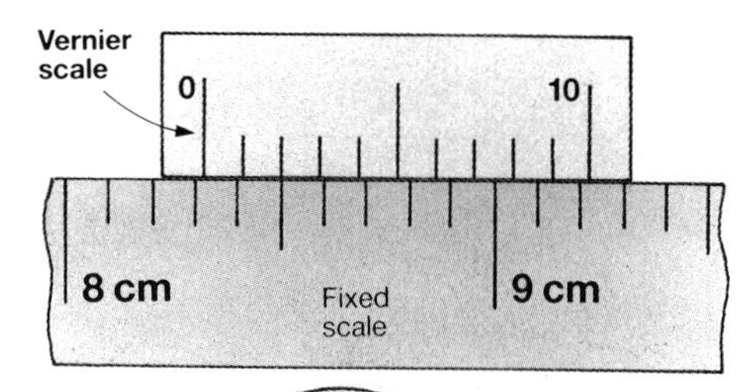

- **Vernier slide callipers**. An instrument containing a **vernier scale**, used to measure lengths in the range 10 to 100 mm.

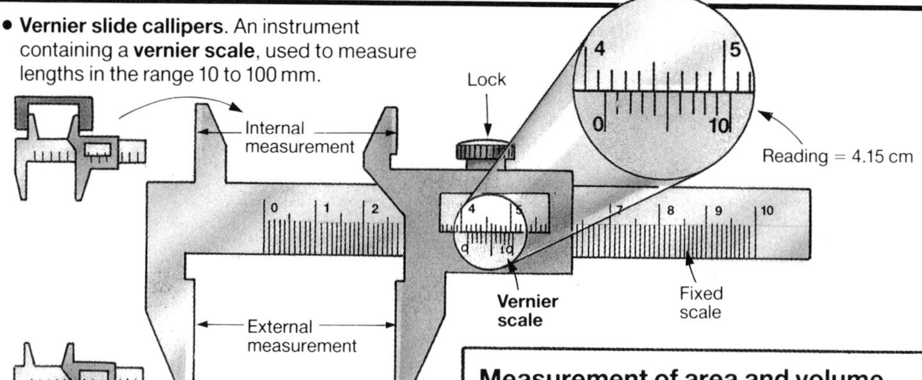

Method of measurement:

1. Close the jaws and check that the zero on the **vernier scale** coincides with the zero on the fixed scale. If not, note the reading (this is the **zero error***).

2. Close or open the jaws onto the object to be measured.

3. Lock the sliding jaw into position.

4. Record the reading on the scale.

5. Add or subtract the zero error (see 1) to get the correct reading.

Measurement of area and volume

The volume of a liquid is calculated from the space it takes up in its containing vessel. The internal volume of the containing vessel is called its **capacity**. The **SI unit*** of capacity is the **litre (l)**, equal to 10^{-3} m^3. Note that 1 ml = 1 cm^3. The volume of a liquid is measured using a graduated vessel.

Examples of graduated vessels for measuring volume

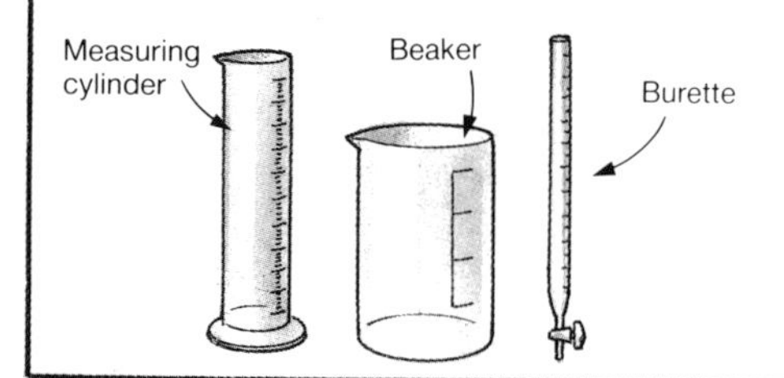

* **SI units**, 96; **Zero error**, 102.

- **Micrometer screw gauge**. An instrument used for accurate measurements up to about 30 mm.

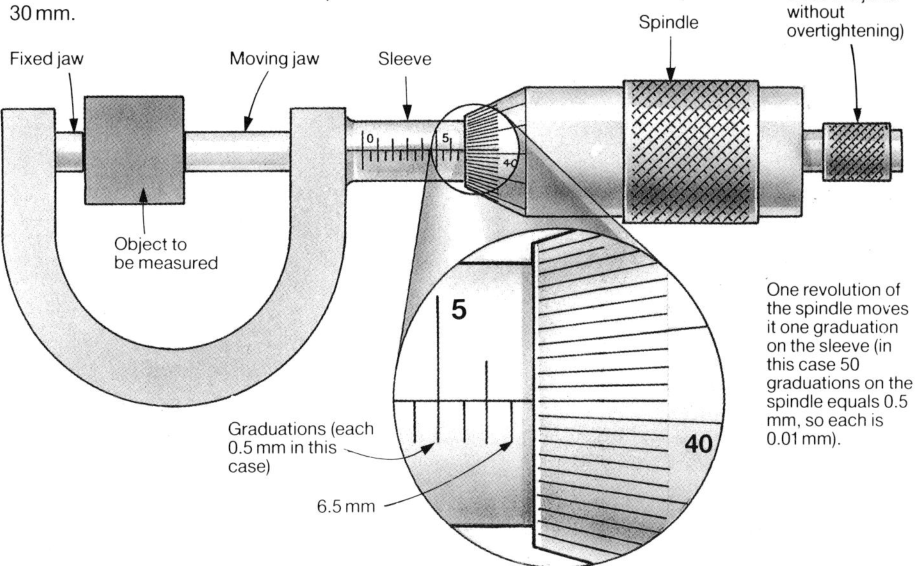

Method of measurement

1. Work out the value of a division on the spindle scale (see diagram).

2. Using the ratchet, close the jaws of the instrument fully. The zero on the spindle scale should coincide with the horizontal reference line. If not, note the **zero error***.

3. Using the ratchet, close the jaws on the object to be measured until it is gripped.

4. Note the reading of the highest visible mark on the sleeve scale (in this case 6.5 mm).

5. Note the division on the spindle scale which coincides with the horizontal reference line (in this case 0.41 mm).

6. Add the two readings and add or subtract the zero error (see 2) to get the correct reading (in this case 6.91 mm).

The surface area and volume of a solid of regular shape are calculated from length measurements of the object (see below).

For solids of irregular shape, see **eureka can**, page 24.

Regular shaped solid	Rectangular bar (h, b, l)	Sphere (r)	Cylinder (r, l)
Measurements made using **vernier slide callipers** or **micrometer screw gauge**	h = height b = breadth l = length	r = radius	r = radius l = length
Volume v of solid calculated from	$v = lbh$	$v = 4/3 \pi r^3$	$v = \pi r^2 l$
Surface area a calculated from	$a = 2bl + 2hl + 2hb$ Top (2bl), Sides (2hl), Ends (2hb)	$a = 4\pi r^2$	$a = 2\pi rl + 2\pi r^2$ Curved surface ($2\pi rl$), Ends ($2\pi r^2$)

* **Zero error**, 102.

Accuracy and errors

All experimental measurements are subject to some errors, other than those caused by carelessness (like misreading a scale). The most common errors which occur are **parallax errors**, **zero errors** and **reading errors**. When stating a reading, therefore, a number of **significant figures** should be quoted which give an estimate of the accuracy of the readings.

- **Parallax error**. The error which occurs when the eye is not placed directly opposite a scale when a reading is being taken.

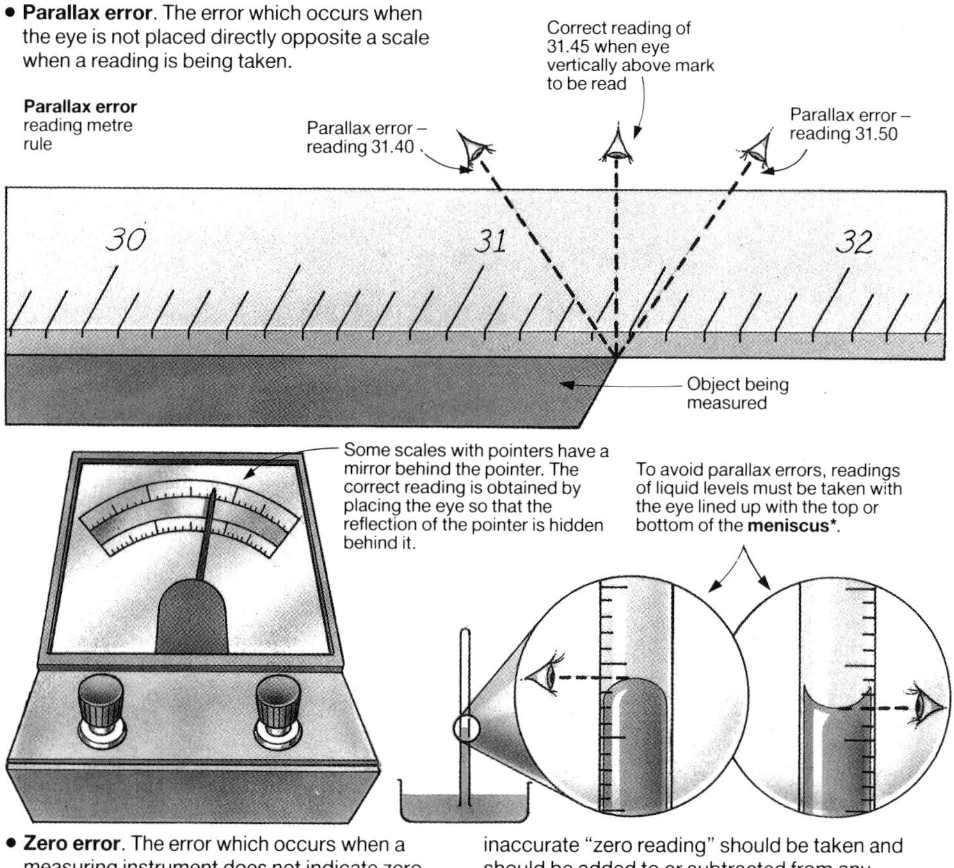

- **Zero error**. The error which occurs when a measuring instrument does not indicate zero when it should. If this happens, the instrument should either be adjusted to read zero or the inaccurate "zero reading" should be taken and should be added to or subtracted from any other reading taken.

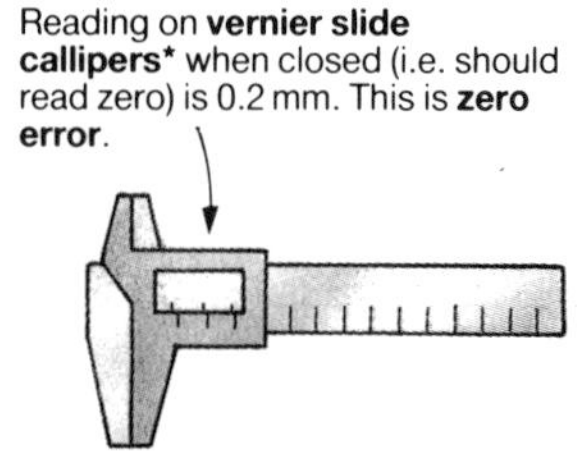

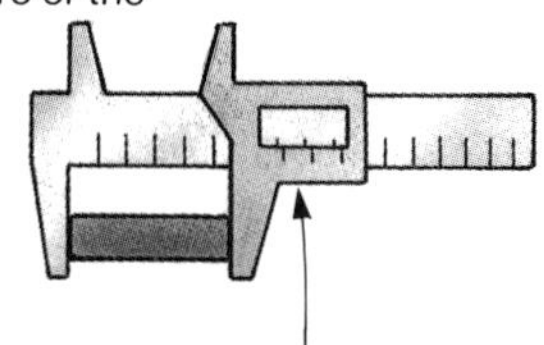

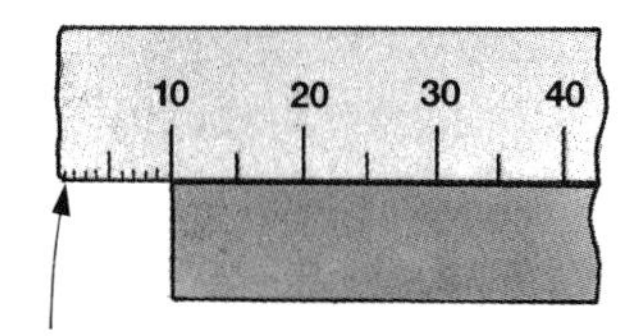

* **Meniscus**, 345; **Vernier slide callipers**, 100.

- **Reading error**. The error due to the guesswork involved in taking a reading from a scale when the reading lies between the scale divisions.

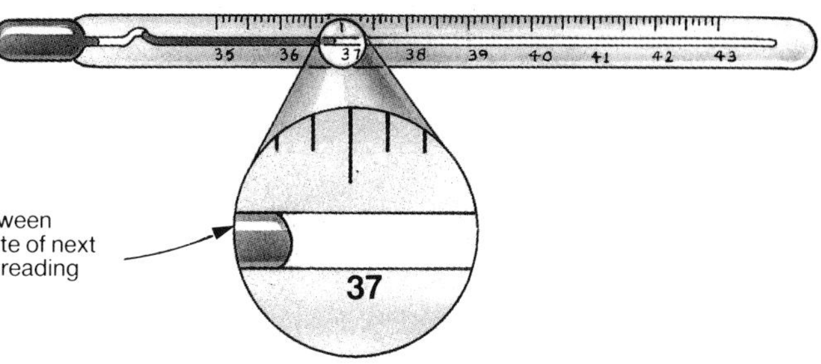

Significant figures

The number of **significant figures** in a value is the number of figures in that value ignoring leading or trailing zeros (but see below) and disregarding the position of the decimal point. They give an indication of the accuracy of a reading.

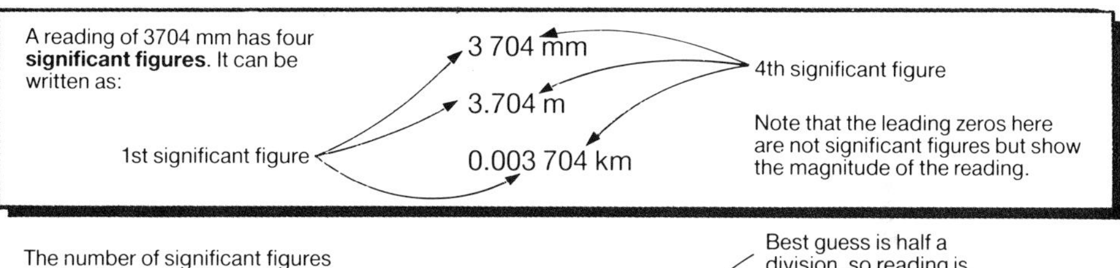

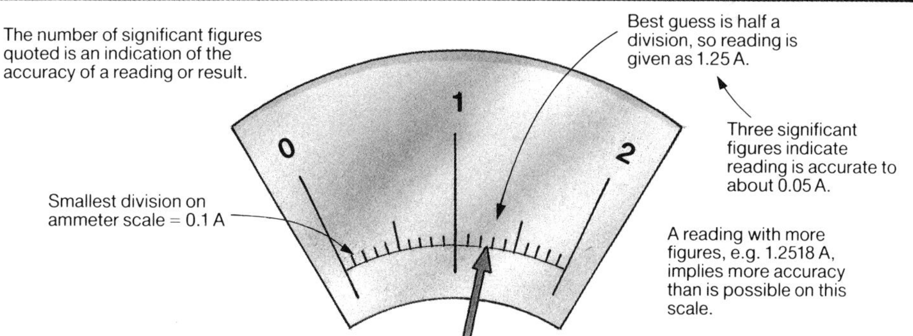

- **Rounding**. The process of reducing the number of figures quoted. The last significant figure is dropped and the new last figure changed depending on the one dropped.

7.3925	(quoted to 5 significant figures)
= 7.393	(rounded to 4 significant figures)
= 7.39	(rounded to 3 significant figures)
= 7.4	(rounded to 2 significant figures)
= 7	(rounded to 1 significant figure)
0.08873	(quoted to 4 significant figures)
= 0.0887	(rounded to 3 significant figures)
= 0.089	(rounded to 2 significant figures)
= 0.09	(rounded to 1 significant figure)

Note that

29.000	is quoted to 5 significant figures
= 29.0	(to 3 significant figures)
= 29	(to 2 significant figures)
= 30	(to 1 significant figure)

In the last case here, the 0 is not a significant figure but must be included (see below).

For large numbers like 283 000 it is impossible to say how many of the figures are significant (the first three must be) because the zeros have to be included to show the magnitude. This ambiguity is removed by using the **index notation** (see page 109).

Fields and forces

This table is a comparison of the three forces normally encountered in physics (excluding the **nuclear force**). In fact, most of the forces dealt with in physics, e.g. the **contact force** between two objects, are examples of the **electromagnetic force** which is a combination of the **magnetic** and **electric forces**. For more about these and all other forces, see pages 6-7.

Gravitational force (see also pages 6 and 18)

Type of force. Note that a force can only exist between two masses, charges or currents and that the size of the force is the same on both of them (see also **Newton's third law**, page 13). Note also that the forces only act between objects which are the same, e.g. there is a force between two masses, but not between a mass and a current.

Force acts between two objects with mass. It is always attractive.

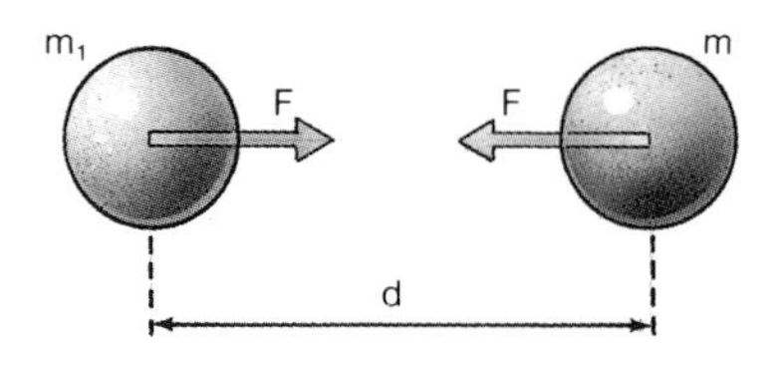

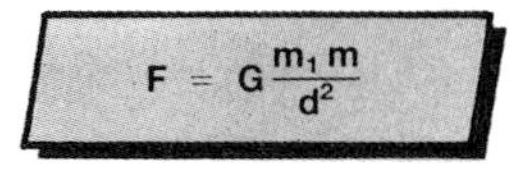

$$F = G\frac{m_1 m}{d^2}$$

$G = 6.7 \times 10^{-11}\ Nm^2kg^{-2}$

G is the **gravitational constant***. Its very small value means that the gravitational force is only noticeable when one of the objects is very large (e.g. a planet).

Description of force in terms of force field. The **force field** is the region around an object (mass, charge or current) in which its effects (gravitational, electric or magnetic) can be detected – see also page 6.

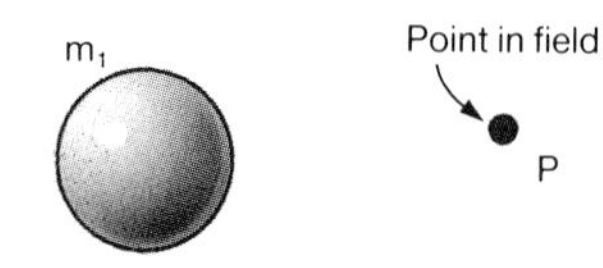

Mass m_1 produces a **gravitational field** in the space around it (see **field intensity**, page 106).

A second mass experiences a gravitational force when placed at any point (e.g. P) in the gravitational field of m_1.

A mass thus produces a gravitational field and is acted upon by a gravitational field.

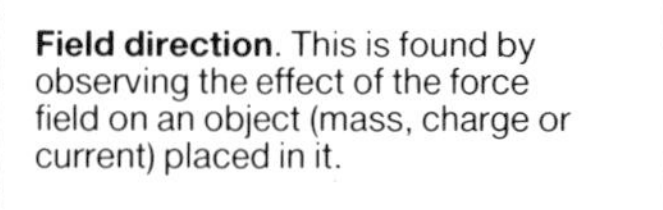

Field direction. This is found by observing the effect of the force field on an object (mass, charge or current) placed in it.

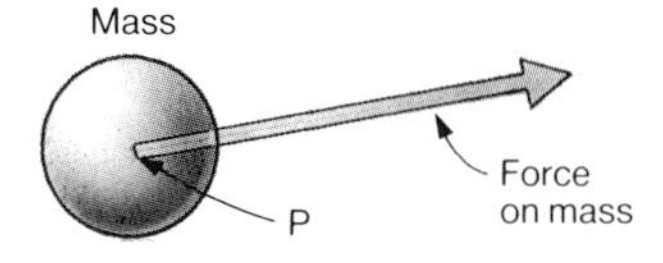

The direction of a gravitational field at a point P is the direction of the force on a mass placed at P.

* **Gravitational constant**, 18 (**Newton's law of gravitation**).

Electric force (see also pages 6 and 58)	Magnetic force (see also pages 6 and 70)

Electric force (see also pages 6 and 58)

Force is between two charges. It is attractive if the charges are of the opposite sign, i.e. one negative and one positive, and repulsive if the charges are of the same sign.

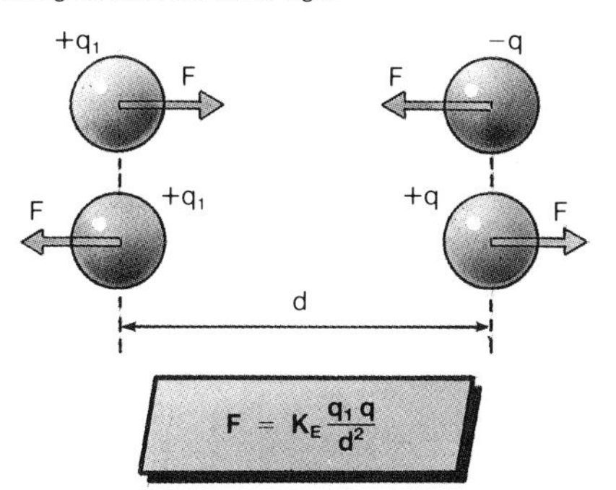

$$F = K_E \frac{q_1 q}{d^2}$$

In air $K_E = 9 \times 10^9\ Nm^2C^{-2}$

This very large value means it is difficult to separate opposite charges.

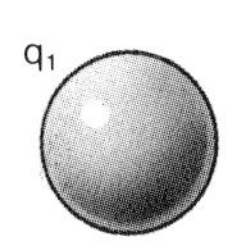

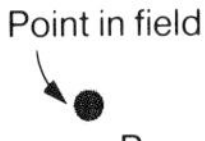

Charge q_1 produces an **electric field*** in the space around it (see **field intensity**, page 106).

A second charge experiences an electric force when placed at any point (e.g. P) in the electric field of q_1.

A charge thus produces an electric field and is acted upon by an electric field.

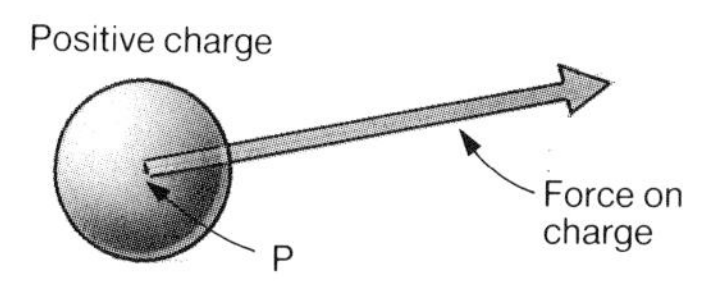

The direction of an electric field intensity at a point P is the direction of the force on a positive charge placed at P.

Magnetic force (see also pages 6 and 70)

Force is between two objects in which current is flowing. If the currents flow in the same direction, the force is attractive. If the currents flow in opposite directions, the force is repulsive.

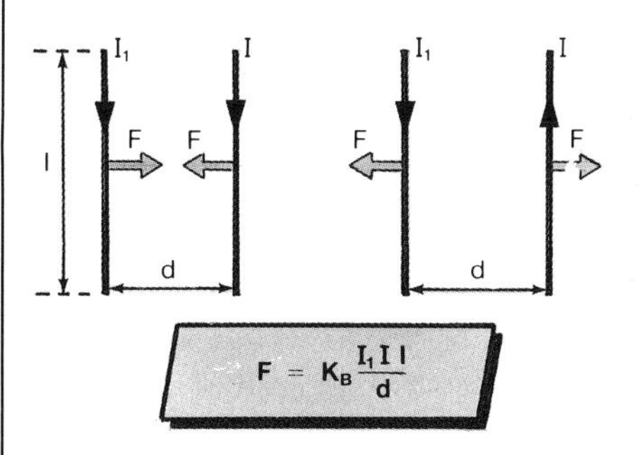

$$F = K_B \frac{I_1 I\, l}{d}$$

In air $K_B = 2.7 \times 10^{-7}\ NA^{-2}$

This small value indicates that the magnetic force is very small in comparison to the electric force.

Point in field

P

Current I_1 produces a **magnetic field*** in the space around it (see **field intensity**, page 106).

A second current experiences a magnetic force when placed at any point P in the magnetic field of I_1.

A current thus produces a magnetic field and is acted upon by a magnetic field.

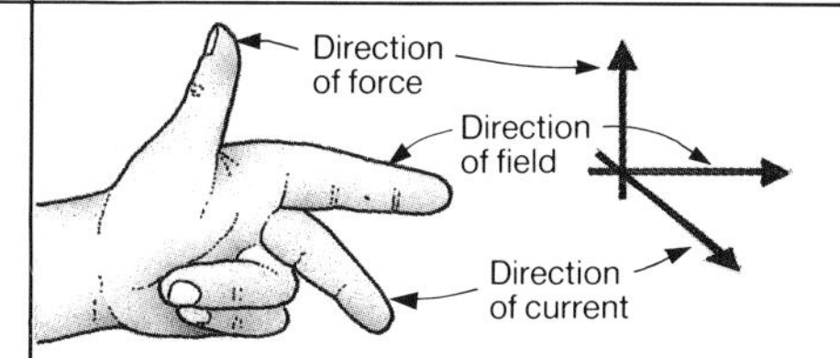

The direction of a magnetic field at a point P is given by **Fleming's left hand rule***.

* **Electric field**, 58; **Fleming's left hand rule**, 76; **Magnetic field**, 72.

Fields and forces (continued)

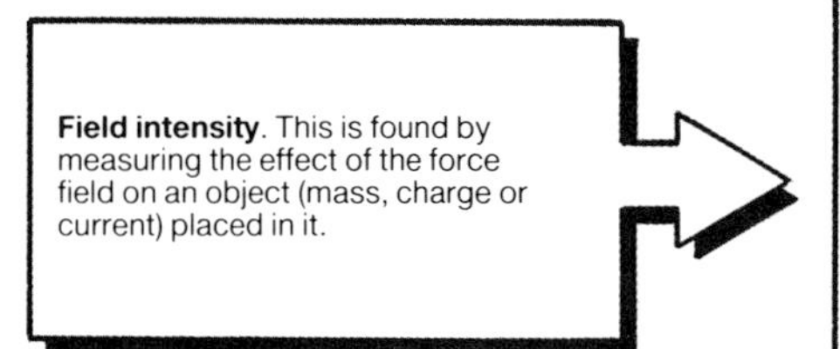
Field intensity. This is found by measuring the effect of the force field on an object (mass, charge or current) placed in it.

Gravitational force

To measure the field intensity g of a gravitational field due to a mass m_1 at a point P, a test mass m is placed at P and the gravitational force F on it is measured. Then:

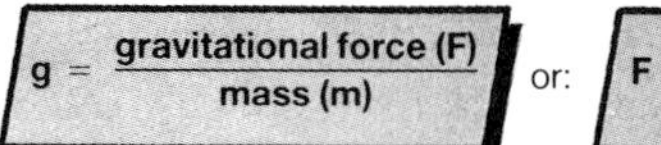

$$g = \frac{\text{gravitational force (F)}}{\text{mass (m)}} \quad \text{or:} \quad F = m g$$

By comparison with the equation above for the gravitational force, the field intensity g at a distance d from a mass m_1 is:

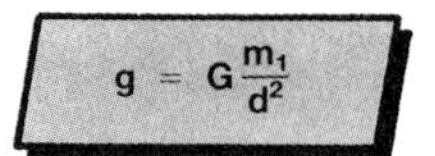

$$g = G\frac{m_1}{d^2}$$

Representation by field lines. Field lines (or **flux lines** or **lines of force** or **flux**) are used in all cases to represent the strength and direction of fields and to visualize them (see panel below). Field lines never cross since the field would then have different directions at one point.

Gravitational field lines always end at a mass

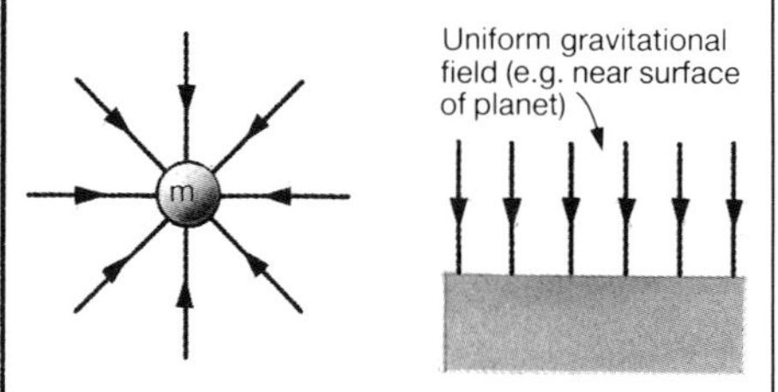

Potential energy (see also page 8). This depends on **field intensity** and the object (its mass in a **gravitational field** or its charge in an **electric field***). The **potential*** at a point in a field is the energy per unit (of mass or charge) and depends upon the field only. Usually the only concern is the difference in potential, or **potential difference***, between two points. Potential can be defined by choosing a reference. The potential at a point is then the potential difference between the point and the reference point.

The gravitational potential difference between two points in a gravitational field is the work done against the forces of the field in moving a unit mass between the points.

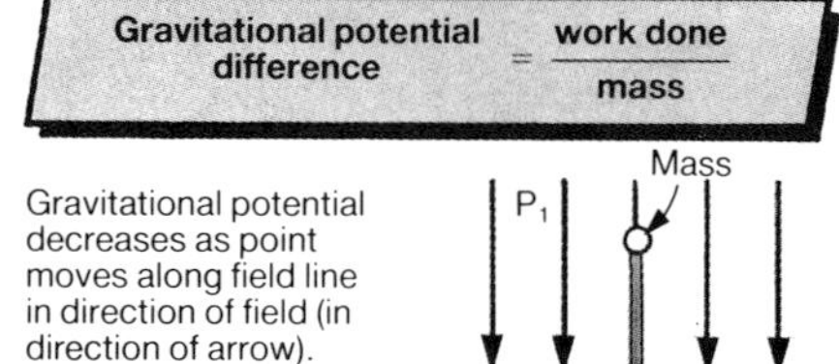

$$\text{Gravitational potential difference} = \frac{\text{work done}}{\text{mass}}$$

Gravitational potential decreases as point moves along field line in direction of field (in direction of arrow).

Gravitational potential higher at P_1 than P_2

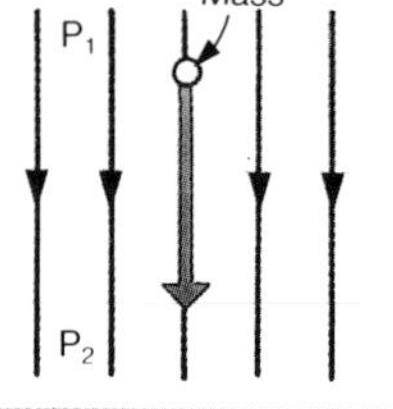

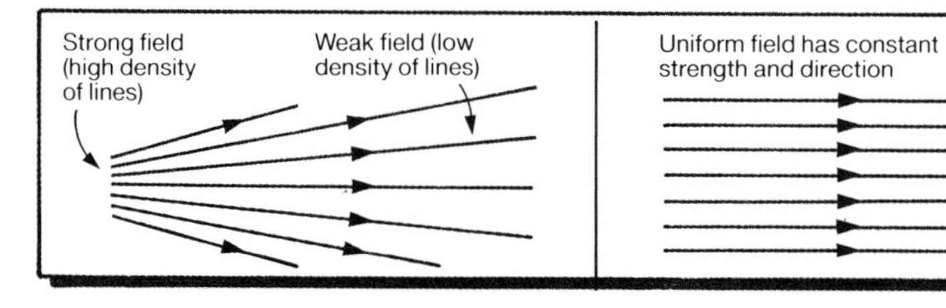

* **Electric field**, **Potential**, **Potential difference**, 58.

Electric force	Magnetic force

Electric force

To measure the field intensity E of an electric field at a point P due to a charge q_1, a test positive charge q is placed at P and the electric force F on it is measured. Then:

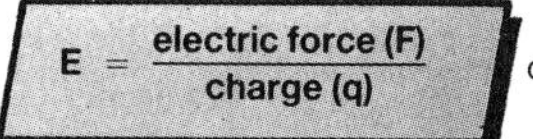

$$E = \frac{\text{electric force (F)}}{\text{charge (q)}}$$

or:

$$F = qE$$

Magnetic force

To measure the field intensity B of a magnetic field due to I_1 at a point P, a conductor of length l carrying a current I is placed at P and the magnetic force F is measured. Then:

$$B = \frac{\text{magnetic force (F)}}{\text{current (I)} \times \text{length (l)}}$$

or:

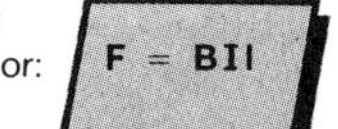

$$F = BIl$$

By comparison with the equation above for the electric force, the field intensity E at a distance d from a charge q_1 is:

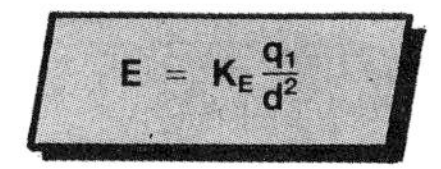

$$E = K_E \frac{q_1}{d^2}$$

By comparison with the equation above for the magnetic force, the field intensity B at a distance d from a current I_1 is:

$$B = K_B \frac{I_1}{d}$$

Electric field lines always begin at a positive charge and end at an equal negative charge.

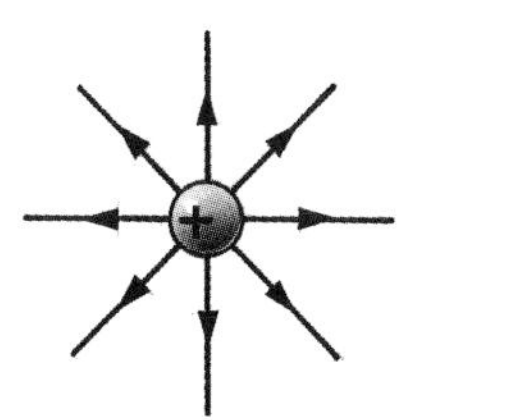

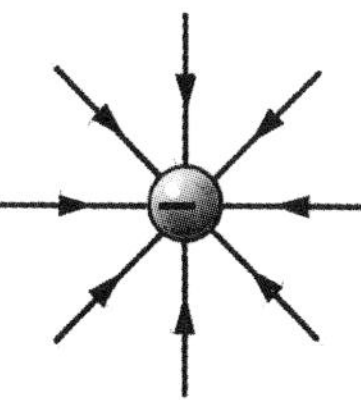

Magnetic field lines* have no beginning or end, but are always closed loops. This is because single north or south poles cannot exist. This is a fundamental difference compared to gravitational and electric fields.

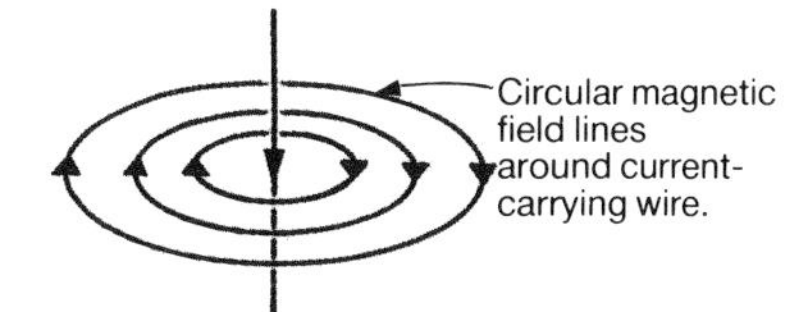

The electric potential difference between two points in an electric field is the work done against the forces of the field in moving a unit positive between them.

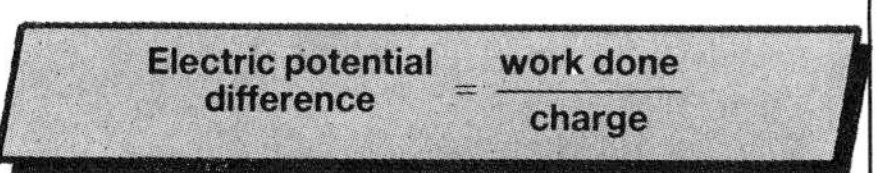

$$\text{Electric potential difference} = \frac{\text{work done}}{\text{charge}}$$

Electric potential decreases as point moves along field line in direction of field (in direction of arrow).

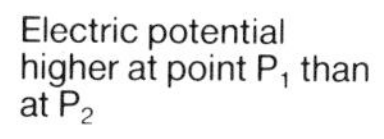

Electric potential higher at point P_1 than at P_2

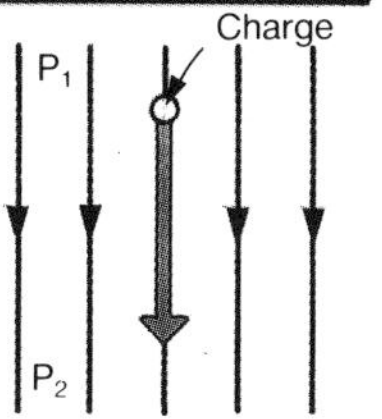

Magnetic potential is much more difficult to define than for gravitational or electric fields because the field lines are circular. Note that if a point moves around a circular line in the diagram above, it returns to the same point, which must have the same potential, although it has moved along a field line. This means that magnetic potential is complicated to calculate.

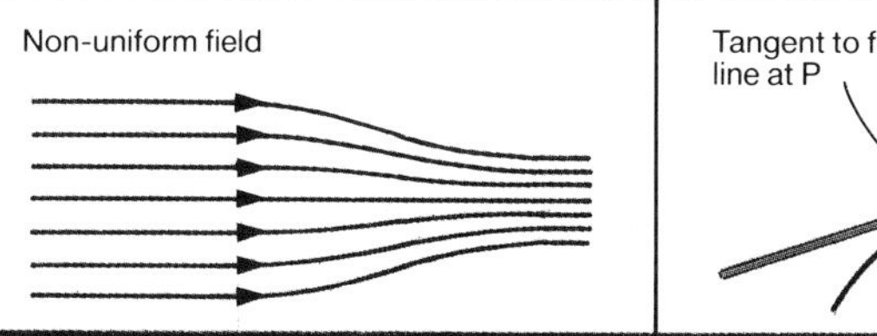

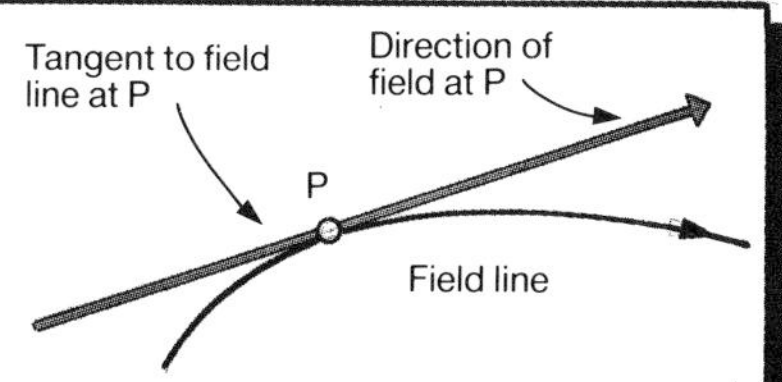

* **Magnetic field lines**, 72.

Vectors and scalars

All quantities in physics are either **scalar** or **vector quantities**, depending on whether the quantity has direction as well as magnitude.

- **Scalar quantity**. Any quantity which has magnitude only, e.g. mass, time, energy, density.

- **Vector quantity**. Any quantity which has both magnitude and direction, e.g. force, displacement, velocity and acceleration. When giving a value to a vector quantity, the direction must be given in some way as well as the magnitude. Usually, the quantity is represented graphically by an arrowed line. The length of the line indicates the magnitude of the quantity (on some chosen scale) and the direction of the arrow the direction of the quantity.

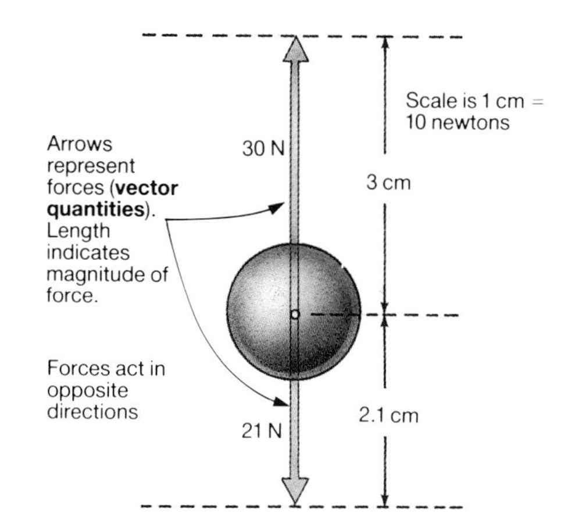

- **Parallelogram rule**. A rule used when adding together two **vector quantities**. The two vectors are drawn from one point to form two sides of a parallelogram which is then completed. The diagonal from the original common point gives the sum of the two vectors (the **resultant**).

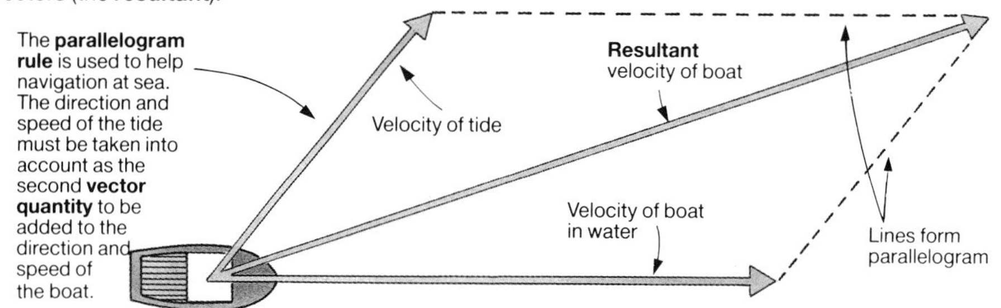

- **Resolution**. The process of splitting one **vector quantity** into two other vectors called its **components**. Normally, the two components are perpendicular to each other. Each component then represents the total effect of the vector in that direction.

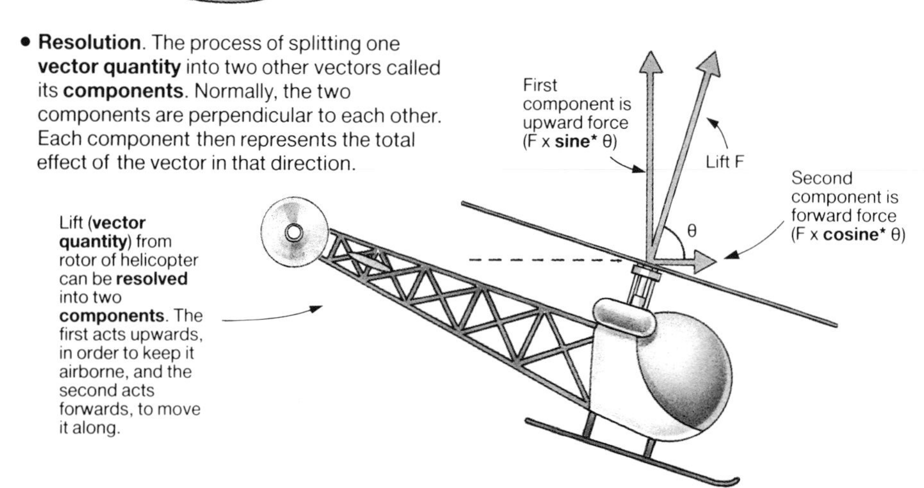

* **Cosine**, 344; **Sine**, 345.

Numbers

Very large or very small numbers (e.g. 10 000 000 or 0.000 001) take a long time to write out and are difficult to read. The **index notation** is therefore used. In this notation, the position of the decimal point is shown by writing the power ten is raised to.

1 000 000	$= 10^6$	or "ten to the six"
100 000	$= 10^5$	or "ten to the five"
10 000	$= 10^4$	or "ten to the four"
1 000	$= 10^3$	or "ten to the three"
100	$= 10^2$	or "ten to the two"
10	$= 10^1$	or "ten to the one"
1	$= 10^0$	any number "to the nought" equals one)
0.1	$= 10^{-1}$	or "ten to the minus one"
0.01	$= 10^{-2}$	or "ten to the minus two"
0.001	$= 10^{-3}$	or "ten to the minus three"
0.000 1	$= 10^{-4}$	or "ten to the minus four"
0.000 01	$= 10^{-5}$	or "ten to the minus five"
0.000 001	$= 10^{-6}$	or "ten to the minus six"

Note that a negative index means "one over" so that $10^{-3} = 1/10^3 = 1/1000$. This also applies to units, e.g. kg m^{-3} means kg/m^3 or kg per m^3.

Indices are added when multiplying numbers, e.g. $10^5 \times 10^{-3}$ (= 100000 x 1/1000) $= 10^{5-3} = 10^2 = 100$

- **Standard form**. A form of expressing numbers in which the number always has one digit before the decimal point and is followed by a power of ten in **index notation** to show its magnitude (see also **significant figures**, page 103).

Examples of numbers written in **standard form**.

56342	5.6342×10^4
4000	4×10^3 (assuming 0s are not significant)
23.3	2.33×10^1
0.98	9.8×10^{-1}
0.00211	2.11×10^{-3}

- **Order of magnitude**. A value which is accurate to within a factor of ten or so. It is important to have an idea of the order of magnitude of some physical quantities so that a figure which has been calculated can be judged. For example, the mass of a person is about 60 kg. Therefore a calculated result of 50 kg or 70 kg is quite reasonable, but a result of 6 kg or 600 kg is obviously not correct.

Typical orders of magnitude

Item	Mass / kg
Earth	5×10^{24}
Car	5×10^3
Human	5×10^1
Bag of sugar	1
Orange	2×10^{-1}
Golfball	5×10^{-2}
Table-tennis ball	2×10^{-3}
Proton	2×10^{-27}
Electron	10^{-30}

Item	Length / m
Radius of galaxy	10^{19}
Radius of solar system	10^{11}
Radius of earth	5×10^6
Height of Mount Everest	10^4
Height of human	2
Thickness of paper	10^{-4}
Wavelength of light	5×10^{-7}
Radius of atom	10^{-10}
Radius of nucleus	10^{-14}

Item	Time / s
Age of earth	2×10^{17}
Time since emergence of man	10^{13}
Human life time	2×10^9
Time span of year	3×10^7
Time span of day	9×10^4
Time between heart beats	1
Camera shutter speed	10^{-2}
Half life of Polonium-214	1.5×10^{-4}
Time for light to travel 1 m	3×10^{-9}

Item	Energy / J
Energy given out by sun per second	10^{26}
Energy released by San Francisco earthquake (1906)	3×10^{17}
Energy released by fission of 1 g of uranium	10^{11}
Energy of lightning discharge	10^9
Energy of 1 KW fire per hour	4×10^6
Kinetic energy of golf ball	20

Circuit symbols

This table shows the main symbols used to represent the various components used in electric circuits (see also pages 60-65).

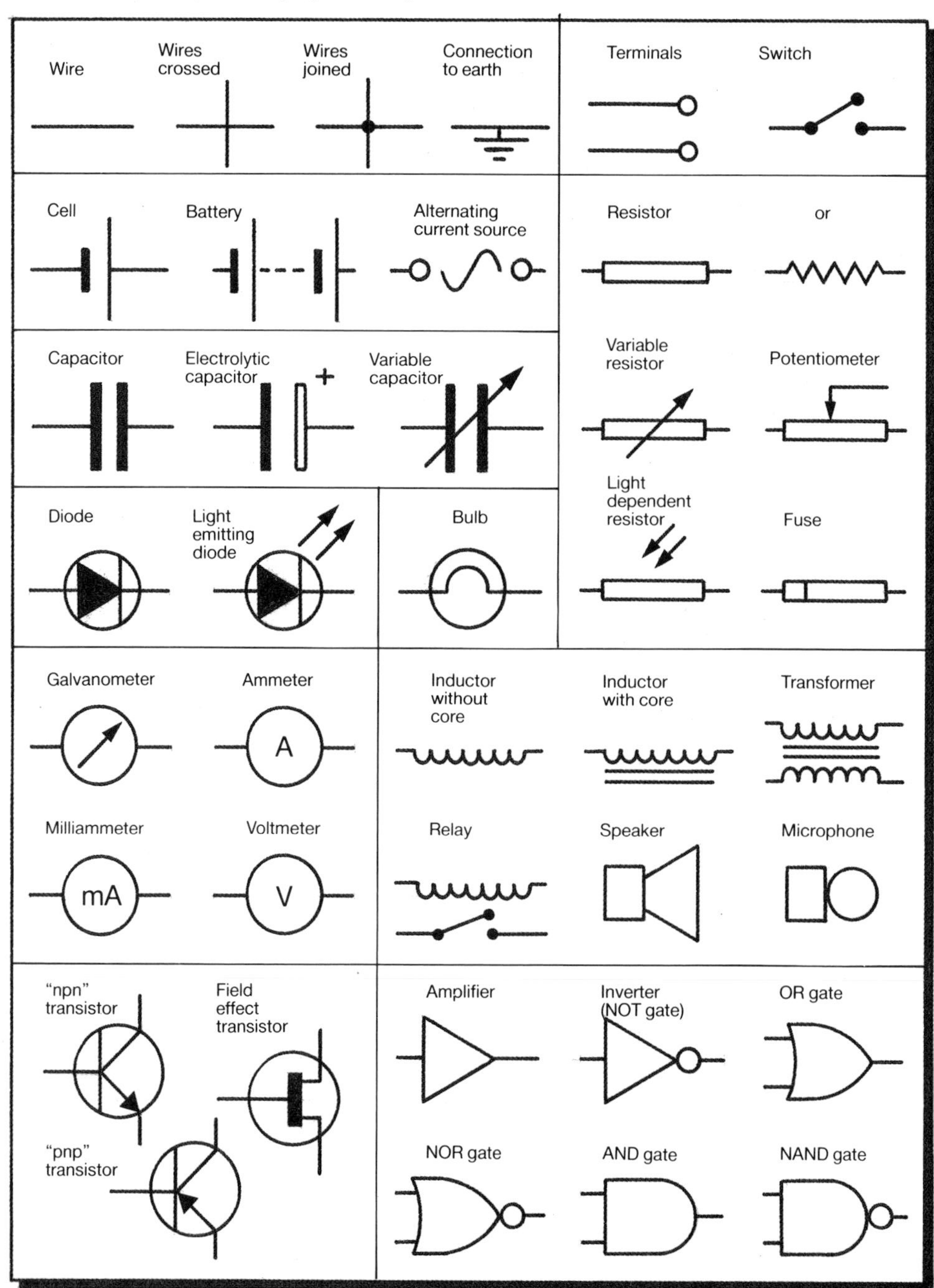

Transistors and gates

Transistors* can be used to amplify electrical signals, such as those from a microphone, and are also used as electronic switches. This has led to their use in complex circuits such as computers. They have replaced the much larger and slower valves and **relays***.

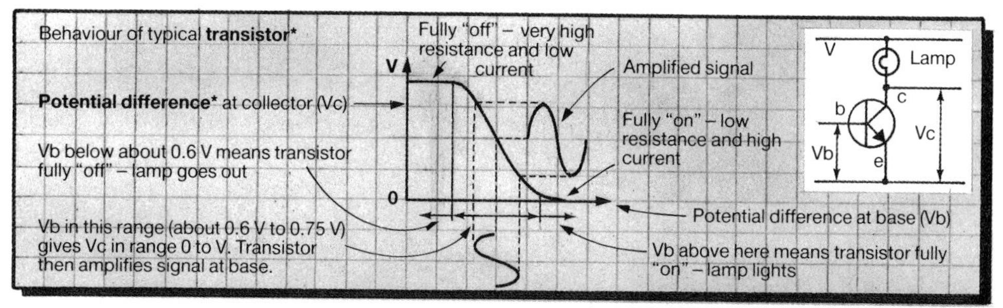

Logic gates

The on and off states of a transistor are used to indicate the numbers 0 and 1. The circuits are therefore known as **digital** (other circuits are called **analogue**). Combinations of transistors with other components are used to make circuits which carry out logical operations.

Truth tables for basic logical operations

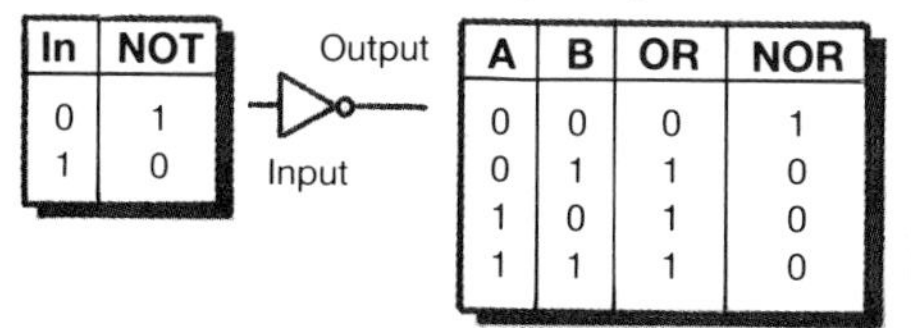

In	NOT
0	1
1	0

A	B	OR	NOR
0	0	0	1
0	1	1	0
1	0	1	0
1	1	1	0

Inputs
A B OR
A B NOR

A	B	AND	NAND
0	0	0	1
0	1	0	1
1	0	0	1
1	1	1	0

Inputs
A B AND
A B NAND

Combinations of these gates and other transistor circuits are used to make complex circuits which can perform mathematical operations, e.g. addition. These are called **integrated circuits**, and may contain many thousands of such components and connections, yet be built into a single slice of silicon.

Computers

Integrated circuits mean that many thousands of logic gates can be put onto a single tiny component called a microchip. The CPU of a computer (see below) can be put on one chip.

Typical computer system

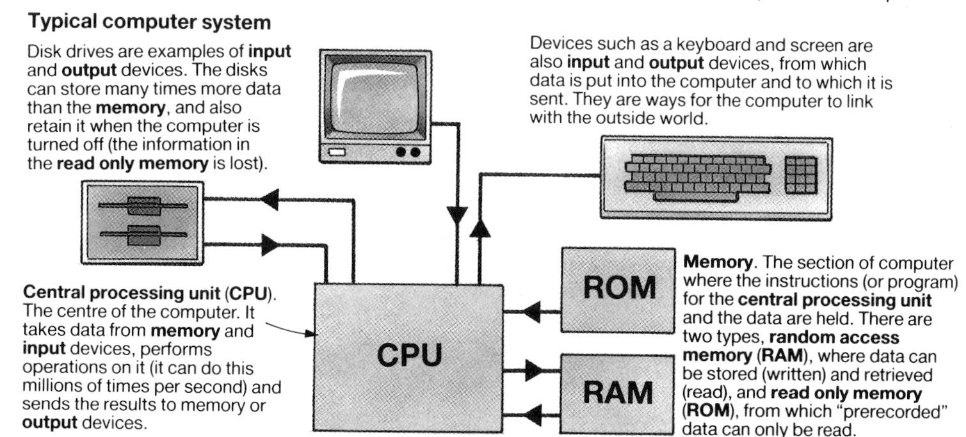

Disk drives are examples of **input** and **output** devices. The disks can store many times more data than the **memory**, and also retain it when the computer is turned off (the information in the **read only memory** is lost).

Devices such as a keyboard and screen are also **input** and **output** devices, from which data is put into the computer and to which it is sent. They are ways for the computer to link with the outside world.

Central processing unit (**CPU**). The centre of the computer. It takes data from **memory** and **input** devices, performs operations on it (it can do this millions of times per second) and sends the results to memory or **output** devices.

Memory. The section of computer where the instructions (or program) for the **central processing unit** and the data are held. There are two types, **random access memory** (**RAM**), where data can be stored (written) and retrieved (read), and **read only memory** (**ROM**), from which "prerecorded" data can only be read.

* **Potential difference**, 58; **Relay**, 75; **Transistor**, 65.

Properties of substances

(Density, specific heat capacity and resistivity all change with temperature. Values quoted here are for room temperature, i.e. 18–22°C.)

Substance	Density / 10^3 kg m^{-3}	Young's modulus / 10^{10} N m^{-2}	Specific heat capacity / J kg^{-1} K^{-1}	Specific latent heat of fusion / 10^4 J kg^{-1}	Linear expansivity / 10^{-6} K^{-1}	Thermal conductivity / W m^{-1} K^{-1}	Resistivity / 10^{-8} ρ m
Aluminium	2.70	7.0	908	40.0	25	242	2.67
Antimony	6.62	7.8	210	16.5	11	19	44
Arsenic	5.73	–	335	–	6.0	–	33.3
Bismuth	9.78	3.2	112	5.5	14	9	117
Brass	8.6 (approx)	9.0	389	–	19	109	8 (approx)
Cadmium	8.65	5.0	230	5.5	30	96	–
Cobalt	8.70	–	435	24.0	12	93	6.4
Constantan	8.90	–	420	–	16	23	49
Copper	8.89	11.0	385	20.0	16	383	1.72
Gallium	5.93	–	377	–	19	34	17.4
Germanium	5.40	–	324	–	5.7	59	4.6 x 10^7
Gold	19.3	8.0	128	6.7	14	300	2.20
Iridium	22.4	–	135	–	6.5	59	5.2
Iron (cast) Iron (wrought)	7.60 7.85	11.0 21.0	460	21.0	12	71	10.3
Lead	11.3	1.6	127	2.5	29	36	20.6
Magnesium	1.74	4.1	1030	30.0	26	154	4.24
Mercury	13.6	–	139	1.2	12	9	95.9
Molybdenum	10.1	–	301	–	5.0	142	5.7
Nickel	8.80	21.0	456	29.0	13	59	6.94
Palladium	12.2	–	247	15.0	12	74	10.7
Platinum	21.5	17.0	135	11.5	9.0	71	10.5
Selenium	4.79	–	324	35.0	26	.24	10^{12} (approx)
Silicon (amorphous)	2.35	11.3	706	–	2.5	175	10^{10} (approx)
Silver	10.5	7.7	234	10.5	19	414	1.63
Steel (mild)	7.80	22.0	450	–	12	46	15 (approx)
Tantalum	16.6	19.0	151	–	6.5	56	13.4
Tellurium	6.2	–	201	–	17	50	1.6 x 10^5
Tin	7.3	5.3	225	5.8	23	63	11.4
Tungsten	19.3	39.0	142	–	4.3	185	5.5
Water	1.00	–	4200	33.4	33.4	.2	–
Zinc	7.10	8.0	387	10.5	11	111	5.92

Useful constants

Quantity	Symbol	Value
Speed of light in vacuum	c	2.998×10^{8} m s^{-1}
Charge on electron	e	1.602×10^{-19} C
Mass of electron	me	9.109×10^{-31} kg
Mass of proton	mp	1.673×10^{-27} kg
Mass of neutron	mn	1.675×10^{-27} kg
Avagadro's number	NA	6.023×10^{23} mol^{-1}
Faraday's constant	F	9.65×10^{4} C mol^{-1}
Gravitational constant	G	6.670×10^{-11} N m^{2} kg^{-2}
Gas constant	R	8.314 J mol^{-1} K^{-1}

Values of common quantities

Quantity	Value
Acceleration due to gravity g (gravitational field strength)	9.81 m s^{-2}
Density of water	1.00×10^{3} kg m^{-3}
Density of mercury	13.6×10^{3} kg m^{-3}
Ice point (standard temperature)	273 K
Steam point	373 K
Standard atmospheric pressure	1.01×10^{5} Pa
Length of earth day	8.64×10^{4} s

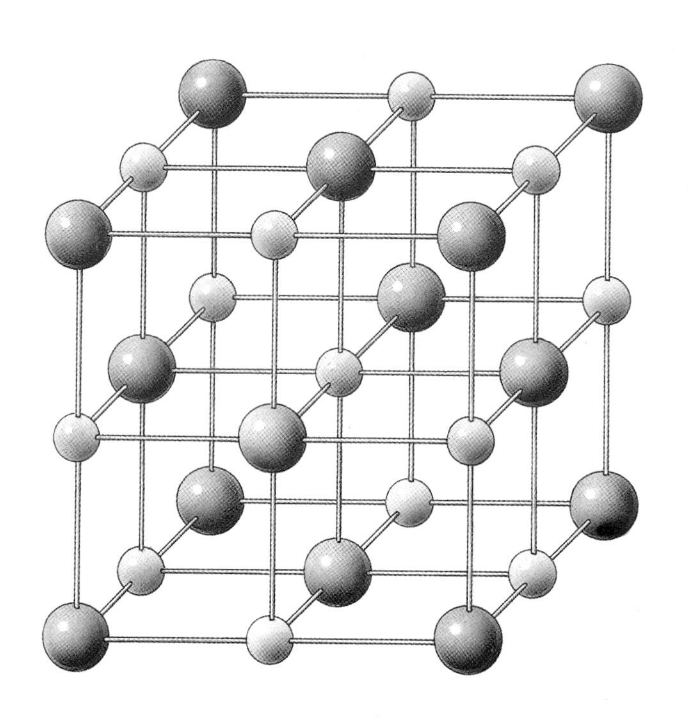

Part two

CHEMISTRY

Designed by Anne Sharples, Roger Berry, Sue Mims and Simon Gooch

Illustrated by Kuo Kang Chen, Kim Blundell, Guy Smith and Chris Lyon

General advisor and co-author: Dr. John Waterhouse

Specialist advisors: John Raffan, Rae Michaelis and Alan Alder

Additional illustrations by Jeremy Gower and Mark Franklin

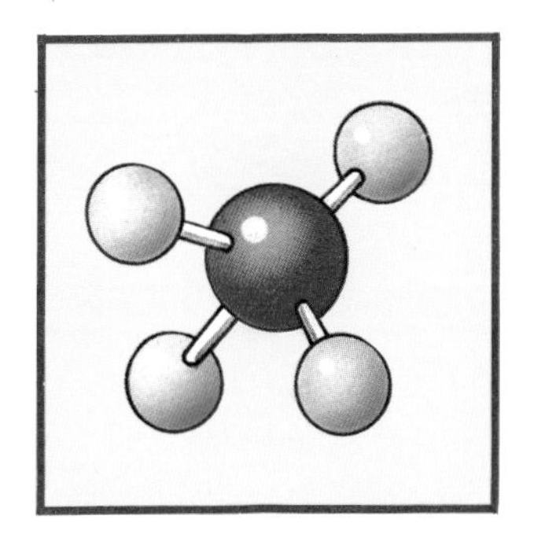

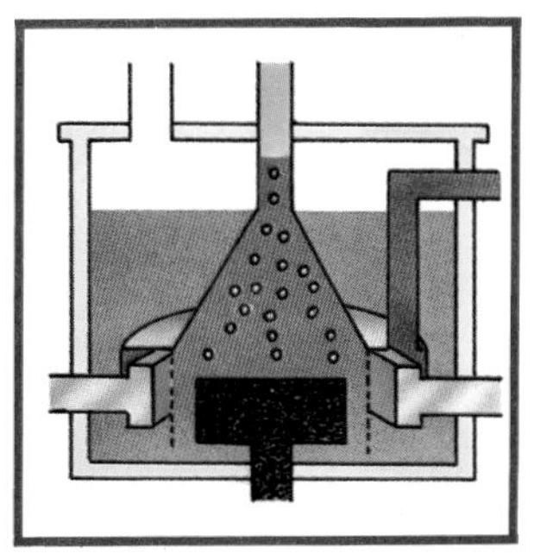

Contents

Physical chemistry

Inorganic chemistry

Organic chemistry

Environmental chemistry

General

We are grateful to the British Standards Institution for permission to reproduce the symbol on page 150.

About chemistry

In this book, chemistry is divided into four main colour-coded sections, followed by a black and white section of general material relating to the whole subject.

Red section

Physical chemistry

Blue section

Inorganic chemistry

Green section

Organic chemistry

Yellow section

Environmental chemistry

Black and white section

General material – charts and tables of properties, symbols and means of identification, plus information on apparatus, preparations, tests and forms of chemical analysis.

Chemistry is the study of the elements which form all existing substances. It covers their structure, how they combine to create other substances and how they react under various conditions. The areas covered by the four different colour-coded sections are explained below.

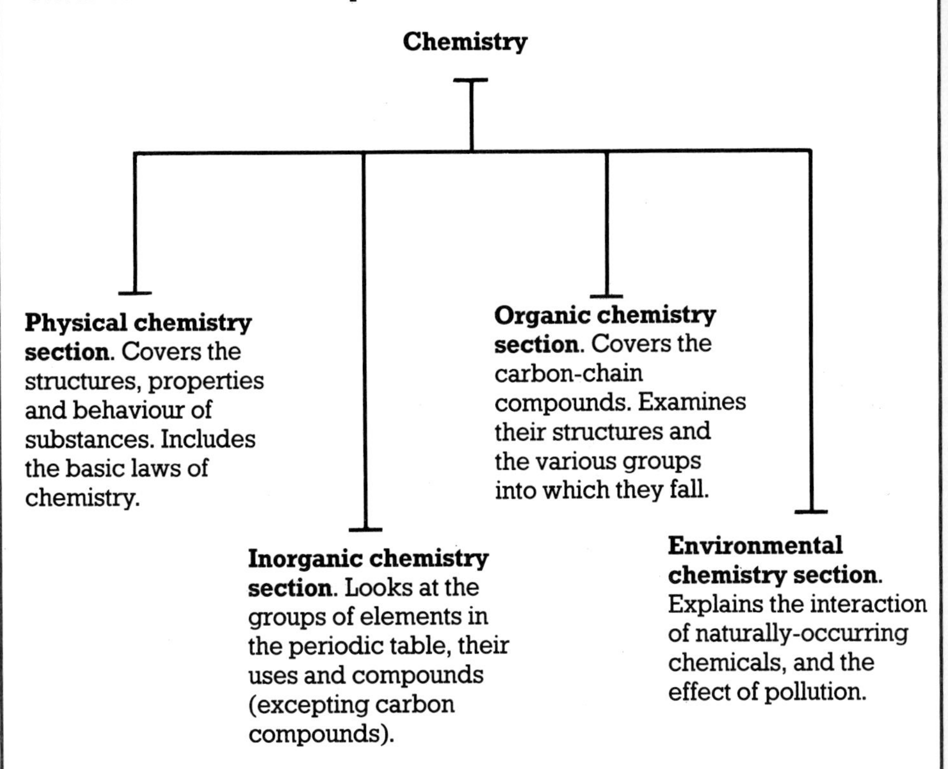

Physical chemistry

Physical chemistry is the study of the patterns of chemical behaviour in **chemical reactions** under various conditions, which result from the **chemical** and **physical properties** of substances. Much of physical chemistry involves measurements of some kind. It covers the following:

1.Solids, liquids and gases, the changes between these states and the reasons for these changes in relation to the structure of a substance (see **states of matter**, pages 120-121, **kinetic theory**, page 123 and **gas laws**, pages 142-143).

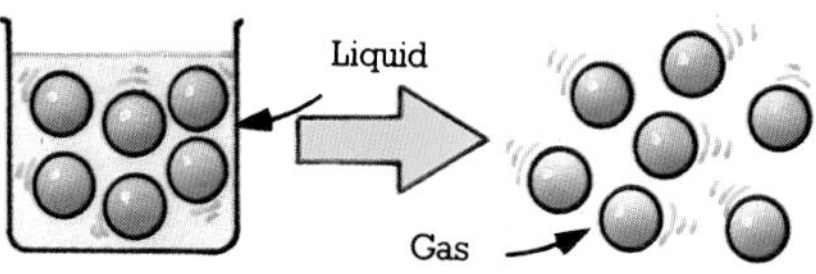

2.The physical and chemical composition of substances – their particles and bonding (see **elements, compounds and mixtures**, pages 122-123, **atoms and molecules**, pages 124-125, **bonding**, pages 130-134 and **crystals**, pages 135-137).

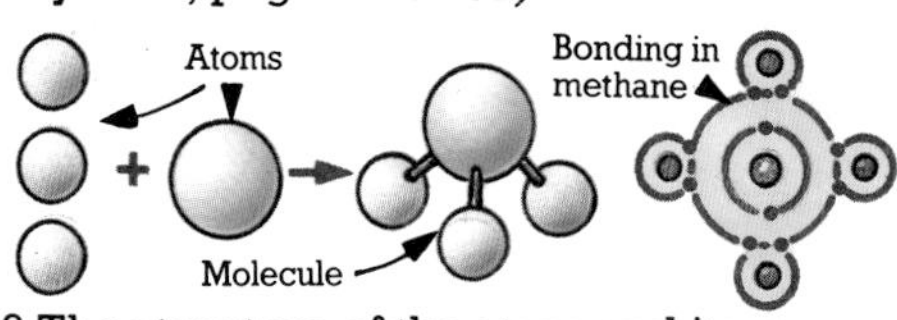

3.The structure of the atom and its importance in the structure of substances (see **atomic structure** and **radioactivity**, pages 126-129).

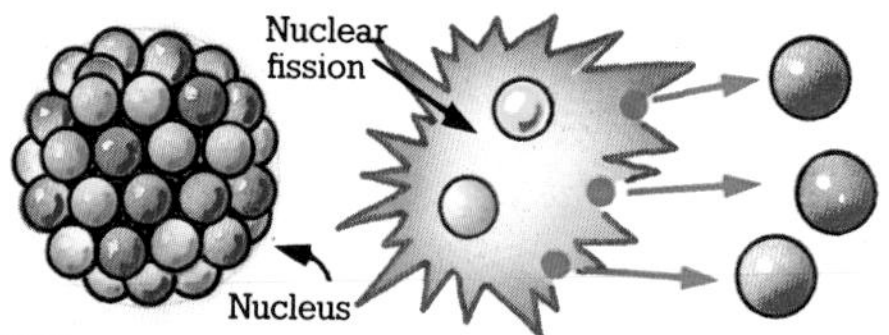

4.The measurement of quantities and the relationship between amounts of gases, liquids and solids (see **measuring atoms**, pages 138-139).

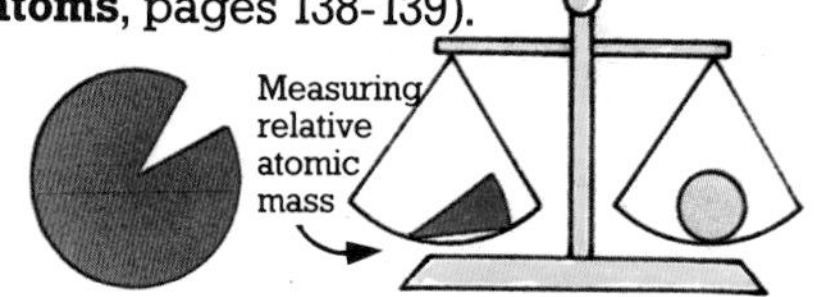

5.Special types of chemical behaviour (see **acids and bases**, pages 150-152 and **salts**, pages 153-155).

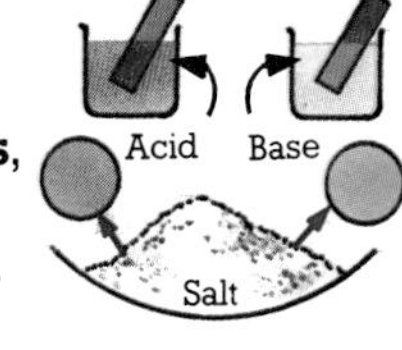

C_2H_4 CH_2 Different formulae Ethene molecule

6.Representing chemicals and chemical reactions (see **representing chemicals**, pages 140-141).

7.How substances mix (see **solutions and solubility**, pages 144-145).

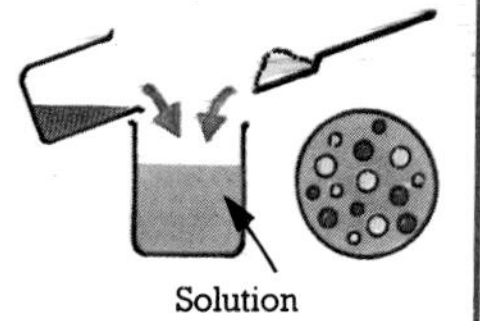

8.Changes during chemical reactions (see **energy and chemical reactions**, pages 146-147 and **rates of reaction**, pages 160-161) and special reactions (see **oxidation and reduction**, pages 148-149 and **reversible reactions**, pages 162-163).

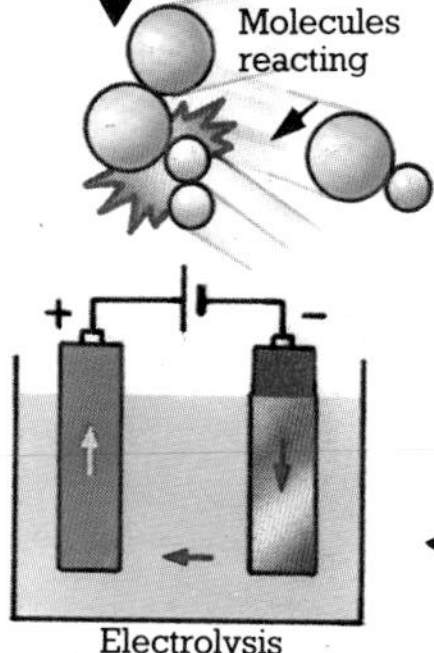

9.The action of electricity on substances and the production of electricity from reactions (see **electrolysis**, pages 156-157 and **reactivity**, pages 158-159).

10.The different levels of reactivity shown by substances and the reasons for this (see **reactivity**, pages 158-159).

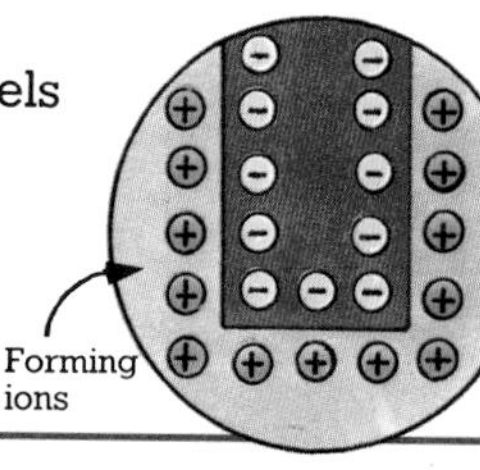

Properties and changes

•**Physical properties**. All the properties of a substance except those which affect its behaviour in **chemical reactions**. There are two main types – **qualitative properties** and **quantitative properties**.

•**Qualitative properties**. Descriptive properties of a substance which cannot be given a mathematical value. They are such things as smell, taste and colour.

Qualitative properties are used in the description of a substance.

•**Quantitative properties.** Properties which can be measured and given a specific mathematical value, e.g. melting point, boiling point, **mass***, hardness and **density***.

Density* (depends on the **mass*** of the particles and how they are packed together)

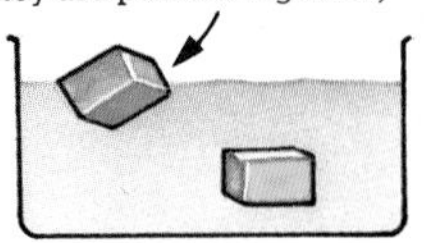

Hardness (depends on **bonding*** and structure)

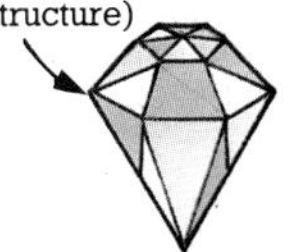

Malleability* and **ductility*** (depend on **bonding*** and structure)

Melting point (depends on **bonding*** and structure)

Boiling point (depends on **bonding*** and structure)

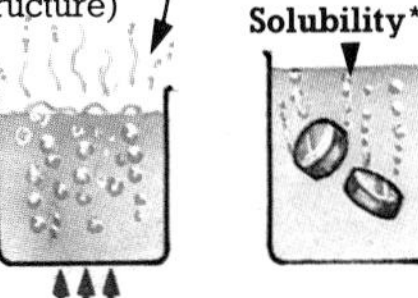

Solubility*

Conductivity* (depends on whether charged particles can move)

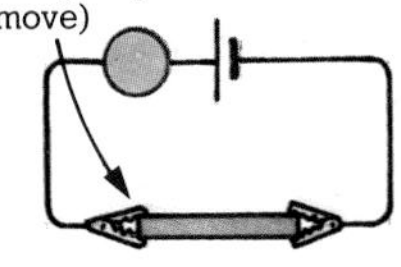

•**Physical change**. A change which occurs when one or more of the **physical properties** of a substance is changed. It is usually easily reversed.

A **physical change** from liquid to solid is caused by removing energy from the particles of the substance (see **kinetic theory**, page 123).

•**Chemical properties**. Properties which cause specific behaviour of substances during **chemical reactions**.

Chemical properties depend on **electron configuration***, **bonding***, structure and energy changes.

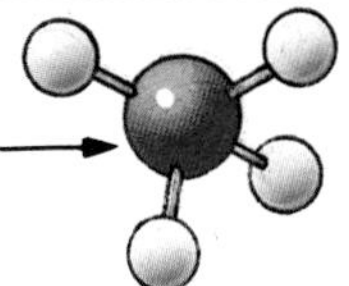

•**Chemical reaction**. Any change which alters the **chemical properties** of a substance or which forms a new substance. During a chemical reaction, **products** are formed from **reactants**.

•**Reactants**. The substances present at the beginning of a **chemical reaction**.

•**Products**. The substances formed in a **chemical reaction**.

The **rusting*** of iron is a **chemical reaction**.

Iron, water and oxygen from the air are the **reactants**.

Rust* is the **product**.

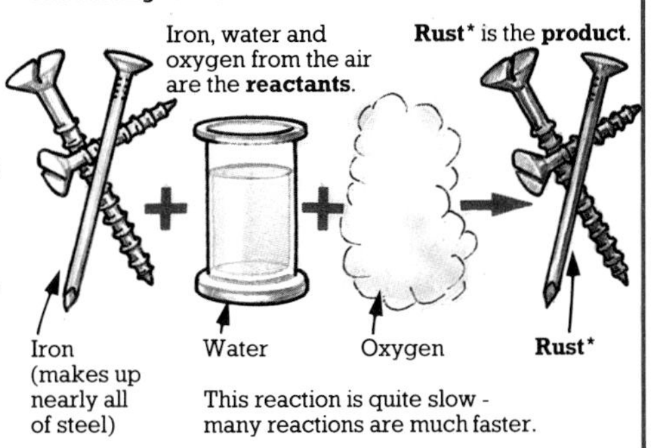

This reaction is quite slow - many reactions are much faster.

•**Reagent**. A substance used to start a **chemical reaction**. It is also one of the **reactants**. Common reagents in the laboratory are hydrochloric acid, sulphuric acid and sodium hydroxide.

* **Bonding**, 130; **Conductivity**, 62; **Density**, 24; **Ductility**, 344 (**Ductile**); **Electron configuration**, 127; **Malleability**, 345 (**Malleable**); **Mass**, 12; **Rust**, 174; **Rusting**, 209 (**Corrosion**); **Solubility**, 145.

States of matter

A substance can be **solid**, **liquid** or **gaseous**. These are the **physical states** or **states of matter** (normally shortened to **states**). Substances can change between states, normally when heated or cooled to increase or decrease the energy of the particles (see **kinetic theory**, page 123).

- **Solid state**. A state in which a substance has a definite volume and shape.

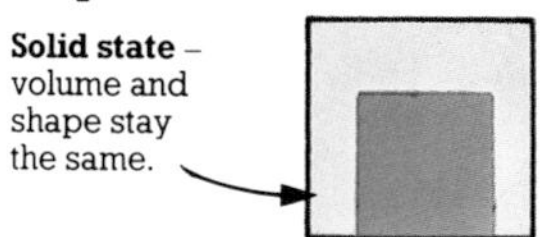

- **Gaseous state**. A state in which a substance has no definite volume or shape. It is either a **vapour** or a **gas**. A vapour can be changed into a liquid by applying pressure alone; a gas must first be turned into a vapour by reducing its temperature to below a level called its **critical temperature**.

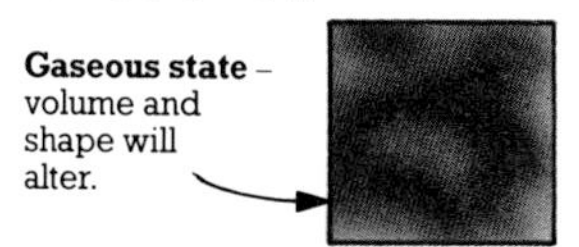

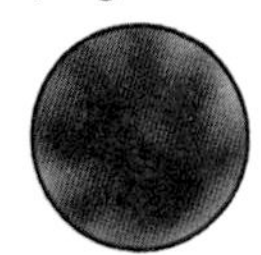

- **Liquid state**. A state in which a substance has a definite volume, but can change shape.

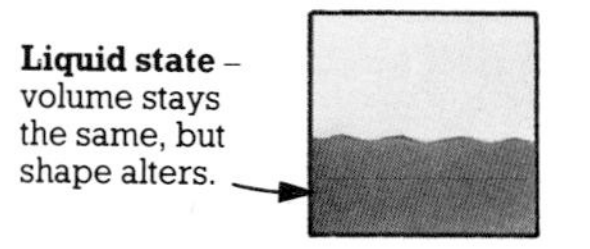

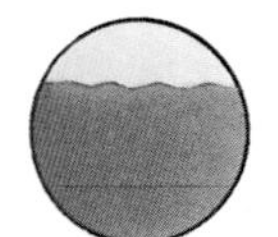

- **Phase**. A separate part of a mixture of substances with two or more states. A mixture of sand and water contains two phases – the **solid** phase (sand) and the **liquid** phase (water).

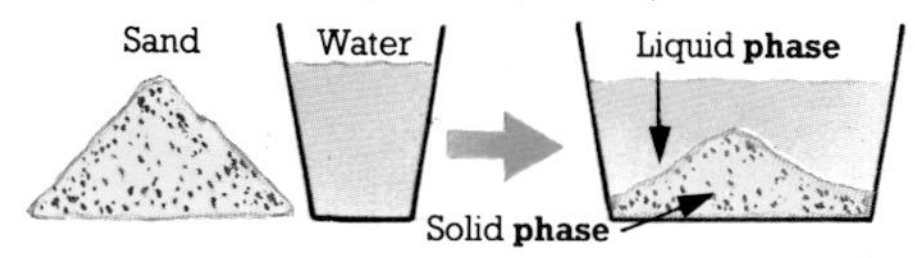

- **Fluid**. A substance that will flow, i.e. is either in the **gaseous** or **liquid state**.

Changes of state

A **change of state** is a **physical change*** of a substance from one state to another. It normally occurs because of a change in the energy of the particles, caused by heating or cooling (see **kinetic theory**, page 123).

- **Melting**. The change of state from **solid** to **liquid**, usually caused by heating. The temperature at which a solid melts is called its **melting point** (see also pages 212-213), which is the same temperature as its **freezing point** (see **freezing**). At the melting point, both solid and liquid states are present. An increase in pressure increases the melting point. All pure samples of a substance at the same pressure have the same melting point.

- **Molten**. Describes the **liquid** state of a substance which is a **solid** at room temperature.

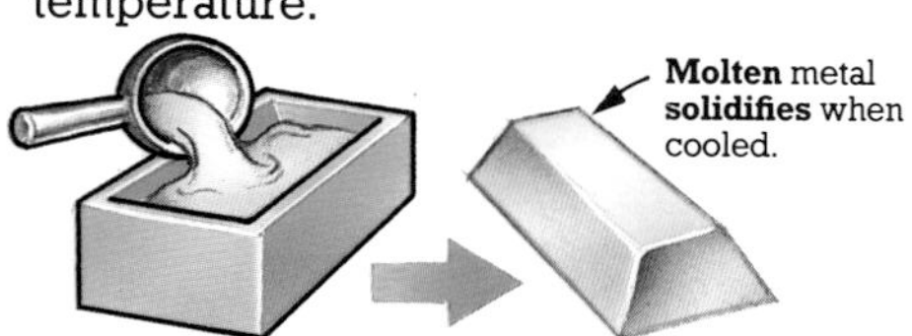

- **Solidification**. The change of state from **liquid** to **solid** of a substance which is a solid at room temperature and atmospheric pressure.

* **Physical change**, 119.

• **Freezing**. The change of state from **liquid** to **solid**, caused by cooling a liquid. The temperature at which a substance freezes is the **freezing point**, which is the same temperature as the **melting point** (see **melting**).

• **Fusion**. The change of state from **solid** to **liquid** of a substance which is solid at room temperature and pressure. The substance is described as **fused** (or **molten**). A solid that has been fused and then solidified into a different form is also described as fused.

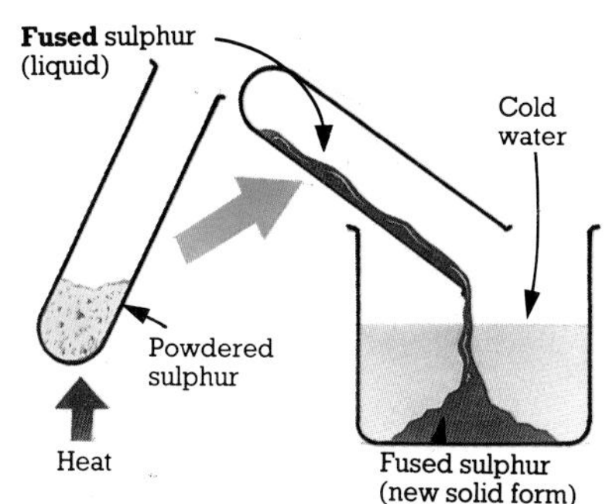

• **Boiling**. A change of state from **liquid** to **gaseous** (**vapour**) at a temperature called the **boiling point** (see also pages 212-213). It occurs by the formation of bubbles throughout the liquid. All pure samples of the same liquid at the same pressure have the same boiling point. An increase in pressure increases the boiling point.

• **Evaporation**. A change of state from **liquid** to **gaseous** (**vapour**), due to the escape of molecules from the surface (for more about this, see page 31). A liquid which readily evaporates is described as **volatile***.

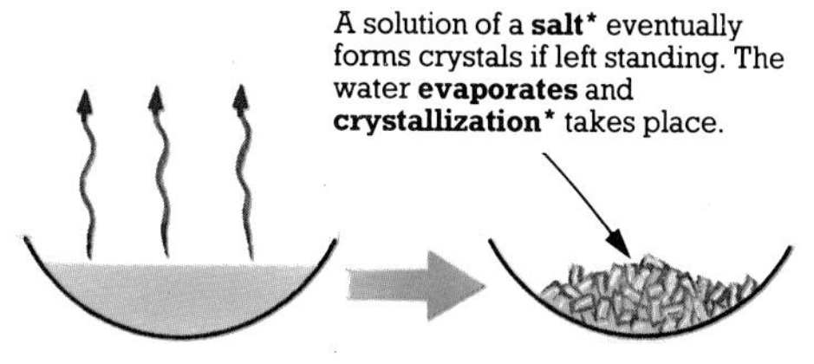

• **Liquefaction**. A change of state from **gaseous** (**gas**) to **liquid**, of a substance which is a gas at room temperature and pressure. It is caused by cooling (to form a **vapour**) and increasing pressure.

• **Condensation**. A change of state from **gaseous** (**gas** or **vapour**) to **liquid**, of a substance which is a liquid at room temperature and pressure. It is normally caused by cooling.

• **Sublimation**. The change of state from **solid** to **gaseous** (**gas**, via **vapour**) on heating, and from gaseous directly to solid on cooling. At no stage is a liquid formed. See picture, page 162.

• **Vaporization**. Any change resulting in a **gaseous state**, i.e. **boiling**, **evaporation** or **sublimation**.

* **Crystallization**, 135; **Salts**, 153; **Volatile**, 345.

Elements, compounds and mixtures

Elements, **compounds** and **mixtures** are the three main types of chemical substance. All substances are made of elements, and most are a combination of two or more elements.

•**Element**. A substance which cannot be split into simpler substances by a chemical reaction. There are just over 100 known elements, classified in the **periodic table***, and most are solids or gases at room temperature. All atoms of the same element have the same number of **protons*** in their **nuclei*** (see **atomic number**, page 127).

•**Chemical symbol**. A shorthand way of representing an **element** in **formulae** and **equations** (see pages 140-141). It represents one atom and usually consists of the first one or two letters of the name of the element, occasionally the Greek or Latin name. See pages 212-213 for a list of elements and their symbols, and pages 226-227 to match symbols to elements.

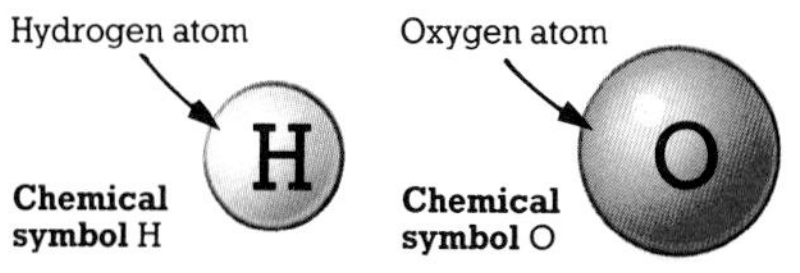

Silver atom

Ag

Chemical symbol Ag – Argentum is Latin for silver

Nitrogen atom

N

Chemical symbol N

•**Compound**. A combination of two or more **elements**, bonded together in some way. It has different physical and chemical properties from the elements it is made of. The proportion of each element in a compound is constant, e.g. water is always formed from two parts hydrogen and one part oxygen. This is shown by its chemical **formula***, H_2O. Compounds are often difficult to split into their elements and can only be separated by chemical reactions.

•**Binary**. Describes a **compound** composed of two **elements** only, e.g. carbon monoxide.

•**Mixture**. A blend of two or more **elements** and/or **compounds** which are not chemically combined. The proportions of each element or compound are not fixed, and each keeps its own properties. A mixture can usually be separated into its elements or compounds fairly easily by physical means.

•**Synthesis**. The process by which a **compound** is built up from its **elements** or from simpler compounds by a sequence of chemical reactions, e.g. iron(III) chloride is made by passing chlorine gas over heated iron.

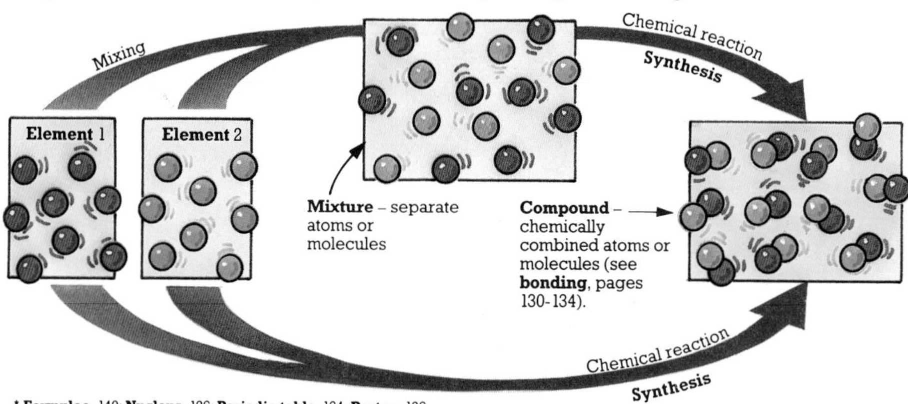

* **Formulae**, 140; **Nucleus**, 126; **Periodic table**, 164; **Proton**, 126.

•**Homogeneous.** Describes a substance which is the same throughout in its properties and composition, e.g. solutions (the particles of the **solute*** and **solvent*** are molecules or ions).

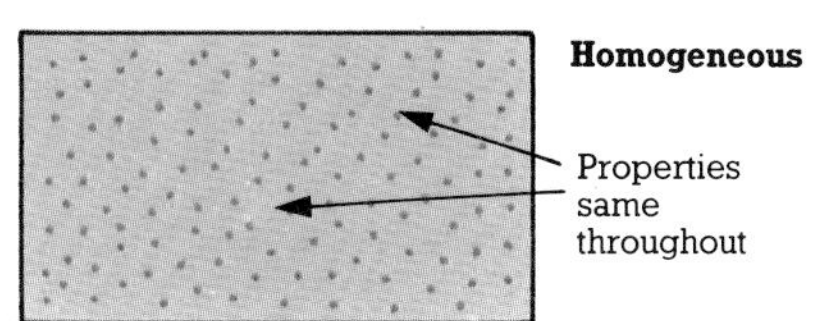

•**Heterogeneous.** Describes a substance which varies in its composition and properties from one part to another, e.g. **suspensions*** (the particles are groups of atoms, molecules or ions).

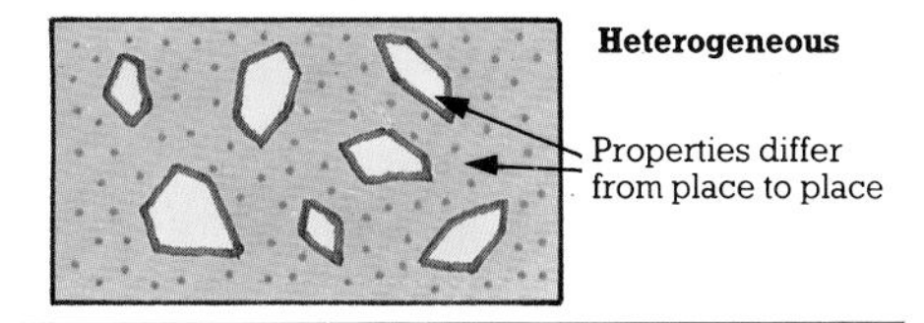

•**Pure**. Describes a sample of a substance which consists only of one **element** or **compound**. It does not contain any other substance in any proportions. If the substance does contain traces of another then it is described as **impure** and the other substance is called an **impurity**.

Kinetic theory

The **kinetic theory** explains the behaviour of solids, liquids and gases, and **changes of state*** between them, in terms of the movement of the particles of which they are made (see diagram below).

•**Brownian motion**. The random motion of small particles in water or air. It supports the kinetic theory, as it is clearly due to unseen impact with the water or air molecules.

•**Diffusion**. The process by which two **fluids*** mix without mechanical help. The process supports the kinetic theory, since the particles must be moving to mix and gases can be seen to diffuse faster than liquids. Only **miscible*** liquids diffuse.

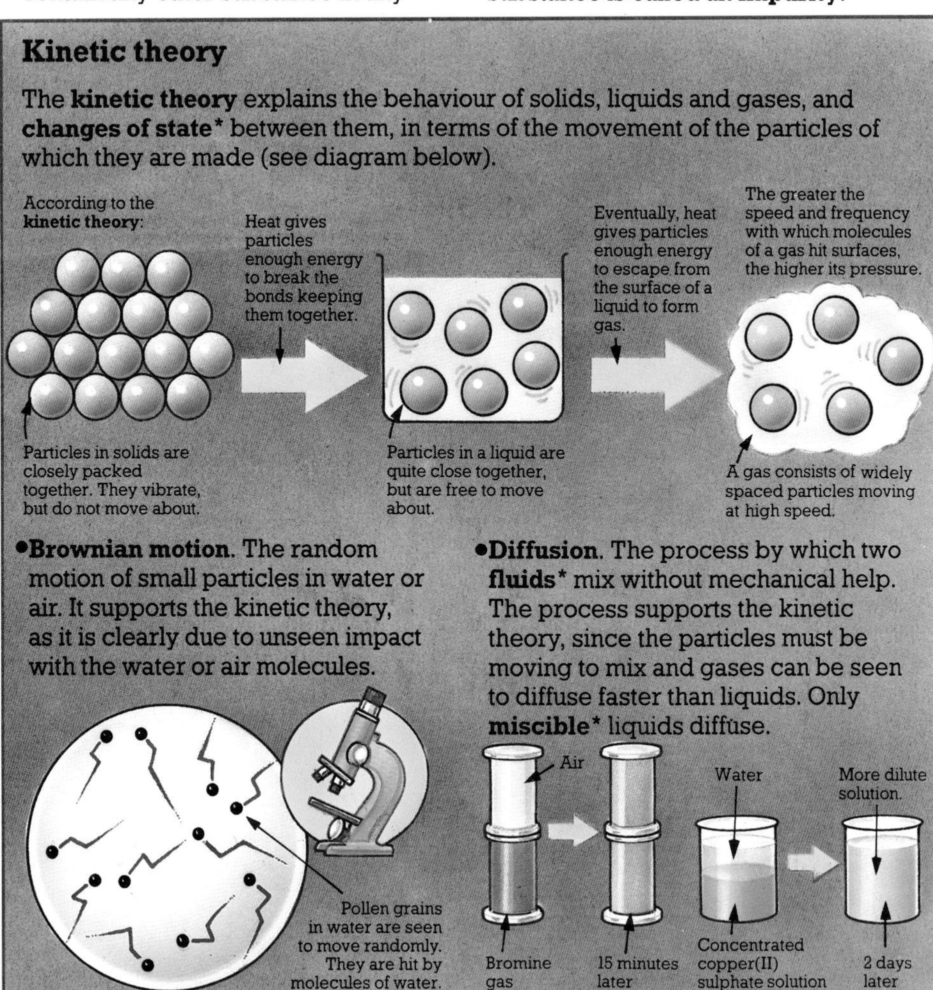

* **Change of state**, **Fluid**, 120; **Miscible**, 145; **Solute**, **Solvent**, 144; **Suspension**, 145.

Atoms and molecules

Over 2000 years ago, the Greeks decided that all substances consisted of small particles which they called **atoms**. Later theories extended this idea to include **molecules** – atoms joined together. **Inorganic*** molecules generally only contain a few atoms, but **organic*** molecules can contain hundreds of atoms.

- **Atom**. The smallest particle of an element that retains the chemical properties of that element. The atoms of many elements are bonded together in groups to form particles called **molecules** (see also **covalent bonding**, page 132). Atoms consist of three main types of smaller particles – see **atomic structure**, page 126.

- **Molecule**. The smallest particle of an element or compound that normally exists on its own and still retains its properties. Molecules normally consist of two or more **atoms** bonded together – some have thousands of atoms. **Ionic compounds*** consist of **ions*** and do not have molecules.

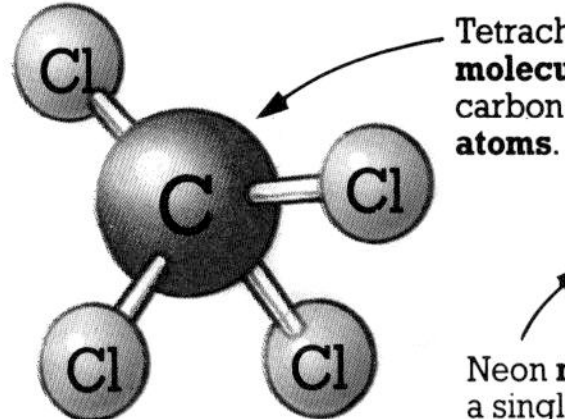

Tetrachloromethane (CCl_4) **molecules** consist of one carbon and four chlorine **atoms**.

Neon **molecules** consist of a single neon **atom**.

- **Atomicity**. The number of **atoms** in a **molecule**, calculated from the **molecular formula*** of the compound.

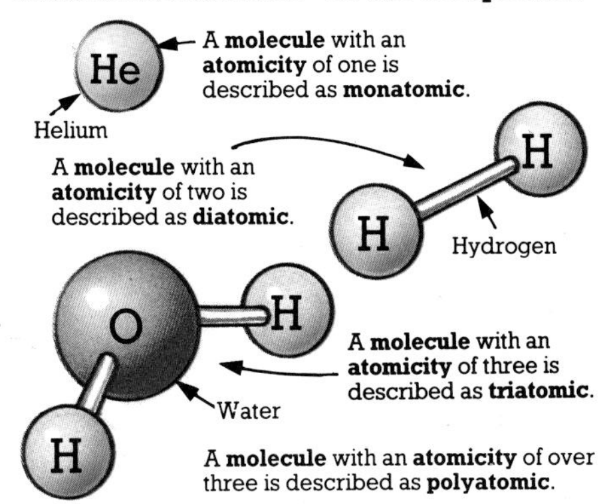

A **molecule** with an **atomicity** of over three is described as **polyatomic**.

- **Dalton's atomic theory**. John Dalton's theory, published in 1808, attempts to explain how **atoms** behave. It is still generally valid. It states that:

1. All matter is made up of tiny particles called **atoms**.

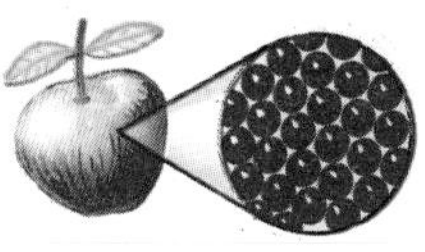

2. Atoms cannot be made, destroyed or divided (now disproved – see **radioactivity**, page 128).

3. All atoms of the same element have the same properties and the same mass (now disproved – see **isotope**, page 127).

4. **Atoms** of different elements have different properties and different masses.

5. When compounds form, the **atoms** of the elements involved combine in simple whole numbers (now known to be untrue for large **organic*** **molecules** containing hundreds of atoms).

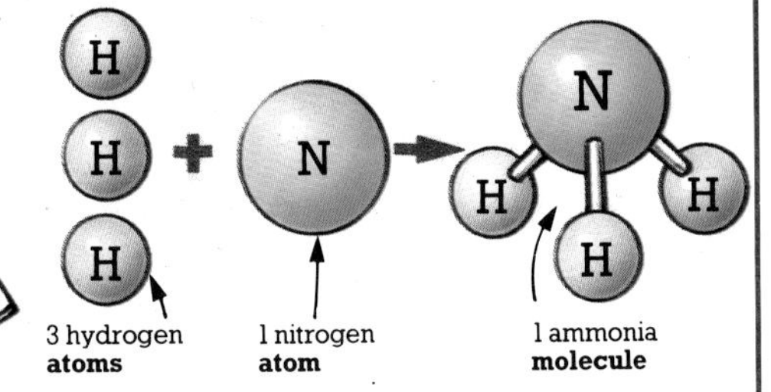

* **Inorganic chemistry**, 166; **Ion**, 130; **Ionic compound**, 131; **Molecular formula**, 140; **Organic chemistry**, 190.

•Dimer. A substance with **molecules** formed from the combination of two molecules of a **monomer***.

Nitrogen dioxide (**monomer***) combines to form dinitrogen tetraoxide (**dimer**).

$$NO_2\,(g) + NO_2\,(g) \rightarrow N_2O_4\,(g)$$

Nitrogen dioxide | Nitrogen dioxide | Dinitrogen tetraoxide

•Trimer. A substance with **molecules** formed from the combination of three molecules of a **monomer***.

•Macromolecule. A **molecule** consisting of a large number of **atoms**. It is normally an **organic*** molecule with a very high **relative molecular mass***.

Basic laws of chemistry

Three laws of chemistry were put forward in the late 18th and early 19th centuries. Two pre-date **Dalton's atomic theory** and the third (the **law of multiple proportions**) was developed from it. These laws were of great importance in the development of the atomic theory.

•Law of conservation of mass. States that matter can neither be created nor destroyed during a chemical reaction. It was developed by a Frenchman, Antoine Lavoisier, in 1774. ▶

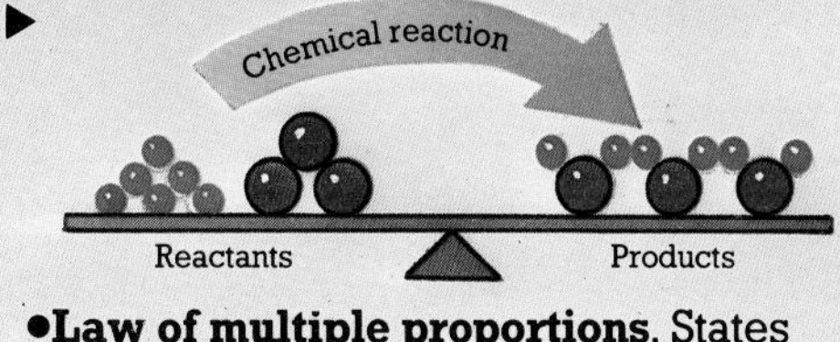

•Law of constant composition. States that all pure samples of the same chemical compound contain the same elements combined in the same proportions by mass. It was developed by a Frenchman, Joseph Proust, in 1799.

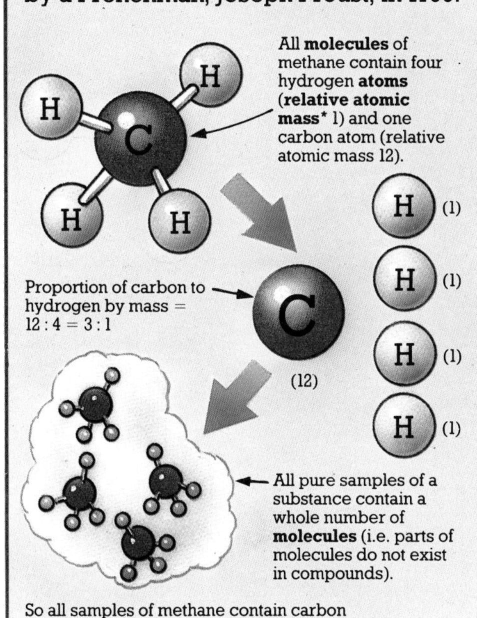

So all samples of methane contain carbon and hydrogen in the ratio 3:1 by mass.

•Law of multiple proportions. States that if two elements, A and B, can combine to form more than one compound, then the different masses of A which combine with a fixed mass of B in each compound are in a simple ratio. It is an extension of **Dalton's atomic theory**.

For one nitrogen **atom**:

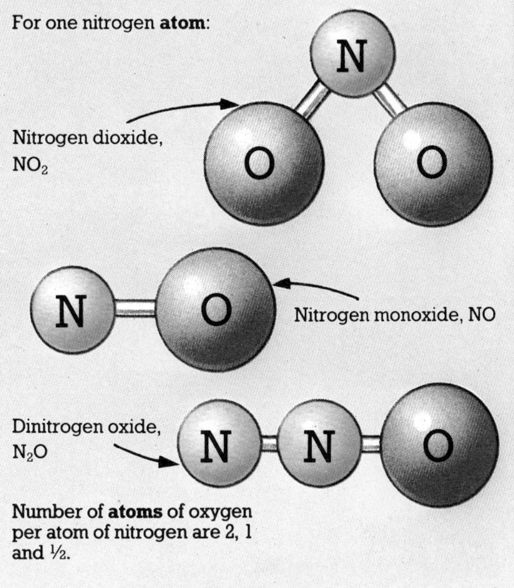

Number of **atoms** of oxygen per atom of nitrogen are 2, 1 and ½.

Masses of oxygen in ratio 4:2:1

* **Monomers**, 200; **Organic chemistry**, 190; **Relative atomic mass**, **Relative molecular mass**, 138.

Atomic structure

Dalton's atomic theory (see page 124) states that the atom is the smallest possible particle. However, experiments have proved that it contains smaller particles, or **subatomic particles**. The three main subatomic particles are **protons** and **neutrons**, which make up the **nucleus**, and **electrons**, which are arranged round the nucleus.

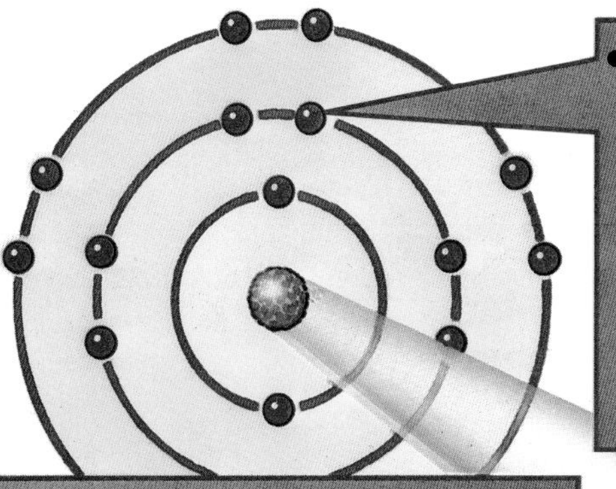

•**Electron**. A **subatomic particle** (see introduction) which moves around the **nucleus** of an atom within an **electron shell**. Its mass is very small, only $\frac{1}{1836}$ that of a **proton**. An electron has a negative electrical charge, equal in size but opposite to that of a proton. There are the same number of electrons as protons in an atom.

•**Neutron**. A **subatomic particle** (see introduction) in the **nucleus** of an atom. It has a **relative atomic mass*** of 1 and no electrical charge. The number of neutrons in atoms of the same element can vary (see **isotope**).

•**Proton**. A **subatomic particle** (see introduction) in the **nucleus** of an atom. It has a **relative atomic mass*** of 1 and a positive electrical charge equal in size but opposite to that of an **electron**. An atom has the same number of protons and electrons, making it electrically neutral.

•**Nucleus** (pl. **nuclei**) or **atomic nucleus**. The structure at the centre of an atom, consisting of **protons** and **neutrons** (usually about the same number of each) packed closely together, around which **electrons** move. The nucleus makes up almost the total mass of the atom, but is very small in relation to the total size.

•**Electron shell**. Also called a **shell** or **energy level**. A region of space in which **electrons** move around the **nucleus** of an atom. An atom can have up to seven shells, increasing in radius, and each can hold up to a certain number of electrons. The model on the right is a simplified one – in fact, the exact positions of electrons cannot be determined at any one time, and each shell consists of **orbitals**.

The first three **electron shells**.

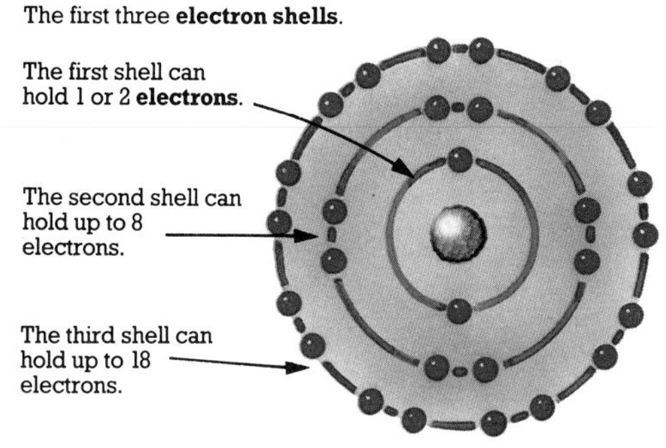

* **Relative atomic mass**, 138.

• **Orbital**. A region in which there can be either one or two **electrons**. Each **electron shell** consists of one or more orbitals of varying shapes.

• **Outer shell**. The last **electron shell** in which there are **electrons**. The number of electrons in the outer shell influences how the element reacts and which **group** it is in (see **periodic table**, pages 164-165).

• **Electron configuration**. A group of numbers which shows the arrangement of the **electrons** in an atom. The numbers are the numbers of electrons in each **electron shell**, starting with the innermost.

A sodium atom has an **electron configuration** of 2.8.1

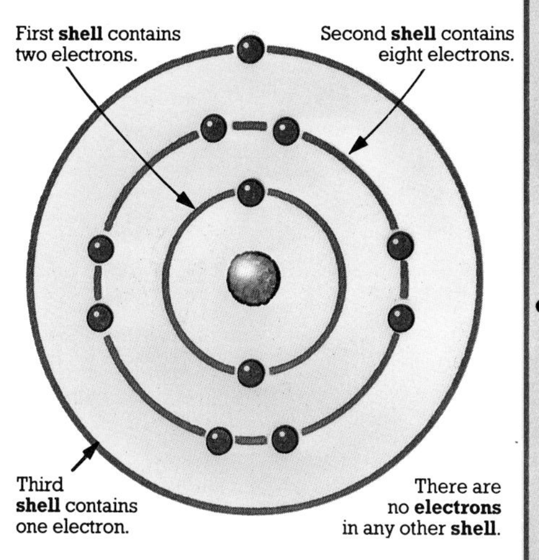

• **Octet**. A group of eight **electrons** in a single **electron shell**. Atoms with an octet for the **outer shell** are very stable and unreactive. All **noble gases*** (except helium) have such an octet. Other atoms can achieve a stable octet (and thus have an **electron configuration** similar to that of the nearest noble gas) either by sharing electrons with other atoms (see **covalent bonding**, page 132) or by gaining or losing electrons (see **ionic bonding**, page 130).

* **Noble gases**, 189.

• **Atomic number**. The number of **protons** in the **nucleus** of an atom. The atomic number determines the element of the atom, e.g. any atom with six protons is carbon, regardless of the number of **neutrons** and **electrons**.

• **Mass number**. The total number of **protons** and **neutrons** in one atom of an element. The mass number of an element can vary because the number of neutrons can change (see **isotope**). The mass number is usually about twice the **atomic number**.

The **atomic number** and **mass number** are often written with the symbol for the element.

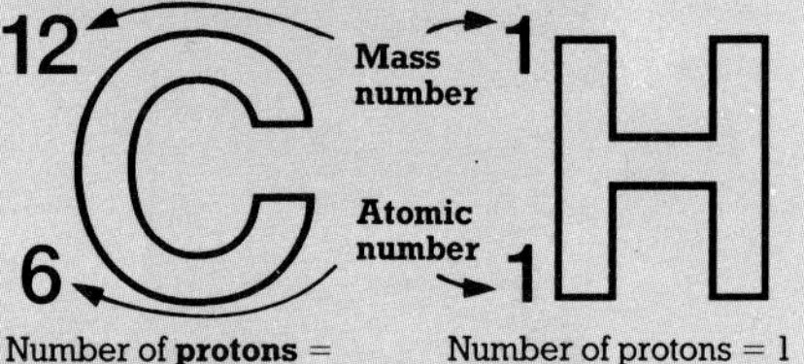

Number of **protons** = **atomic number** = 6

Number of **electrons** = 6 (to balance number of **protons**)

Number of **neutrons** = **mass number** – **atomic number** = 6

Number of protons = 1

Number of neutrons = 0

Mass number and **atomic number** the same. **Nucleus** of hydrogen is a single **proton**.

• **Isotope**. An atom of an element in which the number of **neutrons** is different from that in another atom of the same element. Isotopes of an element have the same **atomic number** but different **mass numbers**. Isotopes are distinguished by writing the mass number by the name or symbol of the element.

Carbon has three **isotopes**.

Carbon-12 $^{12}_{6}C$ → 6 **neutrons**

Carbon-13 $^{13}_{6}C$ → 7 **neutrons**

Carbon-14 $^{14}_{6}C$ → 8 **neutrons**

Radioactivity

Radioactivity is a property of unstable **nuclei*** (for more about the reasons for their instability, see pages 84-85). It involves the nuclei breaking up spontaneously into nuclei of other elements and emitting rays or particles (**radiation**), a process known as **radioactive decay**. A radioactive element is one whose nuclei are gradually splitting up in this way.

- **Radioisotope** or **radioactive isotope**. The general term for a radioactive substance, since all are **isotopes***. There are several naturally-occurring radioisotopes, such as carbon-14, and others are formed in a variety of ways. For more about this, see page 86.

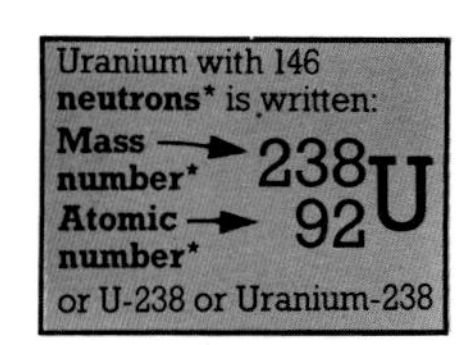

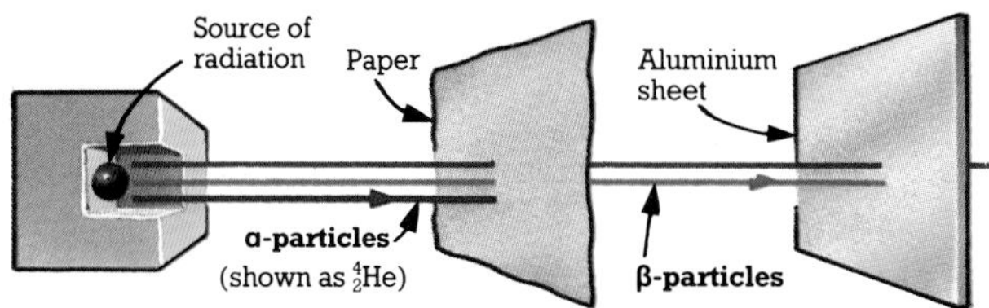

- **Alpha particle** (**α-particle**). One type of particle emitted from the **nucleus*** of a radioactive atom. It is like a helium nucleus, consisting of two **protons*** and two **neutrons***, has a **relative atomic mass*** of 4 and a charge of plus 2. It moves slowly and has a low penetrating power.

- **Beta particle** (**β-particle**). A fast-moving particle emitted from a radioactive **nucleus***. There are two different types of beta particle (for more about this, see page 86). They can penetrate objects which have a low density and/or thickness, for instance paper.

- **Gamma rays** (**γ-rays**). Rays generally emitted after an **alpha** or **beta particle** from a radioactive **nucleus***. They take the form of waves (like light and X-rays) and have a high penetrating power, going through aluminium sheet. They are stopped by lead.

- **Radioactive decay**. The process whereby the **nuclei*** of a radioactive element undergo a series of **disintegrations** (a **decay series**) to become stable. For more about the different types of decay, see page 87.

- **Disintegration**. The splitting of an unstable **nucleus*** into two parts, usually another nucleus and an **alpha** or **beta particle**. The **atomic number*** changes, so an atom of a new element is produced. If this is a stable atom, then no further disintegrations occur. If it is unstable, it disintegrates in turn and the process continues as a **decay series** until a stable atom is formed.

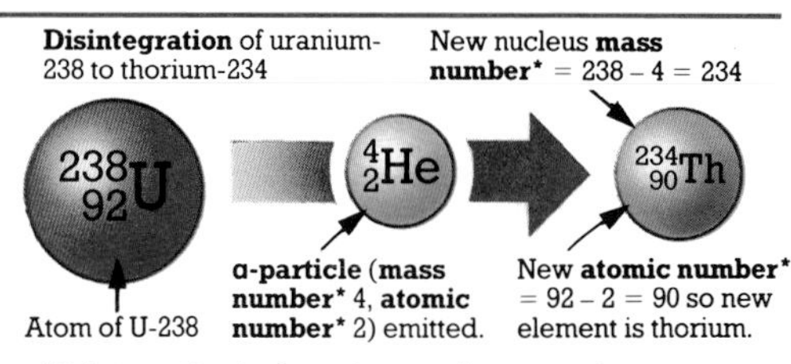

Disintegration is shown by a **nuclear equation**.

$$^{238}_{92}U \rightarrow\ ^{234}_{90}Th +\ ^{4}_{2}He$$

* **Atomic number**, **Isotope**, **Mass number**, 127; **Neutron**, **Nucleus**, **Proton**, 126; **Relative atomic mass**, 138.

- **Decay series** or **radioactive series**. The series of **disintegrations** involved when a radioactive element decays, producing various elements until an element with stable atoms is formed.

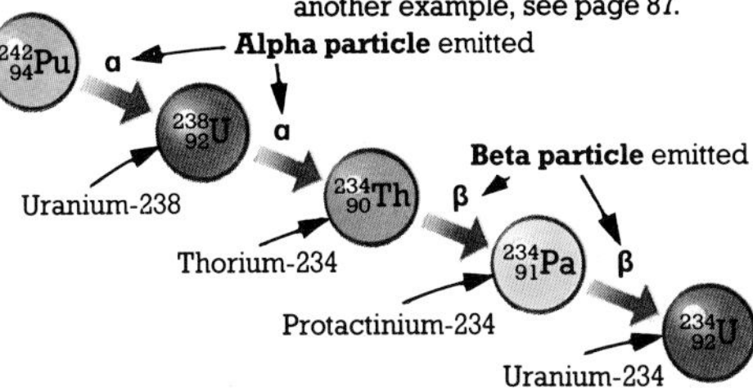

Decay series for plutonium-242 to uranium-234. For another example, see page 87.

- **Becquerel**. A unit of **radioactive decay**. One becquerel is equal to one nuclear **disintegration** per second. A **curie** equals 3.7 x 10^{10} becquerels.

- **Half-life**. The time taken for half of the atoms in a sample of a radioactive element to undergo **radioactive decay**. The amount of radiation emitted is halved. The half-life varies widely, e.g. uranium-238, 4.5 thousand million years; radium-221, 30 seconds.

Radioactive decay curve

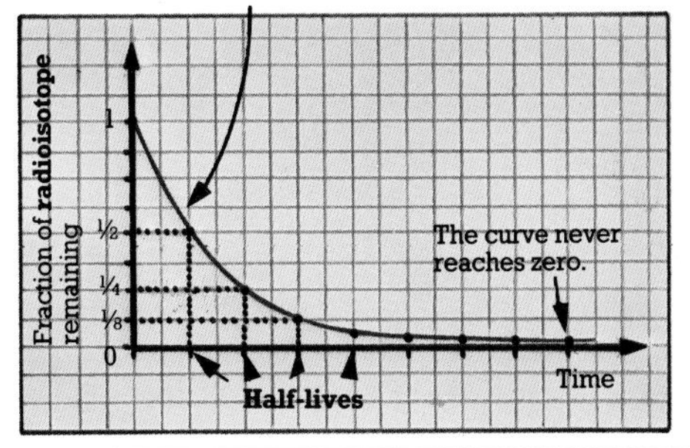

Uses of radioactivity

- **Nuclear fission**. The division of a **nucleus***, caused by bombardment with a **neutron***. The nucleus splits, forming neutrons and nuclei of other elements, and releasing huge amounts of energy. The release of neutrons also causes the fission of other atoms, which in turn produces more neutrons – a **chain reaction**. An element which can undergo fission is described as **fissile**. Controlled nuclear fission is used in **nuclear power stations***, but uncontrolled fission, e.g. in **fission bombs***, is very explosive.

Fission of uranium-235

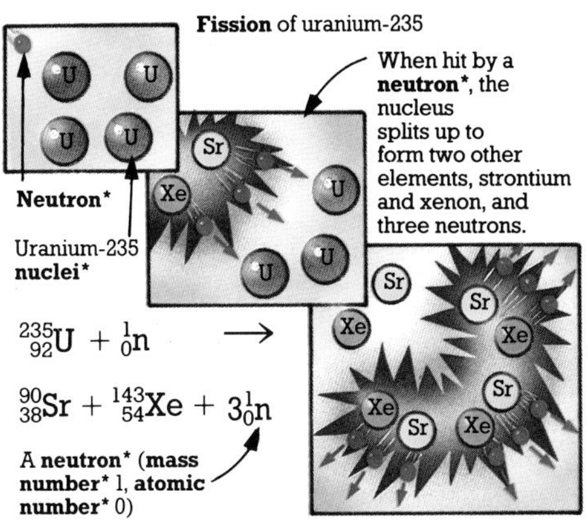

- **Nuclear fusion**. The combination of two **nuclei*** to form a larger one. It will only take place at extremely high temperatures and releases huge amounts of energy. Nuclear fusion takes place in the **fusion bomb***.

- **Radioactive tracing**. A method of following a substance as it moves by tracking radiation from a **radioisotope** introduced into it. The radioisotope used is called a **tracer** and the substance is said to be **labelled**.

- **Radiocarbon dating** or **carbon dating**. A method used to calculate the time elapsed since a living organism died by measuring the radiation it gives off. All living things contain a small amount of carbon-14 (a **radioisotope**) which gradually decreases after death.

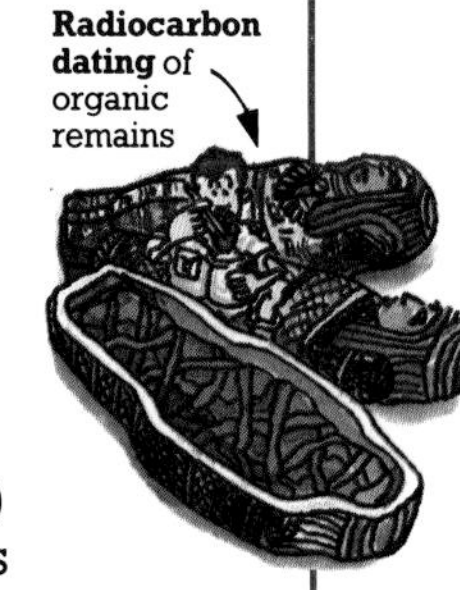

Radiocarbon dating of organic remains

- **Radiology**. The study of radioactivity, especially with regard to its use in medicine (**radiotherapy**). Cancer cells are susceptible to radiation, so cancer can be treated by small doses.

* **Atomic number**, 127; **Fission bomb, Fusion bomb**, 93; **Mass number**, 127; **Neutron**, 126; **Nuclear power station**, 94; **Nucleus**, 126.

Bonding

When substances react together, the tendency is always for their atoms to gain, lose or share electrons so that they each acquire a stable (full) **outer shell*** of electrons. In doing so, these atoms develop some kind of attraction, or **bonding**, between them (they are held together by **bonds**). The three main types of bonding are **ionic bonding**, **covalent bonding** (see pages 132-133) and **metallic bonding** (see page 134). See also **intermolecular forces**, page 134.

• **Valency electron**. An electron, always in the **outer shell*** of an atom, used in forming a bond. It is lost by atoms in **ionic bonding** and **metallic bonding***, but shared with other atoms in **covalent bonding***.

Ions

An **ion** is an electrically charged particle, formed when an atom loses or gains one or more electrons to form a stable **outer shell***. All ions are either **cations** or **anions**.

• **Cation**. An ion with a positive charge, formed when an atom loses electrons in a reaction (it now has more **protons*** than electrons). Hydrogen and metals tend to form cations. Their atoms have one, two or three electrons in their **outer shells***, and it is easier for them to lose electrons (leaving a stable shell underneath) than to gain at least five more.

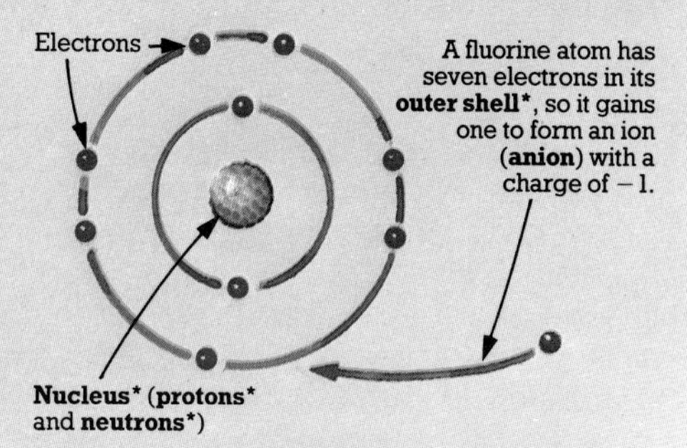

A fluoride ion (**anion**) is written F^-.

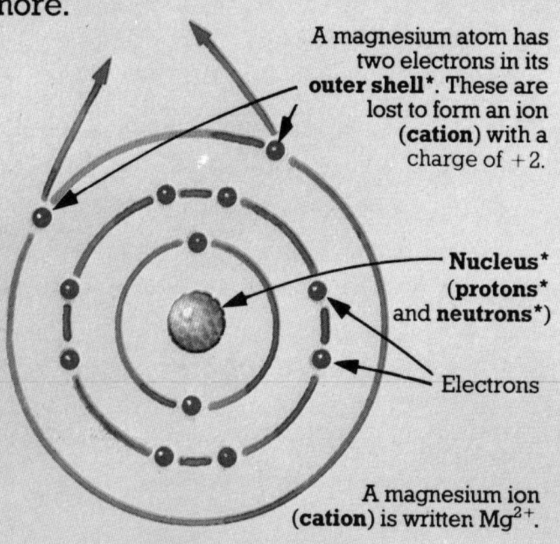

A magnesium ion (**cation**) is written Mg^{2+}.

• **Anion**. An ion with a negative charge, formed when an atom gains electrons in a reaction (it now has more electrons than **protons***). Non-metals tend to form anions. Their atoms have five, six or seven electrons in their **outer shells***, and it is easier for them to gain electrons (and acquire a stable shell) than to lose at least five. Some anions are formed by groups of atoms gaining electrons, e.g. **acid radicals***.

• **Ionization**. The process of forming ions. This either happens when atoms lose or gain electrons or when a compound splits up into ions, e.g. hydrogen chloride forming a solution.

Ionization of hydrogen chloride in water, forming hydrogen ions and chloride ions.

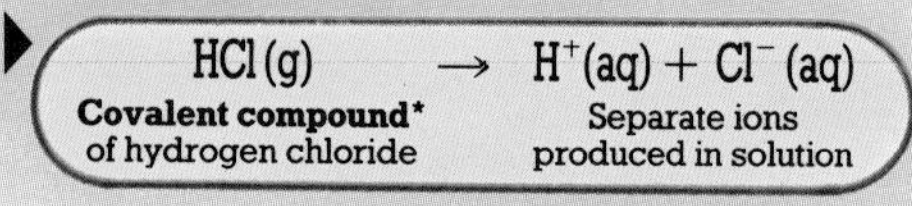

* **Acid radical**, 153; **Covalent bonding**, **Covalent compounds**, 132; **Metallic bonding**, 134; **Neutron, Nucleus**, 126; **Outer shell**, 127; **Proton**, 126.

Ionic bonding

When two elements react together to form ions, the resulting **cations** and **anions**, which have opposite electrical charges, attract each other. They stay together because of this attraction. This type of bonding is known as **ionic bonding** and the electrostatic bonds are called **ionic bonds**. Elements far apart in the periodic table tend to exhibit this kind of bonding, coming together to form **ionic compounds**, e.g. sodium and chlorine (sodium chloride) and magnesium and oxygen (magnesium oxide).

- **Ionic compound**. A compound whose components are held together by ionic bonding. It has no molecules, instead the **cations** and **anions** attract each other to form a **giant ionic lattice***. Ionic compounds have high melting and boiling points (the bonds are strong and hence large amounts of energy are needed to break them). They conduct electricity when **molten*** or in **aqueous solution*** because they contain charged particles (ions) which are free to move.

Sodium and chlorine react to form sodium chloride, an **ionic compound**.

Sodium atom

Chlorine atom

1 electron in **outer shell***

Electron transferred

7 electrons in **outer shell**

Full **outer shell** left

Electrostatic attraction is an **ionic bond**.

Na^+

Cl^-

Full **outer shell** formed

Chloride ion

Sodium ion

Note that there are no molecules – the formula gives the relative numbers of each ion in the **giant ionic lattice*** (in this case one to one).

Formula of sodium chloride is NaCl or Na^+Cl^-.

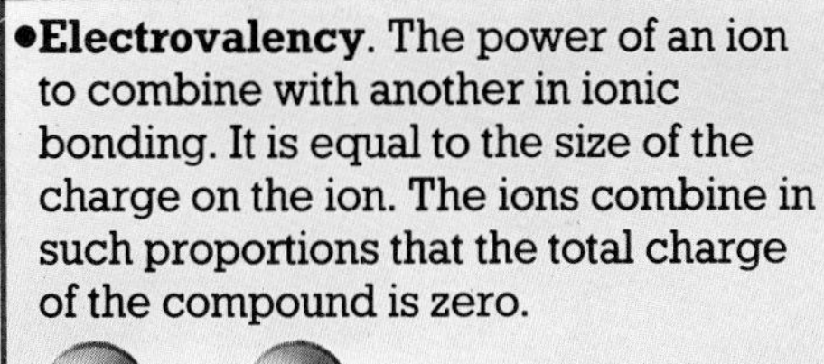

- **Electrovalency**. The power of an ion to combine with another in ionic bonding. It is equal to the size of the charge on the ion. The ions combine in such proportions that the total charge of the compound is zero.

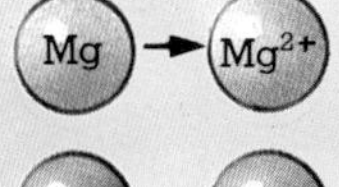

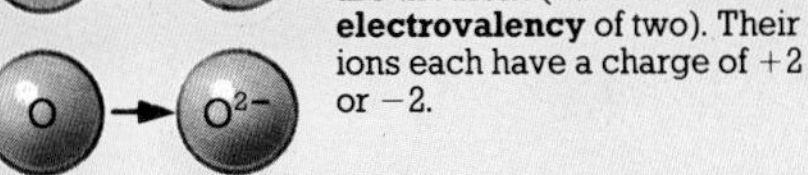

Group 2 and Group 6 elements are **divalent** (have an **electrovalency** of two). Their ions each have a charge of +2 or −2.

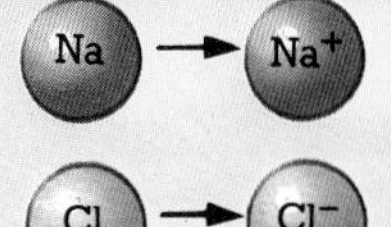

Group 1 and Group 7 elements are **monovalent** (have an **electrovalency** of one). Their ions each have a charge of +1 or −1

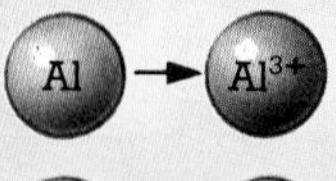

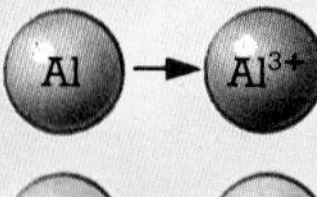

Some Group 3 and Group 5 elements are **trivalent** (have an **electrovalency** of three). Their ions each have a charge of +3 or −3

* **Aqueous solution**, 144; **Giant ionic lattice**, 137; **Molten**, 120; **Outer shell**, 127.

Covalent bonding

Covalent bonding is the sharing of electrons between atoms so that each atom acquires a stable **outer shell***. Electrons are shared in pairs called **electron pairs** (one pair being a **covalent bond**). Covalent bonds between atoms are strong. However, **covalent compounds** (compounds with covalent bonds) are usually liquids or gases at room temperature (see also **molecular lattice**, page 137). The melting and boiling points are low because the attraction between the molecules is small and hence little energy is needed to overcome it. They do not conduct electricity because there are no charged particles (**ions***) present.

• **Single bond**. A covalent bond formed when one pair of electrons is shared between two atoms.

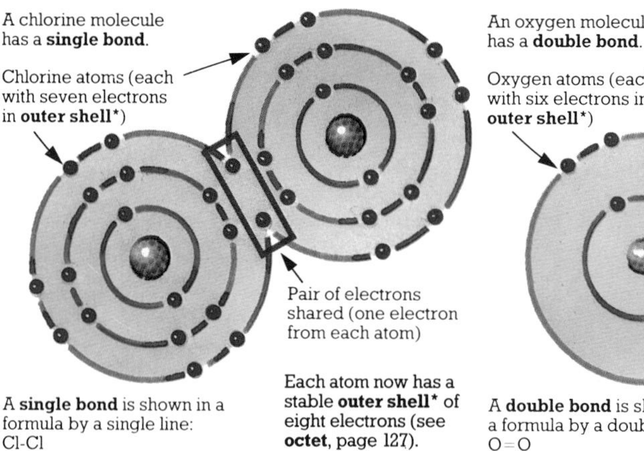

• **Double bond**. A covalent bond formed when two pairs of electrons are shared between two atoms.

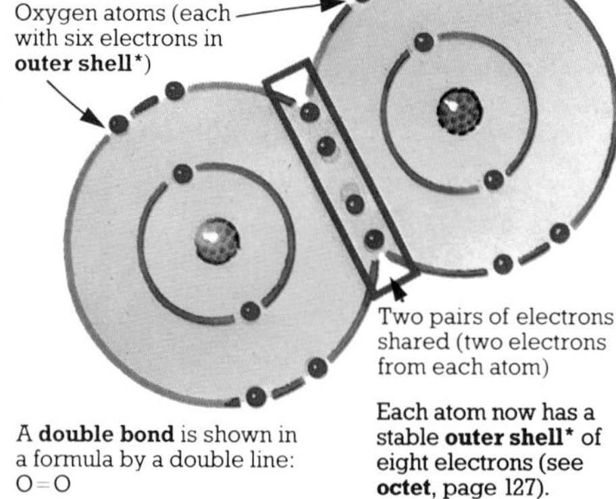

• **Triple bond**. A covalent bond formed when three pairs of electrons are shared between two atoms.

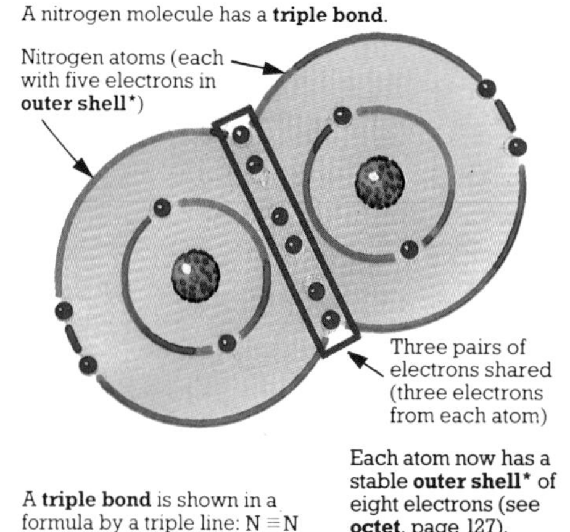

• **Dative covalent bond** or **coordinate bond**. A covalent bond in which both electrons in the bond are provided by the same atom. It donates a **lone pair**.

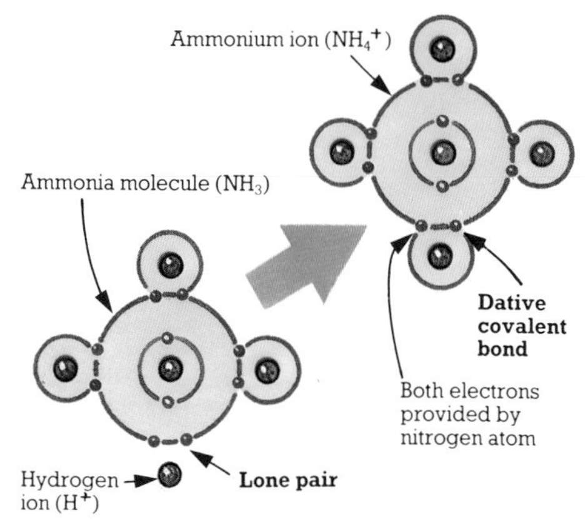

* **Ion**, 130; **Outer shell**, 127.

●**Covalency**. The maximum number of covalent bonds an atom can form. It is equal to the number of hydrogen atoms which will combine with the atom. The covalency of most elements is constant, but that of **transition metals*** varies.

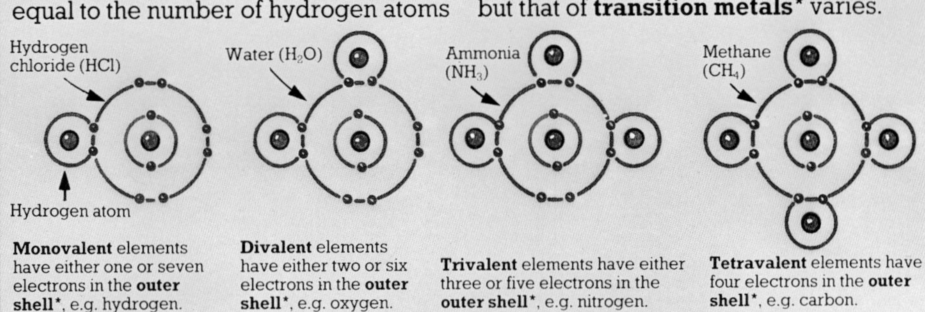

Monovalent elements have either one or seven electrons in the **outer shell***, e.g. hydrogen.

Divalent elements have either two or six electrons in the **outer shell***, e.g. oxygen.

Trivalent elements have either three or five electrons in the **outer shell***, e.g. nitrogen.

Tetravalent elements have four electrons in the **outer shell***, e.g. carbon.

●**Lone pair**. A pair of electrons in the **outer shell*** of an atom which is not part of a covalent bond (see ammonia picture on previous page).

●**Electronegativity**. The power of an atom to attract electrons to itself in a molecule. If two atoms with different electronegativities are joined, a **polar bond** is formed. Weakly electronegative atoms are sometimes called **electropositive** (e.g. sodium) as they form positive ions fairly easily.

●**Polar bond**. A covalent bond in which the electrons are nearer to one atom's **nucleus*** than the other. This effect is called **polarization**. It is caused by a difference in **electronegativity** between the atoms, the electrons being more attracted to one than the other.

●**Polar molecule**. A molecule with a difference in electric charge between its ends, caused by an uneven distribution of **polar bonds**, and sometimes by **lone pairs**. Liquids with polar molecules may be **polar solvents*** and may dissolve **ionic compounds***. A **non-polar molecule** has no difference in charge at its ends.

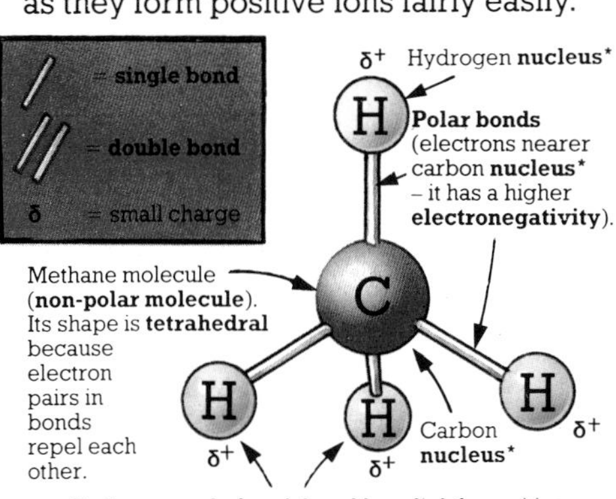

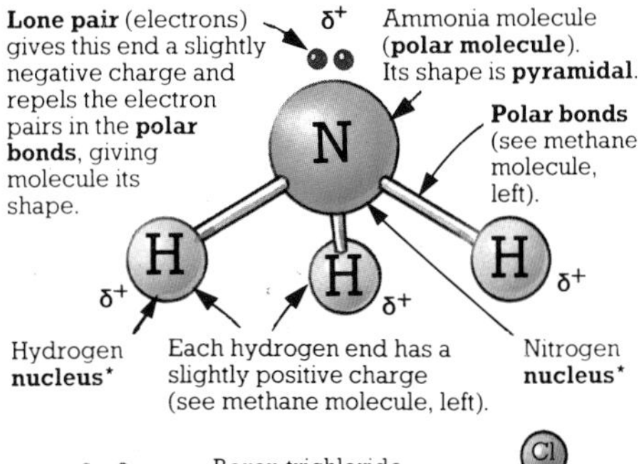

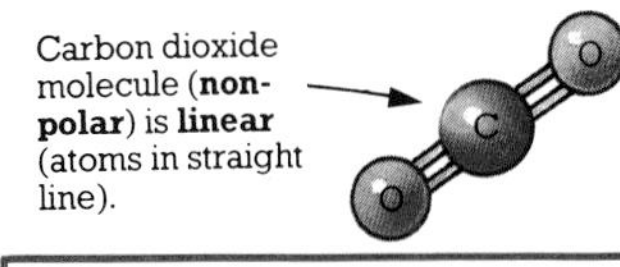

Water molecule (**polar**) is **V-shaped** or **non-linear** – **lone pairs** repel electron pairs in bonds.

Boron trichloride molecule (**non-polar**) is **triagonal** or **trigonal planar** – electrons in bonds repel each other.

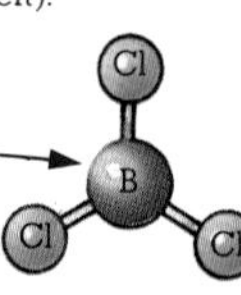

●**Isomerism**. The occurrence of the same atoms forming different arrangements in different molecules. The arrangements are **isomers***. They have the same **molecular formula*** but different **graphic formulae***.

* **Graphic formula**, 140; **Ionic compound**, 131; **Isomers**, 190; **Molecular formula**, 140; **Nucleus**, 126; **Outer shell**, 127; **Polar solvent**, 144; **Transition metals**, 172.

Metallic bonding

Metallic bonding is the attraction between particles in a **giant metallic lattice*** (i.e. in metals). The lattice consists of positive **ions*** of the metal with **valency electrons*** free to move between them. The free or **delocalized** electrons form the bonds between the metal and, because they can move, heat and electricity can be conducted through the metal. The forces between the electrons and ions are strong. This gives metals high melting and boiling points, since relatively large amounts of energy are needed to overcome them. For more about other types of bonding, see pages 130-133.

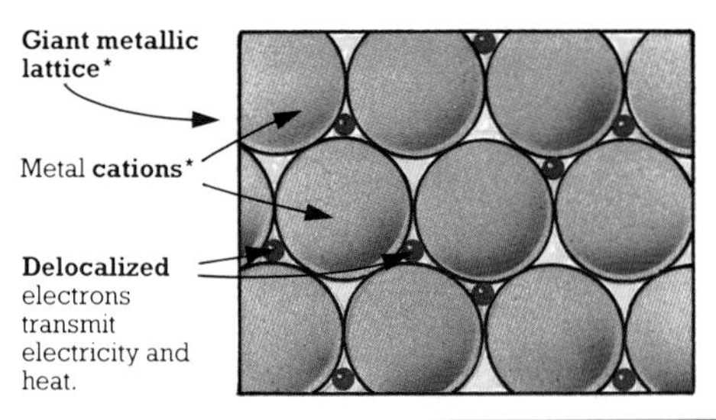

- **Delocalization**. The sharing of **valency electrons*** by all the atoms in a molecule or **giant metallic lattice***. Delocalized electrons can belong to any of the atoms in the lattice and are able to move through the lattice, so the metal can conduct electricity and heat.

Intermolecular forces

- **van der Waals' forces.** Weak attractive forces between molecules (**intermolecular forces***) caused by the uneven distribution and movement of electrons in the atoms of the molecules. The attractive force is approximately twenty times less than in **ionic bonding***. It is the force which holds **molecular lattices*** together, e.g. iodine and solid carbon dioxide. ▶

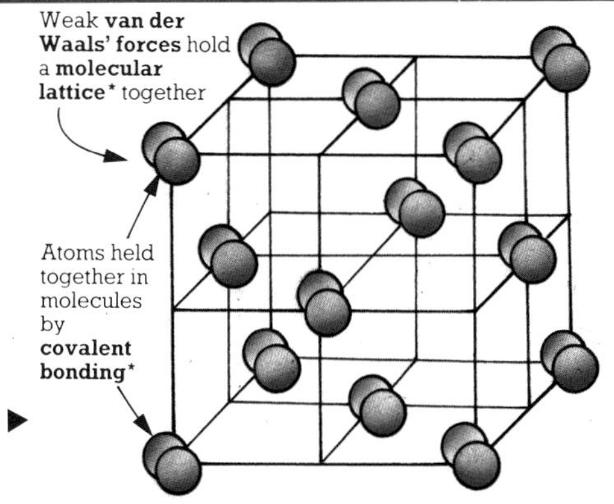

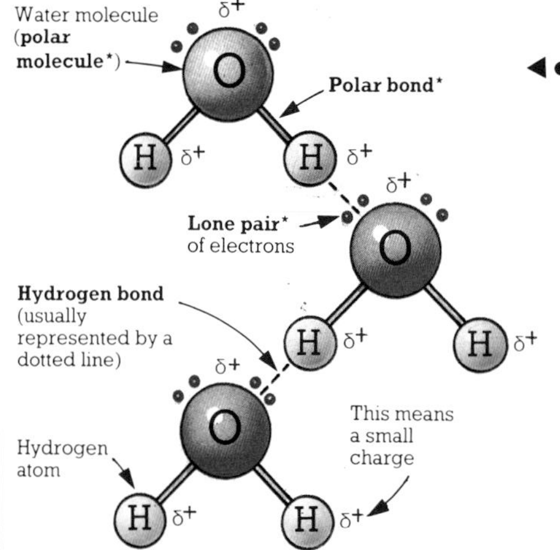

◀ **Hydrogen bond**. An attraction between a **polar molecule*** containing hydrogen and a **lone pair*** of electrons in another molecule. The **polar bonds*** mean that each hydrogen atom has a slightly positive charge and is therefore attracted to the electrons. Hydrogen bonding accounts for high melting and boiling points in water in relation to other substances with small, but **non-polar molecules***. Both the hydrogen bonds and the **van der Waals' forces** must be overcome to separate the molecules.

* **Cation**, 130; **Covalent bonding**, 132; **Giant metallic lattice**, 137; **Intermolecular force**, 7; **Ion**, 130; **Ionic bonding**, 131; **Lone pair**, 133; **Molecular lattice**, 137; **Non-polar molecule**, 133 (**Polar molecule**); **Polar bond**, 133; **Valency electron**, 130.

Crystals

Crystals are solids with regular geometric shapes, formed from regular arrangements of particles. The particles can be atoms, ions or molecules and the bonding of any type or mixture of types. The edges of crystals are straight and the surfaces flat. Substances that form crystals are described as **crystalline**. Solids without a regular shape (i.e. those which do not form crystals) are described as **amorphous**.

• **Crystallization**. The process of forming crystals. It can happen in a number of ways, e.g. cooling **molten*** solids, **subliming*** solids (solid to gas and back), placing a **seed crystal** in a **supersaturated*** solution or placing a seed crystal in a **saturated*** solution and cooling or evaporating the solution. The last method is the most common. Either cooling or evaporating means that the amount of soluble **solute*** decreases, so particles come out of solution and bond to the seed crystal, which is suspended in the solution. Crystallization can be used to purify substances – see page 221.

Methods of **crystallization**

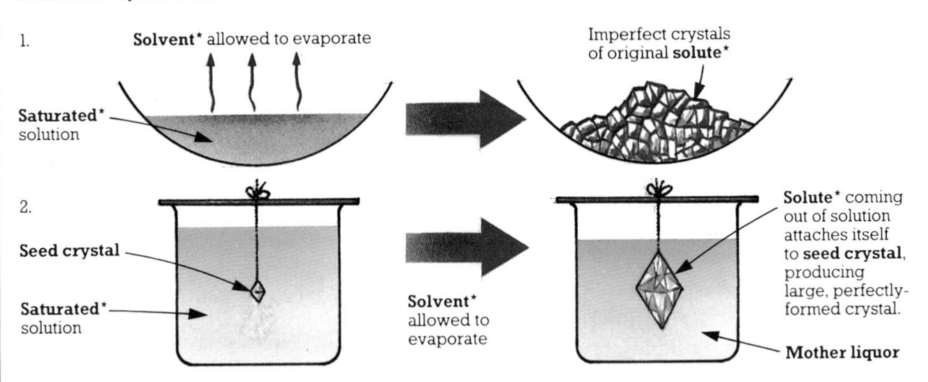

• **Seed crystal**. A small crystal of a substance placed in a solution of the same substance. It acts as a base on which crystals form during **crystallization**. The crystal which grows will take on the same shape as the seed crystal.

• **Water of crystallization**. Water contained in crystals of certain **salts***. The number of molecules of water combined with each pair of ions is usually constant and is often written in the chemical **formula*** for the salt. The water can be driven off by heating. Crystals which contain water of crystallization are **hydrated***.

• **Mother liquor**. The solution left after **crystallization** has taken place in a solution.

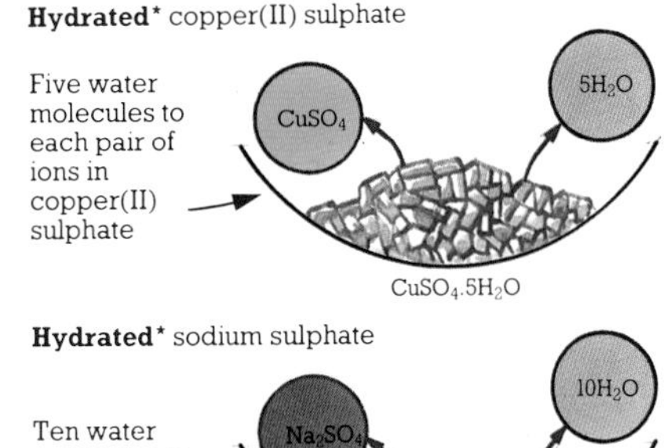

* **Formulae**, 140; **Hydrated**, 155 (**Hydrate**); **Molten**, 120; **Salts**, 153; **Saturated**, **Solute**, **Solvent**, 144; **Sublimation**, 121; **Supersaturated**, 145.

Crystals continued - shapes and structures

Crystals (see page 135) exist in many different shapes and sizes. This is due to the arrangement and bonding of the particles (atoms, molecules or **ions***). The arrangement in space of the particles and the way in which they are joined is called a **crystal lattice**. The shape of a particular crystal depends on its crystal lattice and how this lattice can be split along **cleavage planes**. The main crystal shapes are shown below.

Basic shapes from which large crystals are built.

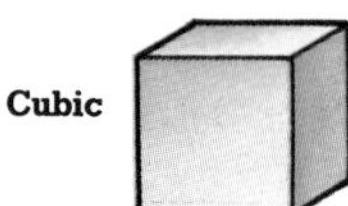

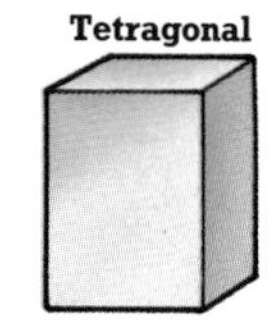

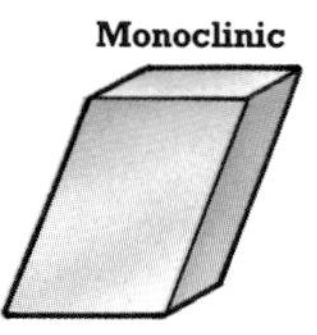

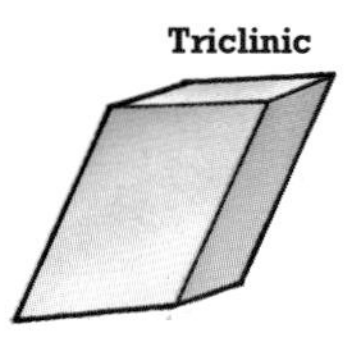

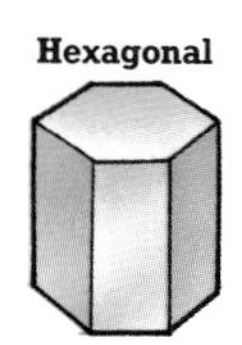

•**Polymorphism**. The occurrence of two or more different crystals of the same substance, differing in shape and appearance. It is caused by different arrangements in the separate types. Changes between types often take place at a certain temperature called the **transition temperature**. Polymorphism in elements is called **allotropy**.

•**Allotropy**. The occurrence of certain elements in more than one crystalline form. It is a specific type of **polymorphism**. The different forms are called **allotropes** and are caused by a change in arrangement of atoms in the crystal.

•**Monotropy**. **Polymorphism** in which there is only one stable form. The other forms are unstable and there is no **transition temperature**.

•**Enantiotropy**. **Polymorphism** in which there are two stable forms of a substance, one above its **transition temperature**, and one below.

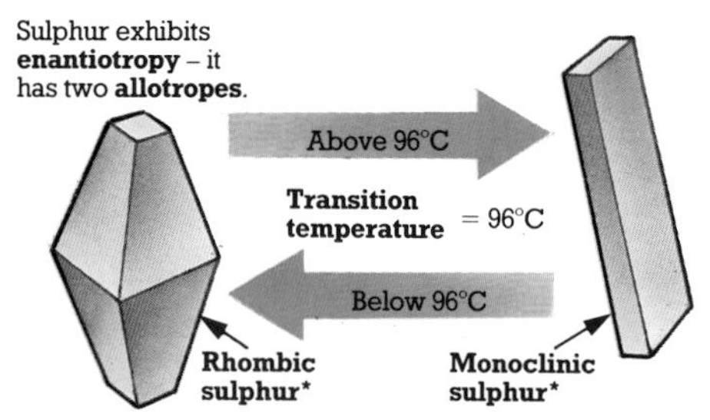

•**Transition temperature**. The temperature at which a substance exhibiting **enantiotropy** changes from one form to another.

•**Isomorphism**. The existence of two or more different substances with the same crystal structure and shape. They are described as **isomorphic**.

•**Cleavage plane**. A plane of particles along which a crystal can be split, leaving a flat surface. If a crystal is not split along the cleavage plane, it shatters.

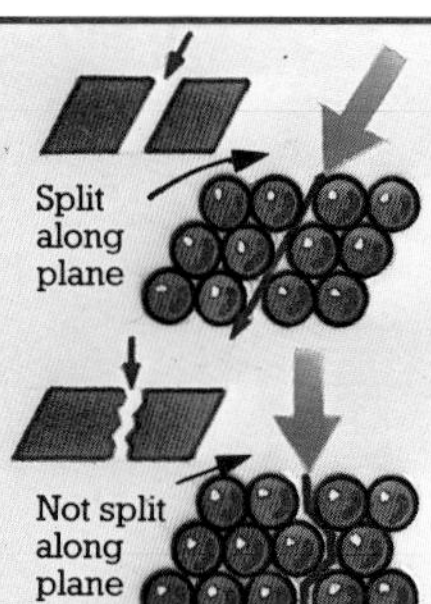

•**X-ray crystallography**. The use of X-rays to work out crystal structure. Deflected X-rays produce a **diffraction pattern** from which the structure is worked out (see below).

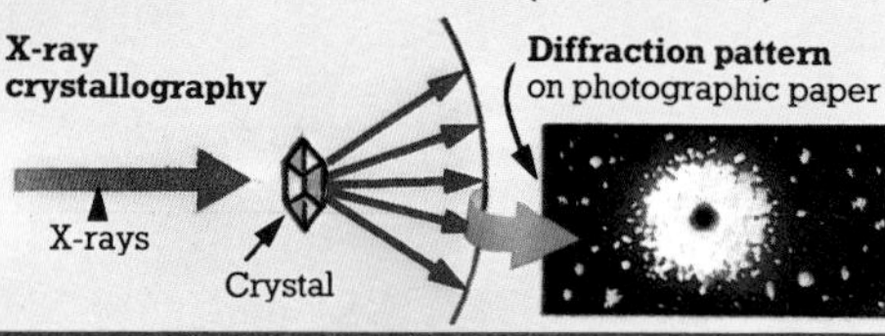

* **Ion**, 130; **Monoclinic sulphur**, **Rhombic sulphur**, 184.

Crystal lattices

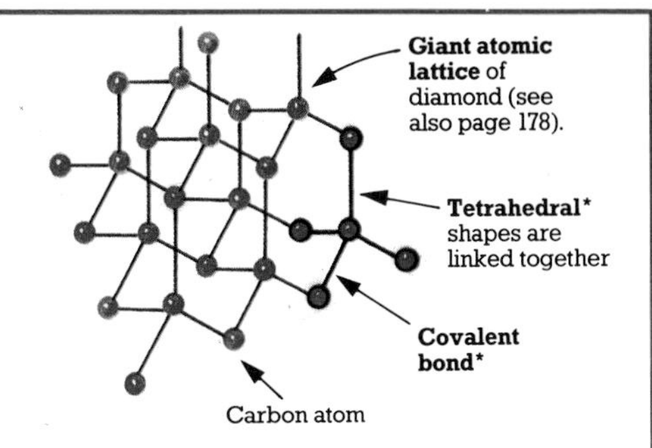

●**Giant atomic lattice**. A **crystal lattice** consisting of atoms held together by **covalent bonding***, e.g. diamond. Substances with giant atomic lattices are extremely strong and have very high melting and boiling points.

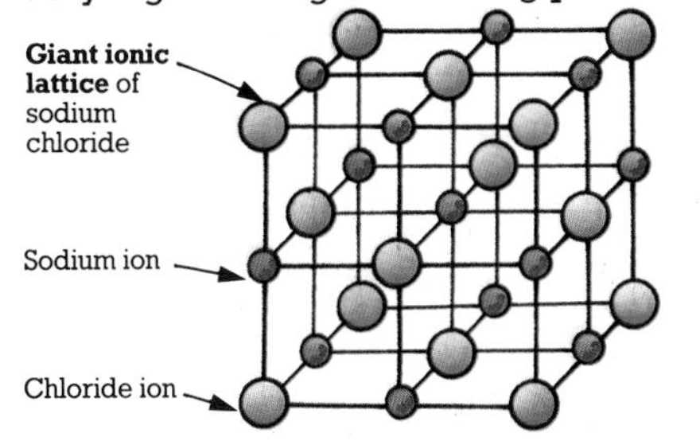

●**Giant ionic lattice**. A **crystal lattice** consisting of **ions*** held together by **ionic bonding***, e.g. sodium chloride. The ionic bonds are strong, which means the substance has high melting and boiling points.

●**Giant metallic lattice**. A **crystal lattice** consisting of metal atoms held together by **metallic bonding***, e.g. zinc. The **delocalized*** electrons are free to move about, making a metal a good conductor of heat and electricity. The layers of atoms can slide over one another, making metals **malleable*** and **ductile***.

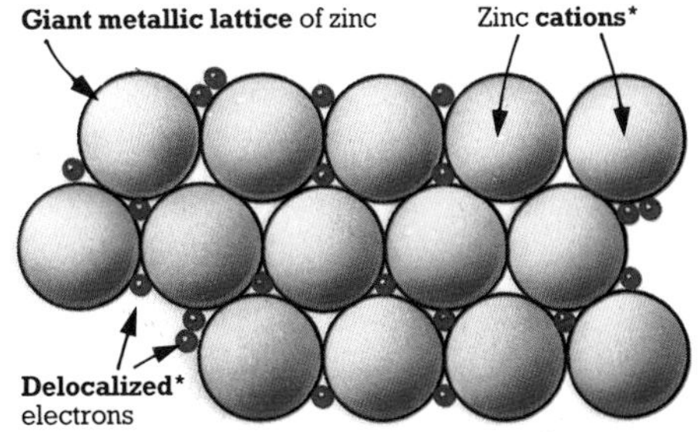

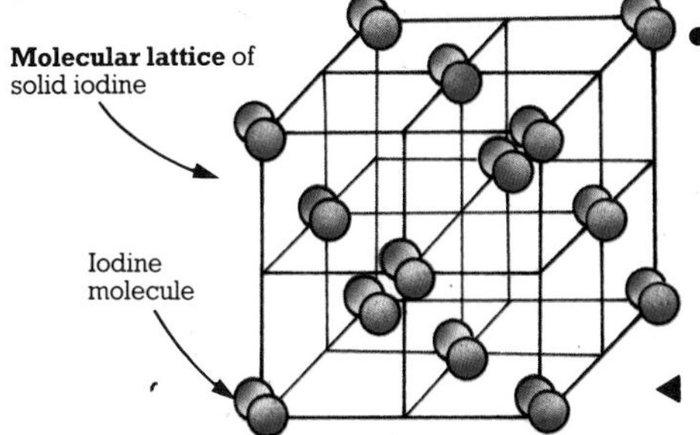

●**Molecular lattice**. A **crystal lattice** consisting of molecules bonded together by weak **intermolecular forces** (see page 134), e.g. iodine. The intermolecular forces are overcome when the crystal is broken, not the **covalent bonds*** in the molecules, so the crystal has low melting and boiling points compared with **ionic compounds***.

In crystals where the particles are all the same size, e.g. in a **giant metallic lattice**, various arrangements of the particles are possible. The most common are shown here.

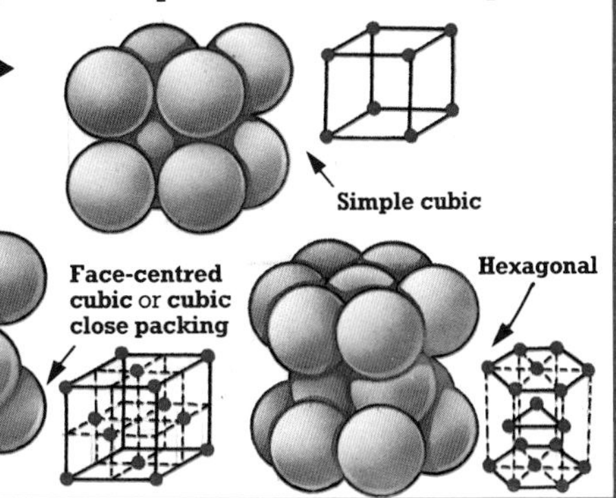

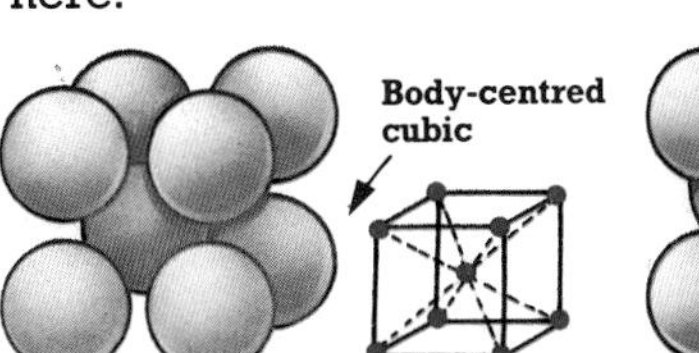

* **Cation**, 130; **Covalent bond, Covalent bonding**, 132; **Delocalization**, 134; **Ductile**, 344; **Ion**, 130; **Ionic bonding, Ionic compound**, 131; **Malleable**, 345; **Metallic bonding**, 134; **Tetrahedral**, 133.

Measuring atoms

With a diameter of about 10^{-7} millimetres and a mass of about 10^{-22} grams, atoms are so small that they are extremely difficult to measure. Their masses are therefore measured in relation to an agreed mass to give them a manageable value. Because there are many millions of atoms in a very small sample of a substance, the **mole** is used for measuring quantities of particles. The masses of atoms and molecules are measured using a machine called a **mass spectrometer**.

• **Relative atomic mass** or **atomic weight**. The average mass (i.e. taking into account **relative isotopic mass** and **isotopic ratio**) of one atom of a substance divided by one twelfth the mass of a carbon-12 atom (see **isotope**, page 127). See page 83 for more about its units, and pages 212-213 for a table of relative atomic masses.

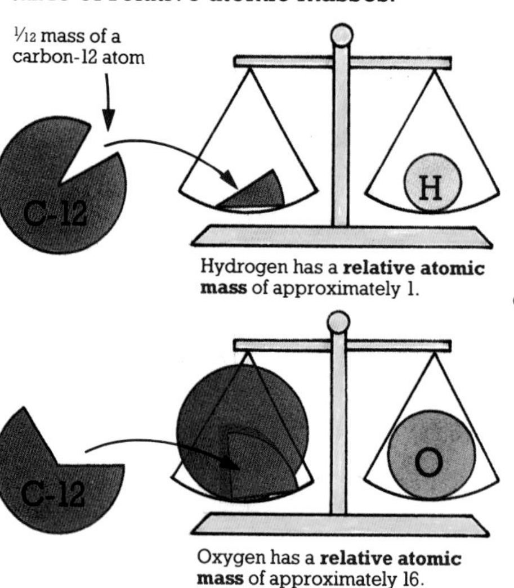

Hydrogen has a **relative atomic mass** of approximately 1.

Oxygen has a **relative atomic mass** of approximately 16.

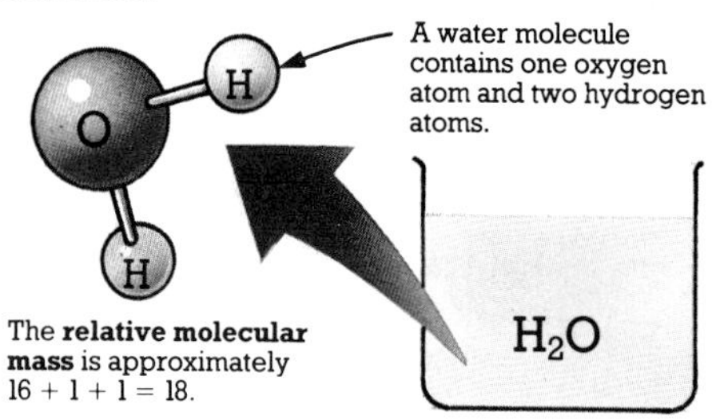

The **relative molecular mass** is approximately 16 + 1 + 1 = 18.

Relative molecular mass also applies to **ionic compounds***, even though they do not have molecules.

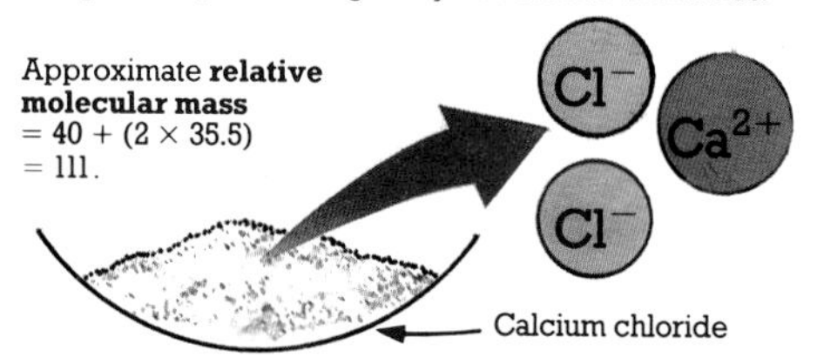

• **Relative molecular mass**. Also called **molecular weight**, **relative formula mass** or **formula weight**. The mass of a molecule of an element or compound divided by one twelfth the mass of a carbon-12 atom (see **isotope**, page 127). It is the sum of the **relative atomic masses** of the atoms in the molecule.

• **Relative isotopic mass**. The mass of an atom of a specific **isotope*** divided by one twelfth the mass of a carbon-12 atom. It is nearly exactly the same as the **mass number*** of the isotope.

• **Isotopic ratio**. The ratio of the number of atoms of each **isotope*** in a sample of an element. It is used with **relative isotopic masses** to calculate the **relative atomic mass** of an element.

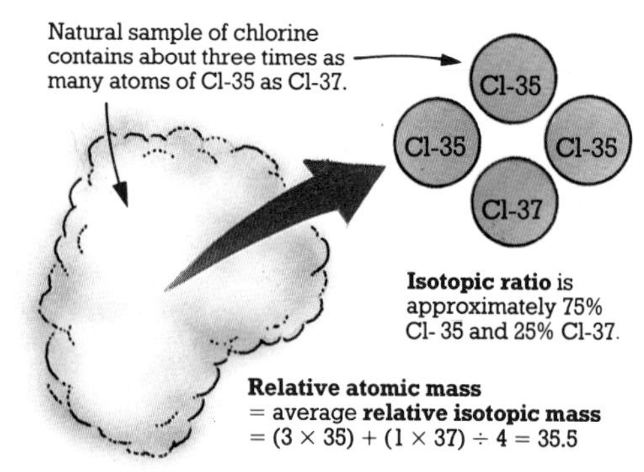

Isotopic ratio is approximately 75% Cl-35 and 25% Cl-37.

Relative atomic mass
= average **relative isotopic mass**
= (3 × 35) + (1 × 37) ÷ 4 = 35.5

* **Ionic compound**, 131; **Isotope**, **Mass number**, 127.

•Mole (mol). The **SI unit*** of the amount of a substance – see also page 96. One mole contains the same number of particles as there are atoms in 12 grams of the carbon-12 **isotope***.

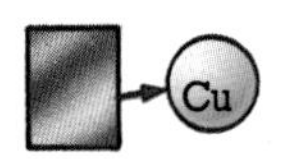

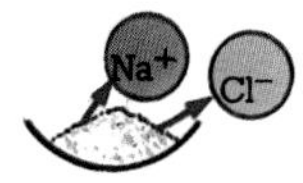

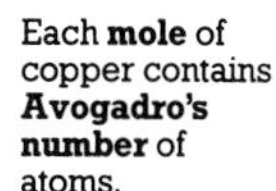

Each **mole** of copper contains **Avogadro's number** of atoms.

Each **mole** of oxygen contains **Avogadro's number** of molecules.

A **mole** of sodium chloride contains 1 mol Na^+ ions and 1 mol Cl^- ions.

•Avogadro's number. The number of particles per **mole**, equal to 6.02 x 10^{23} mol^{-1}.

•Molar mass. The mass of one **mole** of a given substance. It is the **relative atomic** or **molecular mass** of a substance expressed in grams.

Relative atomic mass = 23

Relative molecular mass = 23 + 35.5

Na

NaCl

1 mol of sodium

1 mol of sodium chloride

23 g

58.5 g

Molar mass 23 g **Molar mass** 58.5 g

•Molar volume. The volume of one **mole** of any substance, measured in cubic decimetres (dm^3). Molar volumes of solids and liquids vary, but all gases under the same conditions have the same molar volume. The molar volume of any gas at **s.t.p.*** is 22.4 dm^3 and at **r.t.p.** (**room temperature and pressure**, i.e. 20°C and 101325 **pascals***) it is 24 dm^3.

In solids and liquids, **molar volume** depends on size and arrangement of particles. ▼

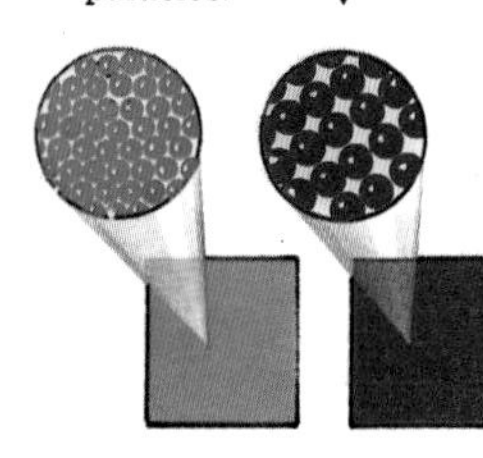

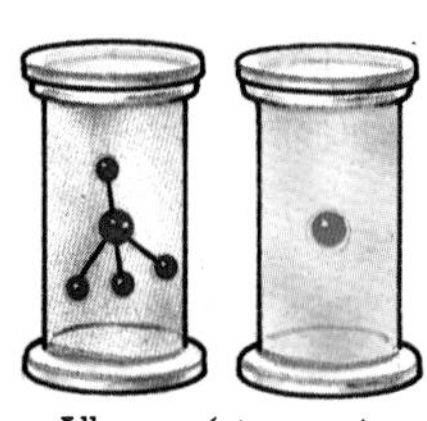

All gases (at same ▲ temperature and pressure) have same **molar volume**. Their particles are not bonded together.

•Concentration. A measurement of the amount of a **solute*** dissolved in a **solvent***, expressed in **moles** per dm^3 ($mol\ dm^{-3}$). **Mass concentration** is the mass of solute per unit volume, e.g. grams per dm^3 ($g\ dm^{-3}$).

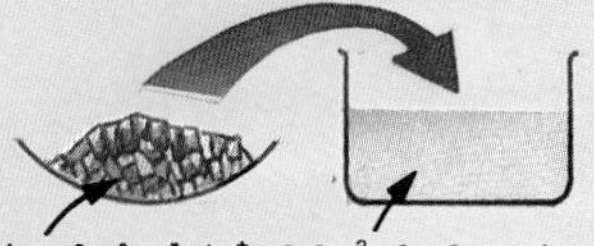
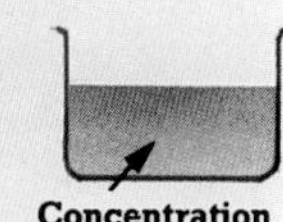

4 **mol** of **solute*** 2 dm^3 of **solvent*** **Concentration** of 2 $mol\ dm^{-3}$

Concentration is the number of **moles** of **solute*** dissolved in each dm^3 of **solvent***.

•Molarity. A term sometimes used to describe the **concentration** when expressed in **moles** of **solute*** per dm^3 of **solvent***. The molarity is also expressed as the **M-value**, e.g. a solution with a concentration of 3 mol dm^{-3} has a molarity of 3 and is described as a 3**M** solution.

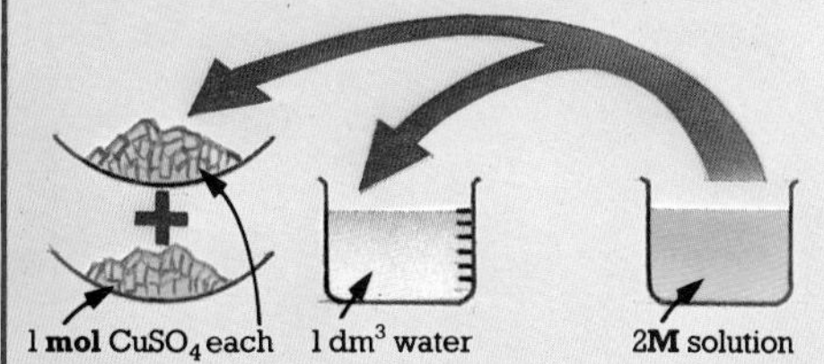

1 **mol** $CuSO_4$ each 1 dm^3 water 2**M** solution

A 2**M** copper(II) sulphate solution contains 2 **mol** of copper(II) sulphate in each dm^3.

•Molar solution. A solution that contains one **mole** of a substance dissolved in every cubic decimetre (dm^3) of solution. It is therefore a **1M** solution (see **molarity**).

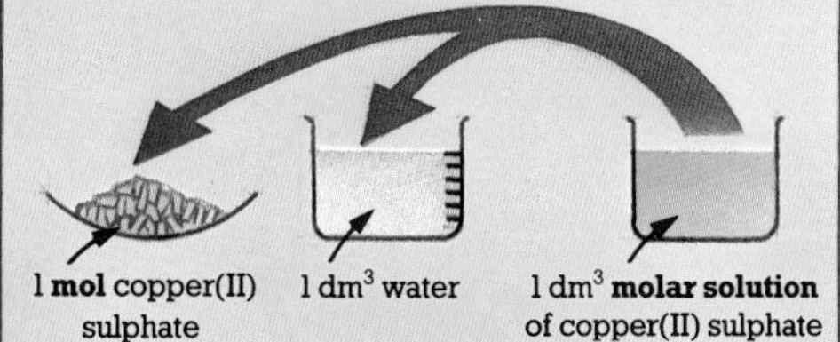

1 **mol** copper(II) sulphate 1 dm^3 water 1 dm^3 **molar solution** of copper(II) sulphate

A **1M** or **molar solution** of copper(II) sulphate contains 1 **mol** of copper(II) sulphate in each dm^3.

•Standard solution. A solution of which the **concentration** is known. It is used for **volumetric analysis***.

* **Isotope**, 127; **Pascal**, 97; **SI units**, 96; **Solute, Solvent**, 144; **s.t.p.**, 143; **Volumetric analysis**, 222.

Representing chemicals

Most chemicals are named according to the predominant elements they contain. Information about the chemical composition and structure of a compound is given by a **formula** (pl. **formulae**), in which the **chemical symbols*** for the elements are used. A chemical **equation** shows the reactants and products of a chemical reaction and gives information about how the reaction happens.

Formulae

- **Empirical formula**. A formula showing the simplest ratio of the atoms of each element in a compound. It does not show the total number of atoms of each element in a **covalent compound***, or the **bonding** in the compound (see pages 130-134).

- **Molecular formula**. A formula representing one molecule of an element or compound. It shows which elements the molecule contains and the number of atoms of each in the molecule, but not the **bonding** of the molecule (see pages 130-134).

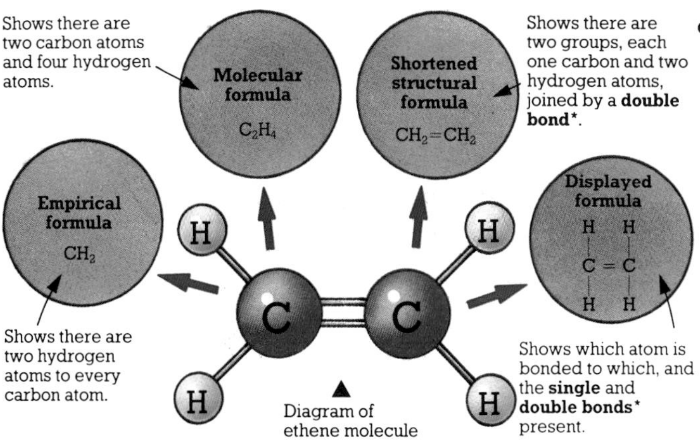

- **Stereochemical formula** or **3-dimensional structural formula**. A formula which shows the 3-dimensional arrangement of the atoms and **bonds*** in a molecule. See page 190 for the stereochemical formula of methane.

- **Shortened structural formula**. A formula which shows the sequence of groups of atoms (e.g. a **carboxyl group***) in a molecule and the **bonding** (see pages 130-134) between the groups of atoms (shown as lines).

- **Displayed formula** or **full structural formula**. A formula which shows the arrangement of the atoms in relation to each other in a molecule. All the bonds in the molecule are shown.

- **Percentage composition**. The composition of a compound expressed in terms of the percentage of its mass taken up by each element.

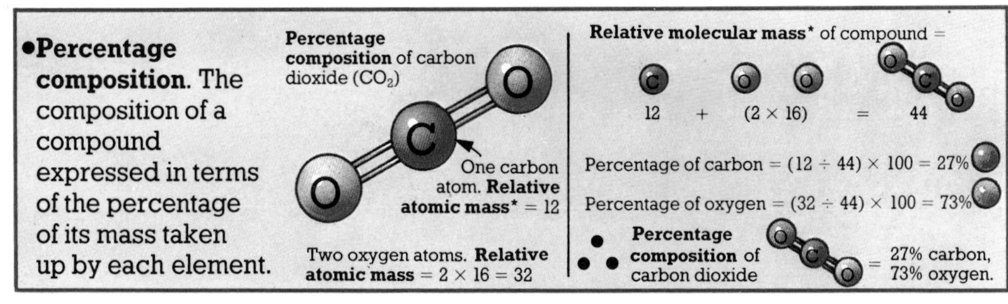

* **Bonds**, 130; **Carboxyl group**, 195 (**Carboxylic acids**); **Chemical symbol**, 122; **Covalent compounds**, **Double bond**, 132; **Relative atomic mass**, **Relative molecular mass**, 138; **Single bond**, 132.

Names

- **Trivial name**. An everyday name given to a compound. It does not usually give any information about the composition or structure of the compound. e.g. salt (sodium chloride), chalk (calcium carbonate).

- **Traditional name**. A name which gives the predominant elements of a substance, without necessarily giving their quantities or showing the structure of the substance. Some traditional names are **systematic names**.

- **Systematic name**. A name which shows the elements a compound contains, the ratio of numbers of atoms of each element and the **oxidation number*** of elements with variable **oxidation states***. The **bonding** (see pages 130-134) can also be worked out from the name. In some cases the systematic name is simplified. Some systematic names are the same as **traditional names**. See also **naming simple organic compounds**, page 214.

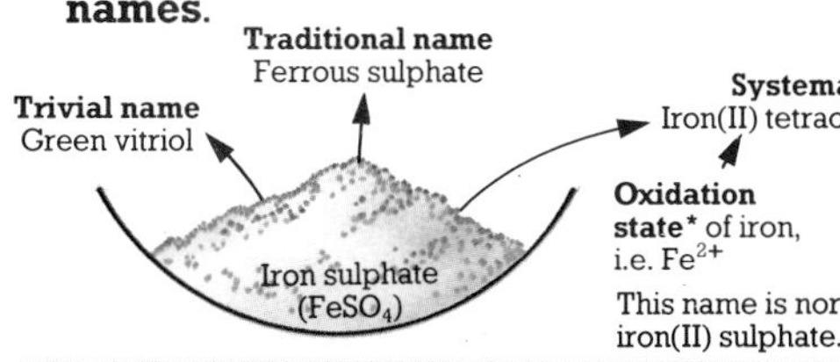

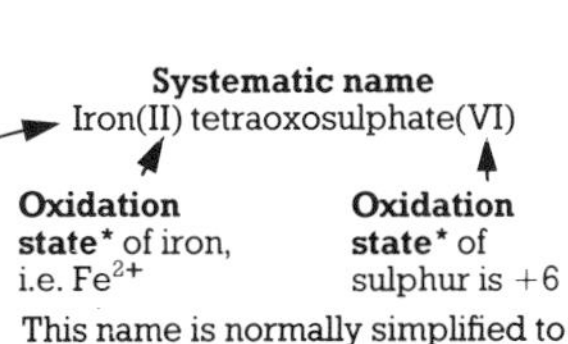

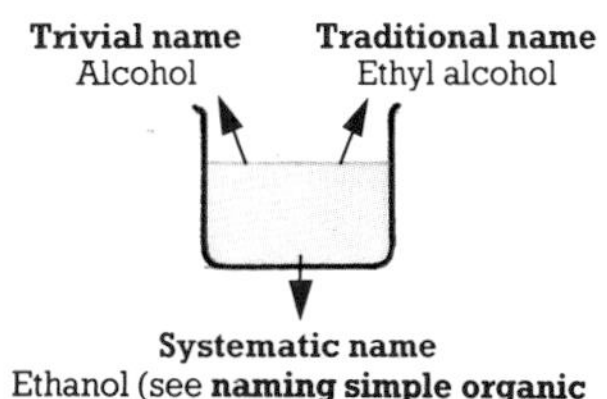

Equations

- **Word equation**. An equation in which the substances involved in a reaction are indicated by their names, e.g:

Sodium + Water → Sodium hydroxide + Hydrogen

However, the names may be replaced by the **formulae** of the substances (see opposite page).

$$Na + H_2O \rightarrow NaOH + H_2$$

- **Balanced equation**. An equation in which the number of atoms of each element involved in the reaction is the same on each side of the equation (i.e. it obeys the **law of conservation of mass***). The numbers of molecules of each substance are shown by the number in front of their **formula**, e.g:

$$2Na + 2H_2O \rightarrow 2NaOH + H_2$$

- **Ionic equation**. An equation which only shows what happens to the ions in a reaction. ▶

$$2Na(s) + 2H_2O(l) \rightarrow 2NaOH(aq) + H_2(g)$$

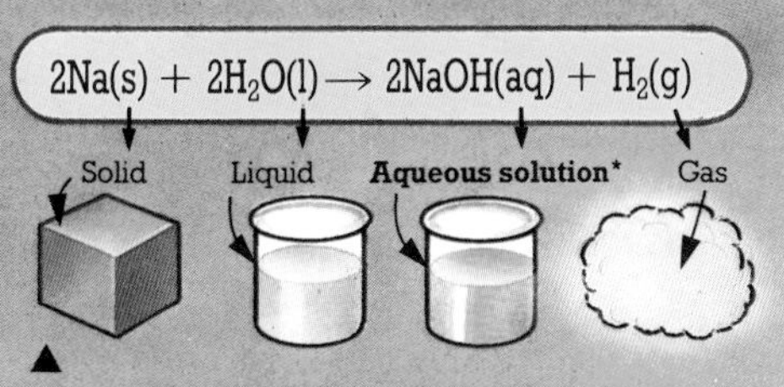

▲

- **State symbols**. Letters written after the **formula** of a substance which show its **physical state*** in a reaction.

- **Spectator ion**. An ion which remains the same after a chemical reaction. It is omitted from **ionic equations**.

In the reaction:

$$NaOH(aq) + HCl(aq) \rightarrow NaCl(aq) + H_2O(l)$$

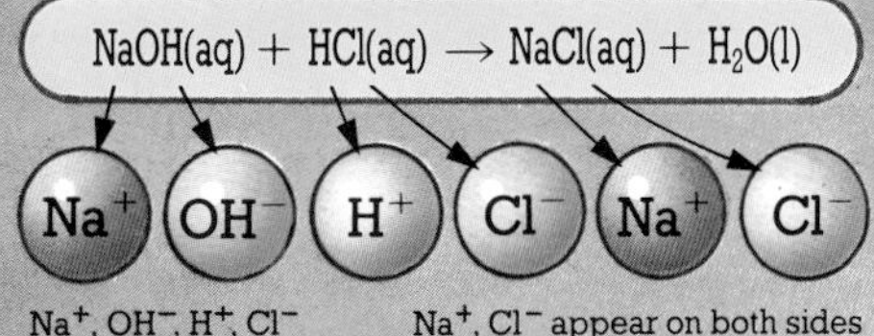

Na^+, OH^-, H^+, Cl^- are all ions

Na^+, Cl^- appear on both sides of the equation – they are **spectator ions**.

Ionic equation is:

$$OH^-(aq) + H^+(aq) \rightarrow H_2O(l)$$

* **Aqueous solution**, 144; **Law of conservation of mass**, 125; **Oxidation number**, **Oxidation state**, 149; **Physical states**, 120.

Gas laws

The molecules in a gas are widely spaced and move about in a rapid, chaotic manner (see **kinetic theory**, page 123). The combined volume of the gas molecules is very much smaller than the volume the gas occupies and the forces of attraction between the molecules are very weak. This is true for all gases, so they all behave in a similar way. There are several **gas laws** that describe this common behaviour (see below).

Symbols used in **gas laws**

P = pressure
T = temperature in **kelvins**
V = volume
k = a **constant***

● **Boyle's law**. At constant temperature, the volume of a gas is inversely proportional to the pressure (the volume decreases as the pressure increases).

● **Charles' law**. At constant pressure, the volume is directly proportional to the temperature on the **absolute temperature scale** (the gas expands as the temperature increases).

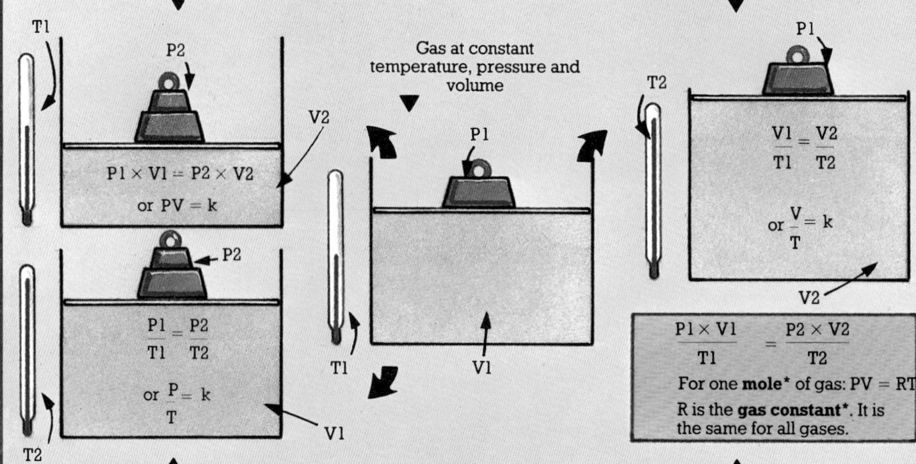

● **Pressure law** or **Third gas law**. At constant volume, the pressure is directly proportional to the temperature on the **absolute temperature scale** (the pressure increases with the temperature).

● **Ideal gas equation** or **General gas equation**. An equation that shows the relationship between the pressure, volume and temperature of a fixed mass of gas.

● **Ideal gas**. A theoretical gas that behaves in an "ideal" way. Its molecules have no volume, do not attract each other, move rapidly in straight lines and lose no energy when they collide. Many real gases behave in approximately the same way as ideal gases when the molecules are small and widely spaced.

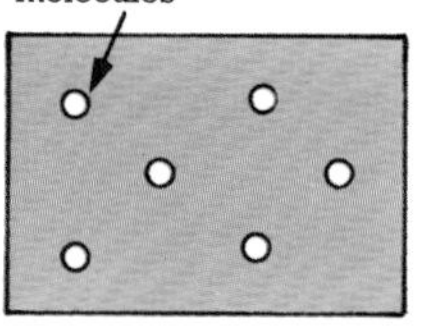

Behaves like an **ideal gas**.

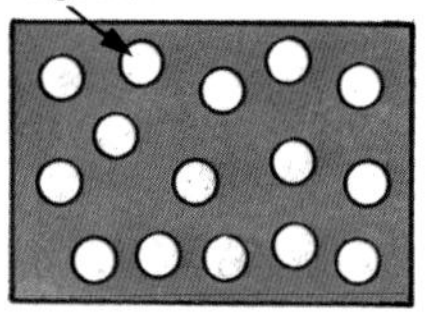

Does not behave like an **ideal gas**.

* **Gas constant**, 113; **Mole**, 139.

- **Partial pressure**. The pressure that each gas in a **mixture*** of gases would exert if it alone filled the volume occupied by the mixture.

- **Dalton's law of partial pressures**. The total pressure exerted by a **mixture*** of gases (which do not react together) is the sum of the **partial pressure** of each gas in the mixture.

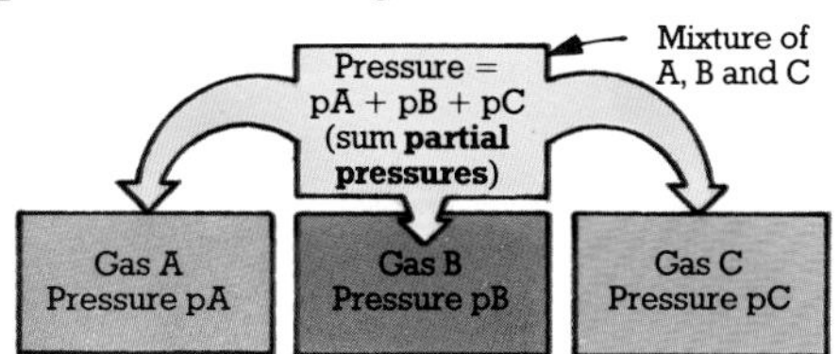

- **Graham's law of diffusion**. If the temperature and pressure are constant, the rate of **diffusion*** of a gas is inversely proportional to the square root of its density. The density of a gas is high if its molecules are heavy and low if its molecules are light. Light molecules move faster than heavy molecules, so a gas with a high density diffuses more slowly than a gas with a low density.

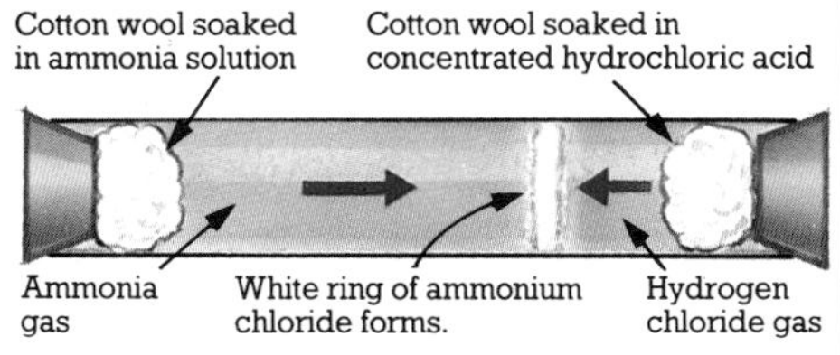

Light ammonia molecules **diffuse*** faster than hydrogen chloride molecules. The two gases meet further towards the right-hand end of the tube.

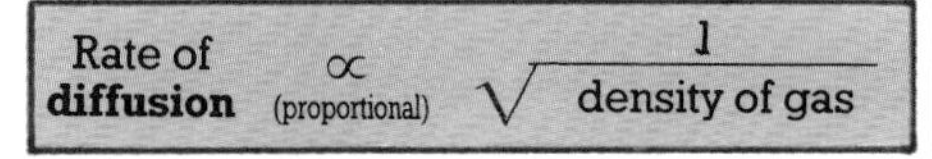

- **Relative vapour density**. The density of a gas relative to the density of hydrogen. It is calculated by dividing the density of a gas by the density of hydrogen. Relative vapour density is a ratio and has no units.

$$\textbf{Relative vapour density} = \frac{\text{Density of the gas}}{\text{Density of hydrogen}}$$

- **Gay-Lussac's law**. When gases react together to produce other gases and all the volumes are measured at the same temperature and pressure, the volumes of the reactants and products are in a ratio of simple whole numbers.

$$2CO(g) + O_2(g) \rightarrow 2CO_2(g)$$

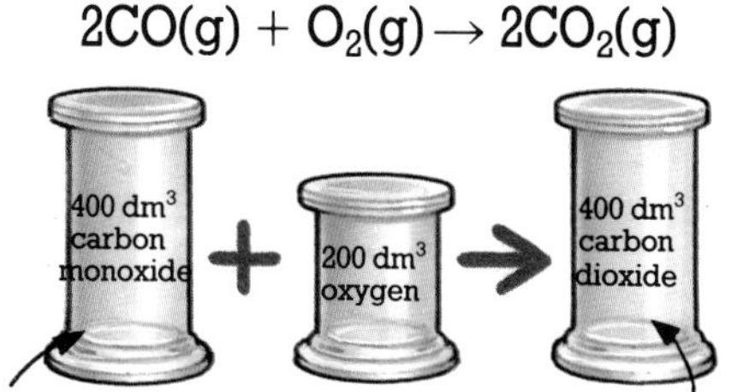

According to **Avogadro's law**, these two jars contain the same number of molecules.

- **Avogadro's law** or **Avogadro's hypothesis**. Equal volumes of all gases at the same temperature and pressure contain the same number of molecules.

- **s.t.p.** An abbreviation for **standard temperature and pressure**. These are internationally agreed standard conditions under which properties such as volume and density of gases are usually measured.

s.t.p. =
temperature: 0°C or 273K (**kelvins**)
pressure: 101325 **pascals***

- **Absolute temperature scale**. A standard temperature scale, using units called **kelvins (K)**. A kelvin is the same size as a degree **Celsius***, but the lowest point on the scale, zero kelvins or **absolute zero**, is equal to −273 degrees Celsius, a theoretical point where an **ideal gas** would occupy zero volume.

To convert degrees **Celsius** to **kelvins**, add 273.

To convert **kelvins** to degrees **Celsius**, subtract 273.

Degrees Celsius | **Kelvins**
100°C | 373K
steam
0°C ice | 273K
Absolute zero −273°C | 0K

* **Celsius scale**, 27; **Diffusion**, 123; **Mixture**, 122; **Pascal**, 97.

Solutions and solubility

When a substance is added to a liquid, several things can happen. If the atoms, molecules or ions of the substance become evenly dispersed (**dissolve**), the **mixture*** is a **solution**. If they do not, the mixture is either a **colloid**, a **suspension** or a **precipitate**. How well a substance dissolves depends on its properties, those of the liquid and other factors such as temperature and pressure.

- **Solvent**. The substance in which the **solute** dissolves to form a solution.
- **Solute**. The substance which dissolves in the **solvent** to form a solution.

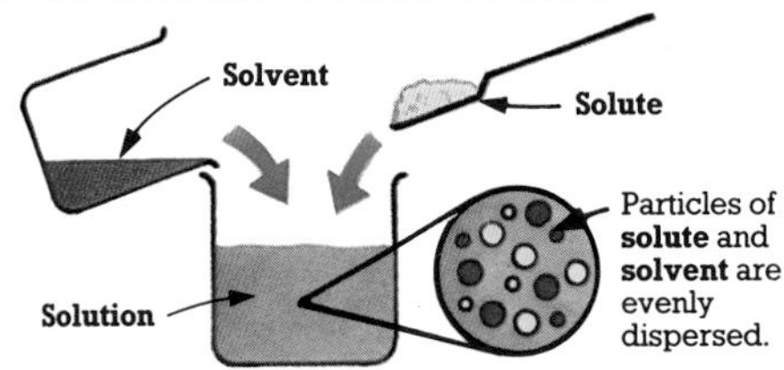

- **Solvation**. The process of **solvent** molecules combining with **solute** molecules as the solute dissolves. When the solvent is water the process is called **hydration**. Whether or not solvation takes place depends on how much the molecules of the solvent and solute attract each other.
- **Polar solvent**. A liquid with **polar molecules***. Polar solvents generally dissolve **ionic compounds***. **Solvation** occurs because the charged ends of the solvent molecules attract the ions of the **giant ionic lattice***. Water is the most common polar solvent.

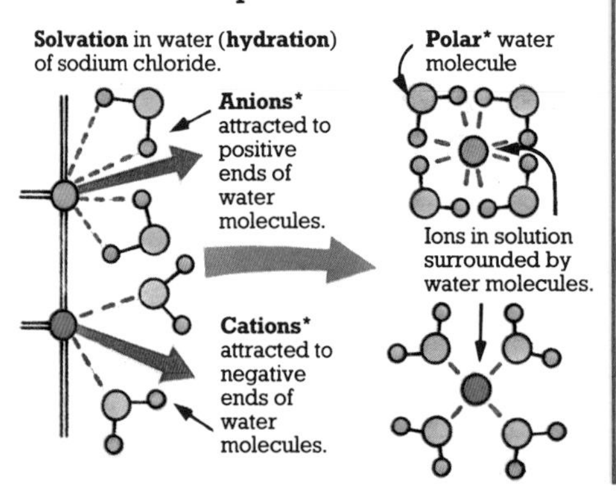

- **Non-polar solvent**. A liquid with **non-polar molecules***. Non-polar solvents dissolve **covalent compounds***. The **solute** molecules are pulled from the **molecular lattice*** by the solvent molecules and **diffuse*** through the solvent. Many organic liquids are non-polar solvents.

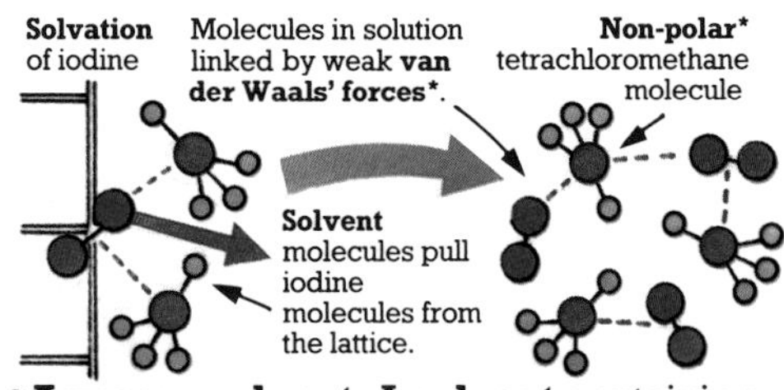

- **Aqueous solvent**. A **solvent** containing water. Water molecules are **polar***, so aqueous solvents are **polar solvents**.
- **Aqueous solution**. A solution formed from an **aqueous solvent**.

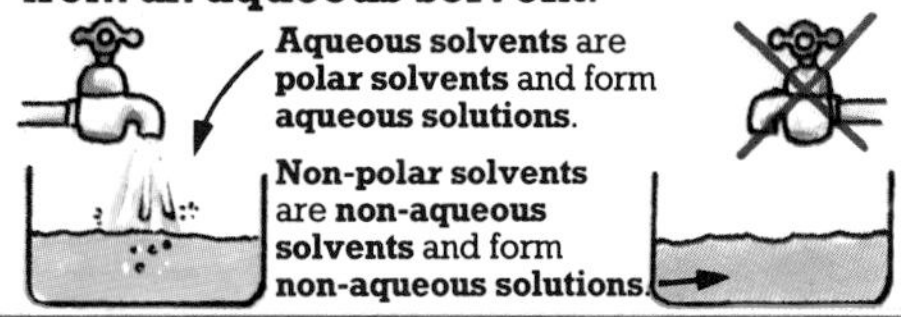

- **Dilute**. Describes a solution with a low **concentration*** of **solute**.
- **Concentrated**. Describes a solution with a high **concentration*** of **solute**.
- **Saturated**. Describes a solution that will not dissolve any more **solute** at a given temperature (any more solute will remain as crystals). If the temperature is raised, more solute may dissolve until the solution becomes saturated again.

* **Anion, Cation**, 130; **Concentration**, 139; **Covalent compounds**, 132; **Diffusion**, 123; **Giant ionic lattice**, 137; **Ionic compound**, 131; **Mixture**, 122; **Molecular lattice**, 137; **Non-polar molecule**, 133 (**Polar molecule**); **van der Waals' forces**, 134.

•Solubility. The amount of a **solute** which dissolves in a particular amount of **solvent** at a known temperature.

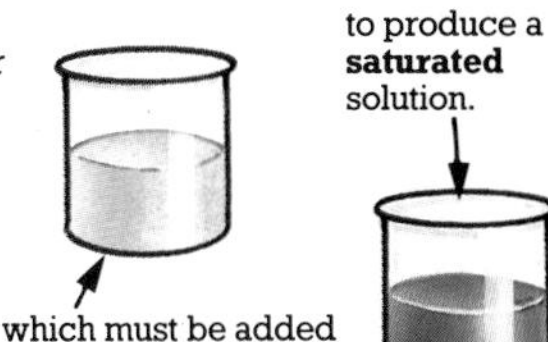
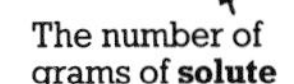

The solubility of a solid usually increases with temperature, while the solubility of a gas decreases.

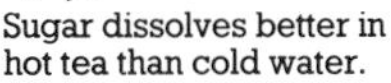

Sugar dissolves better in hot tea than cold water.

Warm soft drinks have more bubbles than cold ones.

The change of solubility with temperature is shown by a **solubility curve**.

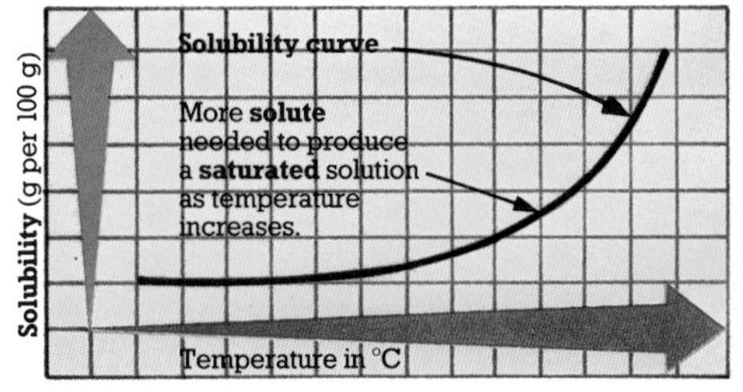

•Soluble. Describes a **solute** which dissolves easily in a **solvent**. The opposite of soluble is **insoluble**.

•Supersaturated. Describes a solution with more dissolved **solute** than a **saturated** solution at the same temperature. It is formed when a solution is cooled below the temperature at which it would be saturated, and there are no particles for the solute to **crystallize*** around, so the "extra" solute remains dissolved. The solution is unstable – if crystals are added or dust enters, the "extra" solute forms crystals.

•Precipitate. An **insoluble** solid (see **soluble**) formed when a reaction occurs in a solution.

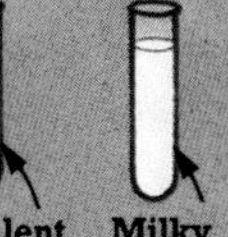
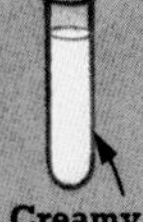
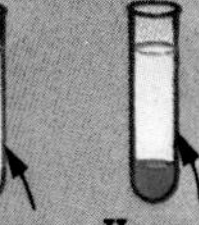

The reaction below forms a dense white **precipitate** of silver chloride.

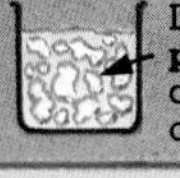

$AgNO_3(aq) + NaCl(aq) \rightarrow AgCl(\downarrow) + NaNO_3(aq)$

Silver nitrate; Sodium chloride; This symbol means **precipitate**

•Miscible. Describes two or more liquids which **diffuse*** together. The opposite is **immiscible**.

•Suspension. Fine particles of a solid (groups of atoms, molecules or ions) suspended in a liquid in which the solid does not dissolve.

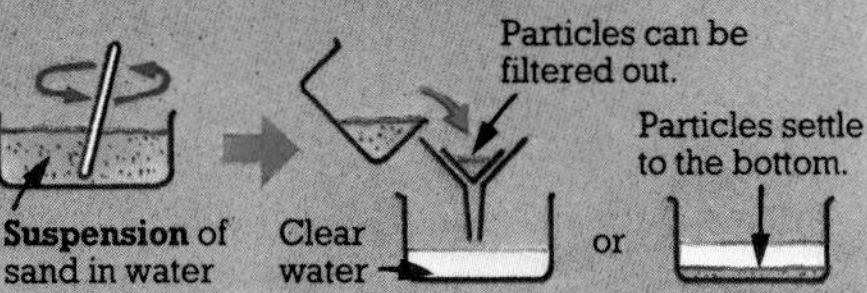

•Colloid. A **mixture*** of extremely small particles of a substance dispersed in another in which it does not dissolve. The particles (groups of atoms, molecules or ions) are smaller than in a **suspension**.

Emulsion. A **colloid** consisting of tiny particles of one liquid dispersed in another liquid, e.g. salad dressing.

Foam. A **colloid** of small bubbles of gas dispersed in a liquid.

Mist. A **colloid** consisting of tiny particles of a liquid dispersed in a gas.

Smoke. A **colloid** consisting of tiny particles of a solid dispersed in a gas.

* **Crystallization**, 135; **Diffusion**, 123; **Mixture**, 122.

Energy and chemical reactions

Nearly all chemical reactions involve a change in energy. Some reactions involve electrical energy or light energy, but almost all involve heat energy. The change in energy in a reaction results from the different amounts of energy involved when bonds are broken and formed. The study of heat energy in chemical reactions is called **thermochemistry**.

● **Enthalpy change of reaction** or **heat of reaction** (**ΔH**). The amount of heat energy given out or absorbed during a chemical reaction. If the reaction is a **change of state***, this amount is also known, particularly in physics, as the **latent heat** (see page 30). Hence the **molar enthalpy changes of fusion** and **vaporization** on page 147 are closely allied to the **specific latent heats** on page 31, though the quantities are different (each being more relevant to its science).

Enthalpy change = total **enthalpy** of products − total **enthalpy** of reactants

$$2H_2(g) + O_2(g) \rightarrow 2H_2O(g) \quad \Delta H = -488kJ$$

The value of △**H** is only true for the number of **moles*** and the **physical states*** of the chemicals in the equation.

J stands for **joule***, a unit of energy. **kJ** stands for **kilojoule** (1000 **joules**).

● **Enthalpy** (**H**). The amount of energy that a substance contains. It is impossible to measure directly, but its change during a reaction can be measured.

● **Energy level diagram**. A diagram which shows the **enthalpy change of reaction** for a reaction.

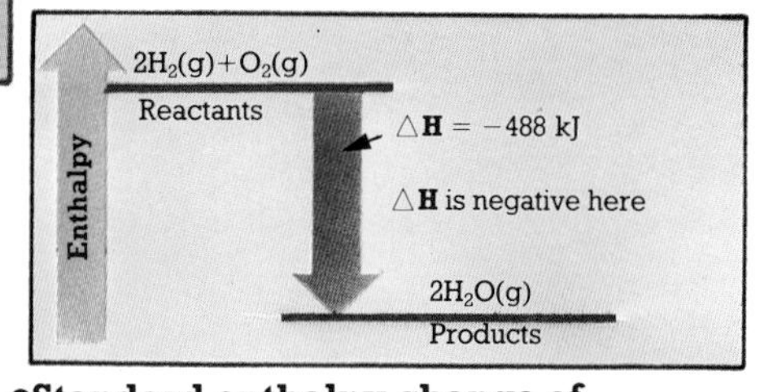

● **Standard enthalpy change of reaction** (△**H**$^{\ominus}$). An **enthalpy change of reaction** measured under standard conditions, i.e. standard temperature and pressure (**s.t.p.***). If solutions are used, their **concentration*** is 1**M***.

Special enthalpy changes

● **Enthalpy change of combustion** or **heat of combustion**. The amount of heat energy given out when one **mole*** of a substance is completely burnt in oxygen. The heat of combustion for a substance is measured using a **bomb calorimeter**.

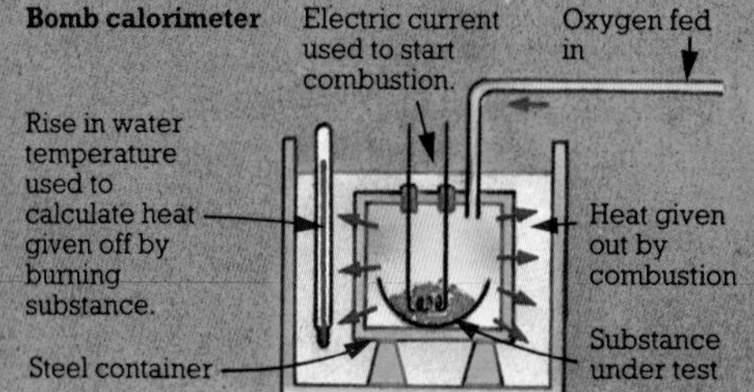

● **Enthalpy change of neutralization** or **heat of neutralization**. The amount of heat energy given out when one **mole*** of hydrogen ions (H^+) is **neutralized*** by one mole of hydroxide ions (OH^-). If the acid and alkali are fully **ionized***, the heat of neutralization is always −57 kJ. The **ionic equation*** for neutralization is:

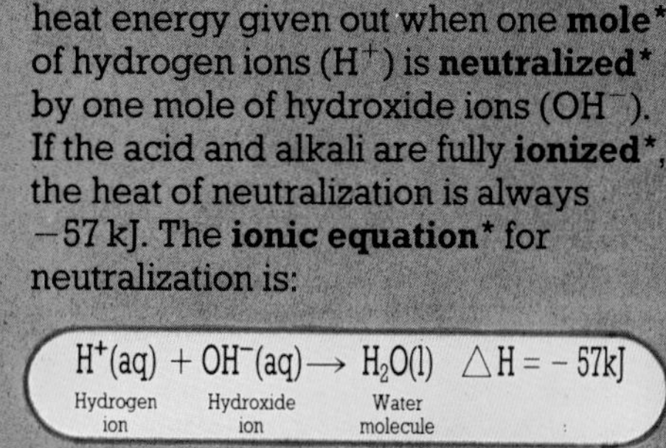

When a **weak acid*** or a **weak base*** is involved, the heat produced is less. Some energy must be supplied to ionize the acid fully.

* **Calorimetry**, 344; **Concentration**, 139; **Ionic equation**, 141; **Ionization**, 130; **Joule**, 97; **M-value**, 139 (**Molarity**); **Mole**, 139; **Neutralization**, 151; **Physical states**, 120; **s.t.p.**, 143; **Weak acid, Weak base**, 152.

•**Exothermic reaction**. A chemical reaction during which heat is transferred to the surroundings.

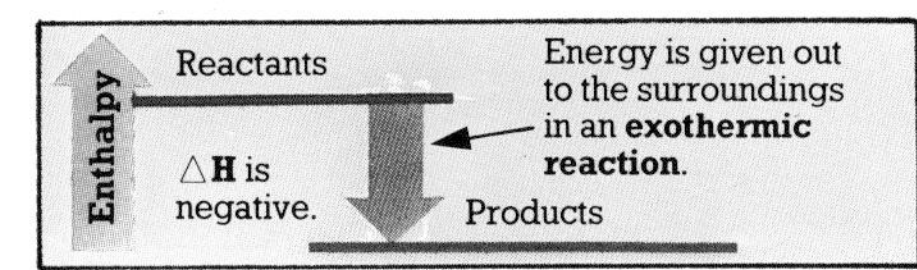

•**Endothermic reaction**. A chemical reaction during which heat is absorbed from the surroundings.

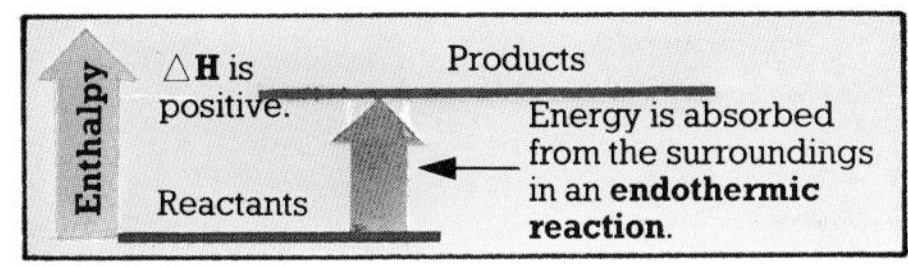

•**Bond energy**. A measure of the strength of a **covalent bond*** formed between two atoms. Energy must be supplied to break bonds and is given out when bonds are formed. A difference in these energies produces a change in energy during a reaction.

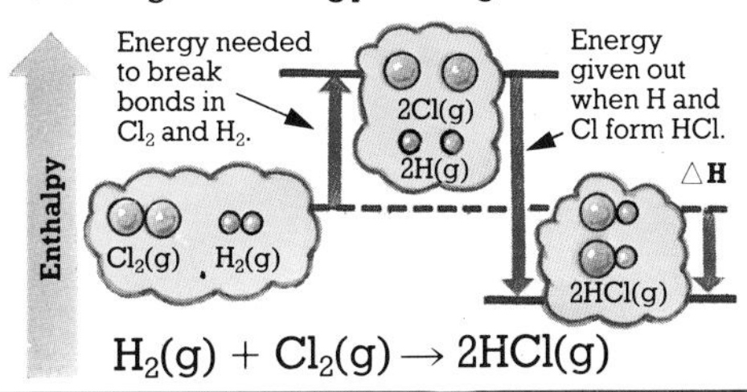

$H_2(g) + Cl_2(g) \rightarrow 2HCl(g)$

•**Law of conservation of energy**. During a chemical reaction, energy cannot be created or destroyed. In a **closed system*** the amount of energy is constant.

•**Hess's law**. This states that the **enthalpy change of reaction** that occurs during a particular chemical reaction is always the same, no matter what route is taken in going from the reactants to the products. Hess's law is illustrated by an **energy cycle**.

Energy cycle

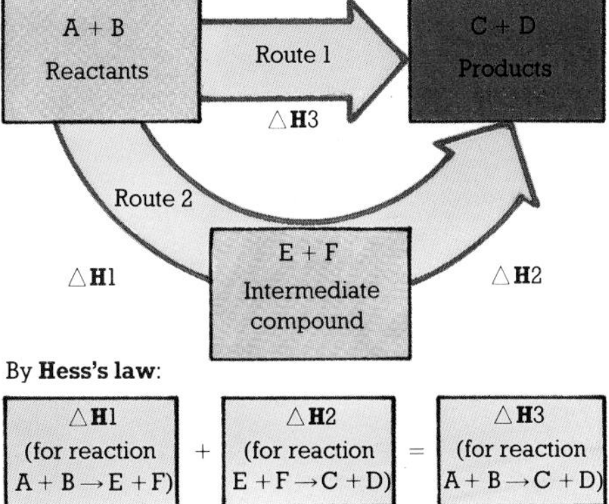

By **Hess's law**:

△**H**1 (for reaction A + B → E + F) + △**H**2 (for reaction E + F → C + D) = △**H**3 (for reaction A + B → C + D)

Hess's law is used to find enthalpy changes of reaction which cannot be measured directly, e.g. the **enthalpy change of formation** of methane.

•**Enthalpy change of solution** or **heat of solution**. The amount of heat energy given out or taken in when one **mole*** of a substance dissolves in such a large volume of **solvent*** that further dilution produces no heat change.

•**Molar enthalpy change of fusion** or **molar heat of fusion**. The amount of heat energy required to change one **mole*** of a solid into a liquid at its melting point. Energy must be supplied to break the bonds in the **crystal lattice*** of the solid.

•**Molar enthalpy change of vaporization** or **molar heat of vaporization**. The heat energy needed to change one **mole*** of a liquid into a vapour at its boiling point.

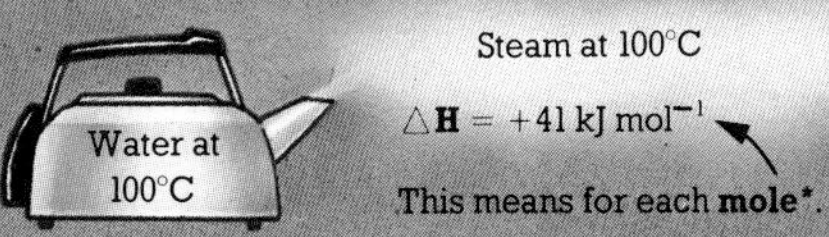

•**Enthalpy change of formation** or **heat of formation**. The heat energy given out or taken in when one **mole*** of a compound is formed from elements.

$C(graphite) + O_2(g) \rightarrow CO_2(g) \quad \Delta H = -394 kJ$

Carbon — Oxygen — Carbon dioxide

* **Closed system**, 162; **Covalent bond**, 132; **Crystal lattice**, 136; **Mole**, 139; **Solvent**, 144.

Oxidation and reduction

The terms **oxidation** and **reduction** originally referred to the gain and loss of oxygen by a substance. They have now been extended to include the gain and loss of hydrogen and electrons. There is always a transfer of electrons in reactions involving oxidation and reduction, that is, the **oxidation state** of one or more of the elements is always changed.

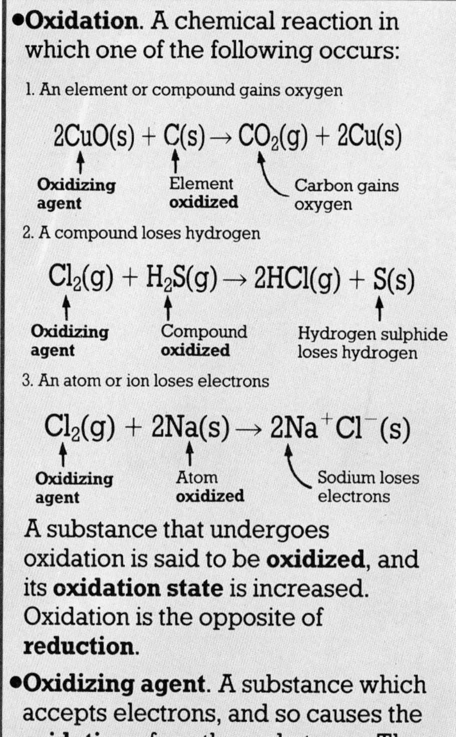

●**Oxidation**. A chemical reaction in which one of the following occurs:

1. An element or compound gains oxygen

$$2CuO(s) + C(s) \rightarrow CO_2(g) + 2Cu(s)$$

2. A compound loses hydrogen

$$Cl_2(g) + H_2S(g) \rightarrow 2HCl(g) + S(s)$$

3. An atom or ion loses electrons

$$Cl_2(g) + 2Na(s) \rightarrow 2Na^+Cl^-(s)$$

A substance that undergoes oxidation is said to be **oxidized**, and its **oxidation state** is increased. Oxidation is the opposite of **reduction**.

●**Oxidizing agent**. A substance which accepts electrons, and so causes the **oxidation** of another substance. The oxidizing agent is always **reduced** in a reaction.

●**Reduction**. A chemical reaction in which one of the following occurs:

1. A compound loses oxygen

$$2CuO(s) + C(s) \rightarrow CO_2(g) + 2Cu(s)$$

Compound **reduced** — **Reducing agent** — Copper(II) oxide loses oxygen

2. A compound or element gains hydrogen

$$Cl_2(g) + H_2S(g) \rightarrow 2HCl(g) + S(s)$$

Element **reduced** — **Reducing agent** — Chlorine gains hydrogen

3. An atom or ion gains electrons.

$$Cl_2(g) + 2Na(s) \rightarrow 2Na^+Cl^-(s)$$

Atom **reduced** — **Reducing agent** — Chlorine gains electron

A substance that undergoes reduction is said to be **reduced**, and its **oxidation state** is decreased. Reduction is the opposite of **oxidation**.

●**Reducing agent**. A substance which donates electrons, and so causes the **reduction** of another substance. The reducing agent is always **oxidized** in a reaction.

●**Redox**. Describes a chemical reaction involving **oxidation** and **reduction**. The two processes always occur together because an **oxidizing agent** is always reduced during oxidation and a **reducing agent** is always oxidized during reduction. The simultaneous oxidation and reduction of the same element in a reaction is called **disproportionation**.

$$Fe(s) + CuSO_4(aq) \rightarrow FeSO_4(aq) + Cu(s)$$

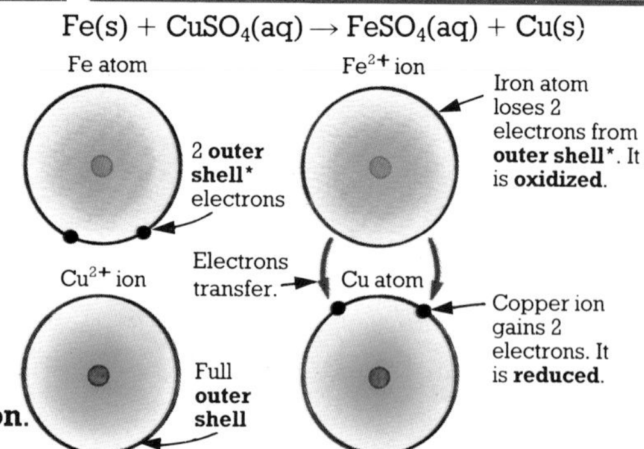

* **Outer shell**, 127.

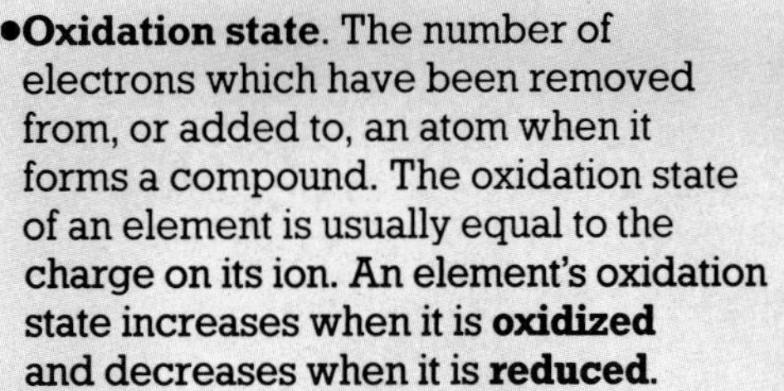

•Oxidation state. The number of electrons which have been removed from, or added to, an atom when it forms a compound. The oxidation state of an element is usually equal to the charge on its ion. An element's oxidation state increases when it is **oxidized** and decreases when it is **reduced.**

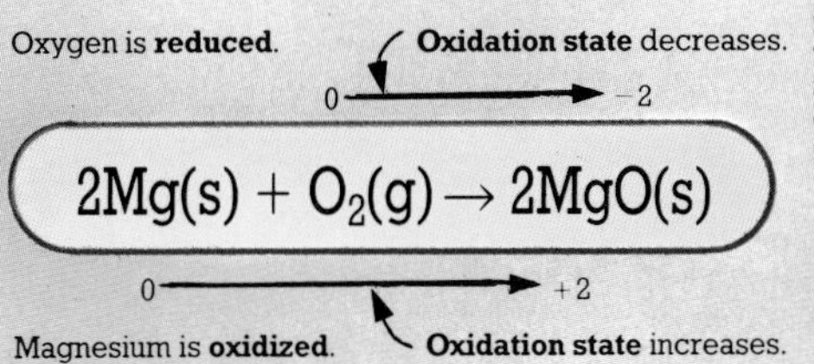

These rules help to work out the oxidation state of an element:

1. The **oxidation state** of a free element (one that is not part of a compound) is zero.

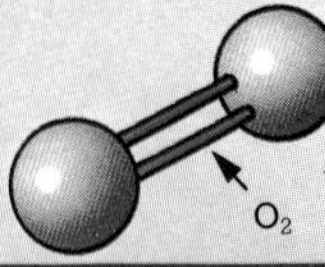

The **oxidation state** of oxygen is 0. No electrons have been lost or gained.

2. The **oxidation state** of an element in an **ionic compound*** is equal to the electrical charge on its ion.

Ca Cl 2

Calcium chloride forms a **giant ionic lattice***.

Oxidation state. 2 electrons removed. +2

−1 **Oxidation state.** 1 electron added.

Oxidation states of elements in **covalent compounds*** are found by assuming the compound is **ionic***, and working out the charge the ions would have.

C H 4

Methane is a **covalent compound***.

−4 **Oxidation state**

+1 **Oxidation state**

3. The sum of the **oxidation states** of all the elements in a compound is zero.

Fe S O 4

Iron(II) sulphate

Sum of **oxidation states** = (+2) + (+6) + (4 × −2) = 0

4. The **oxidation state** of oxygen in a compound is normally −2, but in **peroxides**, e.g. hydrogen peroxide, it is −1.

5. The **oxidation state** of hydrogen is usually +1, except in metal **hydrides**, when it is −1.

•Oxidation number. A number that shows the **oxidation state** of an element in a compound. It is written in Roman numerals and placed in brackets after the name of the element. It is only included in the name of a compound when the element has more than one oxidation state.

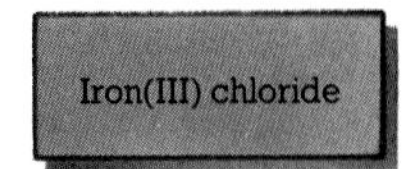

Oxidation number of 3 and **oxidation state** of +3

Lead(IV) oxide

Oxidation number of 4 and **oxidation state** of +4

•Redox series. A list of substances arranged in order of their **redox potentials**, the substance with the most negative redox potential being placed at the top. A substance usually **oxidizes** any substance above it in the series and **reduces** any substance below it. The further apart substances are in the series, the more easily they oxidize or reduce each other. The redox series is an extended version of the **electrochemical series***.

•Redox potential. A measurement of the power of a substance to gain electrons in solution. A strong **reducing agent**, which readily loses electrons (which it can give to another substance), will have a high negative redox potential. A strong **oxidizing agent**, which easily gains electrons, will have a high positive redox potential. Redox potential is the same as **electrode potential***.

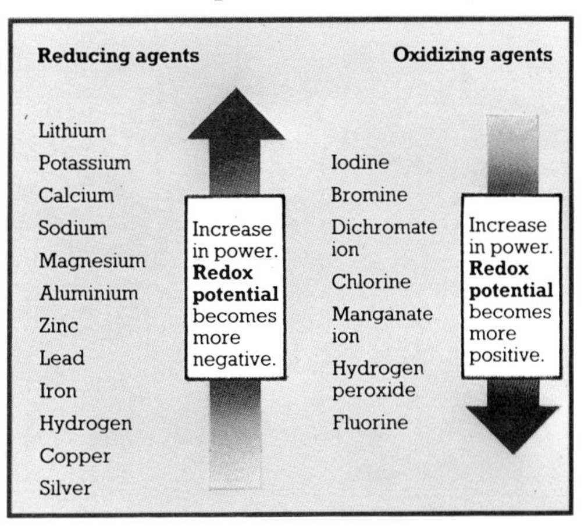

* **Covalent compounds**, 132; **Electrochemical series**, 159; **Electrode potential**, 158; **Giant ionic lattice**, 137; **Ionic compound**, 131.

Acids and bases

All chemicals are either **acidic**, **basic** or **neutral**. In pure water, a small number of molecules **ionize***, each one forming a hydrogen ion (a single **proton***) and a hydroxide ion. The number of hydrogen and hydroxide ions is equal, and the water is described as **neutral**. Some compounds dissolve in, or react with, water to produce hydrogen ions or hydroxide ions, which upset the balance. These compounds are either **acids** or **bases**.

•**Acid**. A compound containing hydrogen which dissolves in water to produce hydrogen ions (H^+ – **protons***) in the solution. Hydrogen ions do not exist on their own in the solution, but join with water molecules to produce **hydroxonium ions**. These ions can only exist in solution, so an acid will only display its properties when it dissolves.

Ascorbic acid Methanoic acid Sulphuric acid in **lead-acid accumulators***

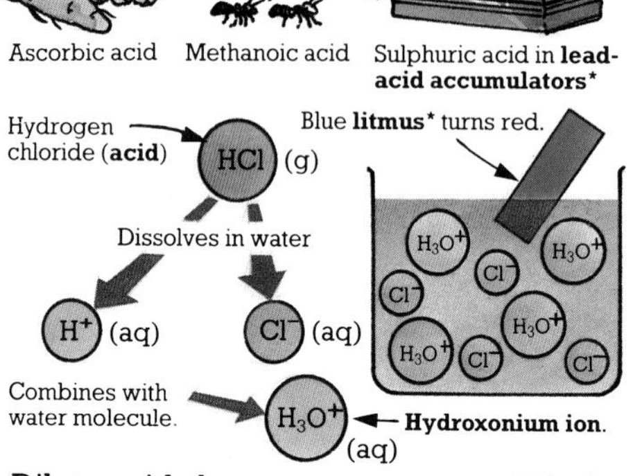

Dilute acids have a sour taste, a **pH*** of less than 7 and turn blue **litmus*** red. They react with metals above hydrogen in the **electrochemical series*** to produce hydrogen gas.

$$H_2SO_4(aq) + Mg(s) \rightarrow MgSO_4(aq) + H_2(g)$$

Acid Metal Salt* Hydrogen

Dilute **strong acids*** react with carbonates or hydrogencarbonates to produce carbon dioxide gas and are **neutralized** by **bases**.

•**Acidic**. Describes any compound with the properties of an **acid**.

•**Hydroxonium ion** (H_3O^+) or **oxonium ion** . An ion formed when a hydrogen ion attaches itself to a water molecule (see **acid**). When a reaction takes place in a solution containing hydroxonium ions, only the hydrogen ion takes part. Hence usually the hydroxonium ion can be considered to be a hydrogen ion.

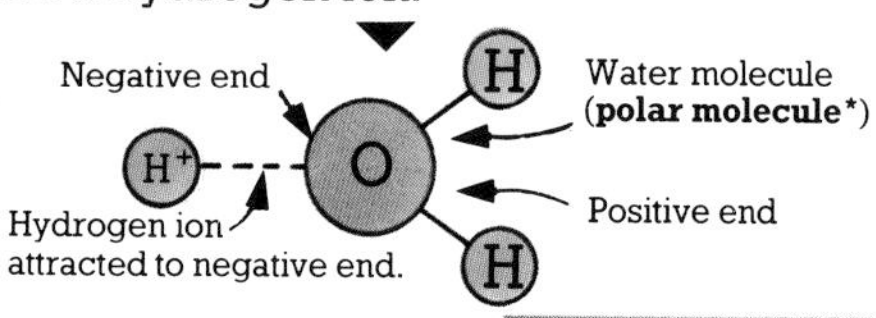

•**Mineral acid**. An **acid** which is produced chemically from a mineral, e.g. hydrochloric acid is produced from sodium chloride and sulphuric acid is produced from sulphur.

Acid	Formula
Hydrochloric	HCl
Sulphuric	H_2SO_4
Sulphurous	H_2SO_3
Nitric	HNO_3
Nitrous	HNO_2
Phosphoric	H_3PO_4

Ethanedioic (Oxalic)	$(COOH)_2$
Methanoic (Formic)	$HCOOH$
Ethanoic (Acetic)	CH_3COOH

•**Organic acid**. An **organic compound*** that is **acidic**. The most common ones are **carboxylic acids***.

* **Carboxylic acids**, 195; **Electrochemical series**, 159; **Ionization**, 130; **Lead-acid accumulator**, 159; **Litmus**, 152; **Organic compounds**, 190; **pH**, 152; **Polar molecule**, 133; **Proton**, 126; **Salts**, 153; **Strong acid**, 152.

•**Base**. A substance that will **neutralize** an **acid** by accepting hydrogen ions. It is the chemical opposite of an acid. Bases are usually metal oxides and hydroxides, although ammonia is also a base. A substance with the properties of a base is described as **basic**. A base which dissolves in water is an **alkali**. Ammonia is produced when a base is heated with an ammonium **salt***.

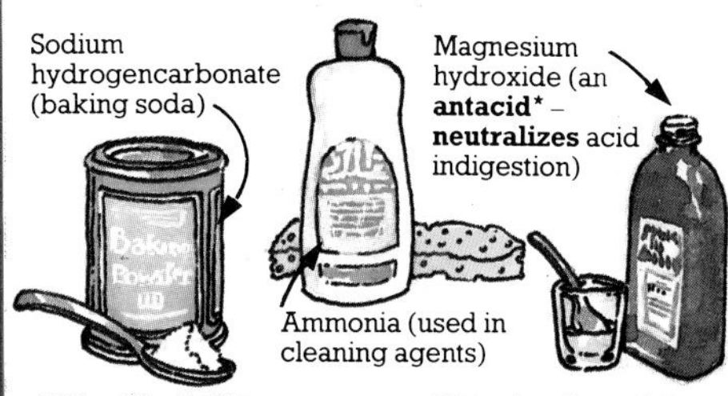

•**Alkali**. A **base**, normally a hydroxide of a Group 1 or Group 2 metal, which is soluble in water and produces hydroxide ions (OH^-) in solution. These make a solution **alkaline**.

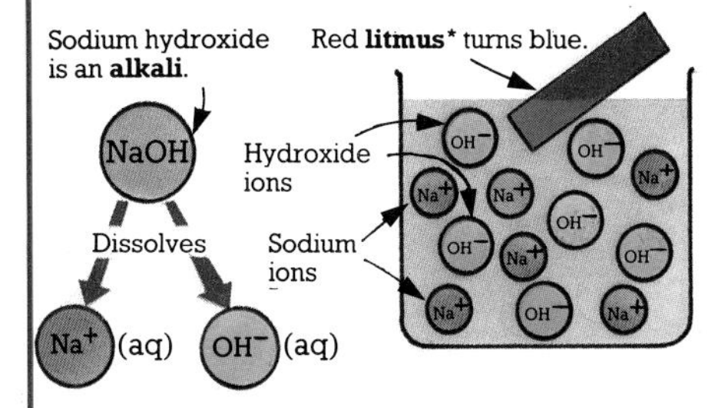

•**Alkaline**. Describes a solution formed when a **base** dissolves in water to form a solution which contains more hydroxide ions than hydrogen ions.

Alkaline solutions have a **pH*** of more than 7, turn red **litmus*** blue, and feel soapy because they react with the skin. Alkaline solutions produced from **strong bases*** react with a few metals, e.g. zinc and aluminium, to give off hydrogen gas.

$$2Al(s) + 2NaOH(aq) + 6H_2O(l) \longrightarrow 2NaAl(OH)_4(aq) + 3H_2(g)$$

Aluminium + Sodium hydroxide + Water → Sodium aluminate + Hydrogen

•**Amphoteric**. Describes a substance that acts as an **acid** in one reaction, but as a **base** in another, e.g. zinc hydroxide.

ACID  **BASE**

•**Anhydride**. A substance that reacts with water to form either an **acidic** or an **alkaline** solution (see **hydrolysis**, page 155). It is usually an oxide.

$$SO_2(g) + H_2O(l) \longrightarrow H_2SO_3(aq)$$

Sulphur dioxide (**anhydride**) + Water → Sulphurous acid

•**Neutral**. Describes a substance that does not have the properties of an **acid** or **base**. A neutral solution has an equal number of hydrogen and hydroxide ions. It has a **pH*** of 7 and does not change the colour of **litmus***.

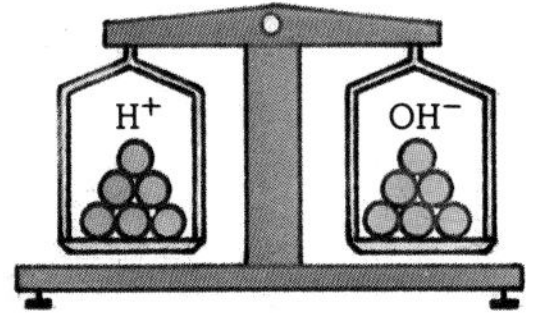

A **neutral** solution contains an equal number of hydrogen and hydroxide ions.

•**Neutralization**. The reaction between an **acid** and a **base** to produce a **salt*** and water only. An equal number of hydrogen and hydroxide ions react together to form a **neutral** solution. The **acid radical*** from the acid and **cation*** from the base form a salt.

Neutralization is:

ACID + BASE → SALT* + WATER

•**Bronsted-Lowry theory**. Another way of describing **acids** and **bases**. It defines an acid as a substance which donates **protons***, and a base as one which accepts them.

Ethanoic acid donates a **proton*** – it is an **acid**.
Water accepts a **proton*** – it is a **base**.

$$CH_3COOH(aq) + H_2O(l) \rightleftharpoons H_3O^+(aq) + CH_3COO^-(aq)$$

Hydroxonium ion donates **proton*** – it is an **acid**.
Ethanoate ion accepts **proton*** – it is a **base**.

* **Acid radical**, 153; **Antacid**, 344; **Cation**, 130; **Litmus**, **pH**, 152; **Proton**, 126; **Salts**, 153; **Strong base**, 152.

Acids and bases continued – strength and concentration.

The **concentration*** of **acids** and **bases** (see previous two pages) depends on how many **moles*** of the acid or base are in a solution, but the strength depends on the proportion of their molecules which **ionize*** to produce **hydroxonium ions*** or hydroxide ions. A dilute **strong acid** can produce more hydrogen ions than a concentrated **weak acid**.

• **Strong acid**. An acid that completely **ionizes*** in water, producing a large number of hydrogen ions in solution.

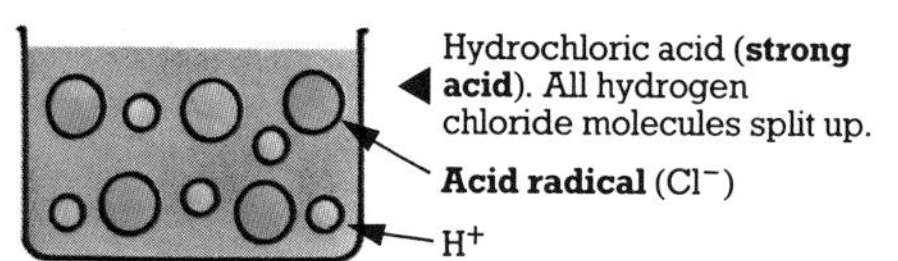

• **Weak acid**. An acid that only partially **ionizes*** in water, i.e. only a small percentage of its molecules split into hydrogen ions and **acid radicals**.

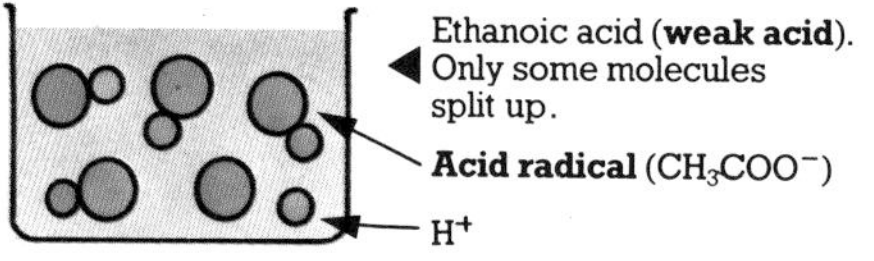

• **Strong base**. A base that is completely **ionized*** in water. A large number of hydroxide ions are released to give a strongly alkaline solution.

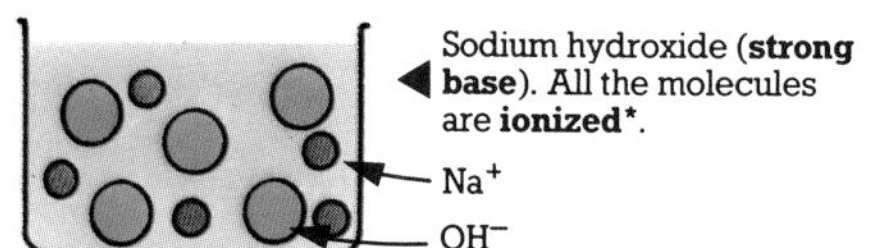

• **Weak base**. A base that is only partially **ionized*** in water. Only some of the molecules of the base split up to produce hydroxide ions, giving a weakly alkaline solution.

Ammonia reacts slightly with water to give a low concentration of hydroxide ions:

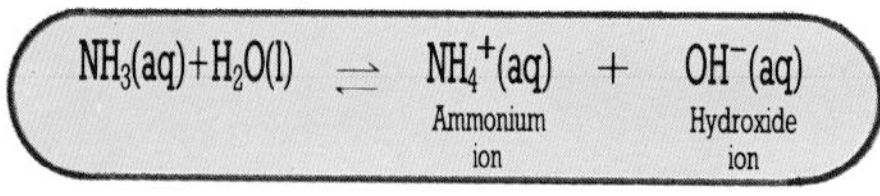

• **pH**. Stands for **power of hydrogen**, a measure of hydrogen ion **concentration*** in a solution.

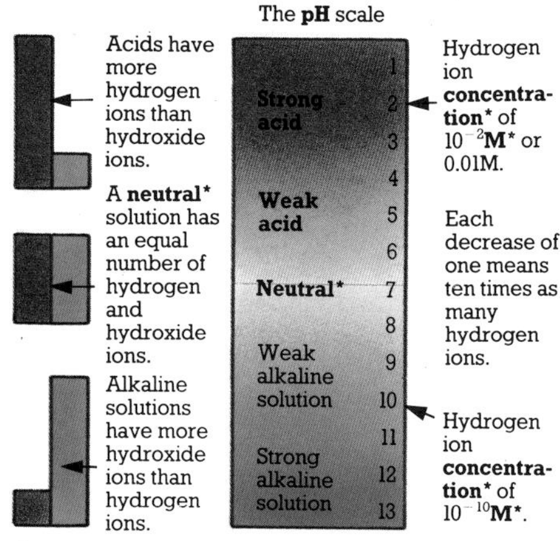

• **Indicator**. A substance whose colour depends on the **pH** of the solution it is in. Indicators can be used in solid or liquid form. Some common ones are shown at the bottom of this column.

• **Litmus**. An **indicator** which shows whether a solution is acidic or alkaline. Acids turn blue litmus paper red, and alkaline solutions turn red litmus paper blue.

• **Universal indicator**. An **indicator**, either in the form of paper or in solution, which shows the **pH** of a solution with a range of colours.

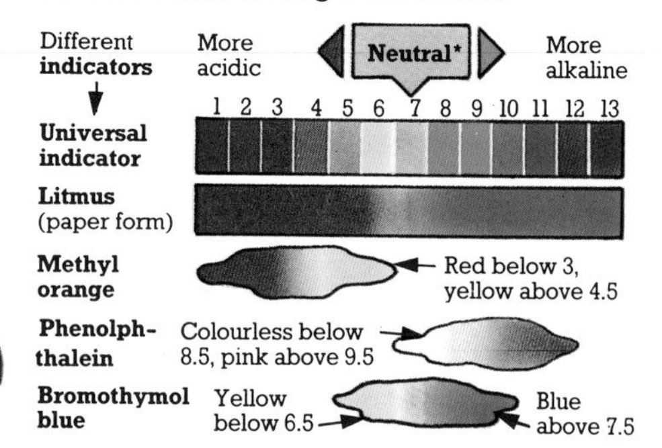

* **Concentration**, 139; **Hydroxonium ion**, 150; **Ionization**, 130; **M-value**, 139 (**Molarity**); **Mole**, 139; **Neutral**, 151.

Salts

All **salts** are **ionic compounds*** which contain at least one **cation*** and one **anion*** (called the **acid radical**). Theoretically, they can all be formed by replacing one or more of the hydrogen ions in an acid by other cations (one or more) e.g. metal ions (see below) or ammonium ions. Salts have many industrial and domestic uses.

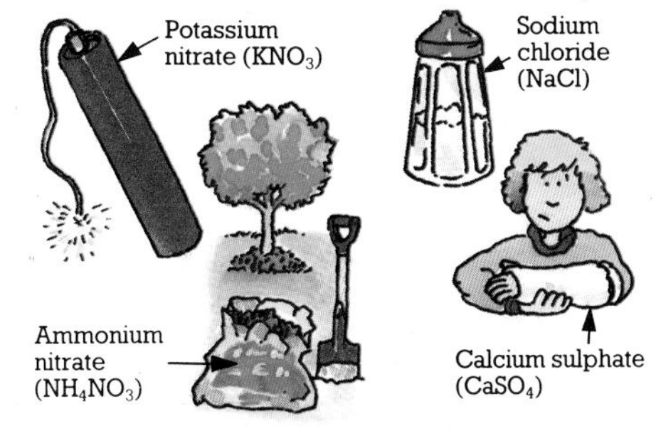

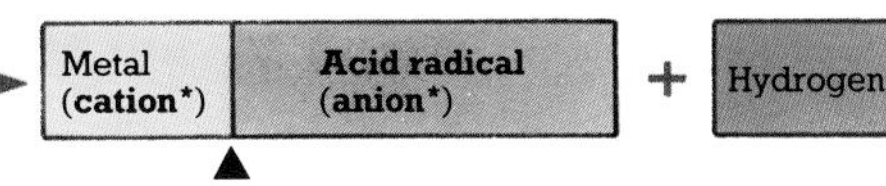

•**Acid radical**. The **anion*** left after the hydrogen ions have been removed from an acid. See table below.

Acid	Radical	Radical name
Hydrochloric	Cl^-	Chloride
Sulphuric	SO_4^{2-}	Sulphate
Sulphurous	SO_3^{2-}	Sulphite
Nitric	NO_3^-	Nitrate
Nitrous	NO_2^-	Nitrite
Carbonic	CO_3^{2-}	Carbonate
Ethanoic	CH_3COO^-	Ethanoate
Phosphoric	PO_4^{3-}	Phosphate

The radical name identifies the salt.

Copper (II) sulphate: Cu^{2+} (Cation*) SO_4^{2-} (Acid radical)

Sodium chloride: Na^+ (Cation*) Cl^- (Acid radical)

•**Basicity**. The number of hydrogen ions in an acid that can be replaced to form a salt. Not all the hydrogen ions are necessarily replaced.

H Cl — Hydrochloric acid is **monobasic**.

CH_3COO H — Ethanoic acid is **monobasic**.

H_2 SO_4 — Sulphuric acid is **dibasic**.

H_3 PO_4 — Phosphoric acid is **tribasic**.

•**Normal salt**. A salt containing only metal ions (or ammonium ions) and the **acid radical**, formed when all the hydrogen ions in an acid are replaced by metal ions (or ammonium ions).

Copper (II) sulphate and ammonium chloride (**normal salts**)

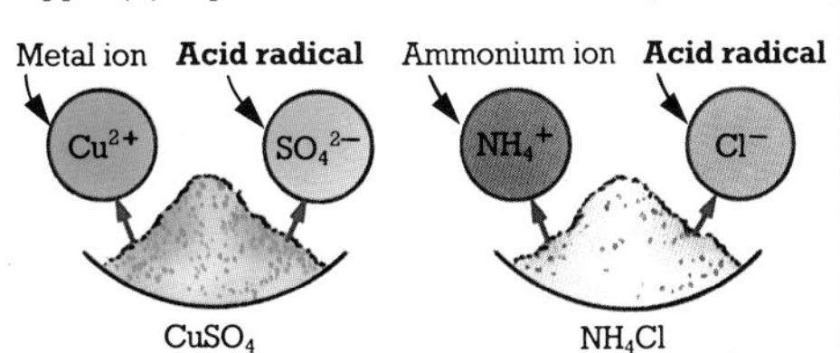

•**Acid salt**. A salt containing hydrogen ions as well as metal ions (or ammonium ions) and the **acid radical**, formed when only some hydrogen ions in an acid are replaced by metal ions (or ammonium ions). Only acids with a **basicity** of two or more can form acid salts. Most acid salts are acidic, but some form alkaline solutions.

Sodium hydrogensulphate (**acid salt**)

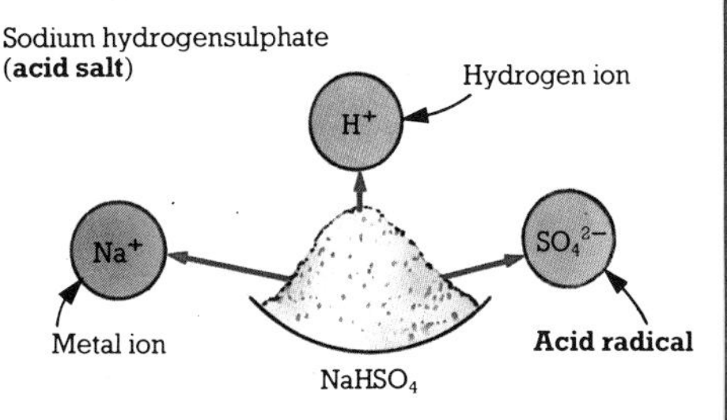

* **Anion**, **Cation**, 130; **Ionic compound**, 131.

Salts (continued)

- **Basic salt**. A salt containing a metal oxide or hydroxide, metal ions and an **acid radical***. It is formed when a **base*** is not completely **neutralized*** by an acid.

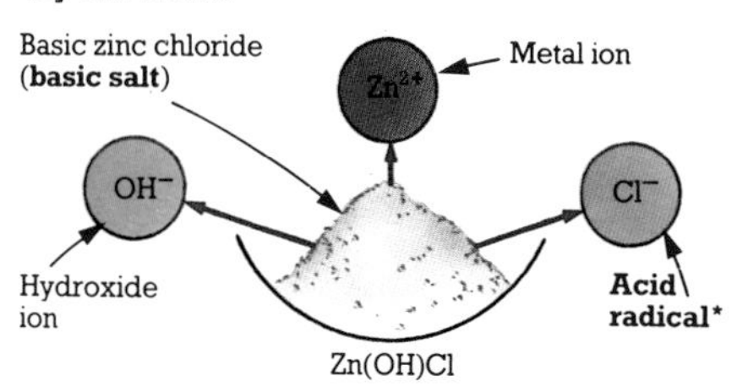

- **Double salt**. A salt formed when solutions of two **normal salts*** react together. It contains two different **cations*** (either two different metal ions or a metal ion and an ammonium ion) and one or more acid **radicals***.

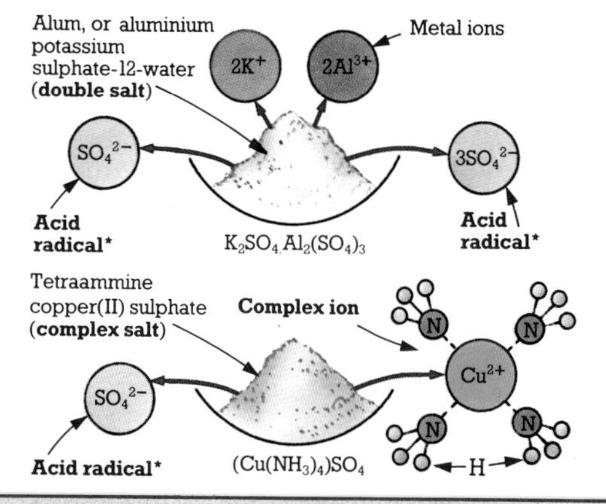

- **Complex salt**. A salt in which one of the ions is a **complex ion**. This is made up of a central **cation*** linked (frequently by **dative covalent bonds***) to several small molecules (usually **polar molecules***) or ions.

Preparation of salts.

Salts can be made in a number of ways, the method depending on whether a salt is soluble or insoluble in water (see table below). Soluble salts are **crystallized*** from solutions of the salts (obtained in various ways – see right) and insoluble salts are obtained in the form of **precipitates***.

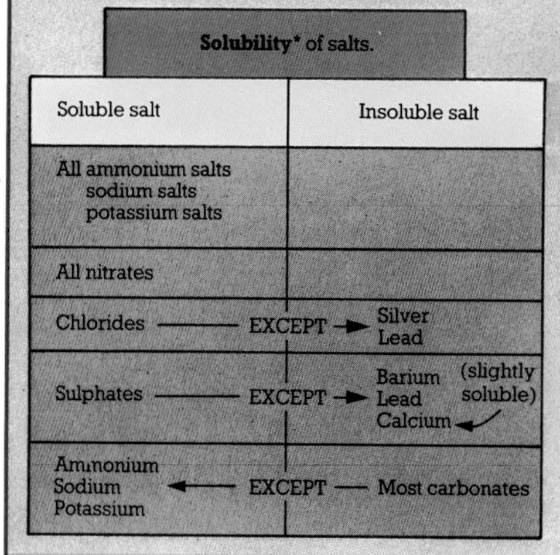

Solubility* of salts.

Soluble salt	Insoluble salt
All ammonium salts, sodium salts, potassium salts	
All nitrates	
Chlorides —— EXCEPT →	Silver, Lead
Sulphates —— EXCEPT →	Barium, Lead, Calcium (slightly soluble)
Ammonium, Sodium, Potassium ← EXCEPT ——	Most carbonates

Soluble salts can be made by the following methods, which all produce a solution of the salt. This is partly evaporated and left to **crystallize***.

1. **Neutralization***, in which an acid is neutralized by an alkali.

$$2NaOH(aq) + H_2SO_4(aq) \rightarrow Na_2SO_4(aq) + 2H_2O(l)$$

Sodium hydroxide	Sulphuric acid	Sodium sulphate	Water
Alkali	Acid	Salt	Water

2. The action of an acid on an insoluble carbonate.

$$MgCO_3(s) + 2HCl(aq) \rightarrow MgCl_2(aq) + H_2O(l) + CO_2(g)$$

Magnesium carbonate	Hydrochloric acid	Magnesium chloride	Water	Carbon dioxide
Insoluble carbonate	Acid	Salt	Water	Carbon dioxide

3. The action of an acid on an insoluble **base***.

$$CuO(s) + H_2SO_4(aq) \rightarrow CuSO_4(aq) + H_2O(l)$$

Copper (II) oxide	Sulphuric acid	Copper (II) sulphate	Water
Insoluble **base***	Acid	Salt	Water

* **Acid radical**, 153; **Base**, 151; **Cation**, 130; **Crystallization**, 135; **Dative covalent bond**, 132; **Neutralization**, 151; **Normal salt**, 153; **Polar molecule**, 133; **Precipitate**, **Solubility**, 145.

•Hydrate. A salt that contains **water of crystallization*** (it is **hydrated**). The salt becomes an **anhydrate** if the water is removed.

•Anhydrate. A salt that does not contain **water of crystallization*** (it is **anhydrous**). The salt becomes a **hydrate** if it absorbs water.

Copper(II) sulphate can be a

hydrate or **anhydrate**.

$CuSO_4.5H_2O$ $CuSO_4 + 5H_2O$

Water driven off by heating

Copper(II) sulphate crystals are **hygroscopic*** (absorb water from the air).

•Dehydration. The removal of water from a substance. It is either removal of hydrogen and oxygen in the correct ratio to give water, or removal of water from a **hydrate** to give an **anhydrate**.

•Hydrolysis. The reaction of a salt with water. The ions of the salt react with water molecules, upsetting the balance of hydrogen and hydroxide ions, and so giving an acidic or alkaline solution. A salt which has been made from the reaction between a **weak acid*** and a **strong base*** dissolves to give an alkaline solution. One which has been made from the reaction between a **strong acid*** and a **weak base*** dissolves to give an acidic solution.

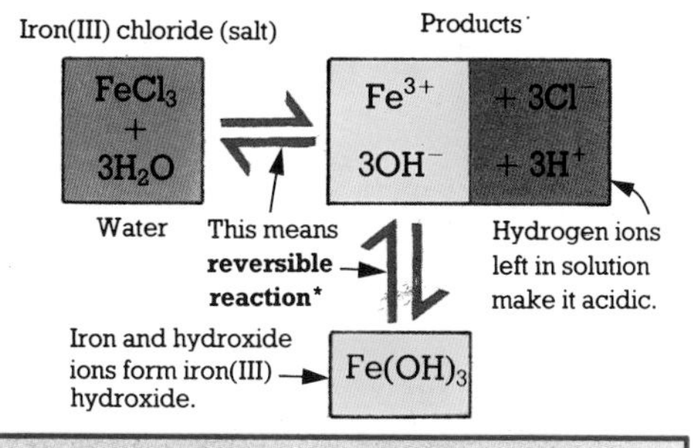

•Double decomposition. A chemical reaction between the solutions of two or more **ionic compounds*** in which ions are exchanged. One of the new compounds formed is an insoluble salt, which forms a **precipitate***. Most insoluble salts and hydroxides are made by this method - the precipitate is filtered out and washed.

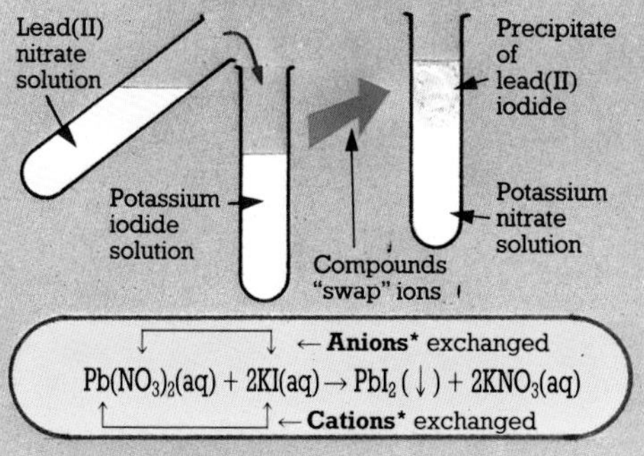

← **Anions*** exchanged

$$Pb(NO_3)_2(aq) + 2KI(aq) \rightarrow PbI_2(\downarrow) + 2KNO_3(aq)$$

← **Cations*** exchanged

•Direct replacement. A reaction in which all or some of the hydrogen in an acid is replaced by another element, usually a metal. It is used to prepare soluble salts, except salts of sodium or potassium, both of which react too violently with the acid.

$$Zn(s) + H_2SO_4(aq) \rightarrow Zn\,SO_4(aq) + H_2(g)$$

•Direct synthesis. A chemical reaction in which a salt is made directly from its elements. This method is used to make salts which react with water and therefore cannot be made by using solutions.

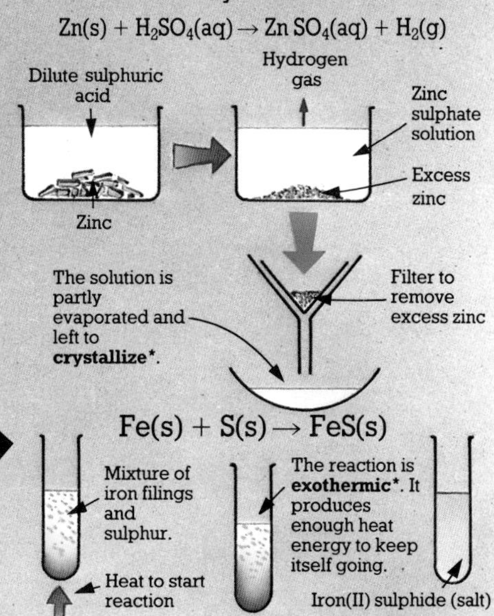

* **Crystallization**, 135; **Exothermic reaction**, 147; **Hygroscopic**, 206; **Ionic compound**, 131; **Precipitate**, 145; **Reversible reaction**, 162; **Strong acid**, **Strong base**, 152; **Water of crystallization**, 135; **Weak acid**, **Weak base**, 152.

Electrolysis

Electrolysis is a term describing the chemical reactions which occur when an electric **current*** is passed through a liquid containing ions. Metals and graphite conduct electric current because some electrons are free to move through the **crystal lattice***, but **molten* ionic compounds*** or compounds which **ionize*** in solution conduct electric current by the movement of ions.

- **Electrolyte**. A compound which conducts electricity when **molten*** or in **aqueous solution*** and decomposes during electrolysis. All **ionic compounds*** are electrolytes. They conduct electricity because when molten or in solution their ions are free to move. **Cations*** carry a positive charge and **anions*** a negative one. The number of ions in an electrolyte determines how well it conducts electricity.

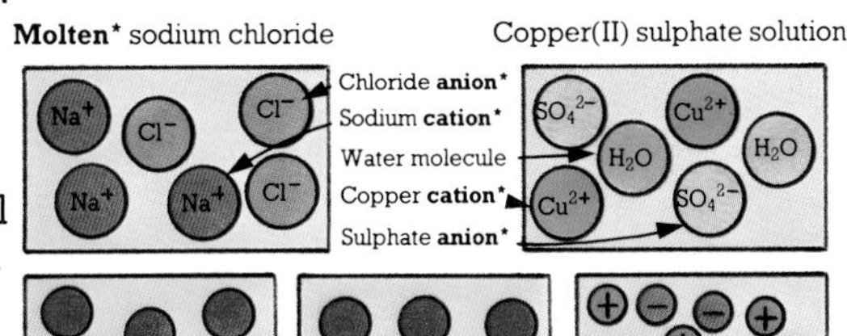

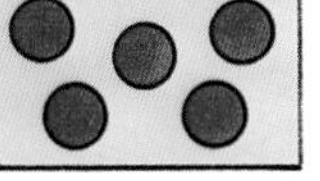

Non-electrolyte. A compound which does not **ionize***.

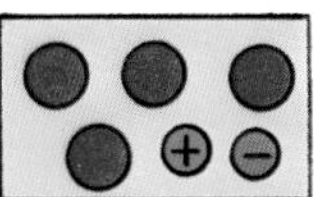

Weak electrolyte. An **electrolyte** which is only partially **ionized***.

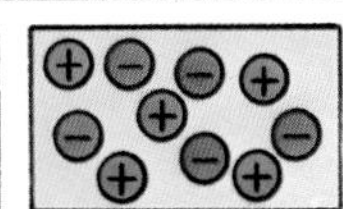

Strong electrolyte. An **electrolyte** which is **ionized*** completely.

- **Electrode**. A piece of metal or graphite placed in an **electrolyte** via which **current*** enters or leaves. There are two electrodes, the **anode** and **cathode**.
- **Inert electrode**. An **electrode** that does not change during electrolysis, e.g. platinum. Some inert electrodes do react with the substances liberated.
- **Active electrode**. An **electrode**, usually a metal, which undergoes chemical change during electrolysis.
- **Electrolytic cell**. A vessel containing the **electrolyte** (either **molten*** or in **aqueous solution***) and the **electrodes**.

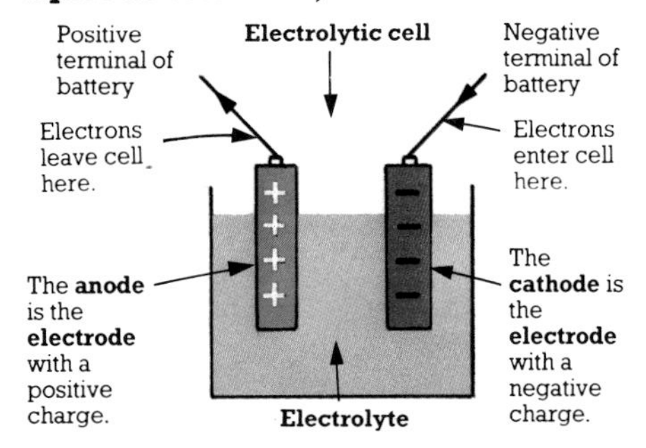

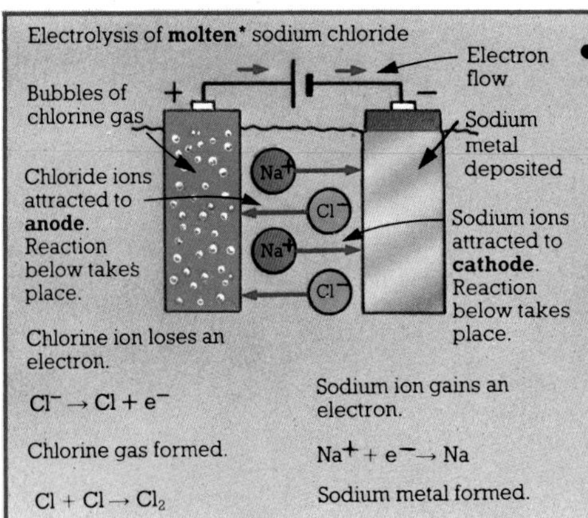

Chlorine ion loses an electron.

$$Cl^- \rightarrow Cl + e^-$$

Chlorine gas formed.

$$Cl + Cl \rightarrow Cl_2$$

Sodium ion gains an electron.

$$Na^+ + e^- \rightarrow Na$$

Sodium metal formed.

- **Ionic theory of electrolysis**. A theory which attempts to explain what happens in an **electrolytic cell** when it is connected to a supply of electricity. It states that **anions*** in the **electrolyte** are attracted to the **anode** (see **electrode**) where they lose electrons. The **cations*** are attracted to the **cathode** where they gain electrons. The ions which react at the electrodes are **discharged**. Electrons flow from the anode to the battery and from the battery to the cathode.

* **Anion**, 130; **Aqueous solution**, 144; **Cation**, 130; **Crystal lattice**, 136; **Current**, 159; **Ionic compound**, 131; **Ionization**, 130; **Molten**, 120.

•Faraday's first law of electrolysis. The mass of a substance produced by chemical reactions at the **electrodes** during electrolysis is proportional to the amount of electricity passed through the **electrolyte**.

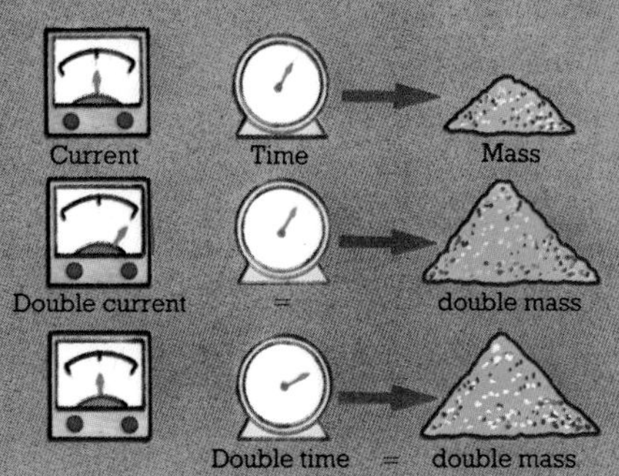

•Voltameter or **coulometer.** A type of **electrolytic cell** used to measure the amount of a substance liberated during electrolysis.

•Coulomb (C). The **SI unit*** of electric charge. One coulomb of electricity passes a point when one **ampere*** flows for one second.

•Faraday's second law of electrolysis. When the same amount of electricity is passed through different **electrolytes**, the number of **moles*** of each element deposited at the **electrodes** is inversely proportional to the size of the charge on its ion.

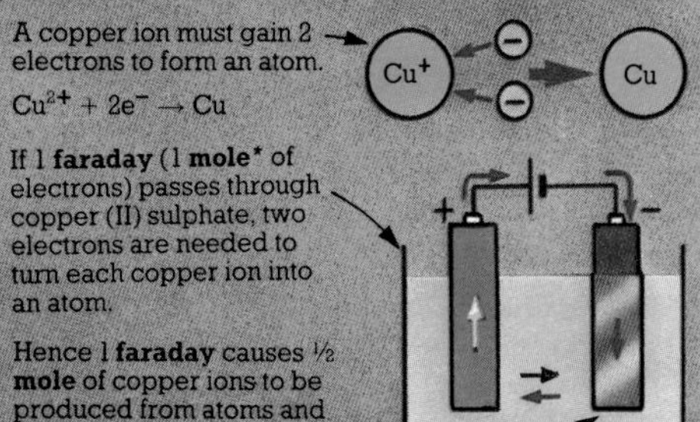

½ is inversely proportional to 2 – charge on ion.

Other examples:	1 **F** produces one **mole*** sodium atoms from sodium ions (Na^+).	1 **F** produces ⅓ **mole*** aluminium atoms from its ions (Al^{3+}).

•Faraday (F). A unit of electric charge equal to 96,500 **coulombs** . It consists of the flow of one **mole*** of electrons and therefore liberates one mole of atoms from singly-charged ions.

Electrolysis in industry

•Electro-refining. A method of purifying metals by electrolysis. Only the metal ions take part in electrolysis, the impurities are lost.

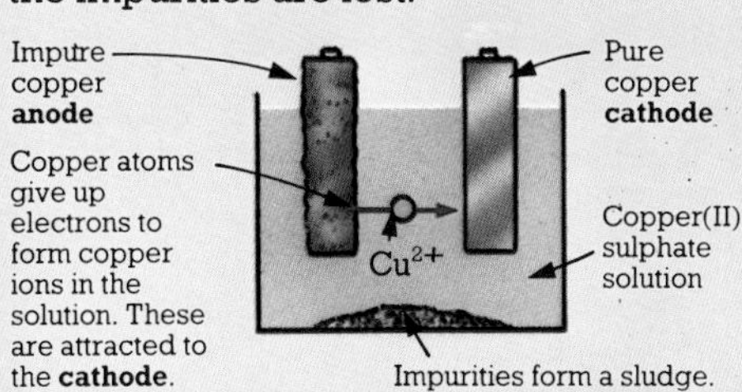

•Metal extraction. A process which produces metals from their **molten*** ores by electrolysis. Metals at the top of the **reactivity series*** are obtained in this way (see **aluminium**, page 176 and **sodium**, page 168).

•Electroplating. The coating of a metal object with a thin layer of another metal by electrolysis. The object forms the **cathode**, on to which metal ions in the **electrolyte** are deposited.

•Anodizing. The coating of a metal object with a thin layer of its oxide. Hydroxide ions are **oxidized*** at the metal **anode** in the electrolysis of dilute sulphuric acid, forming water and oxygen, which oxidizes the metal.

* **Ampere**, 60; **Corrosion**, 209; **Mole**, 139; **Molten**, 120; **Oxidation**, 148; **Reactivity series**, 158; **SI units**, 96.

Reactivity

The **reactivity** of an element depends on its ability to gain or lose the electrons which are used for **bonding** (see pages 130-134). The more reactive an element, the more easily it will combine with others. Some elements are very reactive, others very unreactive. This difference can be used to produce electricity and protect metals from **corrosion***.

- **Reactivity series** or **activity series**. A list of elements (usually metals), placed in order of their reactivity. The series is constructed by comparing the reactions of the metals with other substances, e.g. acids and oxygen (for a summary of reactions, see page 211).

- **Displacement**. A reaction in which one element replaces another in a compound. An element will only displace another lower than itself in the **reactivity series**.

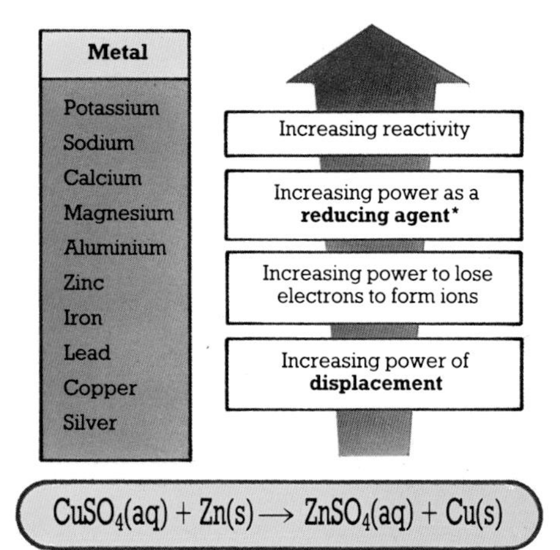

Zinc **displaces** copper from copper(II) sulphate solution. ▶

$$CuSO_4(aq) + Zn(s) \longrightarrow ZnSO_4(aq) + Cu(s)$$

- **Half cell**. An element in contact with water or an **aqueous solution*** of one of its compounds. Atoms on the surface form **cations***, which are released into the solution, leaving electrons behind. The solution has a positive charge and the metal a negative charge, so there is a **potential difference** between them.

- **Electrode potential (E)**. The **potential difference** in a **half cell**. It is impossible to measure directly, so is measured relative to that of another half cell, normally a **hydrogen electrode** (see diagram). Electrode potentials show the ability to **ionize*** in **aqueous solution*** and are used to construct the **electrochemical series**.

Measuring the **electrode potential** of a metal. ▼

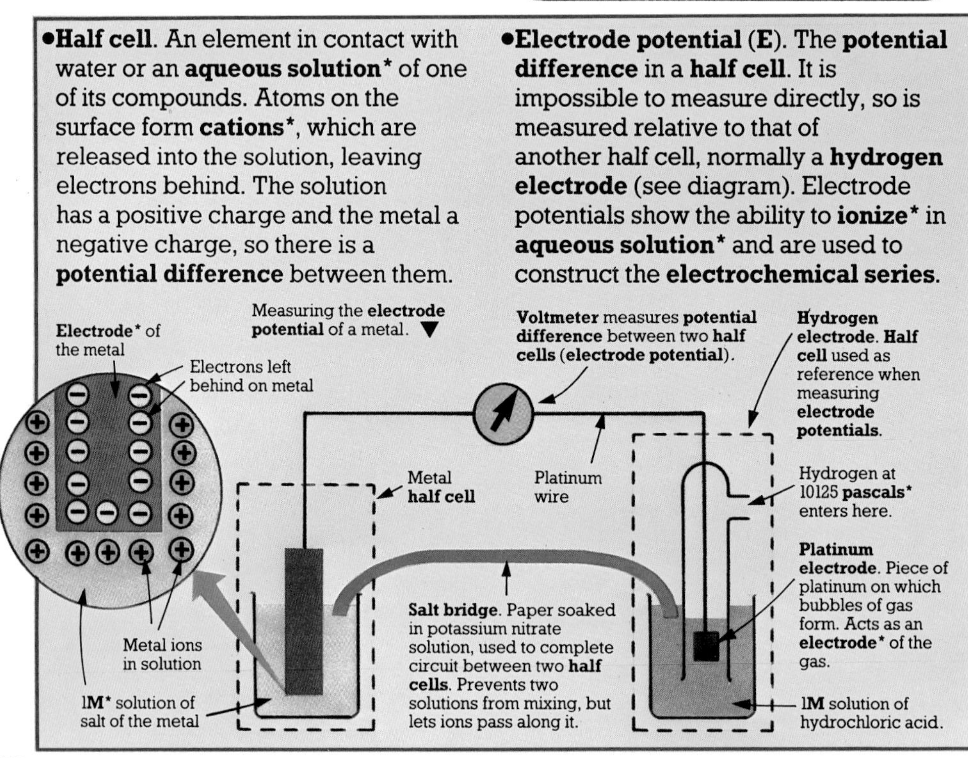

* **Aqueous solution**, 144; **Cation**, 130; **Corrosion**, 209; **Electrode**, 156; **Ionization**, 130; **M-value**, 139 (**Molarity**); **Pascal**, 97; **Reducing agent**, 148.

Electrochemical series. A list of the elements in order of their **electrode potentials**. The element with the most negative electrode potential is placed at the top. The position of an element in the series shows how readily it forms ions in **aqueous solution***, and is thus an indication of how reactive it is likely to be.

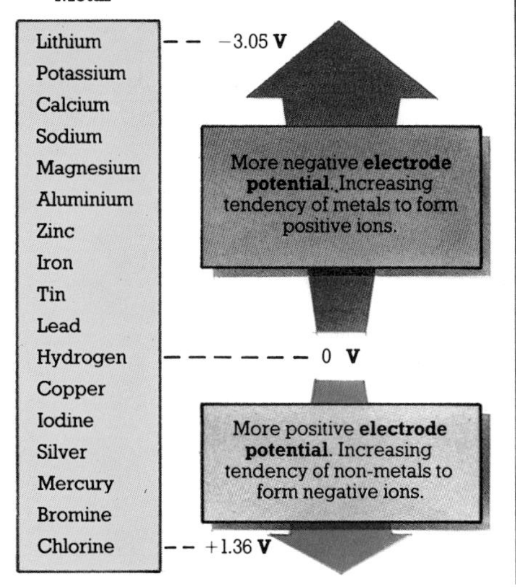

Cell or **electrochemical cell**. An arrangement of two **half cells** of different elements. The half cell with the most negative **electrode potential** forms the **negative terminal** and the other forms the **positive terminal**. When these are connected, a **current** flows between them. There are two types of cell – **primary cells**, which cannot be recharged, and **secondary cells**, which can be reacharged. A **battery** is a number of linked cells.

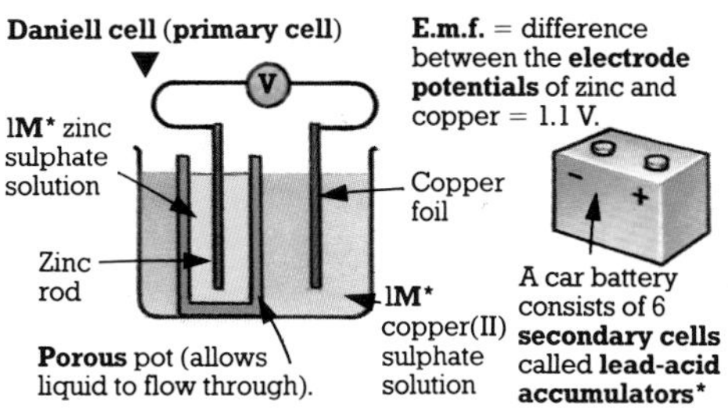

Potential difference or **voltage**. A difference in electric charge between two points, measured in **volts (V)** by an instrument called a **voltmeter**. If two points with a potential difference are joined, an electric **current**, proportional to the potential difference, flows between them.

Current. A flow of electrons (negatively-charged particles) through a material. The **SI unit*** of current is the **ampere*** (**A**), and current is measured using an **ammeter**. A current will flow in a loop, or **circuit**, between two points if there is a **potential difference** between them.

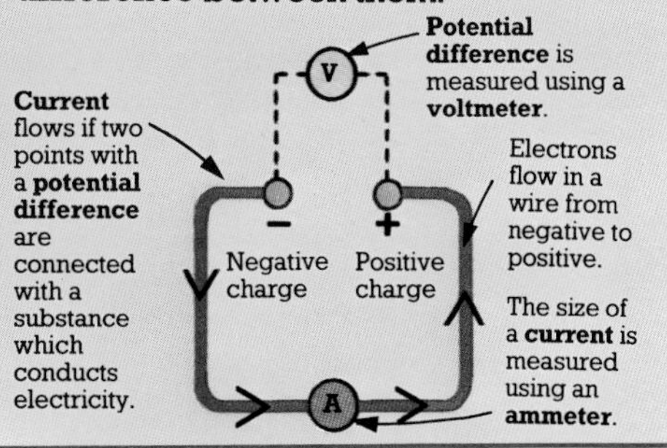

Electromotive force (e.m.f.). The name given to the **potential difference** between the two terminals of a **cell** (i.e. the difference between the **electrode potentials** of the two **half cells**).

Sacrificial protection. Also known as **cathodic protection** or **electrical protection**. A method of preventing iron from **rusting*** by attaching a metal higher in the **electrochemical series** to it, which rusts instead.

Iron pipes are protected by attaching scrap aluminium to them.

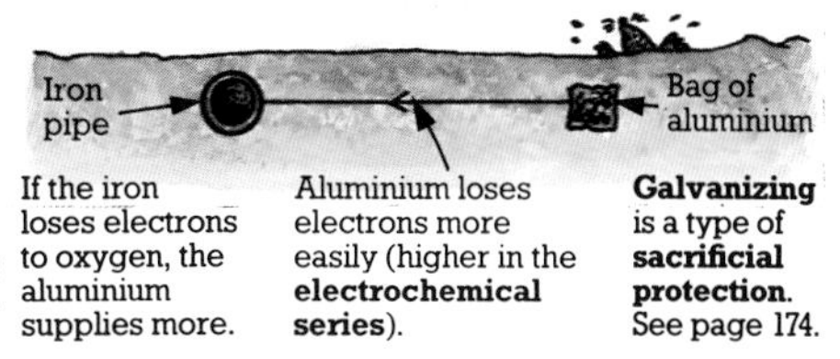

* **Ampere**, 60; **Aqueous solution**, 144; **Lead-acid accumulator**, 69; **M-value**, 139 (**Molarity**); **Rusting**, 209 (**Corrosion**); **SI units**, 96.

Rates of reaction

The time it takes for a chemical reaction to finish varies from less than a millionth of a second to weeks or even years. It is possible to predict how long a particular reaction will take and how to speed it up or slow it down by altering the conditions under which it takes place. The efficiency of many industrial processes is improved by increasing the **rate of reaction**, e.g. by using high temperature and pressure, or a **catalyst**.

•**Rate of reaction**. A measurement of the speed of a reaction. It is calculated by measuring how quickly reactants are used up or products are formed. The experimental method used to measure the rate of reaction depends on the **physical states*** of the reactants and products, and the data from such an experiment is plotted on a **rate curve**. The speed of a reaction varies as it proceeds. The rate at any time during the reaction is the **instantaneous rate**. The instantaneous rate at the start of the reaction is the **initial rate**. The **average rate** is calculated by dividing the total change in the amount of products or reactants by the time the reaction took to finish.

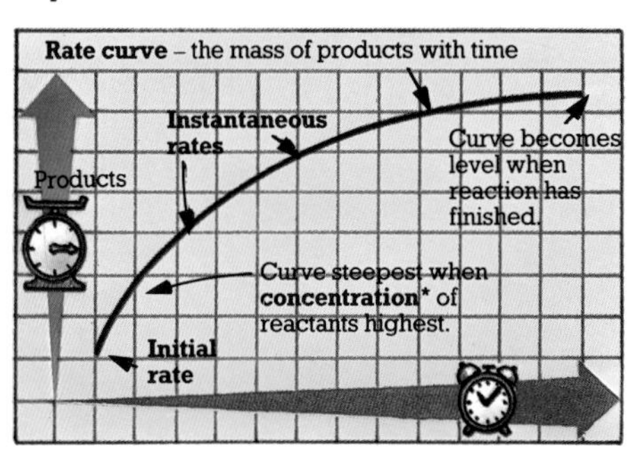

•**Collision theory**. Explains why altering the conditions under which a reaction takes place affects its rate. For a reaction to take place between two particles, they must collide, so if more collisions occur, the **rate of reaction** increases. However, only some collisions cause a reaction, since not all particles have enough energy to react (see **activation energy**).

•**Photochemical reaction**. A reaction whose speed is affected by the intensity of light, e.g. **photosynthesis***. Light gives reacting particles more energy and so increases the **rate of reaction**.

Photochemical reactions occur in photography.

$$2AgCl(s) \rightarrow 2Ag(s) + Cl_2(g)$$

Silver crystals form where light falls on the film, recording the picture.

•**Activation energy (E)**. The minimum energy that the particles of reactants must have for them to react when they collide (see **collision theory**). The **rate of reaction** depends on how many reacting particles have this minimum energy. In many reactions, the particles already have this energy and react straight away. In others, energy has to be supplied for the particles to reach the activation energy.

Friction produces heat, giving particles in match **activation energy**.

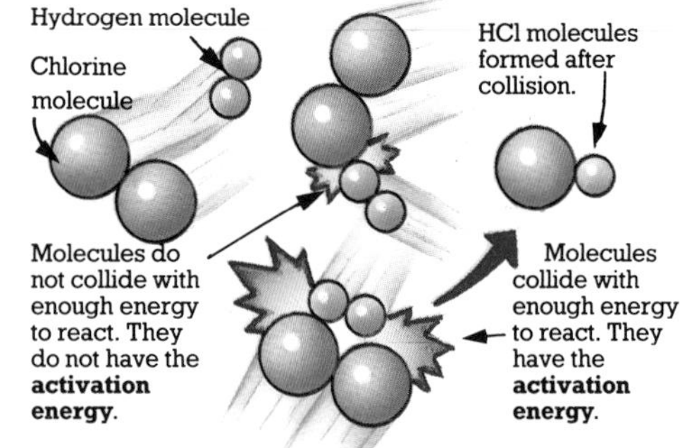

* **Concentration**, 139; **Physical states**, 120; **Photosynthesis**, 209.

Changing rates of reaction

The **rate of reaction** will increase if the temperature is increased. The heat energy gives more particles an energy greater than the **activation energy**.

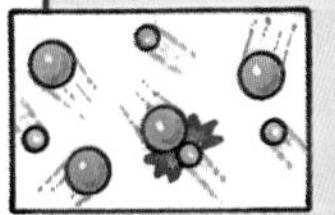

Particles collide with greater energy, so more react.

For reactions involving gases, the **rate of reaction** will increase if the pressure is increased. An increase in the pressure of a gas increases the temperature and decreases the volume (i.e. increases the **concentration*** – see also **gas laws**, page 142). The particles collide more often and with greater energy.

The **rate of reaction** will increase if the **concentration*** of one or more of the reactants is increased.

More molecules in the same space means more collisions.

Low **concentration***

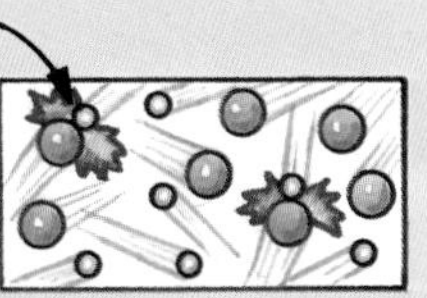

High **concentration***

The **rate of reaction** will increase if the surface area of a solid reactant is increased. Reactions in which one reactant is a solid can only take place at the surface of the solid.

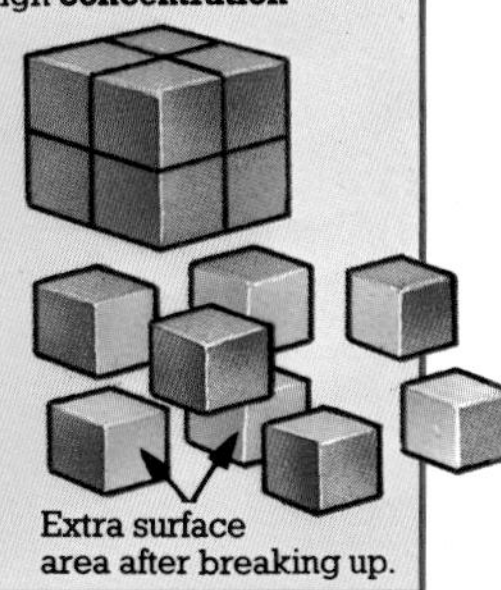

Extra surface area after breaking up.

- **Catalyst**. A substance that increases the rate of a chemical reaction, but is chemically unchanged itself at the end of the reaction. This process is known as **catalysis**. Catalysts work by lowering the **activation energy** of a reaction. The catalyst used in a reaction is written over the arrow in the equation (see page 182). A catalyst which increases the rate of one reaction may have no effect on another.

Catalysts provide an alternative chemical route for a reaction to take.

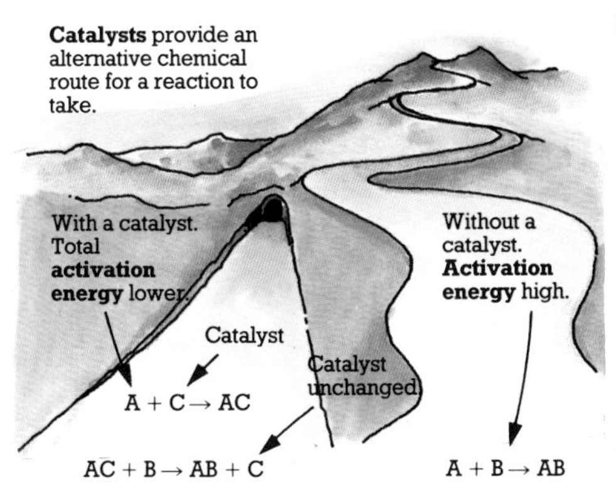

- **Autocatalysis**. A process in which one of the products of a reaction acts as a **catalyst** for the reaction.

- **Surface catalyst**. A **catalyst** which attracts the reactants to itself. It holds them close to each other on its surface, so they react easily.

- **Promoter**. A substance which increases the power of a **catalyst**, so speeding up the reaction.

- **Inhibitor**. A substance that slows a reaction. Some work by reducing the power of a **catalyst**.

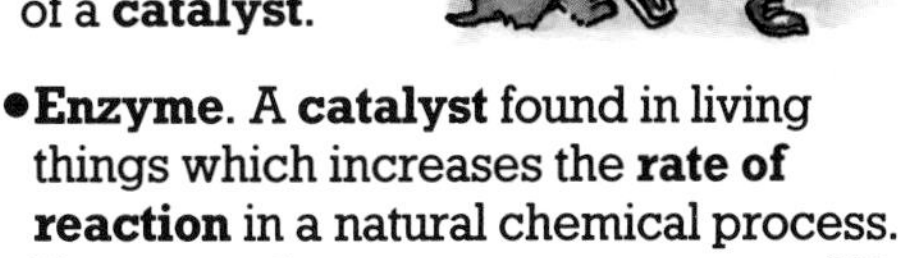

- **Enzyme**. A **catalyst** found in living things which increases the **rate of reaction** in a natural chemical process. For more about enzymes, see page 331.

- **Homogenous catalyst**. A **catalyst** in the same **physical state*** as the reactants.

- **Heterogenous catalyst**. A **catalyst** in a different **physical state*** to the reactants.

* **Concentration**, 139; **Physical states**, 120.

Reversible reactions

Many chemical reactions continue until one or all of the reactants are used up, and their products do not react together. This is known as **completion**. Other reactions, however, never reach this stage. They are called **reversible reactions**.

•**Reversible reaction**. A chemical reaction in which the products react together to reform the original reactants. These react again to form the products and so on. The two reactions are simultaneous, and the process will not come to **completion** (see introduction) if it takes place in a **closed system**. At some stage during a reversible reaction, **chemical equilibrium** is reached.

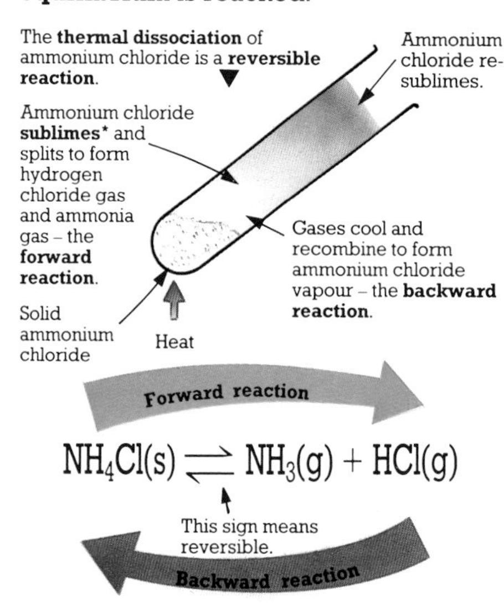

•**Forward reaction**. The reaction in which products are formed from the original reactants in a **reversible reaction**. It goes from left to right in the equation.

•**Backward reaction** or **reverse reaction**. The reaction in which the original reactants are reformed from their products in a **reversible reaction**. It goes from right to left in the equation.

•**Dissociation**. A type of **reversible reaction** in which a compound is divided into other compounds or elements. **Thermal dissociation** is dissociation caused by heating (the products formed recombine when cooled). Dissociation should not be confused with **decomposition**, in which a compound is irreversibly split up.

Nitrogen dioxide undergoes **thermal dissociation** into nitrogen monoxide and oxygen.

Brown nitrogen dioxide gas.

Increase temperature – colour fades.

Heat

Cool

Decrease temperature – gases recombine.

Nitrogen monoxide and oxygen are colourless.

$$2NO_2(g) \rightleftharpoons 2NO(g) + O_2(g)$$

•**Closed system**. A **system*** in which no chemicals can escape or enter. If a product of a **reversible reaction** escapes, for example into the atmosphere, the reaction can no longer move back the other way. A system from which chemicals can escape is an **open system**.

•**Equilibrium**. The cancelling out of two equal but opposite movements. **Chemical equilibrium** is an example of this – it occurs when the **forward** and **backward reactions** are taking place, but are cancelling each other out.

A person walking up an escalator at the same speed as the escalator is moving down is in **equilibrium**.

* **Sublimation**, 121; **System**, 345.

• **Chemical equilibrium**. A stage reached in a **reversible reaction** in a **closed system** when the **forward** and **backward reactions** take place at the same rate. Their effects cancel each other out, and the **concentrations*** of the reactants and products no longer change. Chemical equilibrium is a form of **equilibrium**.

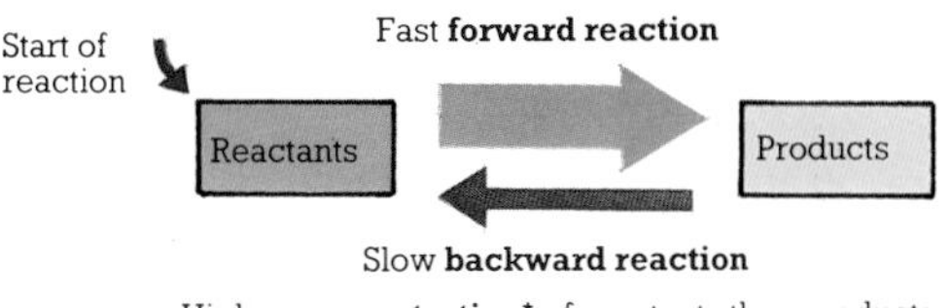

Higher **concentration*** of reactants than products

At **chemical equilibrium**

Products and reactants formed at same rate.

The position of chemical equilibrium

Any change of conditions (temperature, **concentration*** or pressure) during a **reversible reaction** alters the rate of either the **forward** or **backward reaction**, destroying the **chemical equilibrium**. This is eventually restored, but with a different proportion of reactants and products. The **equilibrium position** is said to have changed.

First **equilibrium position** ▶

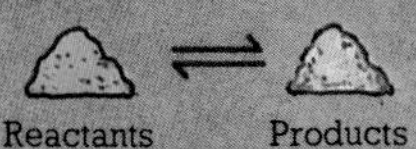

Alter conditions to favour **forward reaction** – **equilibrium position** said to move right.

More products formed.

Alter conditions to favour **backward reaction** – **equilibrium position** is said to move left.

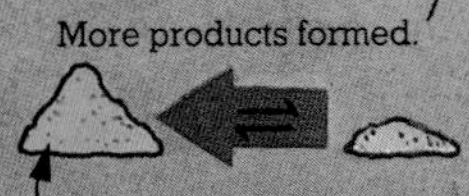

• **Le Chatelier's principle**. A law stating that if changes are made to a **system*** in equilibrium, the system adjusts itself to reduce the effects of the change.

Changing the pressure in **reversible reactions** involving gases may alter the equilibrium position.

In the reaction $A(g) + B(g) \rightleftharpoons AB(g)$:

Molecule of A

Molecule of B

Molecule of AB

Raise pressure – position moves right – more AB formed – i.e. number of molecules decreases to lower pressure again.

Lower pressure – position moves left – more A and B formed – i.e. number of molecules increases to raise pressure again.

Changing the temperature in a **reversible reaction** also alters the equilibrium position. This depends on whether the reaction is **exothermic*** or **endothermic***. A reversible reaction which is exothermic in one direction is endothermic in the other.

Ammonia is made by the **Haber process***.

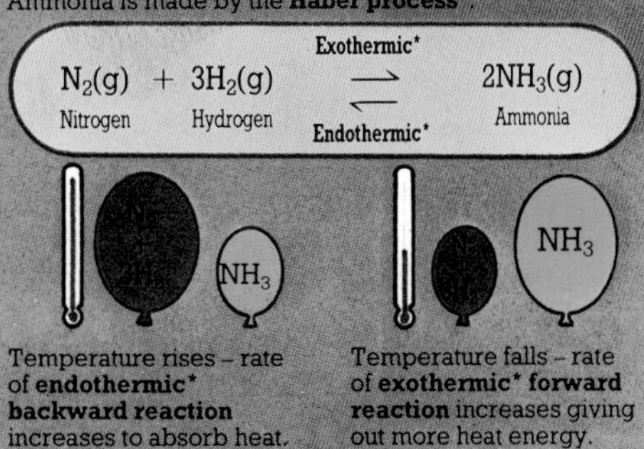

Temperature rises – rate of **endothermic* backward reaction** increases to absorb heat. Less ammonia formed – position moves left.

Temperature falls – rate of **exothermic* forward reaction** increases giving out more heat energy. More ammonia produced – position moves right.

Changing the **concentration*** of the reactants or products in a **reversible reaction** also changes the equilibrium position.

Raise **concentration*** of reactants – increases rate of **forward reaction**. OR Lower **concentration** of products – decreases rate of **backward reaction**.

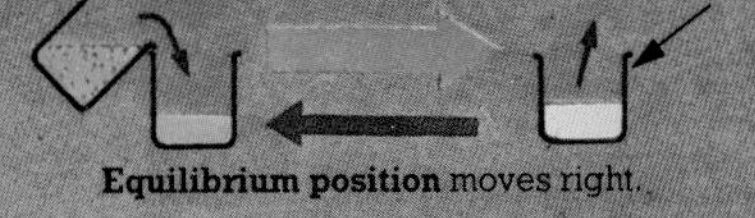

Equilibrium position moves right.

Lower **concentration** of reactants – decreases rate of **forward reaction**. OR Raise **concentration** of products – increases rate of **backward reaction**.

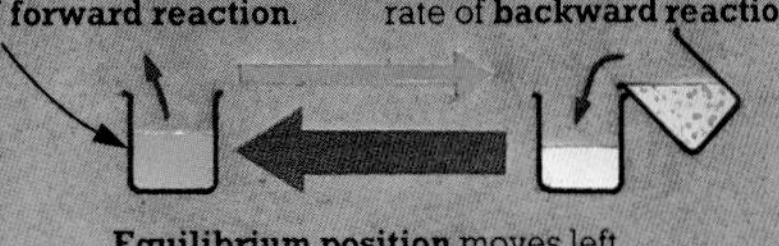

Equilibrium position moves left.

* **Concentration**, 139; **Endothermic reaction**, **Exothermic reaction**, 147; **Haber process**, 180; **System**, 345.

The periodic table

During the 19th century, many chemists tried to arrange the elements in an order which related to the size of their atoms and also showed regular repeating patterns in their behaviour or properties. The most successful attempt was published by the Russian, Dimitri Mendeléev, in 1869, and still forms the basis of the modern **periodic table**.

•**Periodic table**. An arrangement of the elements in order of their **atomic numbers***. Both the physical properties and chemical properties of an element and its compounds are related to the position of the element in the periodic table. This relationship has led to the table being divided into **groups** and **periods**. The arrangement of elements starts on the left of period 1 with hydrogen and moves in order of increasing atomic number from left to right across each period in turn (see picture on the right).

Periodic table

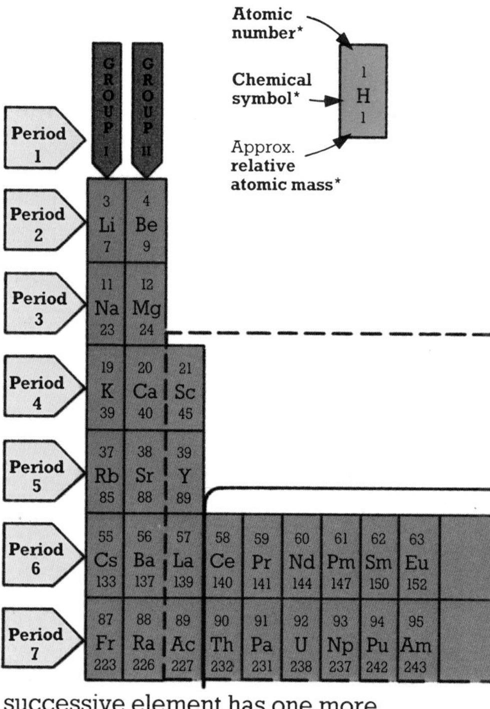

•**Period**. A horizontal row of elements in the **periodic table**. There are seven periods in all. Period 1 has only two elements – hydrogen and helium. Periods 2 and 3 each contain eight elements and are called the **short periods**. Periods 4, 5, 6 and 7 each contain between 18 and 32 elements. They are called the **long periods**. Moving from left to right across a period, the **atomic number*** increases by one from one element to the next. Each successive element has one more electron in the **outer shell*** of its atoms. All elements in the same period have the same number of shells, and the regular change in the number of electrons from one element to the next leads to a fairly regular pattern of change in the chemical properties of the elements across a period. For an example of such a gradual change in property, see below.

Electron configuration* of elements across **Period** 2

All elements have the same **outer shell***, but each successive element, going from left to right, has one more electron added to that shell.

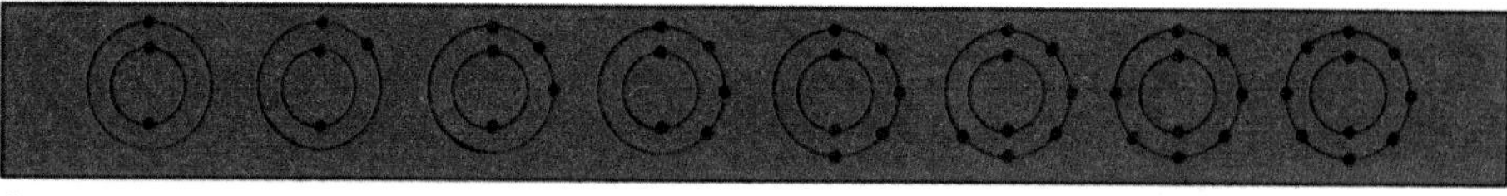

Strong **reducing agents*** ⟶ Weak **reducing agents** ⟶ Strong **oxidizing agents***

This shows a regular pattern of change across **Period 2** in the ability of elements to **reduce*** or **oxidize*** other elements and compounds (see also page 166). Neon is the exception – it is unreactive.

* **Atomic number**, 127; **Chemical symbol**, 122; **Electron configuration**, **Outer shell**, 127; **Oxidation**, **Oxidizing agent**, **Reducing agent**, **Reduction**, 148; **Relative atomic mass**, 138.

•**Group**. A vertical column of elements in the **periodic table**. All groups are numbered (except for **transition metal*** groups) using roman numerals and some have names. Elements in the same group have the same number of electrons in their **outer shell***, and so have similar chemical properties.

Groups with alternative names:

Group number	Group name
Group I	The **alkali metals** (see pages 168-169)
Group II	The **alkaline-earth metals** (see pages 170-171)
Group VII	The **halogens** (see pages 186-188)
Group VIII (or **Group 0**)	The **noble gases** (see page 189)

Colour-coding used in table.

Metals | Metalloids | Non-metals

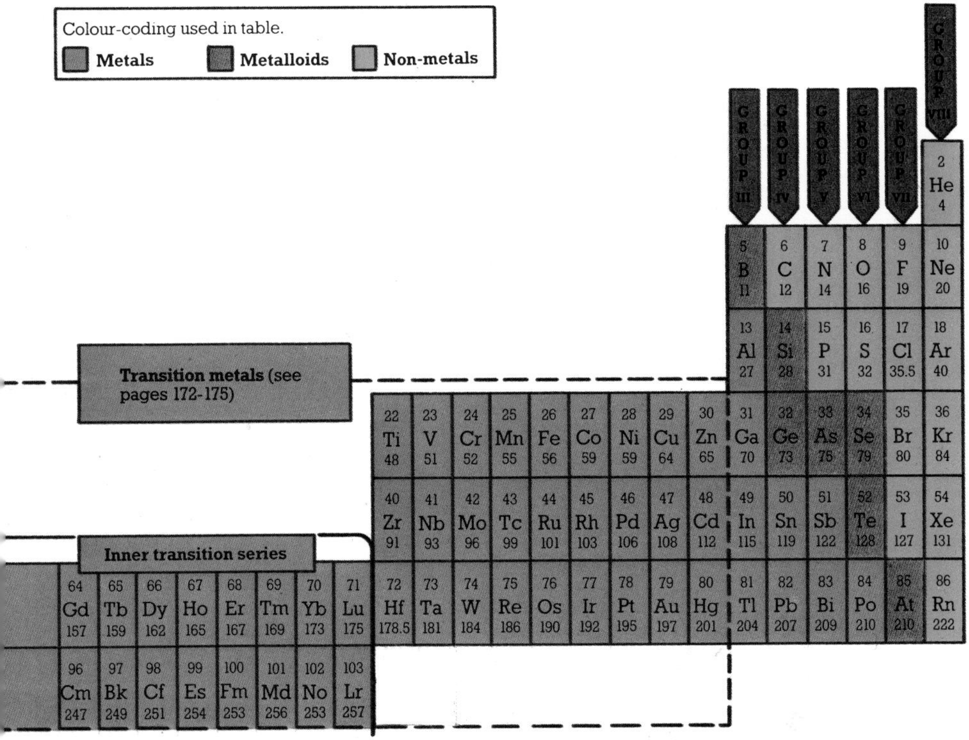

Transition metals (see pages 172-175)

Inner transition series

																	GROUP III	GROUP IV	GROUP V	GROUP VI	GROUP VII	GROUP VIII
																						2 He 4
																	5 B 11	6 C 12	7 N 14	8 O 16	9 F 19	10 Ne 20
																	13 Al 27	14 Si 28	15 P 31	16 S 32	17 Cl 35.5	18 Ar 40
								22 Ti 48	23 V 51	24 Cr 52	25 Mn 55	26 Fe 56	27 Co 59	28 Ni 59	29 Cu 64	30 Zn 65	31 Ga 70	32 Ge 73	33 As 75	34 Se 79	35 Br 80	36 Kr 84
								40 Zr 91	41 Nb 93	42 Mo 96	43 Tc 99	44 Ru 101	45 Rh 103	46 Pd 106	47 Ag 108	48 Cd 112	49 In 115	50 Sn 119	51 Sb 122	52 Te 128	53 I 127	54 Xe 131
64 Gd 157	65 Tb 159	66 Dy 162	67 Ho 165	68 Er 167	69 Tm 169	70 Yb 173	71 Lu 175	72 Hf 178.5	73 Ta 181	74 W 184	75 Re 186	76 Os 190	77 Ir 192	78 Pt 195	79 Au 197	80 Hg 201	81 Tl 204	82 Pb 207	83 Bi 209	84 Po 210	85 At 210	86 Rn 222
96 Cm 247	97 Bk 249	98 Cf 251	99 Es 254	100 Fm 253	101 Md 256	102 No 253	103 Lr 257															

Metals and non-metals

•**Metal**. An element with characteristic physical properties that distinguish it from a **non-metal**. Elements on the left of a **period** have metallic properties. Moving to the right, the elements gradually become less metallic. Elements that are not distinctly metal or non-metal, but have a mixture of properties, are called **metalloids**. Elements to the right of metalloids are non-metals.

Property	Metal	Non-metal
Physical state*	Solids (except mercury)	Solid, liquid or gas (bromine is the only liquid).
Appearance	Shiny	Mainly non-shiny (iodine is one of the exceptions).
Conductivity*	Good	Poor (except graphite)
Malleability*	Good	Poor
Ductility*	Good	Poor
Melting point	Generally high	Generally low
Boiling point	Generally high	Generally low

* **Conductivity**, 62; **Ductility**, 344 (**Ductile**); **Malleability**, 345 (**Malleable**); **Outer shell**, 127; **Physical states**, 120; **Transition metals**, 172.

Inorganic chemistry

Inorganic chemistry is the study of all the elements and their compounds except those compounds made of chains of carbon atoms (see **organic chemistry**, pages 190-205). The properties and reactions of inorganic elements and compounds follow certain patterns, or **trends**, in the **periodic table***. By looking up and down **groups*** and across **periods*** of the table, it is possible to predict the reactions of elements.

Major periodic table trends

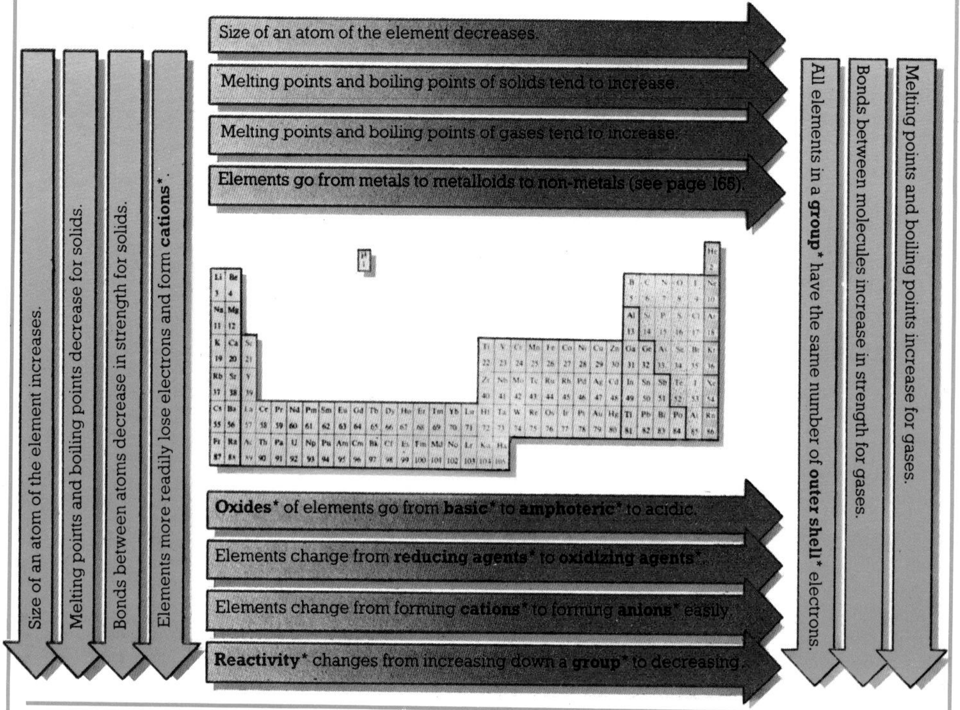

Predicting reactions

Throughout the inorganic section of this book, each **group*** of elements has an introduction and chart which summarizes some of the properties of the group's elements. Below the charts are blue boxes which highlight trends going down the group. After the introduction more common group members are defined. Information on the other members of the group can often be worked out from trends in **reactivity*** going down the group.

The following steps show how to predict the **reactivity*** of caesium with cold water. Caesium is in Group I (see pages 168-169).

1. The chart introducing Group I shows that the reactivity of the elements increases going down the group.

2. From the definitions of lithium, sodium and potassium it is found that all three elements react with water with increasing violence going down the group – lithium reacts gently, sodium reacts violently and potassium reacts very violently.

It is predicted that caesium, as it comes after potassium going down the group, will react extremely violently with water.

* **Amphoteric**, 151; **Anion**, 130; **Basic**, 151 (**Base**); **Cation**, 130; **Group**, 165; **Outer shell**, 127; **Oxides**, 183; **Oxidizing agent**, 148; **Period**, 164; **Periodic table**, 164; **Reactivity**, 158; **Reducing agent**, 148.

Hydrogen

Hydrogen (**H_2**), with an **atomic number*** of one, is the first and lightest element in the periodic table, and the commonest in the universe. It is a **diatomic***, odourless, inflammable gas which only occurs naturally on Earth in compounds. It is made by the reaction of natural gas and steam at high temperatures, or the reaction of **water gas*** and steam over a **catalyst***. It is a **reducing agent***, burns in air with a light blue flame and, when heated, reacts with many substances, e.g. with sodium to form sodium hydride (all compounds of hydrogen and one other element are **hydrides**). Hydrogen is used, for example, to make margarines (see **hydrogenation**, page 193) and ammonia (see **Haber process**, page 180), and as a rocket fuel. See also pages 217 and 219.

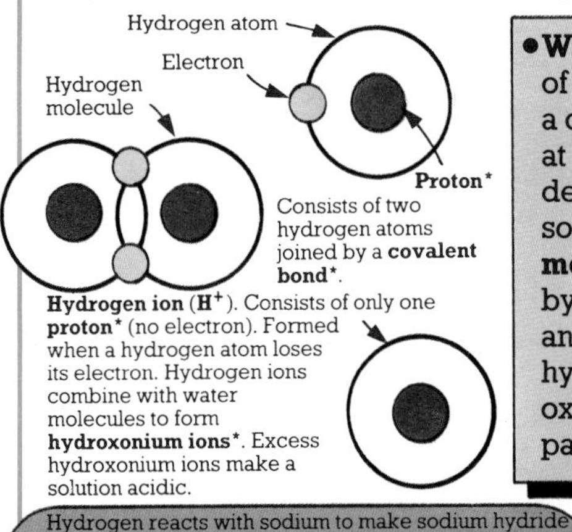

Consists of two hydrogen atoms joined by a **covalent bond***.

Hydrogen ion (H^+). Consists of only one **proton*** (no electron). Formed when a hydrogen atom loses its electron. Hydrogen ions combine with water molecules to form **hydroxonium ions***. Excess hydroxonium ions make a solution acidic.

• **Water** (**H_2O**). An oxide of hydrogen and one of the commonest compounds on Earth. It is a colourless, odourless liquid which freezes at 0°C, boils at 100°C, has its greatest density ($1\ g\ cm^{-3}$) at 4°C and is the best solvent known. It is made of **polar molecules*** linked by **hydrogen bonds*** and is formed when hydrogen burns in oxygen. See also pages 206 and 218.

Ice (frozen **water**) floats on liquid water as it is less dense (see also page 206).

Hydrogen reacts with sodium to make sodium hydride

$$2Na(s) + H_2(g) \rightarrow 2NaH(s)$$

Sodium + Hydrogen → Sodium hydride

Hydrogen is a **reducing agent***

$$CuO(s) + H_2(g) \longrightarrow Cu(s) + H_2O(l)$$

Copper(II) oxide + Hydrogen → Copper + Water

• **Hydroxide**. A compound made of a **hydroxide ion** (**OH^-**) and a **cation***. Solutions containing more OH^- ions than H^+ ions are alkaline. Many hydroxides are not water-soluble, e.g. **lead(II) hydroxide ($Pb(OH)_2$)**. However, the hydroxides of Group I elements and some others are water-soluble.

• **Deuterium** (**D** or **^2_1H**). An **isotope*** of hydrogen with one **proton*** and one **neutron***. It makes up 0.0156% of natural hydrogen. Water molecules containing deuterium are called **deuterium oxide** (**D_2O**) or **heavy water** molecules. Heavy water is used in nuclear reactors to slow the fast moving neutrons.

• **Hydrogen peroxide** (**H_2O_2**). A syrupy liquid. It is an oxide of hydrogen and a strong **oxidizing agent***. It is sold in solution as disinfectant and bleach.

• **Tritium** (**T** or **^3_1H**). An **isotope*** of hydrogen with one **proton*** and two **neutrons***. It is rare in hydrogen but is produced by nuclear reactors. It is **radioactive**, emitting **beta particles** (see pages 128-129)

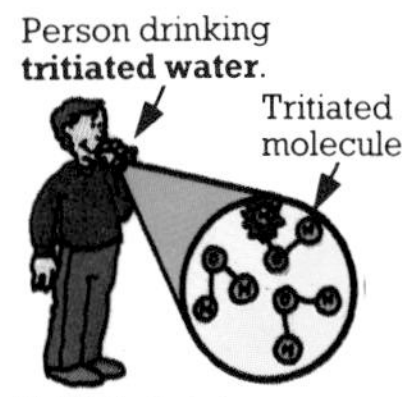

Tritiated water contains some water molecules in which a hydrogen atom has been replaced by a **tritium** atom. It is used by doctors to help find out how much fluid a patient passes.

* **Atomic number**, 127; **Catalyst**, 161; **Cation**, 130; **Covalent bond**, 132; **Diatomic**, 124; **Hydrogen bond**, 134; **Hydroxonium ion**, 150; **Isotope**, 127; **Neutron**, 126; **Oxidizing agent**, 148; **Polar molecule**, 133; **Proton**, 126; **Reducing agent**, 148; **Water gas**, 179 (**Carbon monoxide**).

Group I, the alkali metals

The elements in **Group I** of the periodic table are called **alkali metals** as they are all metals which react with water to form alkaline solutions. They all have similar chemical properties and their physical properties follow certain patterns. The chart below shows some of their properties.

Some properties of Group I elements

Name of element	Chemical symbol	Relative atomic mass*	Electron configuration*	Reactivity	Appearance	Uses
Lithium	Li	6.94	2,1	INCREASING ↓	Silver-white metal	See below.
Sodium	Na	22.99	2,8,1		Soft silver-white metal	See below.
Potassium	K	39.10	2,8,8,1		Soft silver-white metal	See page 169
Rubidium	Rb	85.47	Complex configuration, but still one outer electron		Soft silver-white metal	To make special glass.
Caesium	Cs	132.90			Soft metal with gold sheen	In **photocells*** and as a **catalyst***.
Francium	Fr	No known stable isotope*				

The atoms of all Group I elements have one electron in their **outer shell***, hence the elements are powerful **reducing agents*** because this electron is easily lost in reactions. The resulting ion has a charge of +1 and is more stable because its new outer shell is complete (see **octet**, page 127). All Group I elements react in this way to form **ionic compounds***.

Going down the group, the reaction of the elements with water gets more violent. The first three tarnish in air and **rubidium** and **caesium** catch fire. All members are stored under oil because of their reactivity.

These two pages contain more information on **lithium**, **sodium**, **potassium** and their compounds. They are typical Group I elements.

•**Lithium** (**Li**). The least reactive element in Group I of the periodic table and the lightest solid element. Lithium is rare and is only found in a few compounds, from which it is extracted by **electrolysis***. It burns in air with a pinkish flame. A piece of lithium placed on water glides across the surface, fizzing gently.

$$2Li(s) + 2H_2O(l) \rightarrow 2LiOH(aq) + H_2(g)$$

Lithium + Water → Lithium hydroxide + Hydrogen

After the reaction, the solution is strongly alkaline, due to the **lithium hydroxide** formed.

Lithium reacts vigorously with chlorine to form **lithium chloride** (**LiCl**) which is used in welding flux and air conditioners.

•**Sodium** (**Na**). A member of Group I of the periodic table, found in many compounds. Its main ore is **rock salt** (containing **sodium chloride** – see also **potassium**). It is extracted from molten sodium chloride by **electrolysis***, using a **Downs' cell**. Sodium burns in air with an orange flame and reacts violently with non-metals and water (see equation for **lithium**, and substitute Na for Li). It is used in sodium vapour lamps and as a coolant in nuclear power stations.

Downs' cell. Used to extract **sodium** from molten **sodium chloride** by **electrolysis***.

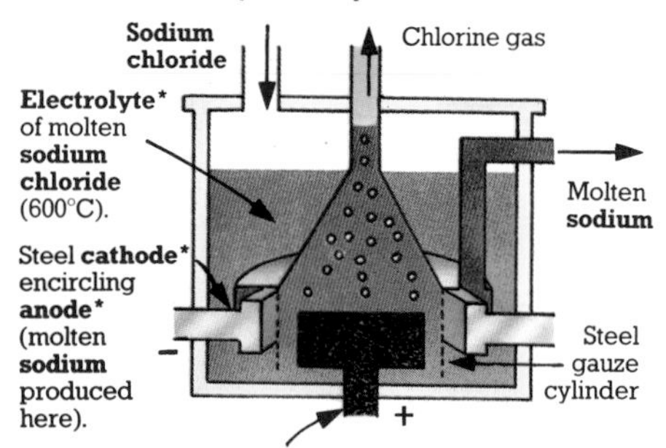

* **Anode**, 156 (**Electrode**); **Catalyst**, 161; **Cathode**, 156 (**Electrode**); **Electrolysis, Electrolyte**, 156; **Electron configuration**, 127; **Ionic compound**, 131; **Isotope, Outer shell**, 127; **Photocell**, 345; **Reducing agent**, 148; **Relative atomic mass**, 138.

•**Sodium hydroxide** ($NaOH$) or **caustic soda**. A white, **deliquescent*** solid, produced by **electrolysis*** of brine. A **strong base***, it reacts with acids to form a sodium **salt*** and water. It is used to make soaps and paper.

•**Sodium carbonate** (Na_2CO_3). A white solid that dissolves in water to form an alkaline solution. Its **hydrate***, called **washing soda** (**$Na_2CO_3.10H_2O$** – see also page 207), has white, **efflorescent*** crystals and is made when ammonia, water and **sodium chloride** react with carbon dioxide in the **Solvay process**.

Washing soda is used in the making of glass, as a **water softener*** and in bath crystals.

•**Sodium hydrogencarbonate** (**$NaHCO_3$**). Also called **sodium bicarbonate** or **bicarbonate of soda**. A white solid made by the **Solvay process** (see **sodium carbonate**). In water it forms a weak alkaline solution.

Sodium hydrogencarbonate is used in baking. The carbon dioxide gas it gives off when heated makes dough rise. It is also used as an **antacid*** to relieve indigestion.

$$2NaHCO_3(s) \rightarrow Na_2CO_3(s) + H_2O(l) + CO_2(g)$$

Sodium hydrogencarbonate → Sodium carbonate + Water + Carbon dioxide

•**Sodium chloride** (**$NaCl$**) or **salt**. A white solid which occurs in sea water and **rock salt** (see **sodium**). It forms **brine** when dissolved in water and is used to make **sodium hydroxide** and **sodium carbonate**.

Sodium chloride is used to preserve and flavour food.

•**Sodium nitrate** (**$NaNO_3$**) or **Chile saltpetre**. A white solid used as a fertilizer and also to preserve meat.

•**Potassium** (**K**). A member of Group I of the periodic table. Potassium compounds are found in sea water and **rock salt** (containing **potassium chloride** – see also **sodium**). Potassium is extracted from molten potassium chloride by **electrolysis***. It is very reactive, reacting violently with chlorine and also with water (see equation for **lithium**, and substitute K for Li). It has few uses, but some of its compounds are important.

Potassium reacting with water. It whizzes across the water giving off so much heat that the hydrogen produced bursts into flames.

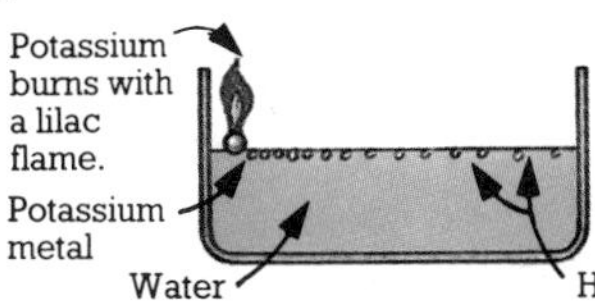

•**Potassium hydroxide** (**KOH**) or **caustic potash**. A white, **deliquescent*** solid. It is a **strong base*** which reacts with acids to form a potassium **salt*** and water. It is used to make toilet soap (see page 202).

•**Potassium carbonate** (**K_2CO_3**). A white solid which is very water-soluble, forming an alkaline solution. It is used to make glass, dyes and soap.

•**Potassium chloride** (**KCl**). A white, water-soluble solid. Large amounts are found in sea water and **rock salt** (see **potassium**). It is used in fertilizer and to produce **potassium hydroxide**.

•**Potassium nitrate** (**KNO_3**), or **saltpetre**. A white solid which dissolves in water to form a **neutral*** solution.

Potassium nitrate is used:

To preserve meats.

In explosives

In fertilizers

•**Potassium sulphate** (**K_2SO_4**). A white solid, forming a **neutral*** solution in water. It is an important fertilizer.

* **Antacid**, 344; **Deliquescent**, **Efflorescent**, 206; **Electrolysis**, 156; **Hydrate**, 155; **Neutral**, 151; **Salts**, 153; **Strong base**, 152; **Water softeners**, 207.

Group II, the alkaline-earth metals

The elements in **Group II** of the periodic table are called the **alkaline-earth metals**. The physical properties of the members of Group II follow certain trends, and, except **beryllium**, they all have similar chemical properties. They are very reactive, though less reactive than Group I elements. The chart below shows some of their properties.

Some properties of Group II elements

Name of element	Chemical symbol	Relative atomic mass*	Electron configuration*	Reactivity	Appearance	Uses
Beryllium	**Be**	9.01	2,2	INCREASING ↓	Hard, white metal	In light, corrosion-resistant **alloys***
Magnesium	**Mg**	24.31	2,8,2		Silver-white metal	See below
Calcium	**Ca**	40.31	2,8,8,2		Soft, silver-white metal	See below
Strontium	**Sr**	87.62	Complex configuration, but still 2 outer electrons		Soft, silver-white metal	In fireworks
Barium	**Ba**	137.34			Soft, silver-white metal	In fireworks and medicine
Radium	**Ra**	Rare **radioactive*** metal			Soft, silver-white metal	An **isotope*** is used to treat cancer.

The atoms of all Group II elements have two electrons in their **outer shell***, hence the elements are good **reducing agents*** because these electrons are fairly easily lost in reactions. Each resulting ion has a charge of +2 and is more stable because its new outer shell is complete (see **octet**, page 127). All Group II elements react this way to form **ionic compounds***, though some **beryllium** compounds have **covalent*** properties.

Going down the group, elements react more readily with both water and oxygen (see **magnesium** and **calcium**). They all **tarnish*** in air but **barium** reacts so violently with both water and oxygen that it is stored under oil.

These two pages contain more information on **magnesium**, **calcium** and their compounds. These metals are typical Group II elements.

•**Magnesium** (**Mg**). A member of Group II of the periodic table. It only occurs naturally in compounds, mainly in either **dolomite** (**$CaCO_3.MgCO_3$** – a rock made of magnesium and **calcium carbonate**) or in **magnesium chloride** (**$MgCl_2$**), found in sea water. Magnesium is produced by the **electrolysis*** of molten magnesium chloride. It burns in air with a bright white flame.

$$2Mg(s) + O_2(g) \rightarrow 2MgO(s)$$
Magnesium + Oxygen → Magnesium oxide

Magnesium reacts rapidly with dilute acids:

$$Mg(s) + 2HCl(aq) \rightarrow MgCl_2(aq) + H_2(g)$$
Magnesium + Hydrochloric acid → Magnesium chloride + Hydrogen

$$Mg(s) + Cl_2(g) \rightarrow MgCl_2(s)$$
Magnesium + Chlorine → Magnesium chloride

Magnesium burns vigorously in chlorine (see above), reacts slowly with cold water and rapidly with steam (see below).

$$Mg(s) + H_2O(g) \rightarrow MgO(s) + H_2(g)$$
Magnesium + Steam → Magnesium oxide + Hydrogen

Magnesium is used to make **alloys***, e.g. for building aircraft. It is also needed for plant **photosynthesis*** (it is found in **chlorophyll*** – the leaf pigment which absorbs light energy).

* **Alloy**, 344; **Chlorophyll**, 255; **Covalent compounds**, 132; **Electrolysis**, 156; **Electron configuration**, 127; **Ionic compound**, 131; **Isotope**, **Outer shell**, 127; **Photosynthesis**, 254; **Radioactivity**, 128; **Reducing agent**, 148; **Relative atomic mass**, 138; **Tarnish**, 345.

• **Magnesium hydroxide** (**$Mg(OH)_2$**). A white solid that is only slightly soluble in water. It is a **base*** and therefore **neutralizes*** acids.

Magnesium hydroxide is used in **antacids*** for treating stomach upsets, particularly indigestion.

• **Magnesium sulphate** (**$MgSO_4$**). A white solid used in medicines for treating constipation, in leather processing and in fire-proofing.

• **Magnesium oxide** (**MgO**). A white solid which is slightly water-soluble. It is a **base***, forming magnesium **salts*** when it reacts with acids. It has a very high melting point and is used to line some furnaces.

Magnesium oxide is added to cocoa powder to stop the particles sticking together.

$$\underset{\text{Magnesium oxide}}{MgO(s)} + \underset{\text{Hydrochloric acid}}{2HCl(aq)} \rightarrow \underset{\text{Magnesium chloride}}{MgCl_2(aq)} + \underset{\text{Water}}{H_2O(l)}$$

• **Calcium** (**Ca**). A member of Group II of the periodic table. It occurs naturally in many compounds, e.g. those found in the Earth's crust, milk and bones. Calcium is extracted from its compounds by **electrolysis***. It burns in oxygen with a red flame and reacts readily with cold water and very rapidly with dilute acids (for equations see **magnesium** and substitute Ca for Mg). Calcium is used to make high-grade steel and in the production of uranium.

• **Calcium hydroxide** (**$Ca(OH)_2$**) or **slaked lime**. A white solid which dissolves slightly in water to form **limewater**. This is weakly alkaline and is used to test for carbon dioxide (see page 219). Calcium hydroxide is used in mortars and to remove excess acidity in soils.

• **Calcium sulphate**. A white solid that occurs both as **anhydrite calcium sulphate** (**$CaSO_4$**) and **gypsum** (**$CaSO_4.2H_2O$**). When heated, gypsum forms plaster of Paris.

Calcium sulphate is used in:

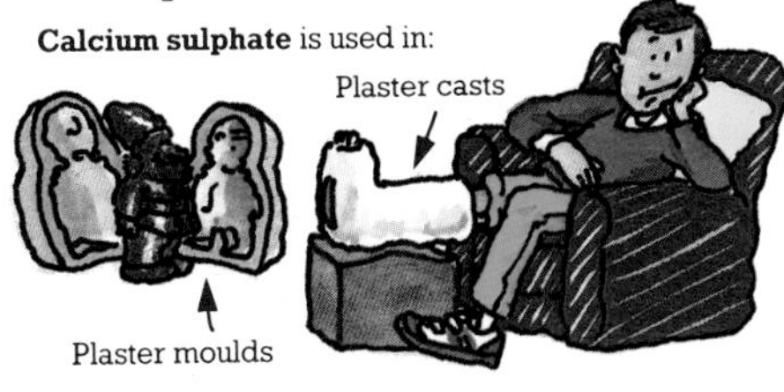

• **Calcium oxide** (**CaO**) or **quicklime**. A white solid. It is a **base*** which is made by heating **calcium carbonate** in a lime kiln.

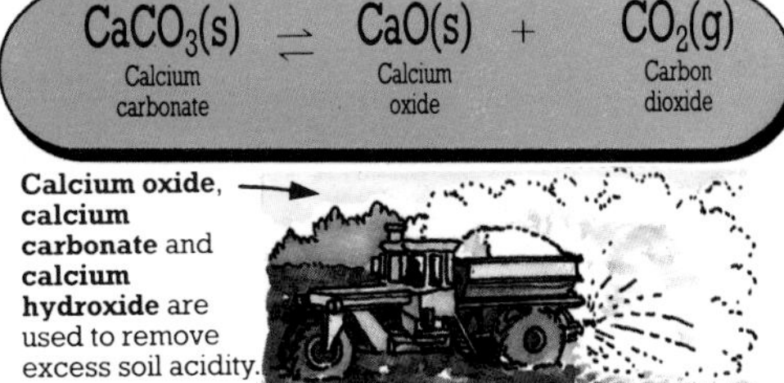

$$\underset{\text{Calcium carbonate}}{CaCO_3(s)} \rightleftharpoons \underset{\text{Calcium oxide}}{CaO(s)} + \underset{\text{Carbon dioxide}}{CO_2(g)}$$

Calcium oxide, **calcium carbonate** and **calcium hydroxide** are used to remove excess soil acidity.

• **Calcium carbonate** (**$CaCO_3$**). A white insoluble solid that occurs naturally as **limestone**, **chalk**, **marble** and **calcite**. It dissolves in dilute acids.

Limestone rock is corroded because rain water containing dissolved carbon dioxide reacts with the limestone to form **calcium hydrogencarbonate** which dissolves slightly in water.

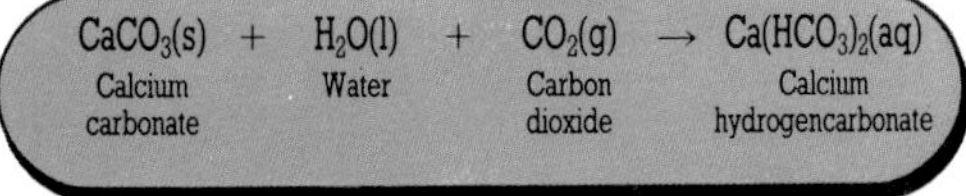

$$\underset{\text{Calcium carbonate}}{CaCO_3(s)} + \underset{\text{Water}}{H_2O(l)} + \underset{\text{Carbon dioxide}}{CO_2(g)} \rightarrow \underset{\text{Calcium hydrogencarbonate}}{Ca(HCO_3)_2(aq)}$$

Calcium hydrogencarbonate formed when **limestone** dissolves in water causes **temporary hardness***.

Calcium carbonate is used to obtain **calcium oxide**, make cement and as building stone.

• **Calcium chloride** (**$CaCl_2$**). A white, **deliquescent***, water-soluble solid which is used as a **drying agent***.

* **Antacid**, 344; **Base**, 151; **Deliquescent**, 206; **Drying agent**, 344; **Electrolysis**, 156; **Neutralization**, 151; **Salts**, 153; **Temporary hardness**, 207.

Transition metals

Transition metals have certain properties in common – they are hard, tough, shiny, **malleable*** and **ductile***. They **conduct*** heat and electricity, and have high melting points, boiling points and densities. Transition metals form **complex ions*** which are coloured in solution. They also have more than one possible charge, e.g. Fe^{2+} and Fe^{3+}.

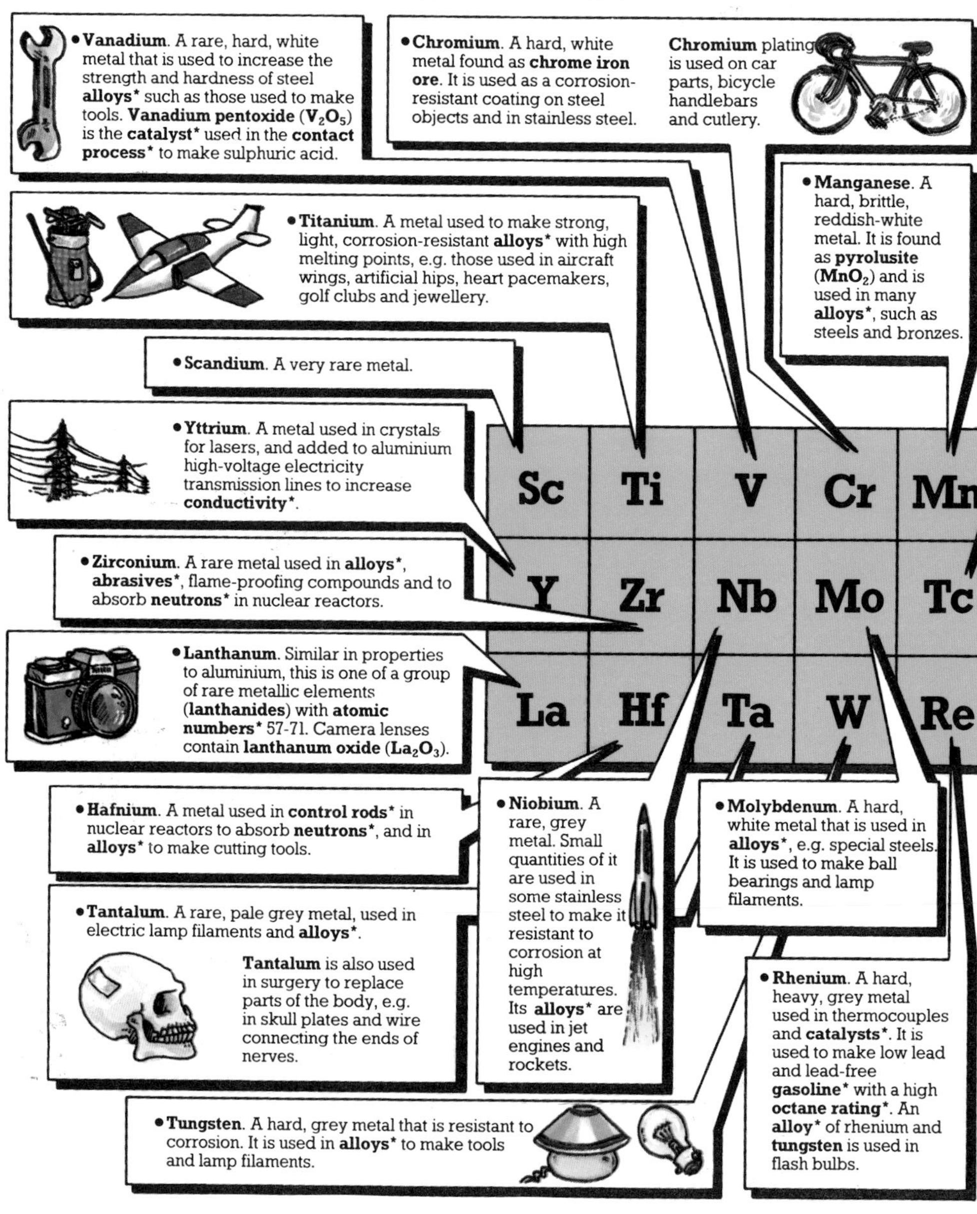

* **Abrasive, Alloy**, 344; **Atomic number**, 127; **Catalyst**, 161; **Complex ion**, 154 (**Complex salt**); **Conductivity**, 62; **Contact process**, 185; **Control rods**, 94; **Ductile**, 344; **Gasoline**, 199; **Malleable**, 345; **Neutron**, 126; **Octane rating**, 199.

Transition metals have many uses, some of which are shown below. There is more information on iron, copper and zinc on pages 174-175. The members of the **inner transition series** (see page 165) are not shown below, as they are very rare and often unstable.

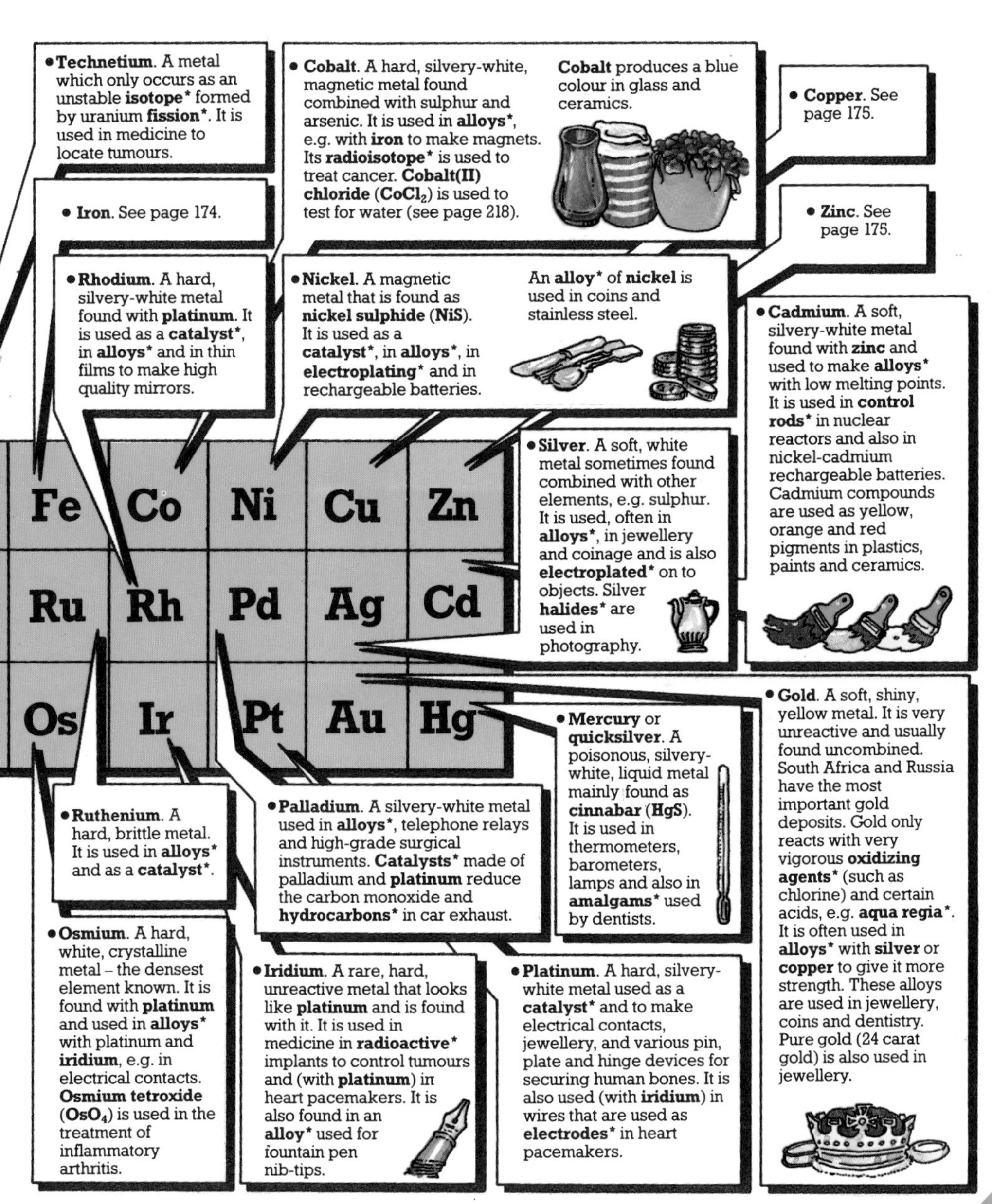

* **Alloy**, **Amalgam**, 344; **Aqua regia**, 182 (**Nitric acid**); **Catalyst**, 161; **Control rods**, 94; **Electrode**, 156; **Electroplating**, 157; **Halides**, 187; **Hydrocarbons**, 191; **Isotope**, 127; **Neutron**, 126; **Nuclear fission**, 129; **Oxidizing agent**, 148; **Radioactivity**, **Radioisotope**, 128.

Iron, copper and zinc

•**Iron (Fe)**. A **transition metal*** in Period 4 of the periodic table. It is a fairly soft, white, magnetic metal which only occurs naturally in compounds. One of its main ores is **haematite** (**Fe_2O_3**), or **iron(III) oxide**, from which it is extracted in a **blast furnace**. Iron reacts to form both **ionic** and **covalent compounds*** and reacts with moist air to form **rust**. It burns in air when cut very finely into iron filings and also reacts with dilute acids. It is above hydrogen in the **electrochemical series***.

Extracting **iron** using a **blast furnace**.

Raw materials fed in here are iron ore (Fe_2O_3), coke (C) and limestone ($CaCO_3$).

Skip

Gas outlet

Heat-resistant brick lining

Blast furnace

Melting zone

Blast of hot air

Molten iron tapped off here.

Reactions in the furnace

Heated limestone forms calcium oxide and gives off carbon dioxide. Oxygen from hot air reacts with coke and also forms carbon dioxide.

Carbon dioxide reacts with coke to form carbon monoxide.

$$CO_2 + C \rightarrow 2CO$$

Iron ore is **reduced*** to iron by carbon monoxide.

$$Fe_2O_3 + 3CO \rightleftarrows 2Fe + 3CO_2$$

Slag removed (impurities plus calcium oxide).

Iron made in the blast furnace is called **pig iron**. It contains about 5% carbon and 4% other impurities, such as sulphur. Most pig iron is converted to **steel**, although some is converted to **wrought iron** (by **oxidizing*** impurities) and some is melted down again along with scrap steel to make **cast iron**.

Wrought iron is used to make crane hooks and anchor chains.

Cast iron is used to make drain covers.

Iron is a vital mineral in the human diet, as it is needed for respiration.

•**Steel**. An **alloy*** of **iron** and carbon which usually contains below 1.5% carbon. The carbon gives the alloy strength and hardness but reduces **malleability*** and **ductility***. Measured amounts of one or more **transition metals*** are often added to steel to give it specific properties, such as corrosion-resistance in the case of **stainless steel** (which contains 11-14% chromium). Steel is often made by the **basic oxygen process**. Scrap steel, molten iron and lime are put into a furnace, and oxygen is blasted on to the metal to **oxidize*** impurities.

Steel is used to make many objects, e.g:

•**Iron(II)** or **ferrous compounds**. Iron compounds containing Fe^{2+} ions, e.g. **iron(II) chloride** (**$FeCl_2$**). Their solutions are green.

•**Iron(III)** or **ferric compounds**. Iron compounds containing Fe^{3+} ions, e.g. **iron(III) chloride** (**$FeCl_3$**). Their solutions are yellow or orange.

Rust

Rust (**$Fe_2O_3.xH_2O$**) or **hydrated iron(III) oxide**. A brown solid formed when **iron**, water and air react together (see **corrosion**, page 209). The "x" in the formula shows that the number of water molecules varies.

Iron objects are given a protective coating to prevent rust, e.g. cars are painted and grease is put on engine parts. If a car rusts, **phosphoric acid*** can be put on to stop the rust spreading.

Iron and **steel** can be protected from rust by **galvanizing** – coating with a layer of zinc (see also **sacrificial protection**, page 159). The surface zinc **oxidizes*** in air, stopping the zinc and iron below being oxidized.

Even when the zinc coating is scratched, exposing iron, it is zinc that reacts with oxygen and water. Galvanized cars remain rust-free longer than others.

* **Alloy**, 344; **Covalent compounds**, 132; **Ductility**, 344 (**Ductile**); **Electrochemical series**, 159; **Ionic compound**, 131; **Malleability**, 345 (**Malleable**); **Oxidation**, 148; **Phosphoric acid**, 182 (**Phosphorus pentoxide**); **Reduction**, 148; **Transition metals**, 172.

•**Copper** (**Cu**). A **transition metal*** in Period 4 of the periodic table. It is a red-brown, soft but tough metal found naturally in certain rocks. Its compounds are found in several ores, e.g. **copper pyrites** (($(CuFe)S_2$) and **malachite** (**$CuCO_3.Cu(OH)_2$**). Copper is extracted from the former by crushing and removing sand and then roasting in a limited supply of air and with silica. The iron combines with silica and forms **slag**. The sulphur is removed by burning to form sulphur dioxide. The copper produced is purified further by **electro-refining***. It is an unreactive metal and only **tarnishes*** very slowly in air to form a thin, green surface film of **basic copper sulphate** (**$CuSO_4.3Cu(OH)_2$**). Copper is below hydrogen in the **electrochemical series***. It does not react with water, dilute acids or alkalis. However, it does react with concentrated nitric or concentrated sulphuric acid. (See also page 219).

Copper is a very good conductor of electricity (although silver is better), so it is used to make wires for electric circuits.

It is used in **alloys*** such as **brass** (copper plus zinc) and **bronze** (copper plus tin) to make "copper" coins, and in an alloy with nickel to make "silver" coins.

Because it is soft but tough, it is used to make pipes for plumbing and central-heating systems.

An **alloy*** of copper and gold is used to make jewellery. The greater the amount of copper, the less the number of carats the gold will be, i.e. less than 24 carats (pure gold).

Copper(II) sulphate (**$CuSO_4$**) has many uses, e.g. in dyeing and **electroplating***. It is also used in **Bordeaux mixture**, which kills moulds growing on fruit and vegetables. (See also test for water, page 218)

Copper(II) chloride (**$CuCl_2$**) is used in fireworks to give a green colour. It is also used to remove sulphur from **petroleum***.

Copper(I) oxide (**Cu_2O**) is used to make glass and paint.

•**Copper(I)** or **cuprous compounds**. Compounds containing Cu^+ ions, e.g. **copper(I) oxide** (see left) and **copper(I) chloride** (**CuCl**). Copper(I) compounds do not dissolve in water.

•**Copper(II)** or **cupric compounds**. Compounds that contain Cu^{2+} ions, e.g. **copper(II) sulphate** and **copper(II) chloride** (see left). Copper(II) compounds dissolve in water to form light blue solutions, and are much more common than **copper(I) compounds**.

•**Zinc** (**Zn**). An element in Period 4 of the periodic table. It is a silvery, soft metal which **tarnishes*** in air. It is too reactive to occur naturally, and its main ores are **zinc blende** (**ZnS**), **calamine** (**$ZnCO_3$**) and **zincite** (**ZnO**). The zinc is extracted by roasting the ore to form **zinc oxide** (**ZnO**) and then **reducing*** it by heating it with coke. Zinc is above hydrogen in the **electrochemical series***. It reacts with oxygen, with steam when red-hot, and with acids. It is used to coat iron and steel to prevent rust (**galvanizing** – see also page 174 and **sacrificial protection**, page 159). It is also used in **alloys***, particularly **brass** (copper and zinc).

Zinc oxide is used in a cream as a protection against skin irritation, e.g. nappy rash.

Zinc is used in batteries.

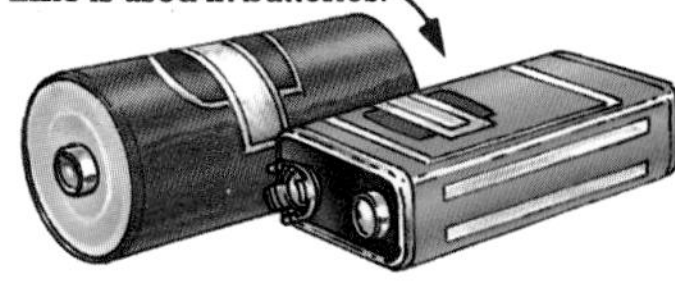

* **Alloy**, 344; **Electrochemical series**, 159; **Electroplating**, **Electro-refining**, 157; **Petroleum**, 198; **Reduction**, 148; **Tarnish**, 345; **Transition metals**, 172.

Group III elements

The elements in **Group III** of the periodic table are generally not as reactive as the elements in Groups I and II. Unlike those elements they show no overall trend in reactivity and the first member of the group is a non-metal. The chart below shows some of their properties.

Some properties of Group III elements

Name of element	Chemical symbol	Relative atomic mass*	Electron configuration*	Reactivity	Appearance	Uses
Boron	**B**	10.81	2,3	NO TREND ↓	Brown powder or yellow crystals	In **control rods***, glass and to harden steel
Aluminium	**Al**	26.98	2,8,3		White metal	See below
Gallium	**Ga**	69.72	Complex configuration but still 3 outer electrons		Silver-white metal	In **semiconductors***
Indium	**In**	114.82			Soft silver-white metal	In **control rods*** and transparent electrodes
Thallium	**Tl**	204.37			Soft silver-white metal	In rat poison

Although all atoms of Group III elements have three outer electrons they react to form different types of compound. Those of **boron**, and some of **aluminium**, are **covalent***. Other members of the group form mostly **ionic compounds***.

Boron oxide (B_2O_3) is added to glass to make special glassware which can be heated or cooled rapidly without cracking.

More information on **aluminium** and its compounds can be found below. Aluminium is the most widely used member of this group.

- **Aluminium (Al)**. A member of Group III of the periodic table. It is the commonest metal found on earth, and occurs naturally in many compounds, e.g. **bauxite** (see **aluminium oxide**) from which it is extracted by **electrolysis***. It is hard, light, **ductile***, **malleable*** and a good conductor of heat and electricity. It reacts with the oxygen in air to form a surface layer of **aluminium oxide** which stops further corrosion. It also reacts with chlorine, dilute acids and alkalis.

Some uses of **aluminium** and its **alloys***:

Powerlines are aluminium as aluminium conducts electricity better for its weight than copper.

Thin sheets of aluminium are used to wrap food, e.g. chocolate bars. It is also used to make fizzy drink cans.

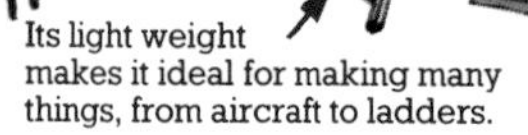

Its light weight makes it ideal for making many things, from aircraft to ladders.

- **Aluminium oxide** (Al_2O_3) or **alumina**. An **amphoteric***, white solid that is almost insoluble in water. It occurs naturally as **bauxite** ($Al_2O_3.2H_2O$ – see also **aluminium**) and as **corundum** (Al_2O_3) – an extremely hard crystalline solid. It is used in some cements, and to line furnaces.

The extraction of **aluminium** from **bauxite** by **electrolysis***.

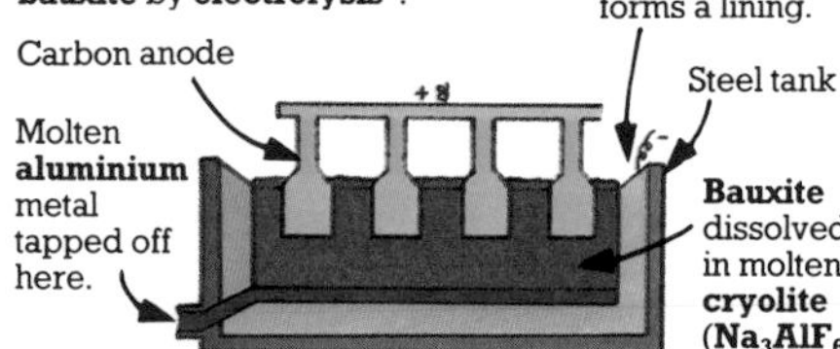

- **Aluminium hydroxide** ($Al(OH)_3$). A white, slightly water-soluble, **amphoteric*** solid, which is used in dyeing cloth, to make ceramics and as an **antacid***.

- **Aluminium sulphate** ($Al_2(SO_4)_3$). A white, water-soluble, crystalline solid used to purify water and make paper.

* **Alloy**, 344; **Amphoteric**, 151; **Antacid**, 344; **Control rods**, 94; **Covalent compounds**, 132; **Ductile**, 344; **Electrolysis**, 156; **Electron configuration**, 127; **Ionic compound**, 131; **Malleable**, 345; **Relative atomic mass**, 138; **Semiconductor**, 65.

Group IV elements

The elements in **Group IV** of the periodic table are generally not very reactive and the members show increasingly metallic properties going down the group. For more about the properties of these elements, see the chart below, **silicon** and **lead** (this page) and **carbon**, pages 178-179.

Some properties of Group IV elements

Name of element	Chemical symbol	Relative atomic mass*	Electron configuration*	Reactivity	Appearance	Uses
Carbon	C	12.01	2,4	NO TREND	Solid non-metal (see page 178)	See page 178
Silicon	Si	20.09	2,8,4		Shiny, grey **metalloid*** solid	See below
Germanium	Ge	72.59	Complex configuration but still 4 outer electrons		Greyish-white **metalloid*** solid	In transistors
Tin	Sn	118.69			Soft silver-white metal	Tin plating, e.g. food containers
Lead	Pb	207.19			Soft silver-grey metal	See below

Although all atoms of Group IV elements have four outer electrons they react to form different compound types. They all form **covalent compounds***, but **tin** and **lead** form **ionic compounds*** as well.

- **Silicon (Si)**. A member of Group IV of the periodic table. It is a hard, shiny, grey **metalloid*** with a high melting point. Silicon is the second most common element in the Earth's crust - it is found in sand and rocks as **silicon dioxide** and **silicates**. When it is ground into a powder it reacts with some alkalis and elements, otherwise it is generally unreactive.

Silicon is a **semiconductor*** and is used to make silicon chips – complete microelectronic circuits.

- **Silicon dioxide** (**SiO_2**). Also called **silicon(IV) oxide**, or **silica**. An insoluble, white, crystalline solid. It occurs in many forms, such as **flint** and **quartz**. It is acidic and reacts with concentrated alkalis.

Silicon dioxide has many uses, e.g. in the making of glass and ceramics.

Quartz crystals are used in watches.

Sand is impure **quartz**.

- **Silicates**. **Silicon** compounds that also contain a metal and oxygen, e.g. **calcium metasilicate** (**$CaSiO_3$**), and make up most of the earth's crust. They are used to make glass and ceramics.

- **Silicones**. Complex, man-made compounds contaning very long chains of **silicon** and oxygen atoms.

Silicones are used in high-performance oils and greases and for non-stick surfaces. They are also used in waxes, polishes and varnishes as they are water-repellant.

- **Lead (Pb)**. A member of Group IV of the periodic table. A soft, **malleable*** metal extracted from **galena** (**lead(II) sulphide**). It is not very reactive, though it **tarnishes*** in air, reacts slightly with **soft water*** and slowly with chlorine and nitric acid. It forms **ionic compounds*** called **lead(II)** or **plumbous compounds**, e.g. **lead(II) oxide (PbO)**, and **covalent compounds*** called **lead(IV)** or **plumbic compounds**, e.g. **lead(IV) oxide** (**PbO_2**).

Lead has many uses, e.g. in car batteries and roofing. It is used in hospitals to protect people from the harmful effects of X-rays.

X-ray machine

The lead inside this rubber apron protects the radiologist.

* **Covalent compounds**, 132; **Electron configuration**, 127; **Ionic compound**, 131; **Malleable**, 345; **Metalloids**, 165 (**Metal**); **Relative atomic mass**, 138; **Semiconductor**, 65; **Soft water**, 207 (**Hard water**); **Tarnish**, 345.

Carbon

Carbon (C) is a member of Group IV of the periodic table (see also chart, page 176). It is a non-metal and has two **allotropes*** – **diamond** and **graphite** – and an **amorphous*** (unstructured) form – **charcoal**. Carbon is not very reactive. It only reacts with steam when heated, and with hot, **concentrated*** sulphuric or nitric acids (see equation below).

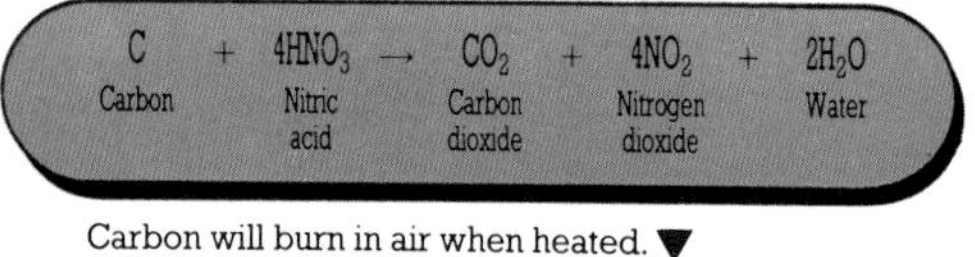

Carbon is a **reducing agent***. It reduces the **oxides*** of any metal below zinc in the **reactivity series*** of metals: ▼

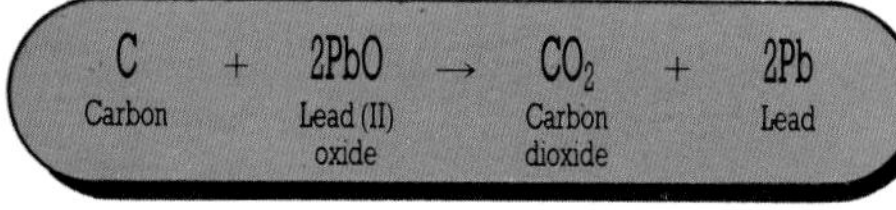

Carbon is used in industry to reduce metal oxide ores to metals (see **iron**, page 174).

Carbon will burn in air when heated. ▼

$$C(s) + O_2(g) \rightarrow CO_2(g)$$

Carbon | Oxygen | Carbon dioxide

If burnt in a limited supply of air, **carbon monoxide** forms.

Carbon atoms can bond with up to four other atoms, including other carbon atoms. As a result, there are a vast number of carbon-based compounds (**organic compounds** – see page 190). Living tissue is made of carbon compounds and animals break down these compounds to liberate energy (see **carbon cycle**, page 209).

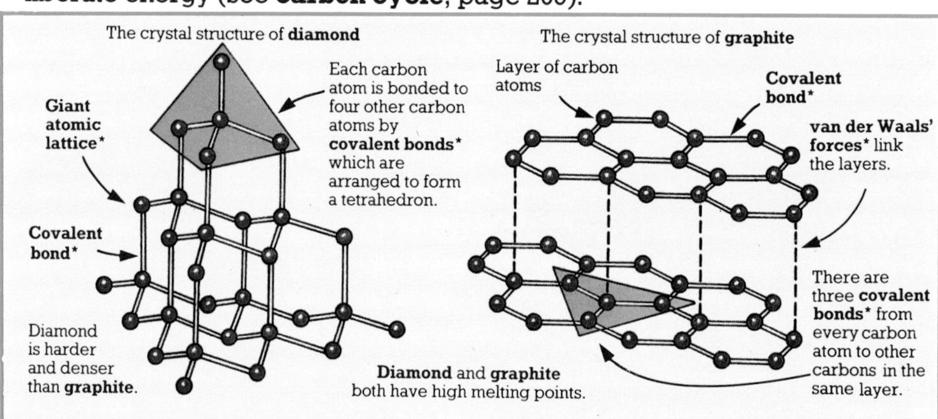

•**Diamond**. A crystalline, transparent form of carbon. It is the hardest naturally-occurring substance. All the carbon atoms are joined by strong **covalent bonds*** – accounting for its hardness and high melting point (3750°C). Diamonds are used as record player styli, **abrasives***, glass cutters, jewellery and on drill bits. **Synthetic diamonds** are made by subjecting **graphite** to high pressure and temperature, a very costly process.

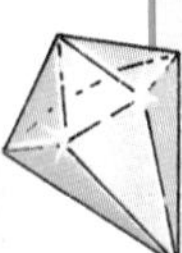

•**Graphite**. A grey, crystalline form of carbon. The atoms in each layer are joined by strong **covalent bonds***, but the layers are only linked by weak **van der Waals' forces*** which allow them to slide over each other, making graphite soft and flaky. Graphite is the only non-metal to conduct electricity well. It also conducts heat. It is used as a lubricant, in **electrolysis*** (as **inert electrodes***), as contacts in electric motors, and in pencil leads.

* **Abrasive**, 344; **Allotropes**, 136 (**Allotropy**); **Amorphous**, 135; **Concentrated**, 144; **Covalent bond**, 132; **Electrolysis**, 156; **Giant atomic lattice**, 137; **Inert electrode**, 156; **Oxides**, 183; **Reactivity series**, 158; **Reducing agent**, 148; **van der Waals' forces**, 134.

•**Coal**. A hard, black solid formed over millions of years from the fossilized remains of plant material. It is mainly carbon, but contains hydrogen, oxygen, nitrogen and sulphur as well. Three types of coal exist – **lignite**, **anthracite** and **bituminous coal**. Coal is used as a fuel in power stations, industry and homes. It also used to be an important source of chemicals (produced by **destructive distillation of coal** – see below) but these are now mostly produced from **petroleum***

This picture shows the products of the **destructive distillation of coal**, achieved by heating coal in the absence of air.

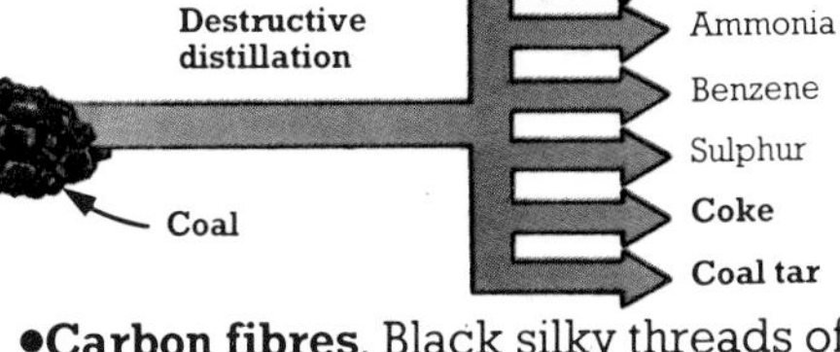

•**Carbon fibres**. Black silky threads of pure carbon made from organic textile fibres. They are stronger and stiffer than other materials of the same weight and are used to make light boats.

•**Coke**. A greyish, porous, brittle solid containing over 80% carbon. It is made by heating **coal** in the absence of air in coke ovens, or as a by-product of making **coal gas** (see **carbon monoxide**). It is a smokeless fuel.

•**Charcoal**. A porous, black, **amorphous***, impure form of carbon. It is made by heating organic material in the absence of air.

Some uses of **charcoal**

Charcoal is used by artists for drawing.

Charcoal is used as a smokeless fuel, e.g. at barbecues.

Activated charcoal absorbs small molecules onto its surface. It is used in shoe lining pads to absorb odours.

•**Carbon dioxide** (**CO_2**). A colourless, odourless gas found in the atmosphere (see **carbon cycle**, page 209). It is made industrially by heating calcium carbonate in a lime kiln (see also page 216 for preparation). It dissolves in water to form **carbonic acid** (**H_2CO_3**).

$$\underset{\text{Carbon dioxide}}{CO_2(aq)} + \underset{\text{Water}}{H_2O(l)} \rightleftharpoons \underset{\text{Carbonic acid}}{2H^+(aq) + CO_3^{2-}(aq)}$$

Carbon dioxide is not very reactive, though it reacts with both sodium and calcium hydroxide solutions (see page 218) and magnesium ribbon burns in it.

Carbon dioxide has many uses.

It is used to make drinks fizzy. Carbon dioxide escapes when the can is opened, as the pressure is released.

It is used in fire extinguishers. It forms a blanket over the flames and does not allow air to reach the fire.

•**Carbon monoxide** (**CO**). A poisonous, colourless, odourless gas, made by passing **carbon dioxide** over hot carbon, and also by burning carbon fuels in a limited supply of air. It is not water-soluble, burns with a blue flame and is a **reducing agent*** (used to reduce metal oxide ores to metal (see **iron**, page 174). It is also used, mixed with other gases, in fuels, e.g. mixed with hydrogen in **water gas**, with nitrogen in **producer gas**, and with hydrogen (50%), methane and other gases in **coal gas**.

If there is not enough oxygen, the **carbon monoxide** produced when fuel is burnt is not changed to **carbon dioxide**. When a car engine runs in a closed garage carbon monoxide accumulates.

•**Carbonates**. Compounds made of a metal **cation*** and a **carbonate anion*** (**CO_3^{2-}**), e.g. **calcium carbonate** (**$CaCO_3$**). Except Group I carbonates, they are insoluble in water and decompose on heating. They all react with acids to give off **carbon dioxide**.

* **Amorphous**, 135; **Anion**, **Cation**, 130; **Petroleum**, 198; **Reducing agent**, 148.

Group V elements

The elements in **Group V** of the periodic table become increasingly metallic going down the group, see chart below.

Some properties of Group V elements

Name of element	Chemical symbol	Relative atomic mass*	Electron configuration*	Reactivity	Appearance	Uses
Nitrogen	N	14.00	2,5	INCREASING ↓	Colourless gas	See below.
Phosphorus	P	30.97	2,8,5		Non-metallic solid (see page 183)	See page 182
Arsenic	As	74.92	Complex configuration but still 5 outer electrons		3 **allotropes***, one of which is metallic.	In **semiconductors*** and **alloys***
Antimony	Sb	121.75			Silver-white metal	In type metal and other **alloys**
Bismuth	Bi	208.98			White metal with reddish tinge	In low melting point **alloys** and medicines

More information on **nitrogen**, **phosphorus** and their compounds can be found below and on pages 181-182. They are the two most abundant members of the group.

All the atoms of Group V elements have five electrons in their **outer shell***. They all react to form **covalent compounds*** in which they share three of these electrons with three from another atom, or atoms (see **octet**, page 127). **Antimony**, **bismuth** and **nitrogen** also form **ionic compounds***.

●**Nitrogen** (**N_2**). A member of Group V of the periodic table. A colourless, odourless, **diatomic*** gas that makes up 78% of the atmosphere. It can be produced by **fractional distillation of liquid air*** (but see also page 217). Its **oxidation state*** in compounds varies from -3 to +5. It reacts with a few reactive metals to form **nitrides**.

$$6Li(s) + N_2(g) \rightarrow 2Li_3N(s)$$

Lithium + Nitrogen → Lithium nitride

Nitrogen is essential for all organisms as it is found in molecules in living cells, e.g. proteins (see also **nitrogen cycle**, page 209). It is used in the manufacture of ammonia (see **Haber process**) and nitric acid. **Liquid nitrogen**, which exists below −196°C, has many uses, including freezing food.

Packets of crisps are filled with **nitrogen** gas to keep them fresh longer (when air is left in the packet the crisps go stale). The gas in the packet also cushions the crisps against damage in transport.

●**Haber process**. This process is used to make **ammonia** from **nitrogen** and hydrogen which are reacted in a ratio of 1:3. Ammonia is produced as fast and economically as possible by using a suitable temperature, pressure and **catalyst*** (see below). The reaction is **exothermic*** and **reversible***.

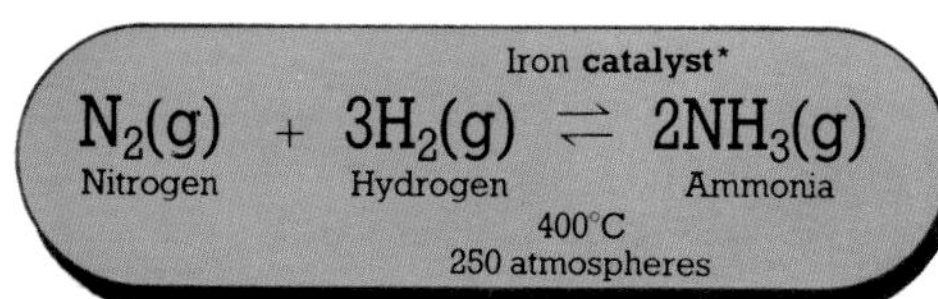

(Under these conditions 15% of the reactants combine to form **ammonia**.)

Haber process

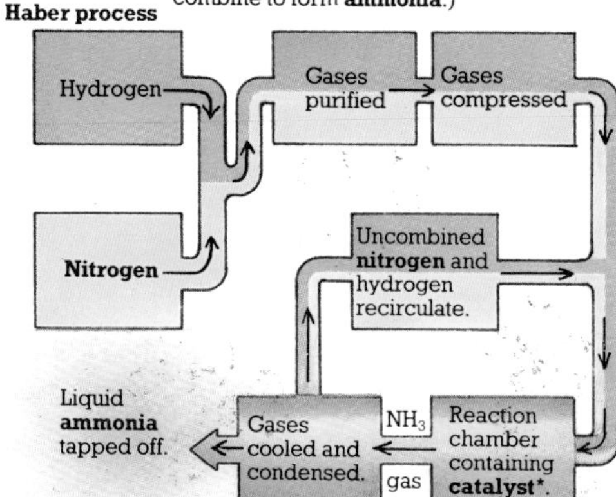

* **Allotropes**, 136 (**Allotropy**); **Alloy**, 344; **Catalyst**, 151; **Covalent compounds**, 132; **Diatomic**, 124; **Electron configuration**, 127; **Exothermic reaction**, 147; **Fractional distillation of liquid air**, 183; **Ionic compound**, 131; **Outer shell**, 127; **Oxidation state**, 149; **Relative atomic mass**, 138; **Reversible reaction**, 162; **Semiconductor**, 65.

• **Ammonia** (NH_3). A colourless, strong-smelling gas that is less dense than air and is a **covalent compound*** made by the **Haber process**. It is a **reducing agent*** and the only common gas to form an alkaline solution in water – this solution is known as **ammonia solution** (**NH_4OH**) or **ammonium hydroxide**. Ammonia burns in pure oxygen to give nitrogen and water, and reacts with chlorine to give **ammonium chloride**.

Ammonia is used to make nitric acid, fertilizers, plastics, explosives and household cleaners.

• **Ammonium chloride** (**NH_4Cl**) or **sal ammoniac**. A white, water-soluble, crystalline solid made when **ammonia solution** (see **ammonia**) reacts with dilute hydrochloric acid. When heated it **sublimes*** and **dissociates*** (see equation below and page 162). It is used in the dry batteries which run many electrical appliances.

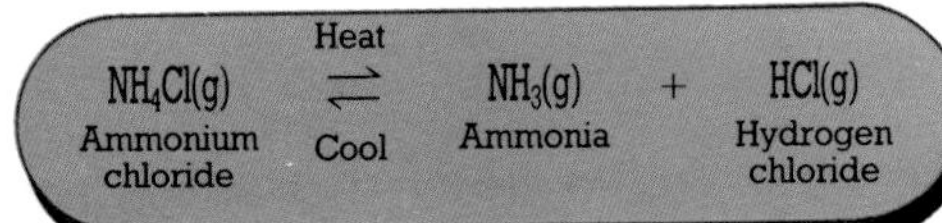

• **Ammonium sulphate** (**$(NH_4)_2SO_4$**). A white, water-soluble, crystalline solid produced by the reaction of **ammonia** and sulphuric acid. It is a fertilizer.

• **Ammonium nitrate** (**NH_4NO_3**). A white, water-soluble, crystalline solid formed when **ammonia solution** (see **ammonia**) reacts with dilute nitric acid. It gives off dinitrogen oxide when heated.

Ammonium nitrate is used in explosives and fertilizers. It is also found in mixtures used to feed pot plants.

• **Dinitrogen oxide** (**N_2O**). Also called **nitrous oxide** or **laughing gas**. A colourless, slightly sweet-smelling, water-soluble gas. It is a **covalent compound*** formed by gently heating **ammonium nitrate**. It supports the combustion of some reactive substances and relights a glowing splint.

Dinitrogen oxide is used as an anaesthetic by both dentists (see **halothane***) and hospital anaesthetists.

• **Nitrogen monoxide** (**NO**). Also called **nitric oxide** or **nitrogen oxide**. A colourless gas that is insoluble in water. It is a **covalent compound*** made when copper reacts with 50% concentrated nitric acid. It reacts with oxygen to form **nitrogen dioxide** and also supports the combustion of reactive elements.

• **Nitrogen dioxide** (**NO_2**). A very dark brown gas with a choking smell. It is a **covalent compound***.

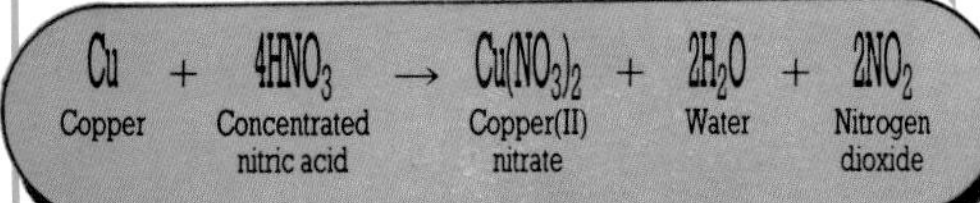

Nitrogen dioxide is made when copper reacts with concentrated nitric acid and when some **nitrates*** are heated. It supports combustion and dissolves in water to give a mixture of nitric acid and **nitrous acid** (**HNO_2**). It is used as an **oxidizing agent***.

Nitrogen dioxide dimerizes (two molecules of the same substance bond together) below 21.5°C to form **dinitrogen tetraoxide** (**N_2O_4**), a colourless gas.

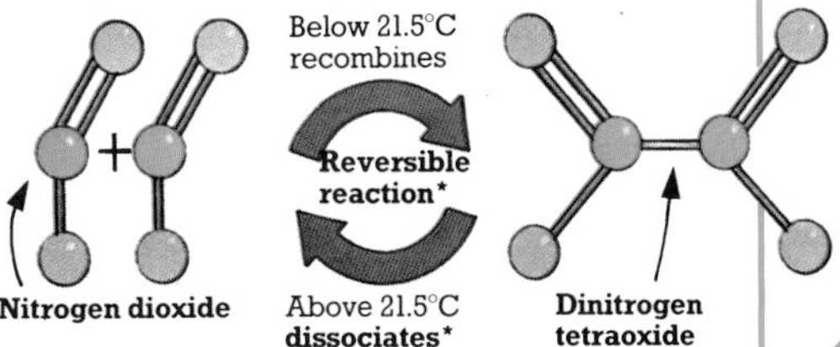

* **Covalent compounds**, 132; **Dissociation**, 162; **Halothane**, 195; **Nitrates**, 182; **Oxidizing agent**, **Reducing agent**, 148; **Reversible reaction**, 162; **Sublimation**, 121.

•**Nitric acid** (**HNO_3**) or **nitric(V) acid**. A light yellow, oily, water-soluble liquid. It is a **covalent compound*** containing nitrogen in **oxidation state*** +5. It is a very strong and corrosive acid which is made industrially by the three-stage **Ostwald process**. This process is shown below:

Stage 1: Ammonia reacts with oxygen.

$$\underset{\text{Ammonia}}{4NH_3} + \underset{\text{Oxygen}}{5O_2} \xrightarrow[\text{900 C}]{\text{Platinum-rhodium catalyst*}} \underset{\text{Nitrogen monoxide}}{4NO} + \underset{\text{Water}}{6H_2O}$$

Stage 2: Nitrogen monoxide cools and reacts with more oxygen to give nitrogen dioxide. Stage 3: Nitrogen dioxide dissolves in water to form **nitric acid**.

$$\underset{\text{Nitrogen dioxide}}{4NO_2} + \underset{\text{Water}}{2H_2O} + \underset{\text{Oxygen}}{O_2} \rightarrow \underset{\text{Nitric acid}}{4HNO_3}$$

Concentrated nitric acid is a mixture of 70% nitric acid and 30% water. It is a powerful **oxidizing agent***. **Dilute nitric acid** is a solution of 10% nitric acid in water. It reacts with **bases*** to give **nitrate salts*** and water. Nitric acid is used to make fertilizers and explosives.

A mixture of concentrated hydrochloric acid and **concentrated nitric acid**, called **aqua regia** (Latin for King's water), will dissolve gold.

•**Nitrates** or **nitrate(V) compounds**. Solid **ionic compounds*** containing the **nitrate anion*** (**NO_3^-**) and a metal **cation*** (see test for nitrate ion, page 218). Nitrogen in a nitrate ion has an **oxidation state*** of +5. Nitrates are **salts*** of **nitric acid** and are made by adding a metal oxide, hydroxide or carbonate to dilute nitric acid. All nitrates are water-soluble and most give off nitrogen dioxide and oxygen on heating (some exceptions are sodium, potassium and ammonium nitrates).

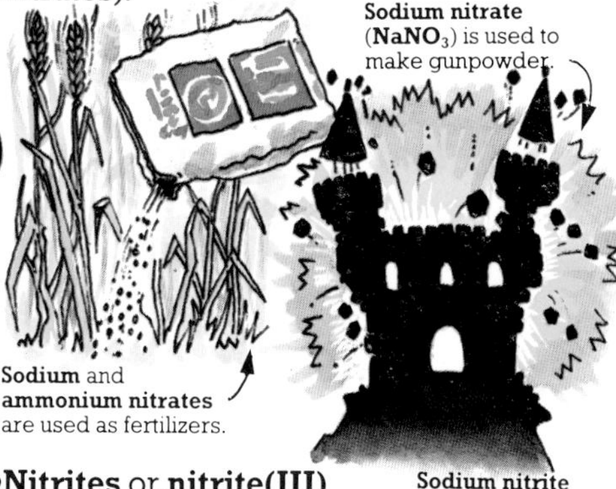

Sodium nitrate (**$NaNO_3$**) is used to make gunpowder.

Sodium and **ammonium nitrates** are used as fertilizers.

•**Nitrites** or **nitrite(III) compounds**. Solid **ionic compounds*** that contain the **nitrite anion*** (**NO_2^-**) and metal **cation***. They are usually **reducing agents***.

Sodium nitrite (**$NaNO_2$**) is used to preserve meat products.

•**Phosphorus** (**P**). A non-metallic member of Group V (see chart, page 180). Phosphorus only occurs naturally in compounds. Its main ore is **apatite** (**$3Ca_3(PO_4)_2.CaF_2$**). It has two common forms. **White phosphorus**, the most reactive form, is a poisonous, waxy, white solid that bursts into flames in air. **Red phosphorus** is a dark red powder that is not poisonous and not very flammable.

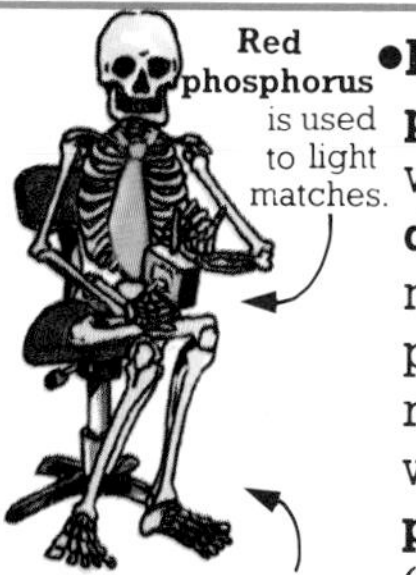

Red phosphorus is used to light matches.

Living organisms contain **phosphorus compounds**, e.g. bones are mainly **calcium phosphate**.

•**Phosphorus pentoxide** (**P_2O_5**). A white solid and **dehydrating agent***, made by burning phosphorus in air. It reacts vigorously with water to form **phosphoric acid** (**H_3PO_4**) and is used to protect against **rust***.

* **Anion**, 130; **Base**, 151; **Catalyst**, 161; **Cation**, 130; **Covalent compounds**, 132; **Dehydrating agent**, 344; **Ionic compound**, 131; **Oxidation state**, 149; **Oxidizing agent**, **Reducing agent**, 148; **Rust**, 174; **Salts**, 153.

Group VI elements

The elements in **Group VI** of the periodic table show increasing metallic properties and decreasing chemical reactivity going down the group. The chart below shows some of the properties of these elements.

Some properties of Group VI elements

Name of element	Chemical symbol	Relative atomic mass*	Electron configuration*	Reactivity	Appearance	Uses
Oxygen	**O**	15.99	2,6	DECREASING ↓	Colourless gas, see below	See below
Sulphur	**S**	32.06	2,8,6		Yellow, non-metallic solid (see page 184)	See page 184
Selenium	**Se**	78.96	Complex configuration but still 6 outer electrons		Several forms, metallic and non-metallic	In **photocells***
Tellurium	**Te**	127.60			Silver-white **metalloid*** solid	In **alloys***, coloured glass, **semiconductors***
Polonium	**Po**	**Radioactive*** element			Metal	

More information on **oxygen**, **sulphur** and their compounds can be found below and on pages 184-185. They are found widely and have many uses.

The atoms of all the elements in Group VI have six electrons in their **outer shell***. They need two electrons to fill their outer shell (see **octet**, page 127) and react with other substances to form both **ionic** and **covalent compounds***. The elements with the smallest atoms are most reactive as the atoms produce the most powerful attraction for the two electrons.

- **Oxygen** (**O_2**). A colourless, odourless, **diatomic*** gas that makes up 21% of the atmosphere. It is the most abundant element in the Earth's crust and is vital for life (see **respiration**, page 209). It supports combustion, dissolves in water to form a **neutral*** solution and is a very reactive **oxidizing agent***, e.g. it oxidizes iron to iron(III) oxide. Plants produce oxygen by photosynthesis and it is obtained industrially by **fractional distillation of liquid air**. It has many uses, e.g. in hospitals and to break down sewage. See test for, and preparation of, oxygen, pages 217 and 219.

- **Ozone** (**O_3**). A poisonous, bluish gas whose molecules contain three **oxygen** atoms. It is an **allotrope*** of oxygen, found in the upper atmosphere where it absorbs most of the sun's harmful ultra-violet radiation. It is produced when electrical sparks pass through air, e.g. when lightning occurs. Ozone is a powerful **oxidizing agent*** and is sometimes used to sterilize water.

Fractional distillation of liquid air (see also page 220).

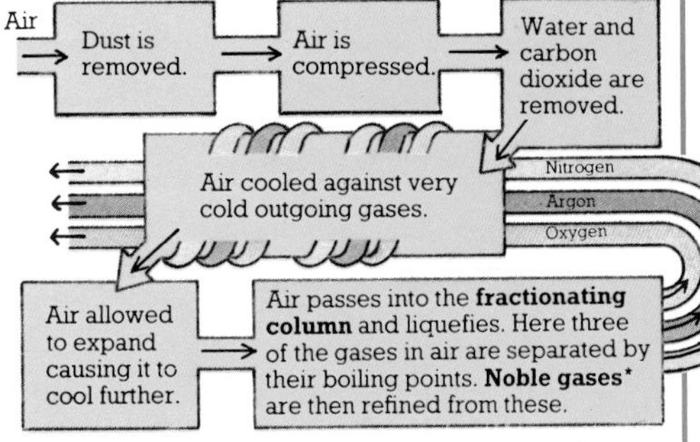

- **Oxides**. Compounds of oxygen and one other element. Metal oxides are mostly **ionic compounds*** and **bases***, e.g. **calcium oxide (CaO)**. Some metal and **metalloid*** oxides are **amphoteric***, e.g. **aluminium oxide** (**Al_2O_3**). Non-metal oxides are **covalent*** and often **acidic***, e.g. **carbon dioxide** (**CO_2**).

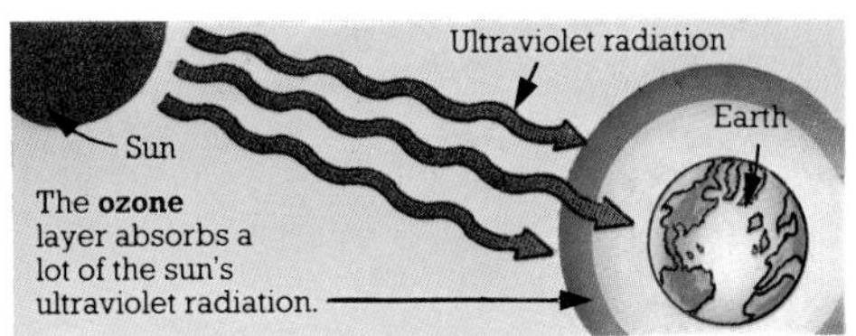

* **Acidic**, 150; **Allotropes**, 136 (**Allotropy**); **Alloy**, 344; **Amphoteric**, **Base**, 151; **Covalent compounds**, 132; **Diatomic**, 124; **Electron configuration**, 127; **Ionic compound**, 131; **Metalloids**, 165 (**Metal**); **Neutral**, 151; **Noble gases**, 189; **Outer shell**, 127; **Oxidizing agent**, 148; **Photocell**, 345; **Radioactivity**, 128; **Relative atomic mass**, 138; **Semiconductor**, 65.

Sulphur

Sulphur (S) is a member of **Group VI** of the periodic table (see chart, page 183). It is a yellow, non-metallic solid that is insoluble in water. It is **polymorphic*** and has two **allotropes*** – **rhombic** and **monoclinic sulphur**. Sulphur is found uncombined in underground deposits (see **Frasch process**) and is also extracted from **petroleum*** and metal **sulphides** (compounds of sulphur and another element), e.g. **iron(II) sulphide (FeS)**. It burns in air with a blue flame to form **sulphur dioxide** and reacts with many metals to form sulphides. It is used to **vulcanize*** rubber, and to make **sulphuric acid**, medicines and **fungicides***.

- **Rhombic sulphur**. Also called **alpha sulphur (α-sulphur)** or **orthorhombic sulphur**. A pale yellow, crystalline **allotrope*** of sulphur, the most stable form at room temperature.

- **Monoclinic sulphur** or **beta sulphur (β-sulphur)**. A yellow, crystalline, **allotrope*** of sulphur. It is more stable than **rhombic sulphur** at temperatures over 96°C.

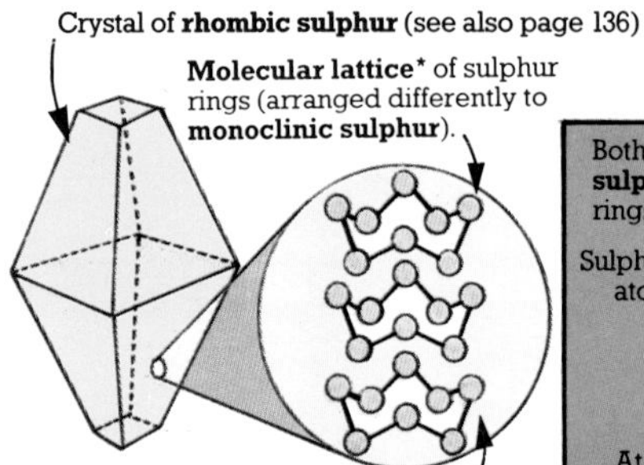

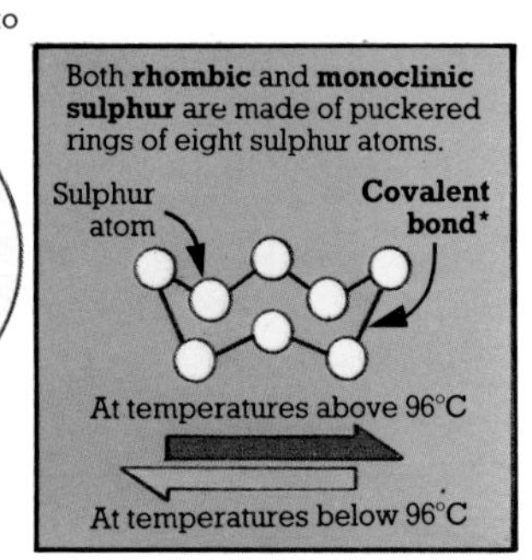

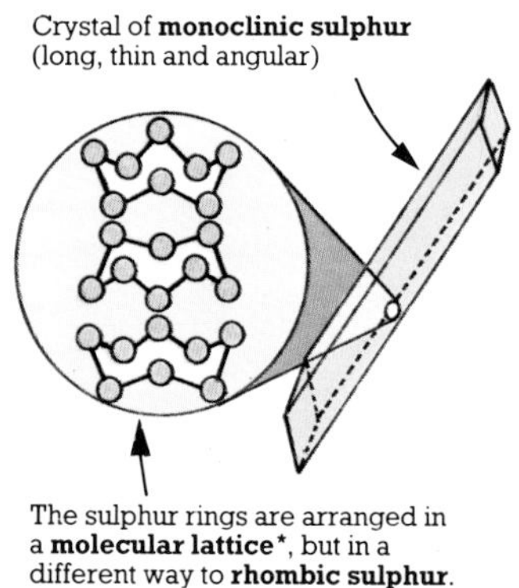

- **Plastic sulphur**. A form of sulphur made when hot liquid sulphur is poured into water to cool it quickly. It can be kneaded and stretched into long fibres. It is not stable and hardens when rings of eight sulphur atoms reform (see above).

- **Flowers of sulphur**. A fine yellow powder formed when sulphur vapour is cooled quickly. The molecules are in rings of eight atoms.

- **Frasch process**. The method used to extract sulphur from underground deposits by melting it. Sulphur produced this way is 99.5% pure.

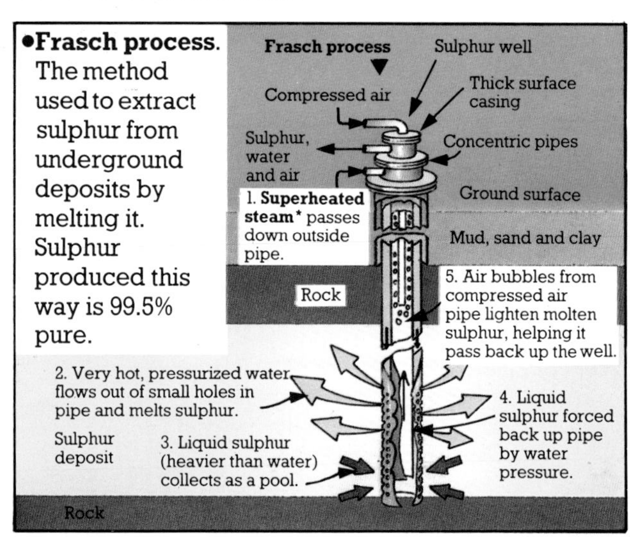

* **Allotropes**, 136 (**Allotropy**); **Covalent bond**, 132; **Fungicide**, 344; **Molecular lattice**, 137; **Petroleum**, 198; **Polymorphism**, 136; **Superheated steam**, **Vulcanization**, 345.

•**Sulphur dioxide** (SO_2) or **sulphur(IV) oxide**. A poisonous, choking gas which forms **sulphurous acid** when dissolved in water. It is a **covalent compound*** made by burning sulphur in air or adding dilute acid to a **sulphite**. It usually acts as a **reducing agent***. It is used to make **sulphuric acid**, in **fumigation*** and as a **bleach***.

Sulphur dioxide is used to preserve many food products containing fruit.

•**Sulphur trioxide** (SO_3) or **sulphur(VI) oxide**. A white **volatile*** solid that is formed by the **contact process**. Sulphur trioxide reacts very vigorously with water to form **sulphuric acid**.

•**Sulphurous acid** (H_2SO_3) or **sulphuric(IV) acid**. A colourless, **weak acid***, formed when **sulphur dioxide** dissolves in water.

•**Hydrogen sulphide** (H_2S). A colourless, poisonous gas, smelling of bad eggs. It dissolves in water to form a **weak acid***. It is given off when organic matter rots and when a dilute acid is added to a metal sulphide.

•**Sulphates** or **sulphate(VI) compounds**. Solid **ionic compounds*** that contain a **sulphate ion** (SO_4^{2-}) and a **cation***. Many occur naturally, e.g. **calcium sulphate** ($CaSO_4$). They are **salts*** of **sulphuric acid**, made by adding **bases*** to dilute sulphuric acid.

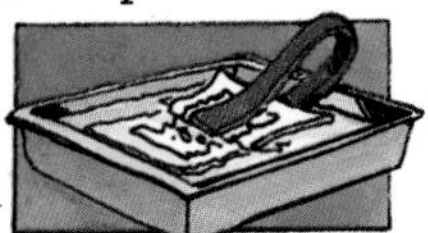

Sodium sulphate solution is used to "fix" photographs. This process stops prints going completely black when exposed to light.

•**Sulphites** or **sulphate(IV) compounds**. **Ionic compounds*** containing a **sulphite ion** (SO_3^{2-}) and a metal **cation*** e.g. **sodium sulphite** (Na_2SO_3). They are **salts*** of **sulphurous acid**, and react with dilute **strong acids*** giving off **sulphur dioxide**.

•**Sulphuric acid** (H_2SO_4) or **sulphuric(VI) acid**. An oily, colourless, corrosive liquid. It is a **dibasic*** acid, made by the **contact process**. **Concentrated sulphuric acid** contains about 2% water, is **hygroscopic*** and a powerful **oxidizing** and **dehydrating agent***. **Dilute sulphuric acid**, a **strong acid***, contains about 90% water. It reacts with metals above hydrogen in the **electrochemical series*** to give the metal **sulphate** and hydrogen.

▲**Concentrated sulphuric acid** is an **oxidizing agent***.

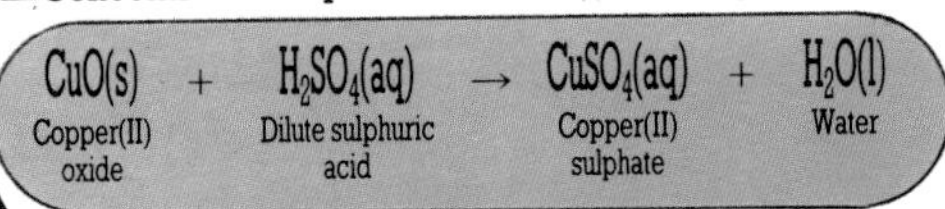

▲**Dilute sulphuric acid** reacts with a **base*** to give a **sulphate**.

Sulphuric acid is used to make many things e.g. fertilizers, man-made fibres, detergents and paints.

The reaction between **concentrated sulphuric acid** and water is very violent. To avoid accidents when the two are mixed, the acid is always added slowly to the water and not vice versa.

•**Contact process**. The industrial process used to make **sulphuric acid**.

Contact process

Dry and pure **sulphur dioxide** and air are passed over a **catalyst*** of vanadium pentoxide at 450°C.

$$2SO_2(g) + O_2(g) \rightarrow 2SO_3(g)$$

Sulphur dioxide + Oxygen → Sulphur trioxide

Sulphur trioxide is formed.

Sulphur trioxide is absorbed by **concentrated sulphuric acid**, and **fuming sulphuric acid**, or **oleum**, is formed.

$$SO_3 + H_2SO_4 \rightarrow H_2S_2O_7$$

Sulphur trioxide + Concentrated sulphuric acid → Fuming sulphuric acid

Fuming sulphuric acid is diluted to form **sulphuric acid**.

$$H_2S_2O_7 + H_2O \rightarrow 2H_2SO_4$$

Fuming sulphuric acid + Water → Sulphuric acid

* **Base**, 151; **Bleach**, 344; **Catalyst**, 151; **Cation**, 130; **Covalent compounds**, 132; **Dehydrating agent**, 344; **Dibasic**, 153; **Electrochemical series**, 159; **Fumigation**, 344; **Hygroscopic**, 206; **Ionic compound**, 131; **Oxidizing agent, Reducing agent**, 148; **Salts**, 153; **Strong acid**, 152; **Volatile**, 345; **Weak acid**, 152.

Group VII, the halogens

The elements in **Group VII** of the periodic table are called the **halogens**, and their compounds and ions are generally known as **halides**. Group VII members are all non-metals and their reactivity decreases going down the group – the chart below shows some of their properties. For further information on group members, see below and pages 187-188.

Name of element	Chemical symbol	Relative atomic mass*	Electron configuration*	Oxidizing power*	Reactivity	Appearance
Fluorine	F	18.99	2,7	DECREASING ↓	DECREASING ↓	Pale yellow -green gas
Chlorine	Cl	35.45	2,8,7			Pale green -yellow gas
Bromine	Br	79.91	2,8,18,7			Dark-red fuming liquid
Iodine	I	126.90	2,8,18,18,7			Non-metallic, black-grey solid
Astatine	At	No stable isotope*				

The atoms of all the elements in Group VII contain seven electrons in their **outer shell*** and they all react to form both **ionic** and **covalent compounds***. The elements at the top of the group form more ionic compounds than those further down the group.

Fluorine is never used in school laboratories as it is very poisonous and attacks glass containers. **Chlorine**, **bromine** and **iodine** do not react with glass, but chlorine is very poisonous, and so are the gases given off by the other two.

The power of Group VII elements as **oxidizing agents*** decreases down the group. They can all oxidize the ions of any members below them in the group. For example, **chlorine** displaces both **bromide** and **iodide anions*** from solution by oxidizing them to **bromine** and **iodine** molecules respectively. Bromine only displaces iodide anions from solution and iodine cannot displace any halide anions from solution.

$$2KI(aq) + Br_2(l) \rightarrow 2KBr(aq) + I_2(s)$$

Bromine displaces **iodide anions*** from potassium iodide. Each iodide anion loses an electron (is **oxidized***) when it is displaced by a **bromide anion***.

•**Fluorine (F_2).** A member of Group VII of the periodic table. It is a **diatomic*** gas, extracted from **fluorospar (CaF_2)** and **cryolite (Na_3AlF_6)**. It is the most reactive member of the group and is a very powerful **oxidizing agent***. It reacts with almost all elements. See pictures for some examples of its uses.

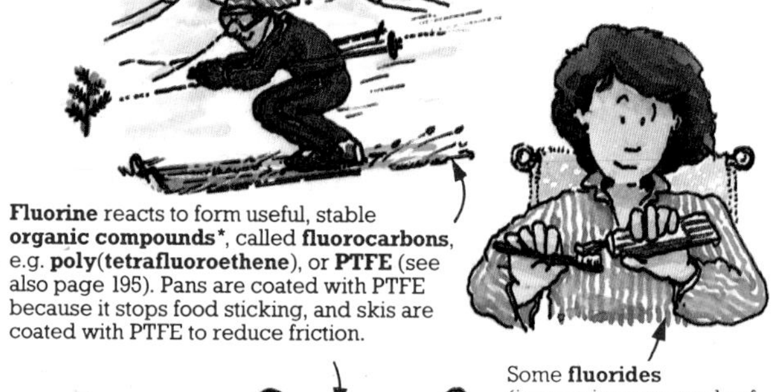

Fluorine reacts to form useful, stable **organic compounds***, called **fluorocarbons**, e.g. **poly(tetrafluoroethene)**, or **PTFE** (see also page 195). Pans are coated with PTFE because it stops food sticking, and skis are coated with PTFE to reduce friction.

Some **fluorides** (inorganic compounds of fluorine) are added to toothpastes, and in some countries to drinking water, to reduce tooth decay.

* **Bromide anion**, 188 (**Bromides**); **Covalent compounds**, 132; **Diatomic**, 124; **Electron configuration**, 127; **Iodide anion**, 188 (**Iodides**); **Ionic compound**, 131; **Isotope**, 127; **Organic compounds**, 190; **Outer shell**, 127; **Oxidation**, **Oxidizing agent**, 148; **Relative atomic mass**, 138.

• **Chlorine** (Cl_2). A member of Group VII of the periodic table. A poisonous, choking **diatomic*** gas which is very reactive and only occurs naturally in compounds. **Sodium chloride** (**NaCl**), its most important compound, is found in rock salt and brine. Chlorine is extracted from brine by **electrolysis***, using the **Downs' cell** (see **sodium**, page 168 and also **chlorine**, page 216). It is a very strong **oxidizing agent***. Many elements react with chlorine to form **chlorides** (see equation below).

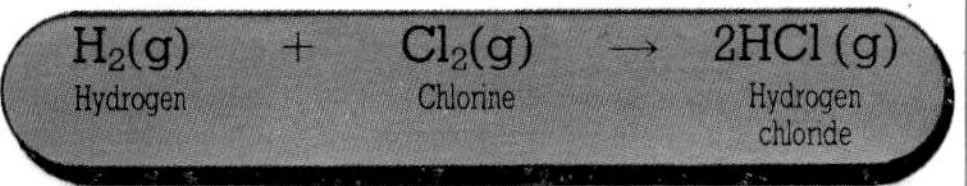

▲ In sunlight this reaction is explosive.

Chlorine has many uses. It is used to make **hydrochloric acid** (see **hydrogen chloride**), some organic solvents and also as a **germicide*** in swimming pools.

It is also used as a germicide in drinking water and disinfectants.

• **Chlorides**. Compounds formed when **chlorine** combines with another element. Chlorides of non-metals (see **hydrogen chloride**) are **covalent compounds***, usually liquids or gases. Chlorides of metals, e.g. **sodium chloride** (**NaCl**), are usually solid, water-soluble, **ionic compounds*** made of a **chloride anion*** (Cl^-) and metal **cation***. See also page 218.

• **Hydrogen chloride** (**HCl**). A colourless, **covalent*** gas that forms ions when dissolved in a **polar solvent***. It is made by burning hydrogen in **chlorine**. It reacts with ammonia and dissolves in water to form **hydrochloric acid**, a **strong acid***. **Concentrated hydrochloric acid**, 35% hydrogen chloride and 65% water, is a fuming, corrosive and colourless solution. **Dilute hydrochloric acid**, about 7% hydrogen chloride and 93% water, is a colourless solution that reacts with **bases***, and with metals above hydrogen in the **electrochemical series***. Concentrated hydrochloric acid is used industrially to remove rust from steel sheets before they are **galvanized***.

Concentrated hydrochloric acid is used to etch metals.

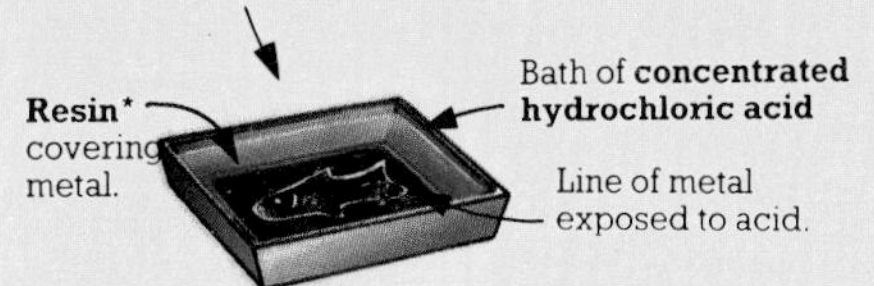

The metal exposed to the acid is eaten away leaving a groove in the surface. When printing a picture the groove is filled with ink.

• **Sodium hypochlorite** (**NaOCl**) or **sodium chlorate(I)**. A crystalline, white solid, stored dissolved in water, and formed when **chlorine** is added to a cold, dilute sodium hydroxide solution. It is used in domestic **bleach*** and also to bleach paper pulp white for writing.

• **Sodium chlorate** ($NaClO_3$) or **sodium chlorate(V)**. A white, crystalline solid, formed when **chlorine** is added to warm concentrated sodium hydroxide, and also when **sodium hypochlorite** is warmed.

Sodium chlorate kills weeds.

* **Anion**, 130; **Base**, 151; **Bleach**, 344; **Cation**, 130; **Covalent compounds**, 132; **Diatomic**, 124; **Electrochemical series**, 159; **Electrolysis**, 156; **Galvanizing**, 174; **Ionic compound**, 131; **Oxidizing agent**, 148; **Polar solvent**, 144; **Resin**, 345; **Strong acid**, 152.

Halogens continued

• **Bromine** (**Br_2**). A member of **Group VII** of the periodic table (the **halogens** – see chart, page 186). It is a **volatile***, **diatomic*** liquid that gives off a poisonous choking vapour. It is very reactive and only occurs naturally in compounds e.g. those found in marine organisms, rocks, sea water and some inland lakes. It is extracted from **sodium bromide** (**NaBr**) in sea water by adding chlorine. Bromine is a strong **oxidizing agent***. It reacts with most elements to form **bromides**, and dissolves slightly in water to give an orange solution of **bromine water**. Bromine compounds are used in medicine, photography, and disinfectants. It is used to make **1,2-dibromoethane** (**CH_2BrCH_2Br**) which is added to petrol to stop lead accumulating in engines.

Silver bromide crystal

Silver bromide is used in photographic film. When exposed to light it decomposes to form silver.

Film

Lens

Sun

Before exposure to light

After exposure to light

In areas of film exposed to light the silver bromide decomposes to form silver which appears black.

In areas of the film not exposed to light the silver bromide is unaffected.

• **Bromides**. Compounds of bromine and one other element. Bromides of non-metals are **covalent compounds*** (see **hydrogen bromide**). Bromides of metals are usually **ionic compounds*** made of **bromide anions*** (**Br^-**) and metal **cations***. Excepting **silver bromide** (**AgBr**), they are all water-soluble. See also page 218.

• **Hydrogen bromide** (**HBr**). A colourless, pungent-smelling gas, made by the reaction of **bromine** with hydrogen. Its chemical properties are similar to those of hydrochloric acid.

• **Iodine** (**I_2**). A member of **Group VII** of the periodic table (the **halogens** – see chart, page 186). A reactive, **diatomic***, crystalline solid. It is extracted from **sodium iodate** (**$NaIO_3$**) and seaweed. It is an **oxidizing agent*** and reacts with many elements to form **iodides**. When heated, it **sublimes***, giving off a purple vapour. Iodine is only slightly soluble in pure water, however, it dissolves well in **potassium iodide** (**KI**) solution and also in some organic solvents.

Lack of **iodine** in the diet means that the thyroid gland cannot produce enough **thyroxin** hormone. Thyroxin is needed to regulate body metabolism. People with a thyroxin deficiency suffer from goitre.

The main food sources of **iodine** are sea food, cod liver oil, fruit and vegetables. Some table salt has iodine added to it.

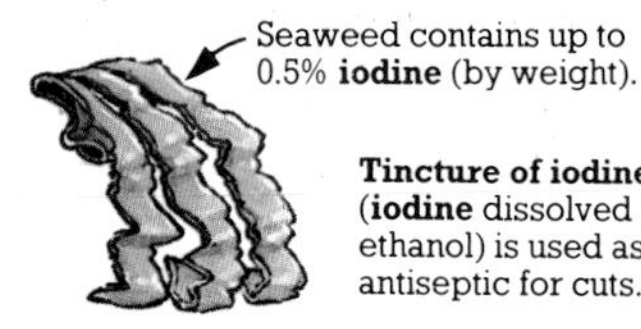

Seaweed contains up to 0.5% **iodine** (by weight).

Tincture of iodine (**iodine** dissolved in ethanol) is used as an antiseptic for cuts.

• **Iodides**. Compounds of **iodine** and one other element. Iodides of non-metals are **covalent compounds*** (see **hydrogen iodide**). Iodides of metals are usually **ionic***, made of **iodide anions*** (**I^-**) and metal **cations***. Except **silver iodide** (**AgI**), ionic iodides are water-soluble. See page 218.

• **Hydrogen iodide** (**HI**). A colourless gas with a pungent smell. It is a **covalent compound***, formed when hydrogen and **iodine** react. It dissolves in water to give a strongly **acidic*** solution called **hydroiodic acid** (its chemical properties are similar to those of hydrochloric acid).

* **Acidic**, 150; **Anion, Cation**, 130; **Covalent compounds**, 132; **Diatomic**, 124; **Ionic compound**, 131; **Oxidizing agent**, 148; **Sublimation**, 121; **Volatile**, 345.

Group VIII, the noble gases

The **noble gases**, also called **inert** or **rare gases**, make up **Group VIII** of the periodic table, also called **Group 0**. They are all **monatomic*** gases, obtained by the **fractional distillation of liquid air***. **Argon** forms 0.9% of the air and the other gases occur in even smaller amounts. They are all unreactive because their atoms' **electron configuration*** is very stable (they all have a full **outer shell***). The lighter members do not form any compounds, but the heavier members form a few.

- **Helium** (**He**). The first member of Group VIII of the periodic table. It is a colourless, odourless, **monatomic*** gas found in the atmosphere (one part in 200,000) and in some natural gases in the USA. It is obtained by the **fractional distillation of liquid air*** and is completely unreactive, having no known compounds. It is used in airships and balloons, as it is eight times less dense than air and not inflammable, and also by deep-sea divers to avoid "the bends".

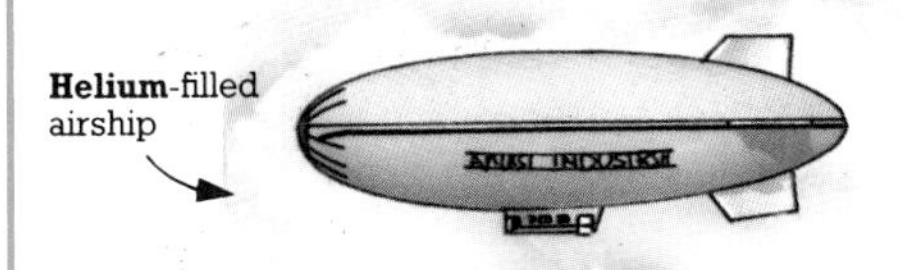

Helium-filled airship

- **Neon** (**Ne**). A member of Group VIII of the periodic table. A colourless, odourless **monatomic*** gas found in the atmosphere (one part in 55,000). It is obtained by the **fractional distillation of liquid air*** and is totally unreactive, having no known compounds. It is used in neon signs and fluorescent lighting as it emits an orange-red glow when an electric discharge passes through it at low pressure.

Neon sign

- **Radon** (**Rn**). The last member of Group VIII of the periodic table. It is **radioactive***, occurring as a result of the **radioactive decay*** of radium.

- **Argon** (**Ar**). The most abundant member of Group VIII of the periodic table. It is a colourless, odourless, **monatomic*** gas that makes up 0.9% of the air. Obtained by the **fractional distillation of liquid air***, it is totally unreactive, having no known compounds. It is used in electric light bulbs and fluorescent tubes.

Electric light bulb

- **Krypton** (**Kr**). A member of Group VIII of the periodic table. It is a colourless, odourless, **monatomic*** gas found in the atmosphere (one part in 670,000). It is obtained by the **fractional distillation of liquid air*** and is unreactive, only forming one known compound, **krypton fluoride** (**KrF_2**). Krypton is used in some lasers and in fluorescent tubes.

Krypton is used in the light bulbs on miners' helmets.

- **Xenon** (**Xe**). A member of Group VIII of the periodic table. A colourless, odourless, **monatomic*** gas found in the atmosphere (0.006 parts per million). Obtained from the **fractional distillation of liquid air***, it is unreactive, forming only a very few compounds, e.g. **xenon tetrafluoride** (**XeF_4**). It is used to fill fluorescent tubes and light bulbs.

Xenon is used in some lighthouse light bulbs.

* **Electron configuration**, 127; **Fractional distillation of liquid air**, 183; **Monatomic**, 124; **Outer shell**, 127; **Radioactive decay**, **Radioactivity**, 128.

Organic chemistry

Originally **organic chemistry** was the study of chemicals found in living organisms. However, it now refers to the study of all carbon-containing compounds, except the **carbonates*** and the **oxides*** of carbon. There are well over two million such compounds (**organic compounds**), more than all the other chemical compounds added together. This vast number of **covalent compounds*** is possible because carbon atoms can bond with each other to make a huge variety of **chains** and **rings**.

- **Aliphatic compounds**. Organic compounds whose molecules contain a **main chain** of carbon atoms. The chain may be **straight**, **branched** or even in **ring** form (though never a **benzene ring** – see **aromatic compounds**).

Straight chain of carbon atoms in a butan-1-ol molecule. No carbon atom is bonded to more than two other carbons.

Main chain – the longest continuous chain of carbon atoms in the molecule.

Side chain – a shorter chain of carbon atoms coming off the main chain.

Branched chain of carbon atoms in a 3-methyl pentane molecule. In a branched chain a carbon atom may be bonded to more than two other carbon atoms.

Cyclohexane molecule. An example of a molecule containing a ring of carbon atoms.

- **Aromatic compounds**. Organic compounds whose molecules contain a **benzene ring**. A benzene ring has six carbon atoms but differs from an **aliphatic** ring because bonds between carbon atoms are neither **single** nor **double bonds*** but midway between both in length and reactivity.

There are two possible ways of representing a **benzene ring**.

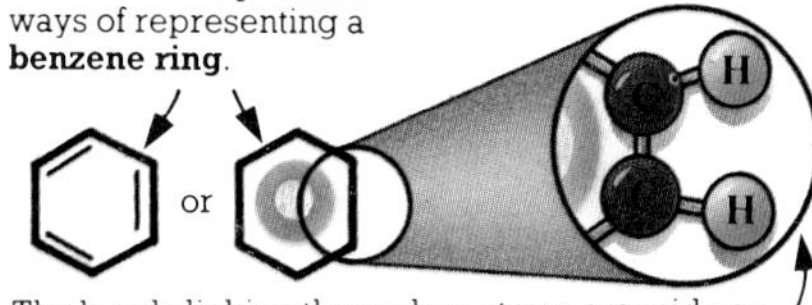

The bonds linking the carbon atoms are midway between **single** and **double bonds*** because some electrons are free to move around the molecule.

- **Stereochemistry**. The study of the 3-dimensional (3-D) structure of molecules. Comparing the 3-D structure of very similar organic molecules, e.g. **stereoisomers**, helps distinguish between them. The 3-D structure of a molecule is often shown by a **stereochemical formula*** – a diagram that shows how atoms are arranged in space.

Structural formula* of methane. This simplified version of the molecule does not show the 3-D arrangement of the atoms.

Stereochemical formula* of methane.

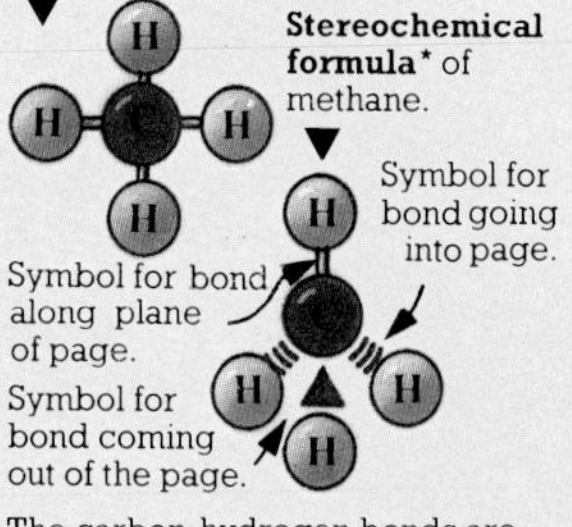

The carbon-hydrogen bonds are arranged to form a tetrahedron.

- **Isomers**. Two or more compounds with the same **molecular formula***, but different arrangements of atoms in their molecules. As a result the compounds have different properties. There are two main types of isomer, **structural isomers** and **stereoisomers**. ▶

* **Carbonates**, 179; **Covalent compounds**, **Double bond**, 132; **Molecular formula**, 140; **Oxides**, 183; **Single bond**, 132; **Stereochemical formula**, **Structural formula (shortened)**, 140.

•Hydrocarbons. Organic compounds that contain only carbon and hydrogen atoms.

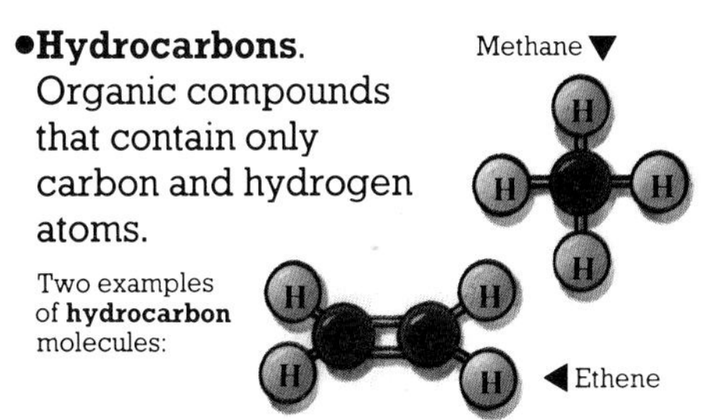

Two examples of **hydrocarbon** molecules:

•Functional group. An atom or group of atoms that gives a molecule most of its chemical properties. Organic molecules can have several such groups (see also pages 194-195).

Most **functional groups** contain at least one atom that is not carbon or hydrogen.

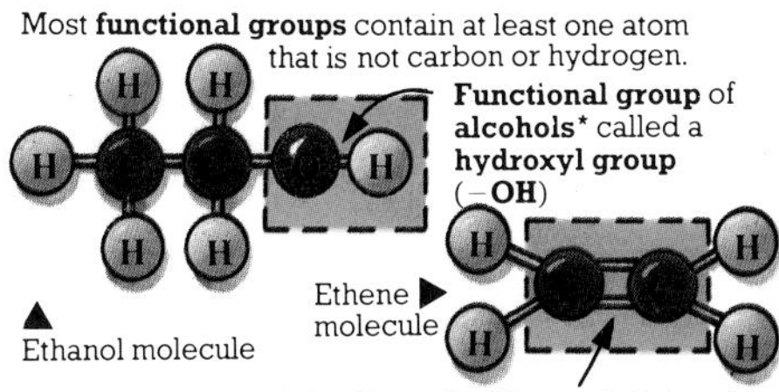

Two carbon atoms joined by a **double** or **triple bond*** are also **functional groups**.

•Saturated compounds. Organic compounds whose molecules only have **single bonds*** between atoms.

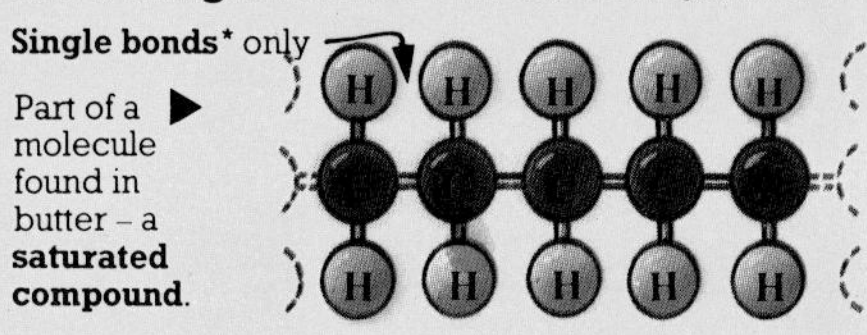

•Unsaturated compounds. Organic compounds whose molecules have at least one **double** or **triple bond***.

•Polyunsaturated compounds. A term used for compounds whose molecules have many **double** or **triple bonds***, e.g. those found in soft margarines.

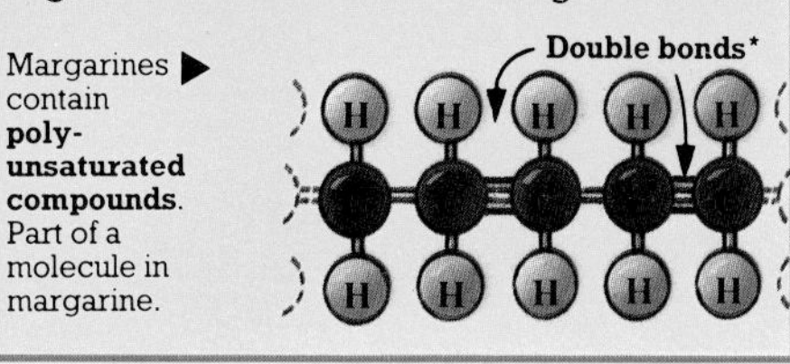

•Homologous series. A group of organic compounds which increase in size by adding a $-CH_2-$ group each time. All series (except the **alkanes***) also have a **functional group**, e.g. the **alcohol* hydroxyl group** (**–OH**). Members of a series have similar chemical properties but their physical properties change as they get larger. A homologous series has a **general formula** for all its members.

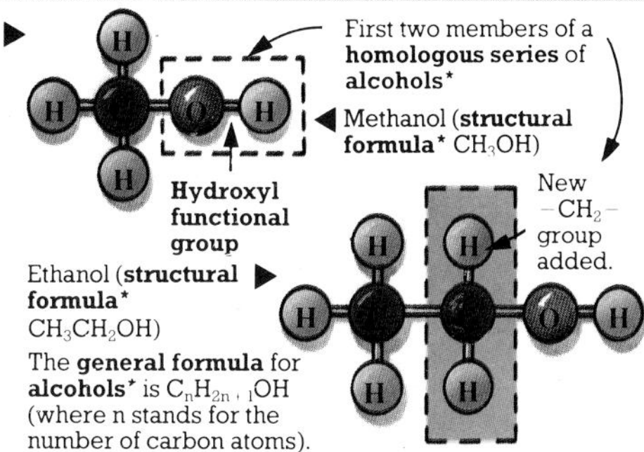

The **general formula** for **alcohols*** is $C_nH_{2n+1}OH$ (where n stands for the number of carbon atoms).

•Structural isomers. Compounds with the same **molecular formula***, but different **structural formulae***, i.e. the atoms are arranged in different ways.

The **molecular formula*** C_2H_6O has two different **structural formulae***

CH_3CH_2OH

CH_3OCH_3

Two **structural isomers**

Ethanol

Methoxymethane

•Stereoisomers. Compounds with the same **molecular formula*** and grouping of atoms but a different 3-D appearance.

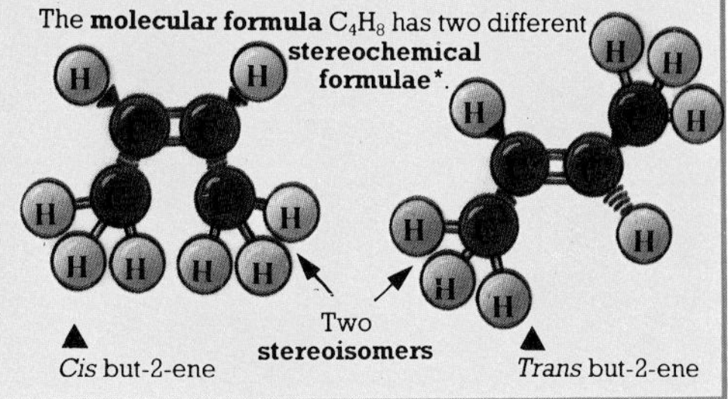

* **Alcohols**, 196; **Alkanes**, 192; **Double bond**, 132; **Molecular formula**, 140; **Single bond**, 132; **Stereochemical formula**, **Structural formula (shortened)**, 140; **Triple bond**, 132.

Alkanes

Alkanes, or **paraffins**, are all **saturated* hydrocarbons*** and **aliphatic compounds***. They form a **homologous series*** which has a **general formula*** of C_nH_{2n+2}. As the molecules in the series increase in size, so the physical properties of the compounds change (see below).

Some properties of alkanes

Name of compound	Molecular formula*	Structural formula*	Physical state at 25°C	Boiling point(°C)
Methane	CH_4	CH_4	Gas	−161.5
Ethane	C_2H_6	CH_3CH_3	Gas	−88.0
Propane	C_3H_8	$CH_3CH_2CH_3$	Gas	−42.2
Butane	C_4H_{10}	$CH_3CH_2CH_2CH_3$	Gas	−0.5
Pentane	C_5H_{12}	$CH_3CH_2CH_2CH_2CH_3$	Liquid	36.0
Hexane	C_6H_{14}	$CH_3CH_2CH_2CH_2CH_2CH_3$	Liquid	69.0

The first part of the name indicates the number of carbon atoms in the molecule. The -ane ending means the molecule is an alkane (see page 214).

The next molecule in the series is always a $-CH_2-$ group longer.

Gradual change of state as molecules get longer.

The boiling points of the alkanes increase regularly as the molecules get longer. Melting points and densities follow the same trend, getting higher as the molecules increase in size.

Alkanes are **non-polar molecules***. They burn in air to form carbon dioxide and water, and react with **halogens***, otherwise they are unreactive. Excepting **methane**, they are obtained from **petroleum***. They are used as fuels and to make other organic substances, e.g. plastics.

- **Methane (CH_4)**. The simplest alkane. It is a colourless, odourless, inflammable gas, which reacts with **halogens*** (see picture below) and is a source of hydrogen. **Natural gas** is composed of 99% methane.
- **Ethane (C_2H_6)**. A member of the alkanes. A gas found in small amounts in **natural gas** (see **methane**), but mostly obtained from **petroleum***. Its properties are similar to methane. It is used to make other organic chemicals.
- **Propane (C_3H_8)**. A member of the alkanes. A gas that is usually obtained from **petroleum***. Its properties are similar to **ethane**. It is bottled and sold as fuel for cooking and heating.
- **Cycloalkanes**. Alkane molecules whose carbon atoms are joined in a ring, e.g. **cyclohexane** (see picture page 190). Their properties are similar to other alkanes.
- **Substitution reaction**. A reaction in which an atom or **functional group*** of a molecule is replaced by a different atom or functional group. The molecules of **saturated compounds***, e.g. alkanes, can undergo substitution reactions, but not **addition reactions**.

Alkanes react with **halogens*** by undergoing a **substitution reaction**. Here is an example:

A chlorine atom is substituted for the hydrogen atom.

Methane | Chlorine | Chloromethane | Hydrogen chloride

* **Aliphatic compounds**, 190; **Functional group**, 191; **General formula**, 191 (**Homologous series**); **Halogens**, 186; **Hydrocarbons**, 191; **Molecular formula**, 140; **Non-polar molecule**, 133; **Petroleum**, 198; **Saturated compounds**, 191; **Structural formula (shortened)**, 140.

Alkenes

Alkenes, or **olefins**, are **unsaturated* hydrocarbons*** and **aliphatic compounds***. Alkene molecules contain one or more **double bonds*** between carbon atoms. Those with only one form a **homologous series*** with the **general formula*** C_nH_{2n}. As the molecules increase in size their physical properties change gradually (see below).

Some properties of alkenes

Name of compound	Molecular formula*	Structural formula*	Physical state at 25°C	Boiling point(°C)
Ethene	C_2H_4	$CH_2{=}CH_2$	Gas	−104.0
Propene	C_3H_6	$CH_3CH{=}CH_2$	Gas	−47.0
But-1-ene	C_4H_8	$CH_3CH_2CH{=}CH_2$	Gas	−6.0
Pent-1-ene	C_5H_{10}	$CH_3CH_2CH_2CH{=}CH_2$	Liquid	30.0

The number denotes the position of the **double bond*** in the molecule. Alkenes are named in the same way as **alkanes**, but end in -ene, not -ane (see page 214).

Each molecule is a $-CH_2-$ group longer. The position of the **double bond*** is shown.

Gradual change from gases to liquids to solids as the molecules get longer.

As the molecules get longer, the boiling points of the alkenes increase regularly. Melting points and densities follow the same trend.

Alkenes are **non-polar molecules***. They burn with a smoky flame and in excess oxygen are completely **oxidized*** to carbon dioxide and water. Alkenes are more reactive than **alkanes**, because of their double bond – they undergo **addition reactions**, and some form **polymers***. Alkenes are made by **cracking*** alkanes and are used to make many products including plastics and antifreeze.

- **Ethene** (C_2H_4) or **ethylene**. The simplest alkene (see chart above) – it is a colourless, sweet-smelling gas which undergoes **addition reactions** to form polymers of **polythene**. It is used to make plastics, ethanol, and many other organic chemicals.

- **Propene** (C_3H_6) or **propylene**. A member of the alkenes. It is a colourless gas used to make propanone and **poly(propene)**, also called **polypropylene**.

Some kitchen tools are made from **poly(propene)**, the **polymer*** of **propene**.

- **Addition reaction**. A reaction in which two molecules react together to produce a single larger molecule. One of the molecules must be **unsaturated*** (have a **double** or **triple bond***).

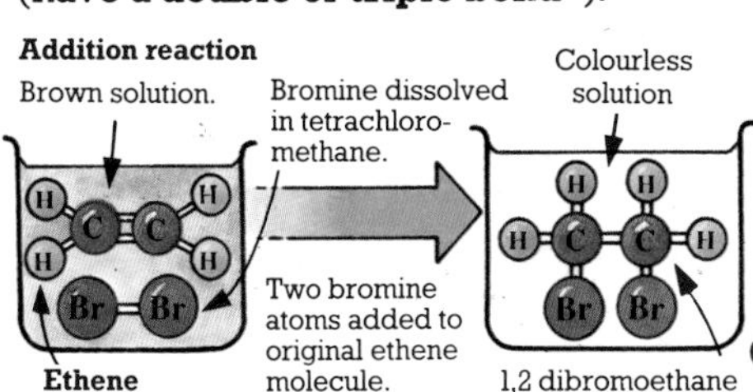

The change of colour is used as a test for **unsaturated compounds*** like alkenes.

- **Hydrogenation**. An **addition reaction** in which hydrogen atoms are added to an **unsaturated compound*** molecule.

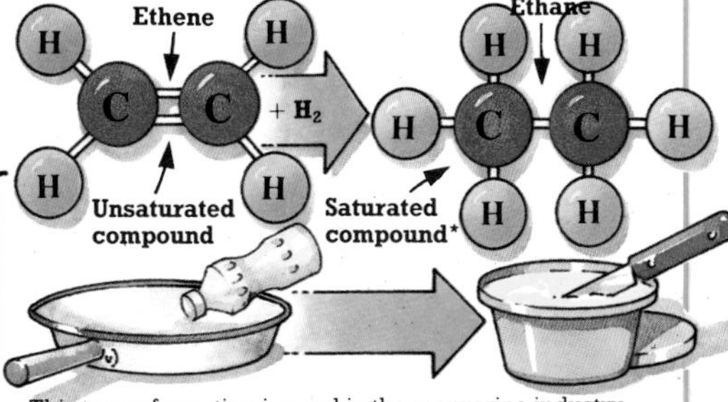

This type of reaction is used in the margarine industry to harden animal and vegetable oils. (These oils are **unsaturated compounds***, but not alkenes.)

* **Aliphatic compounds**, 190; **Cracking**, 198; **Double bond**, 132; **General formula**, 191 (**Homologous series**); **Hydrocarbons**, 191; **Molecular formula**, 140; **Non-polar molecule**, 133 (**Polar molecule**); **Oxidation**, 148; **Polymers**, 200; **Saturated compounds**, 191; **Structural formula (shortened)**, 140; **Triple bond**, 132; **Unsaturated compounds**, 191.

Alkynes

Alkynes, or **acetylenes**, are **unsaturated*** (each molecule has a carbon-carbon **triple bond***) and **aliphatic compounds***. They are **hydrocarbons*** and form a **homologous series*** with a **general formula*** C_nH_{2n-2}. Alkynes are named in the same way as **alkanes***, but end in -yne, not -ane (see page 214). They are **non-polar molecules*** with chemical properties similar to **alkenes***. They burn with a sooty flame in air and a very hot flame in pure oxygen. Alkynes are produced by **cracking***. They are used to make plastics and solvents.

Name of compound	Structural formula*
Ethyne	$CH{\equiv}CH$
Propyne	$CH_3C{\equiv}CH$
Butyne	$CH_3CH_2C{\equiv}CH$

- **Ethyne** (C_2H_2) or **acetylene**. The simplest member of the alkynes. A colourless gas, less dense than air and with a slightly sweet smell. It is the only common alkyne. Ethyne undergoes the same reactions as the other alkynes but more vigorously, e.g. it reacts explosively with chlorine. It is made by **cracking*** and is used to make polyvinyl chloride (PVC) and other vinyl compounds.

Molecule of **ethyne**.

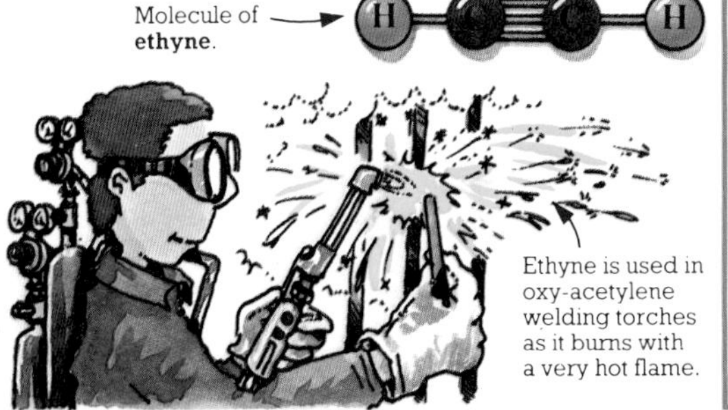

Ethyne is used in oxy-acetylene welding torches as it burns with a very hot flame.

More homologous series

The following groups of organic compounds each form a **homologous series*** of **aliphatic compounds***. Each series has a particular **functional group*** and its members have similar chemical properties.

- **Aldehydes**. Compounds that contain a **-CHO functional group***. They form a **homologous series*** with a **general formula*** $C_nH_{2n+1}CHO$ and are named like **alkanes*** but end in -anal, not -ane (see page 215). They are colourless liquids (except **methanal**) and **reducing agents***, and undergo **addition***, **condensation*** and **polymerization reactions***. When **oxidized***, they form **carboxylic acids**.

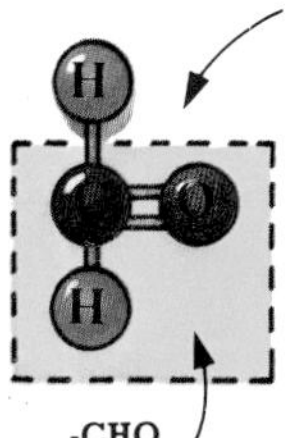

-CHO functional group*

Molecule of **methanal** (**HCHO**) or **formaldehyde**, the simplest **aldehyde**. It is a colourless, poisonous gas with a strong smell. It dissolves in water to make **formalin** – used to preserve biological specimens. It is also used to make **polymers*** and adhesives.

- **Ketones**. Compounds that contain a **carbonyl group** (a **-CO- functional group***). Ketones form a **homologous series*** with a complex **general formula***. They are named like **alkanes*** but end in -anone, not -ane. Most are colourless liquids. They have chemical properties similar to **aldehydes** but are not **reducing agents***.

Carbonyl group (-CO-)

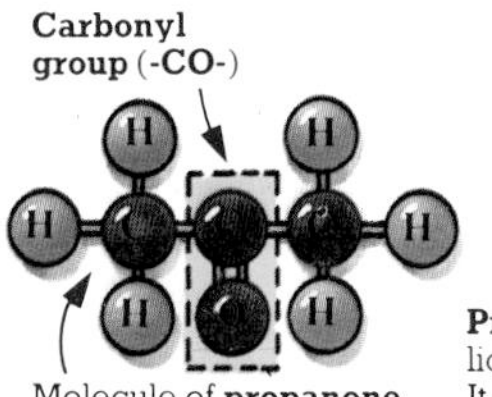

Molecule of **propanone** (CH_3COCH_3) or **acetone**, the simplest **ketone**

Propanone is a colourless liquid that mixes with water. It is used to make **acrylic**, and as an organic solvent, e.g. to remove nail varnish.

* **Addition reaction**, 193; **Aliphatic compounds**, 190; **Alkanes**, 192; **Alkenes**, 193; **Condensation reaction**, 197; **Cracking**, 198; **Functional group**, 191; **General formula**, 191 (**Homologous series**); **Hydrocarbons**, 191; **Non-polar molecule**, 133; **Oxidation**, 148; **Polymerization reaction**, **Polymers**, 200; **Reducing agent**, 148; **Structural formula**, 140; **Triple bond**, 132; **Unsaturated compounds**, 191.

• **Carboxylic acids**. Compounds that contain a **carboxyl group** (a $-COOH$ **functional group***) and form a **homologous series*** with a **general formula*** $C_nH_{2n+1}COOH$. Their names end in -anoic acid (see page 215). Pungent, colourless **weak acids***, they react with **alcohols*** to give **esters** (see **condensation reaction**, page 197).

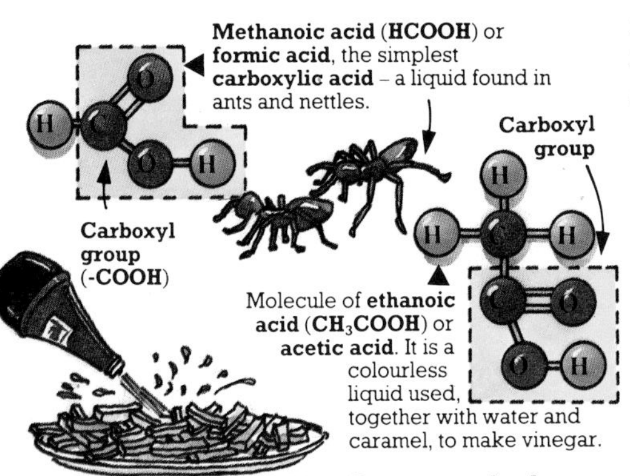

Methanoic acid (**HCOOH**) or **formic acid**, the simplest **carboxylic acid** – a liquid found in ants and nettles.

Molecule of **ethanoic acid** (CH_3COOH) or **acetic acid**. It is a colourless liquid used, together with water and caramel, to make vinegar.

• **Dicarboxylic acids**. Compounds that contain two **carboxyl groups** (see **carboxylic acids**) in each molecule.

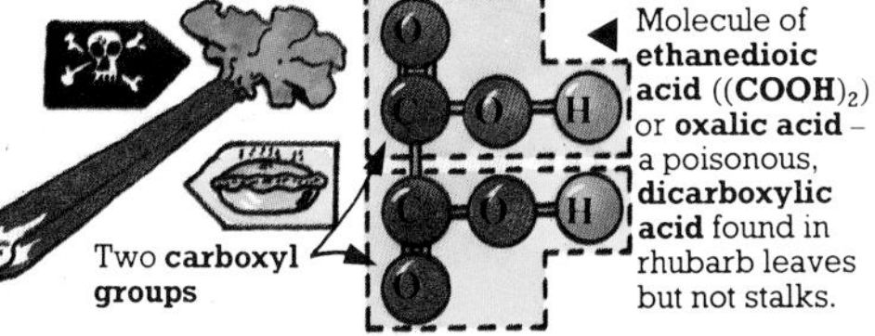

Molecule of **ethanedioic acid** ($(COOH)_2$) or **oxalic acid** – a poisonous, **dicarboxylic acid** found in rhubarb leaves but not stalks.

• **Esters**. A **homologous series*** of compounds containing a $-COO-$ **functional group*** in every molecule. They are unreactive, colourless liquids made by reacting a **carboxylic acid** and **alcohol*** (see **condensation reaction**, page 197). Found in vegetable oils and animal fats, they give fruit and flowers their flavours and smells. They are used in perfumes and flavourings.

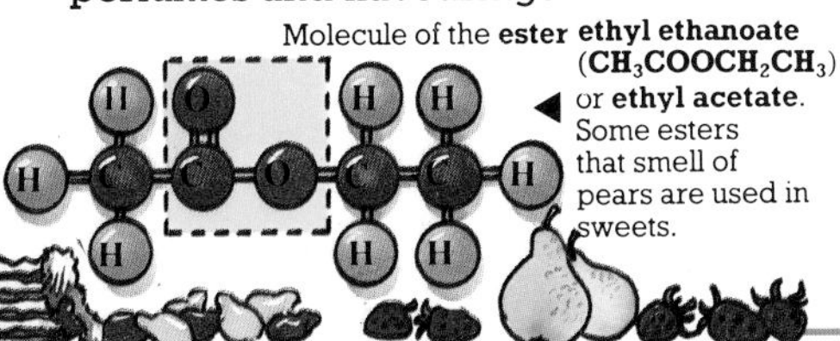

Molecule of the ester **ethyl ethanoate** ($CH_3COOCH_2CH_3$) or **ethyl acetate**. Some esters that smell of pears are used in sweets.

• **Halogenoalkanes** or **alkyl halides**. A **homologous series*** that contains one or more **halogen*** atoms (see also page 215). Most halogenoalkanes are colourless, **volatile*** liquids which do not mix with water. They will undergo **substitution reactions***. The most reactive contain iodine, and the least reactive contain fluorine.

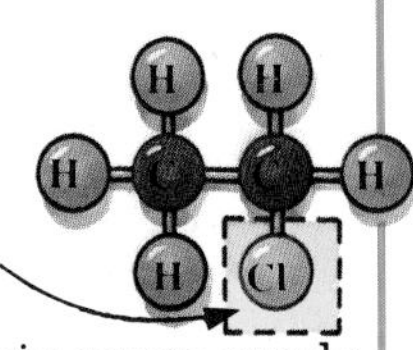

Molecule of **chloroethane** (CH_3CH_2Cl), a **halogenoalkane**. Used to keep fridges cold (see **refrigerant**, page 345).

The chlorine atom is the **halogen* functional group***. It is called a **chloro group** (**−Cl**) (see page 215).

Some important organic compounds have more than one **halogen*** atom in their molecules.

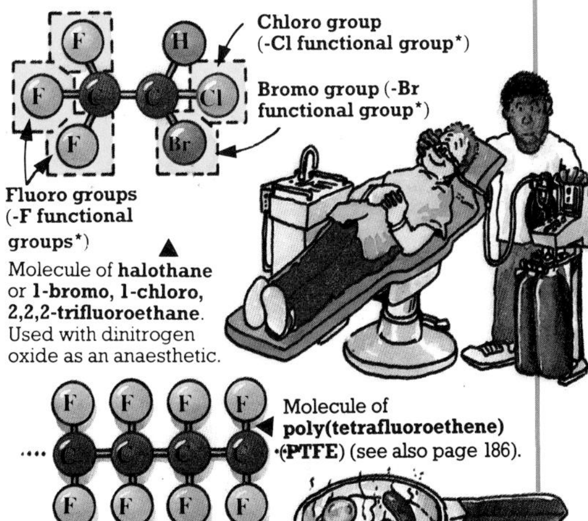

Molecule of **halothane** or **1-bromo, 1-chloro, 2,2,2-trifluoroethane**. Used with dinitrogen oxide as an anaesthetic.

Molecule of **poly(tetrafluoroethene)** (**PTFE**) (see also page 186).

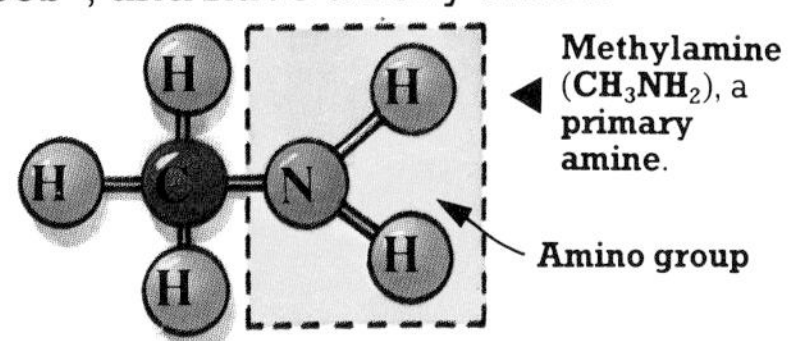

PTFE is used as a non-stick coating on saucepans.

• **Primary amines**. Compounds that contain an **amino group** ($-NH_2$ **functional group***). They are **weak bases***, and have a fishy smell.

Methylamine (CH_3NH_2), a **primary amine**.

Amino group

• **Diamines**. Compounds with two **amino groups** in each molecule.

* **Alcohols**, 196; **Functional group**, 191; **General formula**, 191 (**Homologous series**); **Halogens**, 186; **Substitution reaction**, 192; **Volatile**, 345; **Weak acid, Weak base**, 152.

Alcohols

Alcohols are organic compounds that contain one or more **hydroxyl groups** (**-OH functional groups***) in each molecule. The alcohols shown below in the chart are all members of a **homologous series*** of alcohols which are **aliphatic compounds*** with the **general formula*** $\mathbf{C_nH_{2n+1}OH}$. As the molecules in the series increase in size, their physical properties change steadily. Some of the trends are shown here:

Some properties of alcohols			
Name of compound	**Structural formula***	**Physical state at 25°C**	**Boiling point (°C)**
Methanol	CH_3OH	Liquid	65.6
Ethanol	CH_3CH_2OH	Liquid	78.5
Propan-1-ol	$CH_3CH_2CH_2OH$	Liquid	97.2
Butan-1-ol	$CH_3CH_2CH_2CH_2OH$	Liquid	117.5

Alcohols are named in the same way as **alkanes***, but end in -anol. The number in the name tells you which carbon the **hydroxyl group** is attached to (see opposite and page 214).

The next member of the series (going down) is always a $-CH_2-$ group longer than the last.

The members gradually change to solids as the molecules get longer.

Boiling points of alcohols increase as the molecules get longer. They have high boiling points in relation to their **relative molecular mass***, due to **hydrogen bonding***.

As a result of their **hydroxyl groups**, alcohol molecules are **polar***, and have **hydrogen bonds***. Short chain alcohols mix completely with water, but long chain alcohols do not as their molecules have more $-CH_2-$ groups, making them less polar. Alcohols do not **ionize*** in water and are **neutral***. They burn, giving off carbon dioxide and water.

Alcohols react with sodium:

$$2CH_3CH_2OH + 2Na \rightarrow 2CH_3CH_2ONa + H_2$$

Ethanol + Sodium → Sodium ethoxide + Hydrogen

Alcohols react with phosphorus halides to give **halogenoalkanes** (see page 195), and with **carboxylic acids** to give **esters** (see **condensation reaction** and page 195).

Primary alcohols are **oxidized*** first to **aldehydes*** and then to **carboxylic acids***.

Acidified potassium permanganate **catalyst***

$$CH_3CH_2CH_2OH \rightarrow CH_3CH_2CHO \rightarrow CH_3CH_2COOH$$

Propan-1-ol → Propanal → Propanoic acid

Secondary alcohols are **oxidized*** to **ketones** (see page 194).

Acidified potassium permanganate **catalyst***

$$CH_3CHOHCH_3 \rightarrow CH_3COCH_3$$

Propan-2-ol → Propanone

• **Ethanol** (**$\mathbf{CH_3CH_2OH}$**, often written **$\mathbf{C_2H_5OH}$**). Also called **ethyl alcohol**, or **alcohol**. An alcohol which is a slightly sweet smelling water-soluble liquid with a relatively high boiling point. It burns with an almost colourless flame and is made by ethene reacting with steam. It is also produced by **alcoholic fermentation**.

Ethanol is used as a solvent and in methylated spirits. It has many more uses - some are shown below.

A **breathalyser** is used to see if a person has drunk too much alcohol to drive safely. It contains an electronic device that measures the alcohol concentration in the breath.

* **Aldehydes**, 194; **Aliphatic compounds**, 190; **Alkanes**, 192; **Carboxylic acids**, 195; **Catalyst**, 161; **Functional group**, 191; **General formula**, 191 (**Homologous series**); **Hydrogen bond**, 134; **Ionization**, 130; **Neutral**, 151; **Oxidation**, 148; **Polar molecule**, 133; **Relative molecular mass**, 138; **Structural formula (shortened)**, 140.

•Alcoholic fermentation. The name of the process used to produce **ethanol** (the potent chemical in alcoholic drinks) from fruits or grain. **Glucose*** from fruit or grain is converted into ethanol by **enzymes* (catalysts*** of the reactions in living cells). **Yeast** is used in alcoholic fermentation because it has the enzyme **zymase** which catalyses the change of glucose to ethanol.

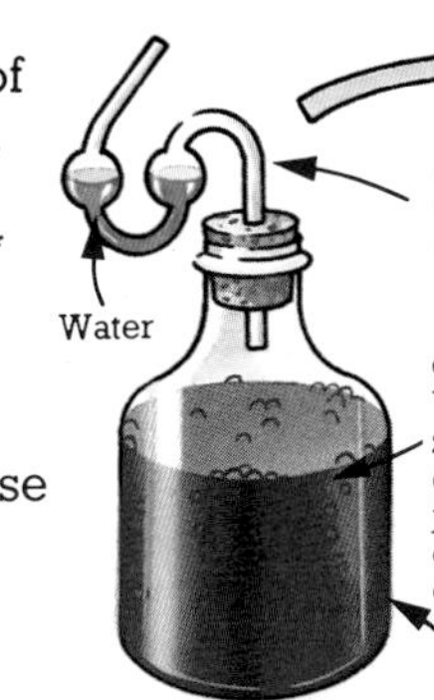

Wine making

Airlock stops air entering (it would **oxidize*** the **ethanol** to ethanoic acid).

Grape juice and yeast. Yeast **enzyme*** **zymase** ferments **glucose*** from the juice, making carbon dioxide gas and ethanol.

Demi-john (special container)

Yeast dies if **ethanol** concentration gets too high. Stronger alcoholic drinks, e.g. whisky, are made by **distilling*** the ethanol solution – this removes water and concentrates the ethanol, making the drink very potent.

$$C_6H_{12}O_6 \xrightarrow{\text{Enzyme*}} 2CH_3CH_2OH + 2CO_2$$

Glucose solution from fruit or barley → Ethanol + Carbon dioxide

Primary, secondary and tertiary alcohols

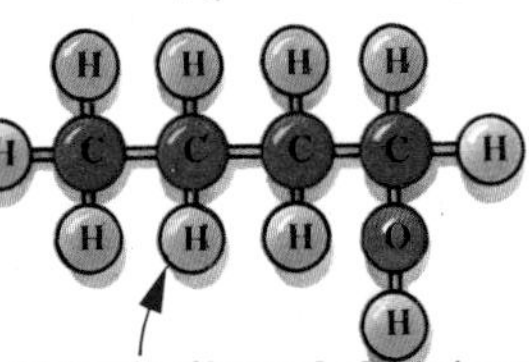

Molecule of **butan-1-ol, a primary alcohol**. The carbon atom attached to the hydroxyl group (see introduction) has two hydrogen atoms attached to it.

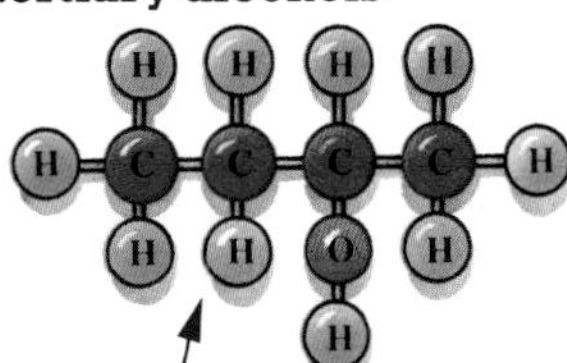

Molecule of **butan-2-ol,** a **secondary alcohol**. The carbon atom attached to the hydroxyl group (see introduction) has one hydrogen atom attached to it.

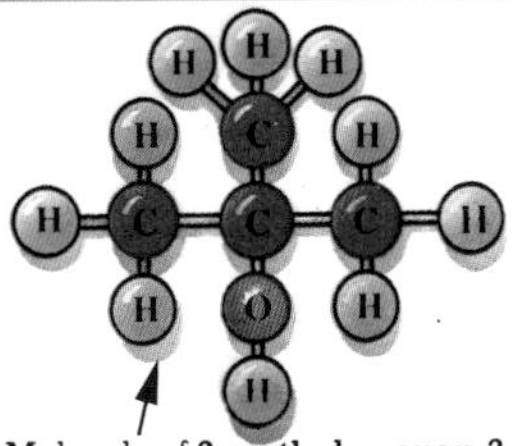

Molecule of **2-methyl propan-2-ol**, a **tertiary alcohol**. The carbon atom attached to the **hydroxyl group** (see introduction) has no hydrogen atoms attached to it.

The numbers in the names of the alcohols give the position of the carbon atom that the **hydroxyl group** is bonded to (see page 214 for more information on naming alcohols).

•Polyhydric alcohols. Alcohols whose molecules contain more than one **hydroxyl group** (see introduction).

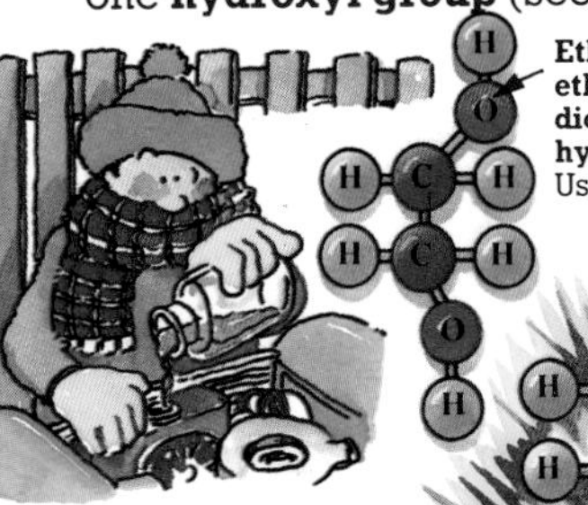

Ethane-1,2-diol, or **ethylene glycol** is a **diol** (contains two **hydroxyl groups**). Used as antifreeze.

Propane-1,2,3-triol, glycerine, or **glycerol**, is a **triol** (contains three **hydroxyl groups**). Used to make explosives.

•Condensation reaction. A type of reaction in which two molecules react together to form one, with the loss of a small molecule, e.g water. (See also **condensation polymerization**, page 200.)

Example of a **condensation reaction**:

$$CH_3CH_2OH \rightarrow CH_3COOH \rightarrow CH_3COOCH_2CH_3 + H_2O$$

Ethanol → Ethanoic acid → Ethyl ethanoate + Water molecule is lost

This reaction is also an **esterification reaction** as the product ethyl ethanoate is an **ester***. An alcohol and an organic acid always react to form an ester.

* **Catalyst**, 161; **Distillation**, 220; **Enzyme**, 161; **Esters**, 195; **Glucose**, 204; **Oxidation**, 148.

Petroleum

Petroleum, or **crude oil**, is a dark, viscous liquid, usually found at great depths beneath the earth or sea-bed. It is often found with **natural gas***, which consists mainly of **methane***. Petroleum is formed over millions of years by the decomposition of animals and plants under pressure. It is a mixture of **alkanes*** which vary greatly in size and structure. Many useful products are produced by **refining** petroleum.

- **Refining**. A set of processes which convert petroleum to more useful products. Refining consists of three main processes – **primary distillation**, **cracking** and **reforming**.

- **Primary distillation** or **fractional distillation of petroleum**. A process used to separate petroleum into **fractions**, according to their boiling points (see also page 220). A **fractionating column** (see diagram) is kept very hot at the bottom, but it gets cooler towards the top. Boiled petroleum passes into the column as vapour, losing heat as it rises. When a fraction reaches a tray at a temperature just below its own boiling point, it condenses onto the tray. It is then drawn off along pipes. Fractions are distilled again to give better separations.

- **Fraction**. A mixture of liquids with similar boiling points, obtained from **primary distillation**. **Light fractions** have low boiling points and short **hydrocarbon*** chains. **Heavy fractions** have higher boiling points and longer chains.

- **Cracking**. A reaction which breaks large **alkanes*** into smaller alkanes and **alkenes***. The smaller alkanes are used as **gasoline**. Cracking occurs at high temperatures, or with a **catalyst*** (**catalytic cracking** or **"cat cracking"**).

$$C_9H_{20} \rightarrow C_7H_{16} + C_2H_4$$

C_9H_{20}	→	C_7H_{16}	+	C_2H_4
Alkane (Nonane)		**Alkane (Heptane)**		**Alkene (Ethene*)**

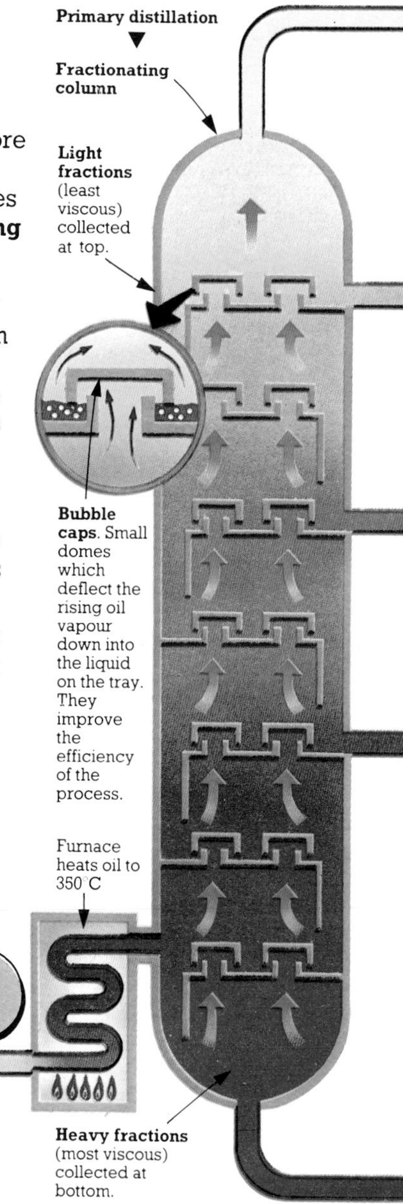

- **Reforming**. A process which produces **gasoline** from lighter **fractions** by breaking up **straight chain*** **alkanes*** and reassembling them as **branched chain*** molecules.

* **Alkanes**, 192; **Alkenes**, 193; **Branched chain**, 190; **Catalyst**, 161; **Ethene**, 193; **Hydrocarbons**, 191; **Natural gas**, 192 (**Methane**); **Straight chain**, 190.

•**Refinery gas**. A gas which consists mainly of **methane***. Other **light fractions** contain **propane** and **butane** (both **alkanes***) and are made into **liquefied petroleum gas** (**LPG**).

Liquefied petroleum gas (see **refinery gas**) is used as bottled gas.

•**Chemical feedstocks**. **Fractions** of petroleum which are used in the production of organic chemicals. These fractions are mainly **refinery gas** and **naptha**, a part of the **gasoline** fraction.

Chemical feedstocks

Refinery gas

•**Gasoline** or **petrol**. A liquid **fraction** obtained from **primary distillation**. It consists of **alkanes*** with 5 to 12 carbon atoms in their molecules and has a boiling point range of 40-150°C. See also **cracking** and **reforming**.

•**Octane rating**. A measure of how well **gasoline** burns, measured on a scale of 0 to 100. It is increased by using an **anti-knock agent** such as tetraethyl-lead ($Pb(OC_2H_5)_4$).

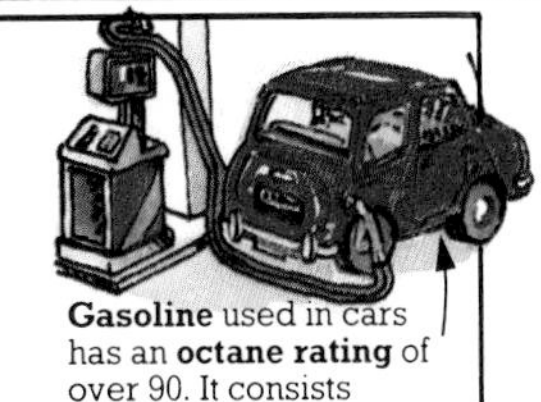

Gasoline used in cars has an **octane rating** of over 90. It consists mainly of **branched chain*** **alkanes***.

Gasoline

•**Kerosene** or **paraffin**. A liquid **fraction** obtained from **primary distillation**. Kerosene consists of **alkanes*** with about 9-15 carbon atoms in their molecules. It has a boiling point range of 150-250°C.

Kerosene is used as a **fuel*** in jet engines and domestic heaters.

Kerosene

•**Diesel oil** or **gas oil**. A liquid **fraction** obtained from **primary distillation**. It consists of **alkanes*** with about 12-25 or more carbon atoms in their molecules. It has a boiling point of 250°C and above.

Diesel oil is used as a **fuel*** in diesel engines.

Diesel oil

•**Residue**. The oil left after **primary distillation**. It consists of **hydrocarbons*** of very high **relative molecular masses***, their molecules containing up to 40 carbon atoms. Its boiling point is greater than 350°C. Some is used as **fuel oil**, the rest is re-distilled to form the substances on the right.

•**Lubricating oil**. A mixture of non-**volatile*** liquids obtained from the distillation of **residue** in a vacuum.

•**Hydrocarbon waxes** or **paraffin waxes**. Soft solids which are separated from **lubricating oil** after the distillation of **residue** in a vacuum.

Candles and polish

•**Bitumen** or **asphalt**. A liquid left after the distillation of **residue** under vacuum. It is a tarry, black semi-solid at room temperature.

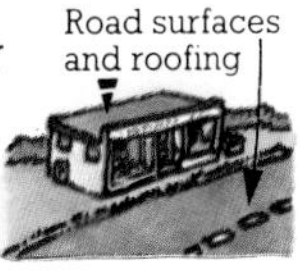

Road surfaces and roofing

Residue

* **Alkanes**, 192; **Branched chain**, 190; **Fuel**, 208; **Hydrocarbons**, 191; **Methane**, 192; **Relative molecular mass**, 138; **Volatile**, 345.

Polymers and plastics

Polymers are substances that consist of many **monomers** (small molecules) bonded together in a repeating sequence. They are very long molecules with a high **relative molecular mass***. Polymers occur naturally, e.g. **proteins***. There are also many **synthetic polymers**, e.g. **plastics**.

• **Monomers**. Relatively small molecules that react to form polymers. For example, **ethene*** molecules are monomers which react together to form **polythene** (see also equation for **homopolymer**). ▶

Simplified picture of a **polymerization reaction** – a reaction in which **monomers** bond to form a polymer.

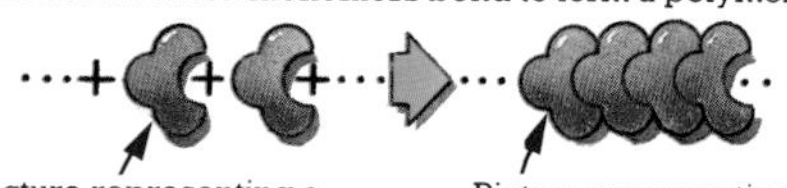

Picture representing a **monomer**.

Picture representing a **polymer**.

• **Addition polymerization**. Polymerization reactions in which **monomers** bond to each other without losing any atoms. The polymer is the only product and has the same **empirical formula*** as the monomer. See also **addition reaction**, page 193.

Example of an **addition polymerization** reaction

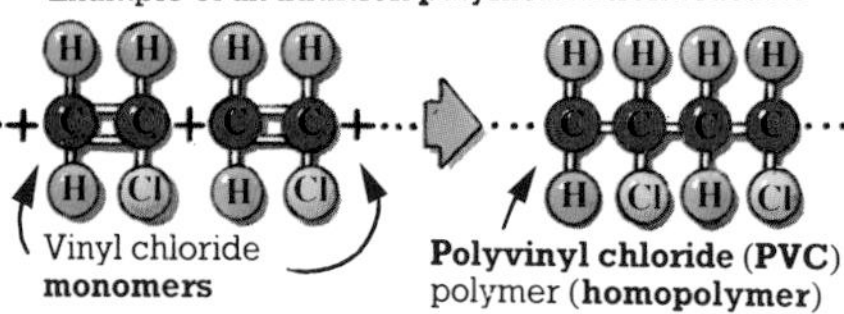

Vinyl chloride **monomers**

Polyvinyl chloride (PVC) polymer (**homopolymer**)

• **Condensation polymerization**. Polymerization reactions in which **monomers** form a polymer with the loss of small molecules such as water. See **condensation reaction**, page 197.

• **Homopolymer**. A polymer made from a single type of **monomer**.

Reaction to produce the **homopolymer polythene**.

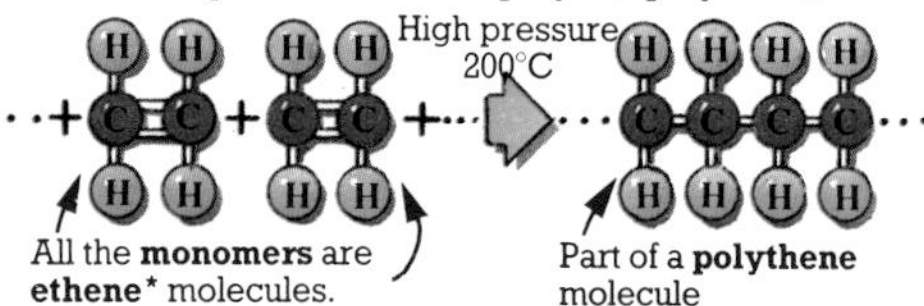

All the **monomers** are **ethene*** molecules.

Part of a **polythene** molecule

This is an **addition polymerization** reaction.

• **Copolymer**. A polymer made from two or more different **monomers**. See **condensation polymerization** example below.

• **Depolymerization**. The breakdown of a polymer into its original **monomers**. It occurs for example when **acrylic** is heated.

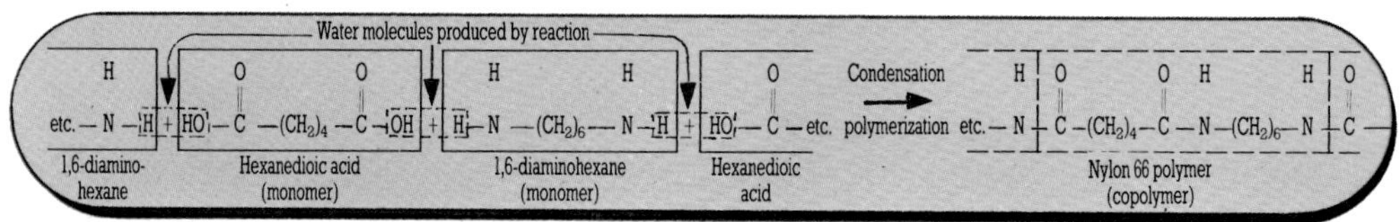

• **Synthetic** or **man-made polymers**. Polymers prepared in the laboratory or in industry (not natural polymers), e.g. **nylons**.

• **Plastics**. **Synthetic polymers** that are easily moulded. They are made from chemicals derived from **petroleum*** and are usually durable, light solids which are thermal and electrical insulators. They are often not **biodegradable*** and give off poisonous fumes when burnt. There are two types of plastic – **thermoplastics** which soften or melt on heating (e.g. **polythene**) and **thermosetting plastics** which harden on heating and do not remelt (e.g. plastic used in worktops).

Some uses of **plastics**

* **Biodegradable**, 210; **Empirical formula**, 140; **Ethene**, 193; **Petroleum**, 198; **Proteins**, 205; **Relative molecular mass**, 138.

• **Polyesters.** **Copolymers**, formed by the **condensation polymerization** of **diol*** and **dicarboxylic acid*** **monomers**. The monomers are linked by **-COO- functional groups***, as found in **esters***.

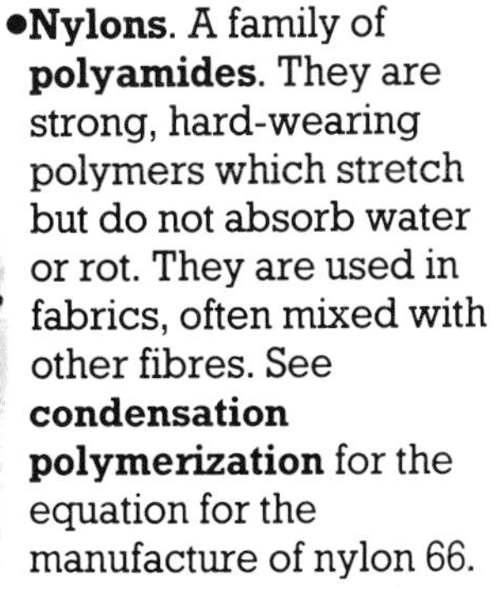

Yachts have sails made of **polyesters**.

Some **polyesters** are produced as fibres which are used in clothing and furnishing materials.

• **Polystyrene** or **poly(phenylethene)**. A **homopolymer** formed by the **addition polymerization** of **styrene** (**phenylethene**).

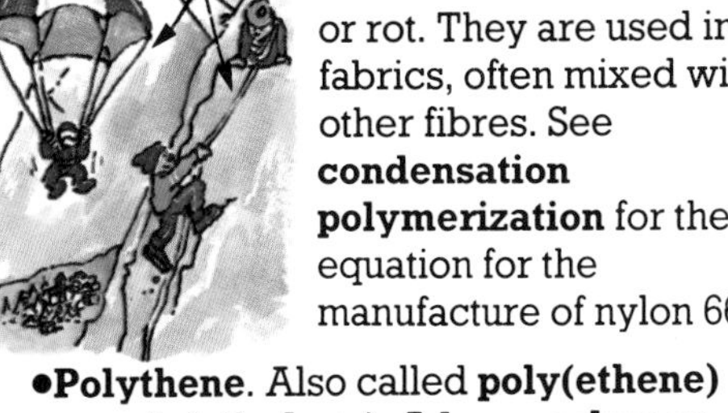

Nylons are used to make many things, e.g. parachutes and climbing ropes.

• **Nylons**. A family of **polyamides**. They are strong, hard-wearing polymers which stretch but do not absorb water or rot. They are used in fabrics, often mixed with other fibres. See **condensation polymerization** for the equation for the manufacture of nylon 66.

• **Polythene**. Also called **poly(ethene)** or **poly(ethylene)**. A **homopolymer** formed by the **addition polymerization** of **ethene*** (see **homopolymer**). Polythene is produced in two forms (depending on the method used) – a soft material of low density, and a hard, more rigid, material of high density. Polythene has a **relative molecular mass*** of between 10,000 and 40,000 and is used to make many things.

Polythene is used to make many kitchen utensils, e.g. washing-up bowls.

• **Polyamides**. **Copolymers** formed by the **condensation polymerization** of a **dicarboxylic acid* monomer** with a **diamine* monomer**, e.g. **nylons**.

Polystyrene is used to make disposable knives, forks and cups. Air-expanded sheets of polystyrene are used in packaging and insulation.

• **Polyvinyl chloride** (**PVC**) or **poly(chloroethene)**. A hard-wearing **homopolymer** used to make many things. Some examples are shown below. (See also **addition polymerization** picture.)

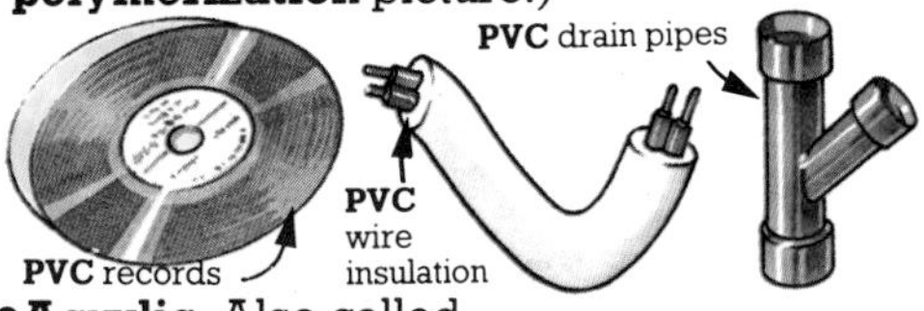

• **Acrylic**. Also called **poly(methylmethacrylate)** or **poly ((1-methoxycarbonyl)-1-methylethene)**. A **homopolymer** formed by **addition polymerization**. It is often used as a glass substitute.

Methyl methacrylate, the acrylic monomer

$$\begin{matrix} H & & & & CH_3 \\ & \diagdown & & \diagup & \\ & & C=C & & \\ & \diagup & & \diagdown & \\ H & & & & COOCH_3 \end{matrix}$$

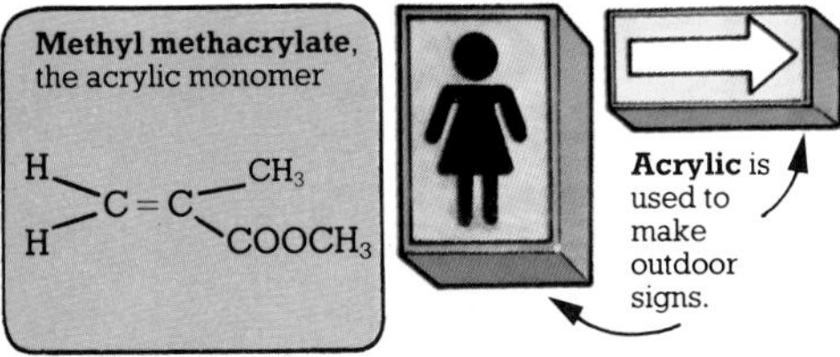

Acrylic is used to make outdoor signs.

• **Natural polymers** or **biopolymers**. Polymers that occur naturally, e.g. **starch** and **rubber**. Starch is made from **monomers** of **glucose***. For a picture of the starch polymer, see **starch**, page 204.

Part of a rubber polymer

$$\begin{matrix} CH_2 & & & CH_2 & & CH_3 & & H & & CH_2 & & H_2C \\ & \diagdown & & \diagup & & & \diagdown & \diagup & & & \diagdown & \diagup \\ & & C=C & & & CH_2 & C=C & & CH_2 & & C=C & \\ & \diagup & & \diagdown & & & & & & \diagup & & \diagdown \\ CH_3 & & & H & & & & & & CH_3 & & H \end{matrix}$$

Rubber polymer is extracted from **latex*** tapped from the rubber tree. It is then **vulcanized*** to produce the rubber used in tyres, hoses etc.

* **Carboxyl group**, 195 (**Carboxylic acids**); **Diamines**, **Dicarboxylic acids**, 195; **Diols**, 197 (**Polyhydric alcohols**); **Esters**, 195; **Ethene**, 193; **Functional group**, 191; **Glucose**, 204; **Latex**, 344; **Relative molecular mass**, 138; **Vulcanization**, 345.

Detergents

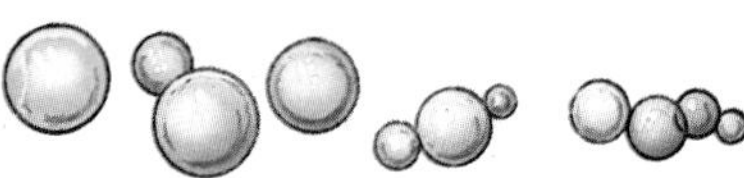

Detergents are substances which, when added to water, enable it to remove dirt. They do this in three ways: by lowering the water's **surface tension*** so that it spreads evenly instead of forming droplets, by enabling grease molecules to dissolve in water, and also by keeping removed dirt suspended in the water. **Soap** is a type of detergent, but there are also many **soapless detergents**.

- **Detergent molecule**. A large molecule consisting of a long **hydrocarbon*** chain with a **functional group*** at one end (making that end **polar***). The **non-polar*** chain is **hydrophobic** (repelled by water) and the polar end is **hydrophilic** (attracted to water). In water, these molecules form **micelles**.

Simple representation of a **detergent molecule**.

Hydrophilic functional group* (head end of molecule)

Hydrophobic hydrocarbon* chain (tail end of molecule)

- **Soap**. A type of detergent. It is the sodium or potassium **salt*** of a long-chain **carboxylic acid*** such as octadecanoic acid, see equation. It is made by reacting animal fats or vegetable oils (**esters***) with sodium hydroxide or potassium hydroxide solution (soap made with potassium hydroxide is softer). The process of making soap is **saponification**. Soap molecules form **micelles** in water. Soap produces a scum in **hard water*** whereas **soapless detergents** do not.

Soap-making factory

Saponification (**soap**-making)

Measured amounts of **fats*** and sodium hydroxide or potassium hydroxide solutions are continuously fed into a large, hollow, column-like structure. The column is at high temperature and pressure. **Soap** and propane-1,2,3-triol are formed.

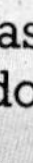

The propane-1,2,3-triol is dissolved in salt water. Then centrifuges used to separate **soap** from propane-1,2,3-triol in salt water.

The final part of the process is **fitting** or **finishing**. Any unreacted long-chain **carboxylic acids*** are **neutralized*** with alkali. Salt concentration is adjusted. Mixture is centrifuged to separate out **soap**.

Saponification equation

$$\begin{array}{c} C_{17}H_{35}COOCH_2 \\ | \\ C_{17}H_{35}COOCH \\ | \\ C_{17}H_{35}COOCH_2 \end{array} + 3NaOH \xrightarrow{\text{Saponification}} 3C_{17}H_{35}COO^-Na^+ + \begin{array}{c} CH_2OH \\ | \\ CHOH \\ | \\ CH_2OH \end{array}$$

Ester (from mutton fat) + Sodium hydroxide → Sodium octadecanoate (sodium stearate) + Propane 1,2,3 triol

All **soap** molecules are sodium or potassium **salts*** of long-chain **carboxylic acids***. In this example, it is a salt of **octadecanoic acid.**

- **Micelle**. A spherical grouping of **detergent molecules** in water. Oils and greases dissolve in the **hydrophobic** centre of the micelle. The picture opposite right shows how micelles remove grease.

Grease stain

Washing powder is one type of detergent used to remove grease.

* **Carboxylic acids**, 195; **Esters**, 194; **Fats**, 205 (**Lipids**); **Functional group**, 191; **Hard water**, 207; **Hydrocarbons**, 191; **Neutralization**, 145; **Non-polar molecule**, 133 (**Polar molecule**); **Salts**, 153; **Surface tension**, 23.

- **Soapless detergents** or **synthetic detergents**. Types of detergent made from by-products of **refining*** crude oil. They are used to make many products, including **washing powders**, shampoos and hair conditioners, and are usually simply referred to as detergents. Soapless detergents do not form a scum in **hard water*** and lather better than **soaps**. If they are not **biodegradable** they pollute rivers.

Example of a **soapless detergent** molecule that does not have an **ionic*** part - used in washing-up liquid.

Non-polar* part of molecule

Polar* part of molecule

CH_3 – CH_2 – CH_2 – CH_2 – CH_2 – CH_2 – CH_2 – CH_2 – (benzene ring) – $(O(CH_2)_2)_n$ OH

Benzene ring*

Example of an **ionic*** **soapless detergent** molecule – used in kitchen soap.

Long **hydrocarbon*** chain (**non-polar*** part of molecule)

Ionic* end (**polar*** part of molecule)

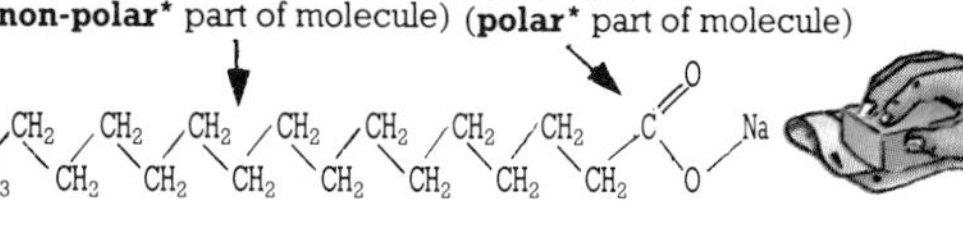

- **Washing powders. Soap** or **soapless detergents** used to wash clothes. They are better for fabrics than water alone, as they make it easier to remove dirt. There are two main types of washing powder - those used when hand-washing clothes (usually soap powders) and those used in washing machines. The latter are mostly **soapless detergents** with other substances added to keep the lather down and to brighten the appearance of the fabric. When they also contain **enzymes***, they are called **biological washing powders**, or **enzyme detergents**. Enzymes help to break down **proteins*** and loosen dirt.

- **Biodegradable detergents. Soapless detergents** that are broken down by bacteria (see **biodegradable**, page 210). Foams from **non-biodegradable detergents** cannot be broken down and cover the water, depriving life of oxygen.

Non-biodegradable foam kills creatures living in water as it stops oxygen dissolving in the water.

- **Surfactants**. Substances which lower the **surface tension*** of water. As a result of this property, detergents have many other uses, as well as removing dirt. They are used in:

Paints, to ensure the pigment is evenly mixed in, and the paint gives a smooth finish and does not drip.

Lubricating greases, to make them gel better.

Cosmetics, to make face powder cover well and evenly, and ensure cosmetic creams mix well with water and thicken properly.

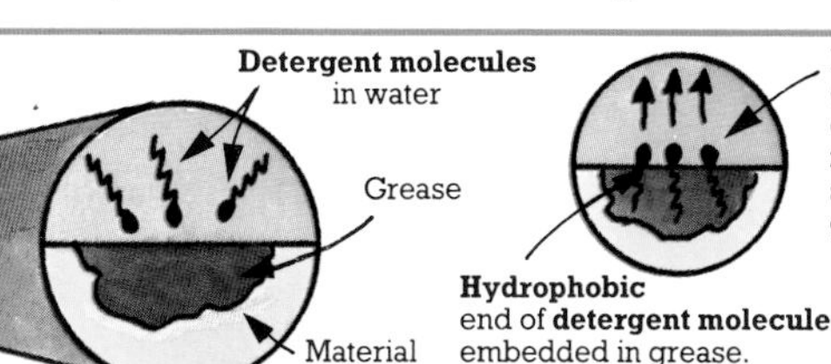

Hydrophobic end of **detergent molecule** embedded in grease.

Hydrophilic end of **detergent molecule** sticking out of grease.

Micelle formed when grease is pulled off.

Constant motion of washing machine, and attraction of head end of **detergent molecules** to water pull off the detergent and grease.

Micelles tend to keep grease suspended in solution.

* **Benzene ring**, 190; **Enzyme**, 161; **Hard water**, 207; **Hydrocarbons**, 191; **Ionic compound**, 131; **Non-polar molecule**, 133 (**Polar molecule**); **Proteins**, 205; **Refining**, 198; **Surface tension**, 23.

Food

In order to survive and grow, living organisms need a number of different substances. These include the **nutrients** – **carbohydrates**, **proteins** and **fats** (see **lipids**) – which are all **organic compounds*** made by plant **photosynthesis*** and taken in by animals. Also important are the **accessory foods** – water and **minerals**, needed by both plants and animals, and **vitamins**, needed by animals only. **Roughage**, or **fibre**, is also needed by many animals to help move food through the gut. Different animals need different amounts of these substances for a healthy diet. For more about minerals and roughage, see page 329.

Some foods containing **carbohydrates**
Cereals
Bread
Sugar
Potatoes

●**Carbohydrates**. **Organic compounds*** of varying complexity – the most complex, made of many individual units, being **polysaccharides** (see **starch**) and the simplest, made of just one unit, being **monosaccharides**. All have the **general formula*** $C_x(H_2O)_y$. Almost all living organisms use the monosaccharide **glucose** for energy.

●**Sucrose**. A **disaccharide**, i.e. a **carbohydrate**, composed of two **monosaccharide** units – in this case **glucose** and **fructose**. It is sweet-tasting, often used to sweeten food, and is commonly known as sugar. It has the **molecular formula** $C_{12}H_{22}O_{11}$ and is obtained from sugar cane and sugar beet.

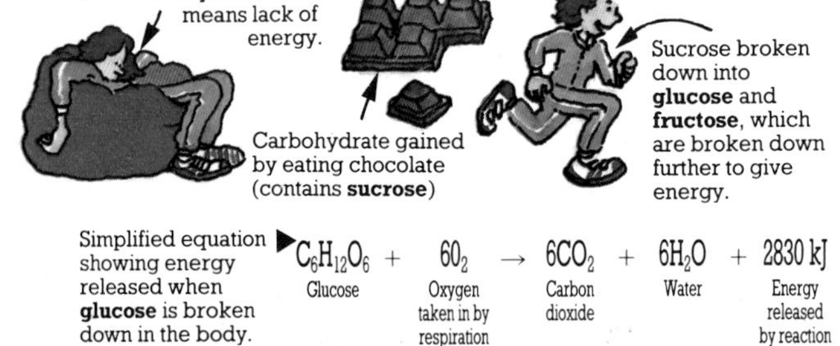

Simplified equation showing energy released when **glucose** is broken down in the body.

$$C_6H_{12}O_6 + 6O_2 \rightarrow 6CO_2 + 6H_2O + 2830\ kJ$$

Glucose + Oxygen taken in by respiration → Carbon dioxide + Water + Energy released by reaction

●**Glucose**. A **monosaccharide** (see **carbohydrates**) with the **molecular formula*** $C_6H_{12}O_6$, the breakdown of which provides energy for plants and animals. Plants make their own by **photosynthesis***, storing it as **starch** until it is needed. Animals take in all forms of carbohydrate, break down the complex ones to glucose, and store this as the **polysaccharide glycogen**. For more about glucose, see pages 328-329, 332-333 and 336-337.

●**Starch**. A **polysaccharide** (see **carbohydrates**) which is the storage form of **glucose** in plants. Like **glycogen** (see **glucose**), it is an example of a **natural polymer*** – the **monomers*** in this case being the glucose **monosaccharides**. Note that when these join, water molecules form at the links (see **condensation polymerization**, page 200).

Part of a **starch** molecule ▼

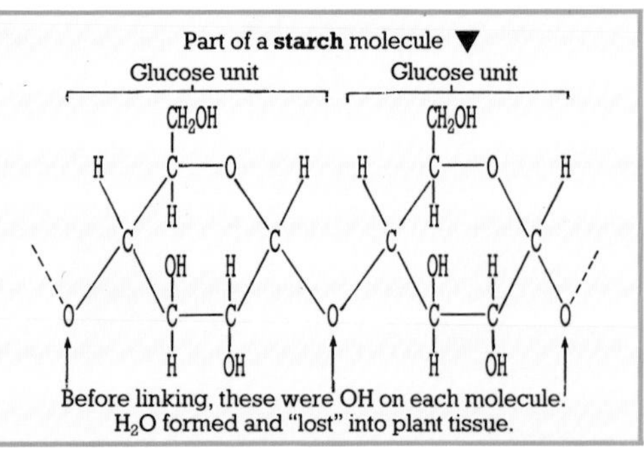

Before linking, these were OH on each molecule. H_2O formed and "lost" into plant tissue.

* **General formula**, 191 (**Homologous series**); **Molecular formula**, 140; **Monomers**, 200; **Natural polymers**, 201; **Organic compounds**, 190; **Photosynthesis**, 209; **Polymer**, 200.

- **Amino acids**. Compounds whose molecules contain a carbon atom joined to a **carboxyl group*** and an **amino group***. **Proteins** are made from amino acids. See also pages 328-329.

There are about 20 different **amino acids**. They all contain an **amino group*** and a **carboxyl group***.

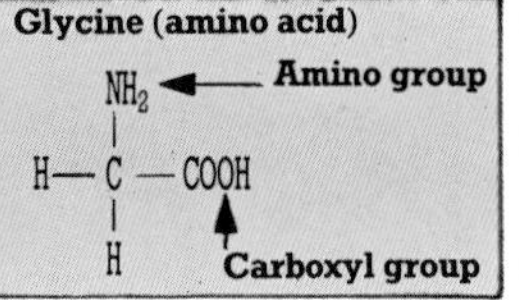

Peanuts contain a lot of **protein**, so they are very nutritious.

1. Chewed peanuts go down the gullet. The **protein** they contain is digested in the stomach and the small intestine.

Gullet

Small intestine

Stomach

2. Each protein consists of many **amino-acid** units (**monomers***) in a different order.

Different coloured squares represent different **amino acids**.

3. An **enzyme*** in the stomach breaks down **protein** molecules (long chains called **polypeptides**) into shorter chains (also called polypeptides).

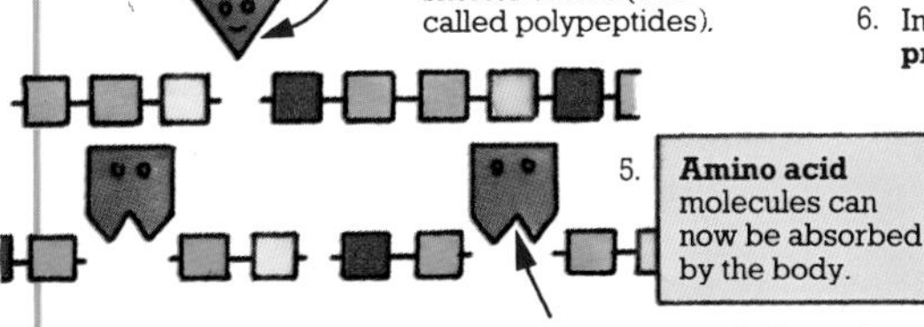

4. An **enzyme*** in the small intestine breaks **polypeptides** into single **amino acids** or molecules made of two amino acids (**dipeptides**).

5. **Amino acid** molecules can now be absorbed by the body.

- **Proteins. Natural polymers*** made from many **amino acid monomers*** joined together. The **relative molecular masses*** of proteins vary from 20,000 to several million. They are found mainly in meat, dairy food, nuts cereal and beans. Animals need proteins for growth and repair of tissue. See also pages 328-329.

Enzymes* are catalysts* that speed up reactions in the body.

6. In the body, certain **enzymes*** make new **proteins** by joining **amino acids** together.

7. The order of the **amino acid monomers*** in the new **protein** chains determines the type of protein. This man needs a lot of the proteins **actin** and **myosin**, found in muscle (see also page 282).

- **Vitamins. Organic compounds*** found in small amounts in food. They are an essential part of the diet of animals. They are needed to help **enzymes* catalyse*** reactions in the body. See page 337 for a list of vitamins.

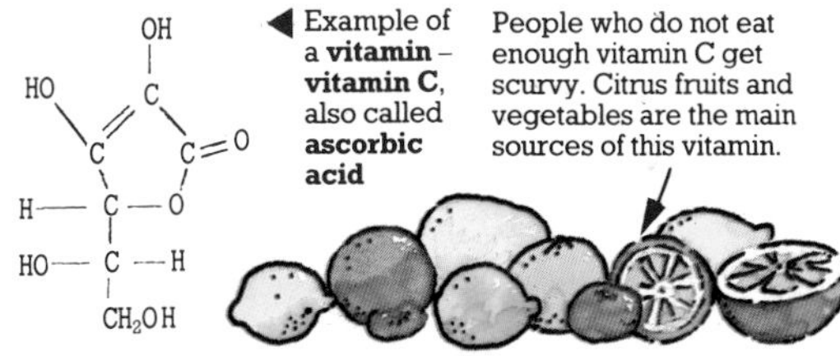

◀ Example of a **vitamin** – **vitamin C**, also called **ascorbic acid**

People who do not eat enough vitamin C get scurvy. Citrus fruits and vegetables are the main sources of this vitamin.

- **Lipids**. A group of **esters***, including **fats** and waxes, found in living tissue (fats form a reserve energy source – see also pages 328-329). Insoluble in water but soluble in **organic solvents***, they are mostly solid or semi-solid and made of **saturated* carboxylic acids***, though a smaller group, the **oils**, are liquids and consist mainly of **unsaturated*** carboxylic acids.

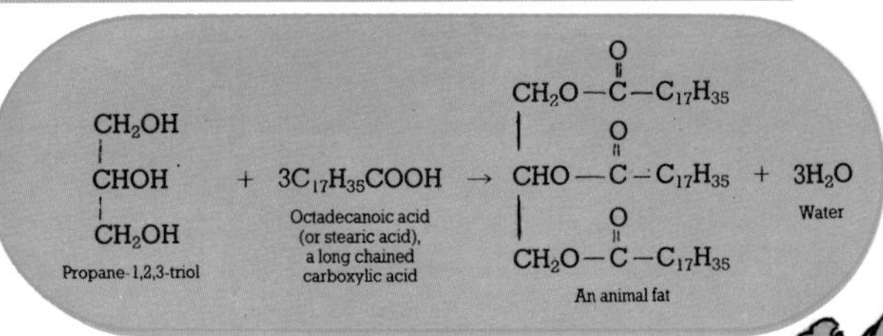

Example of a reaction to make a **fat**: ▲

Coconut **oil** is used to make beauty products.

Palm **oil** is used in soaps.

* **Amino group**, 195; **Carboxyl group**, 195 (**Carboxylic acids**); **Catalysis**, 161 (**Catalyst**); **Enzyme**, 161; **Esters**, 195; **Monomers**, 200; **Natural polymers**, 201; **Organic solvent**, 345; **Relative molecular mass**, 148; **Saturated compounds**, **Unsaturated compounds**, 191.

Water

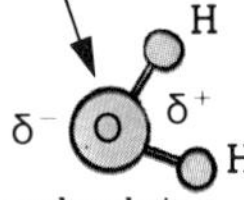

A molecule of water contains one oxygen atom and two hydrogen atoms.

The molecule is a **polar molecule***, which makes water a good **polar solvent***.

Water (H_2O) is the most important compound on Earth. It is found on the surface and in the atmosphere, and is present in animals and plants. Vast amounts of water are used every day in the home and in industry, e.g. for manufacturing processes and the cooling of chemical plants. Water normally contains some dissolved gases, **salts*** and **pollutants***. See also page 167.

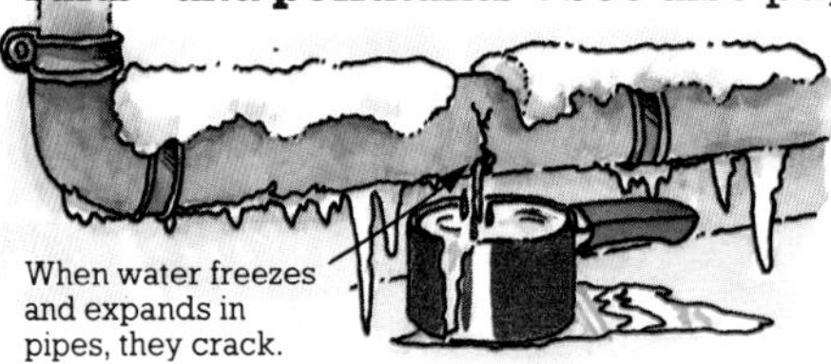
When water freezes and expands in pipes, they crack.

•**Ice**. The solid form of water. It has a **molecular lattice*** in which the molecules are further apart than in water. This is caused by **hydrogen bonds*** and means that ice is less dense than water, and that water expands when it freezes.

•**Water cycle**. The constant circulation of water through the air, rivers and seas.

Rain water is very pure, but does contain some dissolved gases, e.g. carbon dioxide and sulphur dioxide (which produces **acid rain***).

River water is **hard water** if it contains certain **salts***

Sea water contains about 4% dissolved **salts***.

Atmospheric water

•**Humidity**. The amount of water vapour in the air. It depends on the temperature and is higher (up to 4% of the air) in warm air than cold air.

•**Hygroscopic**. Describes a substance which can absorb up to 70% of its own mass of water vapour. Such a substance becomes damp, but does not dissolve.

Sodium chloride absorbs water in a damp atmosphere.

•**Deliquescent**. Describes a substance which absorbs water vapour from the air and dissolves in it, forming a **concentrated*** solution.

Calcium chloride left open to the air forms a **concentrated*** solution

•**Efflorescent**. Describes a crystal which loses part of its **water of crystallization*** to the air. A powdery coating is left on its surface.

A white powder forms on sodium carbonate crystals.

* **Acid rain**, 210; **Concentrated**, 144; **Condensation**, **Evaporation**, 121; **Hydrogen bond**, 134; **Molecular lattice**, 137; **Polar molecule**, 133; **Polar solvent**, 144; **Pollutants**, 210; **Respiration**, 298; **Salts**, 153; **Transpiration**, 252; **Water of crystallization**, 135.

Water supply

- **Distilled water**. Water which has had **salts*** removed by **distillation***. It is very pure, but does contain some dissolved gases.

- **Desalination**. The treatment of sea water to remove dissolved **salts***. It is done by **distillation*** or **ion exchange**.

- **Purification**. The treatment of water to remove bacteria and other harmful substances, and produce pure water.

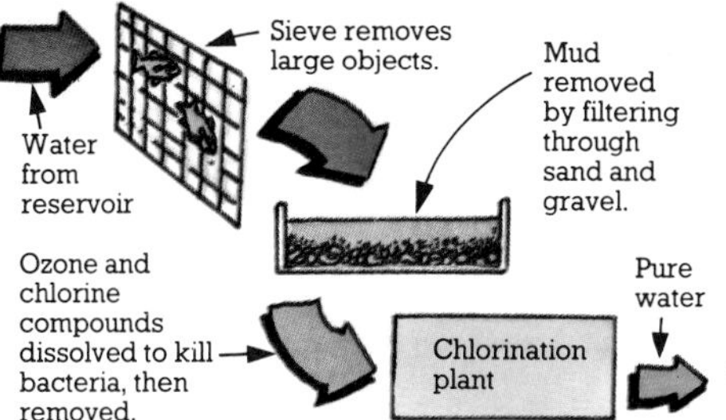

- **Hard water**. Water which contains calcium and magnesium **salts*** that have dissolved from the rocks over which the water has flowed (see **calcium**, page 171). Water that does not contains these salts is called **soft water**. There are two types of hardness – **temporary hardness** (which can be removed relatively easily) and **permanent hardness** (which is more difficult to remove).

Hard water does not lather with soap and forms a **scum**.

Calcium and magnesium ions (in hard water) + Soap (sodium stearate) → Scum (calcium and magnesium stearates) + Sodium ions

Soft water lathers easily because it does not react with soap to form **scum**.

- **Temporary hardness**. One type of water hardness, caused by the **salt*** calcium hydrogencarbonate dissolved in the water. It can be removed by boiling, producing an insoluble white solid (calcium carbonate or "scale").

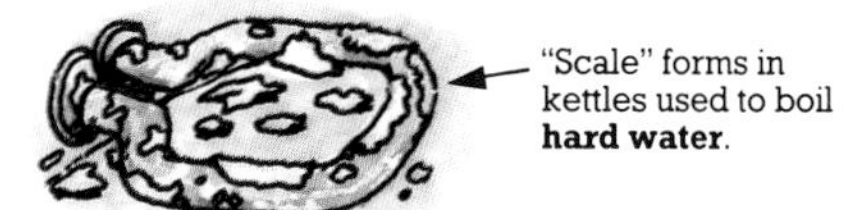

"Scale" forms in kettles used to boil **hard water**.

- **Permanent hardness**. The more severe type of water hardness, caused by calcium and magnesium **salts*** (sulphates and chlorides) dissolved in the water. It cannot be removed by boiling, but can be removed by **distillation*** (producing **distilled water**) or by **water softening** (**ion exchange** or use of **water softeners**).

- **Ion exchange**. A method of **water softening** (see **permanent hardness**). Water is passed over a material such as **zeolite** (sodium aluminium silicate), which removes calcium and magnesium ions and replaces them with sodium ions. Some organic **polymers*** are also used as ion exchange materials.

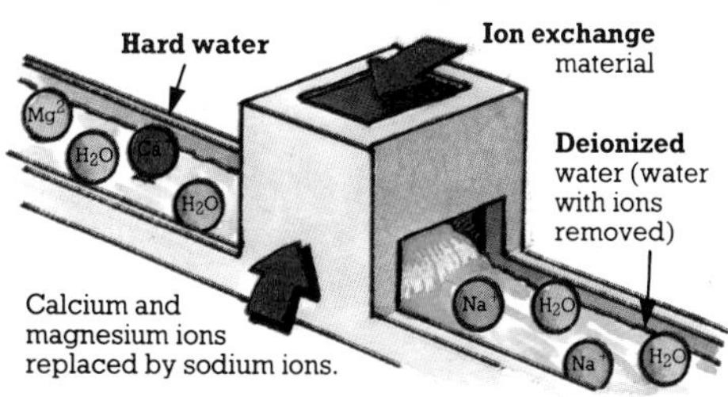

- **Water softeners**. Substances used to remove **permanent hardness**. They react with the calcium and magnesium **salts*** to form compounds which do not react with soap.

- **Washing soda**. The common name for the **hydrate*** of sodium carbonate (see also page 169). It is used as a **water softener** in the home.

* **Distillation**, 220; **Hydrate**, 155; **Polymers**, 200; **Salts**, 153.

Air and burning

Air is a mixture of gases, including oxygen, carbon dioxide and nitrogen, which surrounds the earth and is essential for all forms of life. These gases can be separated by the **fractional distillation of liquid air***, and are used as raw materials in industry. Air also contains some water vapour and may contain **pollutants*** in some areas.

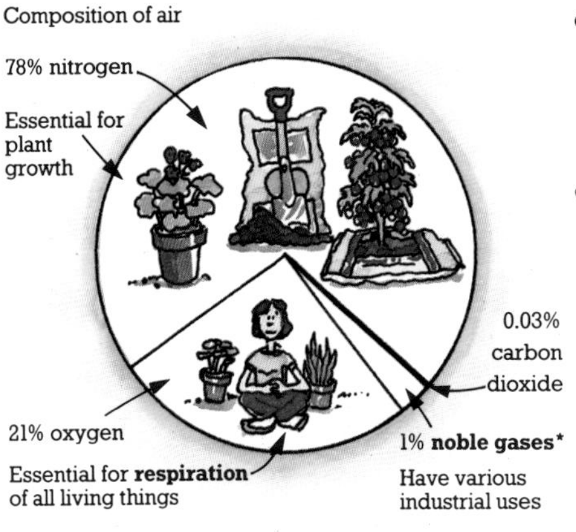

• **Combustion or burning**. An **exothermic reaction*** between a substance and a gas. Combustion usually takes place in air, when the substance which burns combines with oxygen. Substances can also burn in other gases, e.g. chlorine. Combustion does not normally happen spontaneously. It has to be started by heating (see **activation energy**, page 160).

$$CH_4(g) + O_2(g) \rightarrow CO_2(g) + 2H_2O(g) + \text{ENERGY}$$

Methane; Oxygen from air; Carbon dioxide; Water vapour; for cooking

• **Rapid combustion**. **Combustion** in which a large amount of heat and light energy is given out.

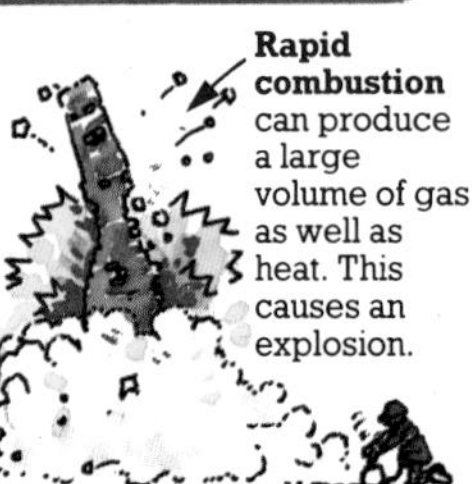

• **Slow combustion**. A form of **combustion** which takes place at low temperature. No **flames** occur. **Respiration** is slow combustion.

• **Flame**. A mixture of heat and light energy produced during **rapid combustion**.

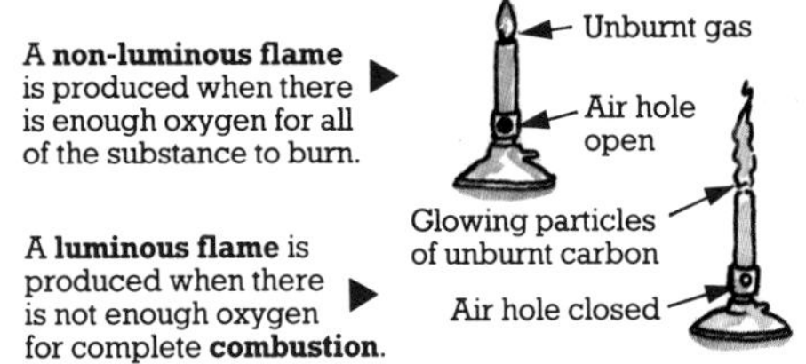

• **Fuel**. A substance which is burned to produce heat energy. Most fuels used today are **fossil fuels**, which were formed from the remains of prehistoric animal and plant life.

• **Calorific value**. A measure of the amount of heat energy produced by a specific amount of a **fuel**. The table below shows the different values for some common fuels.

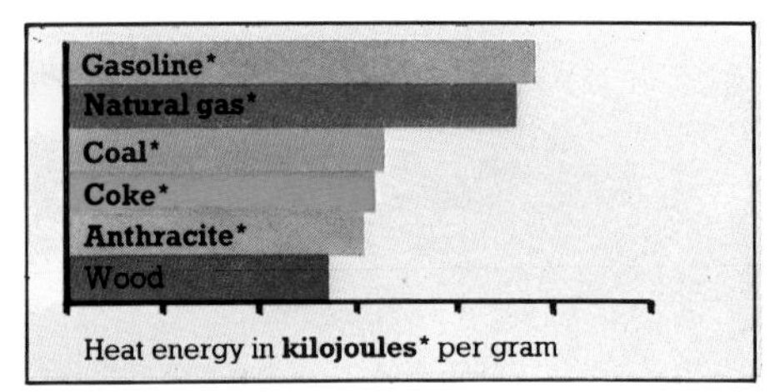

* **Anthracite**, 179 (**Coal**); **Coke**, 179; **Exothermic reaction**, 147; **Fractional distillation of liquid air**, 183; **Gasoline**, 199; **Kilojoule**, 146; **Natural gas**, 192 (**Methane**); **Noble gases**, 189; **Petroleum**, 198; **Pollutants**, 210.

•Corrosion. A reaction between a metal and the gases in air. The metal is **oxidized*** to form an oxide layer on the surface, usually weakening the metal, but sometimes forming a protective coat against further corrosion. Corrosion can be prevented by stopping oxygen reaching the metal or by preventing electrons from leaving it (see **sacrificial protection**, page 159). The corrosion of iron is called **rusting** (see **rust**, page 174).

•Internal respiration. A form of **slow combustion** in animals. It produces energy from the reaction of **glucose*** with oxygen. See also page 332 and **respiration**, page 298.

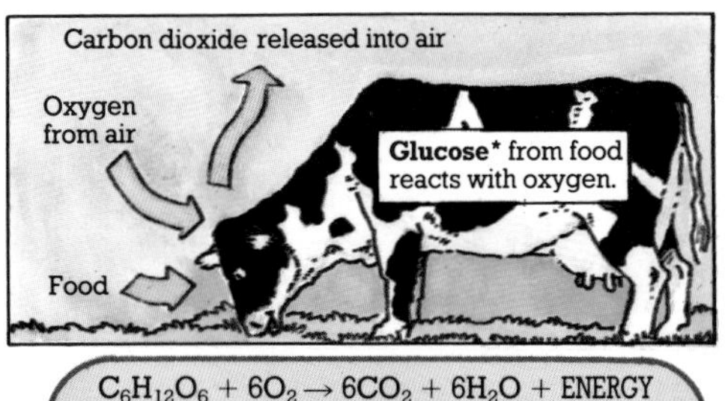

$$C_6H_{12}O_6 + 6O_2 \rightarrow 6CO_2 + 6H_2O + \text{ENERGY}$$

Energy produced by reaction of glucose and oxygen

•Photosynthesis. A **photochemical reaction*** in green plants. It involves the production of **glucose*** from carbon dioxide and water, using the energy from sunlight. Photosynthesis is chemically the opposite of **internal respiration**. See pages 254-255.

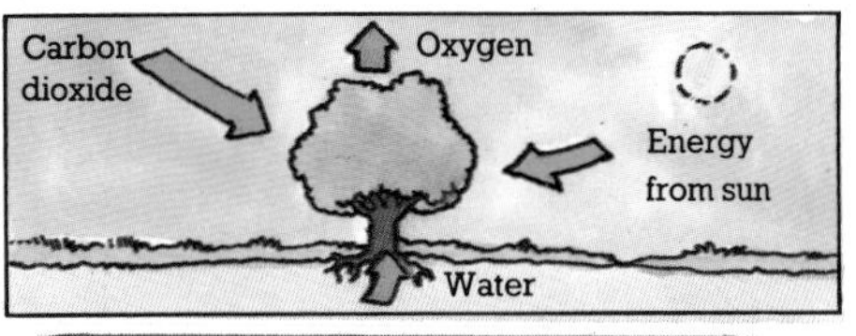

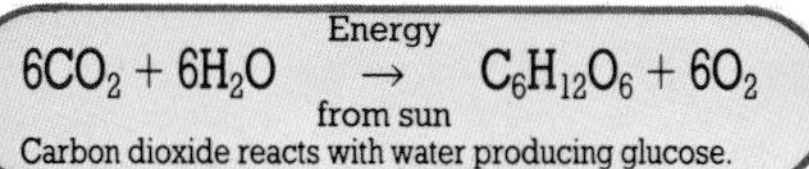

•Nitrogen cycle. The constant circulation of nitrogen through the air, animals, plants and the soil.

•Carbon cycle. The circulation of carbon through the air, animals, plants and the soil.

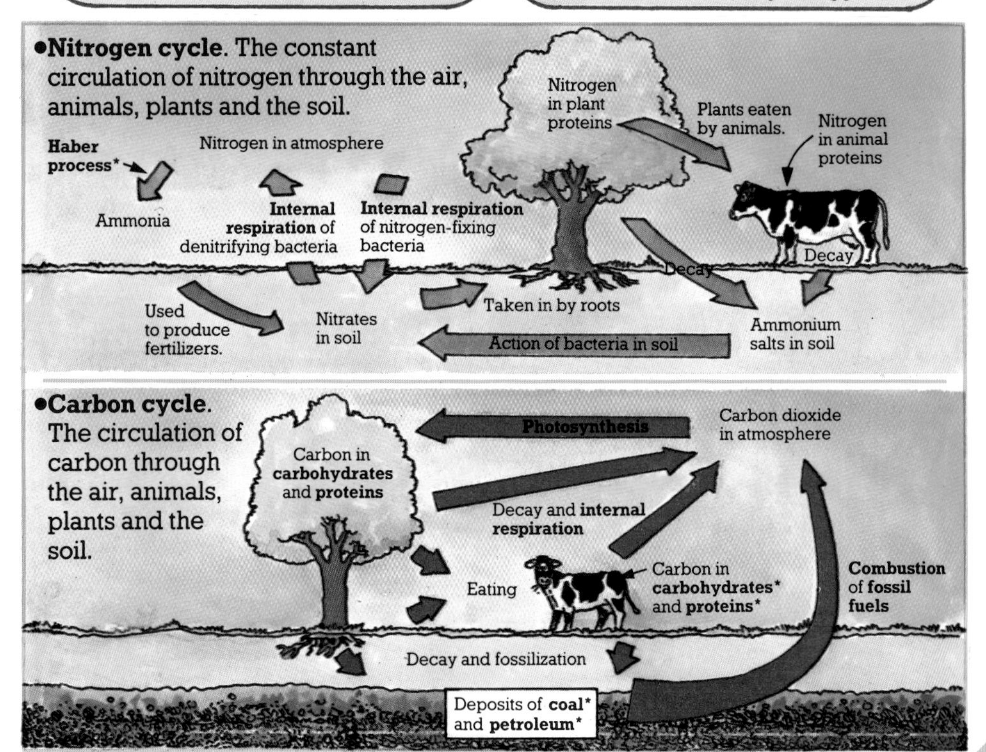

* **Carbohydrates**, 204; **Coal**, 179; **Glucose**, 204; **Haber process**, 180; **Oxidation**, 148; **Petroleum**, 198; **Photochemical reaction**, 160; **Proteins**, 205.

Pollution

Pollution is the release into the land, atmosphere, rivers and oceans, of undesirable substances which upset the natural processes of the Earth. These substances are known as **pollutants**. The major sources and types of pollution are shown below.

- **Biodegradable**. Describes a substance which is converted to simpler compounds by bacteria. Many plastics are not biodegradable (see also **biodegradable detergents**, page 203).
- **Smog**. Fog mixed with dust and soot. It is acidic because of the sulphur dioxide produced when **fuels*** are burnt in industrial cities.
- **Acid rain**. Rain water which is more acidic than usual. Rain water normally has a **pH*** of between 5 and 6, due to dissolved carbon dioxide forming dilute carbonic acid. Sulphur dioxide and oxides of nitrogen, products of the combustion of **fuels***, react with water in the atmosphere to produce sulphuric and nitric acids with a pH of about 3.
- **Eutrophication**. An overgrowth of aquatic plants caused by an excess of nitrates, nitrites and phosphates from fertilizers in rivers. It results in a shortage of oxygen in the water, causing the death of fish.
- **Greenhouse effect**. The trapping of solar energy in the atmosphere by carbon dioxide, causing an increase in temperature. The burning of **fuels*** creates more carbon dioxide, making the problem worse. See also page 29.
- **Thermal pollution**. The effect of releasing warm water from factories and power stations into rivers and lakes. This causes a decrease in the oxygen dissolved in the water and affects aquatic life.

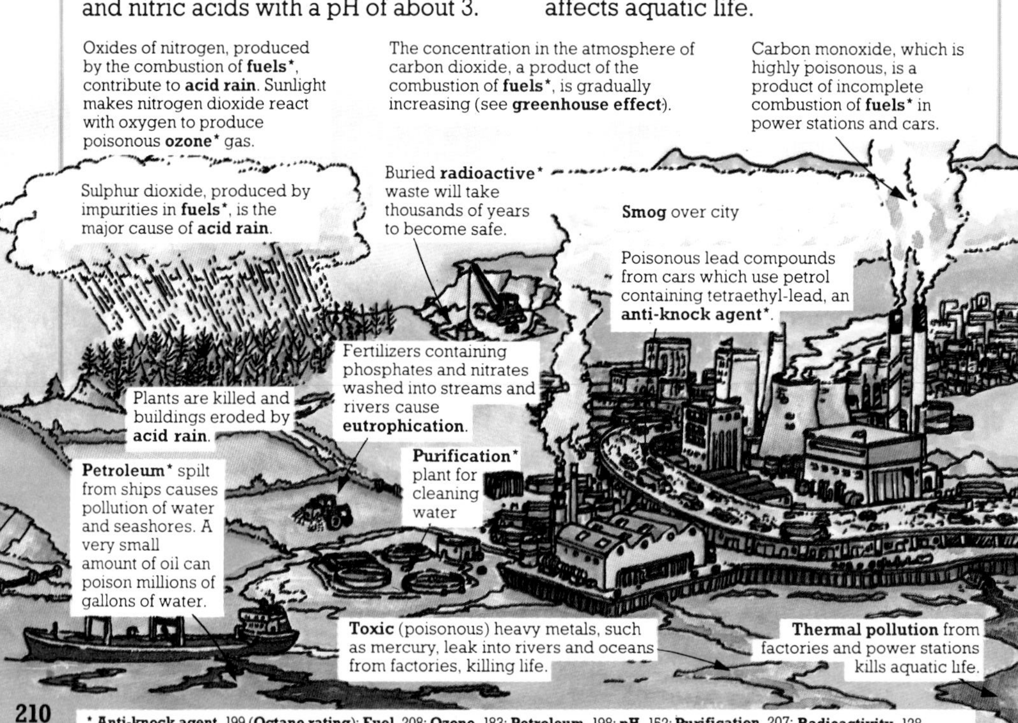

* **Anti-knock agent**, 199 (**Octane rating**); **Fuel**, 208; **Ozone**, 183; **Petroleum**, 198; **pH**, 152; **Purification**, 207; **Radioactivity**, 128.

The reactivity series (showing ten metals – see also page 158)

Metal	Symbol	Reaction with air	Reaction with water	Reaction with dilute strong acids*	Displacement* reactions	Reaction of carbon with oxide	Reaction of hydrogen with oxide	Action of heat on oxide	Action of heat on carbonate	Action of heat on nitrate	Symbol
Potassium	**K**	Burn strongly to form oxides.	React with cold water to produce hydrogen gas and hydroxide. Hydroxide dissolves in water to form alkaline solution. React with decreasing vigour down the series.	Explosive reaction to give hydrogen gas and **salt*** solution.	All metals displace ions of metals below them from solution.	No reaction	No reaction	No reaction	No reaction	Decompose to form nitrite and oxygen.	**K**
Sodium	**Na**										**Na**
Calcium	**Ca**	Burn, when heated, to form oxides. Burn with decreasing vigour down the series.		React to give hydrogen gas and **salt*** solution with decreasing vigour down the series.					Decompose to form oxide and carbon dioxide with increasing ease down the series.	Decompose to form oxide, oxygen and nitrogen dioxide with increasing ease down the series.	**Ca**
Magnesium	**Mg**										**Mg**
Aluminium	**Al**		No reaction with cold water. React with steam to form hydrogen gas and oxide. React with decreasing vigour down the series.								**Al**
Zinc	**Zn**					Oxide **reduced*** to metal with increasing ease down the series. Carbon dioxide formed.					**Zn**
Iron	**Fe**										**Fe**
Lead	**Pb**	Do not burn when heated, but form an oxide layer on surface.	No reaction				Oxide **reduced*** to metal with increasing ease down the series. Water is formed.				**Pb**
Copper	**Cu**			No reaction							**Cu**
Silver	**Ag**	No reaction						Decomposes to form metal and oxygen only.	and carbon dioxide.	and nitrogen dioxide.	**Ag**

* **Displacement**, 158; **Reduction**, 148; **Salts**, 153; **Strong acid**, 152.

The properties of the elements

Below is a chart giving information on the physical properties of the elements in the periodic table (see pages 164-165). The last eight elements (**atomic numbers*** 96-103 – see pages 165 and 226-227 for symbols and names) are not listed, as there is very little known about them – they all have to be made under special laboratory conditions and only exist for a fraction of a second.

All the density measurements below are taken at room temperature except those of gases (marked with a †), which are measured at their boiling points. A dash (-) at any place on the chart indicates that there is no known value.

Element	Symbol	Atomic number	Approx. relative atomic mass	Density (g cm^{-3})	Melting point (°C)	Boiling point (°C)
					(brackets indicate approximations)	
Actinium	Ac	89	227	10.1	1050	3200
Aluminium	Al	13	27	2.7	660	2470
Americium	Am	95	243	11.7	(1200)	(2600)
Antimony	Sb	51	122	6.62	630	1380
Argon	Ar	18	40	1.4 †	−189	−186
Arsenic	As	33	75	5.73	–	613 (**sublimes***)
Astatine	At	85	210	–	(302)	–
Barium	Ba	56	137	3.51	714	1640
Beryllium	Be	4	9	1.85	1280	2477
Bismuth	Bi	83	209	9.78	271	1560
Boron	B	5	11	2.34	2300	3930
Bromine	Br	35	80	3.12	−7.2	58.8
Cadmium	Cd	48	112	8.65	321	765
Caesium	Cs	55	133	1.9	28.7	690
Calcium	Ca	20	40	1.54	850	1487
Carbon	C	6	12	2.25 (**graphite***)	3730 (**sublimes***)	4830
				3.51 (**diamond***)	3750	–
Cerium	Ce	58	140	6.78	795	3470
Chlorine	Cl	17	35.5	1.56†	−101	−34.7
Chromium	Cr	24	52	7.19	1890	2482
Cobalt	Co	27	59	8.7	1492	2900
Copper	Cu	29	64	8.89	1083	2595
Dysprosium	Dy	66	162	8.56	1410	2600
Erbium	Er	68	167	9.16	1500	2900
Europium	Eu	63	152	5.24	826	1440
Fluorine	F	9	19	1.11†	−220	−188
Francium	Fr	87	223	–	(27)	–
Gadolinium	Gd	64	157	7.95	1310	3000
Gallium	Ga	31	70	5.93	29.8	2400
Germanium	Ge	32	73	5.4	937	2830
Gold	Au	79	197	19.3	1063	2970
Hafnium	Hf	72	178.5	13.3	2220	5400
Helium	He	2	4	0.147†	−270	−269
Holmium	Ho	67	165	8.8	1460	2600
Hydrogen	H	1	1	0.07†	−259	−252
Indium	In	49	115	7.3	157	2000
Iodine	I	53	127	4.93	114	184
Iridium	Ir	77	192	22.4	2440	5300
Iron	Fe	26	56	7.85	1535	3000
Krypton	Kr	36	84	2.16†	−157	−152
Lanthanum	La	57	139	6.19	920	3470
Lead	Pb	82	207	11.3	327	1744
Lithium	Li	3	7	0.53	180	1330
Lutetium	Lu	71	175	9.84	1650	3330

* **Atomic number**, 127; **Diamond**, **Graphite**, 178; **Sublimation**, 121.

Element	Symbol	Atomic number	Approx. relative atomic mass	Density (g cm^{-3})	Melting point (°C)	Boiling point (°C)
Magnesium	**Mg**	12	24	1.74	650	1110
Manganese	**Mn**	25	55	7.2	1240	2100
Mercury	**Hg**	80	201	13.6	−38.9	357
Molybdenum	**Mo**	42	96	10.1	2610	5560
Neodymium	**Nd**	60	144	7.0	1020	3030
Neon	**Ne**	10	20	1.2 †	−249	−246
Neptunium	**Np**	93	237	20.4	640	–
Nickel	**Ni**	28	59	8.8	1453	2730
Niobium	**Nb**	41	93	8.57	2470	3300
Nitrogen	**N**	7	14	0.808†	−210	−196
Osmium	**Os**	76	190	22.5	3000	5000
Oxygen	**O**	8	16	1.15†	−218	−183
Palladium	**Pd**	46	106	12.2	1550	3980
Phosphorus	**P**	15	31	1.82 (**white***) 2.34 (**red***)	44.2 (**white**) 590 (**red**)	280 (**white**) –
Platinum	**Pt**	78	195	21.5	1769	4530
Plutonium	**Pu**	94	242	19.8	640	3240
Polonium	**Po**	84	210	9.4	254	960
Potassium	**K**	19	39	0.86	63.7	774
Praseodymium	**Pr**	59	141	6.78	935	3130
Promethium	**Pm**	61	147	–	1030	2730
Protactinium	**Pa**	91	231	15.4	1230	–
Radium	**Ra**	88	226	5	700	1140
Radon	**Rn**	86	222	4.4†	−71	−61.8
Rhenium	**Re**	75	186	20.5	3180	5630
Rhodium	**Rh**	45	103	12.4	1970	4500
Rubidium	**Rb**	37	85	1.53	38.9	688
Ruthenium	**Ru**	44	101	12.3	2500	4900
Samarium	**Sm**	62	150	7.54	1070	1900
Scandium	**Sc**	21	45	2.99	1540	2730
Selenium	**Se**	34	79	4.79	217	685
Silicon	**Si**	14	28	2.35	1410	2360
Silver	**Ag**	47	108	10.5	961	2210
Sodium	**Na**	11	23	0.97	97.8	890
Strontium	**Sr**	38	88	2.62	768	1380
Sulphur	**S**	16	32	2.07 (**rhombic***) 1.96 (**monoclinic***)	113 (**rhombic**) 119 (**monoclinic**)	444
Tantalum	**Ta**	73	181	16.6	3000	5420
Technetium	**Tc**	43	99	11.5	2200	3500
Tellurium	**Te**	52	128	6.2	450	990
Terbium	**Tb**	65	159	8.27	1360	2800
Thallium	**Tl**	81	204	11.8	304	1460
Thorium	**Th**	90	232	11.7	1750	3850
Thulium	**Tm**	69	169	9.33	1540	1730
Tin	**Sn**	50	119	7.3	232	2270
Titanium	**Ti**	22	48	4.54	1675	3260
Tungsten	**W**	74	184	19.3	3410	5930
Uranium	**U**	92	238	19.1	1130	3820
Vanadium	**V**	23	51	5.96	1900	3000
Xenon	**Xe**	54	131	3.52†	−112	−108
Ytterbium	**Yb**	70	173	6.98	824	1430
Yttrium	**Y**	39	89	4.34	1500	2930
Zinc	**Zn**	30	65	7.1	420	907
Zirconium	**Zr**	40	91	6.49	1850	3580

* **Monoclinic sulphur**, 184; **Red phosphorus**, 182 (**Phosphorus**); **Rhombic sulphur**, 184; **White phosphorus**, 182 (**Phosphorus**).

Naming simple organic compounds

Simple **organic compounds*** (those with one or no **functional group***) can be named by following Stages 1 and 2.

Stage 1 Choose the sentence from a) to i) which describes the unidentified molecule, then go to the Stage 2 number indicated.

a) The molecule contains only carbon and hydrogen atoms and **single bonds***. — Go to 1

b) The molecule contains only carbon and hydrogen atoms and a **double bond***. — Go to 2

c) The molecule contains only carbon and hydrogen atoms and a **triple bond***. — Go to 3

d) The molecule contains carbon, hydrogen and a **hydroxyl group (–OH)**. — Go to 4

e) The molecule contains carbon, hydrogen and a –**CHO group** at one end. — Go to 5

f) The molecule contains carbon, hydrogen and a **carbonyl group (–CO–)** between two carbons in the carbon chain. — Go to 6

g) The molecule contains carbon, hydrogen and a **carboxyl group (–COOH)**. — Go to 7

h) The molecule contains only carbon and hydrogen, but has a **side chain***. — Go to 8

i) The molecule contains carbon, hydrogen and one or more **halogen*** atoms. — Go to 9

Key to atoms

Hydrogen atom

Carbon atom

As named

Stage 2

1 The name of a molecule that contains only carbon and hydrogen atoms joined by **single bonds*** begins with the prefix for the number of carbons (see prefix chart, page 215) and ends in **-ane**. For example:

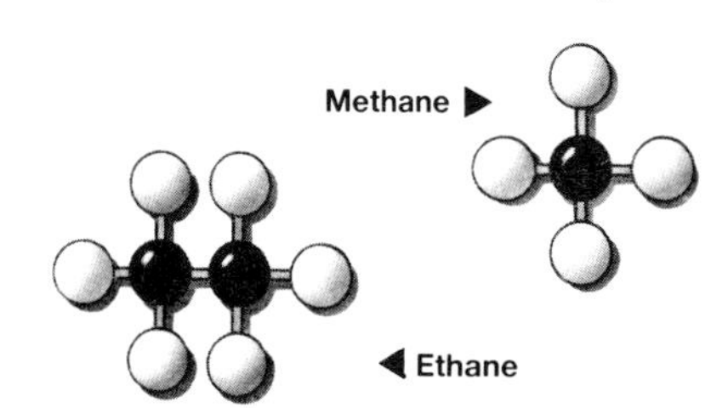

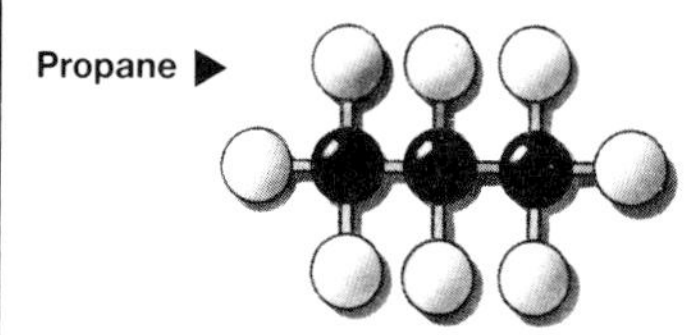

These molecules are all **alkanes***.

2 The name of a molecule that contains only carbon and hydrogen atoms and has one **double bond*** begins with the prefix for the number of carbons (see prefix chart, page 215) and ends in **-ene**. For example:

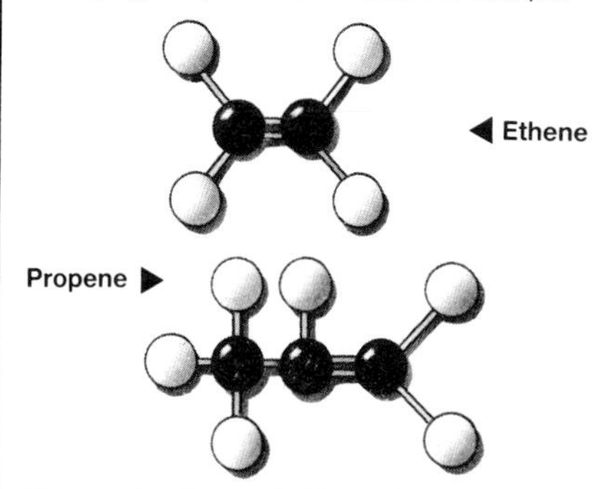

These molecules are all **alkenes***.

3 The name of a molecule that contains only carbon and hydrogen atoms and has one **triple bond*** begins with the prefix for the number of carbons (see prefix chart, page 215) and ends in **-yne**. For example:

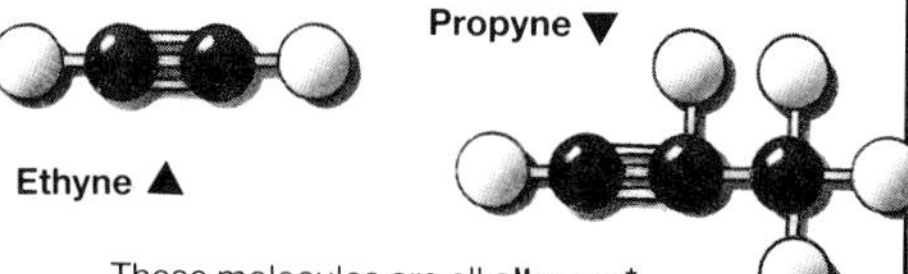

These molecules are all **alkynes***.

4 The name of a molecule that contains only carbon and hydrogen atoms and one **hydroxyl group (-OH)** begins with the prefix for the number of carbons (see prefix chart, page 215) and ends in **-anol**. For example:

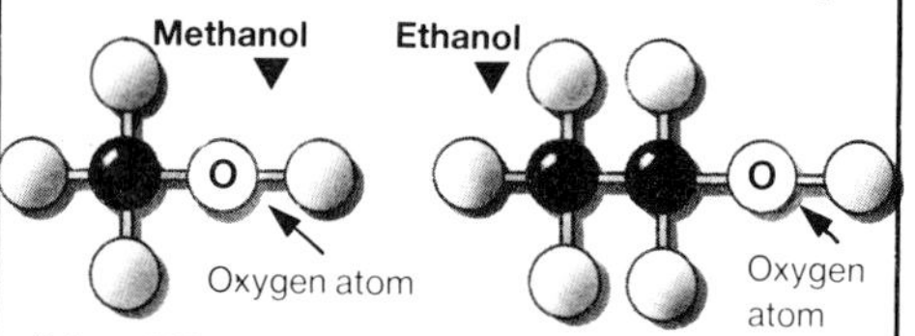

If the –**OH group** is not at one end of the molecule, the number of the carbon to which it is attached is given in front of the name. The carbon atoms are always numbered from the end of the molecule closest to the –OH group. For example:

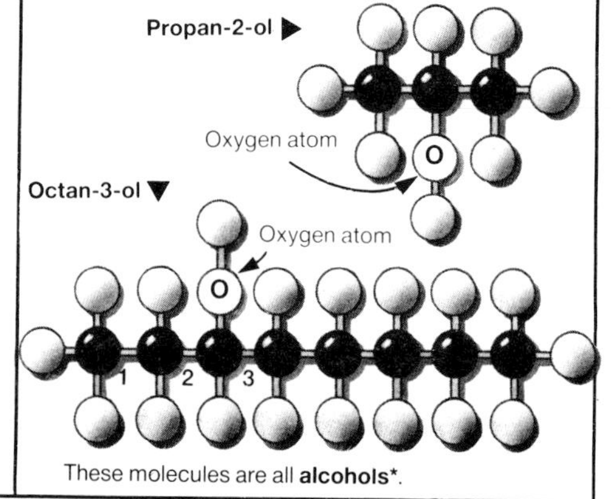

These molecules are all **alcohols***.

* **Alcohols**, 196; **Alkanes**, 192; **Alkenes**, 193; **Alkynes**, 194; **Double bond**, 132; **Functional group**, 191; **Halogens**, 186; **Organic compounds**, **Side chain**, 190; **Single bond**, **Triple bond**, 132.

5 The name of a molecule that contains only carbon and hydrogen atoms and has a –**CHO group** ending the chain begins with the prefix for the number of carbons (see prefix chart below) and ends in **-anal**. For example:

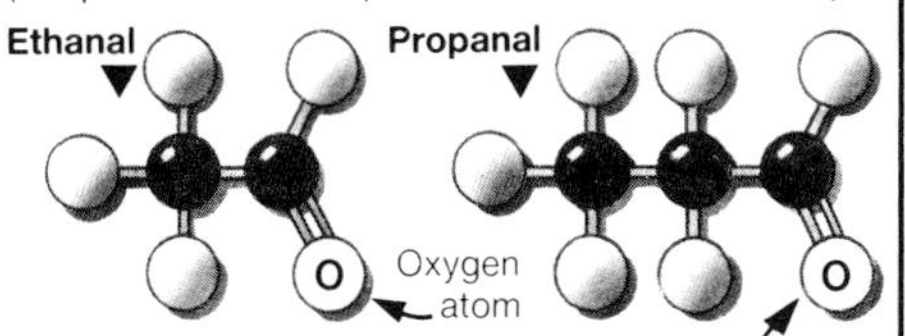

These molecules are all **aldehydes***.

6 The name of a molecule that contains only carbon and hydrogen atoms and has a **carbonyl group (–CO–)** between the ends of the carbon chain begins with the prefix for the number of carbons (see prefix chart below) and ends in **-anone**. For example:

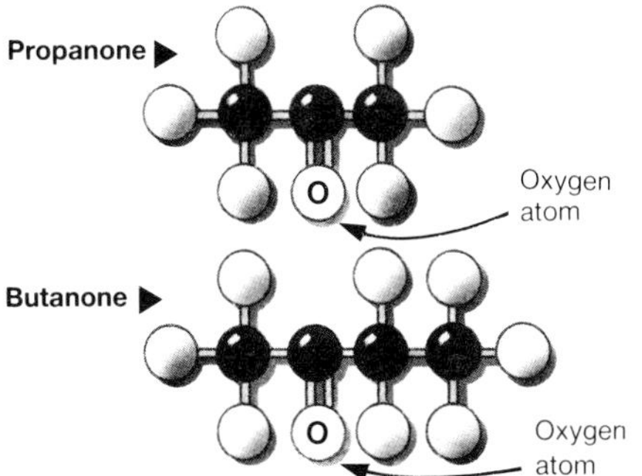

These molecules are all **ketones***.

7 The name of a molecule that contains only carbon and hydrogen atoms and one **carboxyl group (–COOH)** begins with the prefix for the number of carbons (see prefix chart below) and ends in **-anoic acid**. For example:

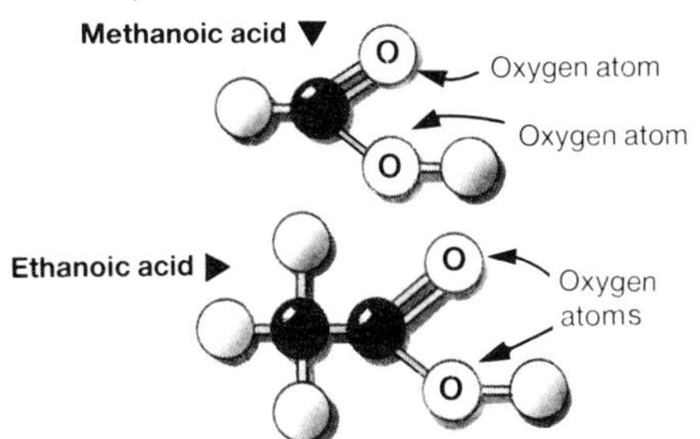

These molecules are all **carboxylic acids***.

8 The name of a branched molecule begins with the name of the branch (**side chain***). If this has only carbon and hydrogen atoms, its name begins with the prefix for the number of carbons in its chain (see prefix chart below) and ends in **-yl**. The main chain is named afterwards in the normal way (see 1). For example:

The **methyl group** is also an example of an **alkyl group**. Alkyl groups are any groups of carbon and hydrogen atoms that have a **general formula*** of C_nH_{2n+1}.

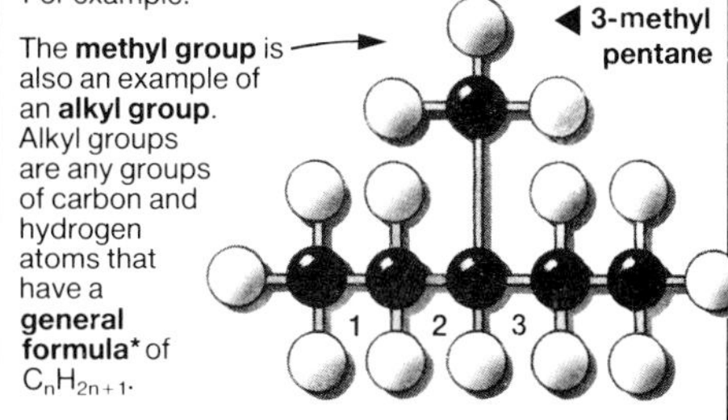

The figure at the beginning of the name gives the number of the carbon atom to which the side chain is joined. The carbon atoms are always numbered from the end of the chain closest to the branch.

9 The name of a molecule that contains carbon and hydrogen atoms and one or more **halogens*** begins with the abbreviation for the halogen(s) – the abbreviations for fluorine, chlorine, bromine and iodine are **fluoro**, **chloro**, **bromo** and **iodo** respectively. For example:

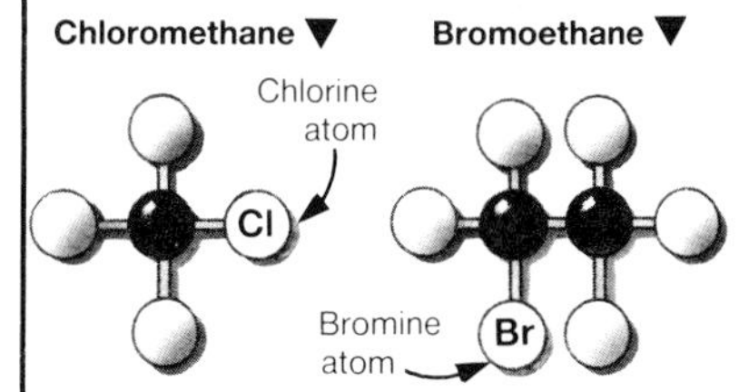

The end of the name is that which the molecule would have had if all the halogen atoms had been replaced by hydrogen atoms (see 1).

If the halogen is not at one end of the molecule, the name includes the number of the carbon atom to which it is attached. The carbon atoms are always numbered from the end of the chain closest to the halogen. For example:

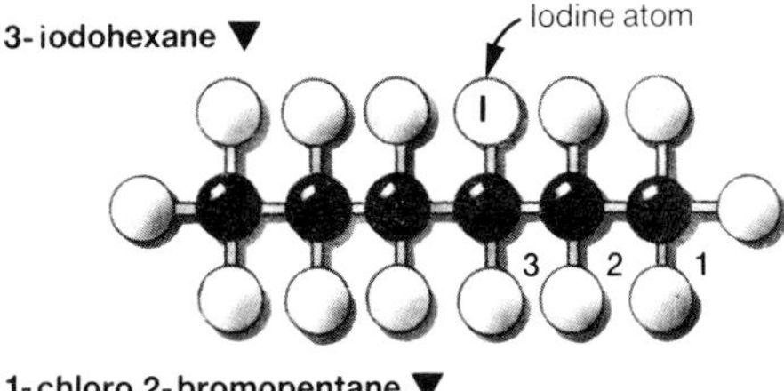

1-chloro,2-bromopentane

Chlorine atom

Cl

Br

Bromine atom

These molecules are all **halogenoalkanes***.

Chart showing prefixes used to denote the number of carbon atoms in a chain.

Number of carbon atoms in chain	Prefix used
One →	**meth-**
Two →	**eth-**
Three →	**prop-**
Four →	**but-**
Five →	**pent-**
Six →	**hex-**
Seven →	**hept-**
Eight →	**oct-**

* **Aldehydes**, 194; **Carboxylic acids**, 195; **General formula**, 191 (**Homologous series**); **Halogenoalkanes**, 195; **Halogens**, 186; **Ketones**, 194; **Side chain**, 190.

The laboratory preparation of six gases

Methods for preparing six gases – **carbon dioxide**, **chlorine**, **hydrogen**, **ethene**, **nitrogen** and **oxygen** – are described below.

Carbon dioxide (see page 179) is obtained from the reaction of hydrochloric acid with calcium carbonate (marble chips).

A gas jar filled with water is placed on the beehive shelf over the mouth of the delivery tube. Gas produced by the reaction comes out of the delivery tube and displaces the water in the gas jar. This method of collecting a gas is called collecting gas **over water**.

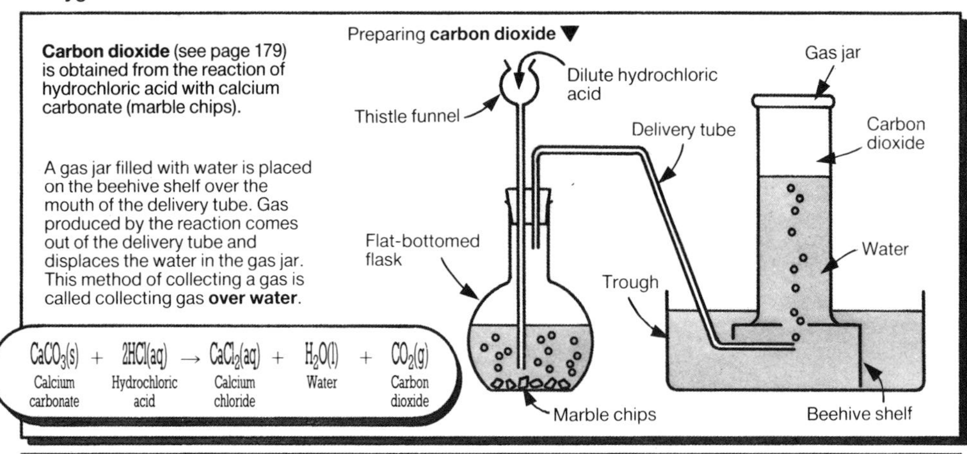

Chlorine (see page 187) is prepared by **oxidizing*** concentrated hydrochloric acid using manganese(IV) oxide. This reaction is always done in a fume cupboard.

The gas produced by the reaction contains some hydrogen chloride and water. The hydrogen chloride is removed by passing the stream of gas through water and the water is removed by passing the gas through concentrated sulphuric acid. Finally the chlorine is collected in a gas jar. It displaces air from the gas jar as it is heavier. This method of gas collection is collecting a gas by **upward displacement of air**.

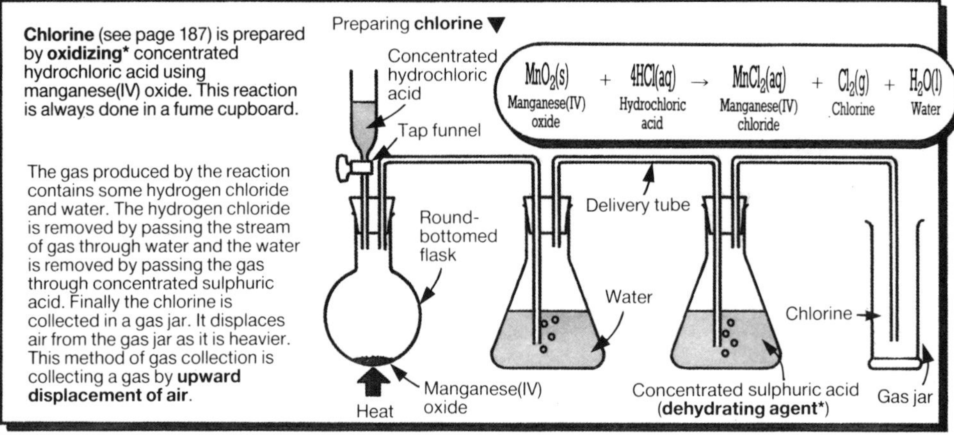

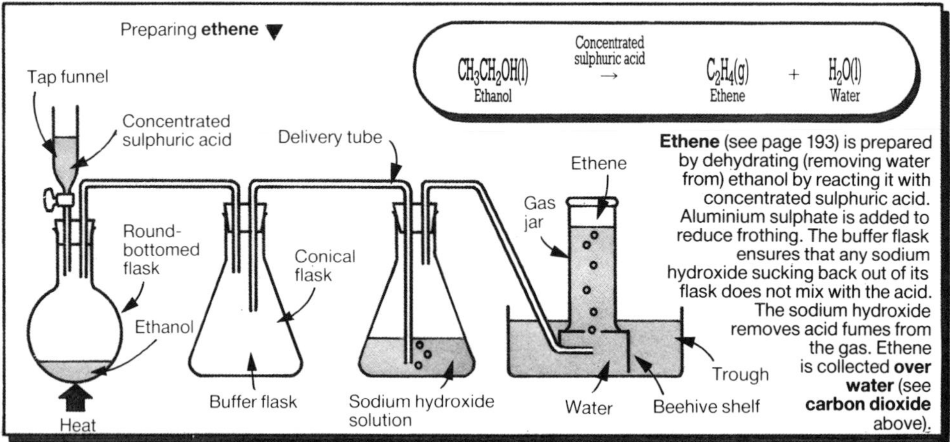

Ethene (see page 193) is prepared by dehydrating (removing water from) ethanol by reacting it with concentrated sulphuric acid. Aluminium sulphate is added to reduce frothing. The buffer flask ensures that any sodium hydroxide sucking back out of its flask does not mix with the acid. The sodium hydroxide removes acid fumes from the gas. Ethene is collected **over water** (see **carbon dioxide** above).

* **Dehydrating agent**, 344; **Oxidation**, 148.

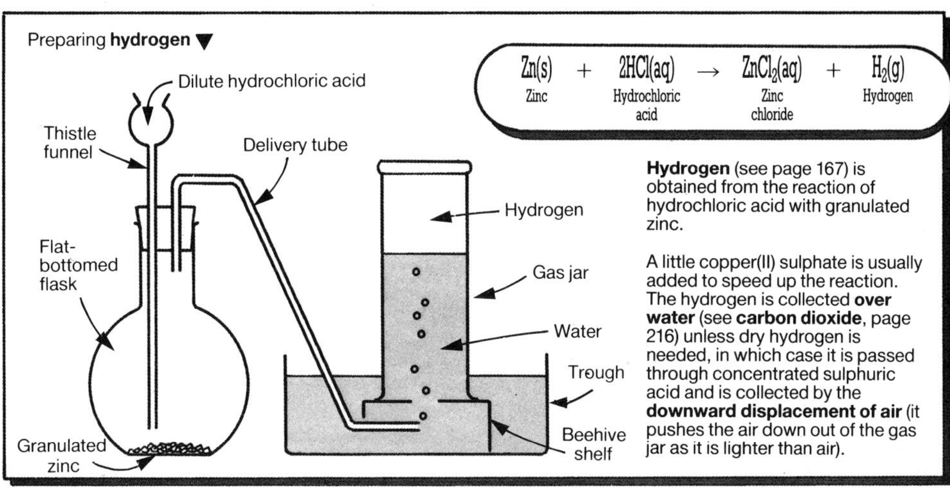

$$Zn(s) + 2HCl(aq) \rightarrow ZnCl_2(aq) + H_2(g)$$

Zinc, Hydrochloric acid, Zinc chloride, Hydrogen

Hydrogen (see page 167) is obtained from the reaction of hydrochloric acid with granulated zinc.

A little copper(II) sulphate is usually added to speed up the reaction. The hydrogen is collected **over water** (see **carbon dioxide**, page 216) unless dry hydrogen is needed, in which case it is passed through concentrated sulphuric acid and is collected by the **downward displacement of air** (it pushes the air down out of the gas jar as it is lighter than air).

Nitrogen (see page 180) is prepared by removing the carbon dioxide and oxygen from air. A residue of **noble gases*** remains in the nitrogen. The carbon dioxide is removed by passing the air through sodium hydroxide solution.

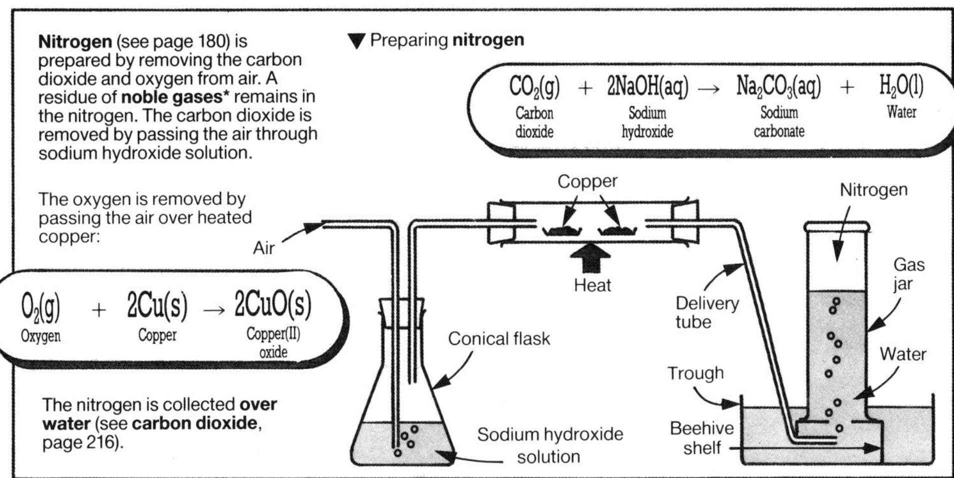

$$CO_2(g) + 2NaOH(aq) \rightarrow Na_2CO_3(aq) + H_2O(l)$$

Carbon dioxide, Sodium hydroxide, Sodium carbonate, Water

The oxygen is removed by passing the air over heated copper:

$$O_2(g) + 2Cu(s) \rightarrow 2CuO(s)$$

Oxygen, Copper, Copper(II) oxide

The nitrogen is collected **over water** (see **carbon dioxide**, page 216).

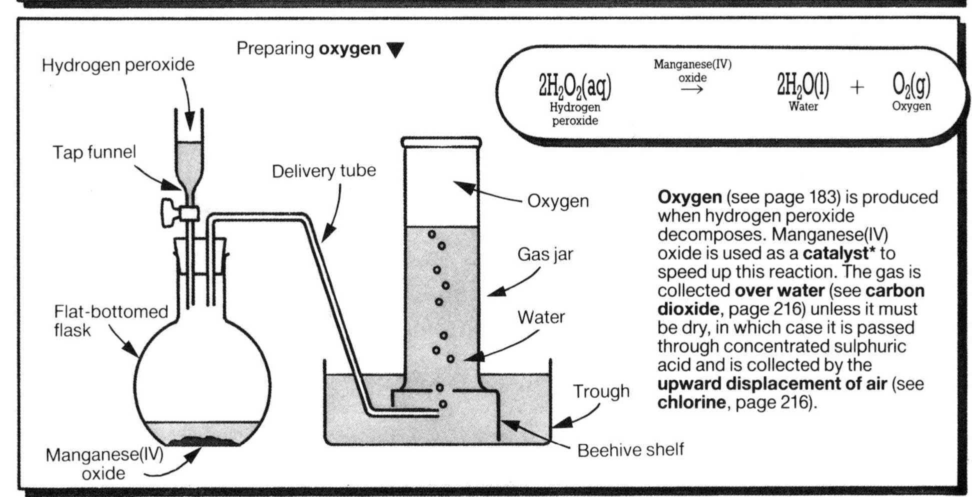

$$2H_2O_2(aq) \xrightarrow{\text{Manganese(IV) oxide}} 2H_2O(l) + O_2(g)$$

Hydrogen peroxide, Water, Oxygen

Oxygen (see page 183) is produced when hydrogen peroxide decomposes. Manganese(IV) oxide is used as a **catalyst*** to speed up this reaction. The gas is collected **over water** (see **carbon dioxide**, page 216) unless it must be dry, in which case it is passed through concentrated sulphuric acid and is collected by the **upward displacement of air** (see **chlorine**, page 216).

* **Catalyst**, 161; **Noble gases**, 189.

Laboratory tests

Various different tests are used to identify substances. Some of the tests involve advanced machinery, others are simple laboratory tests and all are known collectively as **qualitative analysis**. Some of the more advanced tests are shown on page 222; these two pages cover simple laboratory tests leading to the identification of water, common gases, a selection of **anions*** and **cations*** (i.e. components of compounds) and some metals.

The appearance or smell of a substance often gives clues to its identity – these can be confirmed by testing. If there are no such clues, then it is a matter of progressing through the tests, gradually eliminating possibilities (it is often a good idea to start with a **flame test**). Often more than one test is needed to identify an ion (anion or cation), as only one particular combination of results can confirm its presence (compare the tests and results for **lead**, **zinc** and **magnesium**).

Tests for water (H_2O)

Test	Results
Add to **anhydrous*** copper(II) sulphate.	White copper sulphate powder turns blue.
Add to **anhydrous*** cobalt(II) chloride.	Blue cobalt(II) chloride turns pink.

Tests to identify gases

Gas		Test	Results
Carbon dioxide	CO_2	Pass into **limewater** (calcium hydroxide solution).	Turns limewater cloudy.
Hydrogen	H_2	Put a lighted splint into a sample of the gas.	Burns with a "popping" noise.
Oxygen	O_2	Put a glowing splint into a sample of the gas.	Splint relights.

Tests for anions*

These tests are used to identify some of the **anions*** found in compounds.

Anion	Symbol	Test	Results
Bromide	Br^-	Add silver nitrate solution to a solution of substance in dilute nitric acid.	Pale yellow precipitate, dissolves slightly in ammonia solution.
Carbonate	CO_3^{2-}	a) Add dilute hydrochloric acid to the substance. b) Try to dissolve the substance in water containing **universal indicator*** solution.	Carbon dioxide gas given off. Dissolves and turns the indicator purple. (Compare **hydrogencarbonate** test.)
Chloride	Cl^-	Add silver nitrate solution to a solution of substance in dilute nitric acid.	Thick white precipitate dissolves in ammonia solution.
Hydrogen-carbonate	HCO_3^-	a) Add dilute hydrochloric acid to the substance. b) Try to dissolve the substance in water containing **universal indicator*** solution.	Carbon dioxide gas evolved. Dissolves and green indicator solution turns purple when boiled.
Iodide	I^-	Add silver nitrate solution to a solution of substance in dilute nitric acid.	Yellow precipitate, does not dissolve in ammonia solution.
Nitrate	NO_3^-	Add iron(II) sulphate solution followed by concentrated sulphuric acid to the solution.	Brown ring forms at the junction of the two liquids.
Sulphate	SO_4^{2-}	Add barium chloride solution to the solution.	White precipitate, does not dissolve in dilute hydrochloric acid.
Sulphite	SO_3^{2-}	Add barium chloride solution to the solution.	White precipitate, that dissolves in dilute hydrochloric acid.
Sulphide	S^{2-}	Add lead(II) ethanoate solution to the solution.	Black precipitate

* **Anhydrous**, 155 (**Anhydrate**); **Anion**, **Cation**, 130; **Universal indicator**, 152.

Tests for cations*

Most **cations*** in compounds can be identified by the same **flame tests** as those used to identify pure metals (see page 222 for how to carry out a flame test). The chart on the right gives a selection of flame test results. Cations can also be identified by the results of certain reactions. A number of these reactions are listed in the chart below. They cannot be used to identify pure metals, since many metals are insoluble in water and hence cannot form solutions.

Flame tests

Metal	Symbol	Flame colour
Barium	Ba	Yellow-green
Calcium	Ca	Red
Copper	Cu	Green
Lead	Pb	Blue
Lithium	Li	Pink
Potassium	K	Lilac
Sodium	Na	Orange

Cation	Symbol	Test	Results
Aluminium	Al^{3+}	a) Add dilute sodium hydroxide solution to a solution of the substance. b) Add dilute ammonia solution to a solution of the substance. c) Compare with **lead**, see tests below.	White precipitate that dissolves as more sodium hydroxide solution is added White precipitate that does not dissolve as more ammonia solution is added. ————
Ammonium	NH_4^+	Add sodium hydroxide solution to a solution of the substance and heat gently.	Ammonia gas is given off, it has a distinctive choking smell.
Calcium	Ca^{2+}	a) See **flame test** b) Add dilute sulphuric acid to a solution of the substance.	———— White precipitate formed.
Copper(II)	Cu^{2+}	a) See **flame test** b) Add dilute sodium hydroxide solution to a solution of the substance. c) Add dilute ammonia solution to a solution the substance.	———— Pale blue precipitate that dissolves as more sodium hydroxide is added. Pale blue precipitate, changing to deep blue solutions as more ammonia solution is added.
Iron(II)	Fe^{2+}	a) Add dilute sodium hydroxide solution to a solution of the substance. b) Add dilute ammonia solution to a solution of the substance.	Pale green precipitate formed. Pale green precipitate formed.
Iron(III)	Fe^{3+}	a) Add dilute sodium hydroxide solution to a solution of the substance. b) Add dilute ammonia solution to a solution of the substance.	Red-brown precipitate formed. Red-brown precipitate formed.
Lead(II)	Pb^{2+}	a) Add dilute sodium hydroxide solution to a solution of the substance. b) Add dilute ammonia solution to a solution of the substance. c) See also **flame test** to distinguish between lead and **aluminium**	White precipitate that dissolves as more sodium hydroxide solution is added. White precipitate that does not dissolve as more ammonia solution is added. ————
Magnesium	Mg^{2+}	a) Add dilute sodium hydroxide solution to a solution of the substance. b) Add dilute ammonia solution to a solution of the substance.	White precipitate that does not dissolve as more sodium hydroxide solution is added. White precipitate that does not dissolve as more ammonia solution is added.
Zinc	Zn^{2+}	a) Add dilute sodium hydroxide solution to a solution of the substance. b) Add dilute ammonia solution to a solution of the substance.	White precipitate that dissolves as more sodium hydroxide solution is added. White precipitate that dissolves as more ammonia solution is added.

* **Cation**, 130.

Investigating substances

The investigation of chemical substances involves a variety of different techniques. The first step is often to obtain a pure sample of a substance (impurities affect experimental results). Some of the separating and purifying techniques used to achieve this are explained on these two pages. A variety of different methods are then used to find out the chemical composition and the chemical and physical properties of the substance (**qualitative analysis**) and how much of it is present (**quantitative analysis**). For more information, see also pages 218-219 and 222.

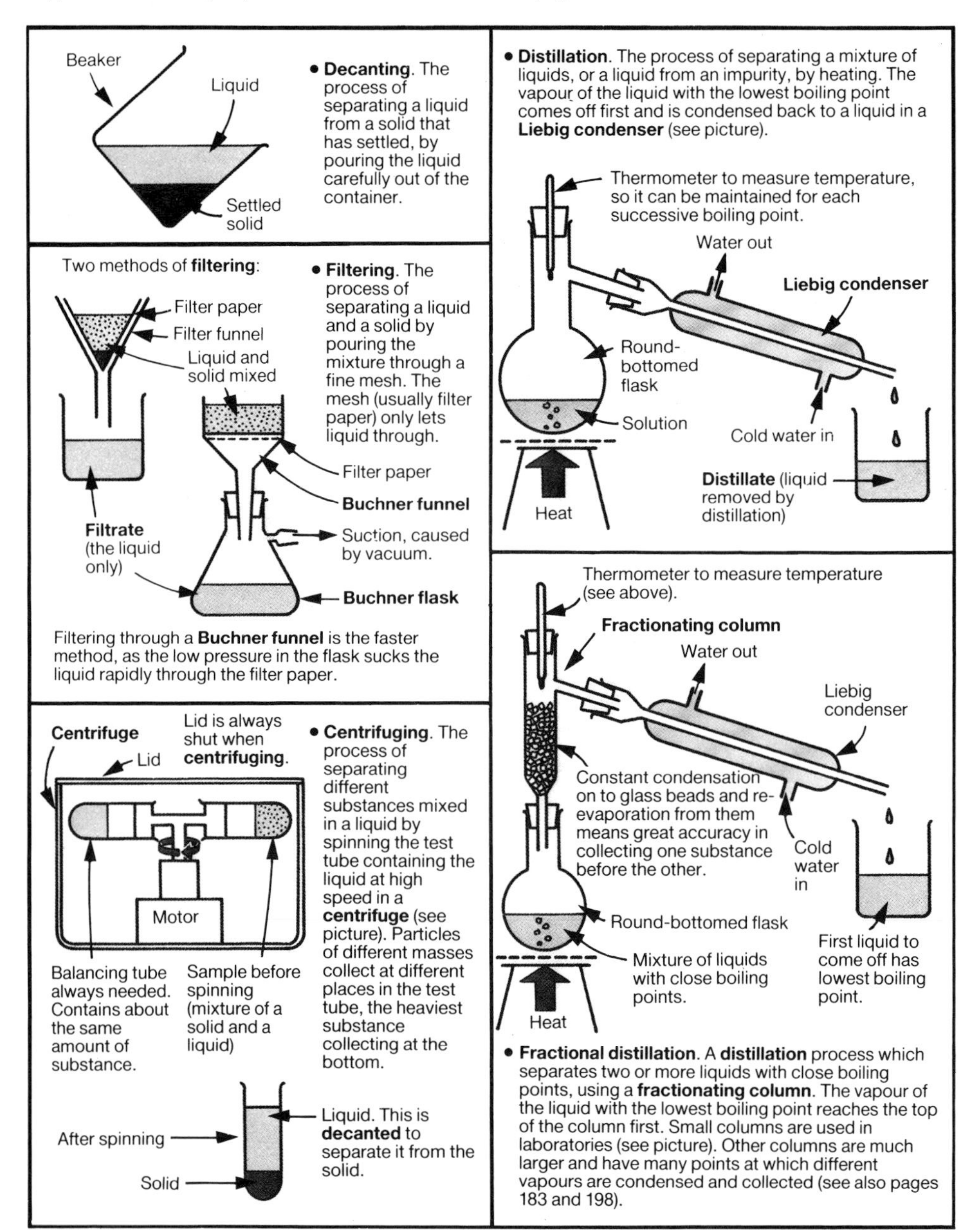

- **Decanting**. The process of separating a liquid from a solid that has settled, by pouring the liquid carefully out of the container.

- **Filtering**. The process of separating a liquid and a solid by pouring the mixture through a fine mesh. The mesh (usually filter paper) only lets liquid through.

Filtering through a **Buchner funnel** is the faster method, as the low pressure in the flask sucks the liquid rapidly through the filter paper.

- **Centrifuging**. The process of separating different substances mixed in a liquid by spinning the test tube containing the liquid at high speed in a **centrifuge** (see picture). Particles of different masses collect at different places in the test tube, the heaviest substance collecting at the bottom.

- **Distillation**. The process of separating a mixture of liquids, or a liquid from an impurity, by heating. The vapour of the liquid with the lowest boiling point comes off first and is condensed back to a liquid in a **Liebig condenser** (see picture).

- **Fractional distillation**. A **distillation** process which separates two or more liquids with close boiling points, using a **fractionating column**. The vapour of the liquid with the lowest boiling point reaches the top of the column first. Small columns are used in laboratories (see picture). Other columns are much larger and have many points at which different vapours are condensed and collected (see also pages 183 and 198).

• **Solvent extraction**. The process of obtaining a **solute*** by transferring it from its original **solvent*** to one in which it is more soluble, and from which it can be easily removed. It is a method of separation often used when the solute cannot be heated, and makes use of a particular property of the solvents, i.e. whether they are **polar** or **non-polar solvents***. **Ether extraction** is an example.

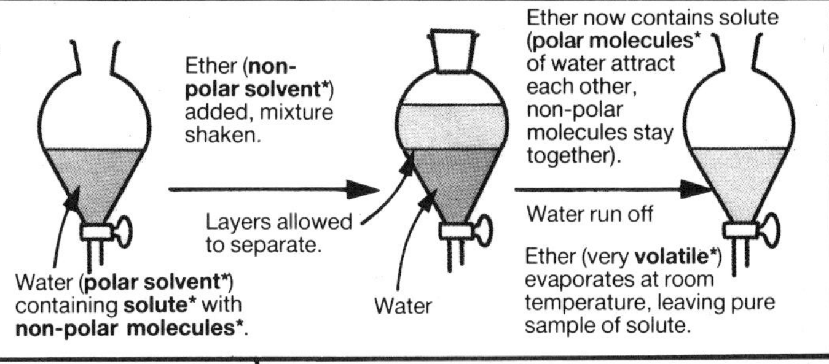

Paper chromatography

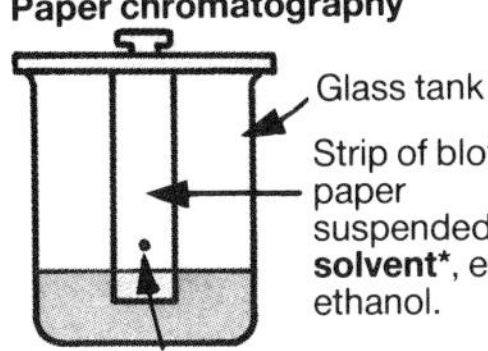

Standard tables identify substances by $\mathbf{R_f}$ **value** - distance moved by substance over distance moved by **solvent**.

After removing from tank

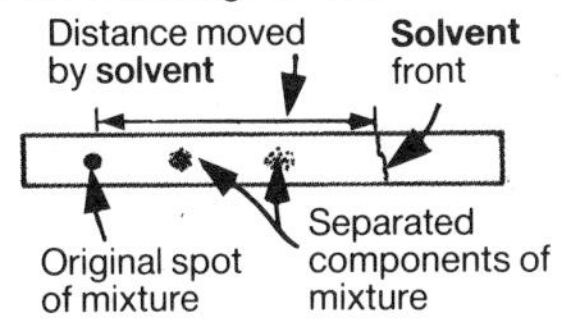

• **Chromatography**. The process of separating small amounts of substances from a mixture by the rates at which they move through or along a medium (the **stationary phase**, e.g. blotting paper). Most methods of chromatography involve dissolving the mixture in a **solvent*** (the **eluent**), though it is vaporized in **gas chromatography**. Substances move at different rates because they vary in their **solubility*** and their attraction to the medium.

There are several methods of chromatography, including **column chromatography** (components in the mixture are separated in a column containing a solvent and a material that attracts molecules) and **gas chromatography** (vaporized mixture is separated as it passes along a heated column in a stream of gas).

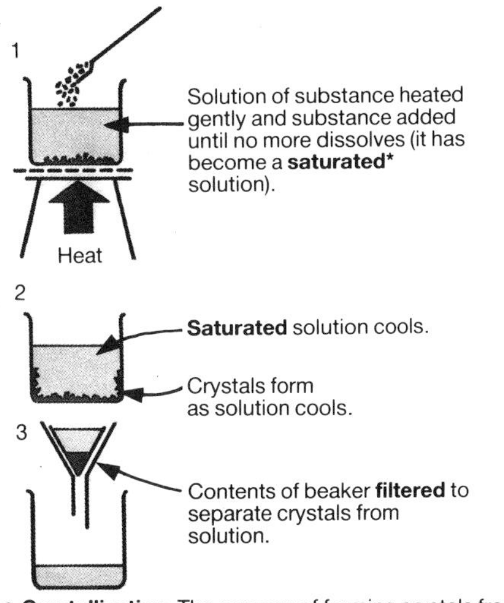

• **Crystallization**. The process of forming crystals from a solution, which can be used to produce a pure sample of a substance, as the impurities will not form crystals. To make pure crystals, a hot, **saturated*** solution of the substance is cooled and the crystals formed on cooling are removed by **filtering**. See also page 135.

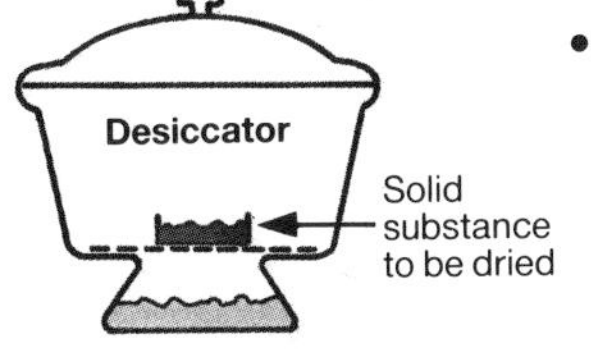

Drying agent* not in contact with solid. Absorbs moisture from air, causing water to evaporate from solid.

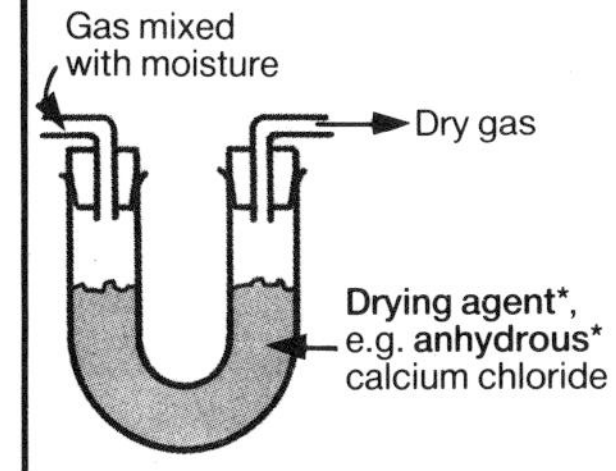

• **Desiccation**. The process of removing water mixed with a substance, or **water of crystallization*** from a substance. Solids are often dried in large glass **desiccators** that contain a **drying agent*** such as silica gel. Water is removed from most gases and liquids by bringing them into direct contact with a drying agent, e.g. **anhydrous*** calcium chloride (which absorbs the water and, in the case of liquids, is then **filtered** off).

Measuring the melting point

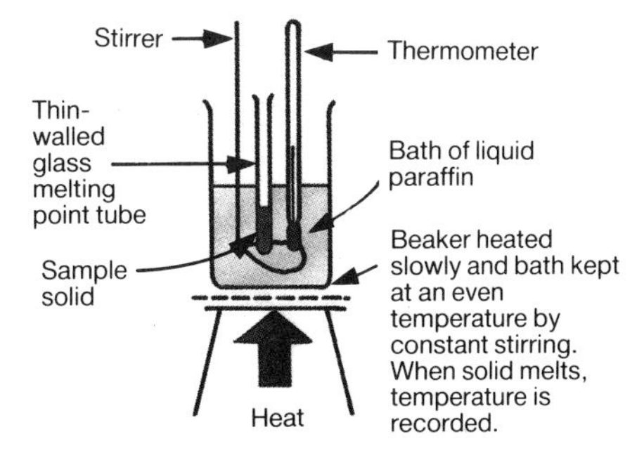

• **Melting point** and **boiling point tests**. Tests used to determine the purity of a sample. A pure sample of a substance has a particular known melting point and boiling point, and any impurities in a sample will alter these measurements.

* **Anhydrous**, 155 (**Anhydrate**); **Drying agent**, 344; **Non-polar molecule**, 133 (**Polar molecule**); **Non-polar solvent**, **Polar solvent**, **Saturated**, 144; **Solubility**, 145; **Solute**, **Solvent**, 144; **Volatile**, 345; **Water of crystallization**, 135.

Qualitative and quantitative analysis

There are two types of analysis used to investigate substances: **qualitative analysis** – any method used to study chemical composition – and **quantitative analysis** – any method used to discover how much of a substance is present in a sample. Below are some examples of both types of analysis.

Qualitative analysis

Below are some example of qualitative analysis. The **flame test** and the tests on pages 218-219 are examples of qualitative analysis used in schools. The other methods described are more advanced.

- **Flame test**. Used to identify metals. A substance is collected on the tip of a clean platinum or nichrome wire. This is held in a flame to observe the colour with which the substance burns (see also page 219). Between tests, the wire is cleaned by dipping it in concentrated hydrochloric acid and then heating strongly.

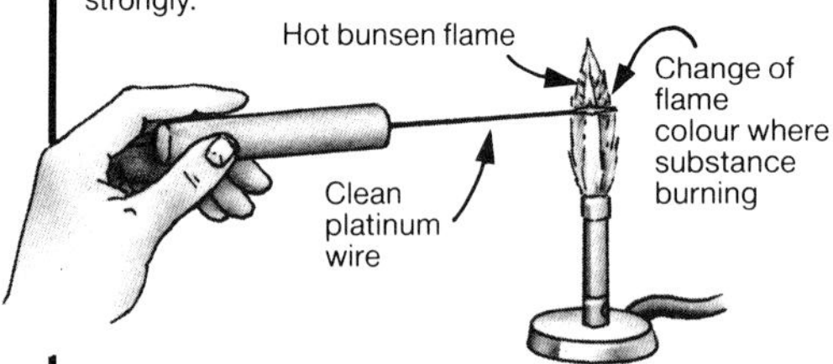

- **Mass spectroscopy**. A method of investigating the composition of a substance, in particular the **isotopes*** it contains. It is also used as a method of quantitative analysis as it involves measuring the relative proportions of isotopes or molecules in the substance. The apparatus used is called a **mass spectrometer**.

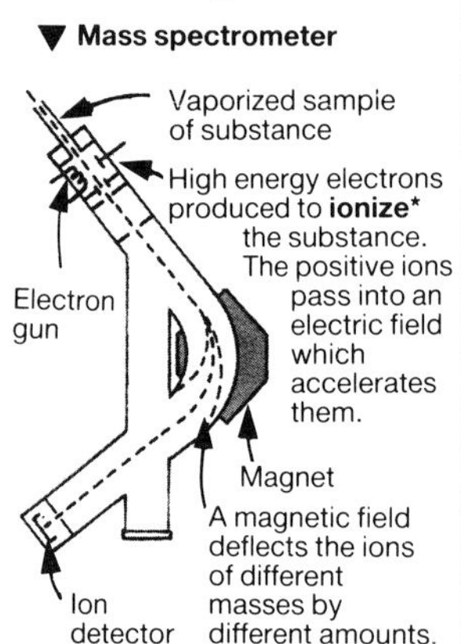

- **Nuclear magnetic resonance (n.m.r.) spectroscopy**. A method used for investigating the position of atoms in a molecule. Radio waves are passed through a sample of a substance held between the poles of a magnet. The amount of absorption reveals the positions of particular atoms within a molecule. This information is presented on a graph called a **nuclear magnetic resonance (n.m.r.) spectrum**. ▼

Nuclear magnetic resonance spectrum of **ethanol*** (CH_3CH_2OH)

Degree of absorption

Peak showing an -OH group | Peaks showing a $-CH_2-$ group | Peaks showing a $-CH_3$ group

Quantitative analysis

Below are some examples of quantitative analysis. See also **mass spectroscopy**.

- **Volumetric analysis**. A method of determining the concentration of a solution using **titration**. This is the addition of one solution into another, using a **burette***. The concentration of one solution is known. The first solution is added from the burette until the **end point**, when all the second solution has reacted. The volume of solution from the burette needed to reach the end point is called the **titre**. This, the volume of solution in the flask and the known concentration of one solution are used to calculate the concentration of the second solution. ▶

Burette*
Solution A
Tap
Conical flask containing measured volume of solution B.
Apparatus used for **titrations**

- **Gravimetric analysis**. A method of determining the amount of a substance present by converting it into another substance of known chemical composition that is easily purified and weighed. ▼

Gravimetric analysis can be used to measure the amount of lead in a sample of water containing a lead salt.

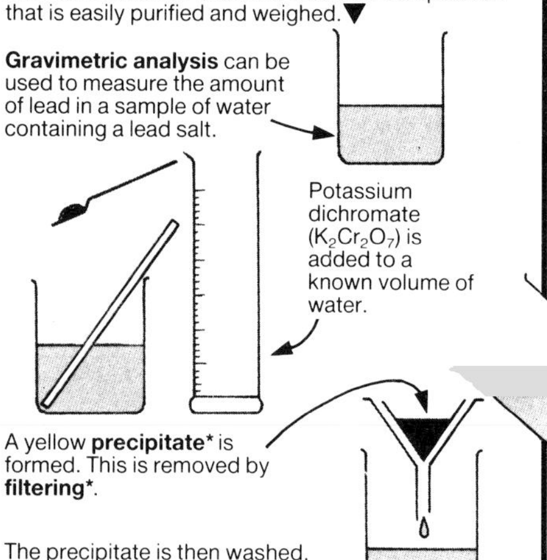

The precipitate is then washed, dried and weighed accurately.

The concentration of the lead in the sample of water is calculated from the volume of water, the weight of lead chromate and the **relative atomic mass*** of lead.

* **Burette**, 223; **Ethanol**, 196; **Filtering**, 220; **Ionization**, 130; **Isotope**, 127; **Precipitate**, 145; **Relative atomic mass**, 138.

Apparatus

Apparatus is chemical equipment. The most common items are described and illustrated below and on pages 224-225. Simple 2-D diagrams used to represent them are also shown, together with approximate ranges of sizes.

- **Beaker**. Used to hold liquids. Shows approximate volume. ▼

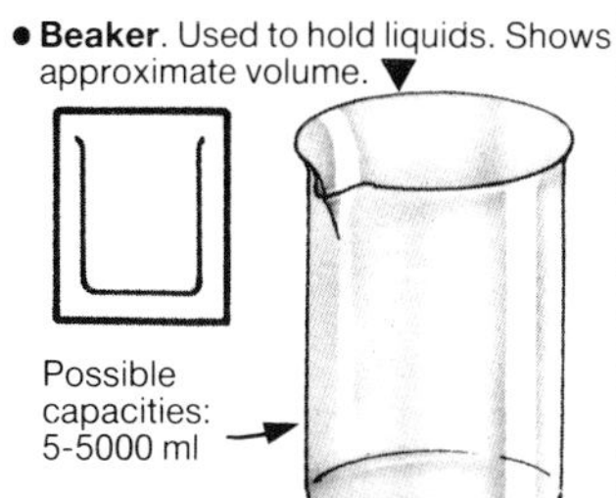

- **Beehive shelf**. Used to support a **gas jar*** while gas is being collected by the displacement of water. For examples of its use, see pages 216-217. ▼

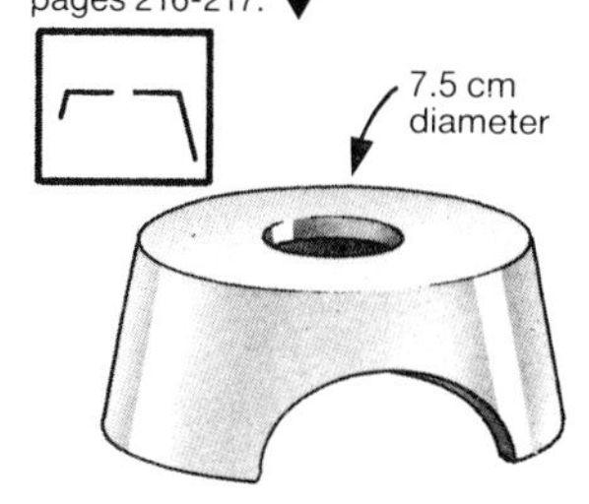

- **Bunsen burner**. Used to provide heat for chemical reactions. Its adjustable air-hole allows some control of the flame temperature. If the hole is closed, the flame is yellow and cooler than the blue flame produced when the hole is open. See picture, page 208. ▶

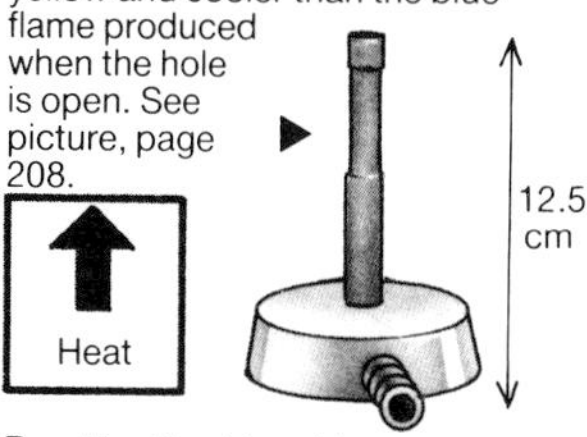

- **Burette**. Used to add accurate volumes of liquid during **titrations** (see **volumetric analysis**, page 222). ▼

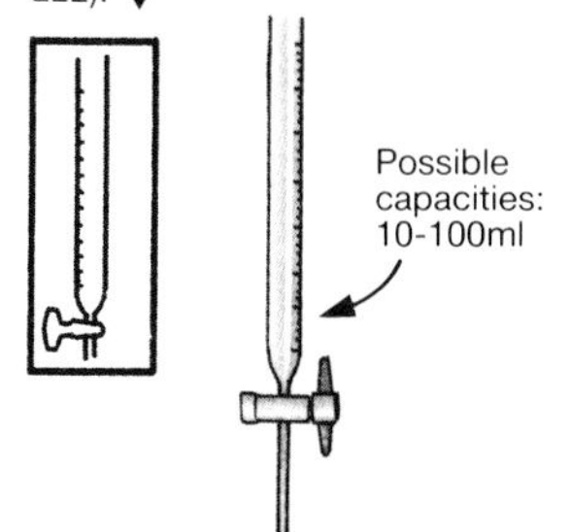

Condensers

- **Liebig condenser**. Used to condense vapours. Vapour passes through the central channel and is cooled by water flowing ◀ through the outer pipe. See **distillation**, page 220.

Possible lengths: 25-50 cm

Water circulates in outer tube.

Vapour condenses in inner tube.

Length: 15.0 cm

- **Reflux condenser**. Used to ▲ return vapour to a liquid to prevent loss by evaporation.

- **Crucible**. Used to hold small quantities of solids which are being heated strongly, either in a furnace or over a **bunsen burner**. They are made of porcelain, silica, fireclay, nickel or steel. ▼

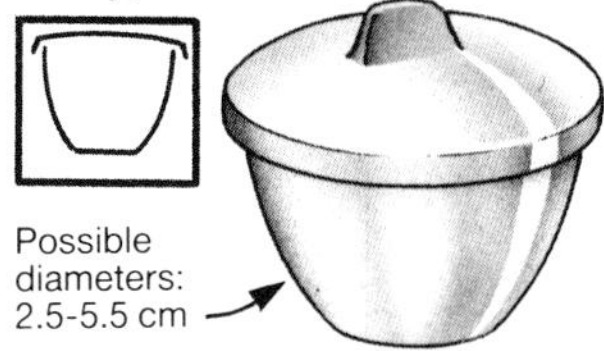

- **Crystallizing dish**. Used to hold solutions which are being evaporated to form crystals. The flat bottom helps to form an even layer of crystals. ▼

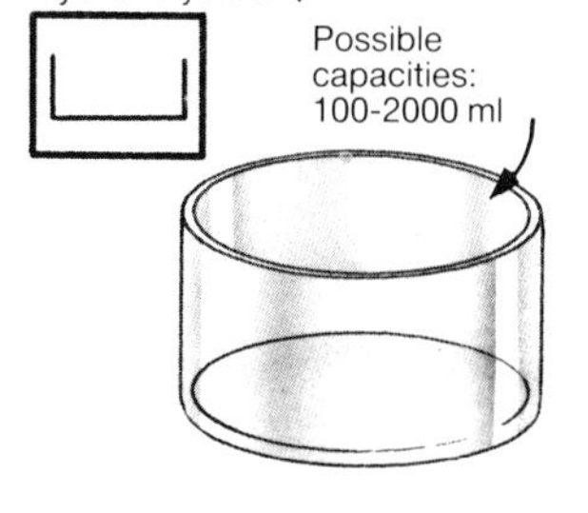

- **Delivery tube**. A tube used to carry gases. ▼

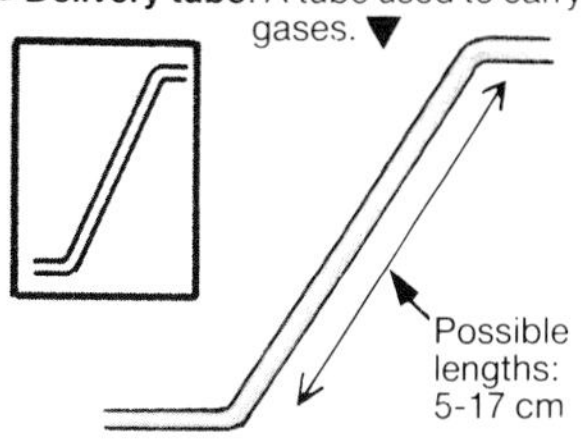

- **Desiccator**. A glass container used to dry solids. It contains a **drying agent***. See **desiccation**, page 221. ▼

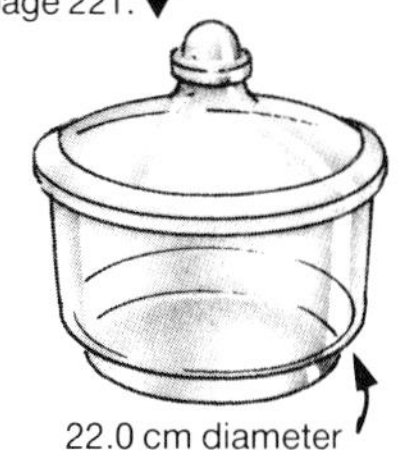

- **Evaporating basin**. Used to hold a solution whose **solvent*** is being separated from the **solute*** by evaporation (often using heat). ▼

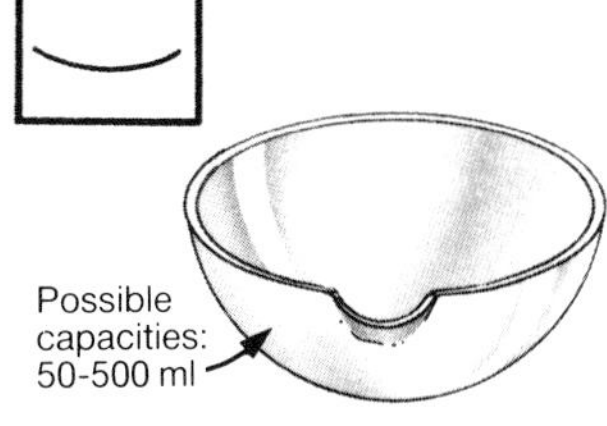

- **Filter paper**. Paper which acts as a sieve, only allowing liquids through, but no solid matter. Filter paper is graded according to how finely it is meshed, i.e. the size of particle it allows through. It is put in a **filter** or **Buchner funnel*** to give support as the liquid passes through, and the solid settles on the paper. See **filtering**, page 220. ▼

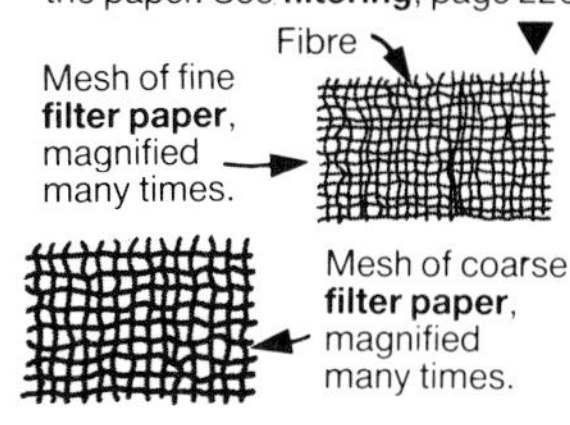

Holes between fibres allow tiny particles to pass through.

* **Buchner funnel**, 224; **Drying agent**, 344; **Filter funnel**, **Gas jar**, 224; **Solute**, **Solvent**, 144.

Flasks

- **Buchner flask**. Used when liquids are filtered by suction. See **filtering**, page 220.

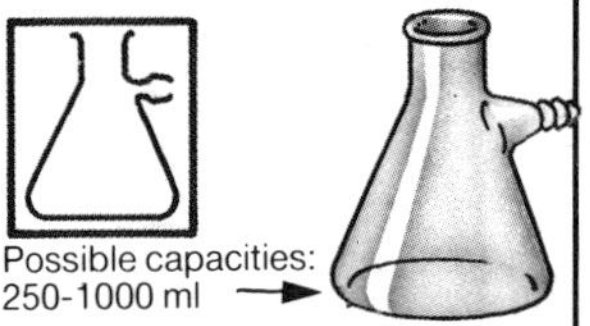

- **Conical flask**. Used to hold liquids when carrying out reactions and preparing solutions of known concentration. They are used in preference to beakers when it is necessary to have a container that can be stoppered. They have some volume markings but these are not as accurate as the markings on a **pipette** or **burette***.

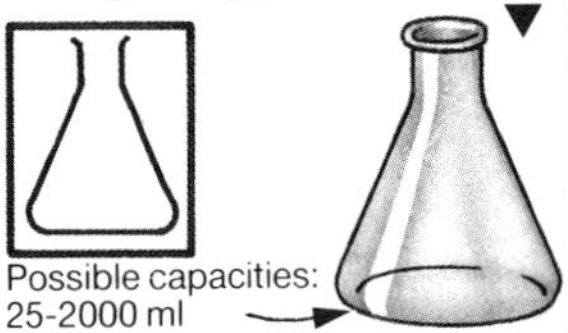

- **Flat-bottomed flask**. Used to hold liquids when carrying out reactions where heating is not required (the flask stands on the work-bench).

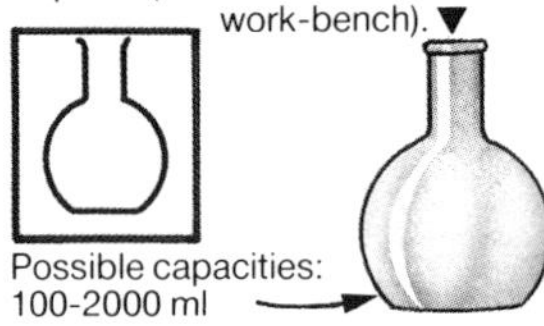

- **Round-bottomed flask**. Used to hold liquids, especially when even heating is needed. Volume markings are approximate. It is held in position above the flame by a clamp.

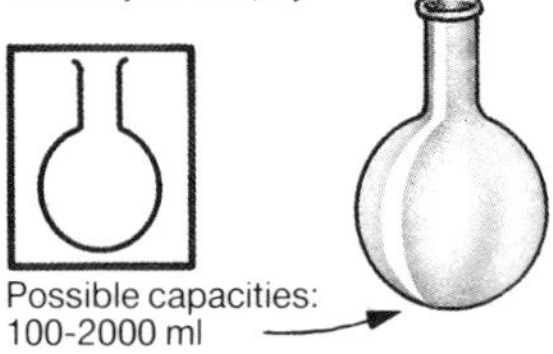

- **Volumetric flask**. Used when mixing accurate concentrations of solutions. Each flask has a volume marking which is very exact and a stopper so that it can be shaken to mix the solution.

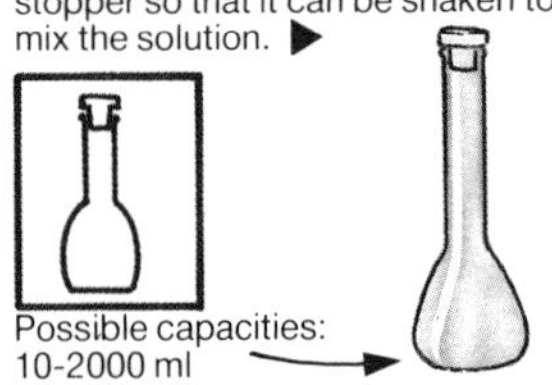

- **Fractionating column**. Used to separate components of a mixture by their boiling points. It contains glass balls or rings that provide a large surface area and thus promote condensation and re-evaporation. See **fractional distillation**, page 220.

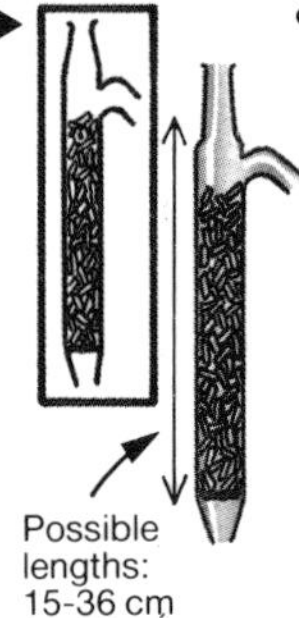

- **Fume cupboard**. A glass panelled cupboard that contains an extractor fan and encloses an area of work-bench. Dangerous experiments are done in a fume cupboard.

Funnels

- **Buchner funnel**. Used when liquids are filtered by suction. It has a flat perforated plate, on which **filter paper*** is placed. See **filtering**, page 220.

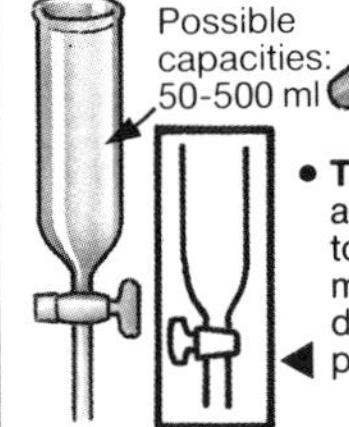

- **Tap funnel**. For adding a liquid to a reaction mixture drop by drop. See pages 216-217.

- **Filter funnel**. Used when separating solids from liquids by **filtering** (see page 220). **Filter paper*** is put inside the funnel.

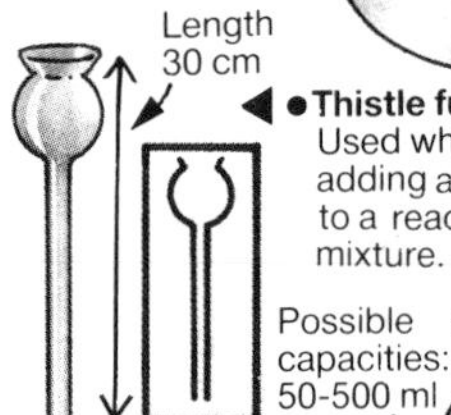

- **Thistle funnel**. Used when adding a liquid to a reaction mixture.

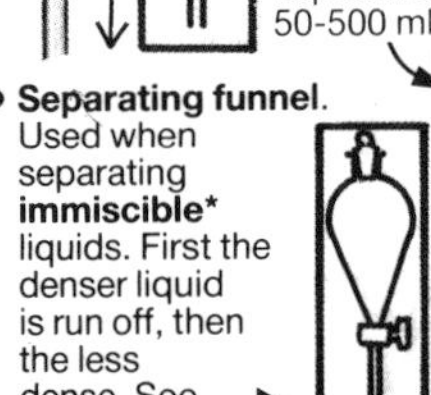

- **Separating funnel**. Used when separating **immiscible*** liquids. First the denser liquid is run off, then the less dense. See **solvent extraction**, page 221.

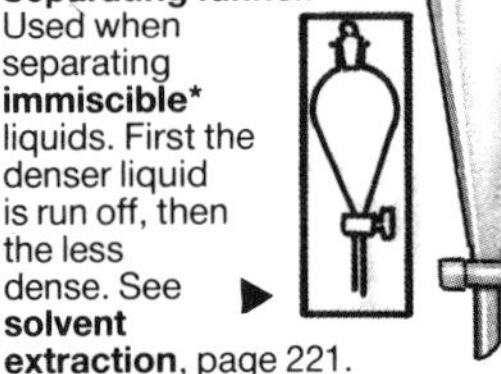

- **Gas jar**. Used when collecting and storing gases. The jar can be sealed, using a glass lid whose rim is coated with a thin layer of grease. See pages 216-217.

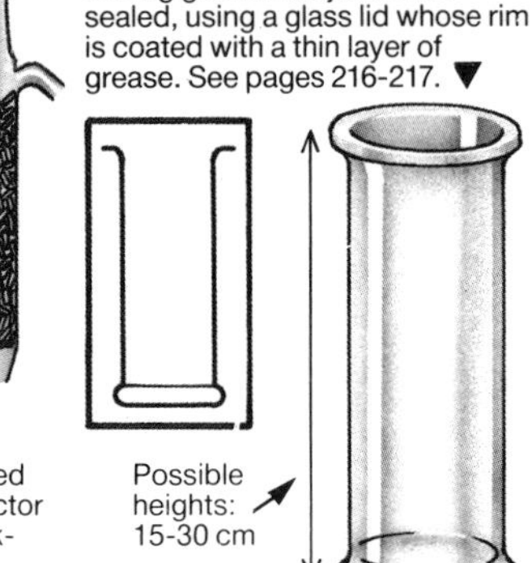

- **Gas syringe**. Used to measure the volume of a gas. It is used both to receive gas and to inject gas into a reaction vessel.

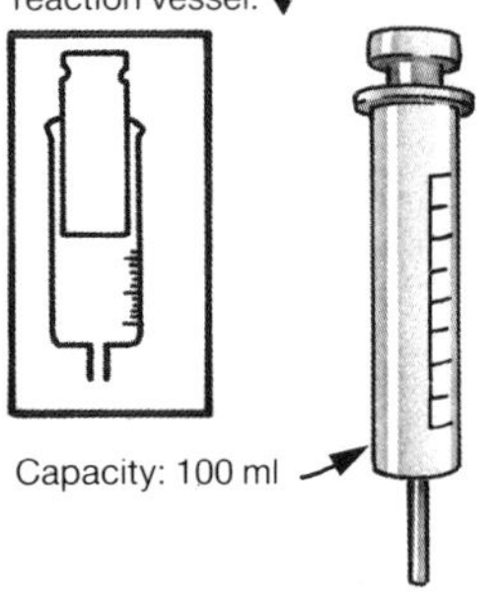

- **Gauze**. Used to spread the heat from a flame evenly over the base of an object being heated. Made of iron, steel, copper or ceramics.

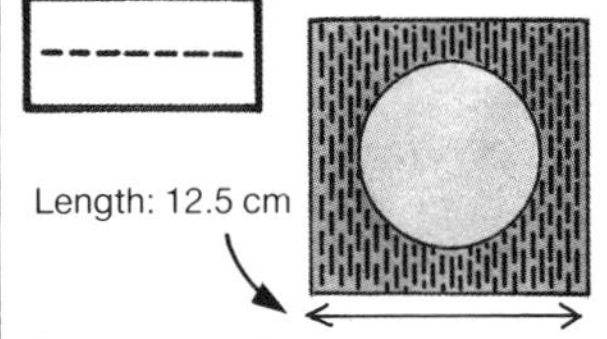

- **Measuring cylinder**. Used to measure the approximate volume of liquids.

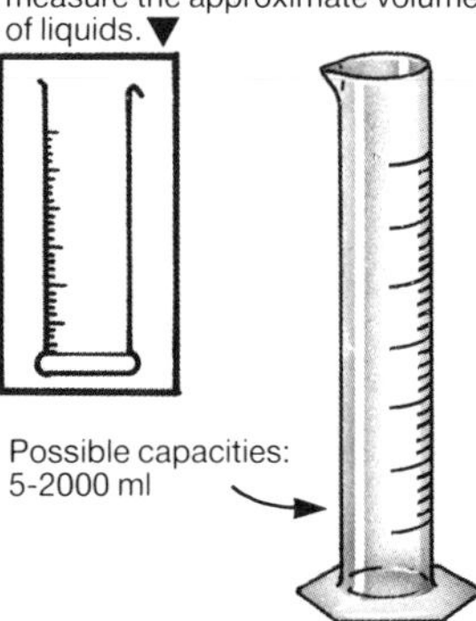

* **Burette**, **Filter paper**, 223; **Immiscible**, 145 (**Miscible**).

- **Pipeclay triangle**. Used to support **crucibles*** on **tripods** when they are being heated. They are made of iron or nickel-chromium wire enclosed in pipeclay tubes. ▼

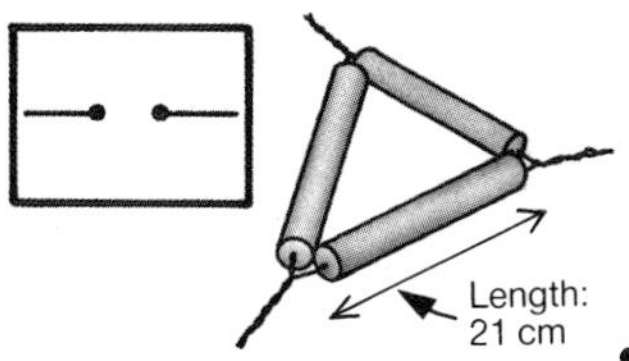

Pipettes

- **Pipette**. Used to dispense accurate volumes of liquid. They come in different sizes for different volumes. The liquid is run out of the pipette until its level has dropped from one volume marking to the next. ▶

Possible capacities: 1-100 ml

Possible capacities: 1-2 ml

- ◀ **Dropping pipette** or **teat pipette**. Used to dispense small volumes or drops of liquid. It does not provide an accurate measurement.

- **Stands and clamps**. Used to hold apparatus in position, e.g. **round-bottomed flasks**. ▼

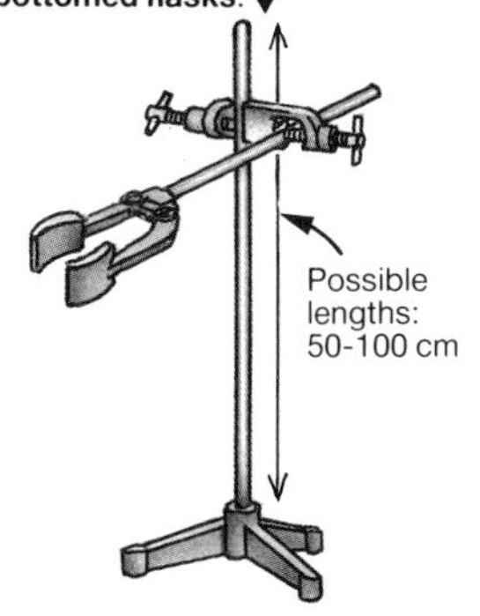

- **Spatula**. Used to pick up small quantities of a solid. ▼

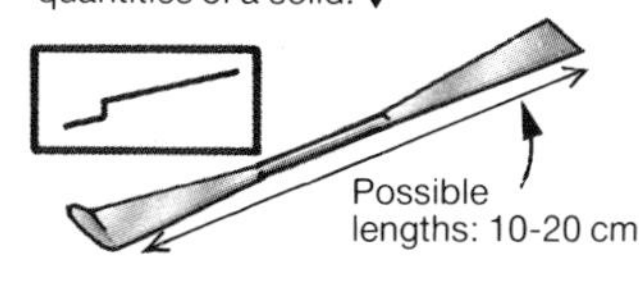

- **Test tube holder**. Used to hold a test tube, e.g. when heating it in a flame, creating a chemical reaction within it, or transferring it from one place to another. ▼

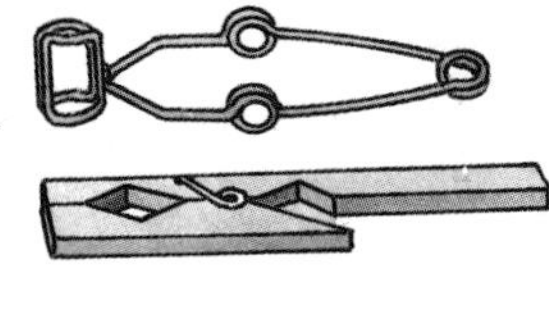

- **Test tube rack**. Used to hold many test tubes upright. ▼

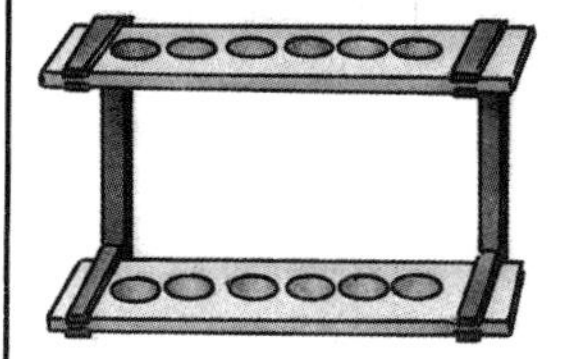

- **Thermometer**. Used to measure temperature. They are filled either with alcohol or with mercury, depending on the temperature range for which they are intended. ▼

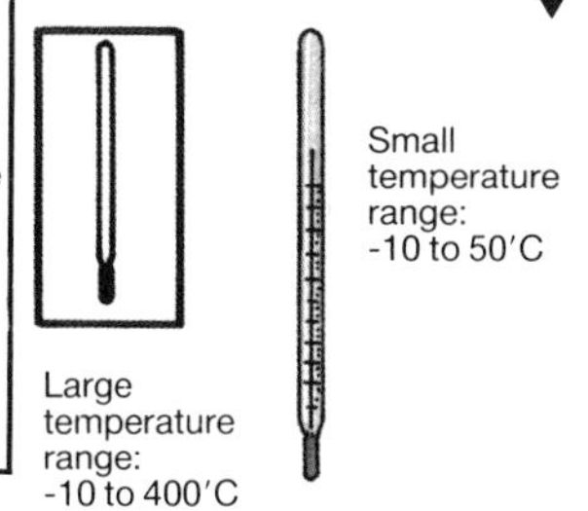

- **Tongs**. Used to move hot objects. ▼

- **Top pan balances**. Used for quick, accurate weighing. ▼

- **Trough**. Used when collecting gas **over water** (see **carbon dioxide**, page 216). The water contained in a gas jar inverted in the trough is displaced into the trough. Troughs are also used when substances such as potassium are reacted (see picture, page 169). ▼

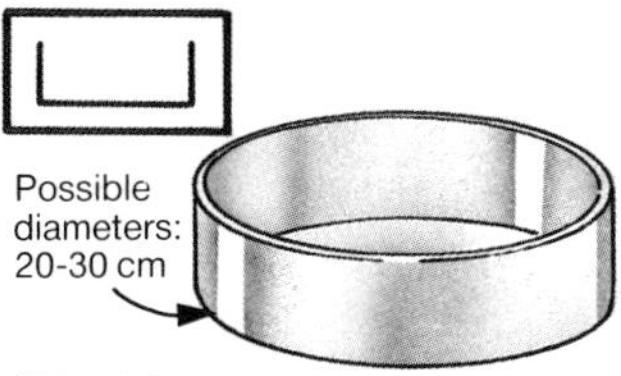

- **Tripod**. Used with a **pipeclay triangle** or **gauze** when heating **crucibles***, **flasks**, etc. ▼

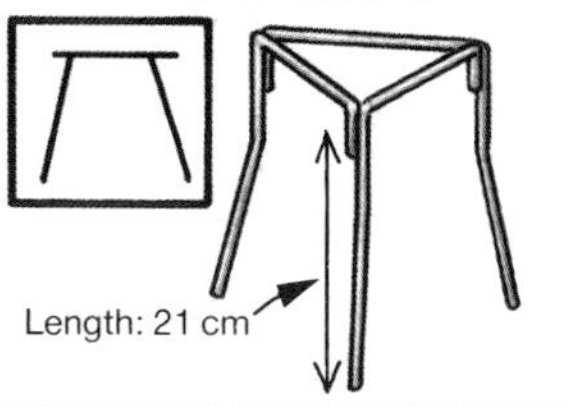

Tubes

- **Boiling tube**. A thick-walled tube used to hold substances being heated strongly. ▶

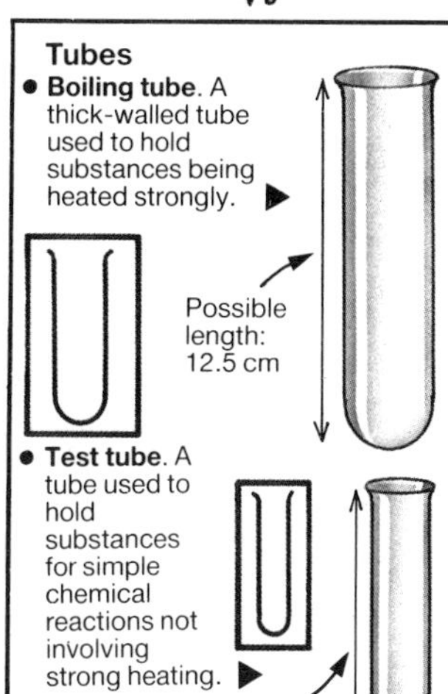

- **Test tube**. A tube used to hold substances for simple chemical reactions not involving strong heating. ▶

Possible length: 7.5 cm

- **Ignition tube**. A disposable tube used to hold small quantities of substances being melted or boiled. ▶

Possible length: 5.0 cm

- **Watch glass**. Used when evaporating small quantities.

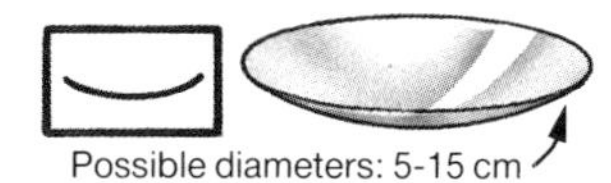

* **Bunsen burner**, **Crucible**, 223.

Chart of substances, symbols and formulae

Below is a comprehensive list of all the symbols and formulae found in the chemistry section of this book. Each one is listed next to the name of the substance it stands for (the substances themselves can be found in the main index on pages 346-384). Capital letters come alphabetically before small ones, i.e. each element is kept together with its compounds. For example, **CH_3OH** (methanol – a carbon compound) is found in an alphabetical list after **C** (carbon), before the **Ca** (calcium) list begins.

Symbol	Substance
$3Ca_3(PO_4)_2.CaF_2$	Apatite
Ac	Actinium
Ag	Silver
AgBr	Silver bromide
AgCl	Silver chloride
AgI	Silver iodide
$AgNO_3$	Silver nitrate
Al	Aluminium
$Al(OH)_3$	Aluminium hydroxide
Al_2O_3	Aluminium oxide
$Al_2O_3.2H_2O$	Bauxite
$Al_2(SO_4)_3$	Aluminium sulphate
Am	Americium
Ar	Argon
As	Arsenic
At	Astatine
Au	Gold
B	Boron
B_2O_3	Boron oxide
BCl_3	Boron trichloride
Ba	Barium
$BaCl_2$	Barium chloride
Be	Beryllium
Bi	Bismuth
Bk	Berkelium
Br/Br_2	Bromine
-Br	Bromo group
C	Carbon
C_2H_2	Ethyne
C_2H_4	Ethene
C_2H_5Br	Bromoethane
C_2H_5CHO	Propanal
C_2H_5Cl	Chloroethane
C_2H_5COOH	Propanoic acid
C_2H_5OH	Ethanol
C_2H_6	Ethane
C_3H_4	Propyne
C_3H_6	Propene
C_3H_6O	Propanone
C_3H_7OH	Propan-1-ol
C_3H_8	Propane
C_4H_6	But-1-yne
C_4H_8	But-1-ene
C_4H_9OH	Butan-1-ol
C_4H_{10}	Butane
C_5H_{10}	Pent-1-ene
C_5H_{12}	Pentane
$C_6H_8O_6$	Ascorbic acid
$C_6H_{12}O_6$	Glucose
C_6H_{14}	Hexane
C_7H_{16}	Heptane
C_8H_{18}	Octane
C_9H_{20}	Nonane
$C_{12}H_{22}O_{11}$	Sucrose
$C_{17}H_{35}COOH$	Octadecanoic acid
CCl_4	Tetrachloromethane

Symbol	Substance
CH_2BrCH_2Br	1,2-dibromoethane
CH_2CHCl	Vinyl chloride
$-CH_3$	Methyl group
CH_3CCH	Propyne
CH_3CH_2CCH	But-1-yne
$CH_3CH_2CH_2CH_2OH$	Butan-1-ol
$CH_3CH_2CH_2OH$	Propan-1-ol
CH_3CH_2CHO	Propanal
CH_3CH_2Cl	Chloroethane
CH_3CH_2COOH	Propanoic acid
CH_3CH_2OH	Ethanol
CH_3CH_2ONa	Sodium ethoxide
CH_3CHO	Ethanal
$CH_3CHOHCH_3$	Propan-2-ol
CH_3Cl	Chloromethane
$CH_3COCH_2CH_3$	Butanone
CH_3COCH_3	Propanone
$CH_3COOCH_2CH_3$	Ethyl ethanoate
CH_3COOH	Ethanoic acid
CH_3NH_2	Methyl amine
CH_3OCH_3	Methoxymethane
CH_3OH	Methanol
CH_4	Methane
CHCH	Ethyne
CO	Carbon monoxide
-CO-	Carbonyl group
CO_2	Carbon dioxide
-COOH	Carboxyl group
$(COOH)_2$	Ethanedioic acid
$COOH(CH_2)_4COOH$	Hexanedioic acid
Ca	Calcium
$Ca_3(PO_4)_2$	Calcium phosphate
$CaCl_2$	Calcium chloride
$CaCO_3$	Calcium carbonate
$CaCO_3.MgCO_3$	Dolomite
CaF_2	Fluorospar
$Ca(HCO_3)_2$	Calcium hydrogencarbonate
CaO	Calcium oxide
$Ca(OH)_2$	Calcium hydroxide
$CaSiO_3$	Calcium metasilicate
$CaSO_4$	Calcium sulphate
$CaSO_4.2H_2O$	Gypsum
Cd	Cadmium
Ce	Cerium
Cf	Californium
Cl/Cl_2	Chlorine
-Cl	Chloro group
Cm	Curium
Co	Cobalt
$CoCl_2$	Cobalt(II) chloride
Cr	Chromium
Cs	Caesium

Symbol	Substance
Cu	Copper
Cu_2O	Copper(I) oxide
CuCl	Copper(I) chloride
$CuCl_2$	Copper(II) chloride
$CuCO_3.Cu(OH)_2$	Malachite
$(CuFe)S_2$	Copper pyrites
$(Cu(NH_3)_4)SO_4$	Tetraammine copper(II) sulphate
$Cu(NO_3)_2$	Copper(II) nitrate
CuO	Copper(II) oxide
$CuSO_4$	Copper(II) sulphate
$CuSO_4.3Cu(OH)_2$	Basic copper sulphate
D	Deuterium
D_2O	Deuterium oxide
Dy	Dysprosium
Er	Erbium
Es	Einsteinium
Eu	Europium
F/F_2	Fluorine
-F	Fluoro group
Fe	Iron
Fe_2O_3	Haematite
$Fe_2O_3.xH_2O$	Rust
$FeCl_2$	Iron(II) chloride
$FeCl_3$	Iron(III) chloride
$Fe(OH)_3$	Iron(III) hydroxide
FeS	Iron(II) sulphide
$FeSO_4$	Iron(II) sulphate
Fm	Fermium
Fr	Francium
Ga	Gallium
Gd	Gadolinium
Ge	Germanium
H/H_2	Hydrogen
H_2CO_3	Carbonic acid
H_2O	Water
H_2O_2	Hydrogen peroxide
H_2S	Hydrogen sulphide
$H_2S_2O_7$	Fuming sulphuric acid
H_2SO_3	Sulphurous acid
H_2SO_4	Sulphuric acid
H_3PO_4	Phosphoric acid
HBr	Hydrogen bromide
HCl	Hydrogen chloride/ Hydrochloric acid
HCHO	Methanal

Symbol	Substance
HCOOH	Methanoic acid
HI	Hydrogen iodide
HNO_2	Nitrous acid
HNO_3	Nitric acid
He	Helium
Hf	Hafnium
Hg	Mercury
HgS	Cinnabar
Ho	Holmium
I/I_2	Iodine
In	Indium
Ir	Iridium
K	Potassium
K_2CO_3	Potassium carbonate
$K_2Cr_2O_7$	Potassium dichromate
K_2SO_4	Potassium sulphate
$K_2SO_4.Al_2(SO_4)_3$	Aluminium potassium sulphate-12-water
KBr	Potassium bromide
KCl	Potassium chloride
KI	Potassium iodide
$KMnO_4$	Potassium permanganate
KNO_3	Potassium nitrate
KOH	Potassium hydroxide
Kr	Krypton
KrF_2	Krypton fluoride
La	Lanthanum
La_2O_3	Lanthanum oxide
Li	Lithium
Li_3N	Lithium nitride
LiCl	Lithium chloride
LiOH	Lithium hydroxide
Lr	Lawrencium
Lu	Lutetium
Lw	Lawrencium
Md	Mendelevium
Mg	Magnesium
$MgCl_2$	Magnesium chloride
$MgCO_3$	Magnesium carbonate
MgO	Magnesium oxide
$Mg(OH)_2$	Magnesium hydroxide
$MgSO_4$	Magnesium sulphate
Mn	Manganese
$MnCl_2$	Manganese(IV) chloride
MnO_2	Pyrolusite/ Manganese(IV) oxide
Mo	Molybdenum
N/N_2	Nitrogen
N_2O	Dinitrogen oxide
N_2O_4	Dinitrogen tetraoxide
$-NH_2$	Amino group
$NH_2(CH_2)_6NH_2$	1,6-diaminohexane
NH_3	Ammonia
$(NH_4)_2SO_4$	Ammonium sulphate
NH_4Cl	Ammonium chloride
NH_4OH	Ammonia solution
NH_4NO_3	Ammonium nitrate
NO	Nitrogen monoxide
NO_2	Nitrogen dioxide
Na	Sodium
Na_2CO_3	Sodium carbonate
$Na_2CO_3.10H_2O$	Washing soda
Na_2SO_3	Sodium sulphite
Na_2SO_4	Sodium sulphate
Na_3AlF_6	Cryolite
$NaAl(OH)_4$	Sodium aluminate
NaBr	Sodium bromide
NaCl	Sodium chloride
$NaClO_3$	Sodium chlorate
$NaHCO_3$	Sodium hydrogencarbonate
$NaHSO_4$	Sodium hydrogensulphate
$NaIO_3$	Sodium iodate
$NaNO_2$	Sodium nitrite
$NaNO_3$	Sodium nitrate
NaOCl	Sodium hypochlorite
NaOH	Sodium hydroxide
Nb	Niobium
Nd	Neodymium
Ne	Neon
Ni	Nickel
NiS	Nickel sulphide
No	Nobelium
Np	Neptunium
O/O_2	Oxygen
O_3	Ozone
-OH	Hydroxyl group
Os	Osmium
OsO_4	Osmium tetroxide
P	Phosphorus
P_2O_5	Phosphorus pentoxide
Pa	Protactinium
Pb	Lead
PbI_2	Lead(II) iodide
$Pb(NO_3)_2$	Lead(II) nitrate
PbO	Lead(II) oxide
PbO_2	Lead(IV) oxide
$Pb(OC_2H_5)_4$	Tetraethyl-lead
$Pb(OH)_2$	Lead(II) hydroxide
PbS	Galena
Pd	Palladium
Pm	Promethium
Po	Polonium
Pr	Praseodymium
Pt	Platinum
Pu	Plutonium
Ra	Radium
Rb	Rubidium
Re	Rhenium
Rh	Rhodium
Rn	Radon
Ru	Ruthenium
S	Sulphur
SO_2	Sulphur dioxide
SO_3	Sulphur trioxide
Sb	Antimony
Sc	Scandium
Se	Selenium
Si	Silicon
SiO_2	Silicon dioxide
Sm	Samarium
Sn	Tin
Sr	Strontium
T	Tritium
Ta	Tantalum
Tb	Terbium
Tc	Technetium
Te	Tellurium
Th	Thorium
Ti	Titanium
Tl	Thallium
Tm	Thulium
U	Uranium
V	Vanadium
V_2O_5	Vanadium pentoxide
W	Tungsten
Xe	Xenon
XeF_4	Xenon tetrafluoride
Y	Yttrium
Yb	Ytterbium
Zn	Zinc
$ZnCl_2$	Zinc chloride
$ZnCO_3$	Calamine
ZnO	Zincite/ Zinc oxide
$Zn(OH)_2$	Zinc hydroxide
Zn(OH)Cl	Basic zinc chloride
ZnS	Zinc blende
$ZnSO_4$	Zinc sulphate
Zr	Zirconium

Part three

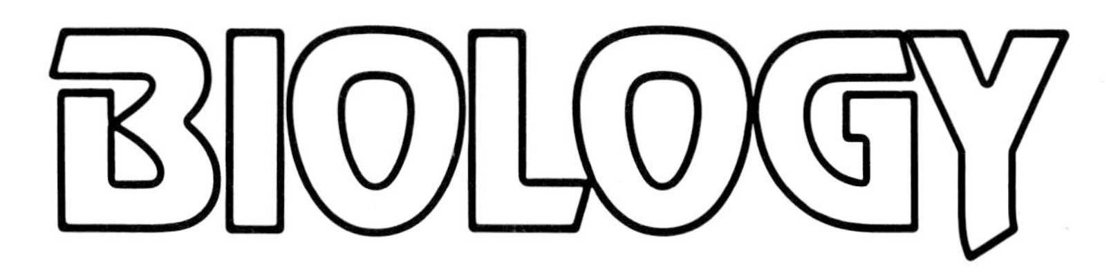

Designed by Nerissa Davies

Illustrated by Kuo Kang Chen

Scientific advisors:
Dr. Margaret Rostron and Dr. John Rostron

Additional illustrations by: Ian Jackson
Chris Lyon, Sue Stitt, Jeremy Banks, Peter Bull,
Chris Shields, Eric Robson, Alan Harris, Gabrielle Smith,
Annabel Milne, Jeane Colville, Sue Walliker and Jayne Goin.

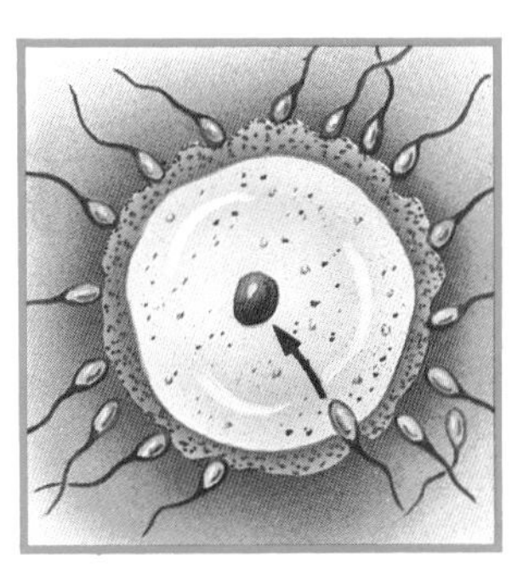

Contents

General

Plants

Animals

Humans

General

We are grateful to the North East London Polytechnic for permission to illustrate their experimental procedure (page 331).

About biology

In this book, biology is divided into four main colour-coded sections, followed by a black and white section of general material relating to the whole subject.

Yellow section
Ecology and living things

Blue section
Botany

Green section
Zoology (animals)

Red section
Zoology (humans)

Black and white section
General material – covers subjects which relate to all living things. Includes tables of general information and classification charts.

Biology is the study of living things. It examines the structures and internal systems of different organisms and how these operate to sustain individual life, as well as looking at the complex web of relationships between organisms which ensure new life is created and maintained. The areas covered by the four different colour-coded sections are explained below.

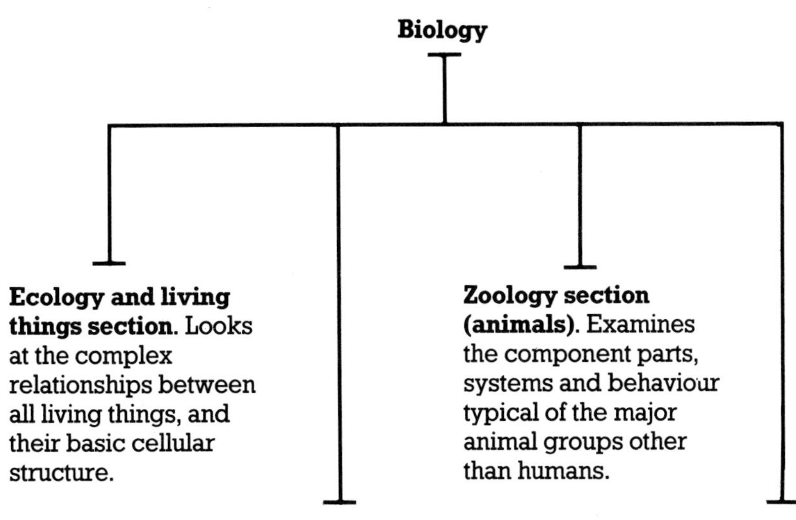

Ecology and living things section. Looks at the complex relationships between all living things, and their basic cellular structure.

Zoology section (animals). Examines the component parts, systems and behaviour typical of the major animal groups other than humans.

Botany section. Covers the plant kingdom – the different types of plant, their characteristics, internal structures and systems.

Zoology section (humans). Covers all the major terms of human biology. In many cases, these also apply to the vertebrates in general (see page 341).

Living things and their environment

The world can be divided into a number of different regions, each with its own characteristic plants and animals. All the plants and animals have become adapted to their own surroundings, or **environment** (see **adaptive radiation**, page 237), and their lives are linked in a complex web of interdependence. The environment is influenced by many different factors, e.g. temperature, water and light (**climatic factors**), the physical and chemical properties of the soil (**edaphic factors**), and the activities of the living things (**biotic factors**). The study of the relationships between plants, animals and the environment is called **ecology**.

- **Biosphere**. The layer of the earth (including the oceans and the atmosphere) which is inhabited by living things, bounded (above) by the upper atmosphere and (below) by the first layers of uninhabited rock.

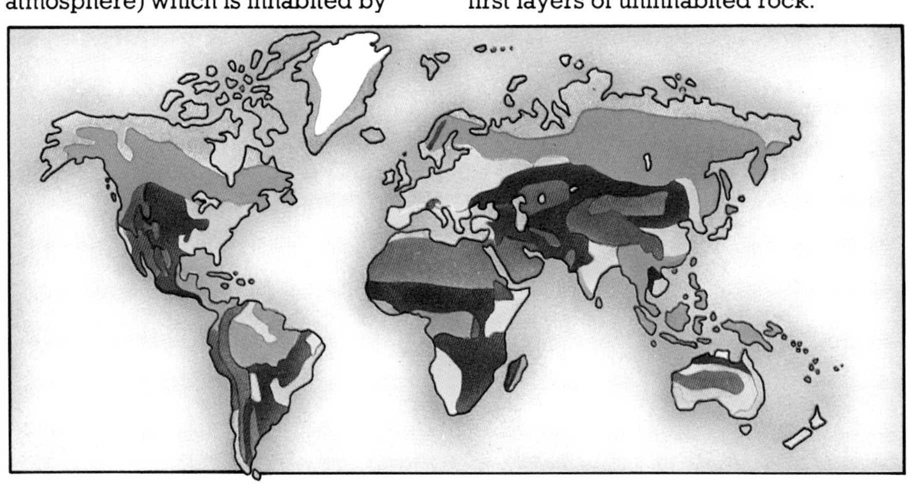

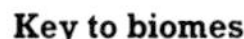

Key to biomes

Tundra. Very cold and windy. Commonest plants: **lichens*** and small shrubs. Animals include musk ox.

Coniferous forest. Low temperatures all year. Dominant plants: **conifers***, e.g. spruce. Commonest large animals: deer.

Deciduous forest. Summers warm, winters cold. Dominant plants: **deciduous*** trees, e.g. beech. Many animals, e.g. foxes.

Tropical forest. High temperatures all year, heavy rainfall. Great variety of plants and animals, e.g. exotic birds.

Grassland / savannah. Main plants: grasses, but savannah (with more rainfall) also has trees. Typical animals: giraffe.

Desert. High temperatures (cold at night), very low rainfall. Dominant plants: cacti. Animals include camels, scorpions.

Other areas

Scrubland (**maquis**)

Ice

Mountains

- **Biomes**. The main ecological regions into which the land surface can be divided. Each has its own characteristic seasons, day length, rainfall pattern and maximum and minimum temperatures. The major biomes are **tundra**, **coniferous forest**, **deciduous forest**, **tropical forest**, **temperate grassland**, **savannah** (tropical grassland) and **desert**. Most are named after the dominant vegetation, since this determines all other living things found there. Each is a giant **habitat** (**macrohabitat**).

* **Conifers**, 339; **Deciduous**, 236; **Lichens**, 342 (**Symbionts**).

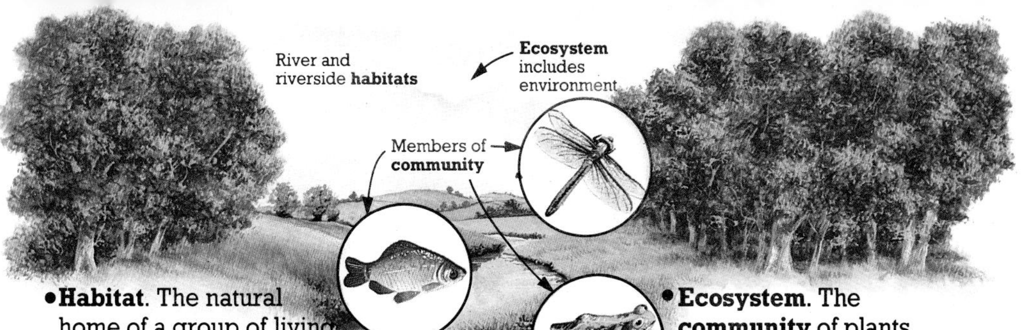

- **Habitat**. The natural home of a group of living things or a single living thing. Small habitats can be found within large habitats, e.g. a river in the **deciduous forest biome**. Very small specialized habitats are called **microhabitats**, e.g. a rotting tree.
- **Community**. The group of plants and animals found in one **habitat**. They all interact with each other and their environment.
- **Ecosystem**. The **community** of plants and animals in a given **habitat**, together with their environment. An ecosystem is a self-contained unit, i.e. the plants and animals interact to produce all the material they need (see also pages 234-235).

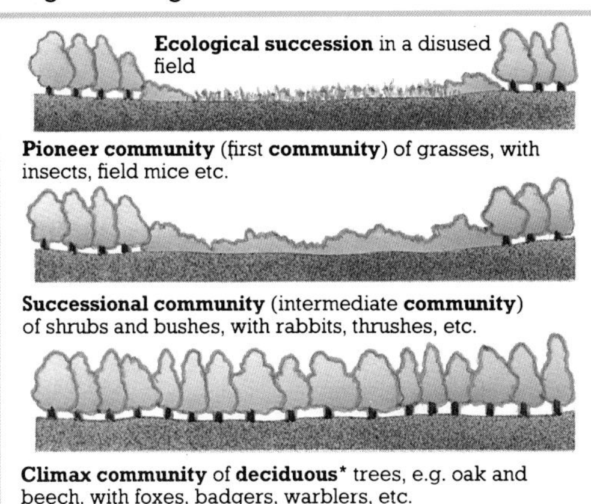

Pioneer community (first **community**) of grasses, with insects, field mice etc.

Successional community (intermediate **community**) of shrubs and bushes, with rabbits, thrushes, etc.

Climax community of **deciduous*** trees, e.g. oak and beech, with foxes, badgers, warblers, etc.

- **Ecological succession**. A process which occurs whenever a new area of land is colonized, e.g. a forest floor after a fire, a farm field which is left uncultivated or a demolition site which remains unused. Over the years, different types of plants (and the animals which go with them) will succeed each other, until a **climax community** is arrived at. This is a very stable **community**, one which will survive without change as long as the same conditions prevail (e.g. the climate).

- **Ecological niche**. The place held in an **ecosystem** by a plant or animal, e.g. what it eats and where it lives. **Gause's principle** states that no two species can occupy the same niche at the same time (if they tried, one species would die out or be driven off). For example, both the curlew and the grey plover can be found (in the winter months) living around the estuaries of Britain, and eating small creatures such as worms and snails. However, they actually occupy different niches. Curlews wade in the shallows, probing the river bed for food with their long beaks. Grey plovers, by contrast, wait on the shore and pick their food off the surface (their beaks are too short for probing). Hence both birds survive in the same general area.

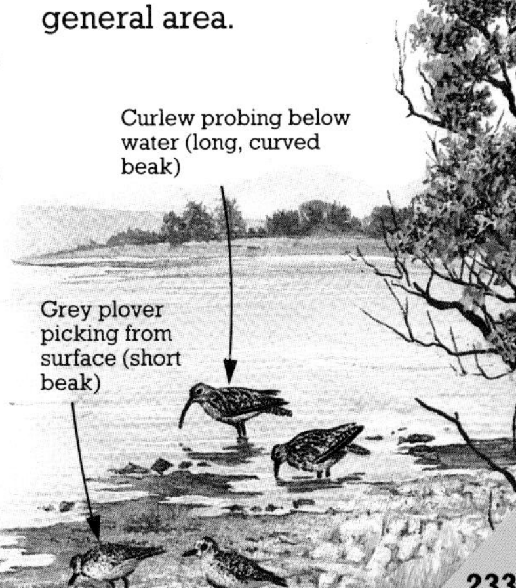

* **Deciduous**, 236.

Within an ecosystem

An **ecosystem** consists of a group (**community***) of animals and plants which interact with each other and with their environment to produce a self-contained ecological unit.

- **Food web**. The complex network of **food chains** in an ecosystem. Each food chain is a linked series of living things, each of which is the food for the next in line. Plants make their food from non-living matter by **photosynthesis*** (they are **autotrophic**) and are always the first members of a chain. Animals cannot make their own food (they are **heterotrophic**) and so rely on the food-making activities of plants.

Simple **food web**

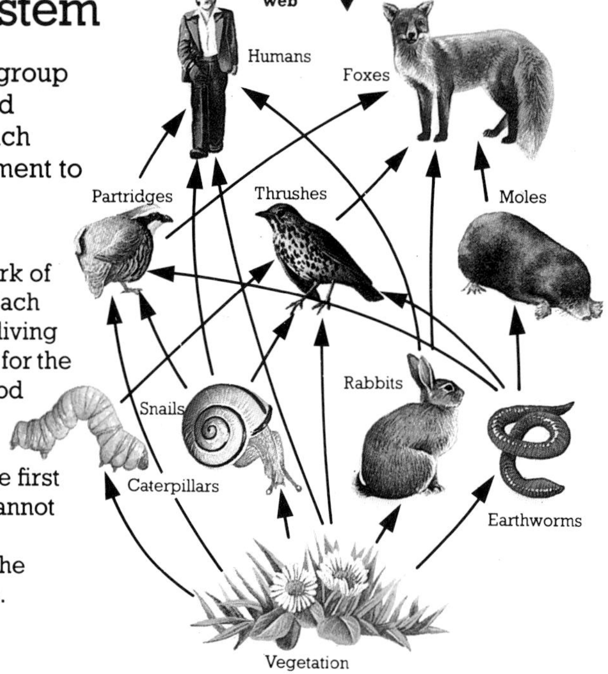

Humans, Foxes, Partridges, Thrushes, Moles, Rabbits, Snails, Caterpillars, Earthworms, Vegetation

Generalized **food chain**, showing **trophic levels**

Producers – green plants, which make their own food. **Trophic level T1**.

→ **Primary consumers** or **first order consumers** – **herbivores** (plant-eating animals), e.g. rabbits. Energy-giving material obtained directly from **producers**. **Trophic level T2**.

→ **Secondary consumers** or **second order consumers** – **carnivores** (flesh-eating animals), e.g. foxes, when they eat herbivores. Energy-giving material obtained from bodies of **primary consumers**. **Trophic level T3**.

→ **Tertiary consumers** or **third order consumers** – carnivores, e.g. foxes, when they eat smaller carnivores. Energy-giving material is obtained by most indirect method – from bodies of secondary consumers, i.e. animals which ate animals which ate **producers**. **Trophic level T4**.

Notes:

1) **Omnivores**, e.g. humans, eat plant and animal matter. They are thus placed on trophic level T2 at some times and on T3 (or T4) at others.

2) Many carnivores, e.g. foxes, will eat both herbivores and smaller carnivores. They are thus on trophic level T3 at some times and on T4 at others.

- **Trophic level** or **energy level**. The level at which living things are positioned within a **food chain** (see **food web**). At each successive level, a great deal of the energy-giving food matter is lost. For example, a cow will break down well over half of the grass it eats (to provide energy). Hence only a small part of the original energy-giving material can be obtained from eating the cow (the part it used to build its own new tissue). This loss of energy means that the higher the trophic level, the fewer the number of animals, since they must eat progressively larger amounts of food to obtain enough energy. This principle is called the **pyramid of numbers**.

Pyramid of numbers

Number of individuals at each **trophic level**

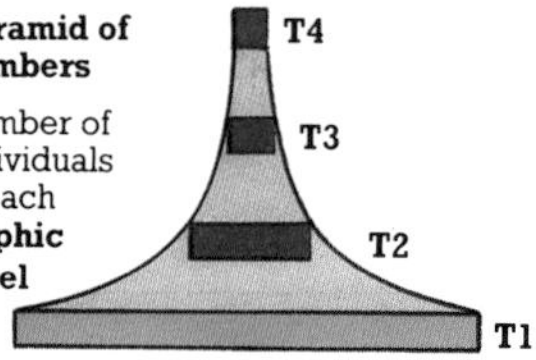

T4, T3, T2, T1

Pyramid of biomass (closely linked to pyramid of numbers)

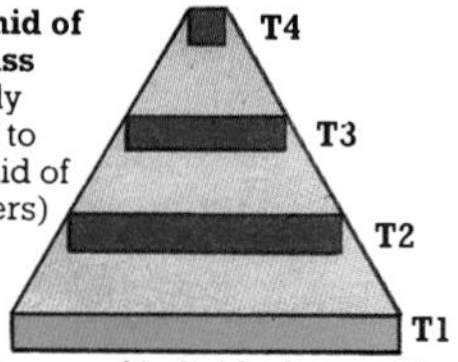

T4, T3, T2, T1

Total mass of individuals at each level (decrease is less extreme than above, since animals at higher levels tend to be larger).

* **Community**, 233; **Photosynthesis**, 254.

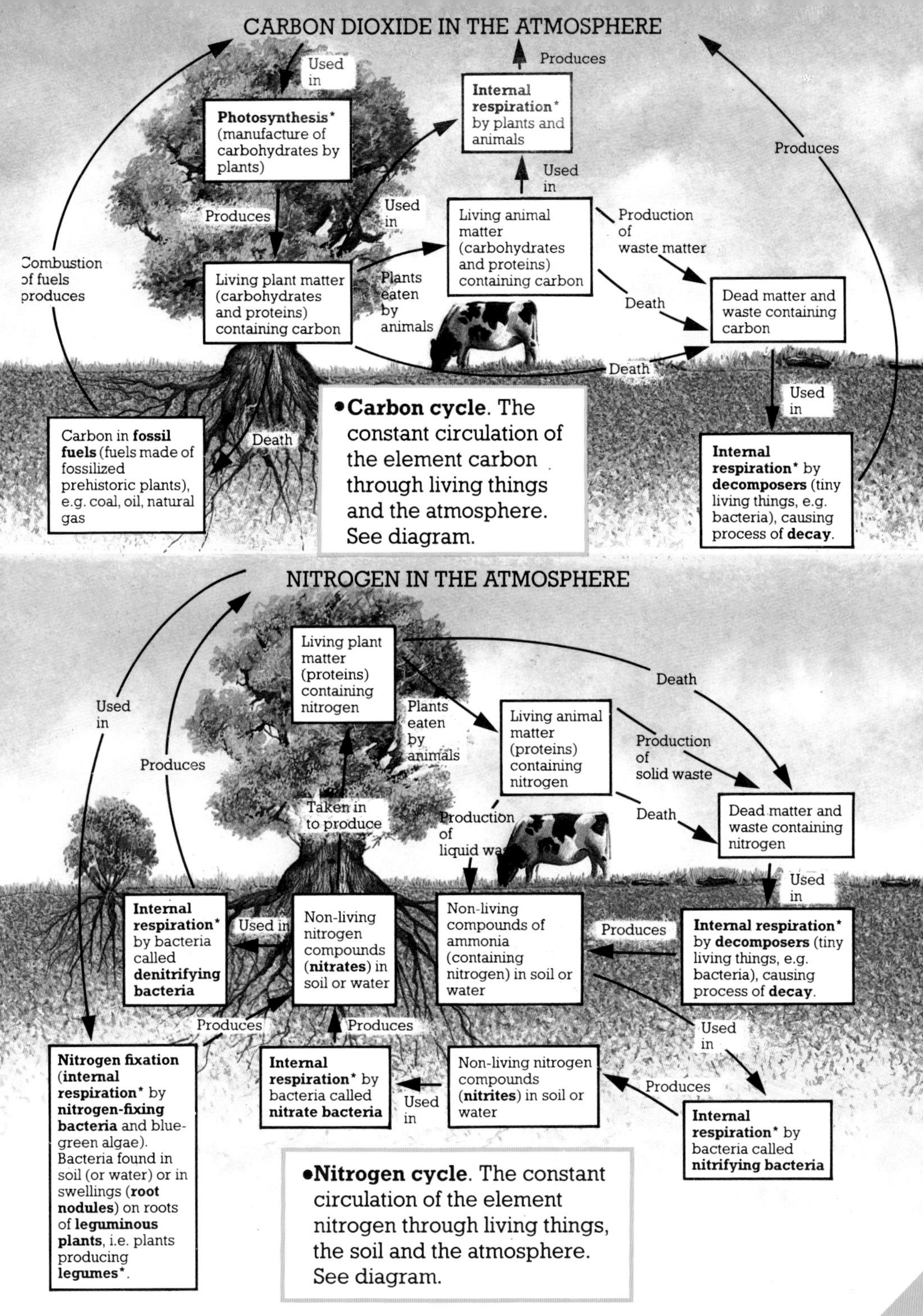

•**Carbon cycle**. The constant circulation of the element carbon through living things and the atmosphere. See diagram.

•**Nitrogen cycle**. The constant circulation of the element nitrogen through living things, the soil and the atmosphere. See diagram.

* **Internal respiration**, 332; **Legume**, 262; **Photosynthesis**, 254.

Life and life cycles

All living things show the same basic **characteristics of life**. These are respiration, feeding, growth, sensitivity (irritability), movement, excretion and reproduction. The **life cycle** of a plant or animal is the progression from its formation to its death, with all the changes this entails (in some cases, these are drastic – see **metamorphosis**, page 277). Below are some terms used to group plants and animals together according to their life cycle, or to describe characteristics of certain life cycles.

- **Perennials**. Plants which live for many years. **Herbaceous perennials**, e.g. delphiniums, lose all the parts above ground at the end of each growing season, and grow new shoots at the start of the next. **Woody perennials**, e.g. trees, produce new growth (**secondary tissue***) each year from permanent stems.

- **Biennials**. Plants which live for two years, e.g. carrots. In the first year they grow and store up food. In the second they produce flowers and seeds, and then die.

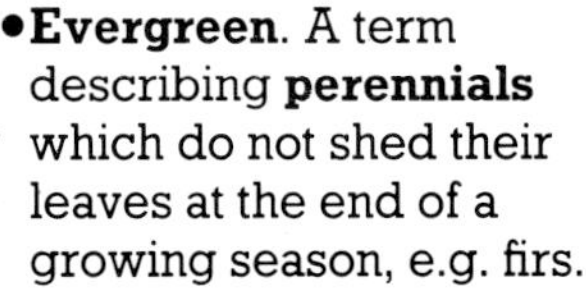

- **Annuals**. Plants which live for one year, e.g. marigolds. In this time they grow from seed, produce flowers and seeds, and then die.

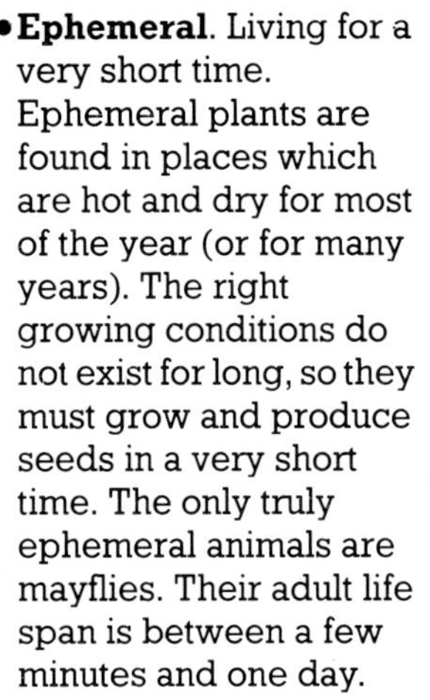

- **Herbaceous**. A term describing plants which do not develop **secondary tissue*** above the ground, i.e. they are "like a herb", as distinct from shrubs and trees (**woody perennials**).

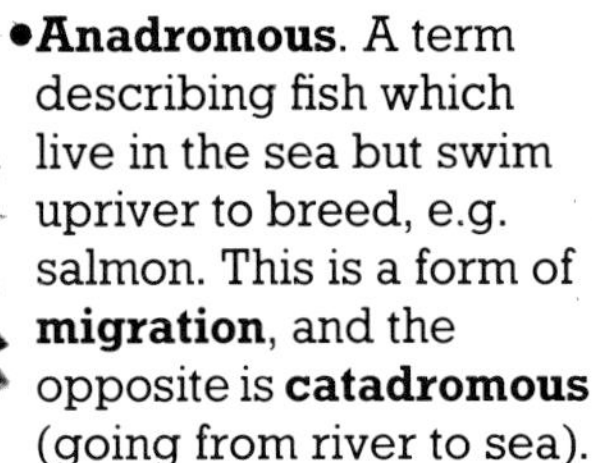

- **Deciduous**. A term describing **perennials** whose leaves lose their **chlorophyll*** and fall off at the end of each growing season, e.g. beeches.

- **Evergreen**. A term describing **perennials** which do not shed their leaves at the end of a growing season, e.g. firs.

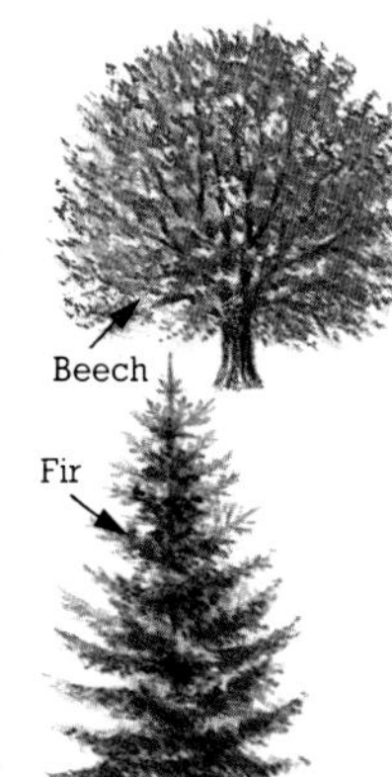

- **Ephemeral**. Living for a very short time. Ephemeral plants are found in places which are hot and dry for most of the year (or for many years). The right growing conditions do not exist for long, so they must grow and produce seeds in a very short time. The only truly ephemeral animals are mayflies. Their adult life span is between a few minutes and one day.

- **Anadromous**. A term describing fish which live in the sea but swim upriver to breed, e.g. salmon. This is a form of **migration**, and the opposite is **catadromous** (going from river to sea).

* **Chlorophyll**, 255 (**Pigments**); **Secondary tissue**, 246.

• **Migration**. Travelling seasonally from one region to another. This normally involves leaving an area in winter to find food elsewhere, and returning in the spring to breed. Migration is part of the life cycle of many animals, especially birds.

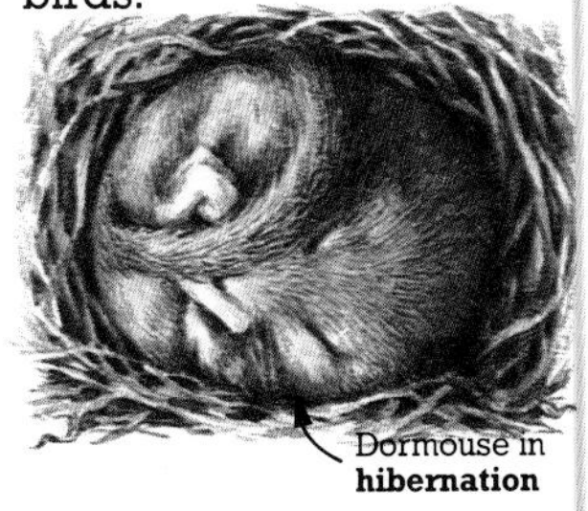

• **Dormancy**. A period, or periods, of suspended activity which is a natural part of the life cycle of many plants and animals. Dormancy in plants occurs when conditions are unfavourable for growth (normally in winter). In animals, dormancy usually occurs because of food scarcity, and is either called **hibernation** or **aestivation**. Hibernation is dormancy in the winter (typical of many animals, e.g. some **mammals***), and aestivation is dormancy in drought conditions (occurs mainly in insects).

Life styles

The world has a vast diversity of living things, each one with its own style of life. This situation is a result of **adaptive radiation**. The living things can be grouped together according to shared characteristics, either by formal classification, based mainly on their structural similarities (see charts, pages 338-341) or by more informal groupings, based on general life styles (see list, page 342).

• **Adaptive radiation** or **evolutionary adaptation**. The gradual process which has produced many different forms of living thing from one prehistoric starting point. Each has become **specialized**, i.e. has evolved the best form to cope with its environment, e.g. streamlined shapes for swimming and flying.

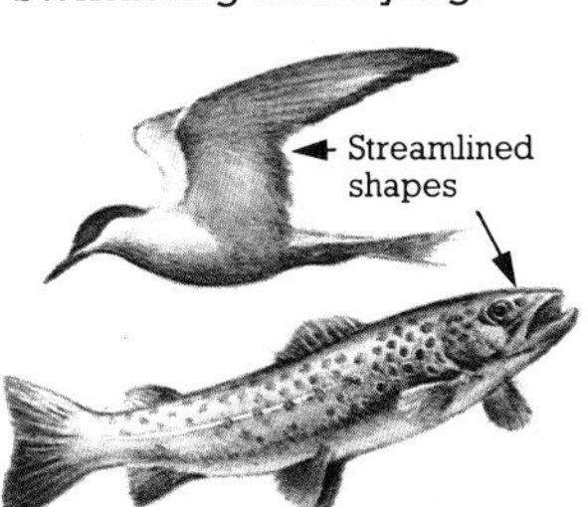

Many living things have also developed **protective adaptations** – protective measures such as thorns or poison

stings. All adaptations become established in successive generations because those creatures with them are the most likely to survive long enough to breed (and perpetuate the adaptation). This is the basis of Darwin's theory of **natural selection**, (also called **Darwinism**), first expounded in the mid-nineteenth century.

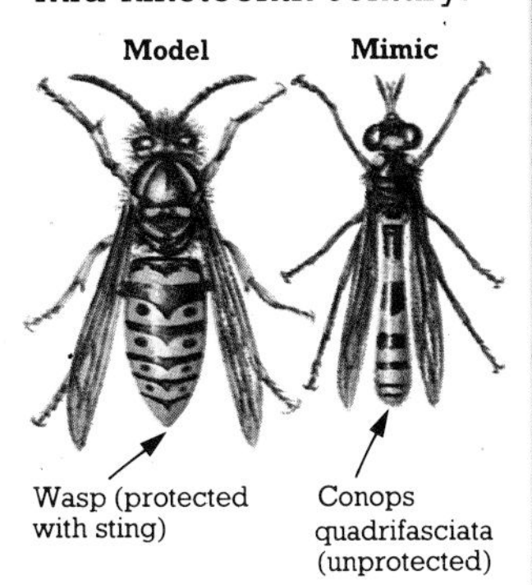

• **Mimicry**. A special type of adaptation, in which a plant or animal (the **mimic**) has developed a resemblance to another plant or animal (the **model**). This is used especially for protection (e.g. many unprotected insects have adopted the colouring of those with stings), but also for other reasons (bee orchids are mimics for reproduction purposes – see page 259).

* **Mammals**, 341.

The structure of living things

A living thing capable of a separate existence is called an **organism**. All organisms are made up of **cells** – the basic units of life, which carry out all the vital chemical processes. The simplest organisms have just one cell (they are **unicellular** or **acellular**), but very complex ones, e.g. humans, have thousands or even millions. They are **multicellular** and their cells are of many different types, each type specially adapted for its own particular job. Groups of cells of the same type (together with non-living material) make up the different **tissues** of the organism, e.g. muscle tissue. Several different types of tissue together form an **organ**, e.g. a stomach, and a number of organs together form a **system**, e.g. a digestive system.

The parts of a cell

Despite the fact that cells can look very different, they are all made up of the same basic parts. Each of these parts has a specific role to play.

- **Cell membrane**. Also called the **plasma membrane** or **plasmalemma**. The outer skin of a cell. It is **semipermeable***, i.e. selective about which substances it allows through.

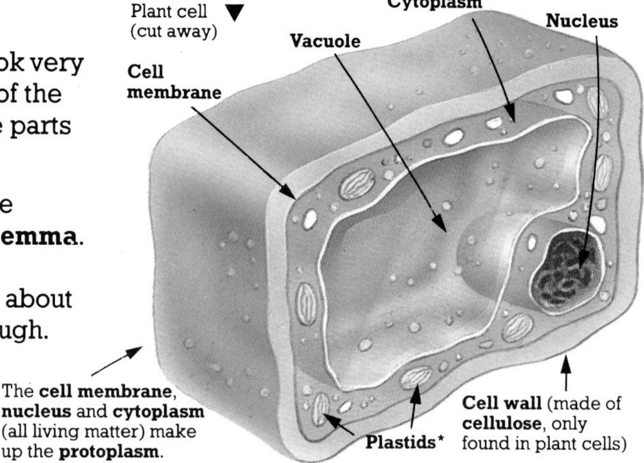

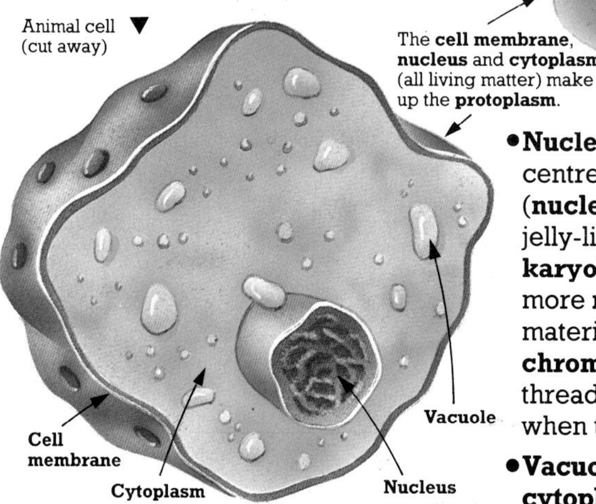

The **cell membrane**, **nucleus** and **cytoplasm** (all living matter) make up the **protoplasm**.

- **Cytoplasm**. The material where all the chemical reactions vital to life occur (see **organelles**). It generally has a jelly-like outer layer and a more liquid inner one (see **ectoplasm** and **endoplasm** – picture, page 268).

- **Nucleus** (pl. **nuclei**). The cell's control centre. Its double-layered outer skin (**nuclear membrane**) encloses a jelly-like fluid (**nucleoplasm** or **karyolymph**), which contains one or more **nucleoli*** and the genetic material **DNA***. This is held in **chromosomes*** – bodies which form a threadlike mass called **chromatin** when the cell is not dividing.

- **Vacuoles**. Fluid-filled sacs in the **cytoplasm**. They are small and temporary in animal cells, and either remove substances (see **Golgi complex**) or contain fluid brought in (see **pinocytosis**, page 327). Most plant cells have one large, permanent vacuole, filled with **cell sap** (dissolved minerals and sugars).

* **Chromosomes**, 324; **DNA**, 324 (**Nucleic acids**); **Nucleoli**, **Plastids**, 240; **Semipermeable**, 327.

Organelles

The **organelles** are tiny bodies in the **cytoplasm**. Each type (listed below and on page 240) has a vital role to play in the chemical reactions within the cell.

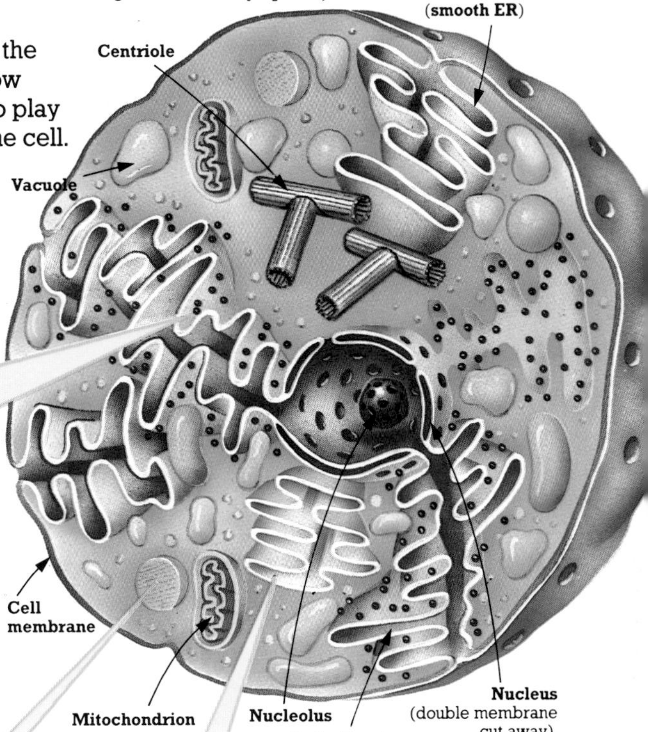

Animal cell (showing the **organelles** in the **cytoplasm**)

Endoplasmic reticulum (**smooth ER**)

Centriole

Vacuole

Cell membrane

Mitochondrion

Nucleolus

Endoplasmic reticulum (**rough ER**)

Nucleus (double membrane cut away). **Nucleoplasm** and **chromosomes** not shown.

The **mitochondria**, **centriole** and **nucleolus** are defined overleaf.

- **Ribosomes**. Tiny round particles (most are attached to the **endoplasmic reticulum**). They are involved in building up proteins from amino acids (see page 328). "Coded" information (held by the **DNA** in the **nucleus**) is sent to the ribosomes in strands of a substance called **messenger RNA** (**mRNA**). These pass on the "codes" so that the ribosomes join the amino acids in the correct way to produce the right proteins. **RNA*** is present in at least two other forms in the cells. The ribosomes are made of **ribosomal RNA** (see **nucleoli***) and molecules of **transfer RNA** (**tRNA**) "carry" the amino acids to the ribosomes.

- **Lysosomes**. Round sacs containing powerful **enzymes***. They take in foreign bodies, e.g. bacteria, to be destroyed by the enzymes. Their outer skins do not usually let the enzymes out into the cell (to break down its contents), but if the cell becomes damaged the skins disappear and the cell digests itself.

- **Golgi complex**. Also called a **Golgi apparatus**, **Golgi body** or **dictyosome**. A special area of **smooth ER**. It collects and distributes the substances made in the cell (e.g. proteins and waste from chemical reactions). The substances fill the sacs, which gradually swell up at their outside edges until pieces "pinch off". These pieces (**vacuoles**) then travel out of the cell via the **cytoplasm** and **cell membrane**.

- **Endoplasmic reticulum** or **ER**. A complex system of flat sacs, folding inwards from the **cell membrane** and joining up with the **nuclear membrane** (see **nucleus**). It provides a large surface area for reactions or fluid storage, and a passageway for fluids passing through. ER with **ribosomes** on its surface is **rough ER**. ER with no ribosomes is **smooth ER**.

* **Enzymes**, 331; **Nucleoli**, 240; **RNA**, 324 (**Nucleic acids**).

Organelles (continued)

- **Centrioles**. Two bodies just outside the **nucleus*** in animal cells. Each lies in a dense area of **cytoplasm*** (**centrosome**) and is made up of two tiny cylinders, forming a T-shape. Each cylinder is made up of nine sets of three tiny tubes (**microtubules**). Centrioles are vital to **cell division**.

- **Mitochondria** (sing. **mitochondrion**) or **chondriosomes**. Rod-shaped bodies with a double layer of outer skin. The inner layer forms a series of folds (**cristae**, sing. **crista**), providing a large surface area for the vital chemical reactions which go on inside the mitochondria (called the "powerhouses" of a cell). They are the places where simple substances taken into the cell are broken down to provide energy. For more about this, see **aerobic respiration**, page 332.

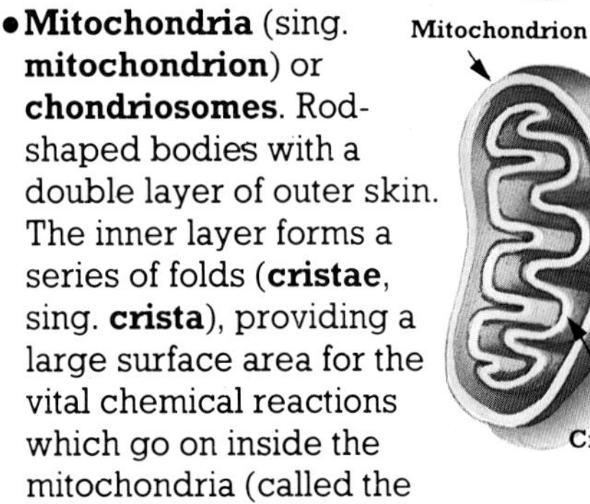

- **Nucleoli** (sing. **nucleolus**). One or more small, round bodies in the **nucleus***. They produce the component parts of the **ribosomes*** (made of **ribosomal RNA**), which are then transported out of the nucleus and assembled in the **cytoplasm***.

- **Plastids**. Tiny bodies in plant cell **cytoplasm***. Some (**leucoplasts**) store starch, oil or proteins. Others – **chloroplasts*** – contain **chlorophyll*** (used in making food).

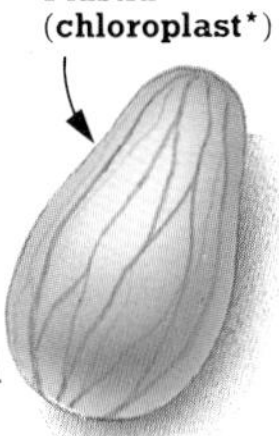

Cell division

Cell division is the splitting up of one cell (the **parent cell**) into two identical **daughter cells**. There are two types of cell division, both involving the division of the **nucleus*** (**karyokinesis**) followed by the division of the **cytoplasm*** (**cytokinesis**). The first type of cell division (**binary fission**) is described on these two pages. It produces new cells for growth and also to replace the millions of cells which die each day (from damage, disease or simply because they are "worn out"). It is also the means of **asexual reproduction*** in many single-celled organisms. The second, special type of cell division produces the **gametes*** (sex cells) which will come together to form a new living thing. For more about this, see pages 322-323.

- **Mitosis**. The division of the **nucleus*** when a plant or animal cell divides for growth or repair (**binary fission**). It ensures that the two new nuclei (**daughter nuclei**) are each given the same number of **chromosomes*** (the bodies which carry the "coded" hereditary information). Each receives the same number as were in the original nucleus, called the **diploid number**. Every living thing has its own characteristic diploid number, i.e. all its cells (with the exception of the **gametes***) contain the same, specific number of chromosomes, grouped in identical pairs called **homologous chromosomes**. Humans have 46 chromosomes, in 23 pairs. Although mitosis is a continuous process, it can be divided for convenience sake into four phases. Before mitosis, however, there is always an **interphase**.

* **Asexual reproduction**, 320; **Chlorophyll**, 255 (**Pigments**); **Chloroplasts**, 255; **Chromosomes**, 324; **Cytoplasm**, 238; **Gametes**, 321, **Nucleus**, 238; **Ribosomes**, 239.

• **Interphase**. The periods between cell divisions. Interphases are very active periods, during which the cells are not only carrying out all the processes needed for life, but are also preparing material to produce "copies" of all their components (so both new cells formed after division will have all they need). Just before **mitosis** begins, the **chromatin*** threads in the **nucleus*** also duplicate, so that, after coiling up, each **chromosome*** will consist of two **chromatids** (see **prophase**).

Phases of **mitosis** (only two **chromosomes*** shown – humans have 46)

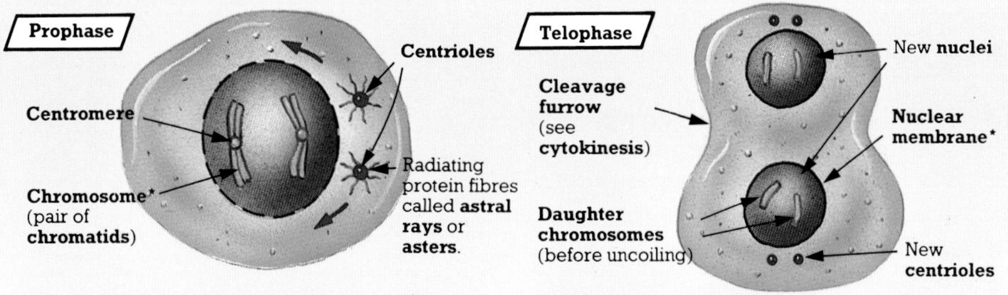

The threads of **chromatin*** in the **nucleus*** coil up to form **chromosomes*** and the **nuclear membrane*** disappears. Each has already duplicated to form two identical coils (**chromatids**), joined by a small sphere (**centromere**). The two **centrioles** move to opposite poles of the cell.

The **spindle fibres** and **astral rays** disappear and a new **nuclear membrane*** forms around each group of **daughter chromosomes**. This creates two new **nuclei*** (**daughter nuclei**), inside which the chromosomes uncoil to once again form a threadlike mass (**chromatin***). The **centrioles** also duplicate, so that a pair will be found in each new cell (after **cytokinesis**).

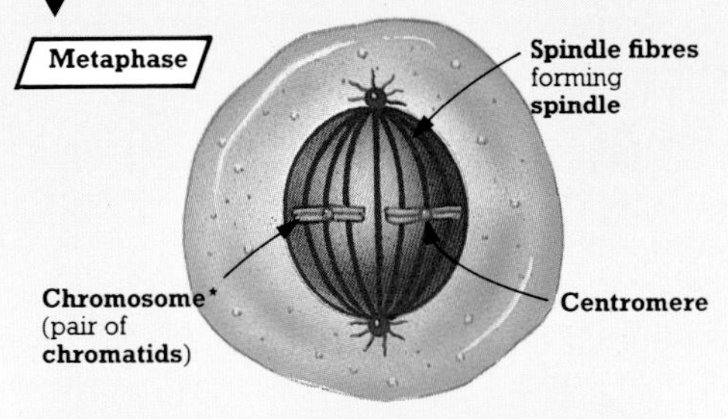

The **centrioles** (at opposite poles) project protein fibres called **spindle fibres** which join together and form a sphere, or **spindle**. The **chromosomes*** (paired **chromatids**) move towards its equator and become attached by their **centromeres** to the spindle fibres.

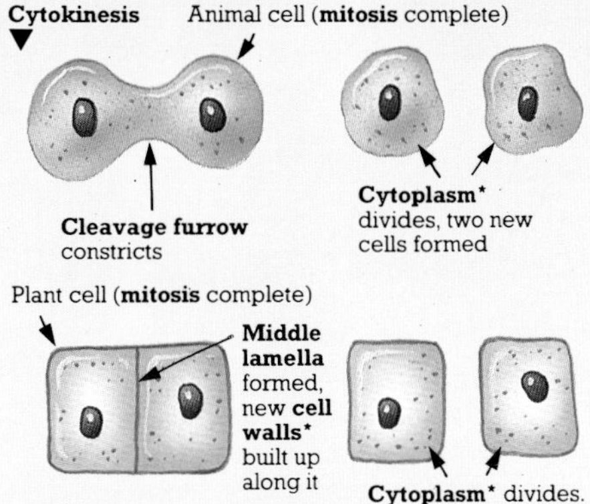

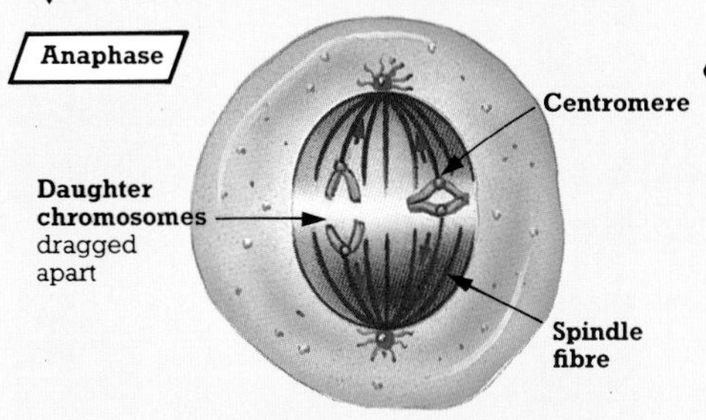

The **centromeres** duplicate and the two **chromatids** from each pair (now called **daughter chromosomes**) move to opposite poles of the **spindle**, seemingly "dragged" there by the contracting **spindle fibres**.

• **Cytokinesis**. The division of the **cytoplasm*** of a cell, which forms two new cells around the new **nuclei*** created by **mitosis** (or **meiosis***). In animal cells, a **cleavage furrow** forms around the cell's equator and then constricts as a ring until it cuts completely through the cell. In plant cells, a dividing line called the **middle lamella** forms down the centre of the cell, and a new **cell wall*** is built up along each side of it.

* **Cell wall**, 238; **Chromatin**, 238 (**Nucleus**); **Chromosomes**, 324; **Cytoplasm**, 238; **Meiosis**, 322; **Nuclear membrane**, 238 (**Nucleus**).

Vascular plants

With the exception of simple plants such as algae and fungi (see classification chart, pages 338-339), all plants are **vascular plants**. That is, they all have a complex system of special fluid-carrying tissue called **vascular tissue**. For more about how the fluids travel within the vascular tissue, see pages 252-253. All vascular plants are classified within the **Division Tracheophyta** (see page 339).

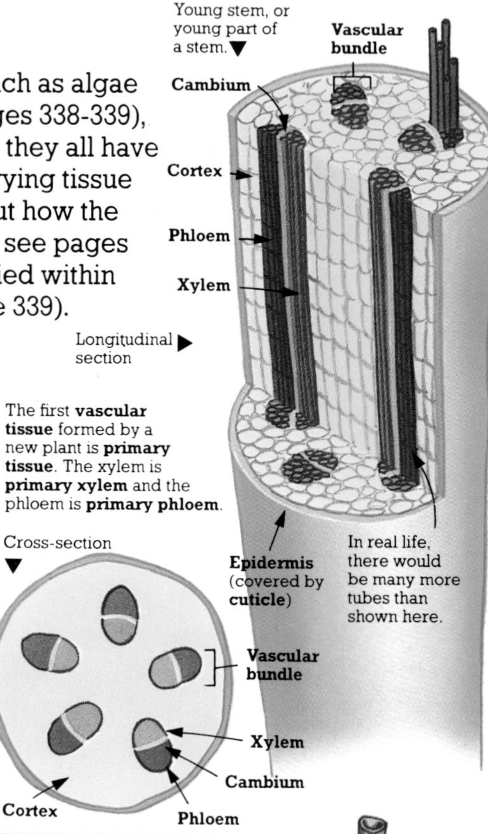

The first **vascular tissue** formed by a new plant is **primary tissue**. The xylem is **primary xylem** and the phloem is **primary phloem**.

- **Vascular tissue**. Special tissue which runs throughout a **vascular plant**, carrying fluids and helping to support the plant. In young stems, it is normally arranged in separate units called **vascular bundles**; in older stems these join up to form a central core (**vascular cylinder***). In young roots, the arrangement of the tissue is slightly different, but a central core is also formed later. For more about the vascular tissue in older plants, see page 246. The vascular tissue is of two different types – **xylem** and **phloem**. They are separated by a layer of tissue called the **cambium**.

Constituents of vascular tissue

- **Xylem**. A tissue which carries water up through a plant. It is made up of **vessels**, with long thin cells (**fibres**) providing support between them. In older stems, the central xylem dies away and its vessels become filled in, forming **heartwood***.

- **Phloem**. A tissue which distributes the food made in the leaves to all parts of the plant. It consists of **sieve tubes**, with special **companion cells** running beside them and other cells packed around them for support. The companion cells are thought to transport fluids.

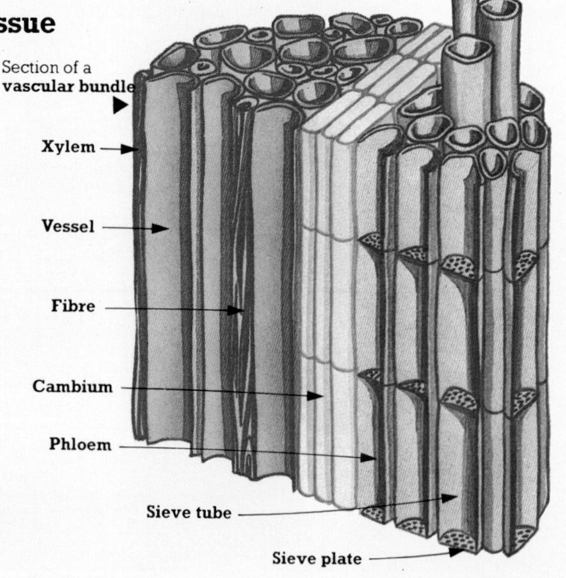

* **Heartwood**, 247; **Vascular cylinder**, 246.

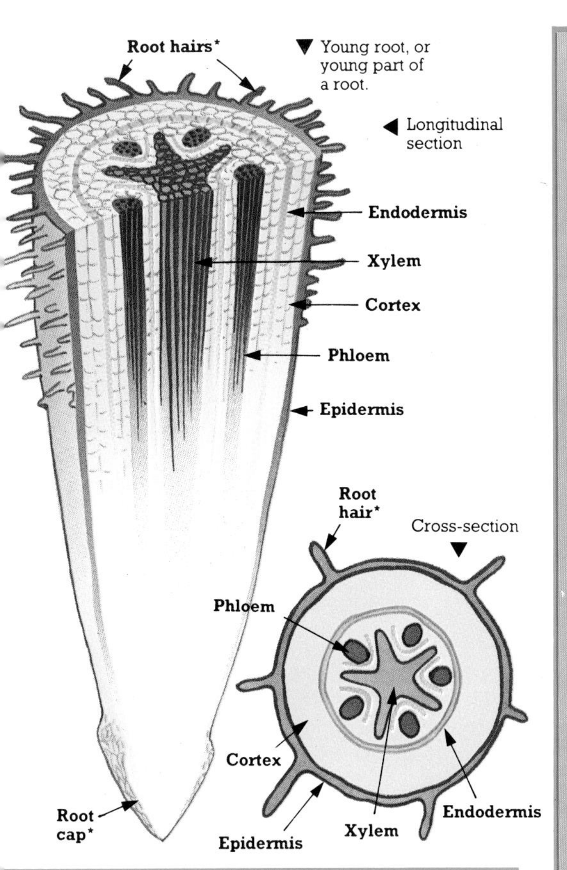

- **Vessels** or **tracheae** (sing. **trachea**). Long tubes in the **xylem** which carry water. Their walls are strengthened with a hard substance called **lignin**. They consist of chains of cells whose walls and **protoplasm*** have died.
- **Sieve tubes**. Long columns of cells in the **phloem**. Their **nuclei*** and **protoplasm*** have been lost but their interconnecting walls have remained. These are called **sieve plates** and have tiny holes in them to allow substances to pass through.
- **Cambium**. A layer of narrow, thin-walled cells between the **xylem** on the inside and the **phloem** on the outside. The cells are able to divide, making more xylem and phloem. Such an area of cells is called a **meristem***.

Other tissues in vascular plants

- **Epidermis**. A thin surface layer of tissue around all parts of a plant. In some areas, especially the leaves, it has many tiny holes, called **stomata***. In older stems, the epidermis is replaced by **phellem***. In older roots, it is first replaced by **exodermis*** and then by phellem.
- **Cuticle**. A thin outer layer of a waxy substance called **cutin**, produced by the **epidermis** above ground. It prevents too much water being lost.
- **Cortex**. A layer of tissue just inside the **epidermis** of stems and roots. It consists mainly of **parenchyma**, a type of tissue with large cells and many air spaces. In some plants there is also some **collenchyma**, a type of supporting tissue with long thick-walled cells. The cortex tends to get compressed and replaced by other tissues as a plant gets older.
- **Endodermis**. The innermost layer of root **cortex**. It contains special **passage cells**. Fluids which have seeped in between the cortex cells, instead of through them, are directed by the passage cells into the central area of **vascular tissue**.
- **Pith** or **medulla**. A central area of tissue found in stems, but not usually in roots. It is generally only called pith once the stem has developed a **vascular cylinder***. It is made up of **parenchyma**, like the **cortex**, and is sometimes used to store food.

* **Exodermis**, 245 (**Piliferous layer**); **Meristem**, 244; **Nucleus**, 238; **Phellem**, 247; **Protoplasm**, 238; **Root cap**, **Root hairs**, 245; **Stomata**, 249; **Vascular cylinder**, 246.

Stems and roots

The **stem** and **roots** of a plant are its main supporting structures, as well as being important in transporting fluids (see pages 242-243 and 252-253). Their various parts are listed here. For more about the development of the stem and roots as a plant gets older, see pages 246-247.

- **Meristem**. Any area from which new growth arises. The cells of a meristem are able to divide, producing new cells. A meristem found at the tip of the root (the **growing point**) or the stem (part of a **terminal bud**) is known as an **apical meristem**.

Stem attachments

- **Shoot**. A new stem growing off the main stem of a plant, or out of a seed.

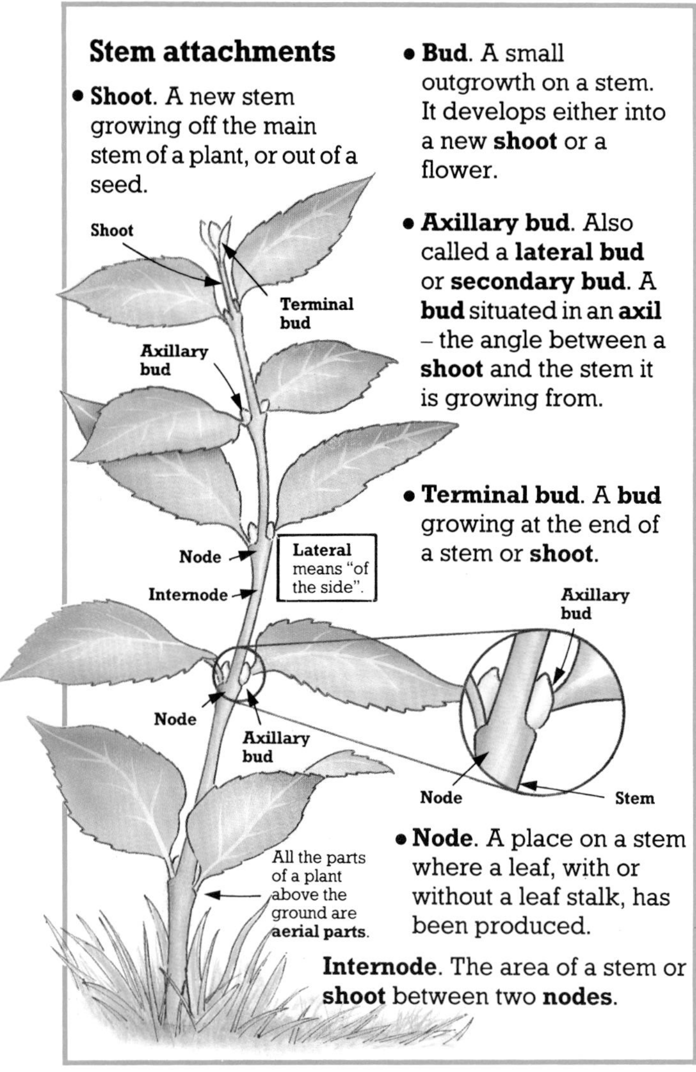

Lateral means "of the side".

- **Bud**. A small outgrowth on a stem. It develops either into a new **shoot** or a flower.
- **Axillary bud**. Also called a **lateral bud** or **secondary bud**. A **bud** situated in an **axil** – the angle between a **shoot** and the stem it is growing from.
- **Terminal bud**. A **bud** growing at the end of a stem or **shoot**.
- **Node**. A place on a stem where a leaf, with or without a leaf stalk, has been produced.

Internode. The area of a stem or **shoot** between two **nodes**.

Parts of a root

- **Growing point**. An area just behind a root tip where the cells divide to produce new growth.
- **Zone of elongation**. The area of new cells produced by the **growing point**, and located just behind it. The cells stretch lengthwise as they take in water, since their **cell walls*** are not yet hard. This elongation pushes the root tip further down into the soil.

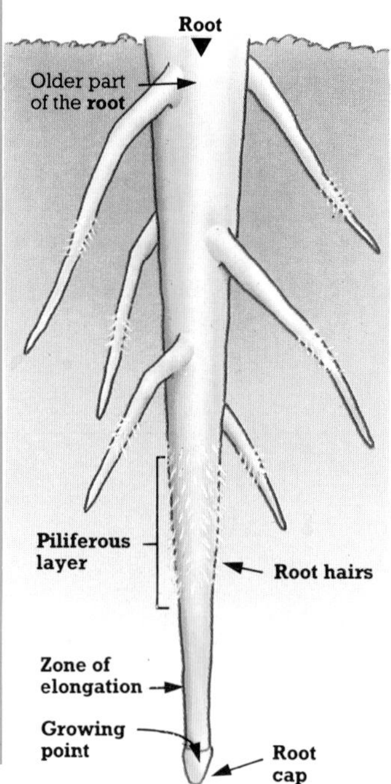

* **Cell wall**, 238.

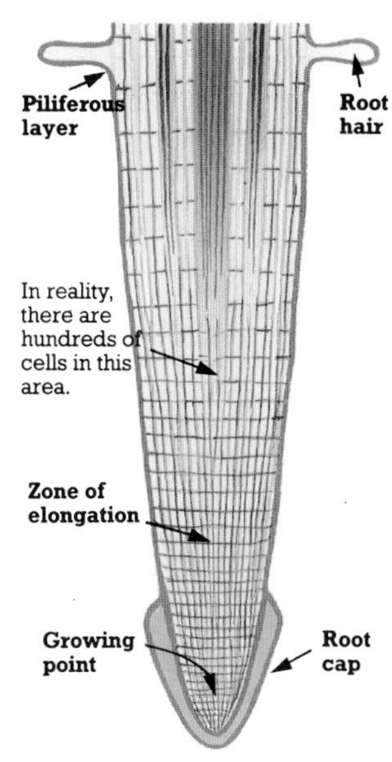

- **Piliferous layer**. The youngest area of the **epidermis***, or outer skin, of a root. It is the area which produces **root hairs**. It is found just behind the **zone of elongation**. As the walls of the elongating cells harden, the outermost cells become the piliferous layer. The older piliferous layer (higher up the root) is slowly worn away, to be replaced by a layer of hardened cells called the **exodermis** (the outermost layer of the **cortex***).

- **Root hairs**. Long outgrowths from the cells of the **piliferous layer**. They take in water and minerals.

- **Root cap**. A layer of cells which protects the root tip as it grows down.

Types of root

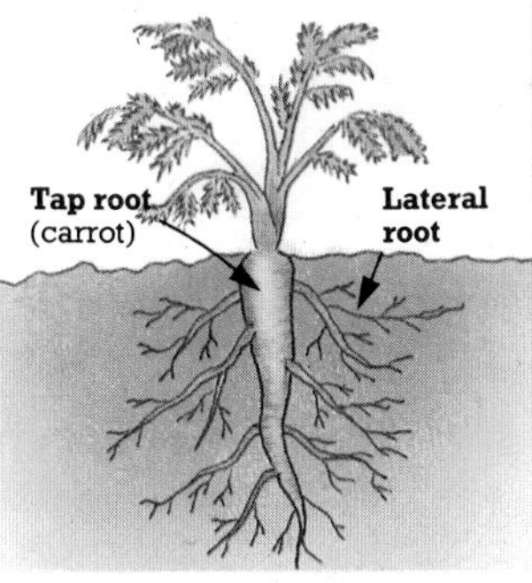

- **Tap root**. A first root, or **primary root**, which is larger than the small roots, called **lateral roots** or **secondary roots**, which grow out of it. Many vegetables are swollen tap roots.

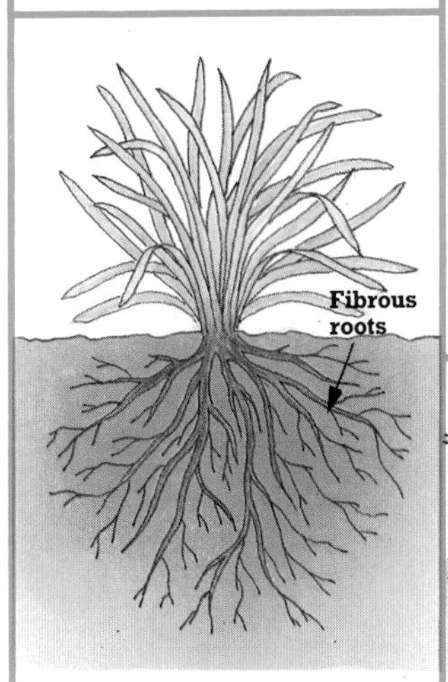

- **Fibrous roots**. A system of fibrous roots is made up of a large number of equal-sized roots, all producing smaller **lateral roots**. The first root is not prominent, as it is in a **tap root** system.

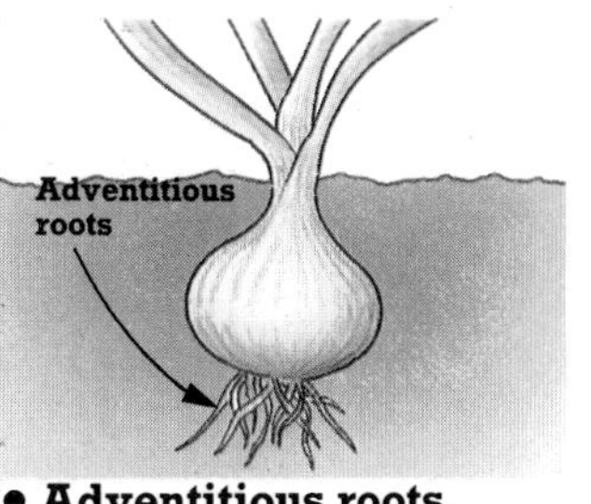

- **Adventitious roots**. Roots which grow directly from a stem. They grow out of **bulbs*** (which are special stems), or from gardeners' cuttings.

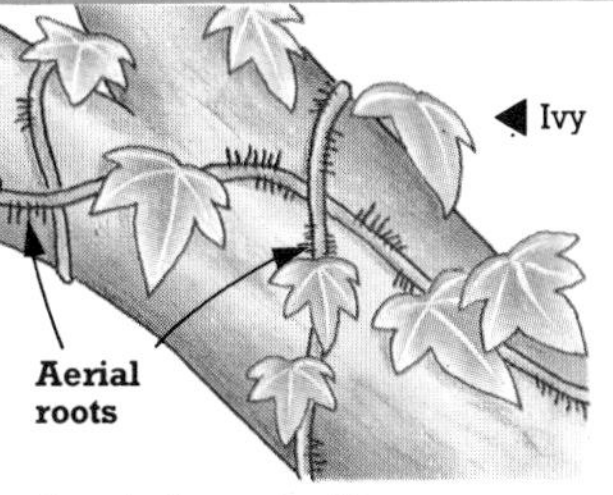

- **Aerial roots**. Roots which grow from stems and do not grow into the ground. They can be used for climbing, e.g. in an ivy. Many absorb moisture from the air.

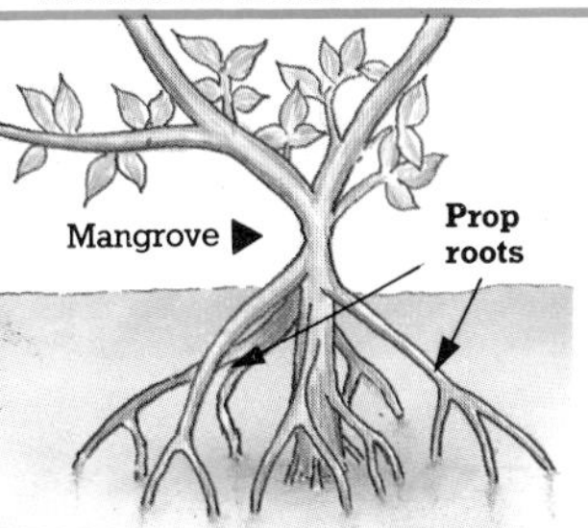

- **Prop roots**. Special types of **aerial root**. They grow out from a stem and then down into the ground, which may be under water. They support a heavy plant, e.g. a mangrove.

* **Bulb**, 262; **Cortex**, **Epidermis**, 243.

Inside an older plant

A plant which lives for many years, such as a tree, forms **secondary tissue** as it gets older. This consists of new layers of tissue to supplement the original tissue, or **primary tissue***. New supportive and fluid-carrying **vascular tissue*** is formed towards the centre of the plant and new protective tissue is produced around the outside. The production of the new vascular tissue is called **secondary thickening**, and results in what is known as a **woody plant**.

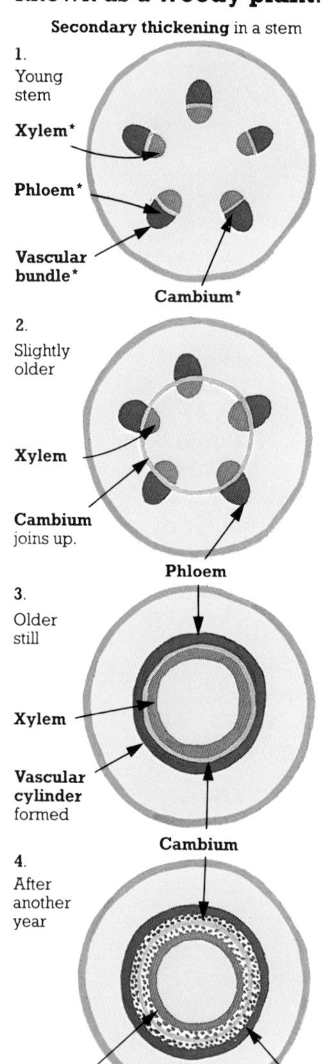

New central tissue

- **Vascular cylinder**. A vascular cylinder develops as the first step of **secondary thickening** in stems. More **cambium*** forms between the **vascular bundles***, and this then gives rise to more **xylem*** and **phloem***, forming a continuous cylinder.

- **Secondary thickening**. The year-by-year production of more fluid-carrying **vascular tissue*** in plants which live for many years, resulting in a gradual increase in the diameter of the stem and roots. Each year, new layers of **xylem*** (**secondary xylem**) and **phloem*** (**secondary phloem**) are produced by the dividing cells of the **cambium*** between them. In stems this happens slightly differently than in roots, but the result throughout the plant is an ever-enlarging core of vascular tissue (which slowly "squeezes out" the **pith*** in stems). Most of this core is xylem, now also known as **wood**. The area of phloem does not widen much at all, because the xylem pushing outwards wears it away.

- **Annual rings**. The concentric circles which can be seen in a cross-section of an older plant. Each ring is one year's new growth of **xylem***, and has two separate areas – the soft **spring wood** (or **early wood**), produced in the early part of the growing season, and the harder **summer wood** (or **late wood**), produced later on.

5. After a number of years

Many **annual rings** (**secondary xylem**)

Cambium

Secondary phloem

Central **pith*** has almost disappeared.

* **Cambium**, 243; **Phloem**, 242; **Pith**, 243; **Primary tissue**, 242; **Vascular bundles**, 242 (**Vascular tissue**); **Xylem**, 242.

Heartwood

- **Heartwood**. The oldest, central part of the **xylem*** in an older plant. The **vessels*** are filled in and no longer carry fluids, but they still provide support.

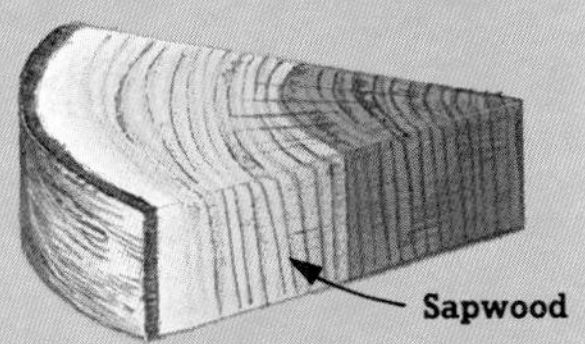

- **Sapwood**. The outer area of **xylem*** in an older plant, whose **vessels*** still carry fluids. It also supports the tree and holds food reserves.

New outer tissue

As well as new **vascular tissue***, an older plant also forms extra areas of tissue around its outside to help protect it. These are called **phelloderm**, **phellogen** and **phellem** respectively (working from the inside). The three areas together are known as the **periderm**.

- **Phellogen** or **cork cambium**. A cell layer which arises towards the outside of the stem and roots of older plants. It is a **meristem***, i.e. an area of cells which keep on dividing. It produces two new layers – **phelloderm** and **phellem**.
- **Phelloderm**. A new cell layer produced by **phellogen** on its inside. It supplements the **cortex*** and is sometimes called **secondary cortex**.
- **Phellem** or **cork**. A new cell layer produced by **phellogen** on its outside. The cells undergo **suberization**, i.e. become impregnated with a waxy substance called **suberin**. This makes the outer layer waterproof. The phellem cells slowly die and replace the previous outer cell layer (**epidermis*** in stems and **exodermis*** in roots). Dead phellem cells are called **bark**.

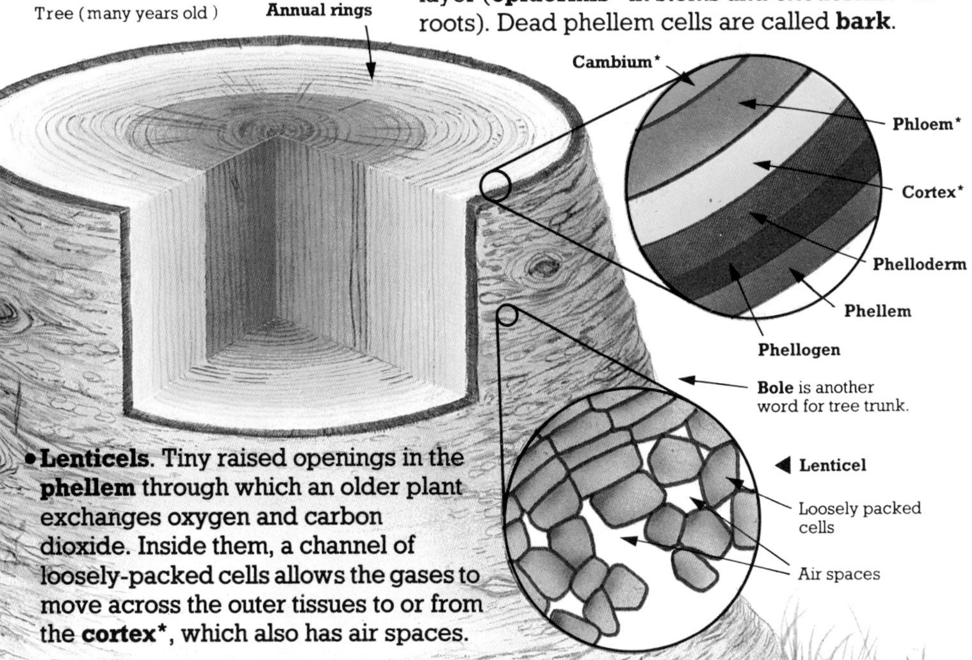

- **Lenticels**. Tiny raised openings in the **phellem** through which an older plant exchanges oxygen and carbon dioxide. Inside them, a channel of loosely-packed cells allows the gases to move across the outer tissues to or from the **cortex***, which also has air spaces.

* **Cambium**, **Cortex**, **Epidermis**, 243; **Exodermis**, 245 (**Piliferous layer**); **Phloem**, 242; **Meristem**, 244; **Vascular tissue**, 242; **Vessels**, 243; **Xylem**, 242.

Leaves

The **leaves** of a plant, collectively known as its **foliage**, are specially adapted to manufacture food. They do this by a special process called **photosynthesis**. For more about this, see pages 254-255. There are many different shapes and sizes of leaves, but only two different types. **Simple leaves** consist of a single leaf blade, or **lamina**, and **compound leaves** are made up of a number of small leaf blades called **leaflets**, all growing from the same leaf stalk. There is a chart showing some of the different leaf shapes on page 250.

Inside a leaf

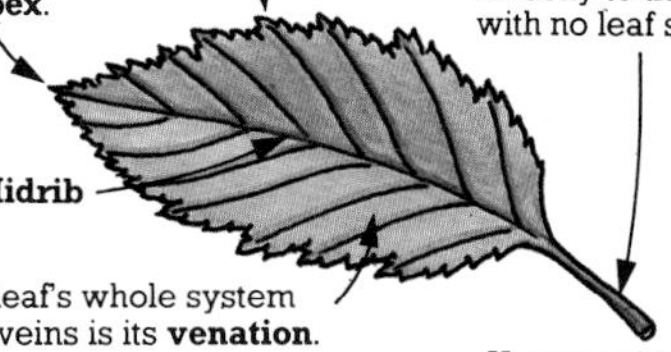

- **Veins**. Long strips of **vascular tissue*** inside a leaf, supplying it with water and minerals and removing the food made inside it. Some leaves have long parallel veins, e.g. those of grasses, but most have a central vein inside a **midrib** (an extension of the leaf stalk), with many smaller branching veins.

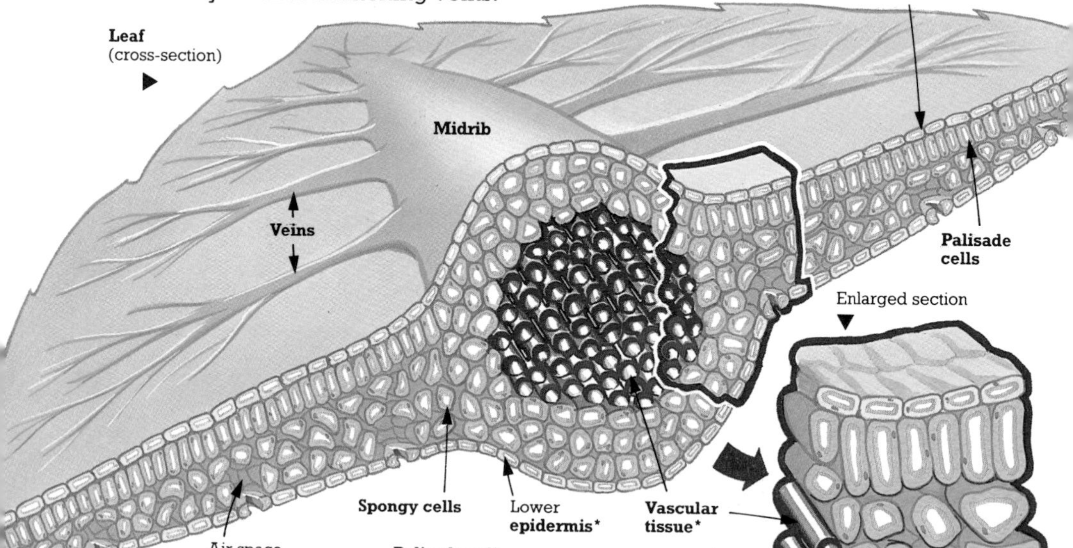

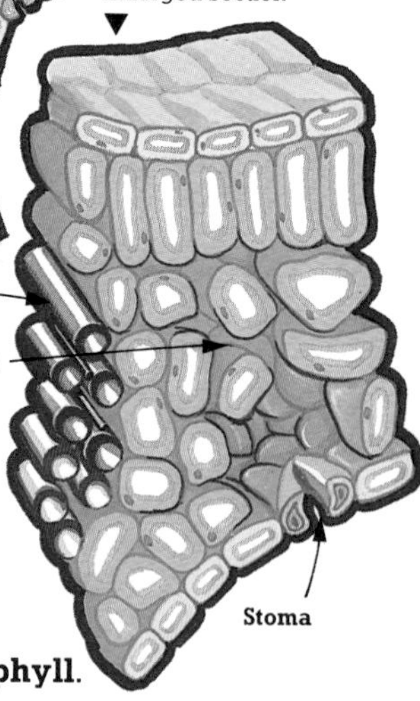

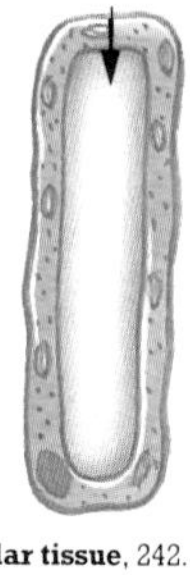

- **Palisade layer**. A cell layer just below the upper surface of a leaf. It is made up of regular, oblong-shaped **palisade cells**. These contain many **chloroplasts***.

- **Spongy layer**. A layer of irregular-shaped **spongy cells** and air spaces where gases circulate. The **spongy** and palisade layers together are the **mesophyll**.

* **Chloroplasts**, 255; **Epidermis**, 243; **Vascular tissue**, 242.

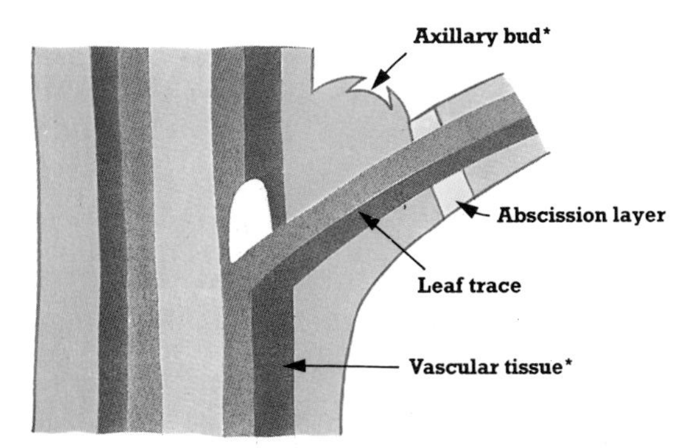

- **Leaf trace**. An area of **vascular tissue*** which branches off that of a stem to become the central **vein** of a leaf.
- **Abscission layer**. A layer of cells at the base of a leaf stalk which separates from the rest of the plant at a certain time of year (stimulated by a **hormone*** called **abscisic acid**). This makes the leaf fall off, forming a **leaf scar** on the stem.
- **Stomata** (sing. **stoma**). Tiny openings in the **epidermis*** (outer skin), through which the exchange of water (**transpiration***) and gases takes place. They are mainly found on the underside of leaves.
- **Guard cells**. Pairs of crescent-shaped cells. The members of each pair are found on either side of a **stoma**, which they open and close by changing shape. This controls water and gas exchange. They are the only surface cells with **chloroplasts***.

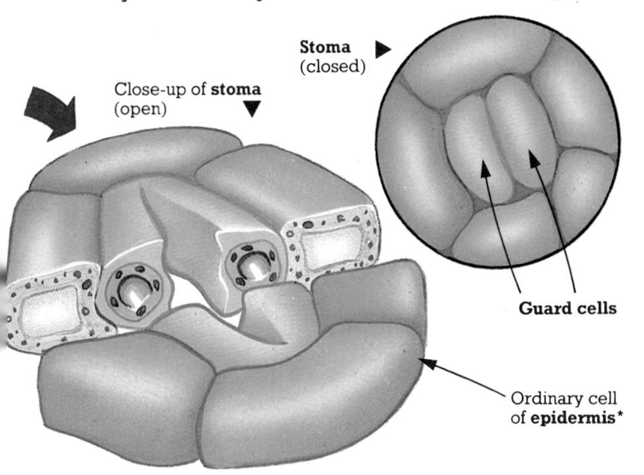

Special leaves

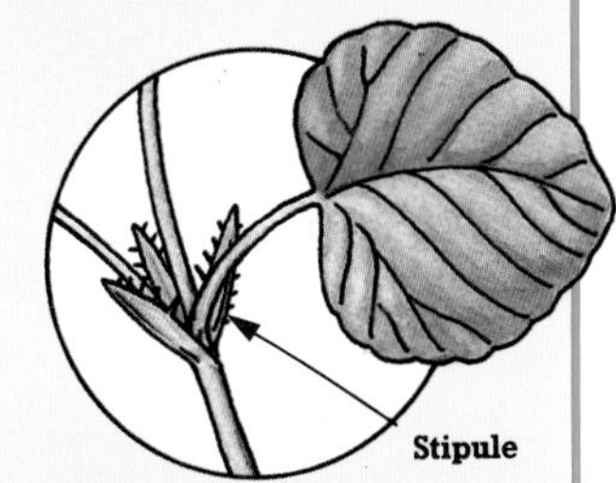

- **Stipule**. A small stalkless leaf at the base of a leaf stalk in many plants.

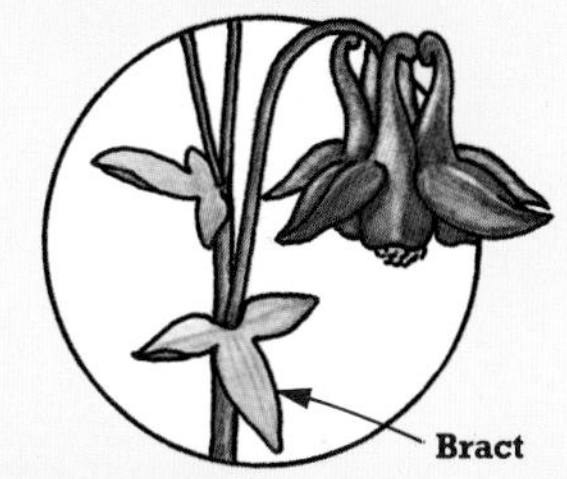

- **Bract**. A leaf at the base of a flower stalk in many plants.

- **Tendril**. A special threadlike leaf (or stem) which either twines round or sticks to a support.

- **Spine**. A specially modified leaf of a cactus. It has a reduced surface area to avoid losing much water.

* **Axillary bud**, 244; **Chloroplasts**, 255; **Epidermis**, 243; **Hormones**, 334; **Transpiration**, 252; **Vascular tissue**, 242.

Types of compound leaf

Shown here are some types of **compound leaf** (leaves made up of **leaflets***), as well as some common leaf arrangements and leaf edges, or **margins**. The pictures are not to scale.

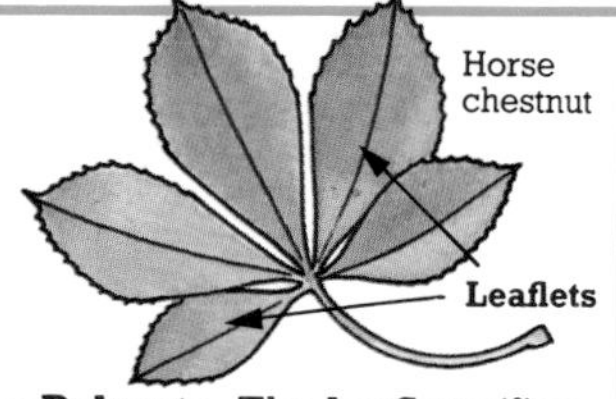

- **Palmate**. The **leaflets** (five or more) radiate from one common point.

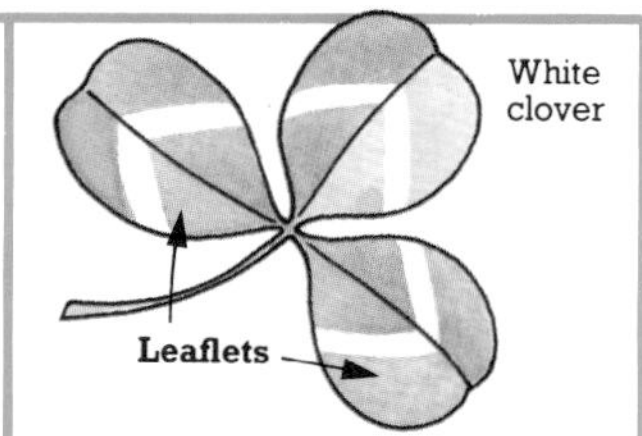

- **Trifoliate**. Three **leaflets** grow from the same point.

- **Ternate**. Special type of **trifoliate** leaf. Each **leaflet** has three **lobes**.

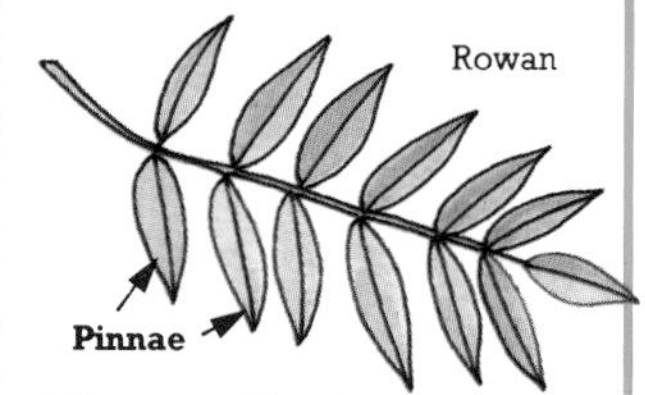

- **Pinnate**. The **leaflets, or pinnae** (sing. **pinna**), are in **opposite** pairs.

- **Bipinnate / tripinnate**. A **pinnate** leaf with pinnate **leaflets**.

Leaf arrangements

- **Spiral**. Leaves growing out from points forming a spiral around the stem.

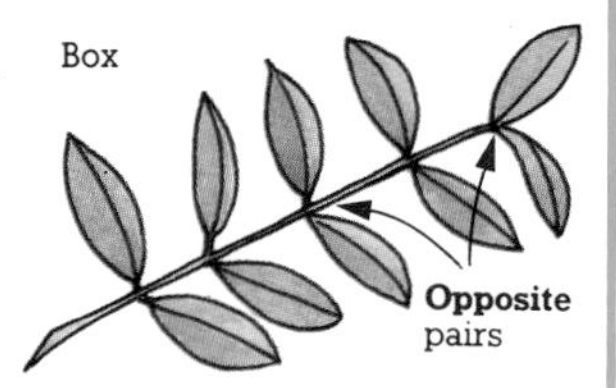

- **Opposite**. Leaf pairs whose members grow from opposite stem sides.

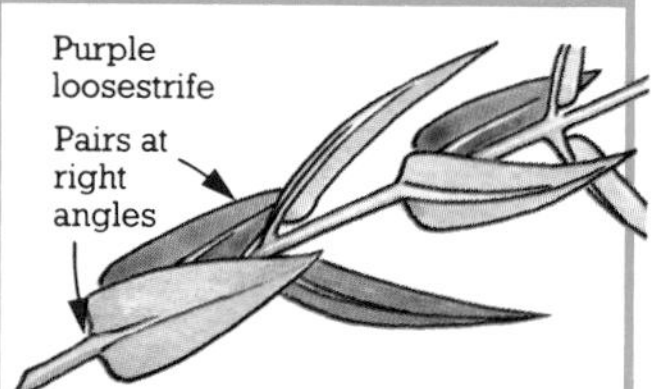

- **Decussate**. **Opposite** pairs, each pair at right angles to the one before.

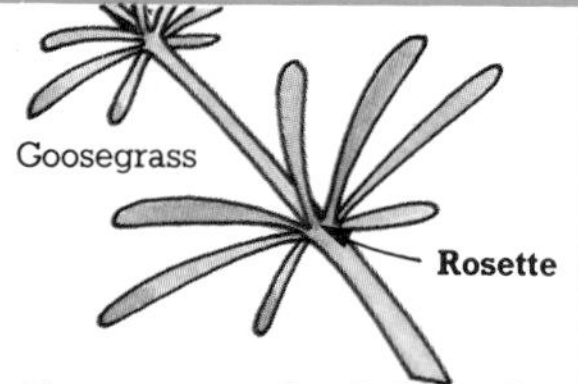

- **Rosette** or **whorl**. A circle of leaves growing from one point.

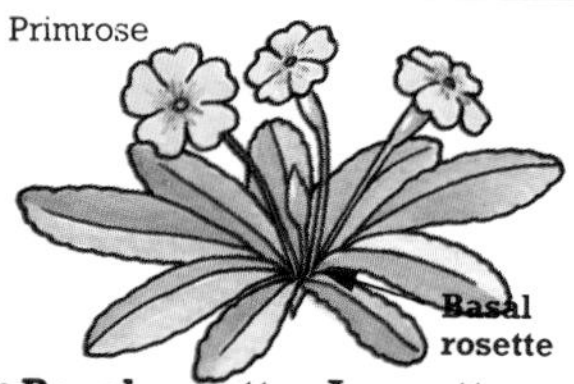

- **Basal rosette**. A **rosette** growing at the base of a stem.

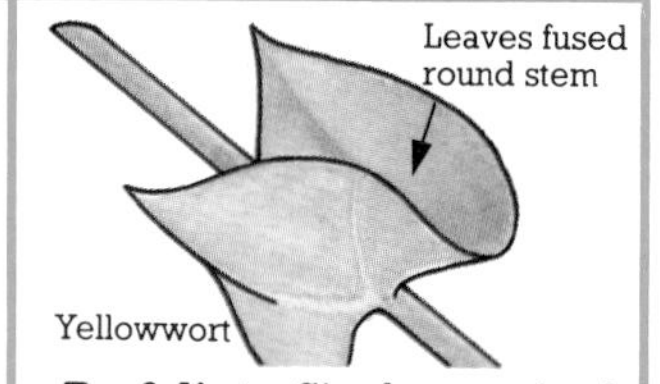

- **Perfoliate**. Single or paired leaves whose bases are fused around the stem.

Leaf margins

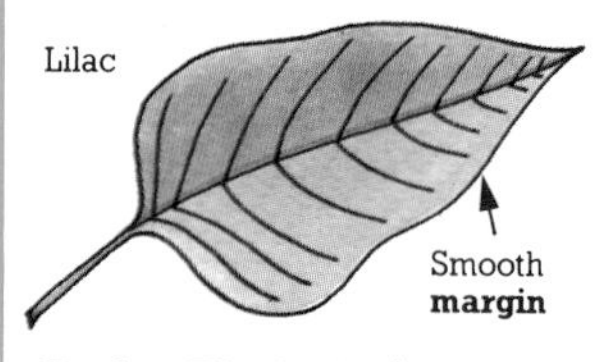

- **Entire**. The leaf **margin** has no indentations of any kind.

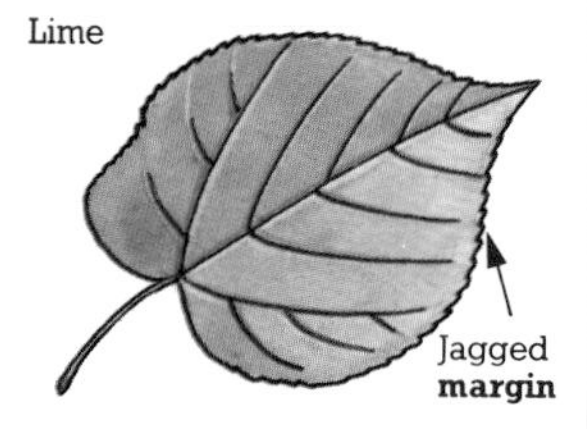

- **Serrate**. The leaf **margin** has tiny jagged "teeth". May also be **lobed**.

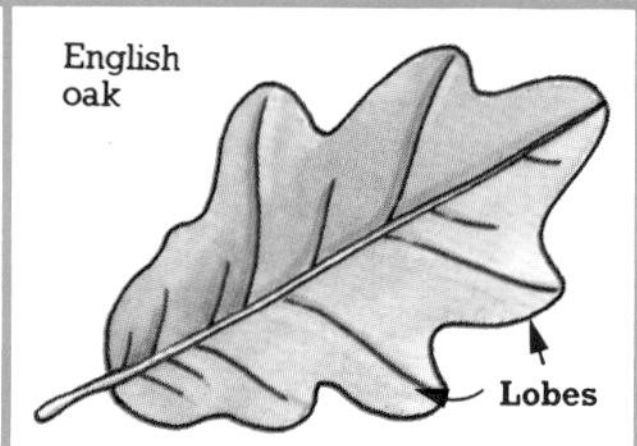

- **Lobed**. The leaf **margin** forms sections, or **lobes**. May also be **serrate**.

* **Leaflets**, 248.

Plant sensitivity

Plants have no nervous system, but they do still show **sensitivity**, i.e. they react to certain forms of stimulation. They do this by moving specific parts or by growing. This is called **tropism**. **Positive tropism** is movement or growth towards the stimulus and **negative tropism** is movement or growth away from it.

- **Phototropism**. Response to light. When the light is sunlight, the response is called **heliotropism**. Most leaves and stems show this by curving around to grow towards the light.

- **Haptotropism** or **thigmotropism**. Response to touch or contact. For example, the sticky hairs of a sundew plant curl around an insect when it comes into contact with them.

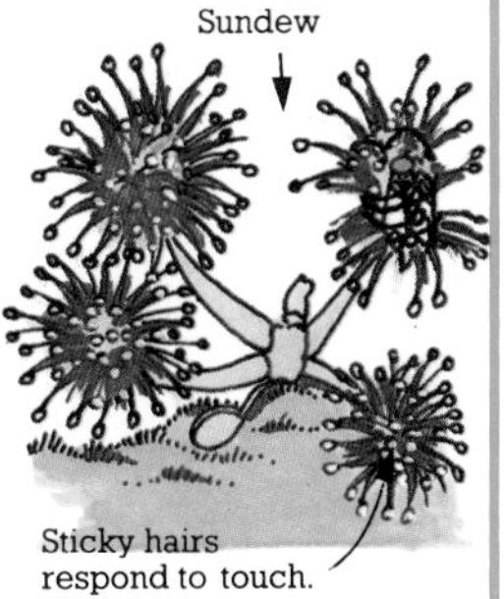

- **Hydrotropism**. Response to water. For example, some roots may grow out sideways if there is more water in that direction.
- **Geotropism**. Response to the pull of gravity. This is shown by all roots, i.e. they all grow down through the soil.

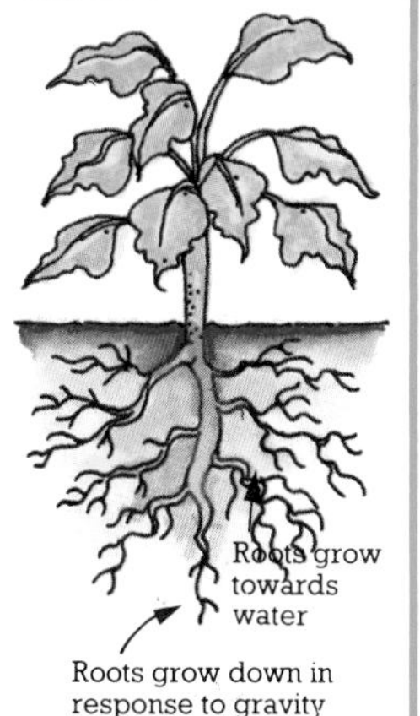

- **Photoperiodism**. The response of plants to the length of day or night (**photoperiods**), especially with regard to the production of flowers. It depends on a number of things, e.g. the plant's age and the temperature of its environment. **Short-day plants** only produce flowers if the length of the day is shorter than a certain length (called its **critical length**), **long-day plants** only if it is longer. It is thought that a "message" to produce flowers is carried to the relevant area by a **hormone***, produced in the leaves when conditions are right. This hormone has been called **florigen**. Some plants are **day-neutral plants**, i.e. their flowering does not depend on the length of day.

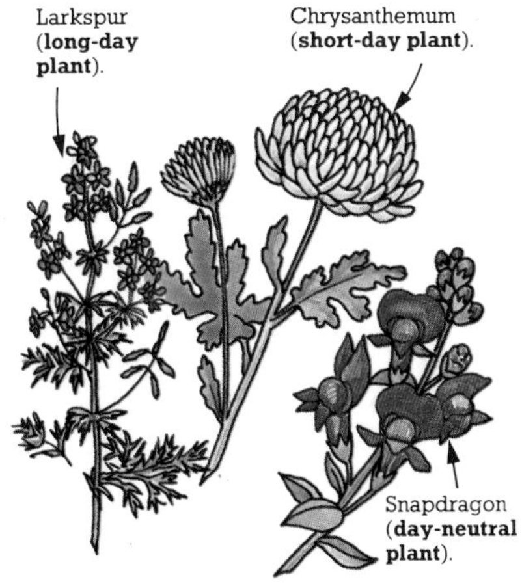

- **Growth hormones** or **growth regulators**. Substances which promote and regulate plant growth. They are produced in **meristems*** (areas where cells are constantly dividing). **Auxins**, **cytokinins** and **gibberellins** are types of growth hormone.

* **Hormones**, 334; **Meristem**, 244.

Plant fluid transportation

The transportation of fluids in a plant is called **translocation**. The fluids travel within the **vascular tissue***, made up of **xylem*** and **phloem***. The xylem carries water (with dissolved minerals) from the roots to the leaves. The phloem carries food from the leaves to areas where it is needed.

- **Transpiration**. The loss of water by evaporation, mainly through tiny holes called **stomata*** on the undersides of leaves.
- **Transpiration stream**. A constant chain of events inside a plant. As the outer leaf cells lose water by **transpiration**, the concentration of minerals and sugars in their **vacuoles*** becomes higher than that of the cells further in. Water then passes outwards by **osmosis***, causing more water to be "pulled" up through the tubes of the **xylem*** in the stem and roots (helped by **capillary action**). The roots then take in more water.
- **Capillary action**. The way that fluids travel up narrow tubes (see also page 23). The molecules of the fluid are "pulled" upwards by the attraction between them and the molecules of the tube.
- **Root pressure**. A pressure which builds up in the roots of some plants. In all plants, water travels in from the soil and on through the layers of root cells by **osmosis***. In plants which develop root pressure, the pressure of this water movement is enough to force the water some way up into the tubes of the **xylem***. It is then "pulled" on upwards by the **transpiration stream**. In other plants, the movement of water through root cells is all due to the "pull" of the transpiration stream.

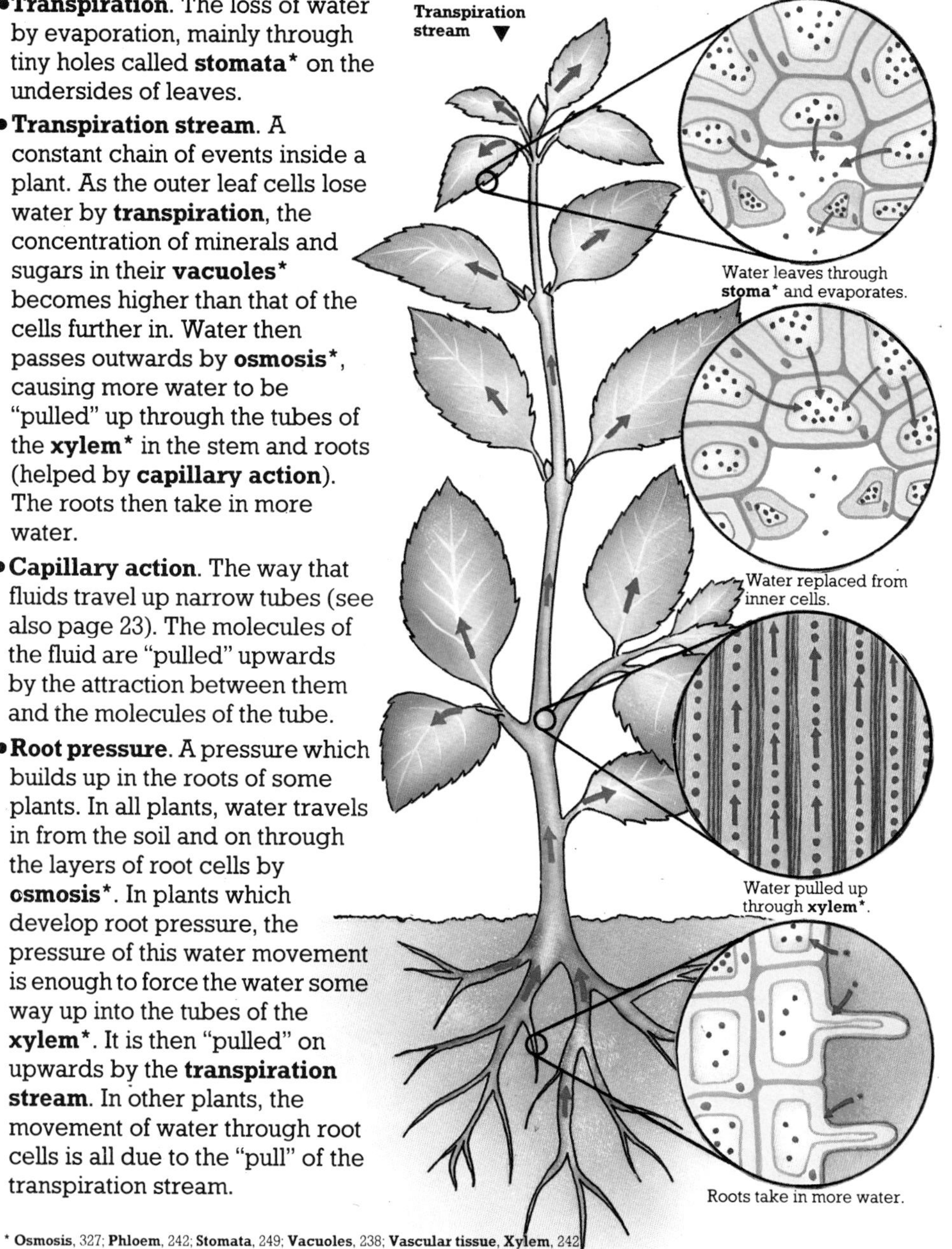

* **Osmosis**, 327; **Phloem**, 242; **Stomata**, 249; **Vacuoles**, 238; **Vascular tissue**, **Xylem**, 242

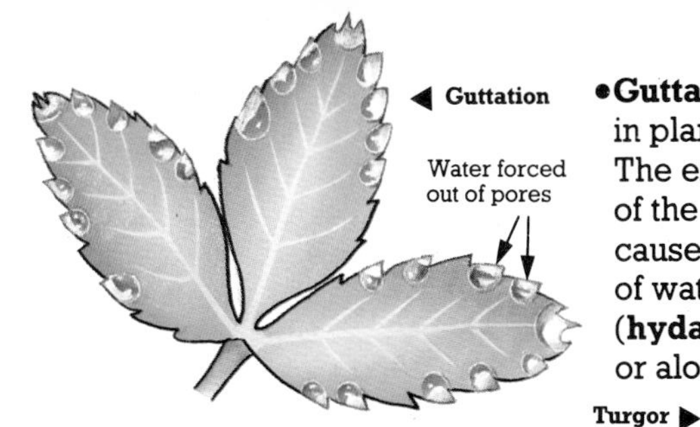

•**Guttation**. A phenomenon occurring in plants which show **root pressure**. The extra pressure, added to the "pull" of the **transpiration stream**, may cause drops of water to be forced out of water-secreting areas of cells (**hydathodes**) via tiny pores at the tips or along the edges of the leaves.

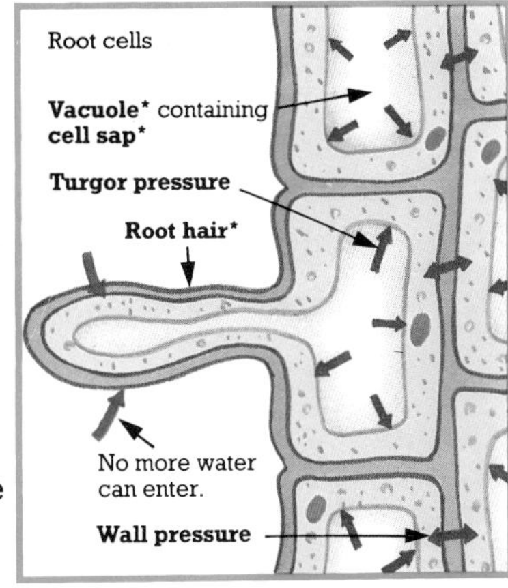

•**Turgor**. The state of the cells in a healthy plant. Each cell can take in no more water (it is **turgid**). Water has passed by **osmosis*** into the **cell sap*** (dissolved minerals and sugars) in its large central **vacuole***, and the vacuole has pushed as far out as it can go. It can go no further because its outward pressure (**turgor pressure**) is equalled by the opposing force of the rigid **cell wall*** (**wall pressure**). Such cells enable a plant to stand firm and upright.

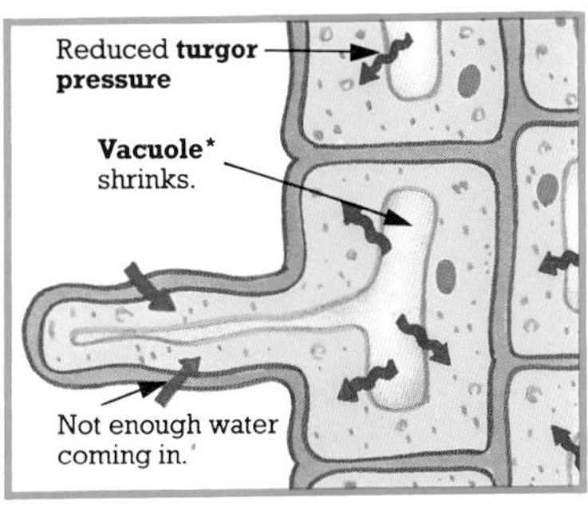

•**Wilting**. A state of drooping, found in a plant subjected to certain conditions, such as excess heat. The plant is losing more water (by **transpiration**) than it can take in, and the **turgor pressure** (see **turgor**) of its cell **vacuoles*** drops. The cells become limp and can no longer support the plant, so it will droop.

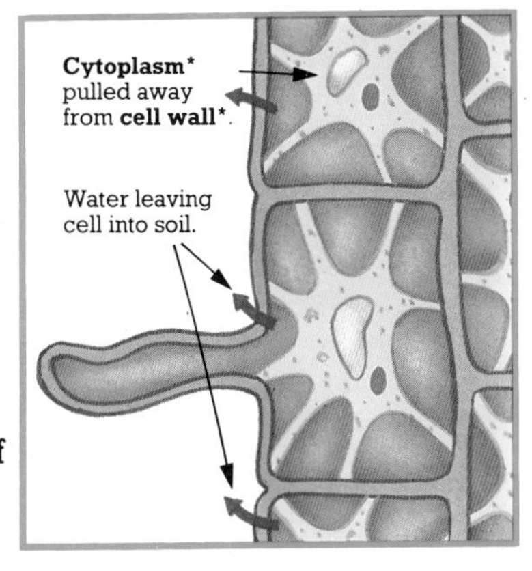

•**Plasmolysis**. An extreme state in a plant, which may cause it to die. Such a plant is losing a large amount of water, often not only by **transpiration** in excess heat (see **wilting**), but also by **osmosis*** into very dry soil or soil with a very high concentration of minerals. The **vacuoles*** of the plant cells then shrink so much that they pull the **cytoplasm*** away from the **cell walls***.

* **Cell sap**, 238 (**Vacuoles**); **Cell wall**, **Cytoplasm**, 238; **Osmosis**, 327; **Root hairs**, 245.

Plant food production

Most plants have the ability to make the food they need for growth and energy (unlike animals, which must take it in). The manufacturing process by which they make their complex food substances from other, simpler substances is called **photosynthesis**.

- **Photosynthesis**. The series of chemical reactions (for basic equation, see page 209) by which green plants make their food. It occurs mainly in the **palisade cells***. Carbon dioxide is combined with water, using energy taken in from sunlight by **chloroplasts**. This produces oxygen as well as the plant's food. ▶

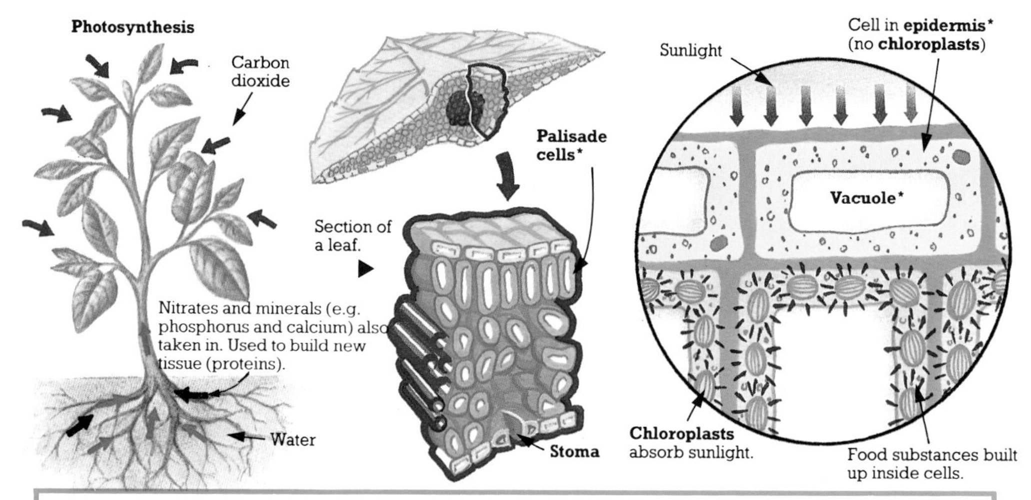

- **Compensation points**. Two points in a 24-hour period (normally around dawn and around dusk) when the two processes of **photosynthesis** and **internal respiration*** (see top of next page) are exactly balanced. Photosynthesis is producing just the right amounts of carbohydrates and oxygen for internal respiration, and this is producing just the right amounts of carbon dioxide and water for photosynthesis.

1. Around dawn (**compensation point**)

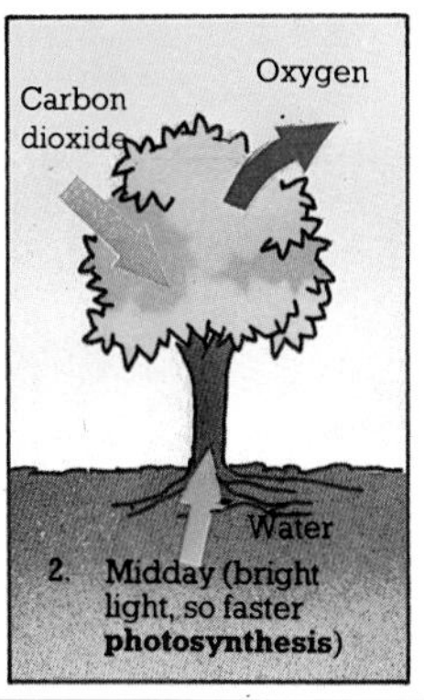

2. Midday (bright light, so faster **photosynthesis**)

3. Around dusk (**compensation point**)

4. Midnight (no light, so no **photosynthesis**)

* **Epidermis**, 243; **Internal respiration**, 332; **Palisade cells**, 248 (**Palisade layer**); **Vacuoles**, 238.

▶ The process of photosynthesis works in co-ordination with that of **internal respiration***, the breakdown of food for energy. Photosynthesis produces oxygen and carbohydrates (needed by internal respiration) and internal respiration produces carbon dioxide and water (needed by photosynthesis). At most times,

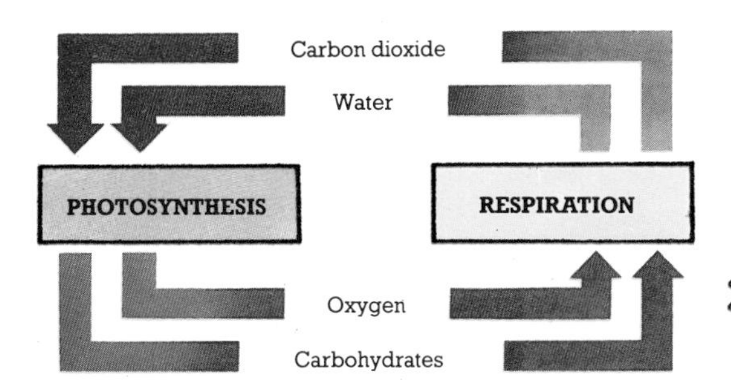

one of the two processes is occurring at a faster rate than the other. This means that excess amounts of its products are being produced, and not enough of the substances it needs are being made in the plant. In this case, extra amounts must be taken in and excess amounts given off or stored (see pictures 2 and 4 on the opposite page).

•**Chloroplasts**. Tiny bodies in plant cells (mainly in the leaves) which contain a green **pigment** called **chlorophyll**. This absorbs the sun's light energy and uses it to "power" **photosynthesis**. Chloroplasts can move around inside a cell, according to light intensity and direction. See also page 240.

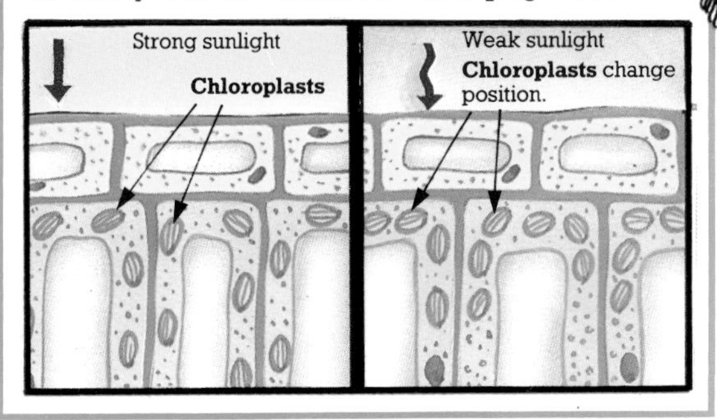

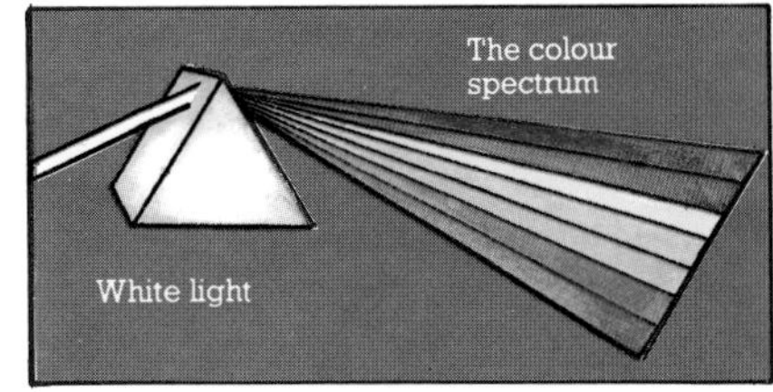

•**Pigments**. Substances which absorb light. White light is actually made up of a spectrum of many different colours. Each pigment absorbs some colours and reflects others.

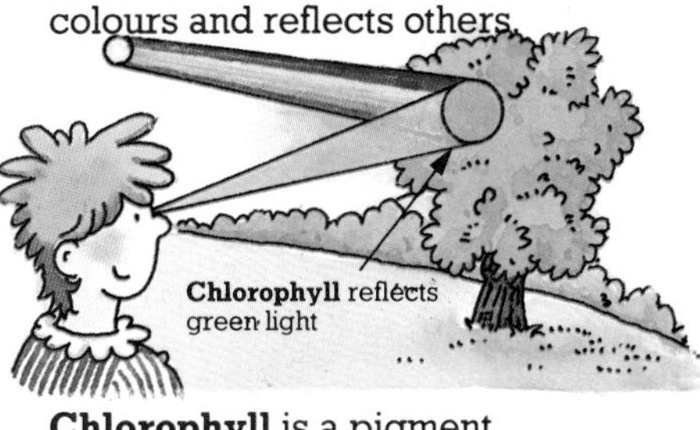

Chlorophyll is a pigment found in all leaves. It absorbs blue, violet and red light, and reflects green light. This is why leaves look green. Other pigments, such as

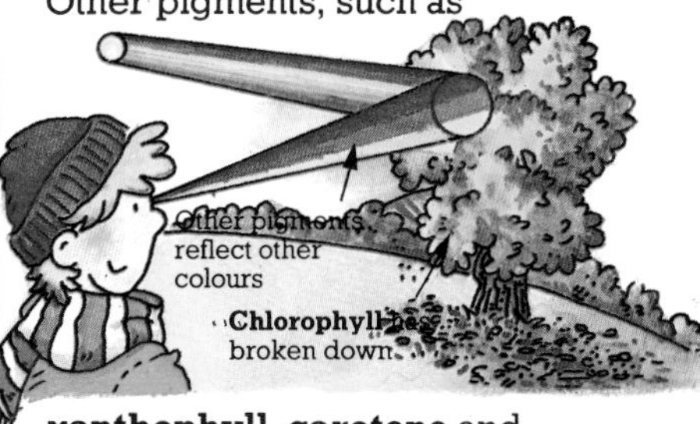

xanthophyll, **carotene** and **tannin** are also present in leaves. They reflect orange, yellow and red light, but are masked by chlorophyll during the growing season. In autumn, the chlorophyll breaks down, and so the autumn colours appear. Pigments are used to give colour to many things, e.g. paints and plastics.

* **Internal respiration**, 332.

Flowers

The **flowers** of a plant contain its organs of **reproduction** (producing new life – see also page 258). In **hermaphrodite** plants, e.g. the buttercup and poppy below, each flower has both male and female organs. **Monoecious** plants, e.g. maize, have two types of flower on one plant – **staminate** flowers, which have just male organs, and **pistillate** flowers, which have just female organs. **Dioecious** plants, e.g. holly, have staminate flowers on one plant and pistillate flowers on a separate plant.

- **Receptacle**. The expanded tip of the flower stalk, or **peduncle**, from which the flower grows.
- **Petals**. The delicate, usually brightly coloured structures around the reproductive organs. They are often scented (to attract insects) and are known collectively as the **corolla**.

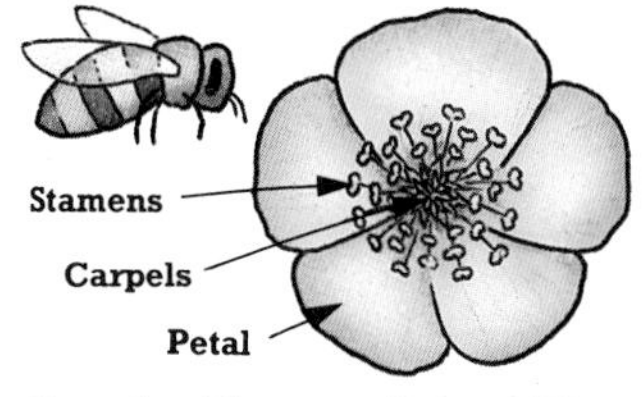

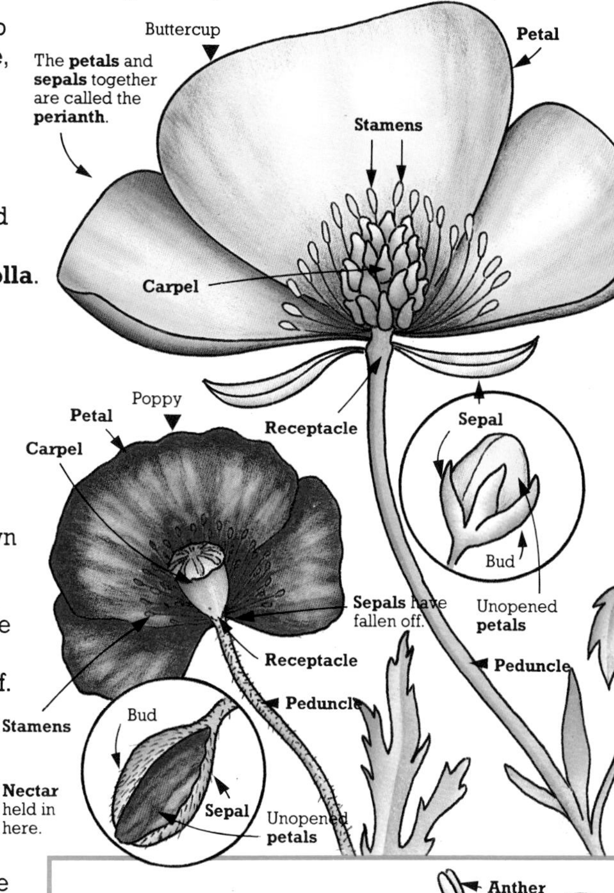

- **Sepals**. The small, leaf-like structures around a bud, known collectively as the **calyx**. In some flowers, e.g. buttercups, they remain as a ring around the opened **petals**; in others, e.g. poppies, they wither and fall off.

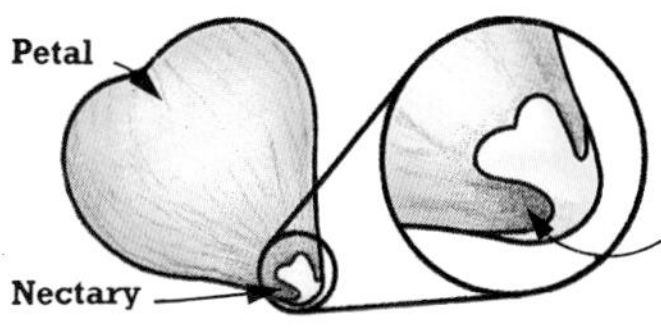

- **Nectaries**. Areas of cells at the base of the **petals** which produce a sugary liquid called **nectar**. This attracts insects needed for **pollination***. It is thought that the dark lines down many petals are there to direct an insect to the nectar, and they are known as **honey guides**.

The male organs

- **Stamens**. The male reproductive organs. Each has a thin stalk, or **filament**, with an **anther** at the tip. Each anther is made up of **pollen sacs**, which contain grains of **pollen***.

Anther
Stamen
Filament

* **Pollen**, **Pollination**, 258.

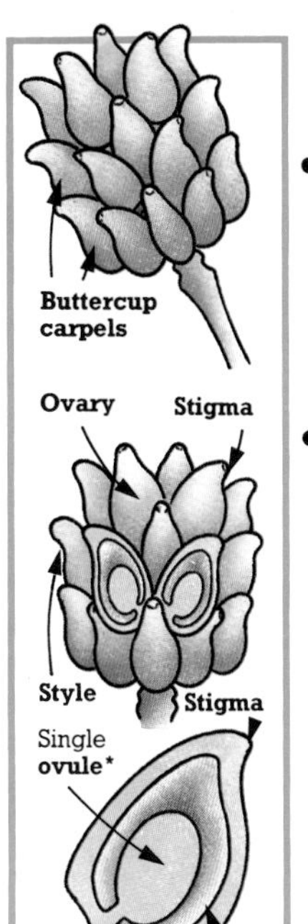

The female organs

- **Carpel** or **pistil**. A female reproductive organ, consisting of an **ovary**, **stigma** and **style**. Some flowers have only one carpel, others have several clustered together.
- **Ovaries**. Female reproductive structures. Each is the main part of a **carpel** and contains one or more tiny bodies called **ovules***, each of which contains a female sex cell. An ovule is fixed by a stalk (**funicle**) to an area of the ovary's inside wall called the **placenta**. The stalk is attached to the ovule at a point called a **chalaza**.
- **Stigma**. The uppermost part of a **carpel**, with a sticky surface to which grains of **pollen*** become attached during **pollination***.
- **Style**. The part of a **carpel** which joins the **stigma** to the **ovary**. Many flowers have an obvious style, e.g. daffodils, but in others it is very short (e.g. buttercups) or almost non-existent (e.g. poppies).
- **Gynaecium**. The whole female reproductive structure, made up of one or more **carpels**.

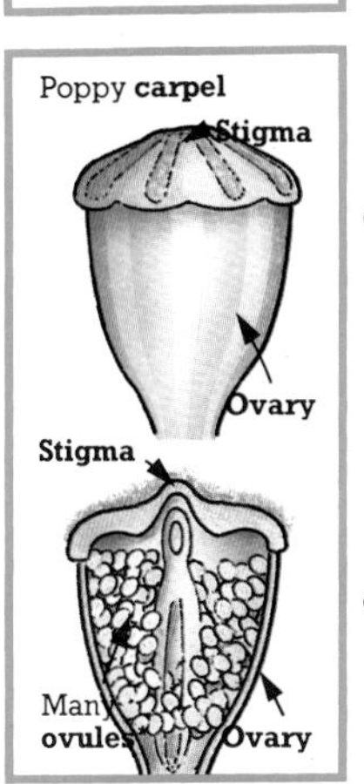

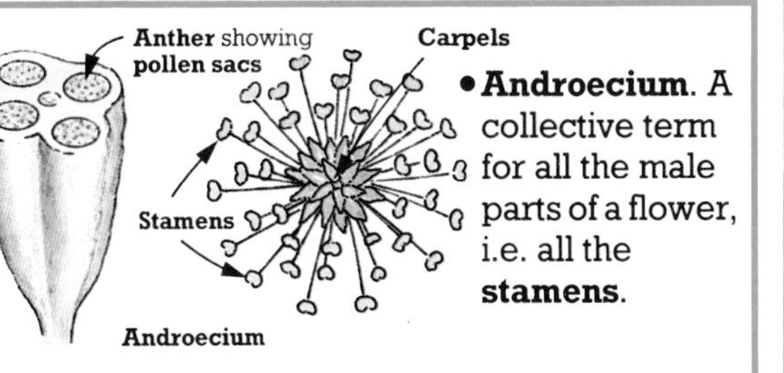

- **Androecium**. A collective term for all the male parts of a flower, i.e. all the **stamens**.

How the parts are arranged

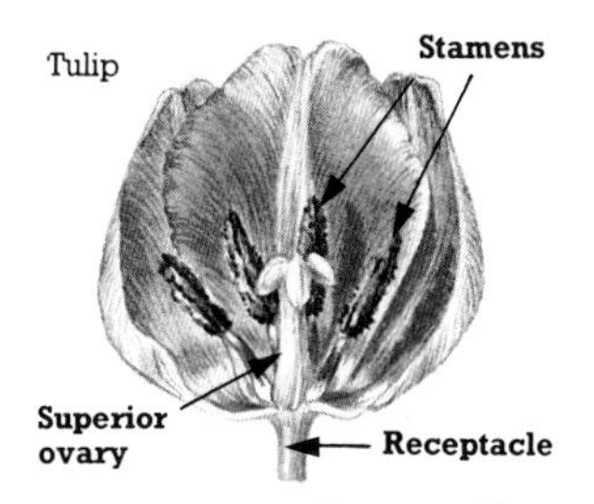

- **Hypogynous flower**. The **carpel** (or carpels) sit on top of the **receptacle**; all the other parts grow out from around its base. The position of the carpel is described as **superior**.

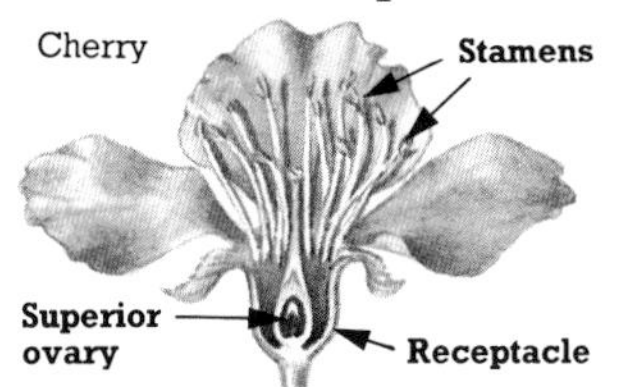

- **Perigynous flower**. The **carpel** (or carpels) rest in a cup-shaped **receptacle**; all the other parts grow out from around its rim. The position of the carpel is described as **superior**.

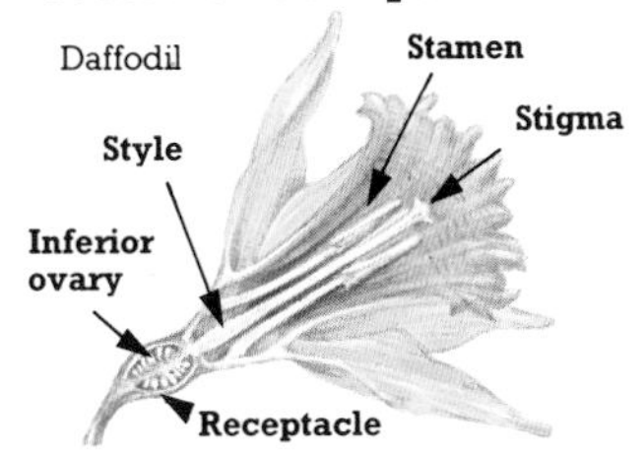

- **Epigynous flower**. The flower parts grow from the top of a **receptacle** which completely encloses the **ovary** (or ovaries), but not the **stigma** and **style**. The position of the ovary is described as **inferior**.

* **Ovules**, **Pollen**, **Pollination**, 258.

Reproduction in a flowering plant

Reproduction is the creation of new life. All flowering plants reproduce by **sexual reproduction***, when a male **gamete*** (sex cell) joins with a female gamete. In flowering plants, the male gametes (strictly speaking only **male nuclei***) are held in **pollen** and the female gametes in **ovules**.

- **Pollen**. Tiny grains formed by the **stamens*** (male parts) of flowers. Each grain is a special cell which has two **nuclei***. When a pollen grain lands on an **ovary*** (female body), one nucleus (the **generative nucleus**) splits in half, forming two **male nuclei** (reproductive bodies – see introduction).

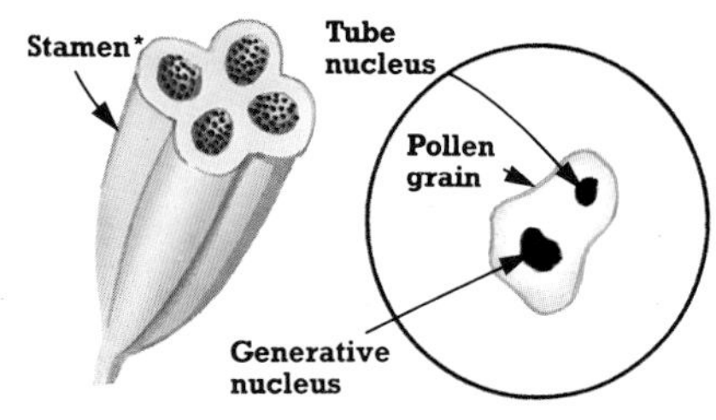

- **Ovules**. The tiny structures inside a flower's female body, or **ovary***. They become seeds after **fertilization**. Each consists of an oval cell (the **embryo sac**), surrounded by layers of tissue called **integuments**, except at one point where there is a tiny hole (**micropyle**). Before fertilization, the embryo sac **nucleus*** undergoes several divisions (looked at in more detail on page 323 – under **gamete production, female**). This results in a number of new cells (some of which become part of the seed's food store) and two naked nuclei which fuse together. One of the new cells is the female **gamete*** (sex cell), or **egg cell**.

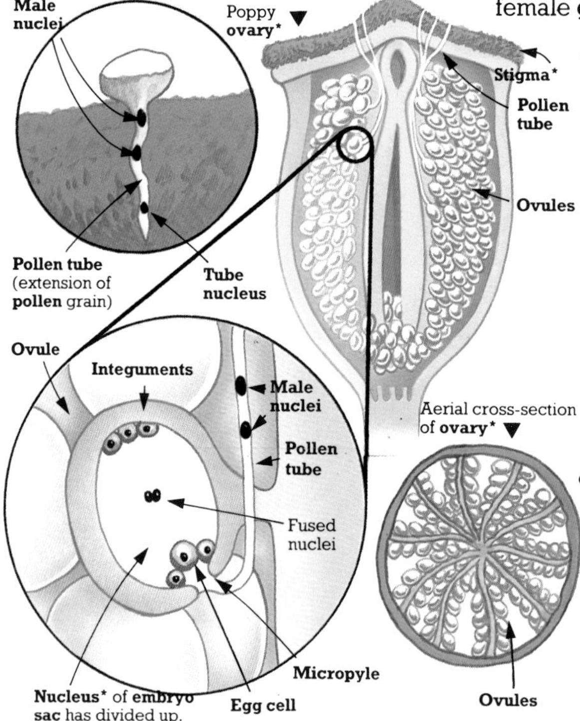

- **Pollination**. The process by which a grain of **pollen** transfers its **male nuclei** (see **pollen**) into the **ovary*** of a flower. The grain lands on the **stigma***, and forms a **pollen tube**, under the control of the **tube nucleus** (the one which did not divide – see **pollen**). The tube grows down through the ovary tissue and enters an **ovule** via its **micropyle**. The two male nuclei then travel along it.

- **Fertilization**. After **pollination**, one **male nucleus** (see **pollen**) fuses with the **egg cell** in the **ovule** to form a **zygote*** (the first cell of a new plant). The other joins with the two fused female nuclei to form a cell which develops into the **endosperm***.

* **Endosperm**, 261; **Male nuclei**, 321 (**Gametes**); **Nucleus**, 238; **Ovaries**, 257; **Sexual reproduction**, 320; **Stamens**, 256; **Stigma**, 257; **Zygote**, 321.

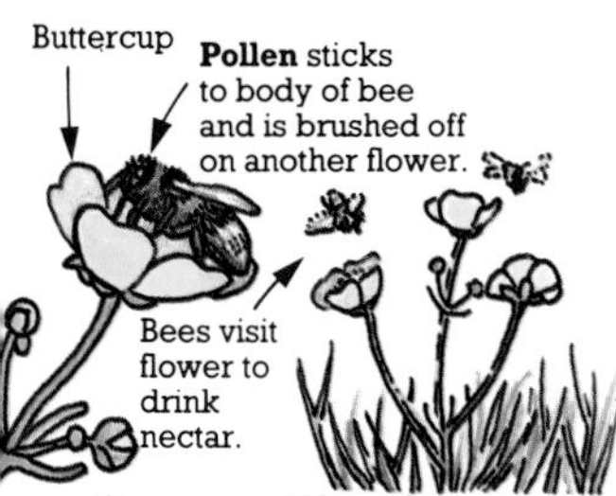

- **Cross pollination**. The **pollination** of one plant by **pollen** grains from another plant of the same type (if the grains land on a different type of plant, they do not develop further, i.e. they do not produce **pollen tubes**). The pollen may be carried by the wind, or by insects which drink the **nectar***.

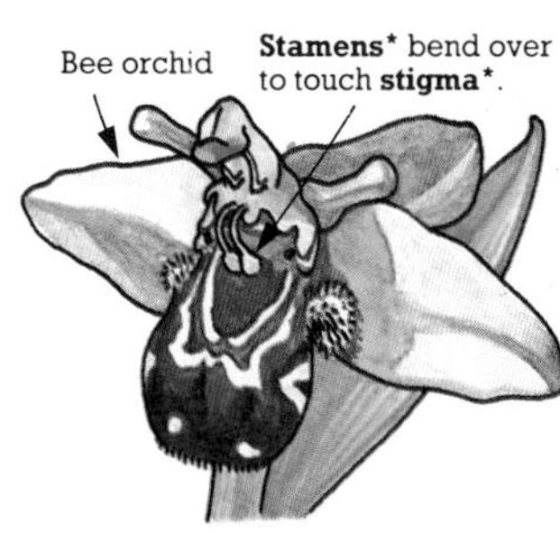

- **Self pollination**. The **pollination** of a plant by its own **pollen** grains. For example, a bee orchid tries to attract male Eucera bees (for **cross pollination**) by looking and smelling like a female bee. But if it is not visited, its **stamens*** (male parts) bend over and transfer pollen to the **stigma*** of its **ovary*** (female body).

Types and arrangements of flowers

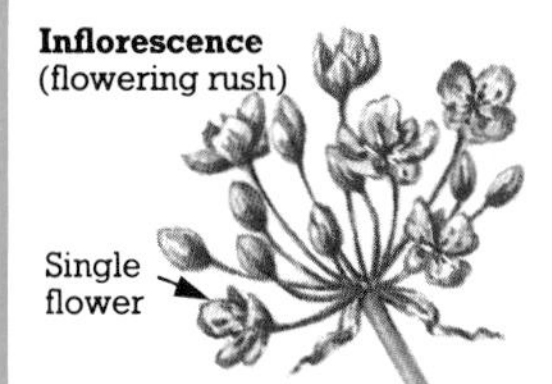

- **Inflorescence**. A group of flowers or **flowerheads** growing from one point.

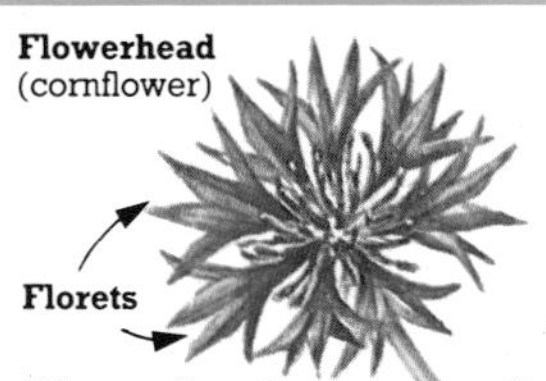

- **Flowerhead** or **composite flower**. A cluster of tiny flowers, or **florets**.

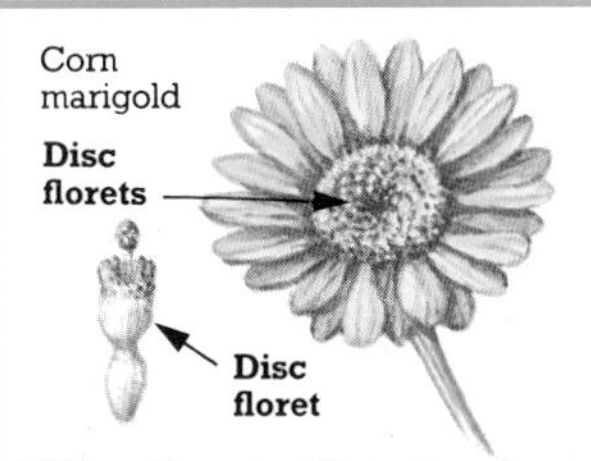

- **Disc florets**. **Florets** whose petals are all the same size.

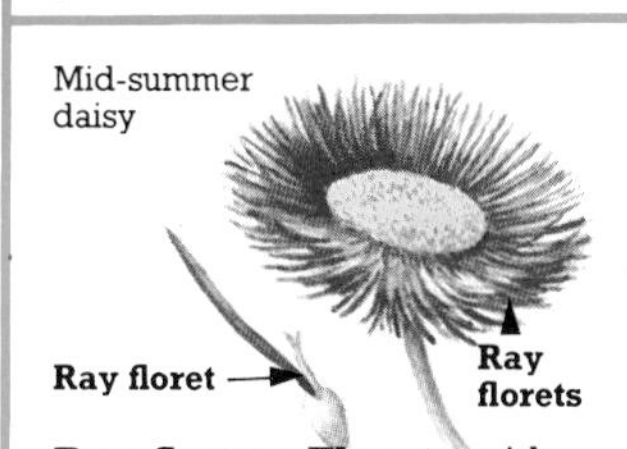

- **Ray florets**. **Florets** with one long petal.

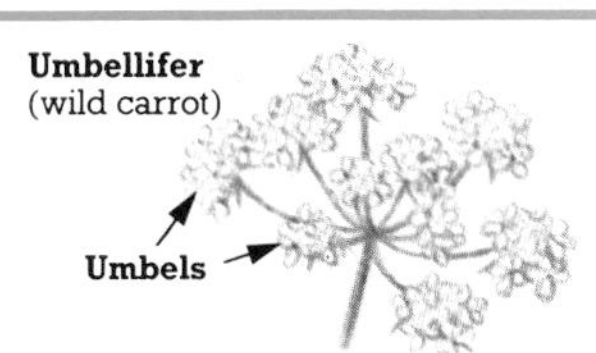

- **Umbellifer**. An **inflorescence** with umbrella-shaped **flowerheads** (**umbels**).

- **Bell flower**. Also called a **tubular** or **campanulate flower**. Its petals are joined to make a bell shape.

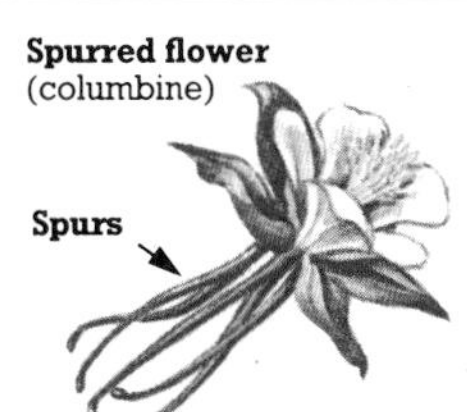

- **Spurred flower**. A flower with one or more petals extended backwards to form **spurs**.

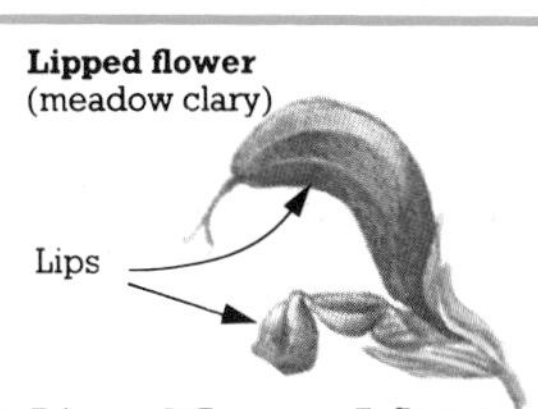

- **Lipped flower**. A flower with two "lips" – an upper and lower one. The upper one often has a hood.

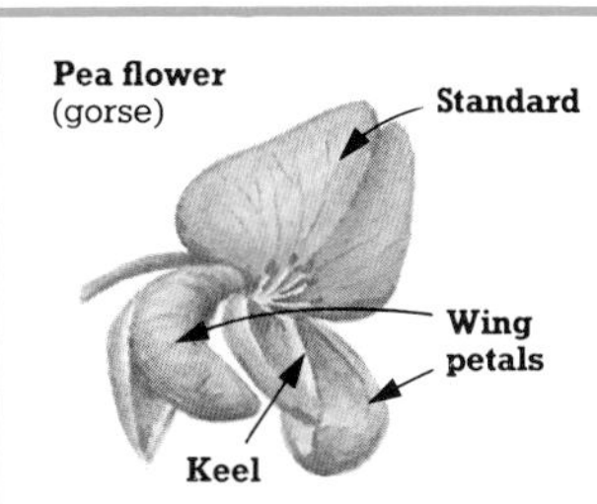

- **Pea flower**. A flower with an upper petal (the **standard**), two side petals (**wing petals**) and two lower petals forming the **keel** (which encloses the reproductive parts).

* **Nectar**, 256 (**Nectaries**); **Ovaries**, 257; **Stamens**, 256; **Stigma**, 257.

Seeds and germination

After **fertilization*** in a flowering plant, an **ovule*** develops into a **seed**. This contains an **embryo**, i.e. a new developing plant, and a store of food. The **ovary*** ripens into a fruit, carrying the seed or seeds. There is a chart of different fruits on page 262.

• **Dispersal** or **dissemination**. The shedding of ripe seeds from the fruit of a parent plant. This happens in one of two main ways, depending on whether a fruit is **dehiscent** or **indehiscent**.

• **Dehiscent**. A word describing a fruit from which the seeds are expelled before the fruit itself disintegrates. For

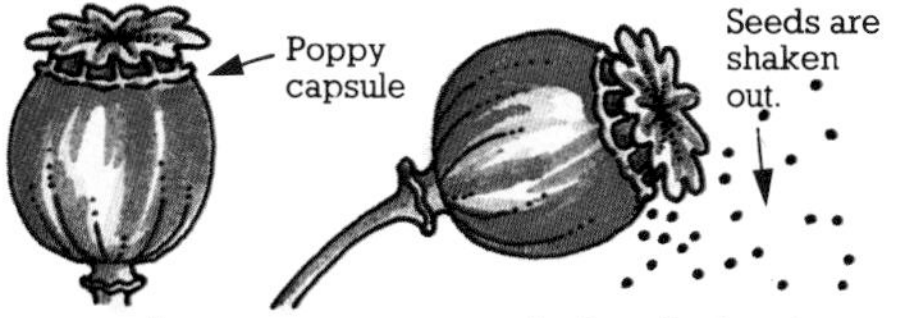

example, a poppy capsule has holes in, and the seeds are shaken out by the wind. Other fruit, e.g. broom pods,

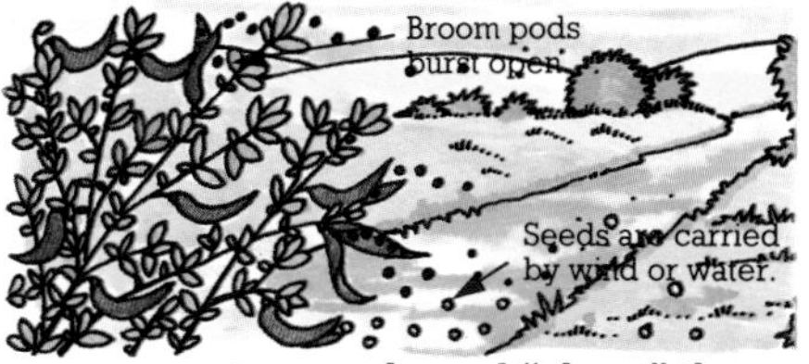

open spontaneously and "shoot" the seeds out. In all cases, the seeds may then be carried by wind, water or other means.

Indehiscent. A word describing a fruit which becomes detached from the plant and disintegrates to free the

seeds. For example, the "keys" of sycamores or the "parachutes" of dandelions are carried by the air, and hooked burrs catch on animal fur. The

fruit then rot away in the ground to expose the seeds. Edible fruit may be eaten by animals, which then expel the seeds in their droppings.

Germination

When conditions are right, a seed will **germinate**. The **plumule** and **radicle** emerge from the seed coat, and begin to grow into the new plant, or **seedling**.

• **Hypogeal**. A type of **germination**, e.g. in pea plants, in which the **cotyledons** remain below the ground within the **testa**, and the **plumule** is the only part to come above the ground.

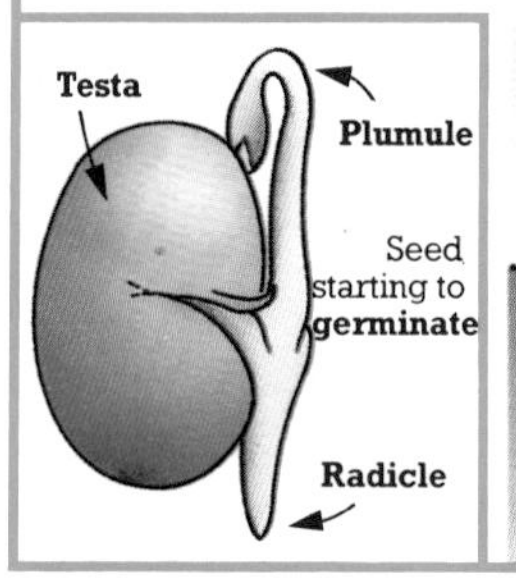

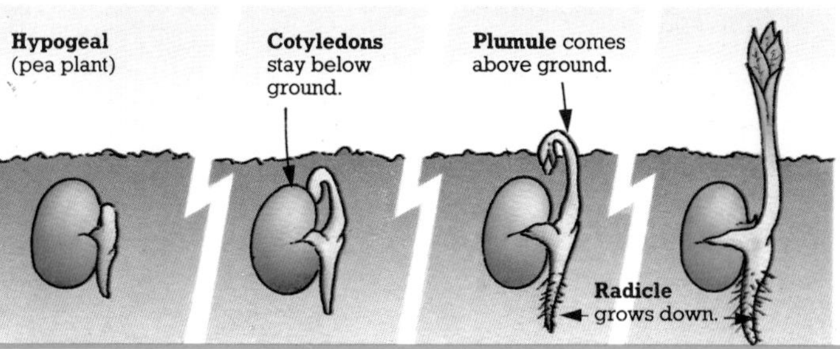

* **Fertilization**, 258; **Ovaries**, 257; **Ovules**, 258.

Parts of a seed

- **Hilum**. A mark on a seed, showing where the **ovule*** was attached to the **ovary***.
- **Testa**. The seed coat. It develops from the **integuments***.

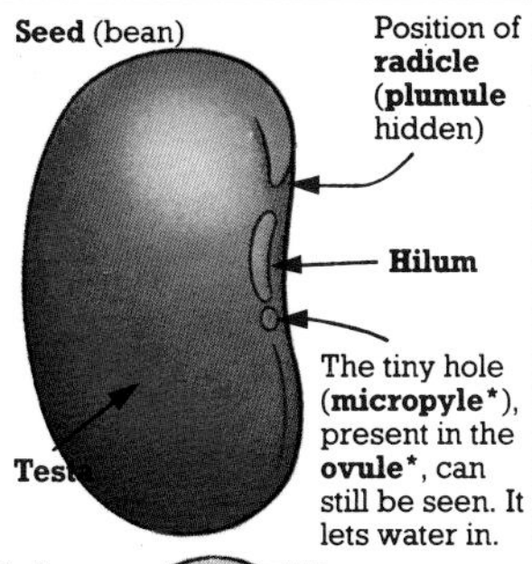

- **Plumule**. The first bud, or **primary bud**, formed inside a seed. It will develop into the first shoot of the new plant.
- **Radicle**. The first root, or **primary root**, of a new plant. It is formed inside a seed.

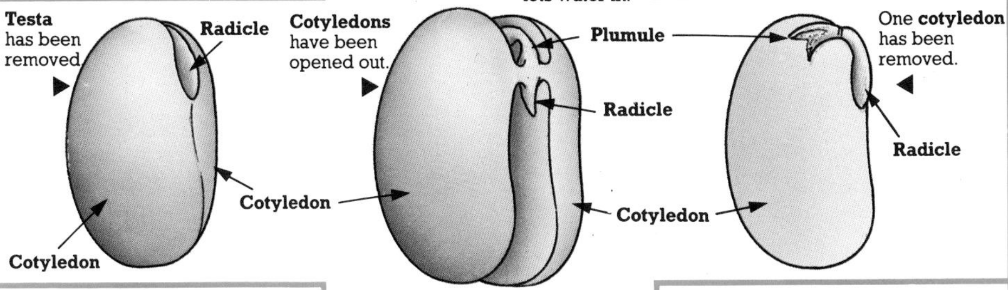

- **Endosperm**. A layer of tissue inside a seed which surrounds the developing plant and gives it nourishment. In some plants, e.g. pea, the **cotyledons** absorb and store all the endosperm before the seed is ripe, in others, e.g. grasses, it is not fully absorbed until after the seed **germinates**.

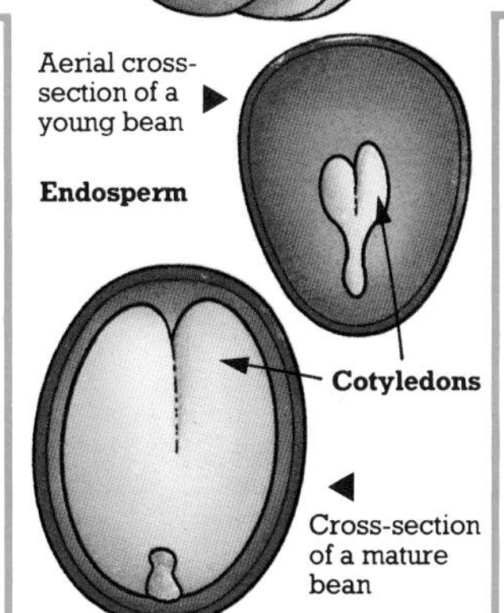

- **Cotyledon** or **seed-leaf**. A simple leaf which forms part of the developing plant. In some seeds, e.g. bean seeds, it absorbs and stores all the food from the **endosperm**. **Monocotyledons** are plants with one cotyledon, e.g. grasses, in **dicotyledons**, e.g. peas, there are two.

- **Epigeal**. A type of **germination**, e.g. in tomato plants, in which the **cotyledons** appear above the ground, below the first leaves – the true leaves.
- **Coleoptile**. The first leaf of many **monocotyledons** (see **cotyledon**). It protects the first bud and the first leaves emerge from it.

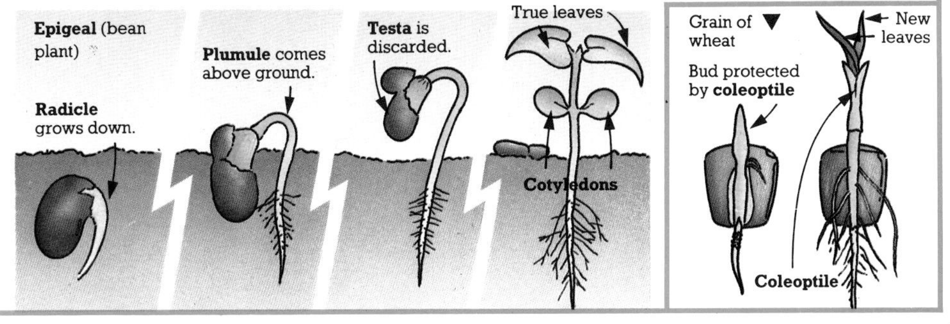

* **Integuments**, **Micropyle**, 258 (**Ovules**); **Ovaries**, 257.

Fruit

A **fruit** contains the seeds of a plant. **True fruit** develop purely from the **ovary***, false fruit from the **receptacle*** as well (e.g. a strawberry). The outer wall of a fruit is called the **pericarp**. In some fruit, it is divided into an outer skin, or **epicarp**, a fleshy part, or **mesocarp**, and an inner layer, or **endocarp**. Listed below are the main types of fruit.

- **Legume** or **pod**. A fruit with seeds attached to its inside wall. It splits along its length to open. E.g. a pea.

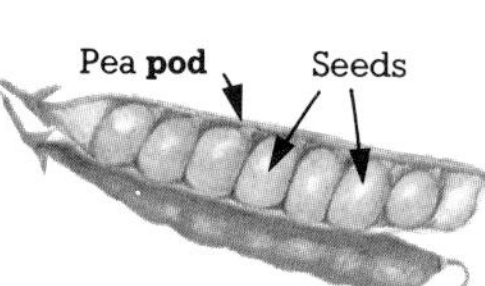

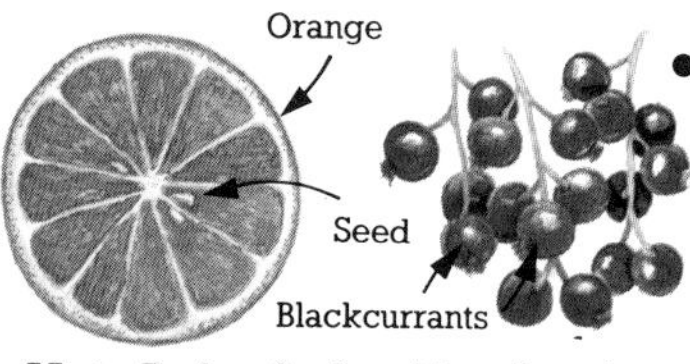

- **Berry**. A fleshy fruit which contains many seeds, e.g. an orange or a blackcurrant.

- **Nut**. A dry fruit with a hard shell, which only contains one seed, e.g. a hazelnut or a walnut.

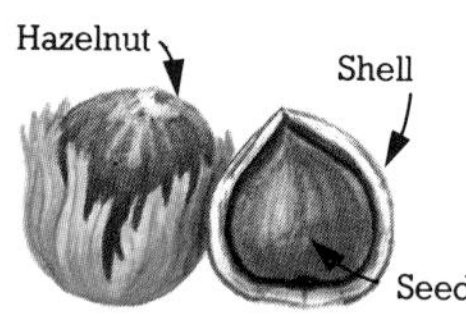

Grains of wheat

- **Grain**. Also called a **caryopsis** or **kernel**. A small fruit whose wall has fused with the seed coat, e.g. wheat.

- **Achene**. A small, dry fruit, with only one seed, e.g. a sycamore or buttercup fruit. A "winged" achene like a sycamore fruit is a **samara** or **key fruit**.

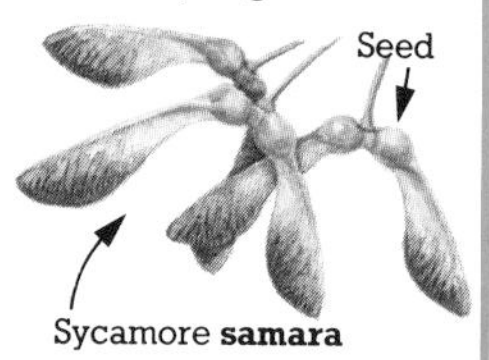

- **Drupe**. A fleshy fruit with a hard seed in the centre, often known as a "stone", e.g. a plum.

- **Pome**. A fruit with a thick, fleshy, outer layer and a core, with the seeds enclosed in a capsule, e.g. an apple. Pomes are examples of **false fruits** (see introduction).

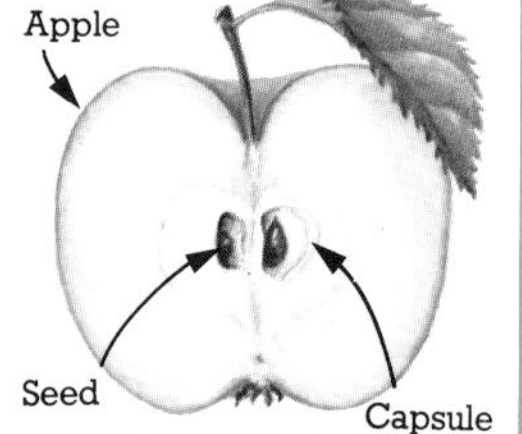

Vegetative reproduction

As well as producing seeds, some plants have developed a special type of **asexual reproduction***, called **vegetative reproduction** or **vegetative propagation**, in which one part of the plant is able to develop unaided into a new plant.

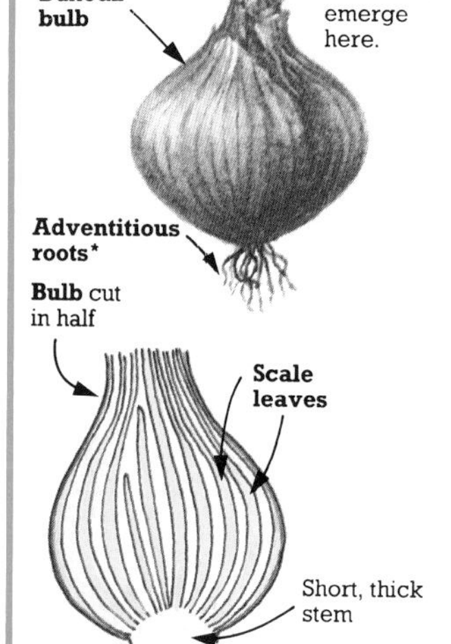

- **Bulb**. A short, thick stem surrounded by scaly leaves (**scale leaves**) which contain stored food material. It is formed underground by an old, dying plant, and represents the first, resting, stage of a new plant, which will emerge as a shoot at the start of the next season. E.g. a daffodil bulb.

* **Adventitious roots**, 245; **Asexual reproduction**, 320; **Ovaries**, 257; **Receptacle**, 256.

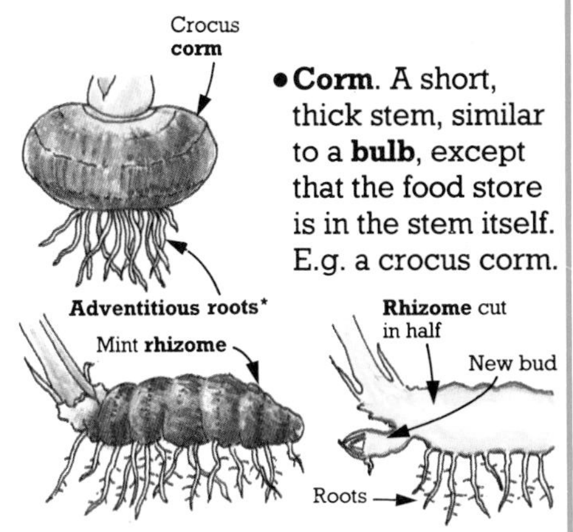

• **Corm**. A short, thick stem, similar to a **bulb**, except that the food store is in the stem itself. E.g. a crocus corm.

• **Rhizome**. A thick stem, which has scaly leaves and grows horizontally underground. It produces roots along its length and also buds from which new shoots grow. Many grasses produce rhizomes, as well as other plants, e.g. ferns and irises.

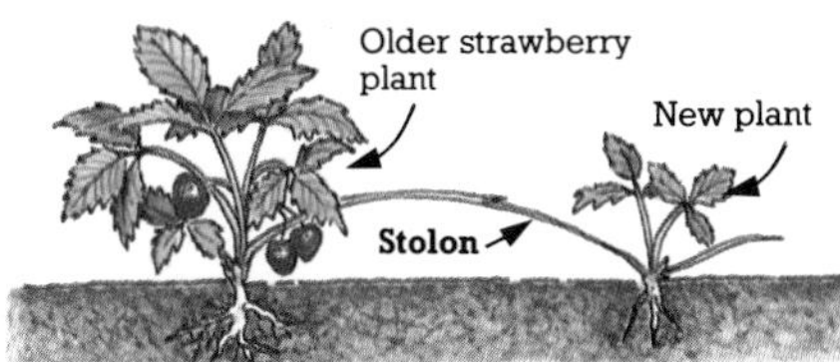

• **Stolon** or **runner**. A stem which grows out horizontally near the base of some plants, e.g. the strawberry. The stolon puts down roots from points at intervals along this stem, and new plants grow at these points.

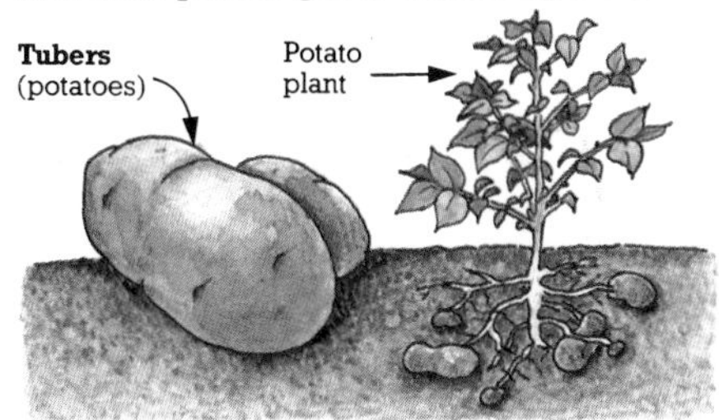

• **Tuber**. A short, swollen, underground stem which contains stored food material and produces buds from which new plants will grow, e.g. a potato.

Artificial propagation

Artificial propagation is the commercial process, in agriculture and market gardening, which makes use of **vegetative reproduction**. The fact that new plants need not always grow from seeds means that many more plants can be produced commercially than would occur naturally.

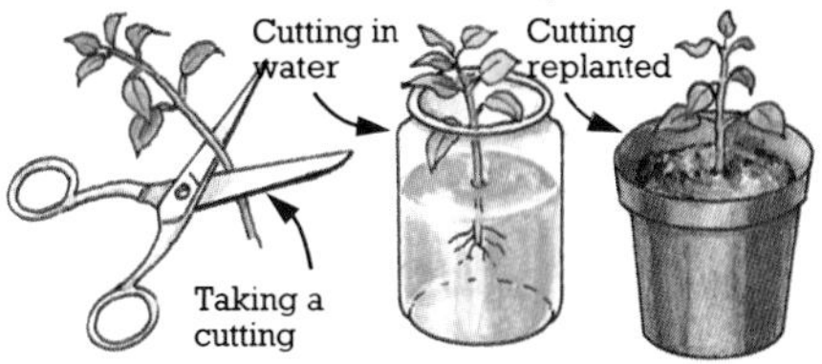

• **Cutting**. A process in which a piece of a plant stem (the cutting) is removed from its parent plant and planted in soil, where it grows into a new plant. In some cases, it is first left in water for a while to develop roots.

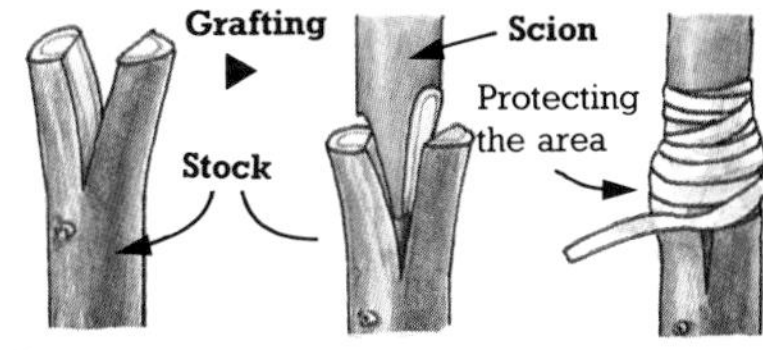

• **Grafting**. The process of removing a piece of a plant stem and re-attaching it elsewhere. This could be to a different part of the same plant (**autografting**), to another plant of the same species (**homografting**), or to a plant of a different species (**heterografting**). The piece removed is called the **scion**, and that to which it is attached is known as the **stock**.

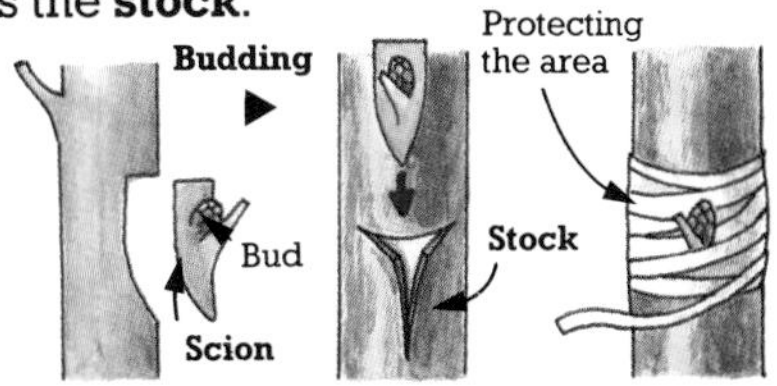

• **Budding**. A type of **grafting** where a bud and its adjacent stem are the parts grafted.

* **Adventitious roots**, 245.

The body structure of animals

Animals exist in a great variety of forms, from single-celled organisms to complex ones made of thousands of cells. The way they are **classified***, or divided into groups, depends to a large extent on how complex their bodies are. The two terms **higher animal** and **lower animal** are often used in this context. The higher an animal is, the more complex its internal organs are. In general, the distinguishing features of higher animals are **segmentation**, body cavities and some kind of skeleton.

- **Segmentation**. The division of a body into separate areas, or **segments**, a step up in complexity from a simple undivided body. Generally, the more complex the animal, the less obvious its segments are. The most primitive form

Metameric segmentation in an earthworm

Metamere

of segmentation is **metameric segmentation**, or **metamerism**. The segments (**metameres**) are very similar, if not identical. Each contains more or less identical parts of the main internal systems, which join up through the internal walls separating the segments. Such segmentation is found in most worms, for example, and in **myriapods***. More complex segmentation is less obvious. In insects, for example, the body has three main parts – the head, **thorax** (upper body region) and **abdomen** (lower body region). Each of these is in fact a group of segments, called a **tagma** (pl. **tagmata**), but the segments are not divided by internal walls. They are simply visible as external markings.

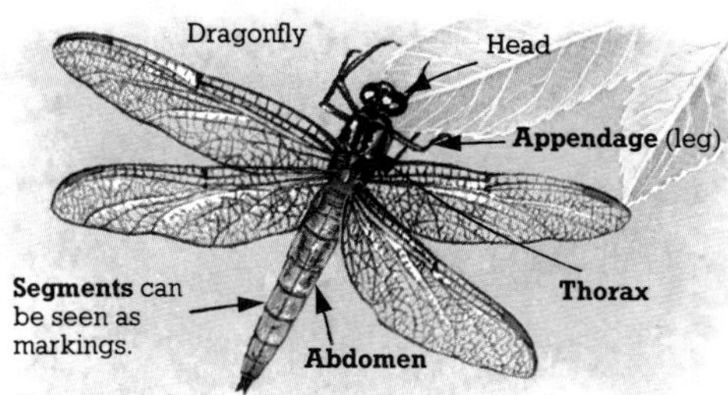

- **Appendage**. A subordinate body part, i.e. one which projects from the body such as an arm, leg, fin or wing.

Arrangement of the parts

- **Bilateral symmetry**. An arrangement of body parts in which there is only one possible body division which will produce two mirror-image halves. It is typical of almost all freely-moving animals. The same state in flowers is called **zygomorphy** (e.g. in snapdragons).

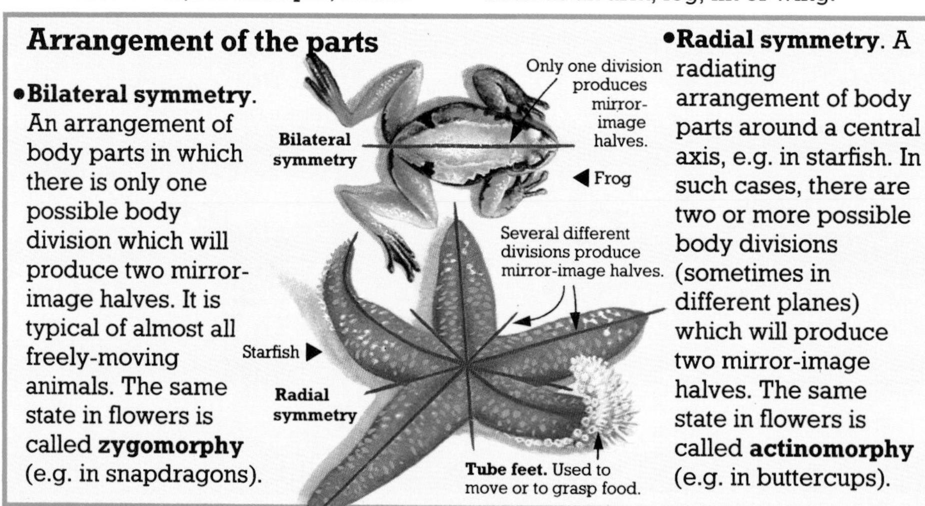

- **Radial symmetry**. A radiating arrangement of body parts around a central axis, e.g. in starfish. In such cases, there are two or more possible body divisions (sometimes in different planes) which will produce two mirror-image halves. The same state in flowers is called **actinomorphy** (e.g. in buttercups).

* **Classification**, 338; **Myriapods**, 341 (Note 5).

Body cavities

Almost all many-celled animals have a main fluid-filled body cavity, or **perivisceral cavity**, to cushion the body organs (very complex animals, e.g. humans, may have other smaller cavities as well). Its exact nature varies, but in most animals it is either a **coelom** or a **haemocoel**. In soft-bodied animals it is vital in movement, providing an incompressible "bag" for their muscles to work against. Such a system is called a **hydrostatic skeleton**.

Simplified cut-away of a peanut worm. Not all body organs shown.

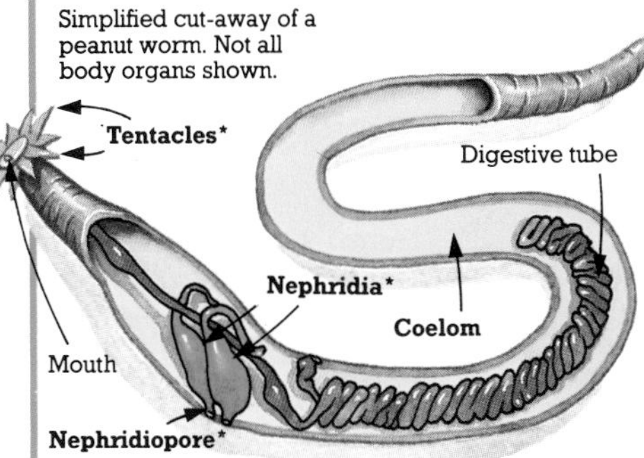

- **Coelom**. The main body cavity (**perivisceral cavity**) of higher worms, **echinoderms***, e.g. starfish, and **vertebrates***, e.g. birds. It is fluid-filled to cushion the organs and is bounded by the **peritoneum**, a thin membrane which lines the body wall. In lower animals, e.g. many worms, the coelom assists in excretion. Their excretory organs, called **nephridia***, project into the coelom and remove fluid waste which has seeped into it. In higher animals, other more complex organs deal with these functions.

Simplified cut-away of a spider. Not all body organs shown.

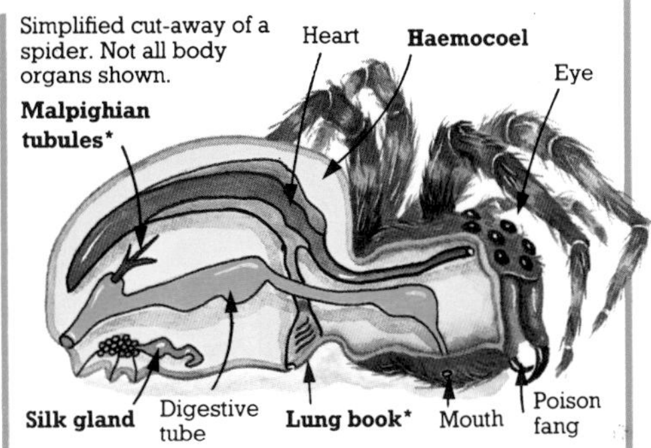

- **Haemocoel**. The fluid-filled main body cavity (**perivisceral cavity**) of **arthropods***, e.g. insects, and **molluscs***, e.g. snails. In molluscs, it is more of a spongy meshwork of tissue than a true cavity. Unlike a **coelom**, a haemocoel contains blood. It is an expanded part of the blood system, through which blood is circulated. In some animals, the haemocoel plays a part in excretion. In insects, for instance, water and fluid waste seep into it, and are then taken up by the **Malpighian tubules*** projecting into it.

- **Mantle cavity**. A body cavity in shelled **molluscs***, e.g. snails. It lies between the **mantle** (a fold of skin lining the shell) and the rest of the body. Digestive and excretory waste is passed into it, for removal from the body. In water-living molluscs, it also holds the **gills***; in land-living snails, it acts as a lung.

Simplified cut-away picture of a whelk. Not all body organs shown.

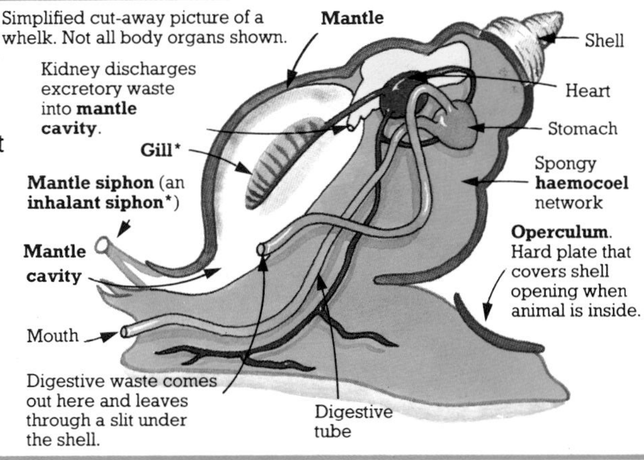

* **Arthropods**, 340; **Echinoderms**, 341; **Gills**, 272; **Inhalant siphon**, 272 (**Siphon**); **Lung books**, 272; **Malpighian tubules**, 273; **Molluscs**, 340; **Nephridiopore**, 273 (**Nephridia**); **Tentacles**, 275; **Vertebrates**, 341 (**Craniata**).

Animal body coverings

All animals have an enclosing outer layer, or "skin", normally with a further covering of some kind. In many cases, the skin is multi-layered, like human skin (see pages 310-311), and in most higher animals its covering is soft, e.g. hair, fur or feathers. Hard coverings, e.g. shells, are found in many lower animals and may form their only supporting framework, if they have no internal skeleton (**endoskeleton**). In such cases, the covering is called an **exoskeleton**. Some of the main body coverings are listed here.

- **Cuticle**. A non-living, waterproof, outer layer in many animals, secreted by the skin. In most soft-bodied animals it hardens to form a supportive outer skeleton, or **exoskeleton**, e.g. the shells of crabs and the tough outer "coat" of insects. The term cuticle is in fact most often used to describe an insect "coat". This consists of a sugar-based substance (**chitin**) and a tough protein (**sclerotin**). It is often made up of **sclerites** – separate pieces joined by flexible, narrow areas. In other animals, e.g. earthworms, the cuticle remains a soft waxy covering.

The term **cuticle** is sometimes used to mean the **stratum corneum*** in humans.

- **Scales**. There are two different types of scales. Those of bony fish (Class **Osteichthyes***), e.g. carp, are small, often bony plates lying within the skin. Those covering the limbs or whole bodies of many **reptiles*** (e.g. the legs of turtles) are thickened areas of skin.

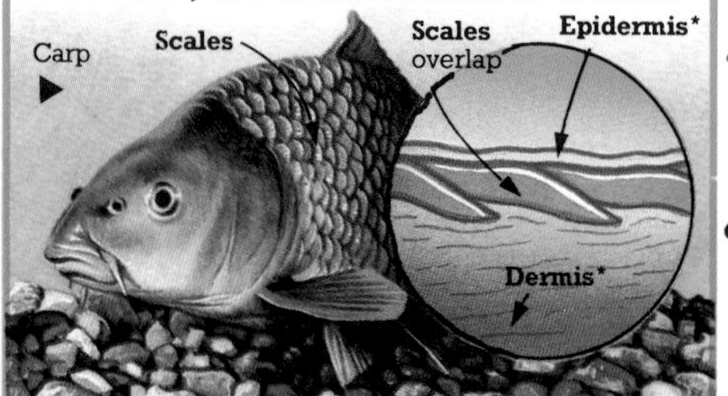

- **Carapace**. The shield-like shell of a crab, tortoise or turtle. In tortoises and turtles, it consists of bony plates fused together under a horny skin, but in crabs, it is a hardened **cuticle**.

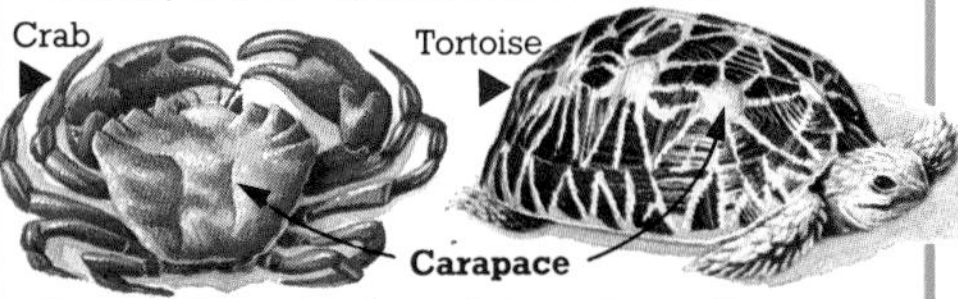

- **Denticles** or **placoid scales**. Sharp backward-pointing plates, covering the bodies of cartilaginous fish (Class **Elasmobranchiomorphi***), e.g. rays. They are similar to teeth, and stick out from the skin, unlike **scales**.

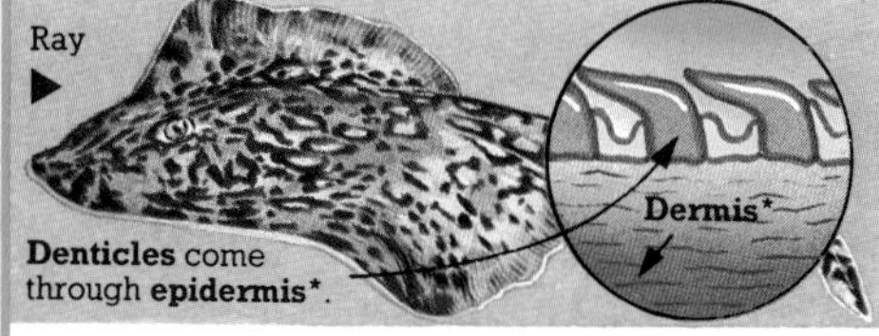

Denticles come through **epidermis***.

- **Elytra** (sing. **elytron**). The front pair of wings of beetles and some bugs. They are modified to form a tough cover for the back pair of wings, used for flying.

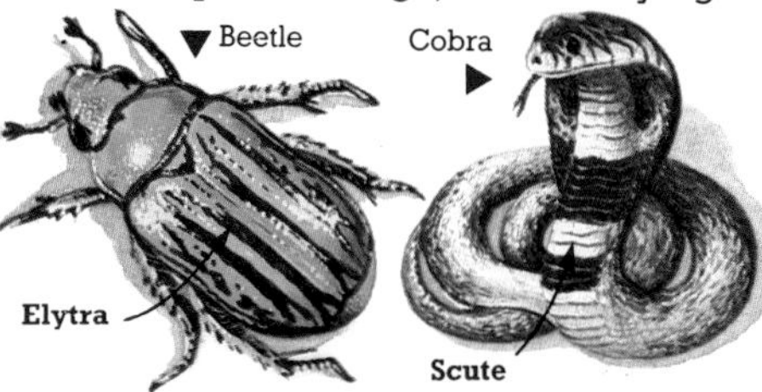

- **Scuta** (sing. **scute** or **scutum**). Any large, hard, external plate, especially those on the underside of a snake, used in movement.

* **Dermis**, 310; **Elasmobranchiomorphi**, 341; **Epidermis**, 310; **Osteichthyes**, **Reptiles**, 341; **Stratum corneum**, 310.

Feathers

The insulating waterproof layer of a bird's body is made up of **feathers**, together known as its **plumage**. Each feather is a light structure made of a fibrous, horny substance called **keratin**. Each has a central **shaft** (or **rachis**) with thin filaments called **barbs**. The barbs of all **contour feathers**, i.e. all the feathers except the **down feathers**, have tiny filaments called **barbules**. Like body hairs, feathers have nerve endings attached to them, as well as muscles which can fluff them up to conserve heat (see **hair erector muscles**, page 310).

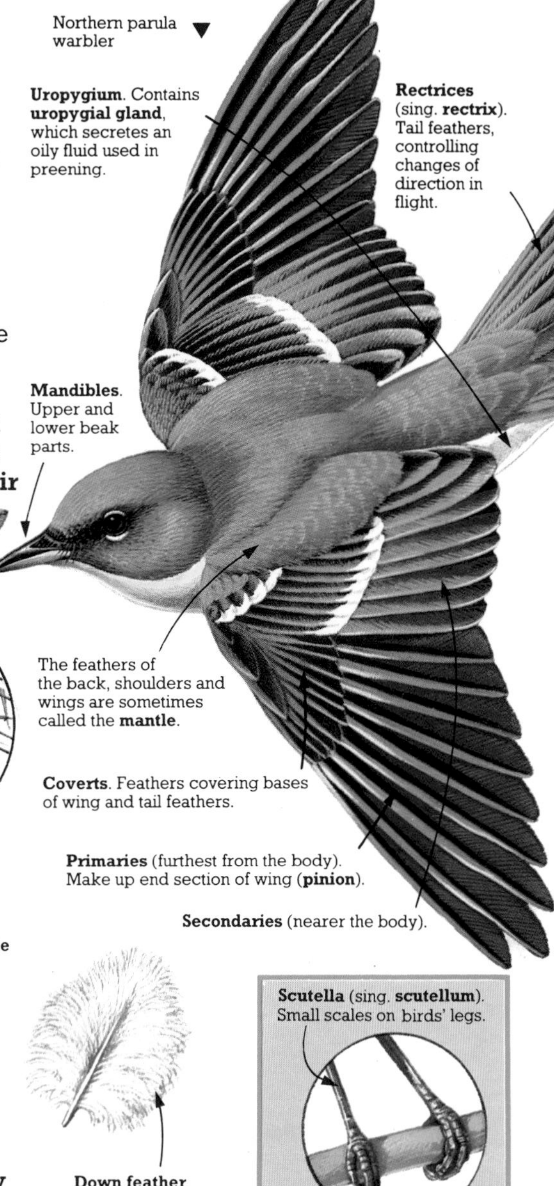

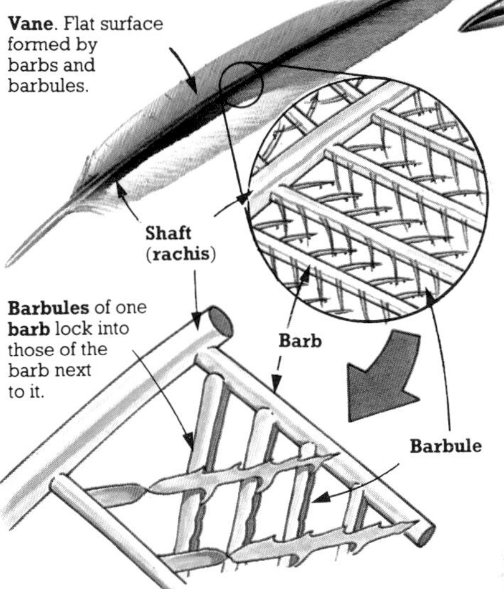

Down feather

Scutella (sing. **scutellum**). Small scales on birds' legs.

- **Remiges** (sing. **remix**) or **flight feathers**. Those feathers of a bird's wings which are used in flight, consisting of the long, strong **primary feathers**, or **primaries**, and the shorter **secondary feathers**, or **secondaries**.
- **Down feathers** or **plumules**. The fluffy, temporary feathers of all young birds, which have flexible **barbs**, but no true **barbules**. The adults of some types of bird keep some down feathers as an insulating layer close to the skin.
- **Feather follicles**. Tiny pits in a bird's skin. Each one has a feather in it, just like a hair in a **hair follicle***. The cells at the base of the follicle grow up and out to form a feather, and then die away, becoming hard and tough.

* **Hair follicle**, 310.

Animal movement

Most animals are capable of movement from place to place (**locomotion**) at least at some stage of their life (plants can only move individual parts – see **tropism**, page 251). The moving parts of animals vary greatly. Many animals have a system of bones and muscles similar to humans (see pages 278-283). Listed here are some of the parts used to move animals.

Movement of simple animals

- **Cilia** (sing. **cilium**). Tiny "hairs" on the outer body surfaces of many small organisms. They flick back and forth to produce movement. Cilia are also found lining the internal passages of more complex animals, e.g. human air passages (they trap foreign particles).

- **Flagella** (sing. **flagellum**). Any long, fine body threads, especially the one or more which project from the surface of many single-celled organisms. These lash backwards and forwards to produce movement. Organisms with flagella are **flagellate**.

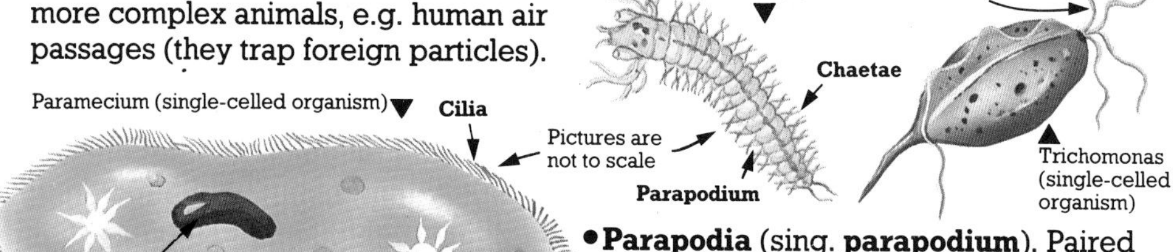

- **Parapodia** (sing. **parapodium**). Paired projections from the sides of many aquatic worms, used to move them along. Each one ends in a bunch of bristles, or **chaetae** (sing. **chaeta**), which may also cover the body in some cases.

- **Pseudopodium** (pl. **pseudopodia**). An extension ("false foot") of the cell matter, or **cytoplasm***, of a single-celled organism. Such extensions are formed either in order for the organism to move or to enable it to engulf a food particle. The latter process is called **phagocytosis**.

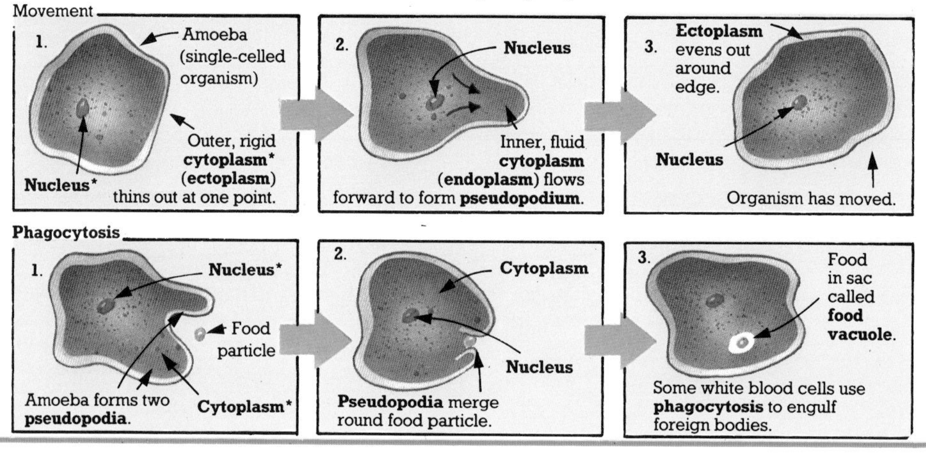

* **Contractile vacuole**, 273; **Cytoplasm**, **Nucleus**, 238.

Swimmers

- **Fins**. Special projections from the body of a fish, which are used as stabilisers and to change direction. They are supported by **rays** – rods of bone or **cartilage*** (depending on the Class of fish – see page 341) radiating out inside them. Fish have two sets of fins, called **median** and **paired fins**.

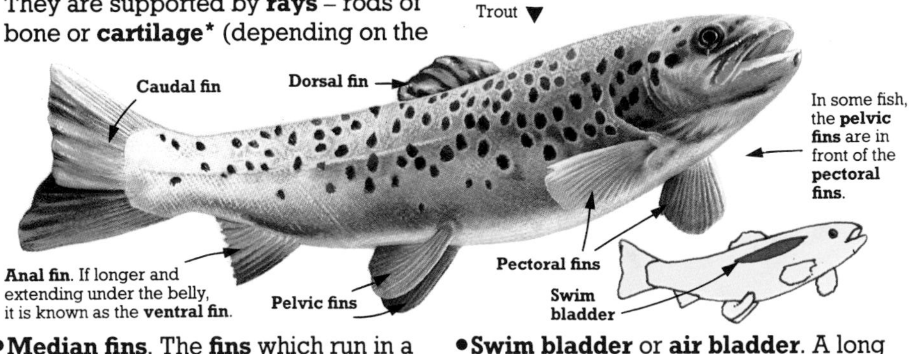

- **Median fins**. The **fins** which run in a line down the centre of the back and the belly. In some fish, e.g. eels, they form one long continuous median fin, but in most they are divided into the **dorsal**, **caudal** (tail) and **anal** (or **ventral**) **fins**. The dorsal fin and the anal fin control changes of direction from side to side. The caudal fin helps to propel the fish through the water.
- **Paired fins**. The **fins** of a fish which stick out from its sides in two pairs: the **pectoral fins** and the **pelvic fins**. They control movement up or down.
- **Swim bladder** or **air bladder**. A long air-filled pouch inside most bony fish (Class **Osteichthyes***). The fish alters the amount of air inside the bladder depending on the depth at which it is swimming. This keeps the density of the fish the same as that of the water, so it will not sink if it stops swimming.

Median or **medial** means "lying on the dividing line between the right and left sides"

Dorsal means "of the back or top surface"

Caudal means "of the tail or hind part"; **caudate** means "having a tail"

Ventral means "of the front or lower surface"

Flyers

- **Pectoralis muscles**. Two large, paired chest muscles, found in many **mammals***, but especially highly developed in birds. Each wing has one **pectoralis major** and one **pectoralis minor**, attached at one end to the **keel**, a large extension of the breastbone. The muscles contract alternately to move the wings.

Keel

Pectoralis minor (pulls wing up)

Bastard wing or **alula**. Short **digit*** with a few feathers. Helps to deal with air turbulence.

Pectoralis major (pulls wing down)

Breastbone, or **sternum**

Coracoid bones

Walkers

- **Unguligrade**. Walking on hooves at the tips of the toes, e.g. horses.

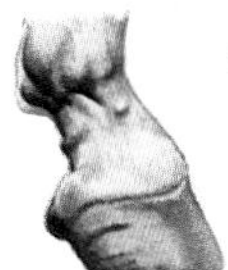

- **Digitigrade**. Walking on the underside of the toes, e.g. dogs and cats.

- **Plantigrade**. Walking on the underside of the whole foot, e.g. man.

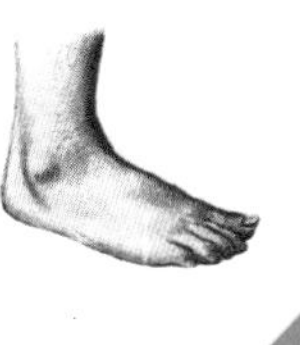

* **Cartilage**, 281; **Digit**, 279 (**Phalanges**); **Mammals**, **Osteichthyes**, 341.

Animal feeding

Different animals take in their food in many different ways, and with many different body parts. Some also have special internal mechanisms for dealing with the food (others have human-like **digestive systems** – see pages 294-295). Listed here are some of the main animal body parts involved in feeding and digestion.

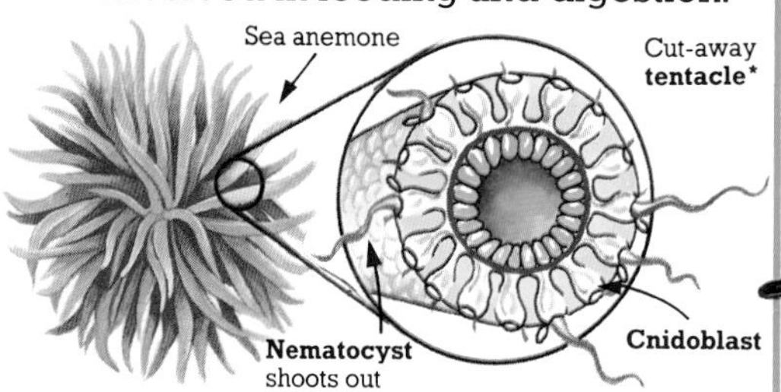

- **Cnidoblasts** or **thread cells**. Special cells found in large numbers on the **tentacles*** of **coelenterates***, e.g. sea anemones, used for seizing food. Each one contains a **nematocyst** – a long thread coiled inside a tiny sac. When a tentacle touches something, the threads shoot out to stick to it or sting it.

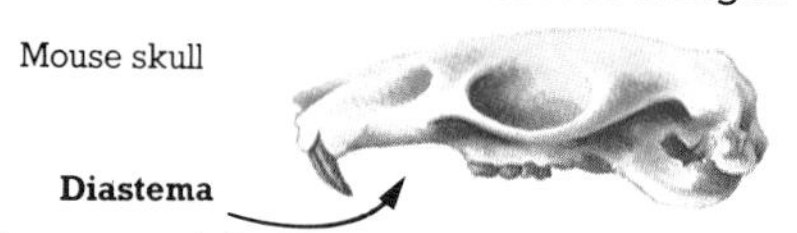

- **Diastema** (pl. **diastemata**). A gap between the front and back teeth of many plant-eaters. It is especially important in rodents, e.g. mice. They can draw their cheeks in through the gaps, so they do not swallow substances they may be gnawing.
- **Carnassial teeth**. The specially adapted second upper **premolar*** and first lower **molar*** of hunters, used for shearing flesh and cracking bones.
- **Radula**. The horny "tongue" of many **molluscs***, e.g. snails. It is covered by tiny teeth, which rasp off food.

Arthropod mouth parts

The mouths of **arthropods***, e.g. insects, are made up of a number of different parts. Depending on the animal's feeding method, these may look very different. The basic mouthparts, found in all insects, are the **mandibles**, **maxillae** (sing. **maxilla**), **labrum** and **labium**. The first two are also found in many other arthropods, e.g. crabs and centipedes (some of these other arthropods have two pairs of maxillae).

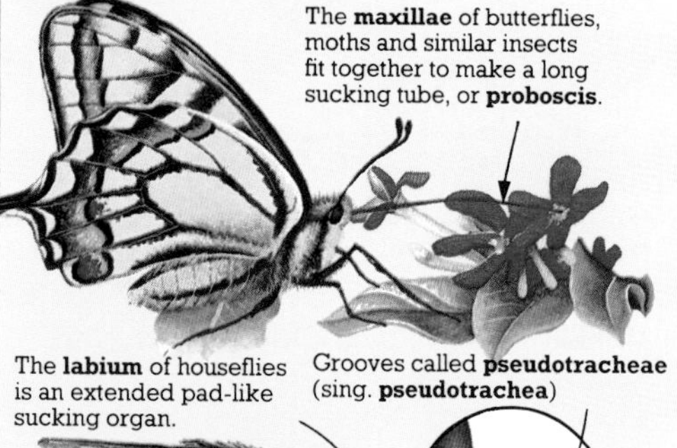

- **Filter-feeding**. The "sieving" of food from water, shown by many aquatic animals. Barnacles, for instance, sieve out microscopic organisms, or **plankton***, with bristly limbs called **cirri** (sing. **cirrus**). Some whales use frayed plates of horny **whalebone**, or **baleen**, hanging down from the top jaw. They sieve out small shrimp-like animals called **krill**.

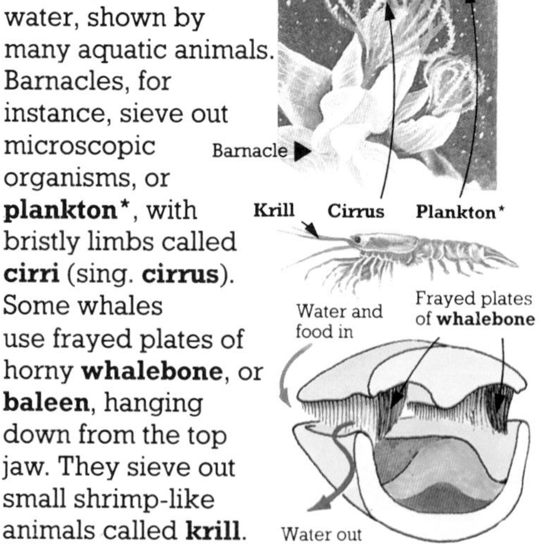

* **Arthropods**, **Coelenterates**, 340; **Molars**, 285; **Molluscs**, 340; **Plankton**, 342; **Premolars**, 285; **Tentacles**, 275.

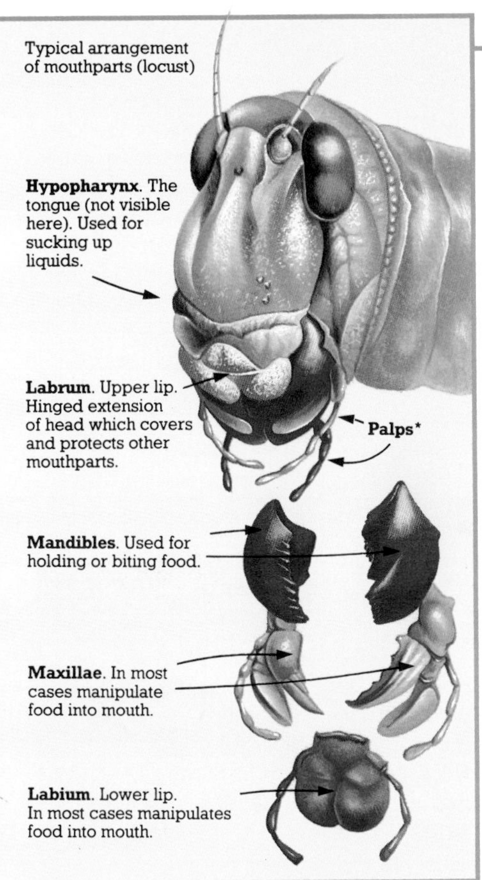

Digestive structures

• **Crop**. A thin-walled pouch, part of the gullet (**oesophagus***) in birds; also a similar structure in some worms, e.g. earthworms, and some insects, e.g. grasshoppers. Food is stored in the crop before it goes into the **gizzard**.

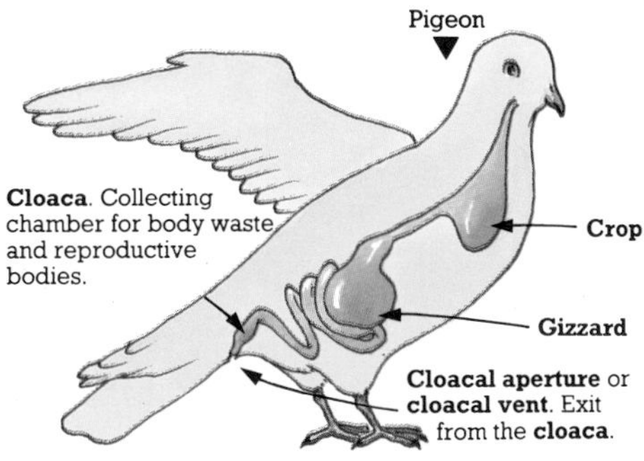

• **Gizzard**. A thick muscular-walled pouch at the base of the gullet (**oesophagus***) in those animals which have **crops**. These animals have no teeth, instead food is ground up in the gizzard. Birds swallow pieces of gravel to act as grindstones; in other animals, the muscular walls of the gizzard do the job, or hard tooth-like structures attached to these walls.

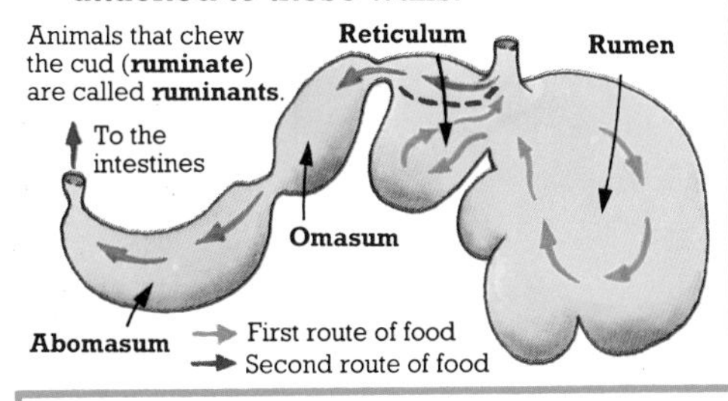

• **Rumen**. The large first chamber of the complex "stomach" of some plant-eating **mammals***, e.g. cows, into which food passes unchewed. It contains bacteria which can break down **cellulose***. Other animals pass this substance as waste, but these animals cannot afford to do this, as it makes up the bulk of their food (grass). The partially-digested food, which has also been processed in the second chamber, or **reticulum**, is then regurgitated to be chewed, and is known as the **cud**. When it is swallowed again it bypasses the first two chambers and is processed further in the third and fourth chambers – the **omasum** and the **abomasum** (the true stomach).

• **Caecum**. Any blind-ended sac inside the body, especially one forming part of a digestive system. In many animals, e.g. rabbits, it is the site of an important stage of digestion (involving bacterial breakdown of **cellulose*** – see **rumen**). In others, e.g. humans (see **large intestine***), it is redundant.

* **Cellulose**, 238 (**Cell wall**); **Large intestine**, 295; **Mammals**, 341; **Oesophagus**, 294; **Palps**, 274.

Animal respiration

The complex process of **respiration** consists of a number of stages (see introduction, page 298). Basically, oxygen is taken in and used by body cells in the breakdown of food, and carbon dioxide is expelled from the cells and the body. Below are some of the main animal respiratory organs.

- **Spiracle**. Any body opening through which oxygen and carbon dioxide are exchanged (e.g. a whale's blowhole). The term is used especially for any of the tiny holes (also called **stigmata**, sing. **stigma**) found in many **arthropods***, e.g. insects.

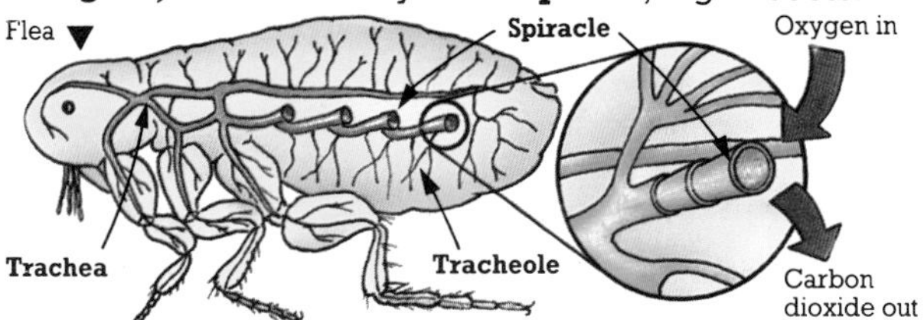

- **Tracheae** (sing. **trachea**). Thin tubes leading in from the **spiracles** of **arthropods***, e.g. all insects and the most advanced spiders. They form an inner network, often branching into narrower tubes called **tracheoles**. Oxygen from the air passes through the tube walls to the body cells. Carbon dioxide leaves via the same route.

- **Lung books** or **book lungs**. Paired breathing organs found in scorpions (which have four pairs) and some (less advanced) spiders (which have one or two). Each one has many blood-filled tissue plates, arranged like book pages. Oxygen comes in through slits (**spiracles**), one by each lung book, and is absorbed into the blood. Carbon dioxide passes out the same way.

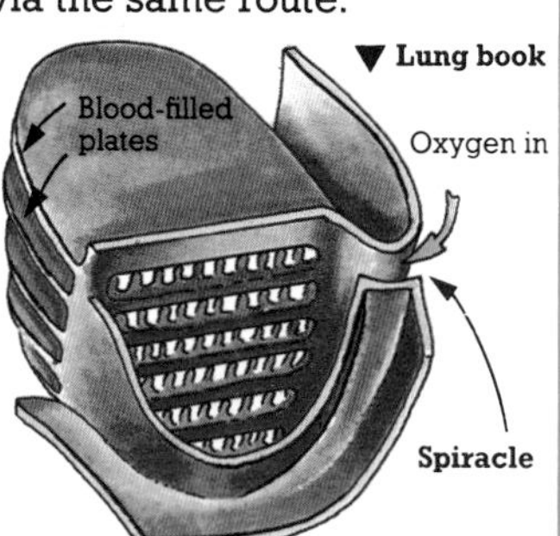

Gills

Gills or **branchiae** (sing. **branchia**), are the breathing organs of most aquatic animals, containing many blood vessels. Oxygen is absorbed into the blood from the water passing over the gills. Carbon dioxide passes out the other way. There are two types of gills – **internal** and **external**.

Breathing with **gills**

1. Water comes in through mouth.

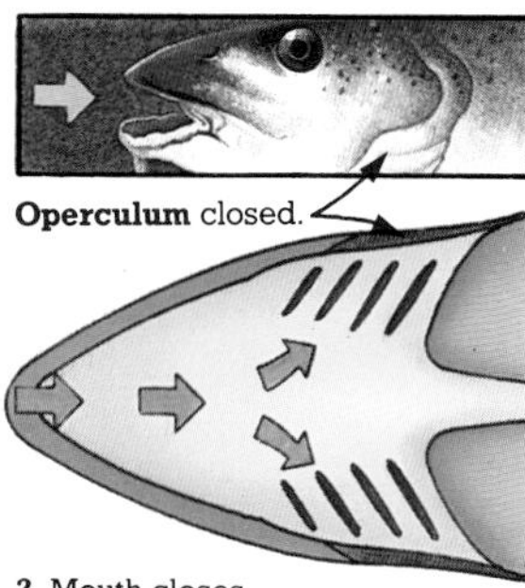

2. Mouth closes, **operculum** opens.

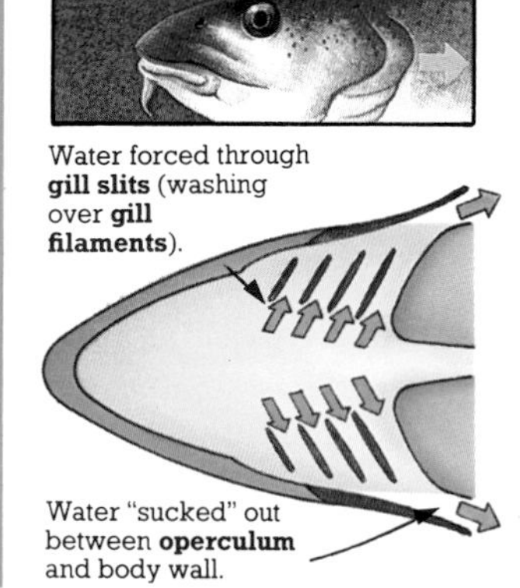

- **Siphon**. A tube carrying water to (**inhalant siphon**) or from (**exhalant siphon**) the **gills** of many lower aquatic animals, e.g. whelks (see picture, page 265). The exhalant siphon of **cephalopods*** (e.g. octopuses) is called the **hyponome***.

* **Arthropods**, **Cephalopods**, 340; **Hyponome**, 275.

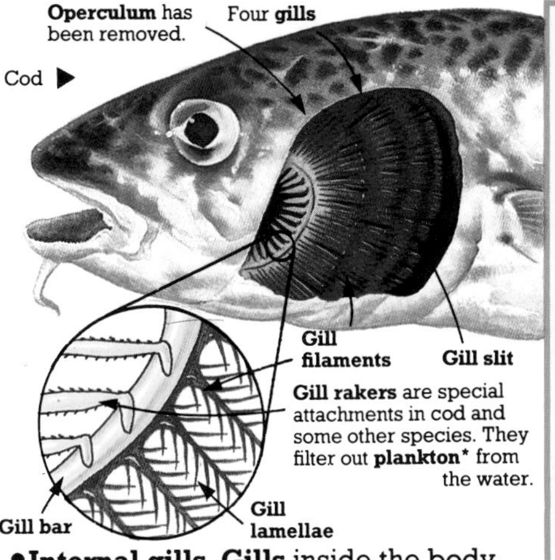

- **Internal gills**. **Gills** inside the body, found in various forms in all fish, most **molluscs***, e.g. limpets, and most **crustaceans***, e.g. crabs. Most fish have four pairs of gills, with channels between them called **gill slits**. In more advanced fish, e.g. cod, they are covered by a flap called the **operculum**. In more primitive fish, e.g. sharks, they end in narrow openings in the skin on the side of the head. Each gill consists of a curved rod, the **gill bar** or **gill arch**, with many fine **gill filaments**, and even finer **gill lamellae** (sing. **lamella**) radiating from it. These all contain blood vessels.
- **External gills**. **Gills** on the outside of the body, found in the young stages of most fish and **amphibians***, some older amphibians and the young aquatic stages of many insects (e.g. caddisfly **larvae*** and mayfly **nymphs***). Their exact form depends on the type of animal, but in many cases they are "frilly" outgrowths from the head, e.g. in young tadpoles or axolotls (types of salamander).

Axolotl

Gill filaments

External gills are soft and "frilly"

Branchial means "of the **gills**"

Animal excretion

Excretion – the expulsion of waste fluid – is vital to life. It gets rid of harmful substances and is also vital to the maintenance of a balanced level of body fluids (see **homeostasis**, page 333).

- **Contractile vacuoles**. Tiny sacs used for water-regulation in single-celled freshwater organisms. Excess water enters a vacuole via several canals arranged around it. When fully expanded, it then contracts and bursts, shooting the water out through the outer membrane.

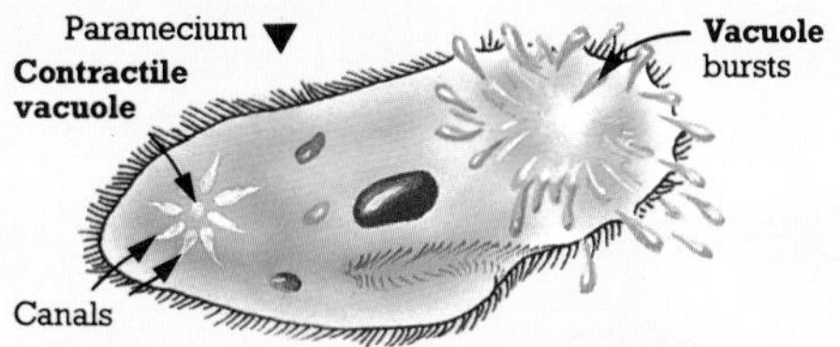

- **Nephridia** (sing. **nephridium**). Waste-collecting tubes in many worms and the **larvae*** of many **molluscs***, e.g. slugs. In higher worms, they collect from the **coelom*** (see picture, page 265). Lower worms and mollusc larvae have more primitive **protonephridia**. The waste fluid enters these via hollow **flame cells (solenocytes)**, which contain hair-like **cilia***. In both a nephridium and a protonephridium, the waste leaves through a tiny hole, or **nephridiopore**.

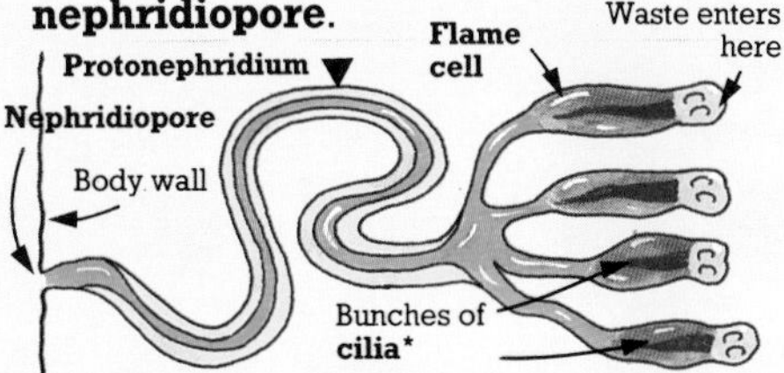

- **Malpighian tubules**. Long tubes found in many **arthropods***, e.g. insects. They carry dissolved waste from the main body cavity (**haemocoel***) into the rear of the gut. See picture, page 265.

* **Amphibians**, 341; **Arthropods**, 340; **Molluscs**, 340; **Nymph**, 277. **Cilia**, 268; **Coelom**, 265; **Crustaceans**, 340; **Haemocoel**, 265; **Larva**, 277; **Plankton**, 342;

Animal senses and communication

All animals show some **sensitivity** (**irritability**), i.e. response to external stimuli such as light and sound vibrations. Humans have a high overall level of sensory development, but individual senses in other animals may be even better developed, e.g. the acute vision of hawks. Listed here are some of the main animal sense organs (and their parts). Their responding parts send "messages" (nervous impulses) to the brain (or more primitive nerve centre), which initiates the response.

Hearing and balance

- **Lateral lines**. Two water-filled tubes lying along each side of the body, just under the skin. They are found in all fish and those **amphibians*** which spend most of their time in water, e.g. some toads. They enable the animals to detect water currents and pressure changes, and they use this information to find their way about.

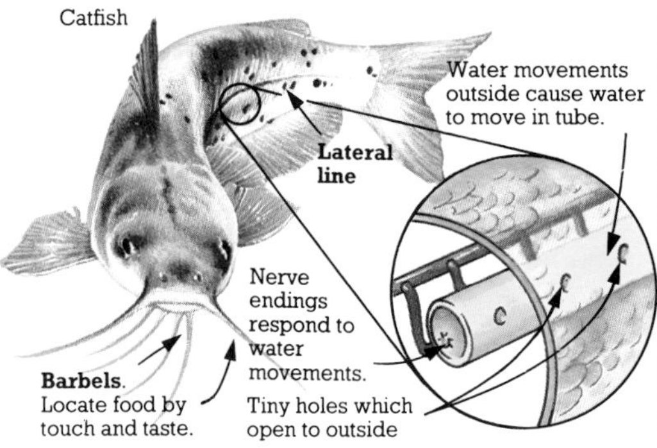

- **Tympanal organs** or **tympani** (sing. **tympanum**). Sound detectors found on the lower body or legs in some insects, e.g. crickets, and on the head in some **amphibians***, e.g. frogs. Each is an air sac covered by a thin layer of tissue. Sensitive fibres in the organs respond to high frequency sound vibrations.

- **Statocysts**. Tiny organs of balance, found in many aquatic **invertebrates***, e.g. jellyfish. Each is a sac with tiny particles called **statoliths** inside, e.g. sand grains. When the animal moves, the grains move, stimulating sensitive cells which set off responses.

Touch, smell and taste

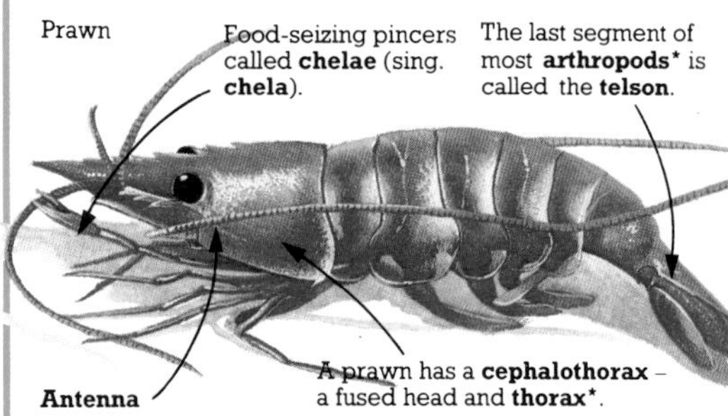

- **Antennae** (sing. **antenna**). Whip-like jointed sense organs on the heads of insects, **myriapods*** (centipedes and millipedes) and **crustaceans***, e.g. prawns. Insects and myriapods have one pair, crustaceans have two. They respond to touch, temperature changes and chemicals (giving "smell" or "taste"). Some crustaceans also use them for swimming or to attach themselves to objects or other animals.

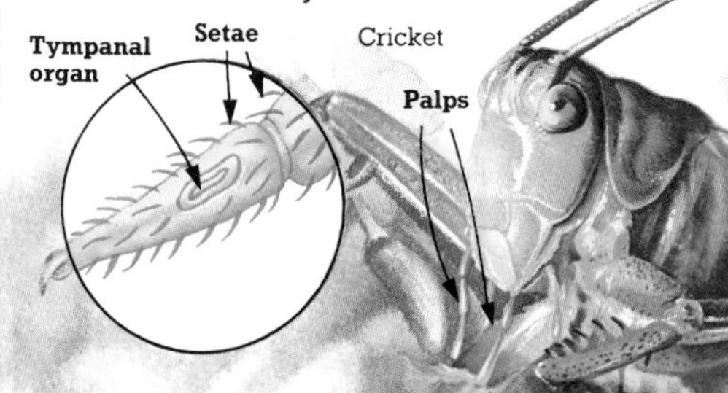

- **Palps**. Projections of the mouthparts of **arthropods***, e.g. insects. They respond to chemicals (giving "smell" or "taste"). The term is also given to various touch-sensitive organs.

* **Amphibians**, 341; **Arthropods**, **Crustaceans**, 340; **Invertebrates**, 341 (Note 8); **Myriapods**, 341 (Note 5); **Thorax**, 264 (Segmentation).

Communication

- **Pheromone**. Any chemical made by an animal that causes responses in other members of the species, e.g. sexual attractants produced by many insects.
- **Syrinx** (pl. **syringes**). The vocal organ of birds, similar to the **larynx***, but found at the base of the windpipe.

- **Vibrissae** (sing. **vibrissa**) or **whiskers**. Stiff hairs standing out from the faces of many **mammals***, e.g. cats, round the nose. They are sensitive to touch.

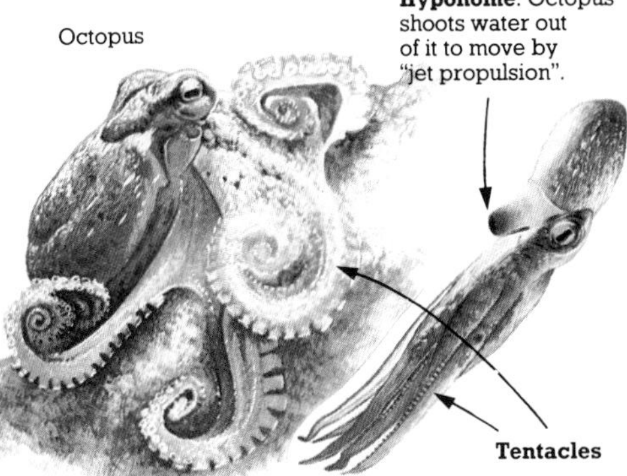

- **Tentacles**. Long, flexible body parts, found in many **molluscs***, e.g. octopuses, and **coelenterates***, e.g. jellyfish. In most cases they are used for grasping food or feeling, though the shorter of the two pairs found in land snails and slugs have eyes on the end.

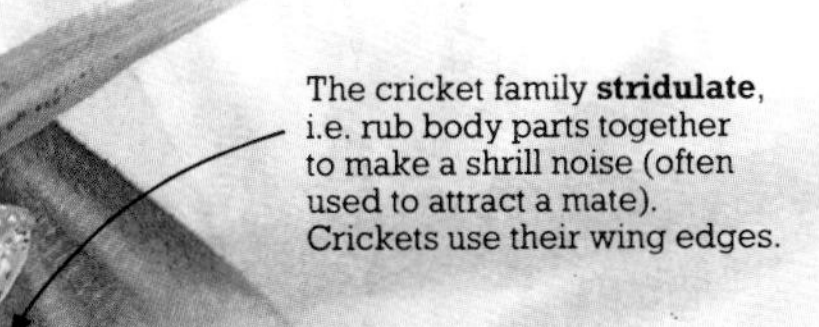

- **Setae** (sing. **seta**). Bristles produced by the skin of many **invertebrates***, e.g. insects. Nerves at their bases respond to movements of air or vibrations.

Sight

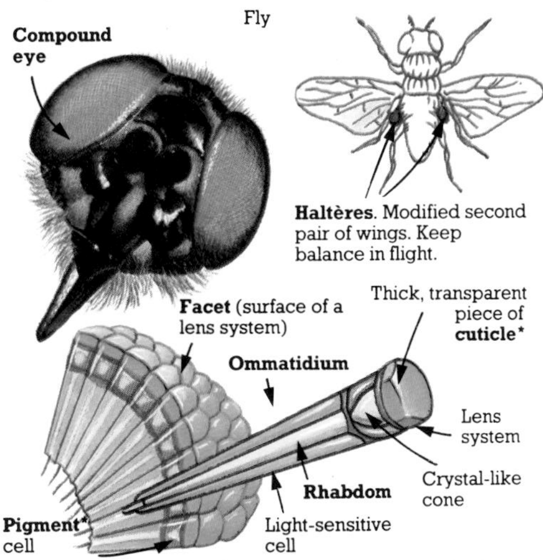

- **Compound eyes**. The special eyes of many insects and some other **arthropods***, e.g. crabs. Each consists of hundreds of separate visual units called **ommatidia** (sing. **ommatidium**). Each of these has an outer lens system which "bends", or refracts, light onto a **rhabdom**, a transparent rod

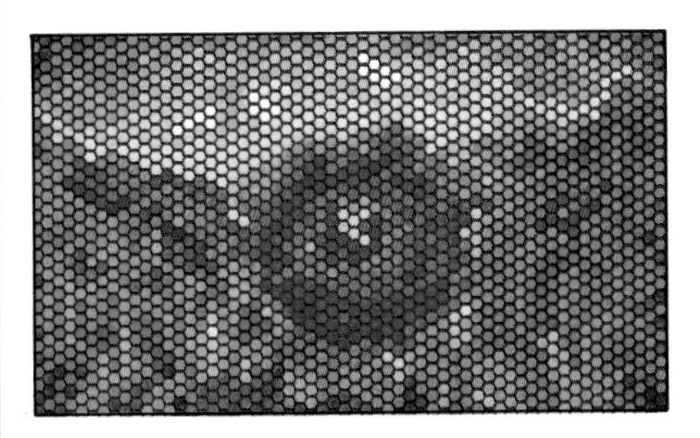

Compound eye view of a flower (**mosaic image**)

surrounded by light-responsive cells.

After receiving information from all the ommatidia (each has a slightly different angle of vision and may record different light intensity or colour), the brain assembles a complete **mosaic image**. This is enough for the animal's needs, but not as well defined as the image produced by the human eye.

* **Arthropods**, **Coelenterates**, 340; **Cuticle**, 266; **Invertebrates**, 341 (Note 8); **Larynx**, 298; **Mammals**, 341; **Molluscs**, 340; **Pigments**, 255.

Animal reproduction

Reproduction is the creation of new life. Most animals reproduce by **sexual reproduction***, the joining of a female sex cell, called an **ovum**, with a male sex cell, or **sperm**. Below are the main terms associated with the reproductive processes of animals.

Harvest mouse

- **Viviparous**. A term describing animals such as humans, in which both the joining of the male and female sex cells (**fertilization**) and the development of the **embryo*** occur inside the female's body (the fertilization is **internal fertilization**), and the baby is born live.

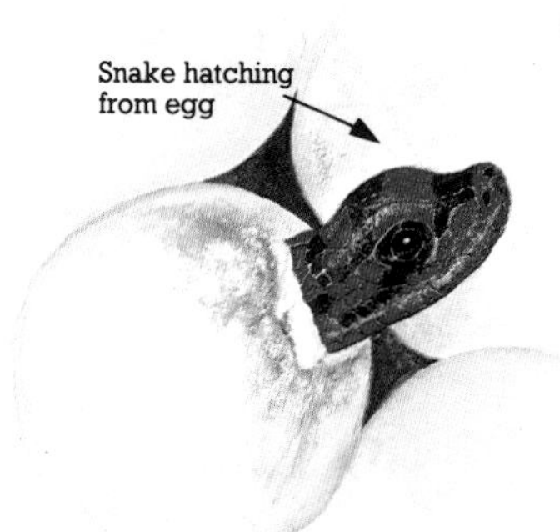

- **Oviparous**. A term describing animals in which the development of the **embryo*** occurs inside an **egg** which has been laid by the mother. In some cases, e.g. in birds, the male and female sex cells join inside the female's body (**internal fertilization**) and the egg already contains the embryo when laid. In other cases, e.g. in many fish, the many eggs each just contain an **ovum** (female sex cell) when laid, and the male then deposits **sperm** (male sex cells) over them (**external fertilization**).

- **Eggs**. There are two main types of egg: **Cleidoic eggs** are produced by most egg-laying animals which live on land, e.g. birds and most **reptiles***, and also by a few aquatic animals, e.g. sharks. Such an egg largely isolates the **embryo*** from its surroundings, allowing only gases to pass through its tough shell (waste matter is stored). It contains enough food (**yolk**) for the complete development of the embryo, and the animal emerges as a tiny version of the adult. The other type of egg, produced by most aquatic animals, e.g. most fish, has a soft outer membrane, through which water and waste matter (as well as gases) can pass. The emerging young are not fully developed.

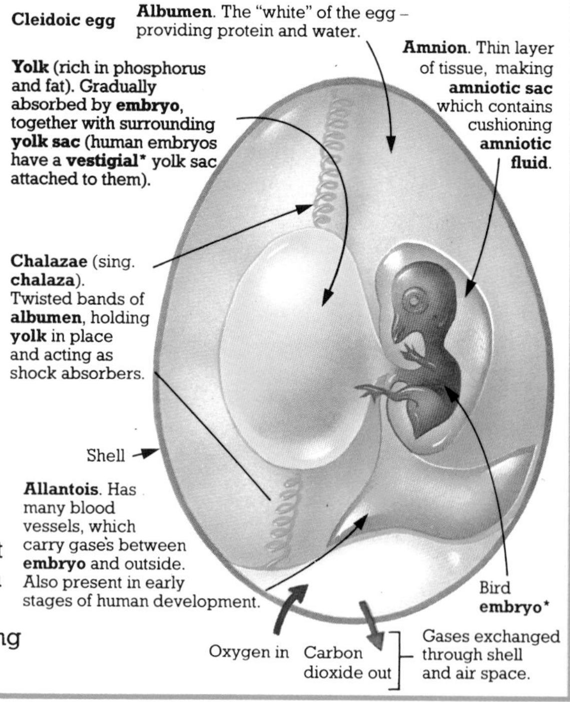

* **Embryo**, 321; **Reptiles**, 341; **Sexual reproduction**, 320; **Vestigial**, 295 (**Appendix**).

• **Oviduct**. Any tube in females through which either **eggs** or **ova** (female sex cells) are discharged. In humans, the **Fallopian tubes***, **uterus*** and **vagina*** form the oviduct.

• **Ovipositor**. An organ extending from the back end of many female insects, through which **eggs** are laid. In many cases, it is long and sharp, and is used to pierce plant or animal tissues before laying.

• **Spermatheca**. A sac for storing **sperm** (male sex cells) in the female of many **invertebrates***, e.g. insects, and some lower **vertebrates***, e.g. newts. The female receives the sperm and stores them until her **ova** (sex cells) are ready to join with them (**fertilization**). Some **hermaphrodite** animals (animals with both male and female organs), e.g. earthworms, have spermathecae. They "swap" sperm when they mate.

Complete metamorphosis (two different forms between **egg** and adult). The many insects which undergo it, e.g. butterflies, are called **endopteryogotes**.

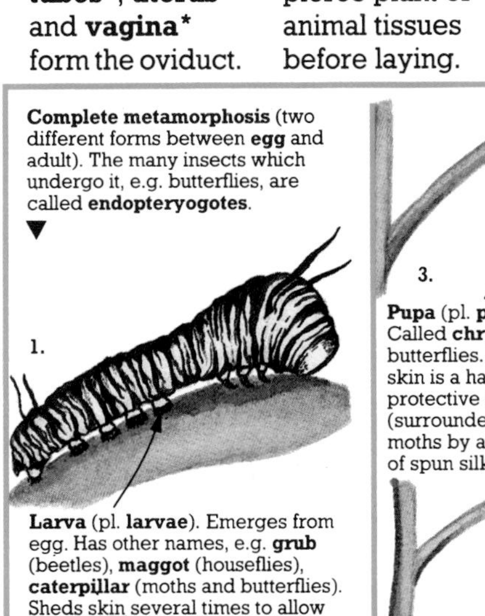

Larva (pl. **larvae**). Emerges from egg. Has other names, e.g. **grub** (beetles), **maggot** (houseflies), **caterpillar** (moths and butterflies). Sheds skin several times to allow for growth (process called **ecdysis**, common to all **arthropods***).

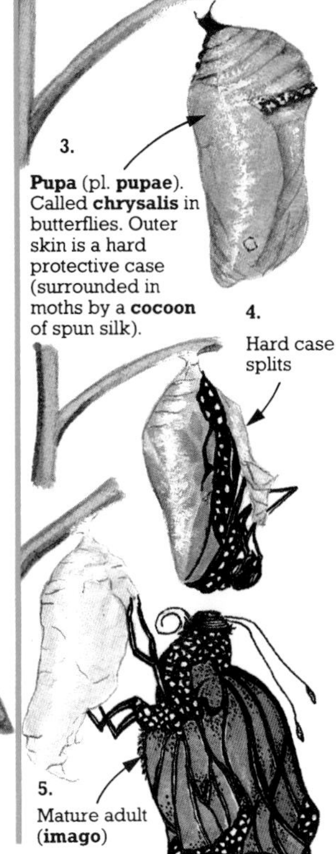

Pupa (pl. **pupae**). Called **chrysalis** in butterflies. Outer skin is a hard protective case (surrounded in moths by a **cocoon** of spun silk).

Sticky pad of silk

2.

Final **ecdysis** (see **larva**) results in **pupa**

Mature adult (**imago**)

Incomplete metamorphosis (gradual development in stages). The insects which undergo it, e.g. locusts, are called **exopterygotes**.

Nymph

Nymph. Emerges from egg. "Mini" version of adult insect, but resemblance only superficial, e.g. wings either in very early stages of development or non-existent, many of inner organs missing. Nymph undergoes several **ecdyses** (see **larva**), with some adult parts emerging each time.

Mature adult (**imago**)

• **Metamorphosis**. The growth and development of some animals involves intermediate forms which are very different from the adult form. Metamorphosis is a series of such changes, producing a complete or partial transformation from the young form to the adult. All insects, most marine **invertebrates***, e.g. lobsters, and most **amphibians***, e.g. frogs, undergo some degree of metamorphosis (intermediate larval forms are common, e.g. legless **tadpoles** in frogs and toads). Above are examples of insect metamorphosis (two different kinds – **complete** and **incomplete metamorphosis**).

* **Amphibians**, 341; **Arthropods**, 340; **Fallopian tubes**, 317; **Invertebrates**, 341 (Note 8); **Uterus**, **Vagina**, 317; **Vertebrates**, 341 (**Craniata**).

The skeleton

The human **skeleton** is a frame of over 200 bones which supports and protects the body organs (the **viscera**) and provides a solid base for the muscles to work against.

- **Cranium** or **skull**. A case protecting the brain and facial organs. It is made of **cranial** and **facial bones**, fused at lines called **sutures**. The upper jaw, for instance, consists of two fused bones called **maxillae** (sing. **maxilla**).

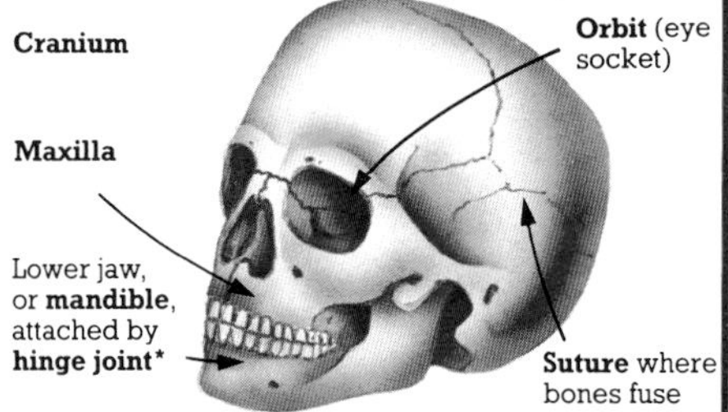

- **Rib cage**. A cage of bones forming the walls of the **thorax** or chest area. It is made up of 12 pairs of **ribs**, the **thoracic vertebrae** and the **sternum**. The ribs are joined to the sternum by bands of **cartilage*** called **costal cartilage**, but only the first seven pairs join it directly. The last five pairs are **false ribs**. The top three of these join the sternum indirectly – their costal cartilage joins that of the seventh pair. The bottom two pairs are **floating ribs**, only attached to the thoracic vertebrae at the back.

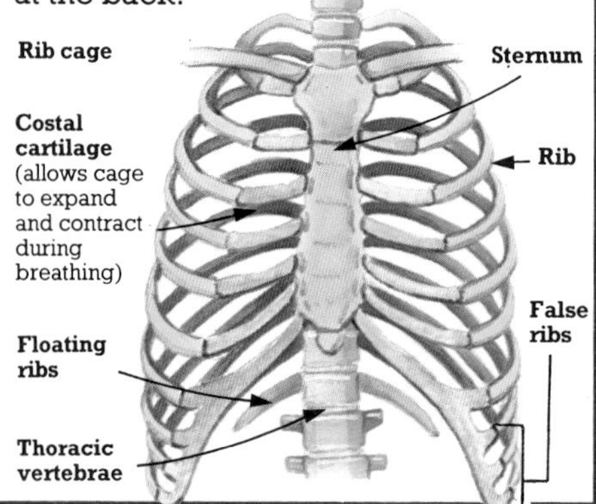

- **Vertebral column**. Also called the **spinal column**, **spine** or **backbone**. A flexible chain of 33 **vertebrae** which protects the **spinal cord***, supports the head and provides points of attachment for the **pelvis** and **rib cage**.

- **Vertebrae** (sing. **vertebra**). The 33 bones of the **vertebral column**. A typical vertebra has a thick "chunk" (the **centrum** or **body**), various projections, or **processes** (named below) and a central hole – the **vertebral foramen** (pl. **foramina**). The foramina together form the **neural**, **spinal** or **vertebral canal**, through which the **spinal cord*** runs.

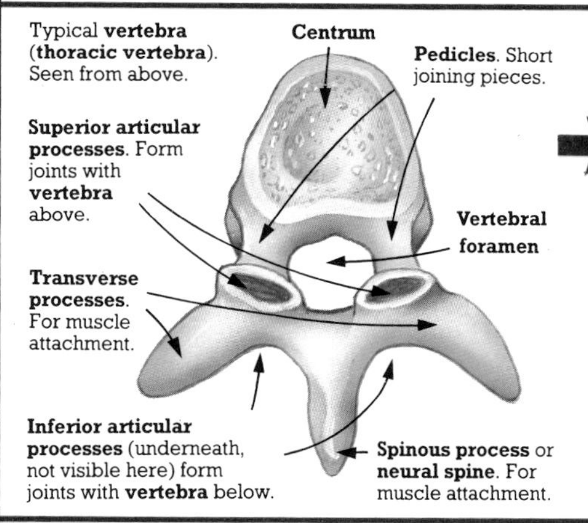

The different vertebrae are named around the skeleton on the next page. The top 24 are movable and linked by **invertebral discs** of **cartilage***. The bottom nine are fused together. They all have the typical structure described above, except for the top two, the **atlas** and **axis**. The atlas (top vertebra) has a special joint with the **skull** which allows the head to nod. The axis has a "peg" (the **dens** or **odontoid process**) which fits into the atlas. This forms a **pivot joint**, a type of joint which allows head rotation.

* **Cartilage**, 281; **Hinge joint**, 280; **Spinal cord**, 302.

The bones of the skeleton

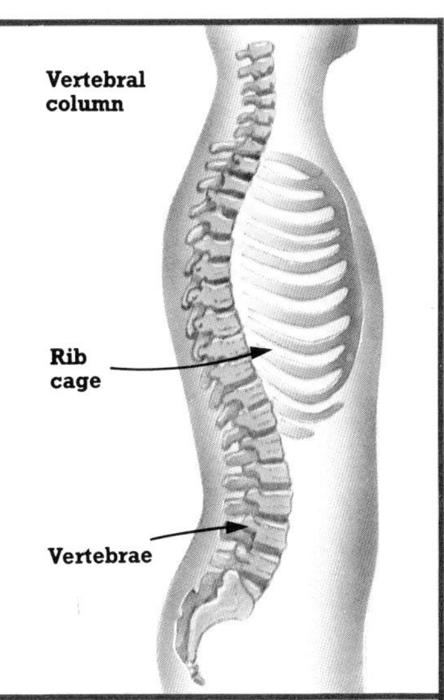

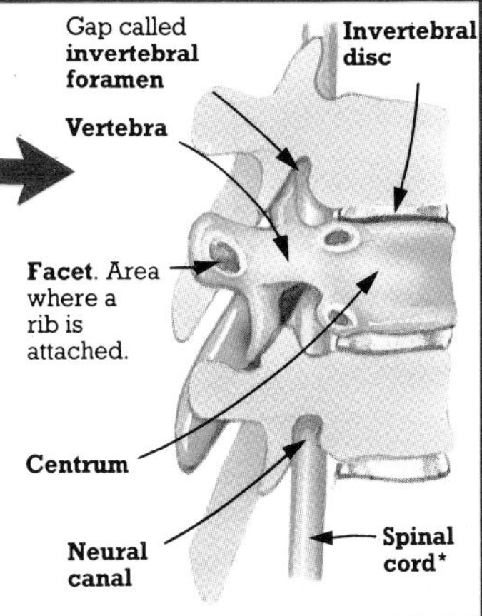

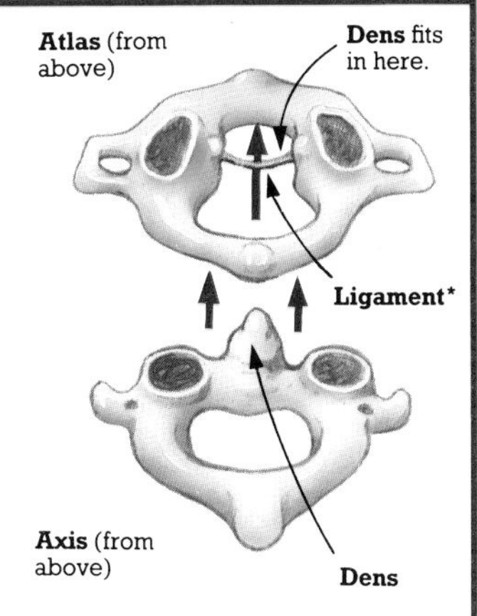

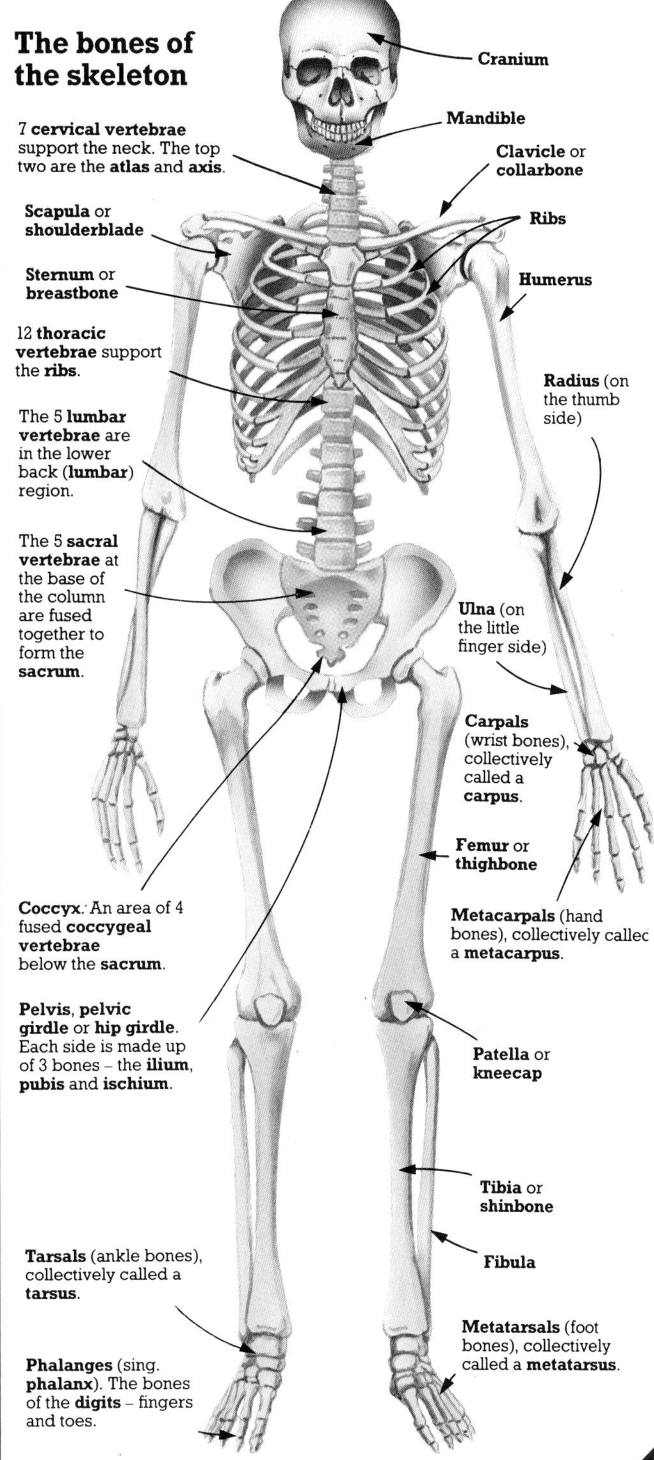

* **Ligament**, 280; **Spinal cord**, 302.

Joints and bone

The bones of the skeleton meet at many **joints**, or **articulations**. Some are **fixed joints**, allowing no movement, e.g. the **sutures*** of the skull. Most, however, are movable, and they give the body great flexibility. The most common are listed below.

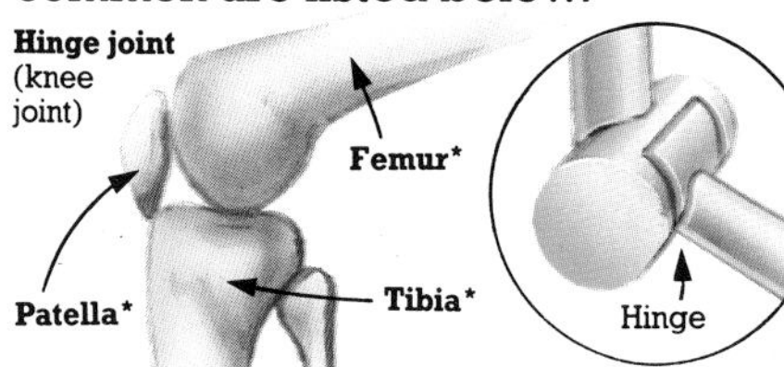

- **Hinge joints**. Joints (e.g. the knee joint) which work like any hinge. That is, the movable part (bone) can only move in one plane, i.e. in either of two opposing directions.

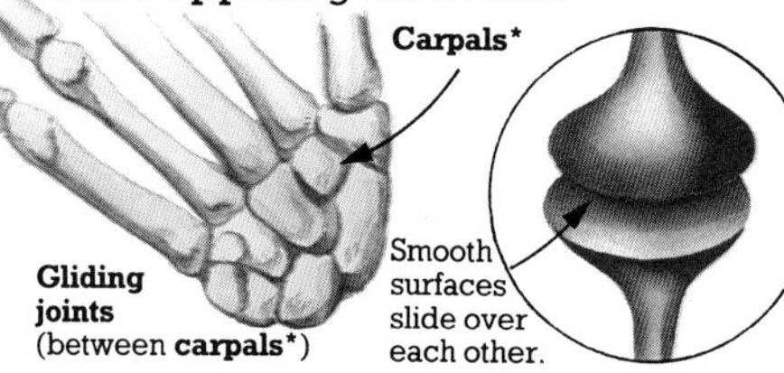

- **Gliding joints**. Also called **sliding** or **plane joints**. Joints in which one or more flat surfaces glide over each other, e.g. those between the **carpals***. They are more flexible than **hinge joints**.

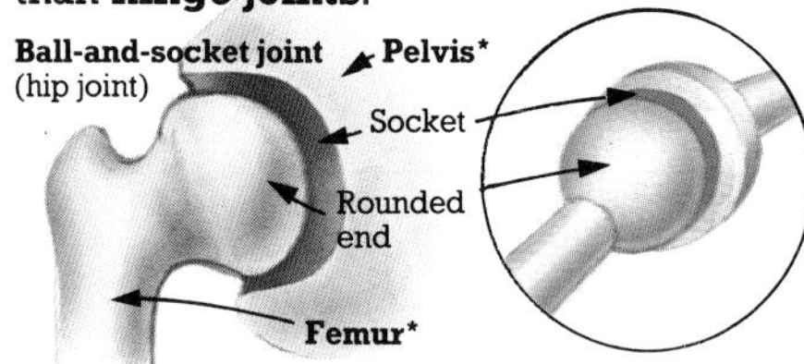

- **Ball-and-socket joints**. The most flexible joints (e.g. the hip joint). The movable bone has a rounded end which fits into a socket in the fixed bone. The movable bone can swivel, or move in many directions.

Connective tissue

There are many different types of **connective tissue** in the body. They all protect and connect cells or organs and have a basis of non-living material (the **matrix**) in which living cells are scattered. The difference between them lies in the nature of this material. The various types of tissue found at a joint, including **bone** itself, are all types of connective tissue. They all contain protein fibres and are either tough (containing **collagen** fibres) or elastic (containing **elastin** fibres).

- **Periosteum**. A thin layer of elastic connective tissue. It surrounds all bones, except at the joints (where **cartilage** takes over), and contains **osteoblasts** – cells which make new bone cells, needed for growth and repair.
- **Ligaments**. Bands of connective tissue which connect the bones of joints (and also hold many organs in place). Most are tough, though some are elastic, e.g. between **vertebrae***.

- **Bone** or **osseous tissue**. A special type of tough connective tissue, made hard and resilient by large deposits of phosphorus and calcium compounds. The living bone cells, or **osteocytes**, are held in tiny spaces (**lacunae**, sing. **lacuna**) within this non-living material.

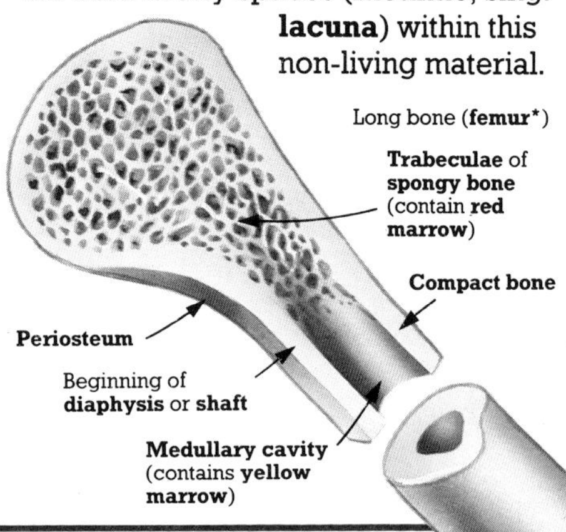

* **Carpals**, **Femur**, **Patella**, **Pelvis**, 279; **Sutures**, 278 (**Cranium**); **Tibia**, 279; **Vertebrae**, 278.

•**Synovial sac** or **synovial capsule**. A cushioning "bag" of lubricating fluid (**synovial fluid**), with an outer skin (**synovial membrane**) of elastic connective tissue. Most movable joints, e.g. the knee, have such a sac lying between the bones. They are known as **synovial joints**.

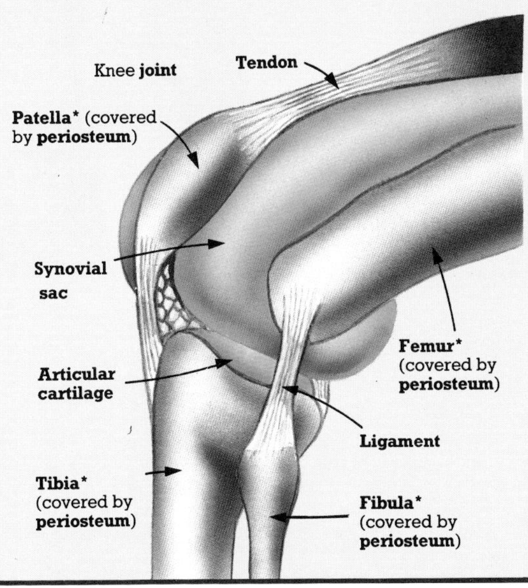

•**Tendons** or **sinews**. Bands of tough connective tissue joining muscles to bones. Each is a continuation of the membrane around the muscle, together with the outer membranes of its bundles of fibres.

•**Cartilage** or **gristle**. A tough connective tissue. In some joints (**cartilaginous joints**) it is the main cushion between the bones (e.g. **vertebrae***). In joints with **synovial sacs**, it covers the ends of the bones and is called **articular cartilage**.

The end of the nose and the outer parts of the ears are made of cartilage, as are young skeletons, though these slowly turn to **bone** as minerals build up (a process called **ossification** or **osteogenesis**).

There are two types of bone. **Spongy bone** is found in short and/or flat bones, e.g. the **sternum***, and fills the ends of long bones, e.g. the **femur***. It consists of a criss-cross network of flat plates called **trabeculae** (sing. **trabecula**), with many large spaces between them, filled by **red marrow** (see **bone marrow**). **Compact bone** forms the outer layer of all bones. It has far fewer spaces, and is laid in concentric layers (**lamellae**, sing. **lamella**) around channels called **Haversian canals**. These link with a complex system of tiny canals carrying blood vessels and nerves to the osteocytes.

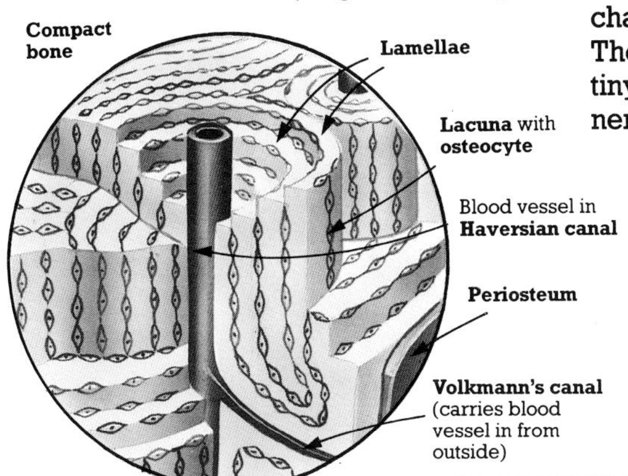

•**Bone marrow**. Two types of soft tissue. **Red marrow**, found in **spongy bone** (see **bone**), is where all new red (and some white) blood cells are made. **Yellow marrow** is a fat store, found in hollow areas (**medullary** or **marrow cavities**) in long bone **shafts**.

* **Femur**, **Fibula**, **Patella**, **Sternum**, **Tibia**, 279; **Vertebrae**, 278.

Muscles

Muscles are areas of special elastic tissue (**muscle**) found all over the body. They may be either **voluntary muscles** (able to be controlled by conscious action) or **involuntary muscles** (not under conscious control). The main types of muscles are listed at the top of the next page.

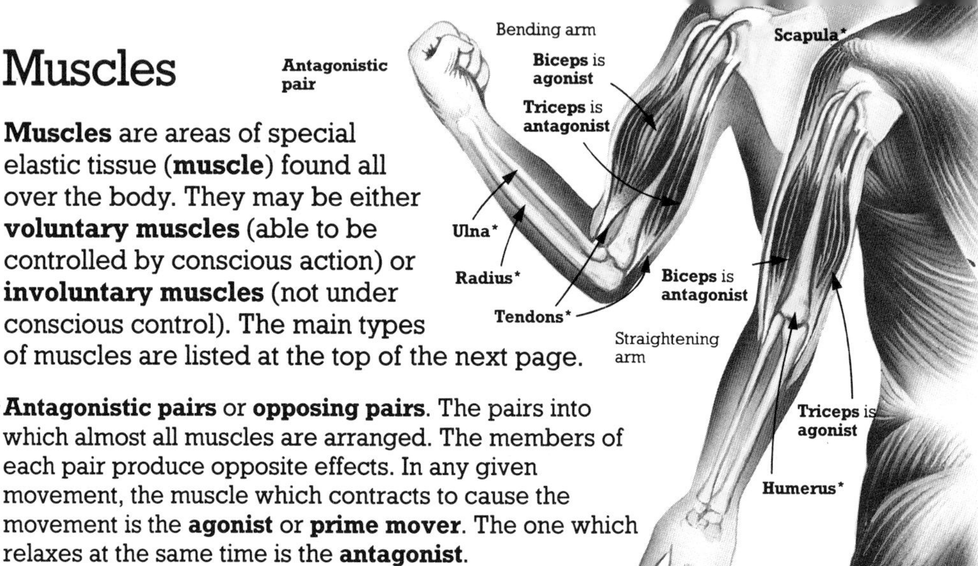

- **Antagonistic pairs** or **opposing pairs**. The pairs into which almost all muscles are arranged. The members of each pair produce opposite effects. In any given movement, the muscle which contracts to cause the movement is the **agonist** or **prime mover**. The one which relaxes at the same time is the **antagonist**.

The structure of muscle tissue

The different types of muscles in the body are made up of different kinds of muscle tissue (groups of cells of different types). The tissue has many blood vessels, bringing food matter to be broken down for energy, and nerves, which stimulate the muscles to act.

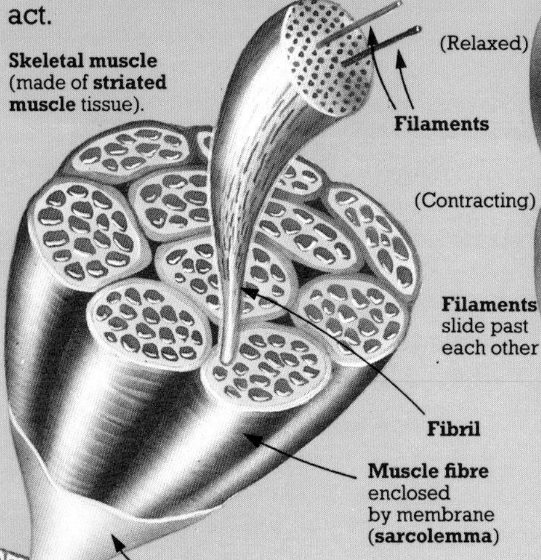

- **Striated** or **striped muscle**. The type of muscle tissue which makes up **skeletal muscles**. It consists of long cells called **muscle fibres**, grouped together in bundles called **fascicles**.

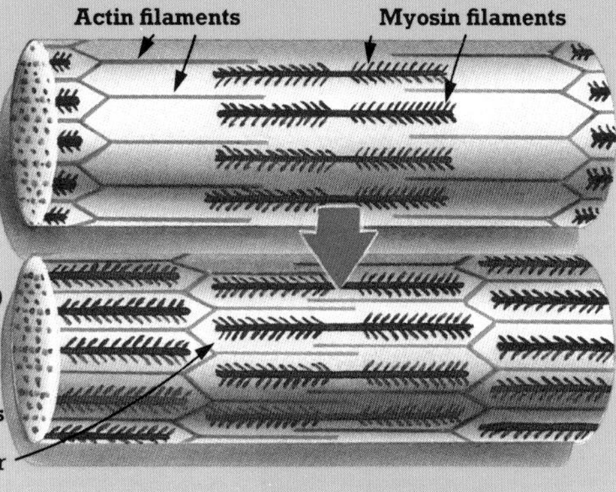

Each fibre has a striped (**striated**) appearance and is made of many smaller cylinders, called **fibrils** or **myofibrils**, which are the parts that contract when a fibre is stimulated by a nerve. The fibrils themselves consist of interlocking **filaments**, or **myofilaments**, of two different types of protein – **actin** (thin filaments) and **myosin** (thicker filaments). These filaments slide past each other as a muscle contracts.

* **Humerus**, **Radius**, **Scapula**, 279; **Tendons**, 281; **Ulna**, 279.

Types of muscles

• **Skeletal muscles**. All the muscles attached to the bones of the skeleton, which contract together or in sequence to move all the body parts. They are all **voluntary muscles** (see introduction) and are made of **striated muscle** tissue. Some are named according to their position, shape or size, others are named after the movement they cause, e.g. **flexors** cause **flexion** (the bending of a limb at a joint), **extensors** straighten a limb.

These are all **skeletal muscles**

• **Cardiac muscle**. The muscle which makes up almost all of the wall of the heart. It is an **involuntary muscle** (see introduction) and is made of **cardiac muscle** tissue.

• **Visceral muscles**. The muscles in the walls of many internal organs, e.g. the intestines and blood vessels. They are all **involuntary muscles** (see introduction) and consist of **smooth muscle** tissue.

• **Cardiac muscle**. A special kind of **striated muscle** tissue, making up the **cardiac muscle** of the heart. Its constant rhythmical contractions are caused by stimulations from special areas of the tissue itself, which produce their own electrical impulses. Any nervous impulses just increase or decrease this heart rate.

• **Smooth muscle** or **visceral muscle**. The type of muscle tissue which makes up the **visceral muscles**. It consists of spindle-shaped cells, much shorter than the complex fibres of **striated muscle**. The way it contracts is not yet fully understood, but it contains **actin** and **myosin**, like striated muscle, and is also stimulated by nerves.

Nervous stimulation

Most muscles are stimulated to move by impulses from nerves running through the body. For more about this, see pages 308-309.

• **Motor end-plate**. The point where the end fibres of an "instruction-carrying" nerve cell (**motor neuron***) meet a **muscle fibre** (see **striated muscle**). The end fibres are branches from one main fibre (**axon***). This carries nervous impulses which make the muscle contract. Each impulse is duplicated and sent down each end branch, hence the whole muscle receives a multiplication of each impulse.

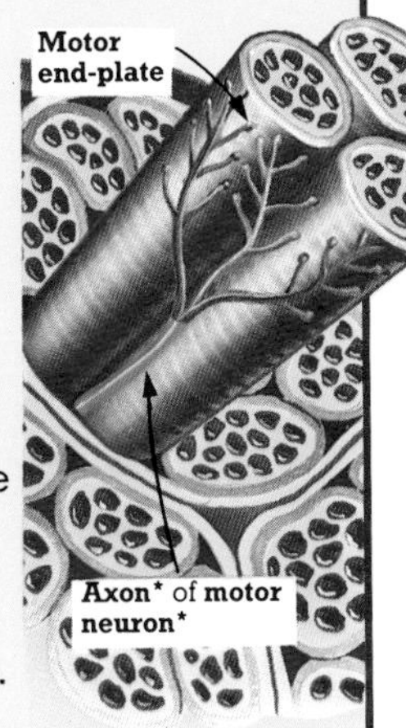

• **Muscle spindle**. A group of **muscle fibres** (see **striated muscle**) which has the end fibres of a sensory nerve cell (**sensory neuron***) wrapped round it. The end fibres are part of one main fibre (**dendron***). When the muscle stretches they are stimulated to send impulses to the brain, "telling" it about the new state of tension. The brain can then work out the changes needed for any further action.

* **Axon**, 304; **Dendron**, 304 (**Dendrites**); **Motor neurons**, **Sensory neurons**, 305.

Teeth

The **teeth** or **dentes** (sing. **dens**) help to prepare food for digestion by cutting and grinding it up. Each tooth is set into the jaw, which has a soft tissue covering called **gum (gingiva)**. During their lives, humans have two sets of teeth (**dentitions**) – a temporary set, or **deciduous dentition**, made up of 20 **deciduous teeth** (also called **milk** or **baby teeth**), and a later **permanent dentition** (32 **permanent teeth**).

Parts of a tooth

- **Crown**. The exposed part of a tooth. It is covered by **enamel**. It is the part most subject to damage or tooth decay.

- **Root**. The part of a tooth that is fixed in a socket in the jaw. **Incisors** and **canines** have one root, **premolars** have one or two and **molars** have two or three. Each root is held in place by the tough fibres of a **ligament*** called the **periodontal ligament**. The fibres are fixed to the jawbone at one end, and to the **cement** at the other. They act as shock absorbers.

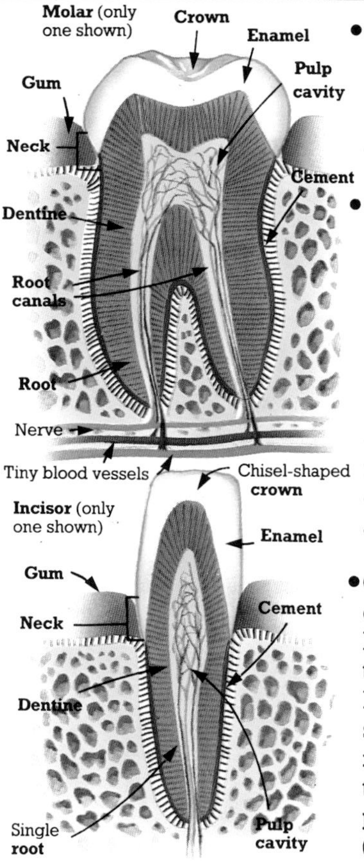

- **Neck** or **cervix**. The part of a tooth just below the surface, lying between the **crown** and the **root**.

- **Enamel**. A substance similar to bone, though it is harder (the hardest substance in the body) and has no living cells. It consists of tightly- packed crystals of **apatite**, a mineral made up of calcium, phosphorus and fluorine.

- **Cement** or **cementum**. A bone-like substance, similar to **enamel** but softer. It forms the thin surface layer of the **root** and is attached to the jaw by the **periodontal ligament** (see **root**).

- **Dentine** or **ivory**. A yellow substance which forms the second layer inside a tooth. Like **enamel**, it has many of the same constituents as bone, but it is softer and it also contains **collagen*** fibres and strands of **cytoplasm***. These run out from the **pulp** cells in the **pulp cavity**.

- **Pulp cavity**. The central area of a tooth, surrounded by **dentine**. It is filled with a soft tissue called **pulp**, which contains blood vessels and nerve fibre endings. These enter at the base of a **root** and run up to the cavity inside **root canals**. The blood vessels supply food and oxygen to the living tissue and the nerve fibre endings are **pain receptors***.

* **Collagen**, 280 (**Connective tissue**); **Cytoplasm**, 238; **Ligaments**, 280; **Pain receptors**, 311.

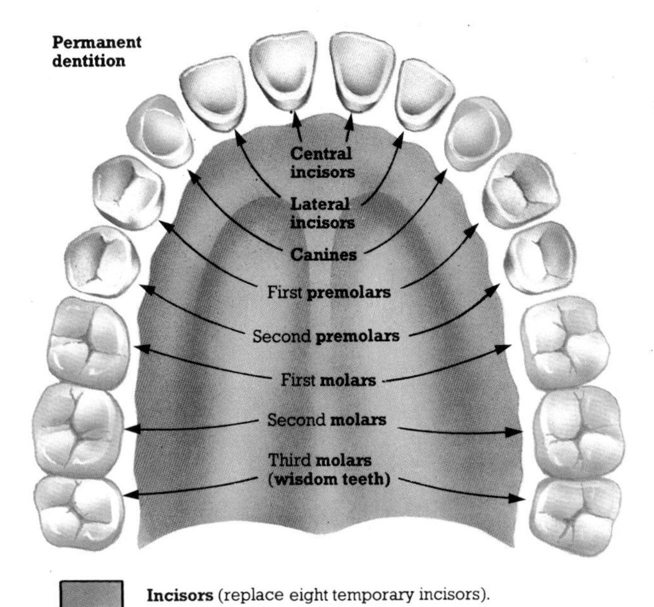

Incisors (replace eight temporary incisors).

Canines (replace four temporary canines).

Premolars (replace eight temporary molars).

Molars (appear behind **premolars** and do not replace any **deciduous teeth**).

Types of teeth

- **Incisors**. Sharp, chisel-shaped teeth, used for biting and cutting. Each has one root, and there are four in each jaw, set at the front of the mouth.
- **Canines** or **cuspids**. Cone-shaped teeth (often called **eye** or **dog teeth**), used to tear food. Each has a sharp point (**cusp**) and one **root**. There are two in each jaw, one each side of the **incisors**. In animals which hunt and kill, they are long and curved.
- **Premolars** or **bicuspids**. Blunt, broad teeth, used for crushing and grinding (found in the permanent set of teeth only). There are four in each jaw, two behind each **canine**. Each has two sharp ridges (**cusps**) and one **root**, except the upper first premolars, which have two.
- **Molars**. Blunt, broad teeth, similar to **premolars** but with a larger surface area. They are also used for crushing and grinding, and each has four surface points (**cusps**). Lower molars have two **roots** each, and upper ones have three. In the permanent set of teeth there are six in each jaw, three behind each pair of **premolars**, and the third ones (at the back) are known as **wisdom teeth**.
- **Wisdom teeth**. Four **molars** (the third ones in line), lying at the end points of the jaws. They appear last of all, when a person is fully mature (hence their name). Often there is no room for them to come through and they get stuck in the jawbone, or **impacted**. A few people never develop wisdom teeth.

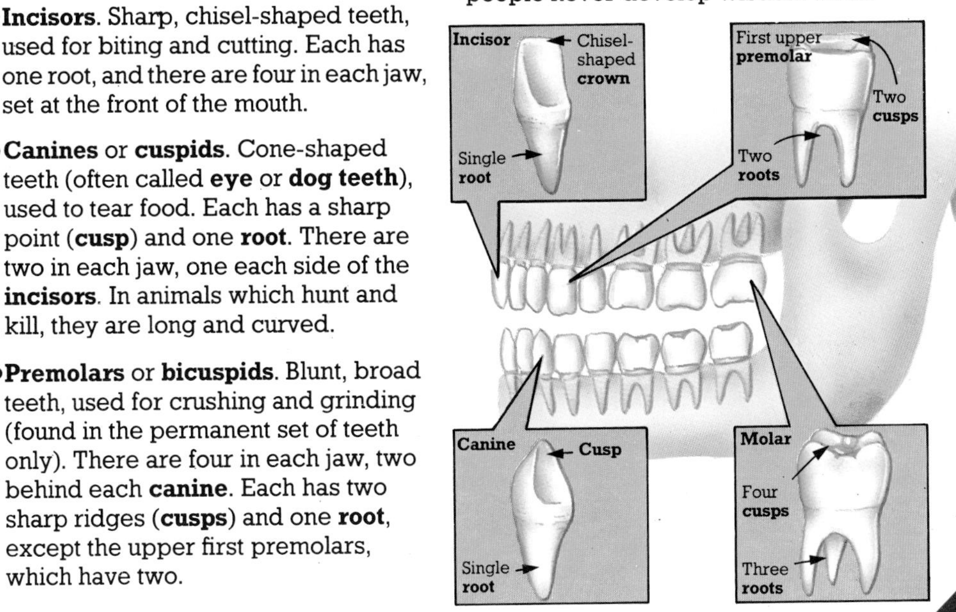

Blood

Blood is a vital body fluid, consisting of **plasma**, **platelets** and **red** and **white blood cells**. An adult human has about 5.5 litres (9.5 pints), which travel around in the **circulatory system*** – a system of tubes called **blood vessels**. The blood distributes heat and carries many important substances in its plasma. Its old, dying blood cells are constantly being replaced by new ones in a process called **haemopoiesis**.

Blood constituents

- **Plasma**. The pale liquid (about 90% water) which contains the blood cells. It carries dissolved food for the body cells, waste matter and carbon dioxide secreted by them, **antibodies** to combat infection, and **enzymes*** and **hormones*** which control body processes.

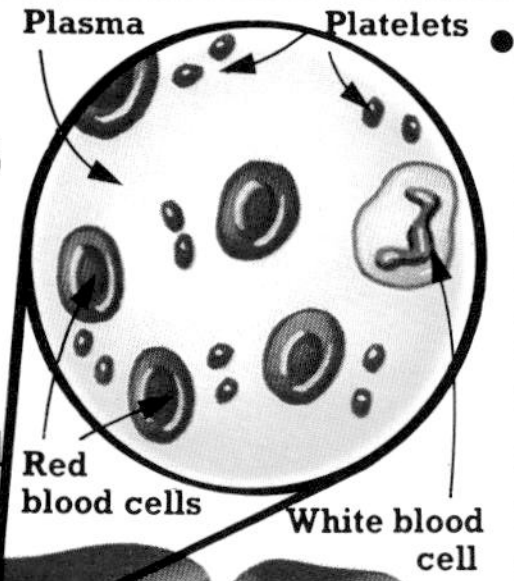

- **Platelets** or **thrombocytes**. Very small, disc-shaped bodies with no **nuclei***, made in the **bone marrow***. They gather particularly at an injured area, where they are important in the **clotting** of blood.

- **White blood cells**. Also called **white corpuscles** or **leucocytes**. Large, opaque blood cells, important in body defence. There are several types. **Lymphocytes**, for example, are made in **lymphoid tissue*** and are found in the **lymphatic system*** as well as blood. They make **antibodies**. Other white cells – **monocytes** – are made in **bone marrow***. They "swallow up" foreign bodies, e.g. bacteria, in a process called **phagocytosis***. Many of them (**macrophages**) leave the blood vessels. They either travel around (**wandering macrophages**) or become fixed in an organ, e.g. a **lymph node*** (**fixed macrophages**).

Lymphocyte

Different types of antibody

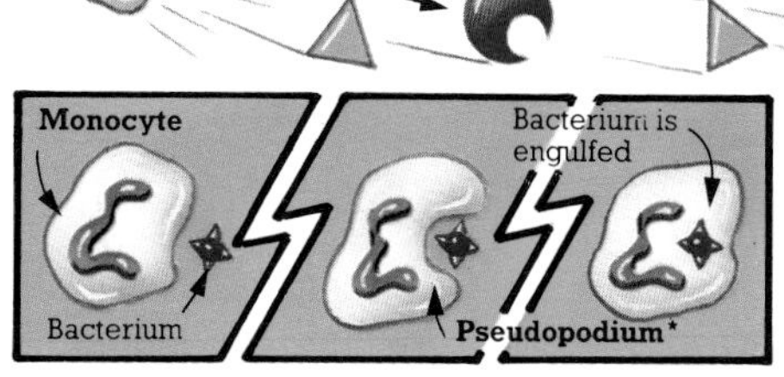

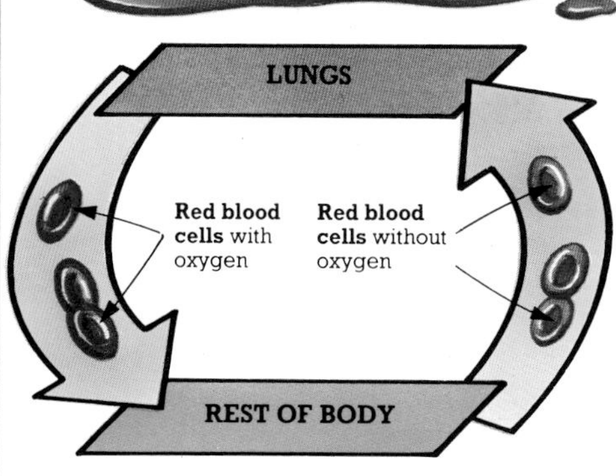

- **Red blood cells**. Also called **red corpuscles** or **erythrocytes**. Red, disc-shaped cells with no **nuclei***. They are made in the **bone marrow*** and contain **haemoglobin** (an iron compound which gives blood a dark red colour). This combines with oxygen in the lungs to form **oxyhaemoglobin**, and the blood becomes bright red. The red cells pass the oxygen to the body cells (by **diffusion***) and then return to the lungs with haemoglobin.

* **Bone marrow**, 281; **Circulatory system**, 288; **Diffusion**, 327; **Enzymes**, 331; **Hormones**, 334; **Lymphatic system**, **Lymph nodes**, 293; **Lymphoid tissue**, 293 (**Lymphoid organs**); **Nucleus**, 238; **Phagocytosis**, 268 (**Pseudopodium**).

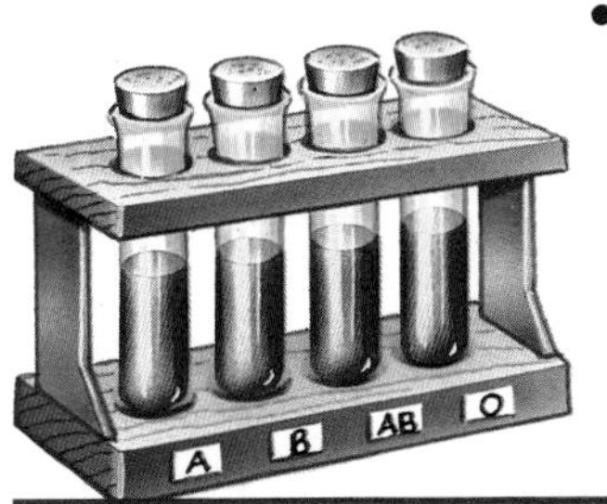

- **Blood groups**. The main way of classifying blood. The group depends on whether the **antigens** A or B are present in the **red blood cells**. Group A blood has A antigen, group B has B, group AB has both and group O has neither.

- **Rhesus factor** or **Rh factor**. A second way of classifying blood (as well as by **blood group**). If it contains the **Rhesus antigen**, it is **Rhesus positive** blood. If not, it is **Rhesus negative**.

Body defence

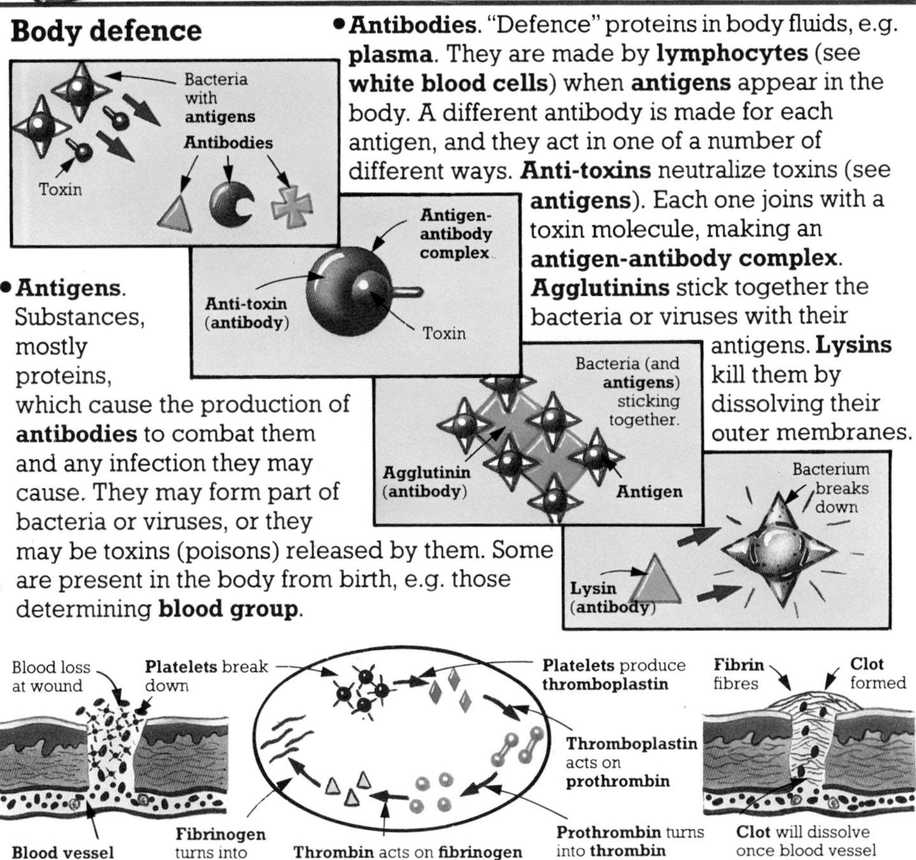

- **Antibodies**. "Defence" proteins in body fluids, e.g. **plasma**. They are made by **lymphocytes** (see **white blood cells**) when **antigens** appear in the body. A different antibody is made for each antigen, and they act in one of a number of different ways. **Anti-toxins** neutralize toxins (see **antigens**). Each one joins with a toxin molecule, making an **antigen-antibody complex**. **Agglutinins** stick together the bacteria or viruses with their antigens. **Lysins** kill them by dissolving their outer membranes.

- **Antigens**. Substances, mostly proteins, which cause the production of **antibodies** to combat them and any infection they may cause. They may form part of bacteria or viruses, or they may be toxins (poisons) released by them. Some are present in the body from birth, e.g. those determining **blood group**.

- **Clotting** or **coagulation**. The thickening of blood into a mass (**clot**) at the site of a wound. First, disintegrating **platelets** and damaged cells release a chemical called **thromboplastin**. This causes **prothrombin** (a **plasma** protein) to turn into **thrombin** (an **enzyme***). This then causes **fibrinogen** (another plasma protein) to harden into **fibrin**, a fibrous substance. A network of its fibres makes up the jelly-like clot.

- **Serum**. A yellowy liquid consisting of the parts of the blood left after **clotting**. It contains many **antibodies** (produced to combat infections). When injected into other people, it can give temporary immunity to the infections.

* **Enzymes**, 331.

The circulatory system

The **circulatory** or **vascular system** is a network of blood-filled tubes, or **blood vessels**, of which there are three main types – **arteries**, **veins** and **capillaries**. A thin tissue layer called the **endothelium** lines arteries and veins, and is the only layer of capillary walls. Blood is kept flowing one way by the pumping of the heart, by muscles in artery and vein walls and by a decrease in pressure through the system (liquids flow from high to low pressure areas).

Passage of main substances in **circulatory system**.

Arteries, arterioles, capillaries
Capillaries, venules, veins

LUNGS
HEART

△ Dissolved food matter
△ Newly-digested food
△ Some food stored
★ Some food used by organ's cells
△● Food and oxygen used by organ's cells
● Oxygen
○ Carbon dioxide
● Oxygen breathed in, carbon dioxide out
□ Waste
■ Waste disposed c

- **Arteries**. Wide, thick-walled blood vessels, making up the **arterial system** and carrying blood away from the heart. Smaller arteries (**arterioles**) branch off the main ones, and **capillaries** branch off the arterioles. Except in the **pulmonary arteries***, the blood contains oxygen (which makes it bright red). In all arteries it also carries dissolved food and waste, brought into the heart by **veins**, and there transferred to the arteries. These carry the food to the cells (via arterioles and capillaries) and the waste to the kidneys.

- **Veins**. Wide, thick-walled blood vessels, making up the **venous system** and carrying blood back to the heart. They contain valves to stop blood flowing backwards due to gravity, and are formed from merging **venules** (small veins). These are formed in turn from merging **capillaries**. The blood contains carbon dioxide (except in the **pulmonary veins***) and waste matter, both picked up from body cells by the capillaries. The blood in the veins leading from the digestive system and liver also carries dissolved food. This is transferred to the arteries in the heart.

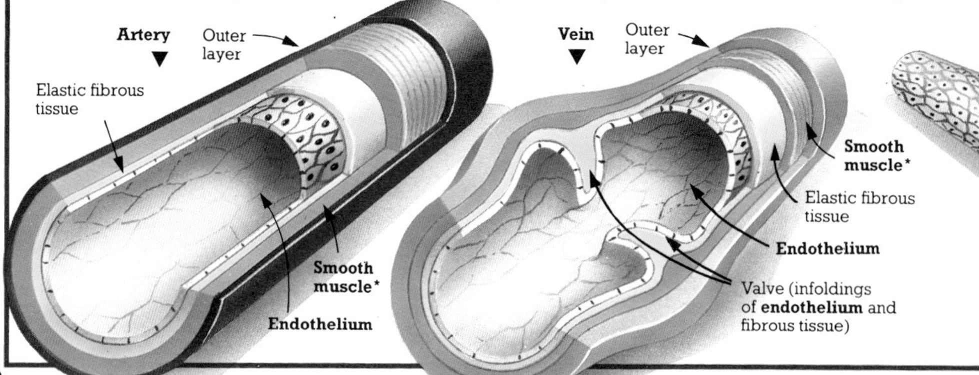

* **Pulmonary arteries**, 291 (**Pulmonary trunk**); **Pulmonary veins**, 291; **Smooth muscle**, 283.

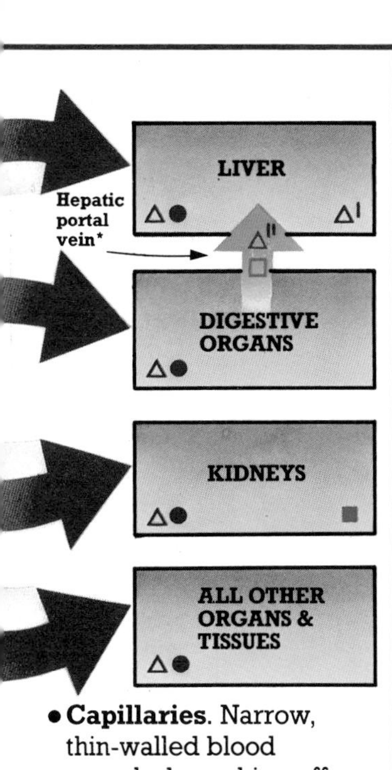

- **Capillaries**. Narrow, thin-walled blood vessels, branching off **arterioles** (see **arteries**) to form a complex network. Oxygen and dissolved food pass out through their walls to the body cells, and carbon dioxide and waste pass in (see **tissue fluid**, page 292). The capillaries of the digestive organs and liver also pick up food. Capillaries finally join up again to form small veins (**venules**).

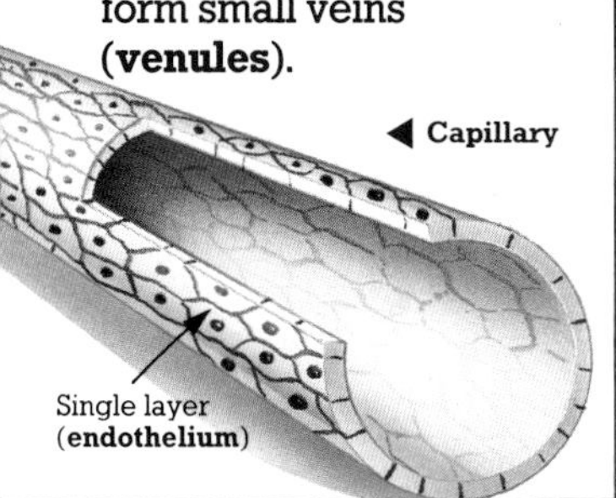

◀ **Capillary**

The main arteries and veins

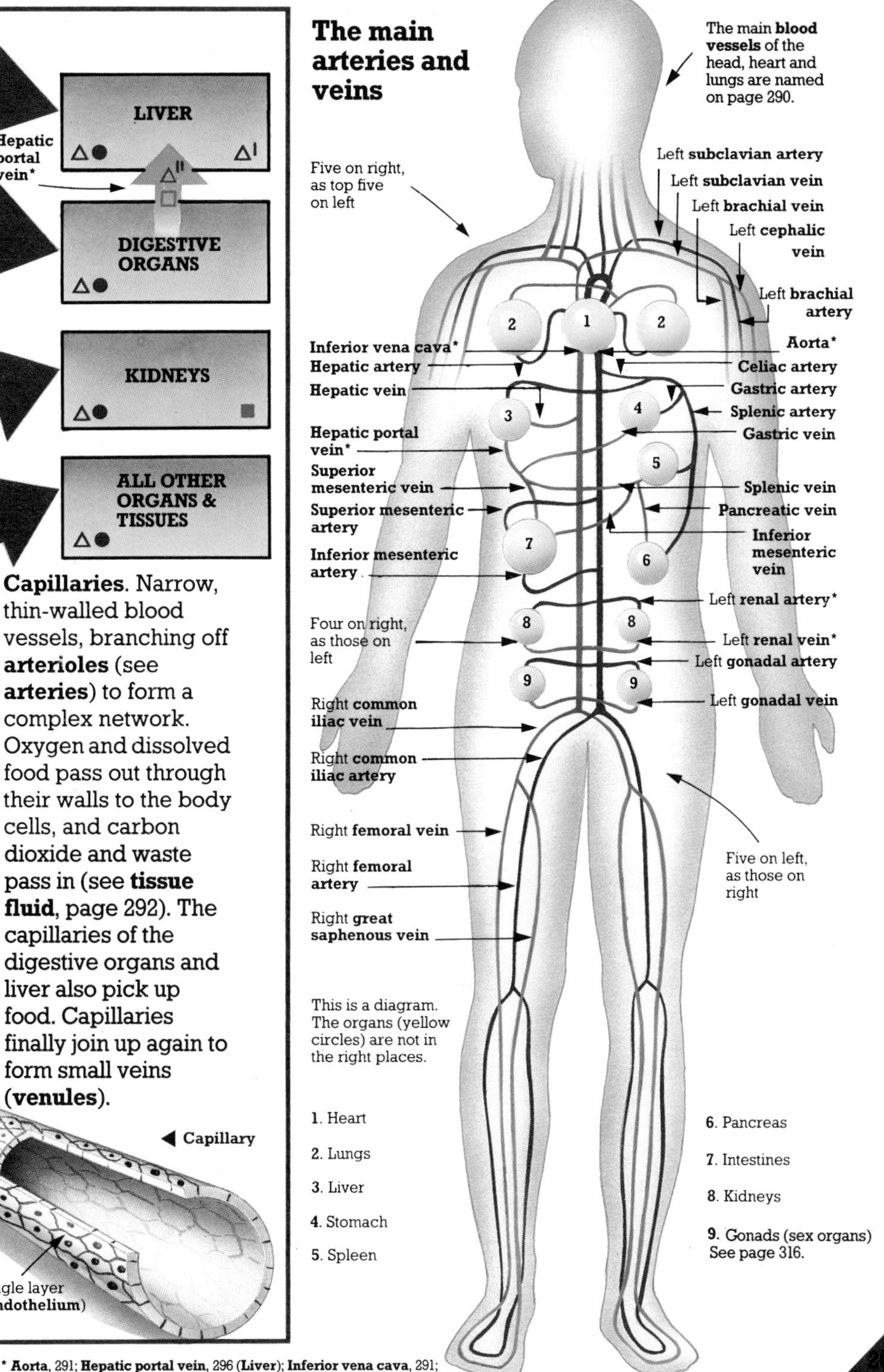

This is a diagram. The organs (yellow circles) are not in the right places.

1. Heart
2. Lungs
3. Liver
4. Stomach
5. Spleen
6. Pancreas
7. Intestines
8. Kidneys
9. Gonads (sex organs) See page 316.

* **Aorta**, 291; **Hepatic portal vein**, 296 (**Liver**); **Inferior vena cava**, 291; **Renal arteries**, **Renal veins**, 300 (**Kidneys**).

The heart

The **heart** is a muscular organ which pumps blood around the blood vessels (the heart and blood vessels together are the **cardiovascular system**). It is surrounded by the **pericardial sac**. This consists of an outer membrane (the **pericardium**) and the cavity (**pericardial cavity**) between it and the heart. This cavity is filled with a cushioning fluid (**pericardial fluid**). The heart has four chambers – two **atria** and two **ventricles**, all lined by a thin tissue layer called the **endocardium**.

Position of **heart**

The cardiac cycle

The **cardiac cycle** is the series of events which make up one complete pumping action of the heart, and which can be heard as the heartbeat (about 70 times a minute). First, both **atria** contract and pump blood into their respective **ventricles**, which relax to receive it. Then the atria relax and take in blood, and the ventricles

The chambers of the heart

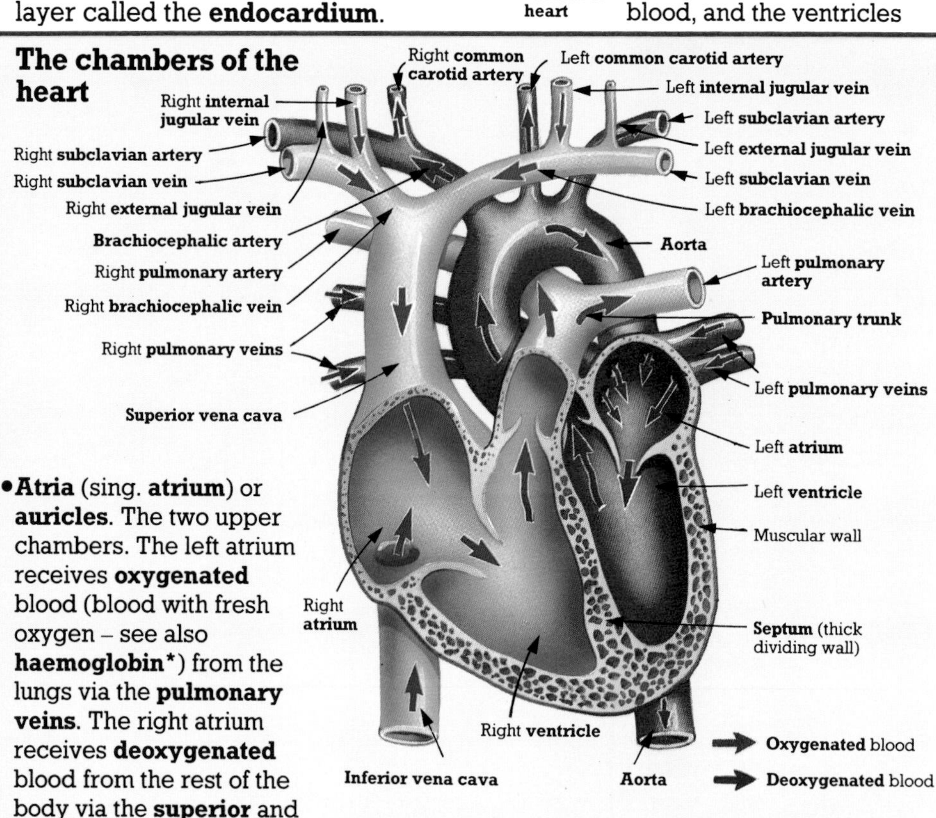

- **Atria** (sing. **atrium**) or **auricles**. The two upper chambers. The left atrium receives **oxygenated** blood (blood with fresh oxygen – see also **haemoglobin***) from the lungs via the **pulmonary veins**. The right atrium receives **deoxygenated** blood from the rest of the body via the **superior** and **inferior vena cavae**. This is blood whose oxygen has been used by the cells and replaced by carbon dioxide.

- **Ventricles**. The two lower chambers. The left ventricle receives blood from the left **atrium** and pumps it into the **aorta**. The right ventricle receives blood from the right atrium and pumps it via the **pulmonary trunk** to the lungs.

* **Haemoglobin**, 286 (**Red blood cells**).

contract to pump it out. The relaxing phase of a chamber is its **diastole phase**; the contracting phase is its **systole phase**. There is a short pause after the systole phase of the ventricles, during which all chambers are in diastole phase (relaxing). The different **valves** which open and close during the cycle are defined below right.

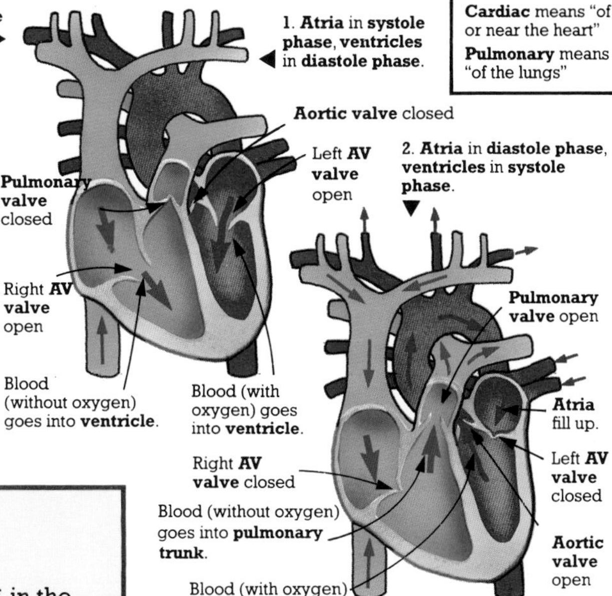

Cardiac means "of or near the heart"
Pulmonary means "of the lungs"

The main arteries and veins

- **Aorta**. The largest **artery*** in the body. It carries blood with fresh oxygen out of the left **ventricle** to begin its journey round the body.
- **Pulmonary trunk**. The **artery*** which carries blood needing fresh oxygen out of the right **ventricle**. After leaving the heart, it splits into the right and left **pulmonary arteries**, one going to each lung.
- **Superior vena cava**. One of the two main **veins***. It carries blood needing fresh oxygen from the upper body to the right **atrium**. All the upper body veins merge into it.
- **Inferior vena cava**. One of the two main **veins***, carrying blood needing fresh oxygen from the lower body to the right **atrium**. All the lower body veins merge into it.
- **Pulmonary veins**. Four **veins*** which carry blood with fresh oxygen to the left **atrium**. Two right pulmonary veins come from the right lung, and two left pulmonary veins come from the left lung.

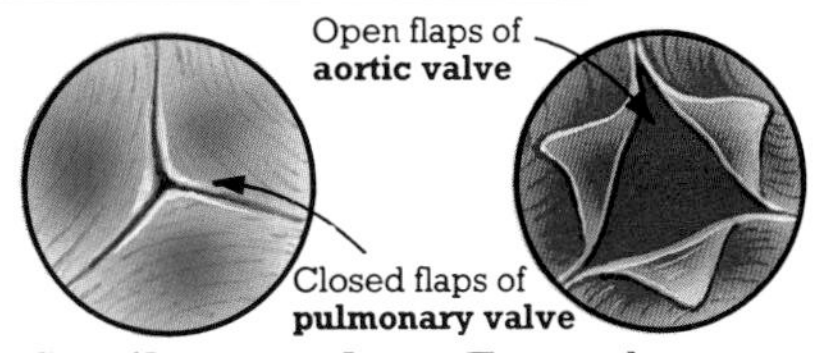

- **Semilunar valves**. Two valves, so called because they have crescent-shaped flaps. One is the **aortic valve** between the left **ventricle** and the **aorta**. The other is the **pulmonary valve** between the right ventricle and the **pulmonary trunk**.

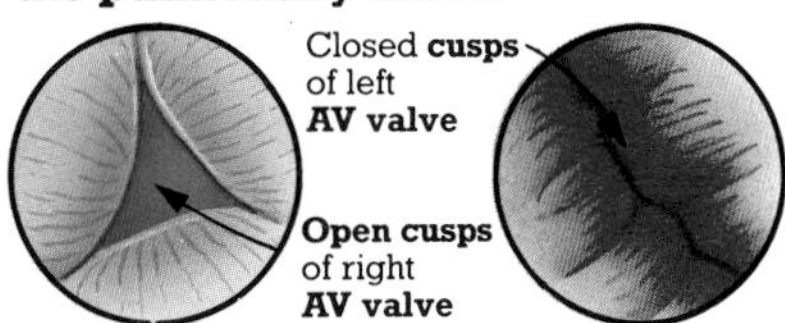

- **Atrioventricular valves** or **AV valves**. Two valves, each between an **atrium** and its corresponding **ventricle**. The left AV valve, or **mitral valve**, is a **bicuspid valve**, i.e. it has two movable flaps, or **cusps**. The right AV valve is a **tricuspid valve**, i.e. it has three cusps.

* **Arteries, Veins**, 288.

Tissue fluid and the lymphatic system

The smallest blood vessels, called **capillaries***, are those in the most direct contact with the individual cells of the body, but even they do not touch the cells. The food and oxygen they carry finally reaches the cells in **tissue fluid**, a substance which forms the link between the **circulatory system*** and the body's drainage system, known as the **lymphatic system**.

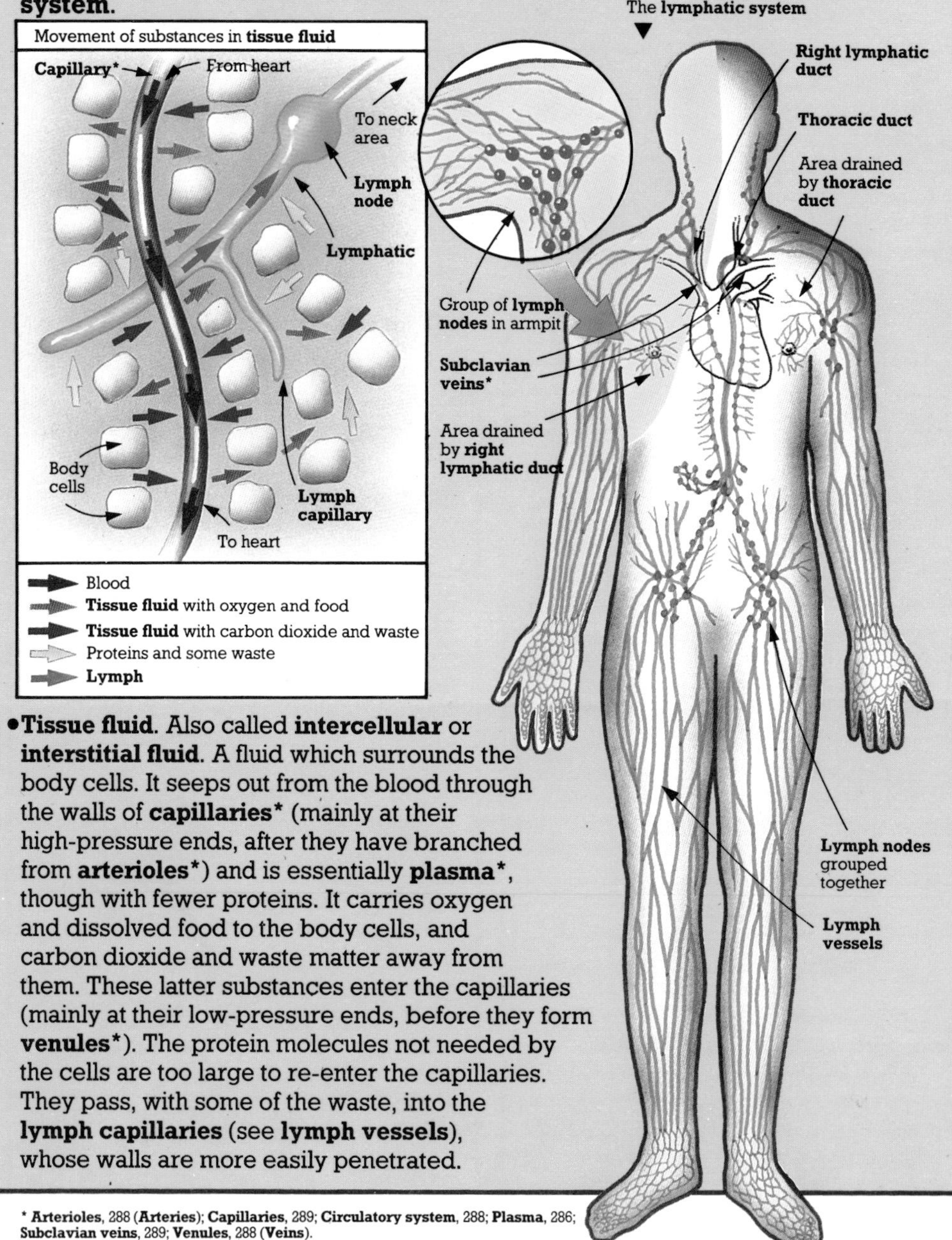

- **Tissue fluid**. Also called **intercellular** or **interstitial fluid**. A fluid which surrounds the body cells. It seeps out from the blood through the walls of **capillaries*** (mainly at their high-pressure ends, after they have branched from **arterioles***) and is essentially **plasma***, though with fewer proteins. It carries oxygen and dissolved food to the body cells, and carbon dioxide and waste matter away from them. These latter substances enter the capillaries (mainly at their low-pressure ends, before they form **venules***). The protein molecules not needed by the cells are too large to re-enter the capillaries. They pass, with some of the waste, into the **lymph capillaries** (see **lymph vessels**), whose walls are more easily penetrated.

* **Arterioles**, 288 (**Arteries**); **Capillaries**, 289; **Circulatory system**, 288; **Plasma**, 286; **Subclavian veins**, 289; **Venules**, 288 (**Veins**).

- **Lymphatic system**. A system of tubes (**lymph vessels**) and small organs (**lymphoid organs**), important in the recycling of body fluids and in the fight against disease. The lymph vessels carry the liquid **lymph** around the body and empty it back into the **veins***, and the lymphoid organs are the source of disease-fighting cells.

- **Lymph vessels** or **lymphatic vessels**. Blind-ended tubes carrying **lymph** from all body areas towards the neck, where it is emptied back into the blood. They are lined with **endothelium***, and have valves to stop the lymph being pulled back by gravity. The thinnest ones are **lymph capillaries**, and include the important **lacteals***, which pick up fat particles (too large to enter the bloodstream directly). The capillaries join to form larger vessels called **lymphatics**, which finally unite to form two tubes – the **right lymphatic duct** (emptying into the right **subclavian vein***) and the **thoracic duct** (emptying into the left **subclavian vein***).

- **Lymph**. The liquid in **lymph vessels**. It contains **lymphocytes** (see **lymphoid organs**), some substances picked up from **tissue fluid** (especially proteins such as **hormones*** and **enzymes***) and also fat particles (see **lymph vessels**).

Lymphoid organs

The **lymphoid organs**, or **lymphatic organs**, are bodies connected to the **lymphatic system**. They are all made of the same type of tissue (**lymphoid** or **lymphatic tissue**) and they all produce **lymphocytes*** – disease-fighting white blood cells.

- **Lymph nodes** or **lymph glands**. Small lymphoid organs found along the course of **lymph vessels**, often in groups, e.g. in the armpits. They are the main sites of **lymphocyte** production (see above) and also contain a filter system which traps bacteria and foreign bodies. These are then engulfed by white blood cells (**fixed macrophages***).

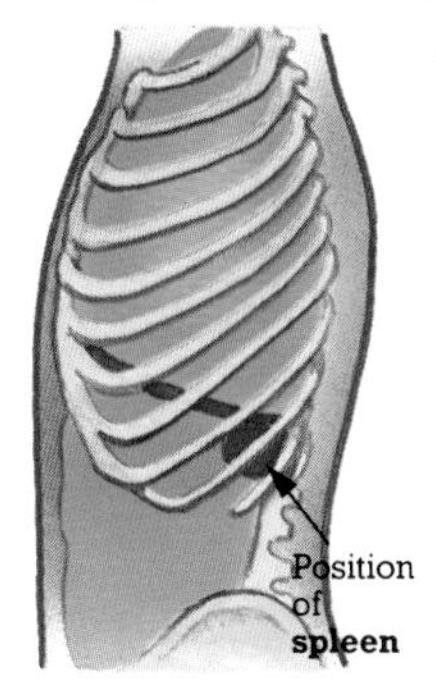

- **Spleen**. The largest lymphoid organ, found just below the **diaphragm*** on the left side of the body. It holds an emergency store of red blood cells and also contains white blood cells (**fixed macrophages***) which destroy foreign bodies, e.g. bacteria, and old blood cells.

- **Tonsils**. Four lymphoid organs: one **pharyngeal tonsil** (the **adenoids**) at the back of the nose, one **lingual tonsil** at the base of the tongue and two **palatine tonsils** at the back of the mouth.

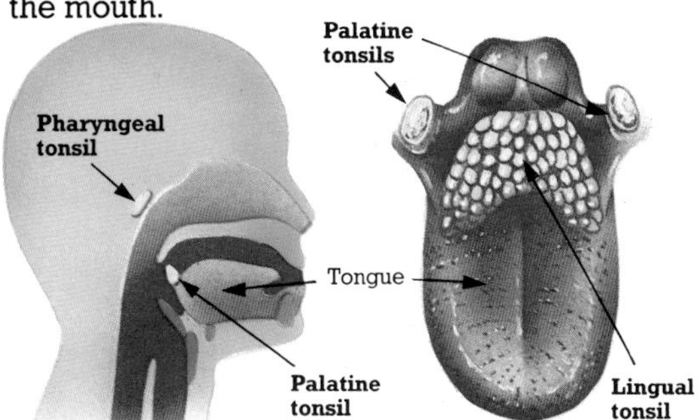

- **Thymus gland**. A lymphoid organ in the upper part of the chest. It is fairly large in children, reaches its maximum size at **puberty*** and then undergoes **atrophy**, i.e. wastes away.

* **Diaphragm**, 298; **Endothelium**, 288; **Enzymes**, 331; **Fixed macrophages**, 286 (**White blood cells**); **Hormones**, 334; **Lacteals**, 295 (**Small intestine**); **Lymphocytes**, 286 (**White blood cells**); **Puberty**, 318; **Subclavian veins**, 289; **Veins**, 288.

The digestive system

After food is taken in, or **ingested**, it passes through the **digestive system**, gradually being broken down into simple soluble substances by a process called **digestion** (see also pages 336-337). The simple substances are absorbed into the blood vessels around the system and transported to the body cells. Here they are used to provide energy and build new tissue. For more about all these different processes, see pages 328-332. The main parts of the digestive system are listed on these two pages. The pancreas and liver (see page 296) also play a vital part in digestion, forming the two main **digestive glands*** (producing **digestive juices***).

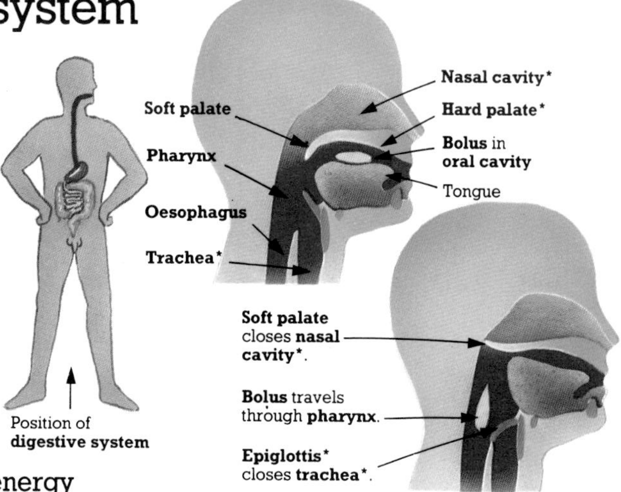

- **Alimentary canal**. Also called the **alimentary tract**, **gastrointestinal** (**GI**) **tract**, **enteric canal** or the **gut**. A collective term for all the parts of the digestive system. It is a long tube running from the mouth to the **anus** (see **large intestine**). Most of its parts are in the lower body, or **abdomen**, inside the main body cavity, or **perivisceral cavity***. They are held in place by **mesenteries** – infoldings of the cavity lining (the **peritoneum**).

- **Pharynx**. A cavity at the back of the mouth, where the mouth cavity (**oral** or **buccal cavity**) and the **nasal cavities*** meet. When food is swallowed, the **soft palate** (a tissue flap at the back of the mouth) closes the nasal cavities and the **epiglottis*** closes the **trachea***.

- **Oesophagus** or **gullet**. The tube down which food travels to the **stomach**. A piece of swallowed food is a **bolus**.

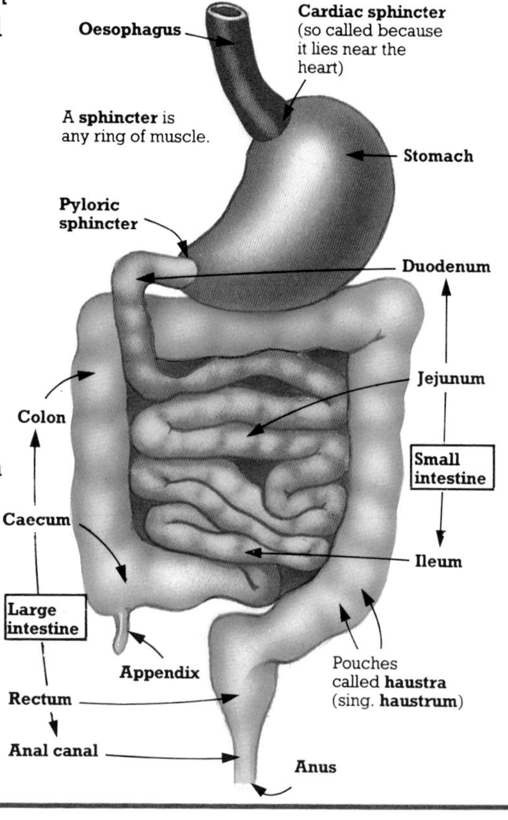

* **Digestive juices**, 296 (**Digestive glands**); **Epiglottis**, 298; **Hard palate**, 307; **Nasal cavities**, 307 (**Nose**); **Perivisceral cavity**, 265; **Trachea**, 298.

• **Cardiac sphincter**. Also called the **gastroesophageal sphincter**. A muscular ring between the **oesophagus** and **stomach**. It relaxes to open and let food through.

• **Stomach**. A large sac in which the first stages of digestion occur. Its lining has many folds (**rugae**, sing. **ruga**) which flatten out to let it expand. Some substances, e.g. water, pass through its wall into nearby blood vessels, but almost all the semi-digested food (**chyme**) goes into the **small intestine** (**duodenum**).

• **Small intestine**. The main site of digestion, a coiled tube with three parts – the **duodenum**, **jejunum** and **ileum**. Many tiny "fingers" called **villi** (sing. **villus**) project inwards from its lining. Each contains **capillaries*** (tiny blood vessels) into which most of the food is absorbed, and a **lymph vessel*** called a **lacteal**, which absorbs recombined fat particles (see **fats**, page 328). The remaining semi-liquid waste mixture passes into the **large intestine**.

• **Large intestine**. A thick tube receiving waste from the **small intestine**. It consists of the **caecum*** (a redundant sac), **colon**, **rectum** and **anal canal**. The colon contains bacteria, which break down any remaining food and make some important vitamins. Most of the water in the waste passes through the colon walls into nearby blood vessels. This leaves a semi-solid mass (**faeces**), which is pushed out of the body (**defaecation**) via the rectum, anal canal and **anus** – a hole surrounded by a muscular ring (the **anal sphincter**).

• **Appendix**. A small blind-ended tube off the **caecum** (see **large intestine**). It is a **vestigial** organ, i.e. one which our ancestors needed, but is now defunct.

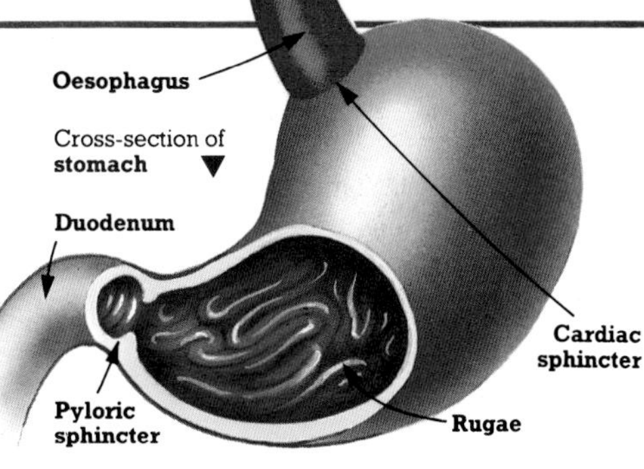

• **Pyloric sphincter**. Also called the **pyloric valve** or **pylorus**. A muscular ring between the **stomach** and the **small intestine**. It relaxes to let food through only after certain digestive changes have occurred.

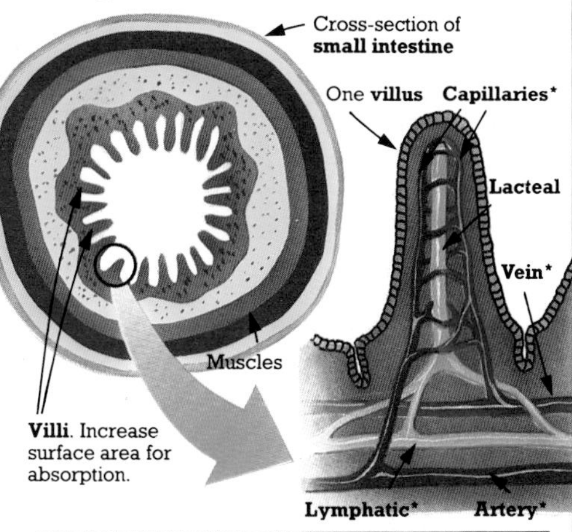

• **Mucous membrane** or **mucosa**. A thin layer of tissue lining all digestive passages (also other passages, e.g. the air passages). It is a special type of **epithelium*** (a surface sheet of cells), containing many single-celled **exocrine glands***, called **mucous glands**. These secrete **mucus** – a lubricating fluid which, in the case of the digestive passages, also protects against the action of **digestive juices***.

• **Peristalsis**. The waves of contraction, produced by muscles in the walls of organs (especially digestive organs), which move substances along.

* **Arteries**, 288; **Caecum**, 271; **Capillaries**, 289; **Digestive juices**, 296 (**Digestive glands**); **Epithelium**, 310 (**Epidermis**); **Exocrine glands**, 296; **Lymphatic**, 293 (**Lymph vessels**); **Veins**, 288.

Glands

Glands are special organs (or sometimes groups of cells or single cells) which produce and secrete a variety of substances vital to life. There are two types of human gland – **exocrine** and **endocrine**.

Exocrine glands

Exocrine glands are glands which secrete substances through tubes, or **ducts**, onto a surface or into a cavity. Most body glands are exocrine, e.g. **sweat glands*** and **digestive glands**.

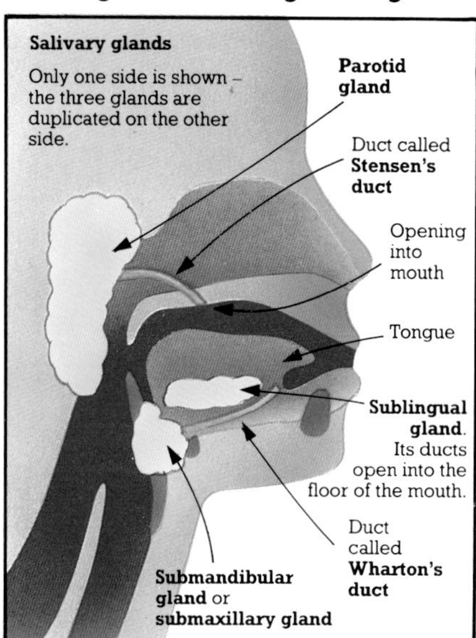

• **Digestive glands**. Exocrine glands which secrete fluids called **digestive juices** into the digestive organs. The juices contain **enzymes*** which cause the breakdown of food (see chart, pages 336-337). Many of the glands are tiny, and set into the walls of the digestive organs, e.g. **gastric glands** in the stomach and **intestinal glands** (or **crypts of Lieberkühn**) in the small intestine. Others are larger and lie more freely, e.g. **salivary glands**. The largest are the **pancreas** and **liver**.

• **Pancreas**. A large gland which is both a **digestive gland** and an **endocrine gland**. It produces **pancreatic juice** (see chart, pages 336-337), which it secretes along the **pancreatic duct**, or **duct of Wirsung**. It also contains groups of cells called the **islets of Langerhans**. These make up the endocrine parts of the organ, and produce the **hormones* insulin*** and **glucagon***.

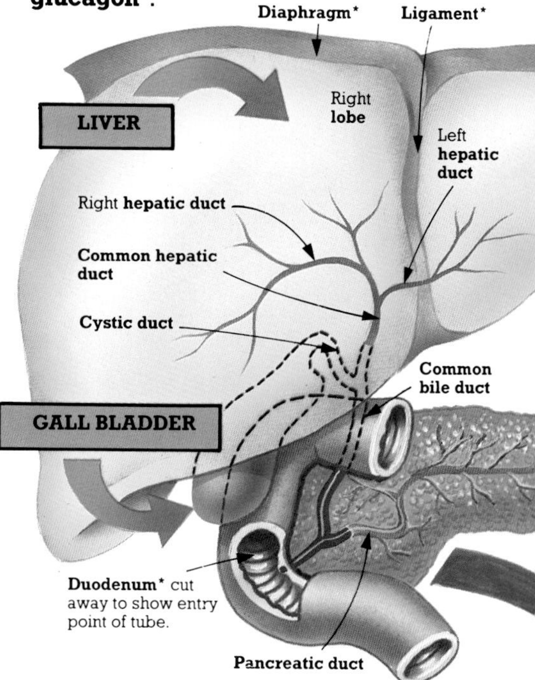

• **Liver**. The largest organ. One of its many roles is that of a **digestive gland**, secreting **bile** (see chart, pages 336-337) along the **common hepatic duct**. Another of its vital jobs is the conversion and storage of newly-digested food matter (see diagram, page 329), which it receives along the **hepatic portal vein** (see picture, page 289), In particular, it regulates the amount of glucose in the blood. It also destroys worn-out red blood cells, stores vitamins and iron, and makes important blood proteins.

* **Diaphragm**, 298; **Duodenum**, 295 (**Small intestine**); **Enzymes**, 331; **Glucagon**, **Hormones**, **Insulin**, 334; **Ligaments**, 280; **Sweat glands**, 311.

- **Gall bladder**. A sac which stores **bile** (made in the **liver**) in a concentrated form until it is needed (i.e. until there is food in the **duodenum***). Its lining has many folds (**rugae**, sing. **ruga**) which flatten out as it expands. When needed, the bile is squeezed along the **cystic duct** and **common bile duct**.

Left lobe

Rugae

Cystic duct

Gall bladder

PANCREAS

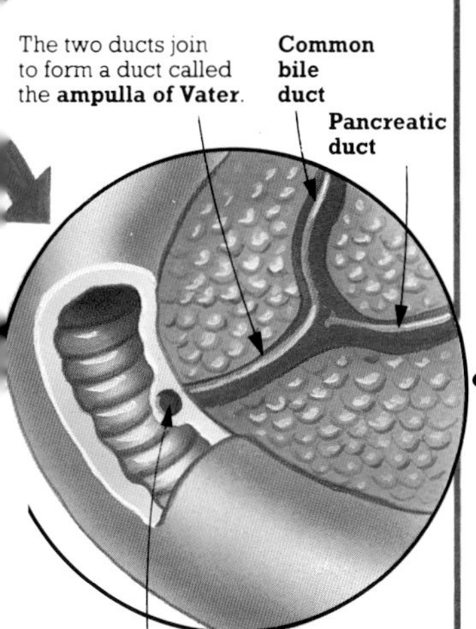

Muscular ring called the **sphincter of Oddi**. If it is closed, **bile** coming from the **liver** is forced back up into the **gall bladder**.

Endocrine glands

Endocrine or **ductless glands** are glands which secrete substances called **hormones** directly into the blood (i.e. blood vessels in the glands). For more about these hormones, and a chart containing all those mentioned below, see pages 334-335. The glands may be separate bodies (e.g. those below) or cells inside organs, e.g. in the sex organs.

- **Pituitary gland**. Also called the **pituitary body** or **hypophysis**. A gland at the base of the brain, directly influenced by the **hypothalamus*** (see also **hormones**, page 334) and made up of an **anterior** (front) **lobe** (**adenohypophysis**) and a **posterior** (back) **lobe** (**neurohypophysis**). Many of its hormones are **tropic hormones**, i.e. they stimulate other glands to secrete hormones. It produces **ACTH**, **TSH**, **STH**, **FSH**, **LH**, **lactogenic hormone**, **oxytocin** and **ADH**.

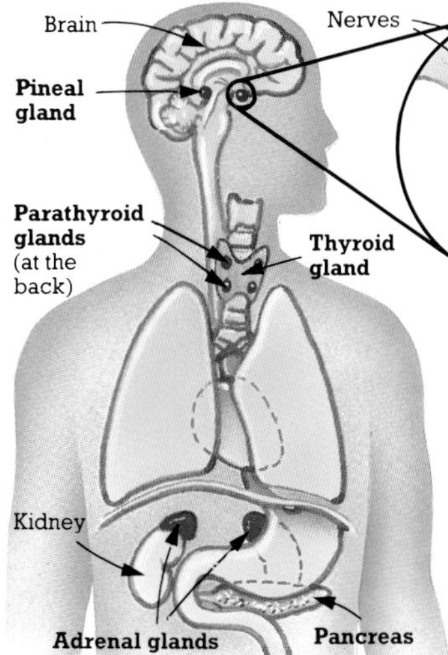

- **Adrenal glands** or **suprarenal glands**. A pair of glands, one gland lying above each kidney. Each has an outer layer (**cortex**), producing **aldosterone**, **cortisone** and **hydrocortisone**, and an inner layer (**medulla**), producing **adrenalin** and **noradrenalin**.

- **Thyroid gland**. A large gland around the **larynx***. It produces **thyroxin** and **thyrocalcitonin**.

- **Parathyroid glands**. Two pairs of small glands embedded in the **thyroid gland**. They produce **PTH**.

- **Pineal gland**. Also called the **pineal body**. A small gland at the front of the brain. Its role is not clear, but it is known to secrete **melatonin**, a hormone thought to influence **sex hormone*** production.

* **Duodenum**, 295 (**Small intestine**); **Hypothalamus**, 303; **Larynx**, 298; **Sex hormones**, 334.

The respiratory system

Position of **respiratory system** ▶

The term **respiration** covers three processes: **ventilation**, or breathing (taking in oxygen and expelling carbon dioxide), **external respiration** (the exchange of gases between the **lungs** and the blood – see also **red blood cells**, page 286) and **internal respiration** (food breakdown, using oxygen and producing carbon dioxide – see pages 332-333). Listed here are the component parts of the human **respiratory system**.

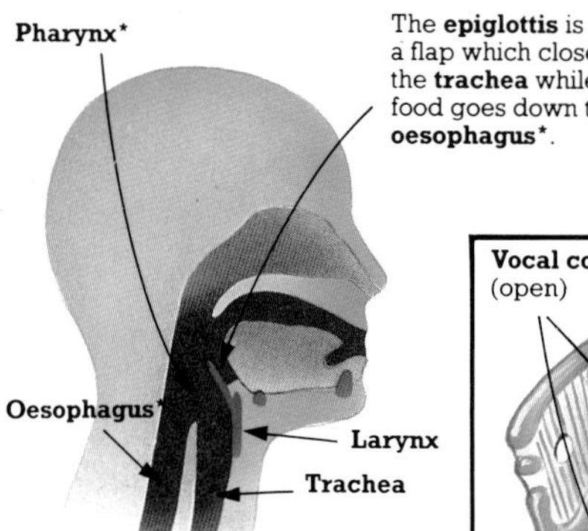

The **epiglottis** is a flap which closes the **trachea** while food goes down the **oesophagus***.

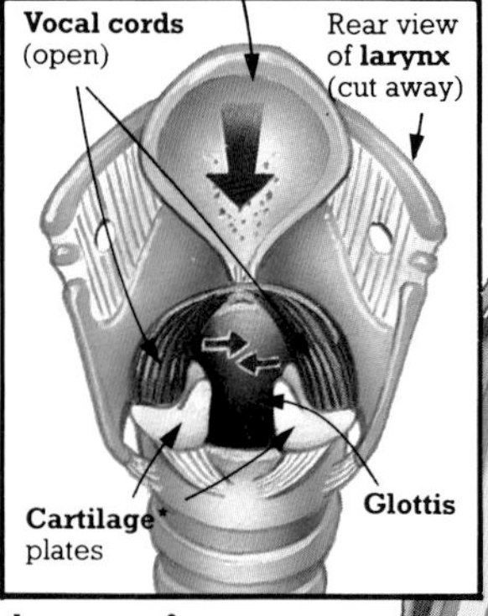

The **lungs** and all the tubes are lined by **mucous membrane*** and **cilia***.

- **Lungs**. The two main breathing organs, inside which gases are exchanged. They contain many tubes (**bronchi** and **bronchioles**) and air sacs (**alveoli**).

- **Trachea** or **windpipe**. The main tube through which air passes on its way to and from the **lungs**.

- **Larynx**. The "voice box" at the top of the **trachea**. It contains the **vocal cords** – two pieces of tissue folding inwards from the trachea lining and attached to plates of **cartilage***. The opening between the cords is called the **glottis**. During speech, muscles pull the cartilage plates (and hence the cords) together, and air passing out through the cords makes them vibrate, producing sounds.

- **Pleura** or **pleural membrane**. A layer of tissue surrounding each **lung** and lining the chest cavity (**thorax**). Between the pleura around a lung and the pleura lining the thorax there is a space (**pleural cavity**). This contains **pleural fluid**. The pleura and fluid-filled cavity make up a cushioning **pleural sac**.

- **Diaphragm** or **midriff**. A sheet of muscular tissue which separates the chest from the lower body, or **abdomen**. At rest, it lies in an arched position, forced up by the abdomen wall below it.

* **Cartilage**, 281; **Cilia**, 268; **Mucous membrane**, 295; **Oesophagus**, **Pharynx**, 294.

- **Bronchi** (sing. **bronchus**). The main tubes into which the **trachea** divides. The first two branches are the right and left **primary bronchi**. Each carries air into a **lung** (via a hole called a **hilum**), alongside a **pulmonary artery*** bringing blood in. They then branch into **secondary bronchi**, **tertiary bronchi** and **bronchioles**, all accompanied by blood vessels, both branching from the pulmonary artery and merging to form **pulmonary veins*** (blood going out).
- **Bronchioles**. The millions of tiny tubes in the **lungs**, all accompanied by blood vessels. They branch off **tertiary bronchi** (see **bronchi**) and have smaller branches called **terminal bronchioles**, each one ending in a cluster of **alveoli**.
- **Alveoli** (sing. **alveolus**). The millions of tiny sacs attached to **terminal bronchioles** (see **bronchioles**). They are surrounded by **capillaries*** (tiny blood vessels) whose blood is rich in carbon dioxide. This passes out through the capillary walls and in through those of the alveoli (to be breathed out). The oxygen which has been breathed into the alveoli passes into the capillaries, which then begin to merge together (eventually forming **pulmonary veins***).

Bronchial means "of the **bronchi** or **bronchioles**"

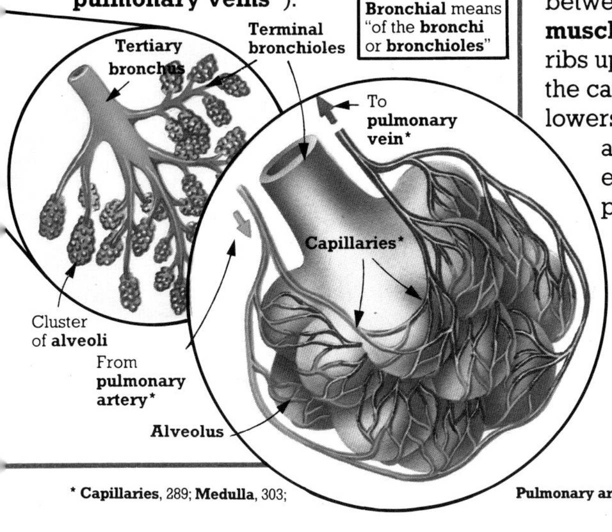

Breathing

Breathing is made up of **inspiration** (breathing in) and **expiration** (breathing out). Both actions are normally automatic, controlled by nerves from the **respiratory centre** in the **medulla*** of the brain. This acts when it detects too high a level of carbon dioxide in the blood.

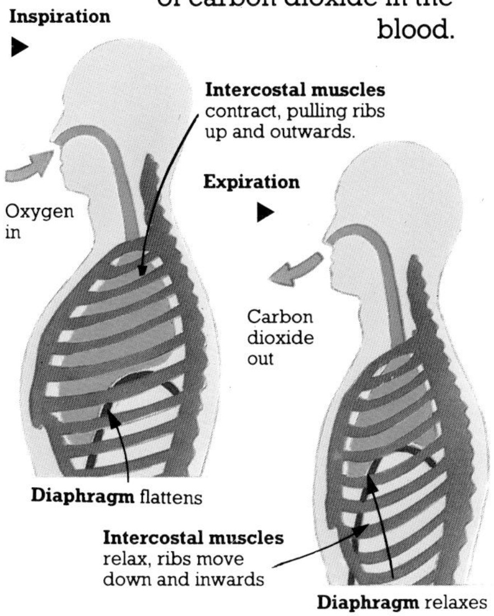

- **Inspiration** or **inhalation**. The act of breathing in. The **diaphragm** contracts and flattens, lengthening the chest cavity. The muscles between the ribs (**intercostal muscles**) also contract, pulling the ribs up and outwards and widening the cavity. The overall expansion lowers the air pressure in the **lungs**, and air rushes in to fill them (i.e. to equalize internal and external pressure).
- **Expiration** or **exhalation**. The act of breathing out. The **diaphragm** and **intercostal muscles** (see **inspiration**) relax, and air is forced out of the **lungs** as the chest cavity becomes smaller.

* **Capillaries**, 289; **Medulla**, 303; **Pulmonary arteries**, 291 (**Pulmonary trunk**); **Pulmonary veins**, 291.

The urinary system

The **urinary system** is the main system of body parts involved in **excretion**, which is the expulsion of unwanted substances. The parts are defined below and right. The lungs and skin are also involved in excretion (expelling carbon dioxide and sweat respectively).

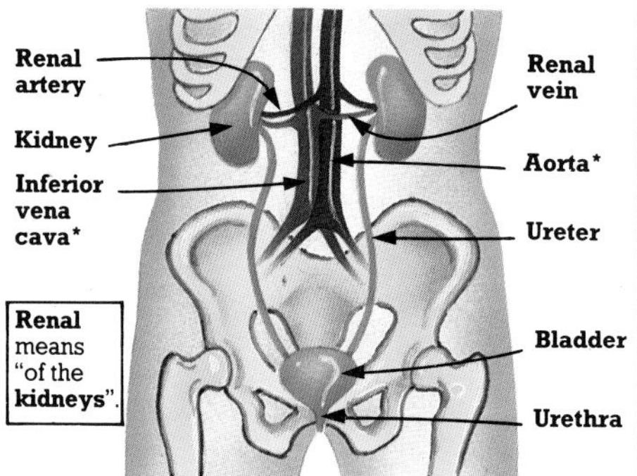

Renal means "of the **kidneys**".

- **Kidneys**. Two organs at the back of the body, just below the ribs. They are the main organs of excretion, filtering out unwanted substances from the blood and regulating the level and contents of body fluids (see also **homeostasis**, page 333). Blood enters a kidney in a **renal artery** and leaves it in a **renal vein**.
- **Ureters**. The two tubes which carry **urine** from the **kidneys** to the **bladder**.

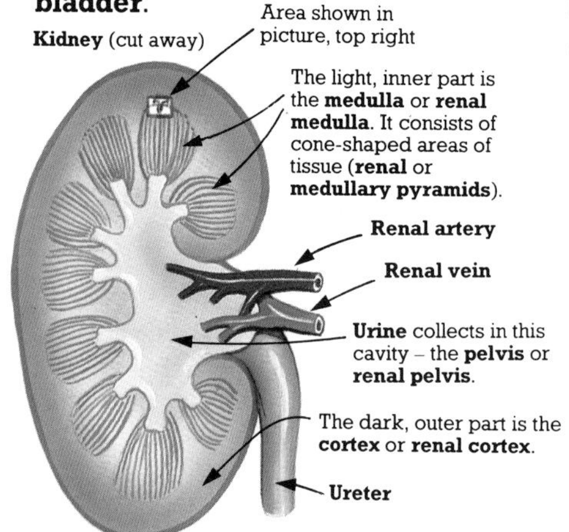

Inside a kidney

1. **Glomerular filtration**. As blood squeezes through the **glomerulus**, most of its water, minerals, vitamins, glucose, **amino acids*** and **urea** are forced into the **Bowman's capsule**, forming **glomerular filtrate**.

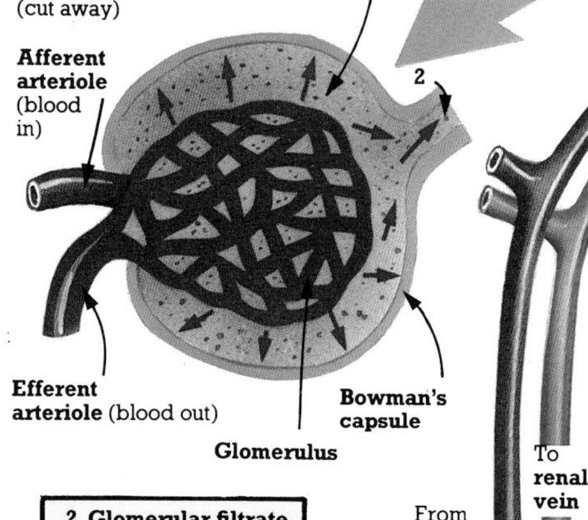

2. **Glomerular filtrate** moves into **proximal convoluted tubule**.

- **Nephrons**. The tiny filtering units of the **kidneys** (there are about a million per kidney). Each consists of a **renal corpuscle** and a **uriniferous tubule**.
- **Renal corpuscles** or **Malpighian corpuscles**. The bodies which filter fluids out of the blood. Each consists of a **glomerulus** and a **Bowman's capsule**.

- **Bladder** or **urinary bladder**. A sac which holds stored **urine**. Its lining has many folds (**rugae**, sing. **ruga**) which flatten out as it fills up, letting it expand. Two muscular rings – the **internal** and **external urinary sphincters** – control the opening from the bladder into the **urethra**. When the volume of urine gets to a certain level, nerves stimulate the internal sphincter to open, but the external sphincter is under conscious control (except in young children), and can be held closed for longer.

* **Amino acids**, 328 (**Proteins**); **Aorta**, **Inferior vena cava**, 291.

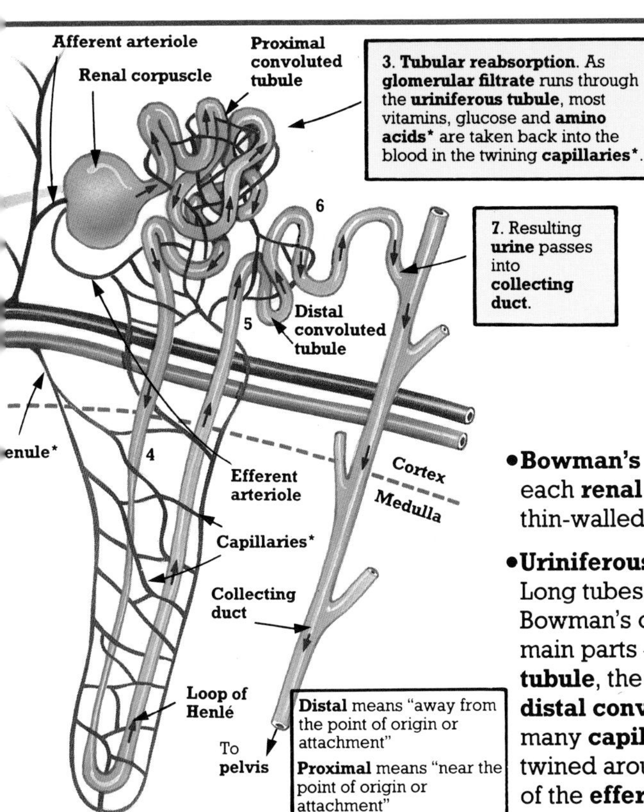

Distal means "away from the point of origin or attachment"

Proximal means "near the point of origin or attachment"

• **Glomerulus**. A ball of coiled-up **capillaries*** (tiny blood vessels) at the centre of each **renal corpuscle**. The capillaries branch from an **arteriole*** entering the corpuscle (an **afferent arteriole)** and re-unite to leave the corpuscle as an **efferent arteriole**.

• **Bowman's capsule**. The outer part of each **renal corpuscle**. It is a thin-walled sac around the **glomerulus**.

• **Uriniferous tubules** or **renal tubules**. Long tubes, each one leading from a Bowman's capsule. Each has three main parts – the **proximal convoluted tubule**, the **loop of Henlé** and the **distal convoluted tubule** – and has many **capillaries*** (tiny blood vessels) twined around it. These are branches of the **efferent arteriole** (see **glomerulus**) and they re-unite to form larger blood vessels carrying blood from the **kidney**.

• **Collecting duct** or **collecting tubule**. A tube which carries **urine** from several **uriniferous tubules** into the **pelvis** of a **kidney**.

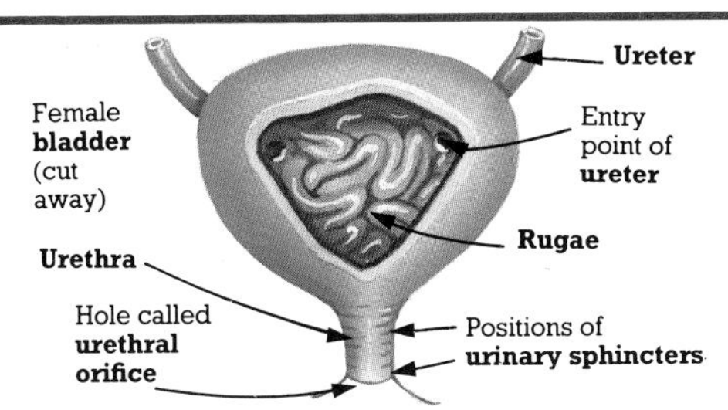

• **Urethra**. The tube carrying **urine** from the **bladder** out of the body (in men, it also carries **sperm*** – see **penis**, page 316). The expulsion of urine is called **urination** or **micturition**.

• **Urea**. A nitrogen-containing (**nitrogenous**) waste substance which is a product of the breakdown of excess **amino acids*** in the liver. It travels in the blood to the **kidneys**, together with smaller amounts of similar substances, e.g. creatinine.

• **Urine**. The liquid which leaves the **kidneys**. Its main constituents are **urea**, excess water and minerals.

* **ADH**, **Aldosterone**, 334; **Amino acids**, 328 (**Proteins**); **Arteriole**, 288 (**Arteries**); **Capillaries**, 289; **Hormones**, 334; **Sperm**, 321 (**Gametes**); **Venule**, 288 (**Veins**).

The central nervous system

The **central nervous system** (**CNS**) is the body's control centre. It co-ordinates all its actions, both mechanical and chemical (working with **hormones***) and is made up of the **brain** and **spinal cord**. The millions of nerves in the body carry "messages" (nervous impulses) to and from these central areas (see pages 306-309).

- **Brain**. The organ which controls most of the body's activities. It is the only organ able to produce "intelligent" action – action based on past experience (stored information), present events and future plans. It is made up of millions of **neurons*** (nerve cells), arranged into **sensory**, **association** and **motor areas**. The sensory areas receive information (nervous impulses) from all body parts and the association areas analyse the impulses and make decisions. The motor areas send impulses (orders) to muscles or glands. The impulses are carried by the fibres of 43 pairs of nerves – 12 pairs of **cranial nerves** serving the head and 31 pairs of **spinal nerves** (see **spinal cord**).

- **Spinal cord**. A long string of nervous tissue running down from the **brain** inside the **vertebral column***. Nervous impulses from all parts of the body pass through it. Some are carried into or away from the brain, some are dealt with in the cord (see **involuntary actions**, page 309). 31 pairs of **spinal nerves** branch out from the cord through the gaps between the **vertebrae***. Each spinal nerve is made up of two groups of fibres: a **dorsal** or **sensory root**, made up of the fibres of **sensory neurons*** (bringing impulses in), and a **ventral** or **motor root**, made up of the fibres of **motor neurons*** (taking impulses out).

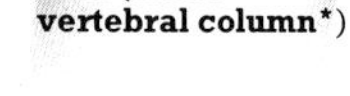

The parts of the brain

- **Cerebrum**. The largest, most highly developed area, with many deep folds. It is composed of two **cerebral hemispheres**, joined by the **corpus callosum** (a band of **nerve fibres***), and its outer layer is called the **cerebral cortex**. This contains all the most important **sensory**, **association** and **motor areas** (see **brain**). It controls most physical activities and is the centre for mental activities such as decision-making, speech, learning, memory and imagination.

- **Cerebellum**. The area which co-ordinates muscle movement and balance, two things under the overall control of the **cerebrum**.

- **Midbrain** or **mesencephalon**. An area joining the **diencephalon** to the **pons**. It carries impulses in towards the **thalamus**, and out from the **cerebrum** towards the **spinal cord**.

- **Pons** or **pons Varolii**. A junction of **nerve fibres*** which forms a link between the parts of the **brain** and the **spinal cord** (via the **medulla**).

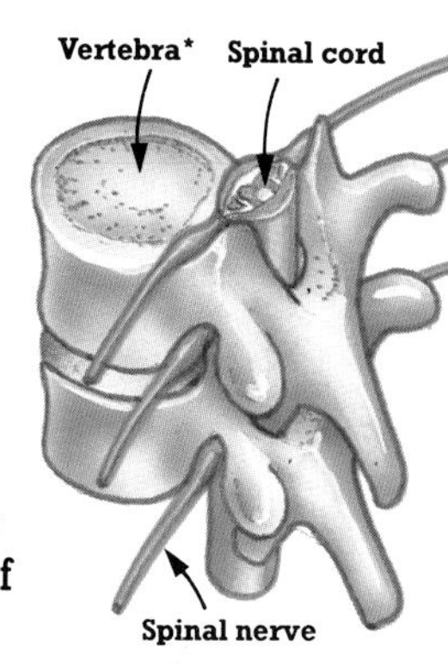

* **Hormones**, 334; **Motor neurons**, 305; **Nerve fibres**, **Neurons**, 304; **Sensory neurons**, 305; **Vertebrae**, **Vertebral column**, 278.

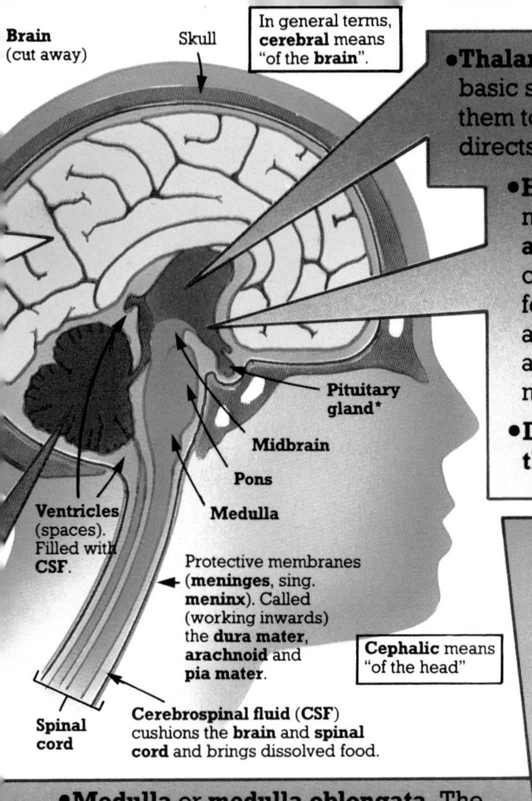

In general terms, **cerebral** means "of the **brain**".

Cephalic means "of the head"

- **Thalamus.** The area which carries out the first, basic sorting of incoming impulses and directs them to different parts of the **cerebrum**. It also directs some outgoing impulses.

- **Hypothalamus.** The master controller of most inner body functions. It controls the **autonomic nervous system*** (the nerve cells causing unconscious action, e.g. food movement through the intestines) and the action of **pituitary gland***. Its activities are vital to **homeostasis*** – the maintenance of stable internal conditions.

- **Diencephalon.** A collective term for the **thalamus** and **hypothalamus**.

- **Medulla** or **medulla oblongata.** The area which controls the "fine tuning" of many unconscious actions (under the overall control of the **hypothalamus**). Different parts of it control different actions, e.g. the **respiratory centre** controls breathing.

- **Brain stem.** A collective term for the **midbrain**, **pons** and **medulla**.

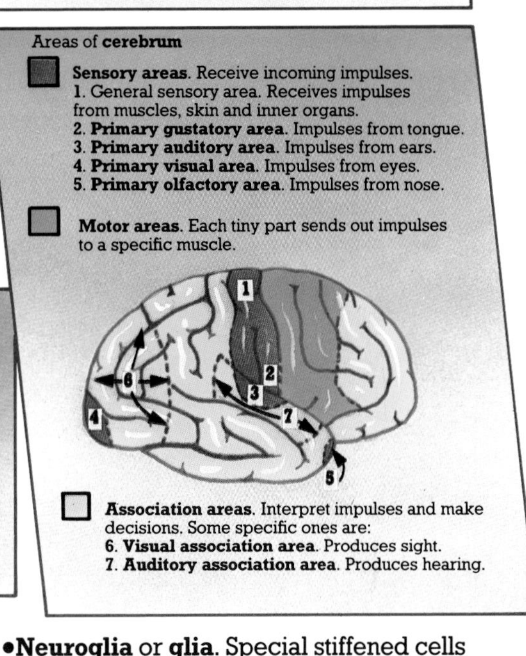

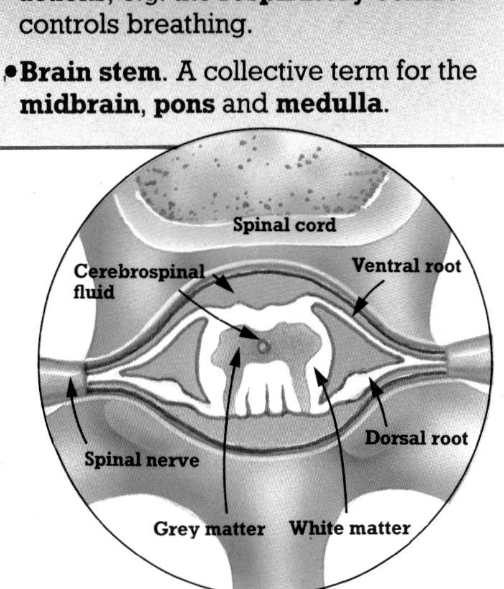

- **Neuroglia** or **glia.** Special stiffened cells which support and protect the nerve cells (**neurons***) of the CNS. Some produce a white fatty substance called **myelin** (see also **Schwann cells**, page 304). This coats the long fibres found in the connective areas of the **brain** and the outer layer of the **spinal cord**, and leads to these areas being known as **white matter. Grey matter**, by contrast, consists mainly of **cell bodies*** and their short fibres, and its neuroglia does not produce myelin.

* **Autonomic nervous system**, 308; **Cell body**, 304; **Homeostasis**, 333; **Neurons**, 304; **Periosteum**, 280; **Pituitary gland**, 297.

The units of the nervous system

The individual units of both the brain and spinal cord (**central nervous system***) and the nerves of the rest of the body (**peripheral nervous system**) are the nerve cells, or **neurons**. They are unique in being able to transmit electrical "messages" (the vital nervous impulses) around the body. Each neuron consists of a **cell body**, an **axon** and one or more **dendrites**, and there are three types of neuron – **sensory**, **association** and **motor neurons**.

The parts of a neuron

- **Cell body** or **perikaryon**. The part of a neuron containing its **nucleus*** and most of its **cytoplasm***. The cell bodies of all **association**, some **sensory** and some **motor neurons** lie in the brain and spinal cord. Those of the other sensory neurons are found in special masses called **ganglia*** or as part of highly specialized **receptors*** in the nose and eyes. Those of the other motor neurons lie in **autonomic ganglia***.

- **Nerve fibres**. The fibres (**axon** and **dendrites**) of a neuron. They are extensions of the **cytoplasm*** of the **cell body** and carry the vital nervous impulses. Most of the long nerve fibres which run out round the body (belonging to **sensory** or **motor neurons**) are accompanied by **neuroglial*** cells. These are called **Schwann cells** and they produce a sheath of **myelin*** around each fibre.

- **Dendrites**. The **nerve fibres** carrying impulses towards a **cell body**. Most neurons have several short dendrites, but one type of **sensory neuron** has just one, elongated dendrite, often called a **dendron**. The endings of these dendrons form **receptors*** all over the body, and the dendrons themselves run inwards to the cell bodies (which are found in **ganglia*** just outside the spinal cord).

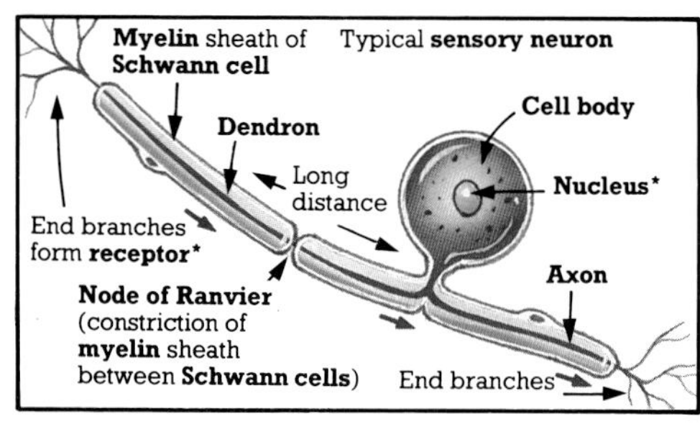

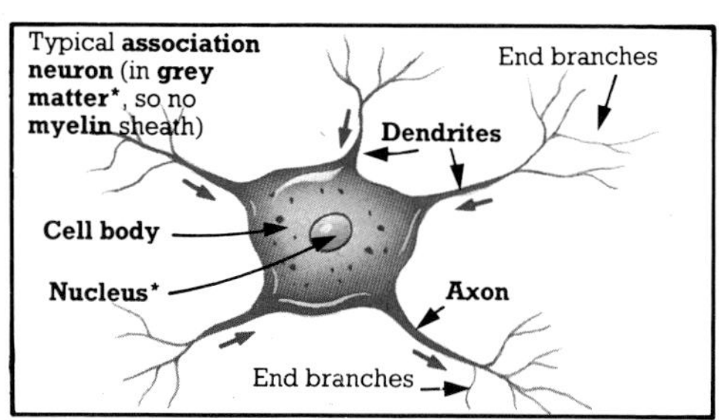

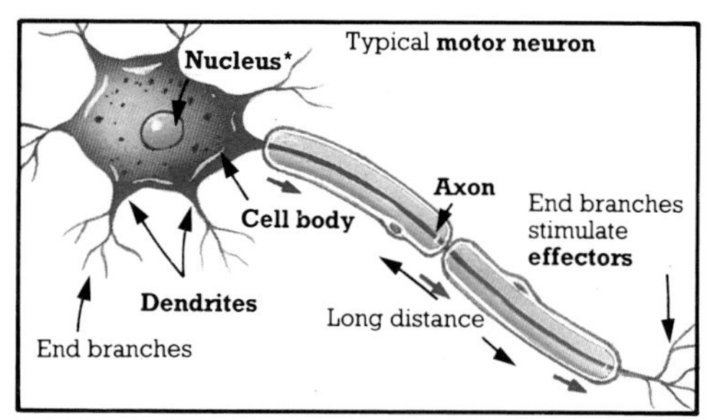

- **Axon**. The single long **nerve fibre** which carries impulses away from a **cell body**. The axons of all **association** and **sensory neurons** and some **motor neurons** lie in the brain and spinal cord. Those of the other motor neurons run out of the spinal cord to **autonomic ganglia***, or further to **effectors** (see **motor neurons**).

* **Autonomic ganglia**, 309; **Central nervous system**, 302; **Cytoplasm**, 238; **Ganglia**, 306; **Grey matter**, **Myelin**, 303 (**Neuroglia**); **Nucleus**, 238; **Receptors**, 307.

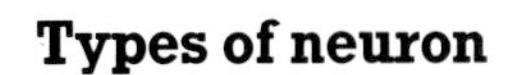

Types of neuron

- **Sensory neurons** or **afferent neurons**. The neurons which carry "information" (nervous impulses) about sensations. The single **dendrites** (**dendrons**) of some sensory neurons run throughout the body, and their endings fire off impulses when stimulated. For more about these endings (**receptors**) and the different sensory neurons, see pages 306-307.

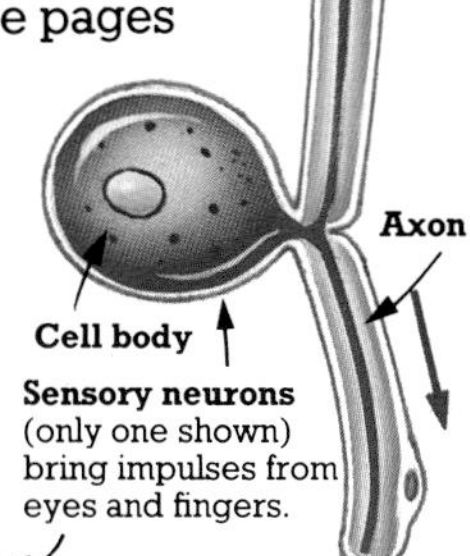

Sensory neurons (only one shown) bring impulses from eyes and fingers.

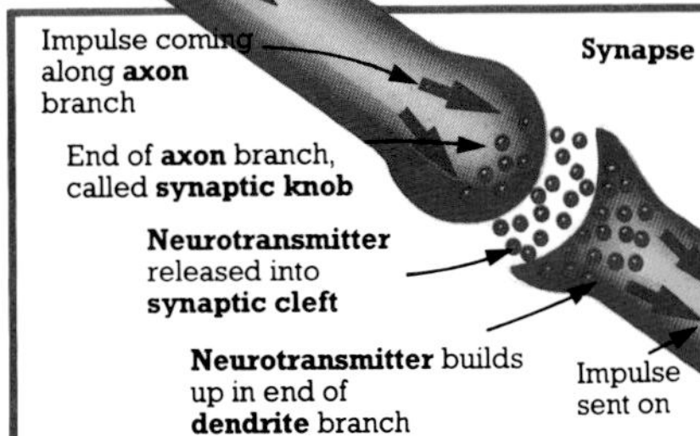

- **Synapses**. The tiny areas where the branching ends of the **axon** of one neuron meet the **dendrites** of the next. When an impulse reaches the end of the axon, a special chemical called a **neurotransmitter** is released into the minute gap (**synaptic cleft**) found at the junction. When enough of this has reached the other side, an impulse is sent on in the dendrites.

Association neurons (only one shown) analyse information and operate in decision-making.

- **Motor neurons** or **efferent neurons**. The neurons which carry "instructions" (nervous impulses) away from the brain and spinal cord. The ends of the **axons** of some motor neurons make connections with muscles or glands (called **effectors**), and the impulses they carry (passed on to them from **association neurons**) stimulate these organs into action. For more about the different motor neurons, see pages 308-309.

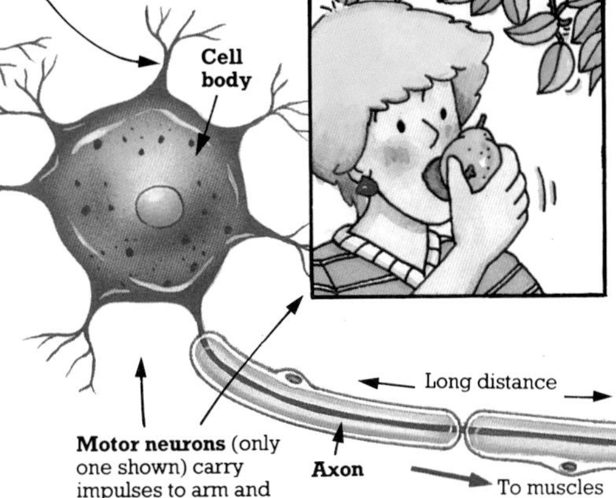

Motor neurons (only one shown) carry impulses to arm and jaw muscles.

- **Association neurons**. Also called **relay**, **internuncial** or **connecting neurons**, or **interneurons**. Special linking neurons, present in vast numbers in the brain and spinal cord. They are involved in picking up impulses (from **sensory neurons**), interpreting the sensory information, and passing impulses to **motor neurons** to initiate actions.

* **Receptors**, 307.

Nerves and nervous pathways

The **sensitivity** (**irritability**) of the body (its ability to respond to stimuli) relies on the transportation of "messages" (nervous impulses) by the fibres of nerve cells (**neurons***). The fibres which bring impulses into the brain and spinal cord are part of the **afferent system**. Those which carry impulses from the brain and cord are part of the **efferent system** (see pages 308-309). The fibres outside the brain and cord make up the **nerves** of the body, known collectively as the **peripheral nervous system** (**PNS**).

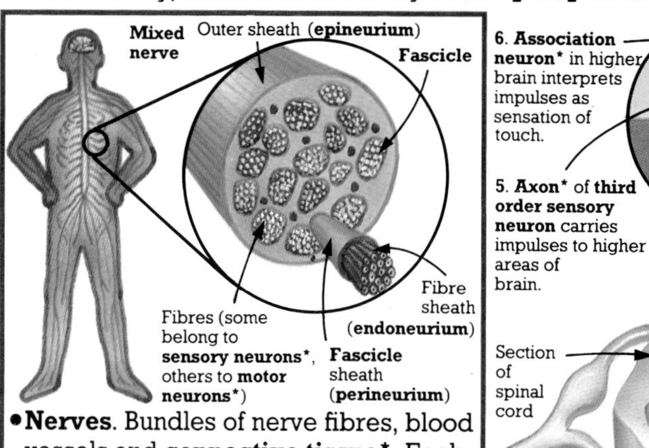

- **Nerves**. Bundles of nerve fibres, blood vessels and **connective tissue***. Each nerve consists of several bundles (**fascicles**) of fibres and each fibre is part of a nerve cell (**neuron***). **Sensory nerves** have just the fibres (**dendrons***) of **sensory** (**afferent**) **neurons***, **motor nerves** have just the fibres (**axons***) of **motor** (**efferent**) **neurons*** and **mixed nerves** have both types of fibre.

The afferent system

The **afferent system** is the system of nerve cells (**neurons***) whose fibres carry sensory information (nervous impulses) towards the spinal cord, up inside it and into the brain. The nerve cells involved are all the **sensory** (**afferent**) **neurons*** of the body. The impulses originate in **receptors** and are interpreted by the brain as sensations.

Tactile means "perceptible by the sense of touch".

6. **Association neuron*** in higher brain interprets impulses as sensation of touch.

BRAIN

5. **Axon*** of **third order sensory neuron** carries impulses to higher areas of brain.

4. **Axon*** of **second order sensory neuron** passes impulses to **dendrites*** of **third order sensory neuron** in lower brain.

Section of spinal cord

Cell bodies* of **first order sensory neurons** lie in bulging masses (**ganglia**, sing. **ganglion**) in **dorsal roots** of **spinal nerves***.

3. **Axon*** of **first order sensory neuron** passes impulses to **dendrites*** of **second order sensory neuron**.

Long distance (inside **nerve**)

Afferent means "leading towards".

2. **Dendron*** of **first order sensory neuron** carries impulses towards spinal cord.

1. **Receptor** in skin (**Meissner's corpuscle***) stimulated by contact.

The routes taken by nervous impulses are **neural pathways**. This is a simplified neural pathway of the **afferent system**. Only one of each type of **neuron*** is shown (in reality, there would be more involved).

* **Association neurons**, 305; **Axon, Cell body**, 304; **Connective tissue**, 280; **Dendron**, 304 (**Dendrites**); **Meissner's corpuscles**, 310; **Motor neurons**, 305; **Neurons**, 304; **Sensory neurons**, 305; **Spinal nerves**, 302 (**Spinal cord**).

• **Receptors**. The parts of the **afferent system** which fire off nervous impulses when they are stimulated. Most are either the single branched ending of the long **dendron*** of a **first order sensory neuron** (see picture) or a group of such endings. They are all embedded in body tissue, and many have some kind of structure formed around them (e.g. a **taste bud** – see **tongue**). They are found all over the body, both near the surface (in the skin, **sense organs**, **skeletal muscles***, etc.) and deeper inside (connected to inner organs, blood vessel walls, etc.).

• **Sense organs**. The highly specialized sensory organs of the body, each with many **receptors**. They are the **nose**, **tongue**, eyes and ears. For more about eyes and ears, see pages 312-315.

Divisions of the afferent system

Central and **peripheral nervous systems**. All nerve cells in body.

- **Afferent system**. Nerve cells bringing impulses in and up.
  - **Somatic afferent system**. Nerve cells bringing impulses from **receptors** near body surface.
  - **Visceral afferent system**. Nerve cells bringing impulses from **receptors** deep inside body.
- **Efferent system**. Nerve cells taking impulses down and out (see pages 308-309).

• **Nose**. The organ of smell. Each of its two nostrils opens into a **nasal cavity** which is lined with **mucous membrane*** and has many **olfactory hairs** extending from its roof. The hairs are the **dendrites*** of special **sensory neurons*** called **olfactory cells**. These are the **receptors** whose impulses are interpreted by the brain as sensations of smell (**olfactory sensations**).

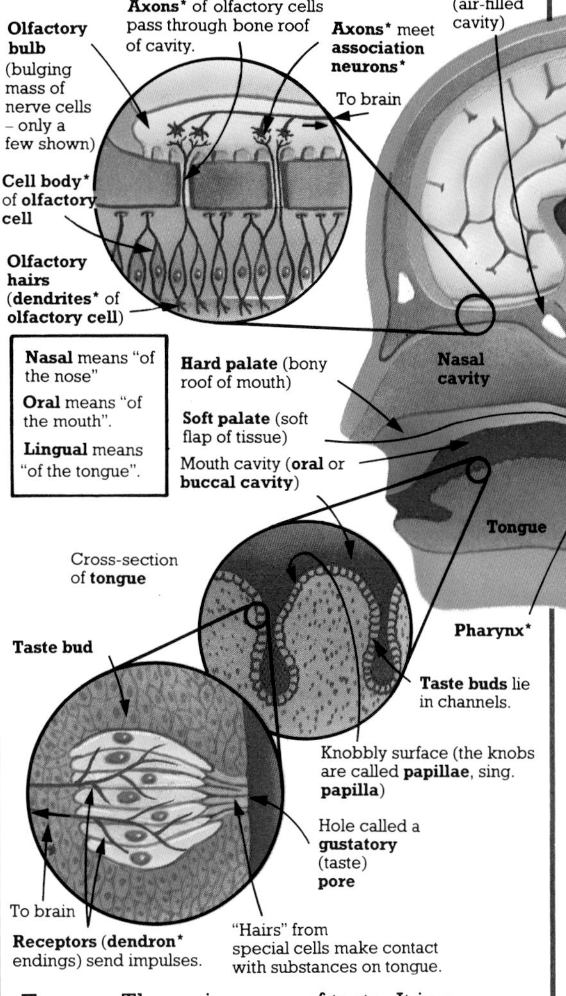

• **Tongue**. The main organ of taste. It is a muscular organ which bears many **taste buds**. These tiny bodies contain the **receptors** whose impulses are interpreted by the brain as taste sensations (**gustatory sensations**).

* **Association neurons**, 305; **Axon**, **Cell body**, 304; **Dendron**, 304 (**Dendrites**); **Mucous membrane**, 295; **Pharynx**, 294; **Sensory neurons**, 305; **Skeletal muscles**, 283.

The efferent system

The **efferent system** is the second system of nerve cells (**neurons***) in the body (see also **afferent system**, pages 306-307). The fibres of its nerve cells carry nervous impulses away from the brain, down through the spinal cord and out around the body. The nerve cells involved are all the **motor** (**efferent**) **neurons*** of the body. The impulses they carry stimulate action in the surface muscles (**skeletal muscles***) or in the glands and internal muscles (in the walls of inner organs and blood vessels). All these organs are known collectively as **effectors**.

Divisions of the efferent system

Central and peripheral nervous systems. All nerve cells in body.

- **Afferent system.** Nerve cells bringing impulses in and up (see pages 306-307).
- **Efferent system.** Nerve cells taking impulses down and out. **Efferent** means "leading away from".
  - **Somatic efferent system**. Nerve cells taking impulses to body surface (**skeletal muscles***). Cause **voluntary actions**.
  - **Autonomic nervous system** (**visceral efferent system**). Nerve cells taking impulses to inner organs. Cause **autonomic actions**.
    - **Sympathetic division**. Nerve cells whose impulses prepare body for action, e.g. increase heart rate.
    - **Parasympathetic division**. Nerve cells whose impulses restore and maintain normal body conditions, e.g. decrease heart rate.

The different actions

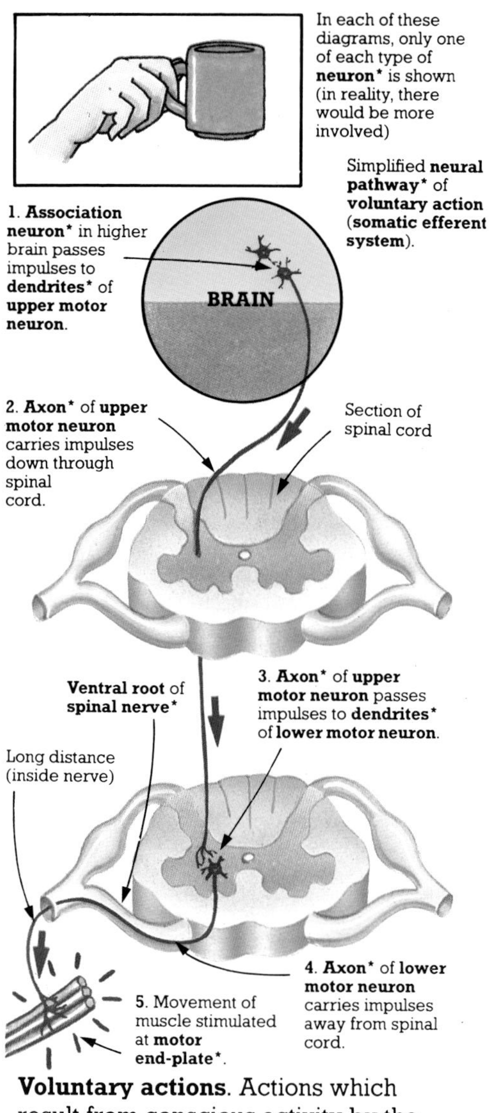

Voluntary actions. Actions which result from conscious activity by the brain, i.e. ones it consciously decides upon, e.g. lifting a cup. We are always aware of these actions, which involve **skeletal muscles*** only. The impulses which cause them originate in higher areas of the brain (especially the **cerebrum***) and are carried by nerve cells of the **somatic efferent system**.

* **Association neurons**, 305; **Axon**, 304; **Cerebrum**, 302; **Dendrites**, 304; **Motor end-plate**, 283; **Motor neurons**, 305; **Neural pathways**, 306; **Neurons**, 304; **Skeletal muscles**, 283; **Spinal nerves**, 302 (**Spinal cord**).

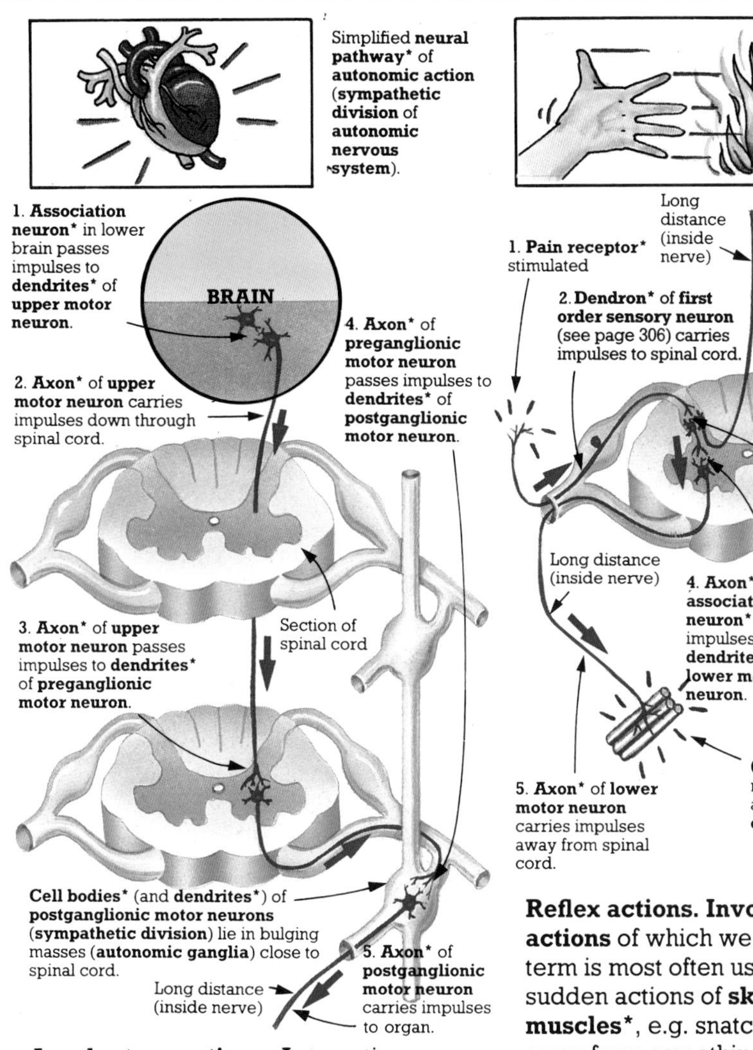

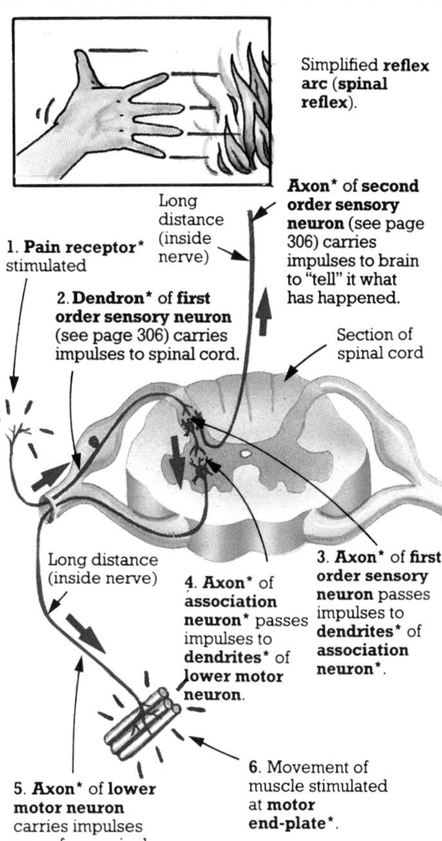

Involuntary actions. Automatic actions (ones the brain does not consciously decide upon). There are two types. Firstly, there are the constant actions of inner organs, e.g. the beating of the heart, of which we are not normally aware. The impulses which cause them originate in the lower brain (especially the **hypothalamus***) and are carried by nerve cells of the **autonomic nervous system**. They are called **autonomic actions**. The other involuntary actions are **reflex actions**.

Reflex actions. Involuntary actions of which we are aware. The term is most often used to refer to sudden actions of **skeletal muscles***, e.g. snatching the hand away from something hot. The impulses which cause such an action are carried by nerve cells of the **somatic efferent system** and the entire **neural pathway*** is a "short-circuited" one, called a **reflex arc**. In the case of **cranial reflexes** (those of the head, e.g. sneezing), this pathway involves a small part of the brain; with **spinal reflexes** (those of the rest of the body), the brain is not actively involved, only the spinal cord.

* **Association neurons**, 305; **Axon**, **Cell body**, 304; **Dendron**, 304 (**Dendrites**); **Hypothalamus**, 303; **Motor end-plate**, 283; **Nerve**, **Neural pathways**, 306; **Pain receptors**, 311; **Skeletal muscles**, 283; **Spinal nerves**, 302 (**Spinal cord**).

The skin

The **skin** or **cutis** is the outer body covering, made up of several tissue layers. It registers external stimulation, protects against damage or infection, prevents drying out, helps regulate body temperature, excretes waste (**sweat**), stores fat and makes **vitamin D***. It contains many tiny structures, each type with a different function. The entire skin (tissue layers and structures), is called the **integumentary system**.

The different layers

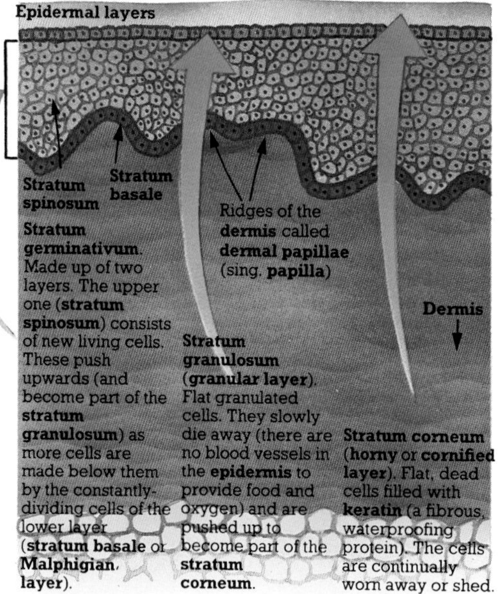

- **Epidermis**. The thin outer layer of the skin which forms its **epithelium** (a term for any sheet of cells which forms a surface covering or a cavity lining). It is made up of several layers (**strata**, sing. **stratum**), shown above.
- **Dermis**. The thick layer of **connective tissue*** under the **epidermis**, containing most of the embedded structures (see introduction). It also contains many **capillaries*** (tiny blood vessels) which supply food and oxygen.
- **Subcutaneous layer** or **superficial fascia**. The layer of fatty tissue (**adipose tissue**) below the **dermis** (it is a fat store). Elastic fibres run through it to connect the dermis to the organs below, e.g. muscles. It forms an insulating layer.

Structures in the skin

- **Meissner's corpuscles**. Special bodies formed around nerve fibre endings. There are especially large numbers on the fingertips and palms. They are touch **receptors***, i.e. they send impulses to the brain when the skin makes contact with an object.

- **Sebaceous glands**. **Exocrine glands*** which open into **hair follicles**. They produce an oil called **sebum** which waterproofs the hairs and **epidermis** and keeps them supple.

- **Hair erector muscles**. Special muscles, each attached to a **hair follicle**. When they contract (in the cold), the hairs straighten. This traps more air and improves insulation (especially in animals with lots of hair, feathers or fur). It also causes "goose-pimples".

- **Hair follicles**. Long narrow tubes, each containing a hair. The hair grows as new cells are added to its base from the cells lining the follicle. Its older cells die as **keratin** forms inside them (see **stratum corneum**).

* **Capillaries**, 289; **Connective tissue**, 280; **Exocrine glands**, 297; **Receptors**, 307; **Vitamin D**, 337.

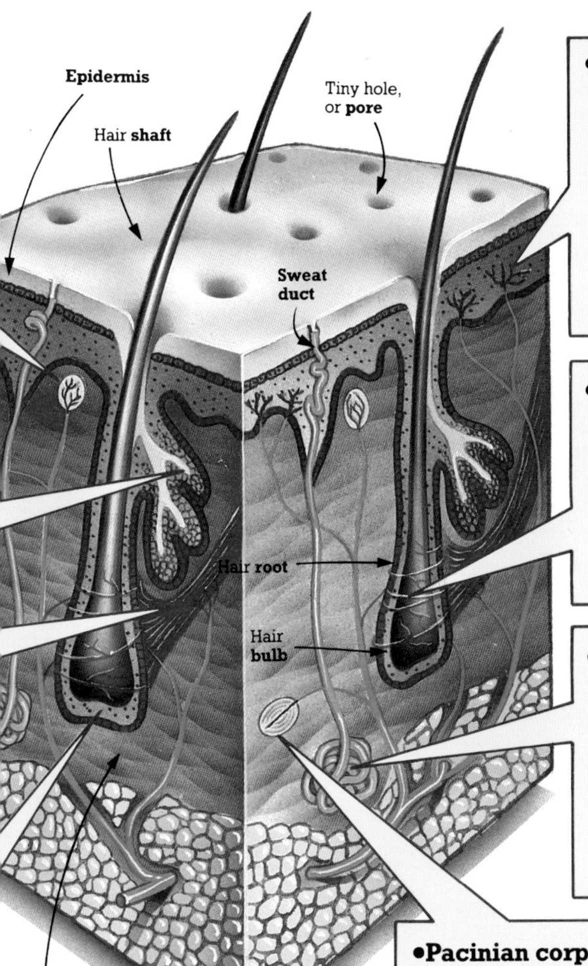

- **Pain receptors**. Nerve fibre endings in the tissue of most inner organs and in the skin (in the **epidermis** and the top of the **dermis**). They are the **receptors*** which send impulses when any stimulation (e.g. pressure. heat, touch) becomes excessive. This is what causes a sensation of pain.

- **Hair plexuses** or **root hair plexuses**. Special groups of nerve fibre endings. Each forms a network around a **hair follicle** and is a **receptor***, i.e. it sends nervous impulses to the brain, in this case when the hair moves.

- **Sweat glands** or **sudoriferous glands**. Coiled **exocrine glands*** which excrete **sweat**. Each has a narrow tube (**sweat duct**) going to the surface. Sweat consists of water, salts and **urea***, which enter the gland from the cells and **capillaries*** (blood vessels).

- **Pacinian corpuscles**. Special bodies formed around single nerve fibre endings, lying in the lower skin layers and the walls of inner organs. They are pressure **receptors***, i.e. they send impulses to the brain when the tissue is receiving deep pressure rather than light touch.

- **Melanin**. A brown **pigment*** which shields against ultra-violet light by absorbing the light energy. It is found in all the layers of the **epidermis** of people from tropical areas, giving them dark skin. Fair-skinned people only have melanin in their lower epidermal layers, but produce more when in direct sunlight, causing a suntan.

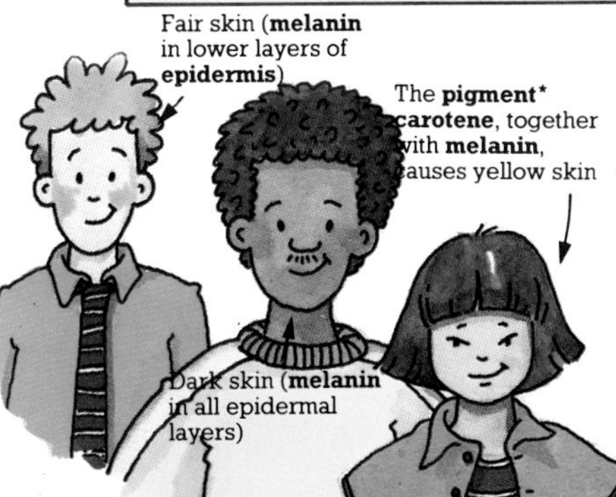

* **Capillaries**, 289; **Exocrine glands**, 297; **Pigment**, 255; **Receptors**, 307; **Urea**, 301.

The eyes

The **eyes** are the organs of **sight**, sending nervous impulses to the brain when stimulated by light rays from external objects. The brain interprets the impulses to produce images. Each eye consists of a hollow, spherical capsule (**eyeball**), made up of several layers and structures. It is set into a socket in the skull (an **orbit**), and is protected by eyelids and eyelashes.

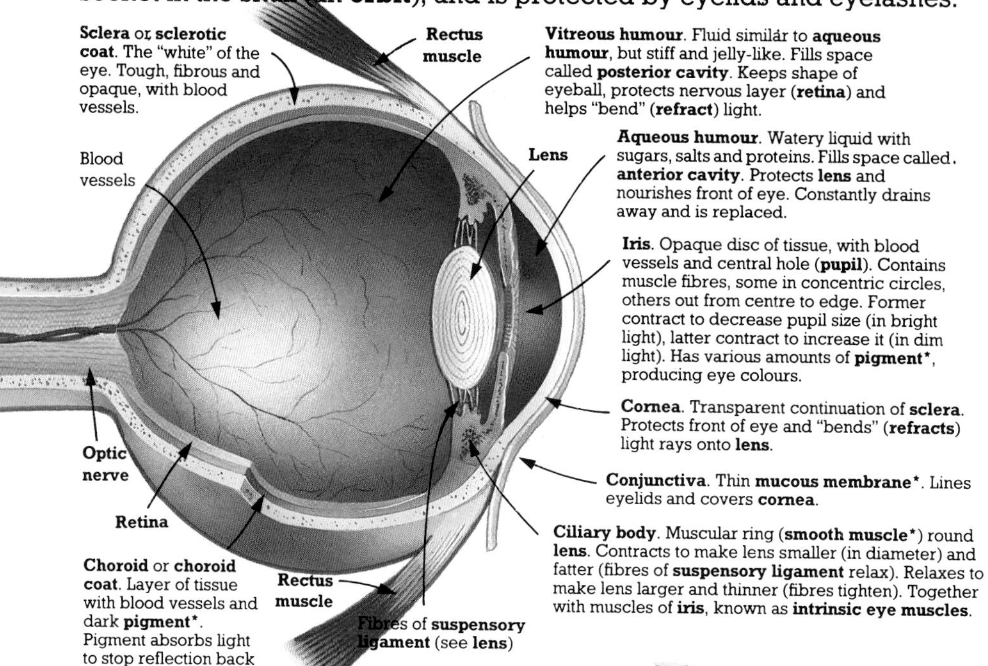

- **Lens**. The transparent body whose role, like that of any lens, is to focus the light rays passing through it, i.e "bend" (**refract**) them so that they come to a point, in this case on the **retina** (for more about lenses and refraction, see pages 50-53). A lens consists of many thin tissue layers and is held in place by the fibres of a **ligament*** called the **suspensory ligament**. These join it to the **ciliary body**, which can alter the lens shape so that light rays are always focused on the retina, whatever the distance of the object being looked at. This is known as **accommodation**. The rays form an upside-down image, but this is corrected by the brain.

Accommodation

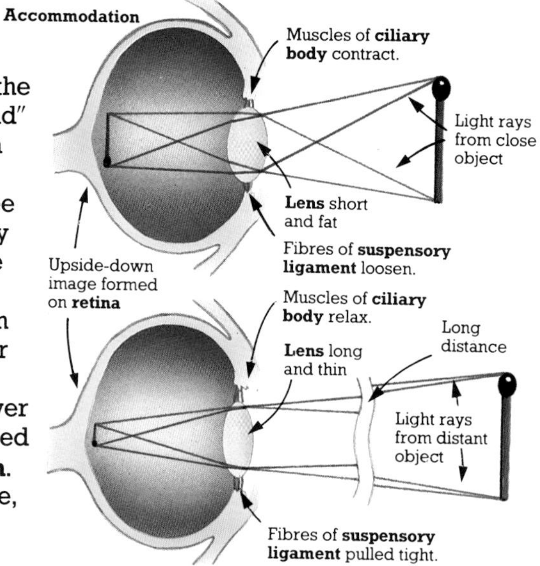

* **Ligaments**, 280; **Mucous membrane**, 295; **Pigments**, 255; **Smooth muscle**, 283.

The inner nervous layer

- **Retina**. The innermost layer of tissue at the back of the eyeball, made up of a layer of **pigment*** and a nervous layer consisting of millions of sensory nerve cells (**sensory neurons***) and their fibres. These lie in chains and carry nervous impulses to the brain. The first cells in the chains are **receptors***, i.e. their end fibres (**dendrons***) fire off the impulses when they are stimulated (by light rays). These fibres are called **rods** and **cones** because of their shapes. The receptors are **photoreceptors** (i.e. stimulated by light).

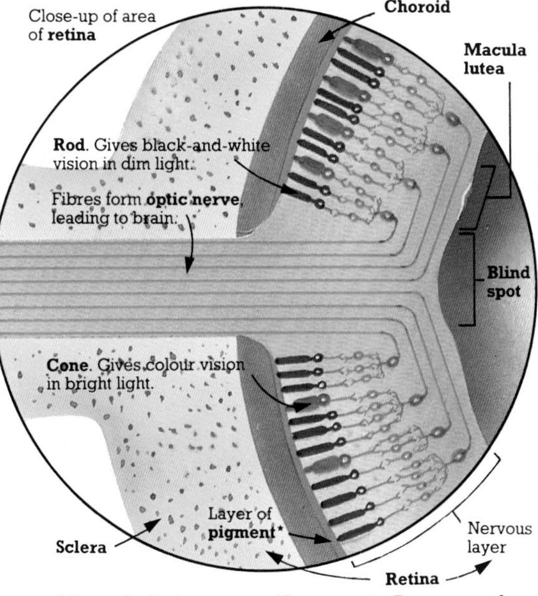

- **Macula lutea** or **yellowspot**. An area of yellowish tissue in the centre of the **retina**. It has a small central dip, called the **fovea** or **fovea centralis**. This has the highest concentration of **cones** (see **retina**) and is the area of acutest vision. If you look directly at a specific object, its light rays are focused on the fovea.

- **Blind spot** or **optic disc**. The point in the retina where the **optic nerve** leaves the eye. It has no **receptors** (see **retina**) and so cannot send any impulses.

Structures around the eyeballs

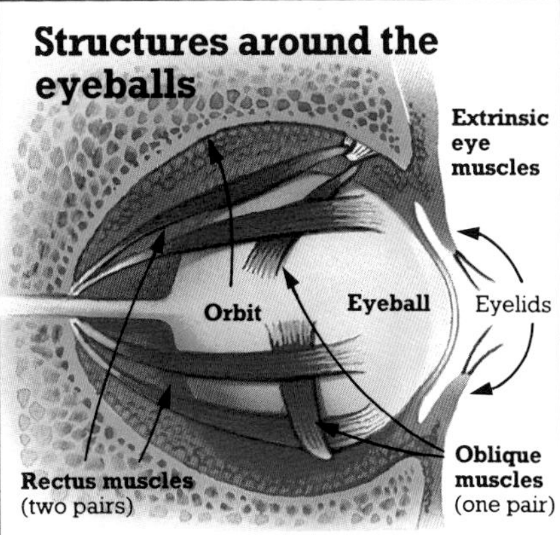

- **Extrinsic eye muscles**. The three pairs of muscles joining the eyeball to the eye socket (**orbit**). They contract to make the eyeball swivel around.

- **Lachrymal glands** or **tear glands**. Two **exocrine glands***, one at the top of each eye socket (**orbit**). They secrete a watery fluid onto the lining of the upper eyelids via tubes called **lachrymal ducts**. The fluid contains salts and an anti-bacterial **enzyme***, and it washes over the surface of the eyes, keeping them moist and clean. It drains away via four **lachrymal canals**, two at the inside corner of each eye, which join to form a **nasolachrymal duct**. This empties into a **nasal cavity***.

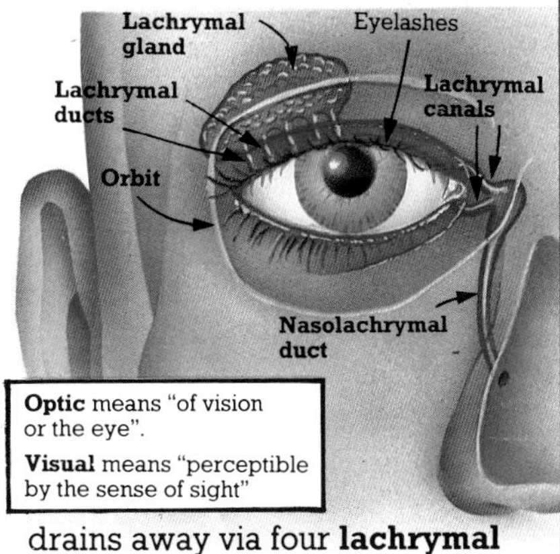

Optic means "of vision or the eye".

Visual means "perceptible by the sense of sight"

* **Dendron**, 304 (**Dendrites**); **Enzymes**, 331; **Exocrine glands**, 297; **Nasal cavity**, 307 (**Nose**); **Pigments**, 255; **Receptors**, 307; **Sensory neurons**, 305.

The ears

The two **ears** are the organs of hearing and balance. Each one is divided into three areas – the **outer ear**, the **middle ear** and the **inner ear**.

- **Outer ear** or **external ear**. An outer "shell" of skin and **cartilage*** (**pinna** or **auricle**), together with a short tube (**ear canal** or **external auditory canal**). The tube lining contains special **sebaceous glands*** (**ceruminous glands**) which secrete **cerumen** (ear wax).

- **Middle ear** or **tympanic cavity**. An air-filled cavity which contains a chain of three tiny bones (**ear ossicles** or **auditory ossicles**) called the **malleus** (or **hammer**), **incus** (or **anvil**) and **stapes** (or **stirrup**).

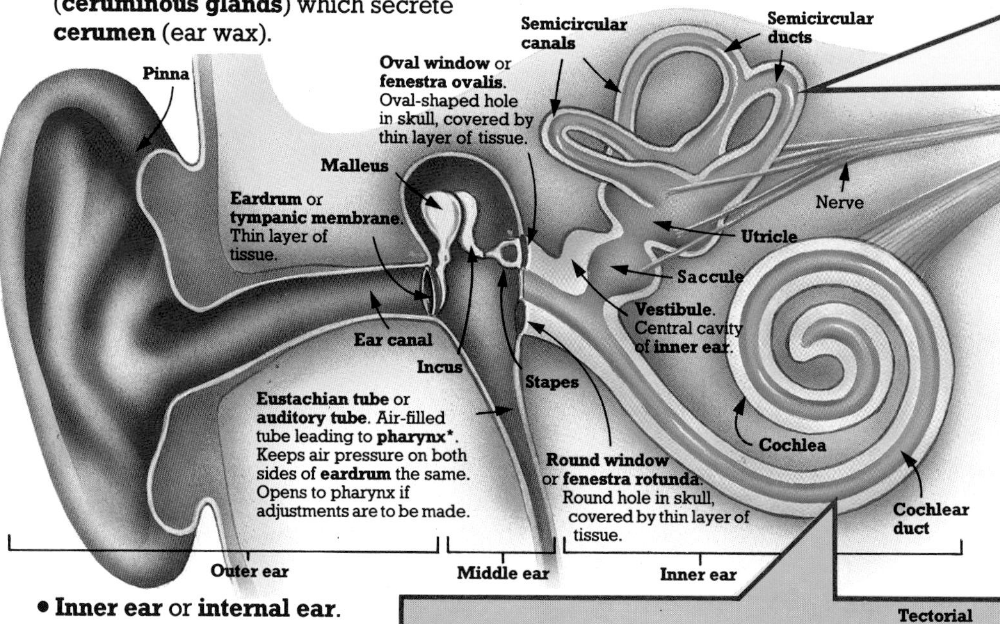

- **Inner ear** or **internal ear**. A connected series of cavities in the skull, with tubes and sacs inside them. The cavities (**cochlea**, **vestibule** and **semicircular canals**) are called the **bony labyrinth** and are filled with one fluid (**perilymph**). The tubes and sacs are filled with another fluid (**endolymph**) and are called the **membranous labyrinth**. They are the **cochlear duct**, **saccule**, **utricle** and **semicircular ducts**.

The inner ear and hearing

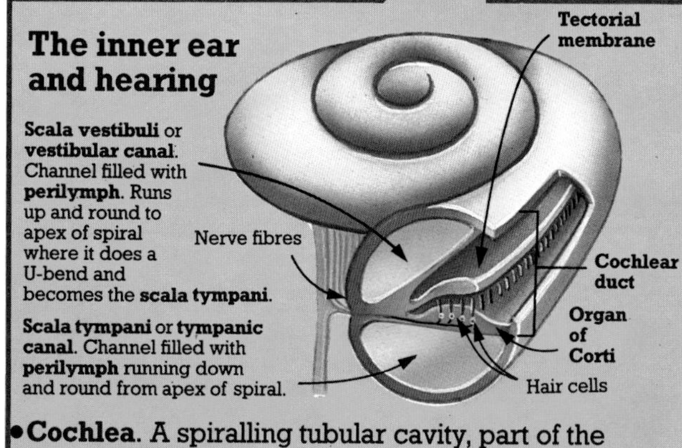

- **Cochlea**. A spiralling tubular cavity, part of the **inner ear**. It contains **perilymph** (see **inner ear**) in two channels (continuous with each other), and also a third channel – the **cochlear duct**.

* **Cartilage**, 281; **Pharynx**, 294; **Sebaceous glands**, 310.

The inner ear and balance

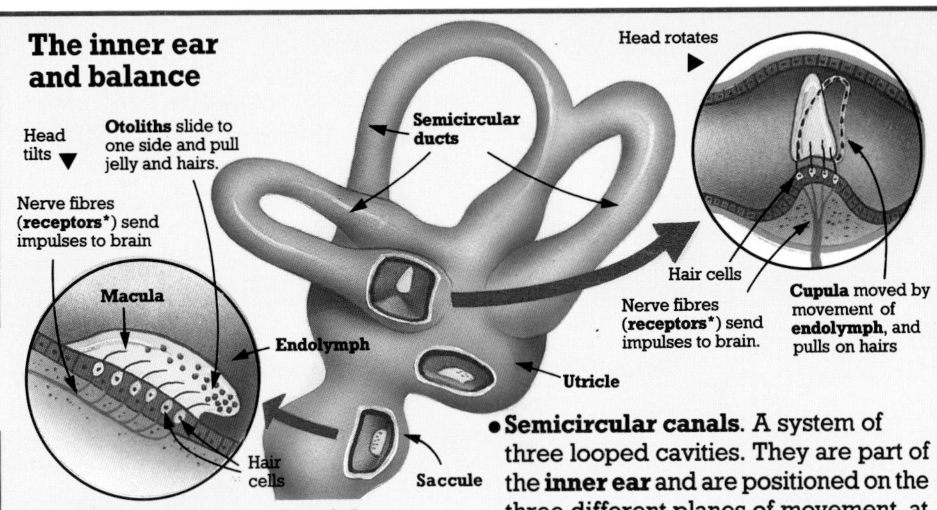

- **Saccule (sacculus)** and **utricle (utriculus)**. Two sacs lying between the **semicircular ducts** and the **cochlear duct**. They contain **endolymph** (see **inner ear**) and have special hair cells in patches in their linings. These cells have nerve fibres (**dendron*** endings) attached to them and hairs embedded in a jelly-like mass called a **macula** (pl. **maculae**). This contains grains of calcium carbonate (**otoliths**). The maculae send the brain information about forward, backward, sideways or tilting motion of the head.

- **Semicircular canals**. A system of three looped cavities. They are part of the **inner ear** and are positioned on the three different planes of movement, at right angles to each other.

- **Semicircular ducts**. Three looped tubes inside the **semicircular canals**. Each contains **endolymph** (see **inner ear**) and a special sensory body, which lies across the basal swelling (**ampulla**, pl. **ampullae**) of the duct. The sensory bodies (**cupulae**, sing. **cupula**) work in a very similar way to **maculae** (see **saccule**) – each consists of a jelly-like mass (without **otoliths**) and hair cells. They send the brain information about rotation and tilting of the head.

- **Cochlear duct**. A spiralling tube within the **cochlea**, connected to the **saccule**. It contains **endolymph** (see **inner ear**) and a long body called the **organ of Corti**. This contains special hair cells whose hairs project into the endolymph and touch a shelf-like tissue layer (**tectorial membrane**). The bases of the cells are attached to nerve fibres (**dendron*** endings).

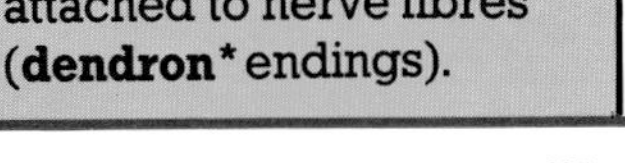

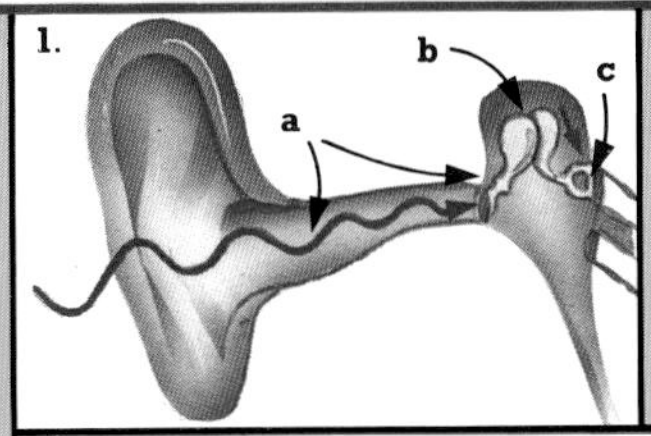

a) Sound waves (air vibrations) come in along **ear canal** and make **eardrum** vibrate.

b) Ear ossicles pick up vibrations and pass them to **oval window** (lever action magnifies vibrations about 20 times).

c) Vibrations of **oval window** cause waves in **perilymph** of **vestibule**.

2.

d) Waves in **perilymph** of **scala vestibuli** cause waves in **endolymph** of **cochlear duct**.

e) Hairs move and cause nerve fibres (**receptors***) to send impulses to brain (which interprets them as sensation of hearing).

f) Waves gradually fade out.

* **Dendron**, 304 (**Dendrites**); **Receptors**, 307.

The reproductive system

Reproduction is the process of producing new life. Humans reproduce by **sexual reproduction*** (described on pages 318-319) and the reproductive organs involved (making up the **reproductive system**) are called the **genital organs** or **genitalia**. They consist of the primary reproductive organs, or **gonads** (two **ovaries** in woman, two **testes** in men) and a number of additional organs. In both women and men, cells in the gonads also act as **endocrine glands***, secreting many important **hormones***.

The male reproductive system

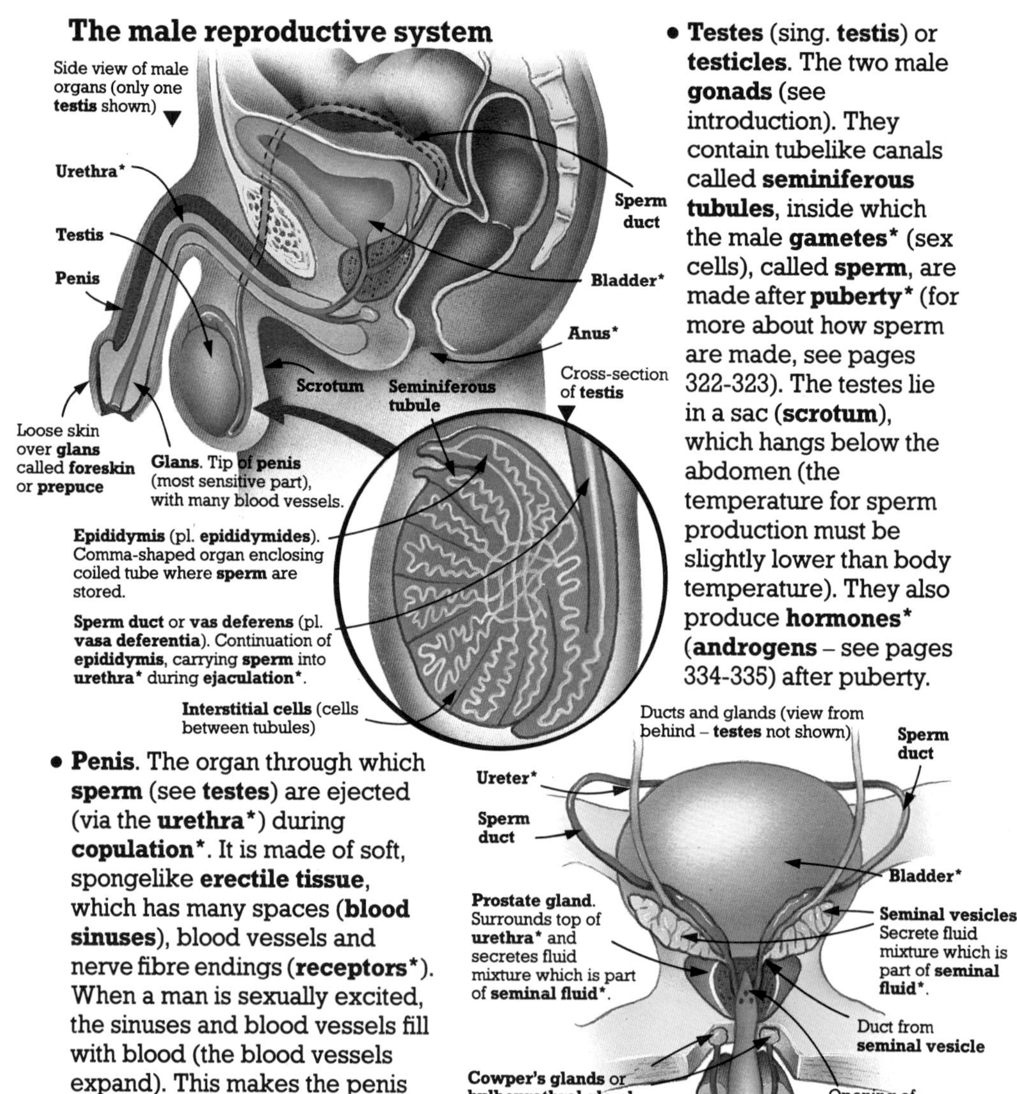

- **Testes** (sing. **testis**) or **testicles**. The two male **gonads** (see introduction). They contain tubelike canals called **seminiferous tubules**, inside which the male **gametes*** (sex cells), called **sperm**, are made after **puberty*** (for more about how sperm are made, see pages 322-323). The testes lie in a sac (**scrotum**), which hangs below the abdomen (the temperature for sperm production must be slightly lower than body temperature). They also produce **hormones*** (**androgens** – see pages 334-335) after puberty.

- **Penis**. The organ through which **sperm** (see **testes**) are ejected (via the **urethra***) during **copulation***. It is made of soft, spongelike **erectile tissue**, which has many spaces (**blood sinuses**), blood vessels and nerve fibre endings (**receptors***). When a man is sexually excited, the sinuses and blood vessels fill with blood (the blood vessels expand). This makes the penis stiff and erect.

* **Anus**, 295 (**Large intestine**); **Bladder**, 300; **Ejaculation**, 319 (**Copulation**); **Endocrine glands**, 297; **Gametes**, 321; **Hormones**, 334; **Mucus**, 295 (**Mucous membrane**); **Puberty**, 318; **Receptors**, 307; **Seminal fluid**, 319 (**Copulation**); **Sexual reproduction**, 320; **Ureters**, 300; **Urethra**, 301.

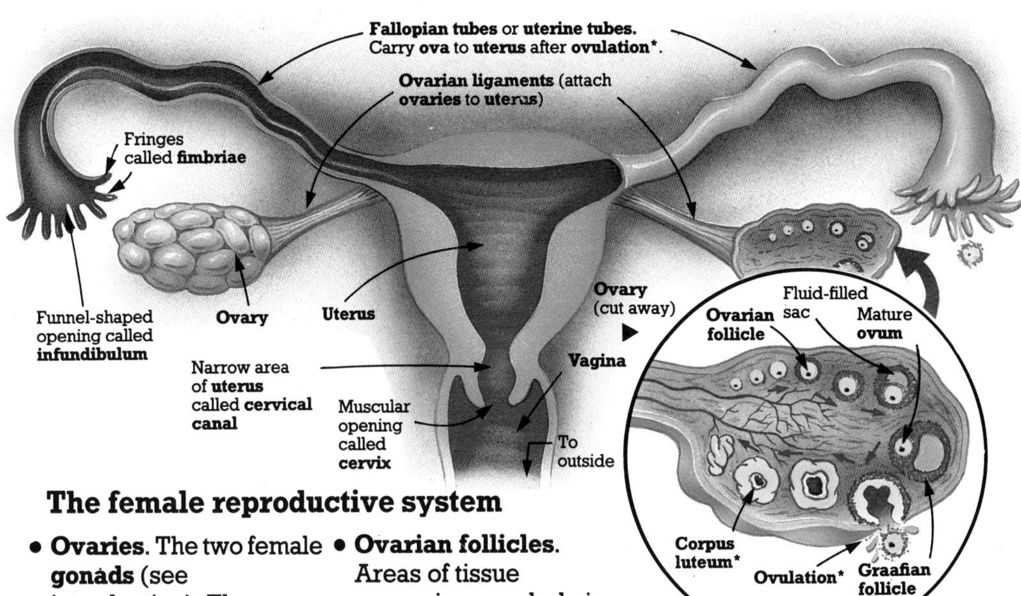

The female reproductive system

- **Ovaries**. The two female **gonads** (see introduction). They are held in place in the lower abdomen (below the kidneys) by **ligaments***. These attach them to the walls of the pelvis. The female **gametes*** (sex cells), called **ova** (sing. **ovum**), are produced regularly in the ovaries (in **ovarian follicles**) after **puberty***. For more about how ova are made, see pages 322-323.

- **Ovarian follicles**. Areas of tissue appearing regularly in the **ovaries** after **puberty***. Each contains a maturing **ovum** (see **ovaries**). The follicles gradually get larger and begin to secrete **hormones*** (see **oestrogen**, page 334). Each round of follicle production results in only one fully mature follicle (**Graafian follicle**).

- **Uterus** or **womb**. The hollow organ, inside which a developing baby (**foetus***) is held, or from which the **ova** (see **ovaries**) are discharged (see **menstrual cycle**, page 318). It has a lining of **mucous membrane*** (the **endometrium**), covering a muscular wall with many blood vessels.

- **Vagina**. The muscular canal leading from the **uterus** out of the body. It carries away the **ova** (see **ovaries**) and inner uterus lining during the **menstrual cycle***, receives the **penis** during **copulation*** and forms the birth canal. Its lining produces a lubricating fluid.

- **Vulva** or **pudendum**. A collective term for the outer parts of the female reproductive system – the **labia** and the **clitoris**. The labia are two folds of skin (one inside the other) which surround the openings from the **vagina** and the **urethra***. The clitoris is the most sensitive part. Like the **penis**, it is made of **erectile tissue** and has many **receptors***.

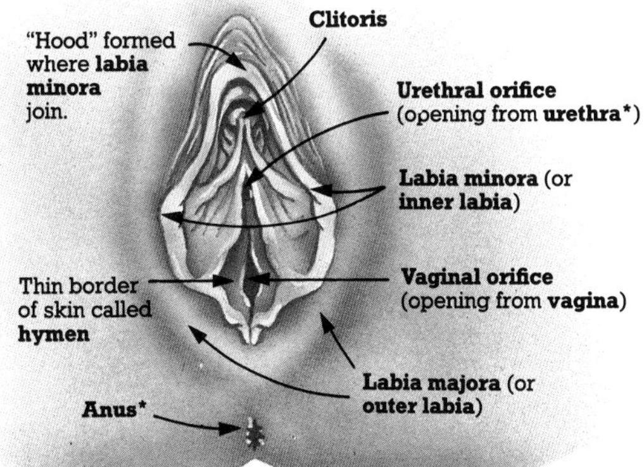

* **Anus**, 295 (**Large intestine**); **Copulation**, 319; **Corpus luteum**, 318 (**Menstrual cycle**); **Foetus**, 319 (**Pregnancy**); **Gametes**, 321; **Hormones**, 334; **Ligaments**, 280; **Mucous membrane**, 295; **Ovulation**, 318 (**Menstrual cycle**); **Puberty**, 318; **Receptors**, 307; **Urethra**, 301.

Development and reproduction

Humans reproduce by **sexual reproduction***. The main processes this involves are described on these two pages, as well as the initial developments which allow it to happen.

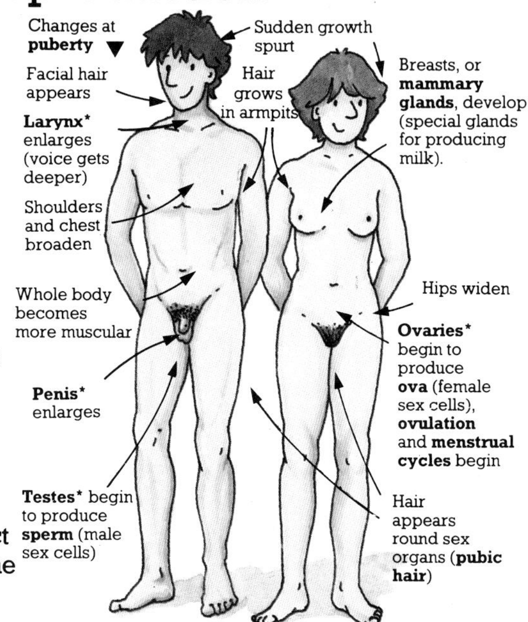

- **Puberty**. The point when the reproductive organs mature, and a person becomes capable of reproducing – roughly between the ages of 11 and 15 in girls, and 13 and 15 in boys. It involves a number of significant changes, all stimulated by **hormones*** (see **oestrogen** and **androgens**, pages 334-335). All the new resulting features are called **secondary sex characters**, as distinct from the **primary sex characters** – the sex organs present from birth (see pages 316-317).

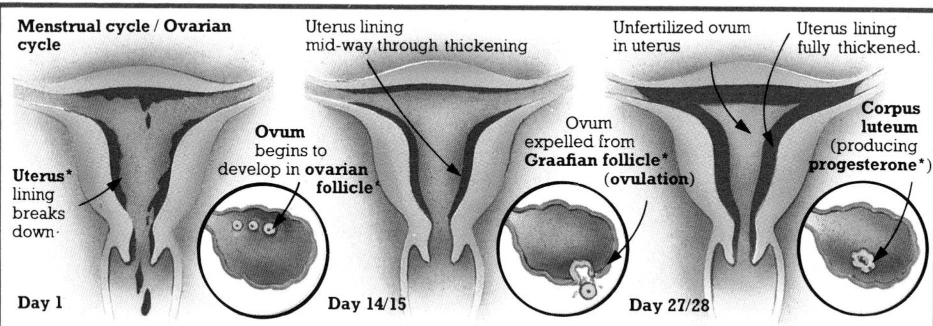

- **Menstrual cycle**. A series of preparatory changes in the **uterus*** lining (**endometrium**), in case of **fertilization**. The lining gradually develops a new inner layer rich in blood vessels. If a fertilized **ovum** (female sex cell) does not appear, this new layer breaks down and leaves the body via the **vagina*** (**menstruation**). Each menstrual cycle lasts about 28 days and they occur continuously from **puberty** to **menopause** (usually between the ages of 45 and 50), when ova production ceases.

  The events of the menstrual cycle run in conjunction with the **ovarian cycle** – the regular maturation of an ovum in an **ovarian follicle***, followed by **ovulation** (the release of the ovum into a **Fallopian tube***), and the breakdown of the **corpus luteum**. This body is formed from the burst **Graafian follicle*** (it does not break down if an ovum is fertilized). Both cycles are controlled by a group of hormones* (see pages 334-335).

* **Fallopian tubes**, 317; **Graafian follicle**, 317 (**Ovarian follicles**); **Hormones**, 334; **Larynx**, 298; **Ovaries**, 317; **Penis**, 316; **Progesterone**, 334; **Sexual reproduction**, 320; **Testes**, 316; **Uterus**, **Vagina**, 317.

- **Copulation**. Also called **coitus** or **sexual intercourse**. The insertion of the **penis*** into the **vagina***, followed by rhythmical movements of the pelvis in one or both sexes. Its culmination in the male is **ejaculation** – the ejection of **semen** from the **urethra*** (in the penis) into the vagina. Semen consists of **sperm** (male sex cells) and a fluid mixture (**seminal fluid**).

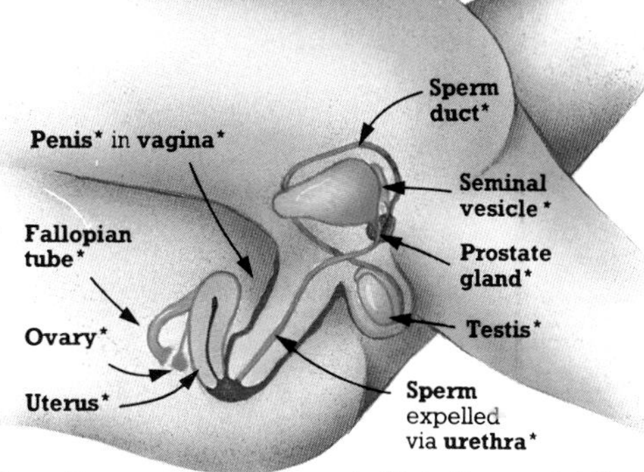

Zona pellucida

Nucleus* will fuse with ovum nucleus

Sperm penetrates ovum

"Tail" left behind

- **Fertilization**. A process which occurs after **ejaculation** if the **sperm** (male sex cells) meet an **ovum** (female sex cell) in a **Fallopian tube***. One sperm penetrates the ovum's outer skin (**zona pellucida**). Its **nucleus*** fuses with that of the ovum, and the first cell of a new baby (**zygote***) is formed. The new cell travels towards the **uterus***, undergoing many cell divisions (**cleavage***) as it does so. The ball of cells formed from these divisions then becomes embedded in the uterus wall (**implantation**), after which it is called an **embryo***.

Pregnancy

- **Pregnancy**, or **gestation**, is the state of carrying young. The time between **fertilization** and giving birth (**parturition**) is the **gestation period** (about 9 months in humans) and the new developing individual in the **uterus*** is called a **foetus**, a term usually used instead of **embryo*** after about 2 months of pregnancy. A series of powerful muscular contractions called **labour** occur just before parturition.

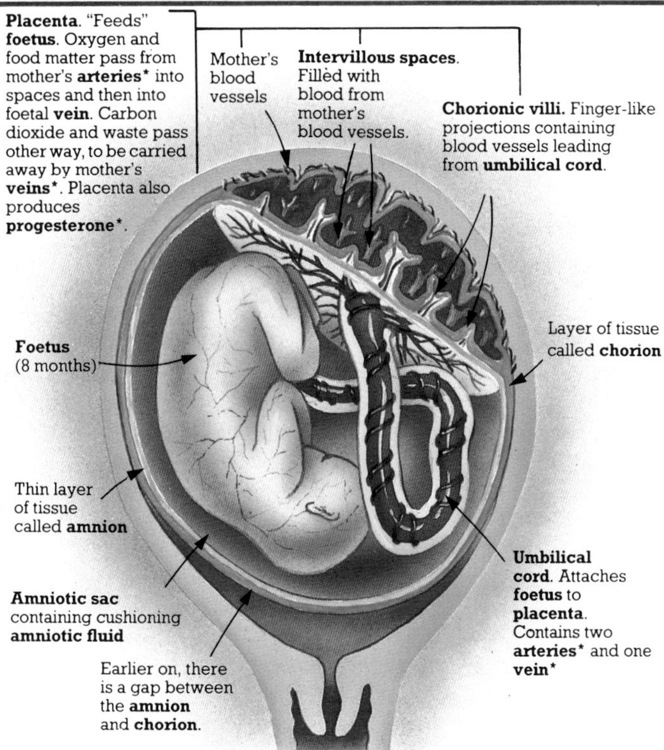

* **Arteries**, 288; **Cleavage**, 321 (**Embryo**); **Fallopian tubes**, 317; **Nucleus**, 238; **Ovaries**, 317; **Penis**, 316; **Progesterone**, 334; **Prostate gland**, **Seminal vesicles**, **Sperm ducts**, **Testes**, 316; **Urethra**, 301; **Uterus**, **Vagina**, 317; **Veins**, 288; **Zygote**, 321.

Types of reproduction

Reproduction is the creation of new life, a process which occurs in all living things. The two main types are **asexual** and **sexual reproduction**, but there is also a special case called **alternation of generations**.

Asexual reproduction

- **Asexual reproduction** is the simplest form of reproduction, occurring in many simple plants and animals. There are a number of different types, e.g. **binary fission***, **vegetative reproduction***, **gemmation** and **sporulation**, but they all share two main features. Firstly, only one parent is needed and secondly, the new individual produced is always genetically identical to this parent.

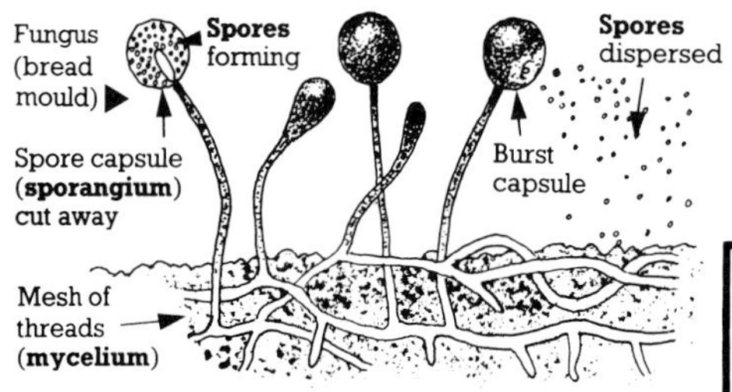

- **Sporulation**. The production of bodies called **spores** by simple plants, e.g. fungi and mosses. After dispersal by wind or water, these develop into new plants. There are two types of spore and, though only one parent is needed in both cases, true **asexual reproduction** really only occurs with one type. These spores are produced in plants such as simple fungi by ordinary cell division (see pages 240-241) and develop into plants which are identical to the parent (an important feature of asexual reproduction). The second kind of spore, however, is produced (e.g. in mosses and ferns) by a special kind of cell division (see pages 322-323) which is a feature of **sexual reproduction**. The new plants are not the same as the parent (see **alternation of generations**).

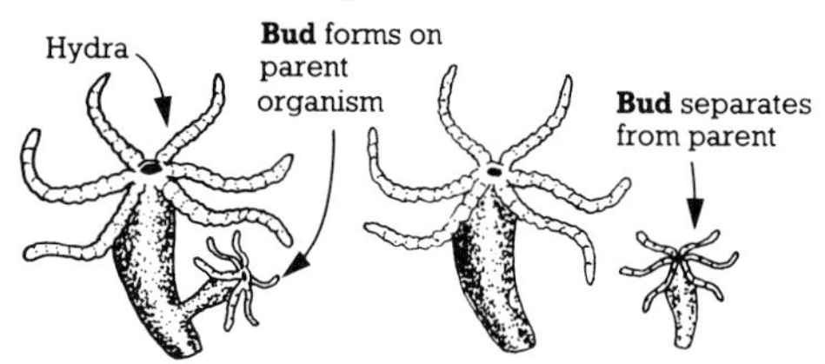

- **Gemmation**. Called **budding** in animals. A type of **asexual reproduction** occurring in many simple plants and animals, e.g. Hydra. It involves the formation of a group of cells which grows out of the organism and develops into a new individual. It either breaks away from the parent or (in **colonial*** animals, e.g. corals) it stays attached (though self-contained).

Sexual reproduction

- **Sexual reproduction** is the type of reproduction shown by all flowering plants and most animals. It involves the joining (**fusion**) of two **gametes** (sex cells), one male and one female. This process is called **fertilization**, and is further described on pages 258 (flowering plants), 319 (humans and similar animals) and 276 (other animals). The two gametes each have only half the number of **chromosomes*** (called the **haploid number***) as the plant or animal which produced them. This is achieved by a special kind of cell division (see pages 322-323) and ensures that when the gametes come together, the new individual produced has the correct, original number of chromosomes (called the **diploid number***).

* **Binary fission**, 240 (**Cell division**); **Chromosomes**, 324; **Colonial**, 342; **Diploid number**, 240 (**Mitosis**); **Haploid number**, 322 (**Meiosis**); **Vegetative reproduction**, 262.

- **Alternation of generations**. A reproductive process found in many simple animals and plants, e.g. jellyfish and mosses. In the animals, a form produced by **sexual reproduction** alternates with one produced **asexually**. In the plants, though, the alternation is really between two stages of sexual reproduction. One plant body (**gametophyte**) produces another (**sporophyte**) by sexual reproduction. This then produces **spores** (see **sporulation**) which grow into new gametophytes. However, the spores are made in the same way as **gametes** (see pages 322-323) and they (and the gametophytes) have only half the original number of **chromosomes***. The gametophytes produce gametes by ordinary cell division (see pages 240-241), as there is no need to halve the chromosomes again.

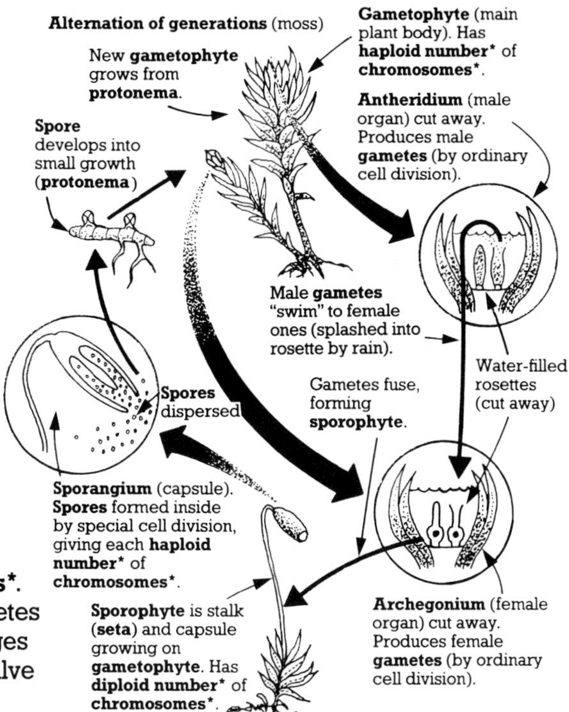

- **Gametes** or **germ cells**. The sex cells which join in **sexual reproduction** to form a new living thing. They are made by a special kind of cell division (see pages 322-323). In animals and simple plants, male gametes are known as **sperm**, short for **spermatozoa** (sing. **spermatozoon**) in animals and **spermatozooids** in simple plants. In flowering plants, they are just **nuclei*** (rather than cells) and are called **male nuclei** (see also pages 258 and 323). Female gametes are called **ova** (sing. **ovum**) or **egg cells** (egg cell is usually used in the case of plants). A sperm is smaller than an ovum and has a "tail" (**flagellum***).

Ovum
Sperm
"Head"
"Tail"
Human **embryo** (eight weeks)

- **Zygote.** The first cell of a new living thing. It is formed when a male and female **gamete** join (see **sexual reproduction**).

- **Embryo**. A new developing individual. It grows from one cell (the **zygote**) by a series of cell divisions (see pages 240-241) called **cleavage**. In humans, this first produces a ball of cells (**morula**) from the one original, and then a larger, hollow ball (**blastocyst**). After **implantation***, this is called the embryo. As it grows, the cells become **differentiated**, i.e. each develops into one kind of cell, e.g. a nerve cell.

* **Chromosomes**, 324; **Diploid number**, 240 (**Mitosis**); **Flagella**, 268; **Haploid number**, 322 (**Meiosis**); **Implantation**, 319 (**Fertilization**); **Nucleus**, 238.

Cell division for reproduction

Many cells within a living thing can divide to produce new cells for growth and repair (see pages 240-241). There is, however, a second type of cell division, which happens specifically to produce the **gametes*** (sex cells) needed for **sexual reproduction*** (and also one of the two types of **spore***). The division of the **nucleus*** in this type of cell division is called **meiosis**. The production of gametes, including both the cell division and the subsequent maturing of the gametes, is called **gametogenesis**.

- **Meiosis**. The division of the **nucleus*** when a cell divides to produce sex cells (see introduction). It can be split into two separate divisions – the **first meiotic division** (or **reduction division**) and the **second meiotic division** (each is followed by division of the **cytoplasm***). These can be divided into different phases (as in **mitosis***). Meiosis in general, and the first meiotic division in particular, ensures that each new **daughter nucleus** receives exactly half the number of **chromosomes*** as the original nucleus. The original number is the **diploid number** (see **mitosis**, page 240); the halved amount is the **haploid number**.

Crossing over (occurs in early **prophase**)

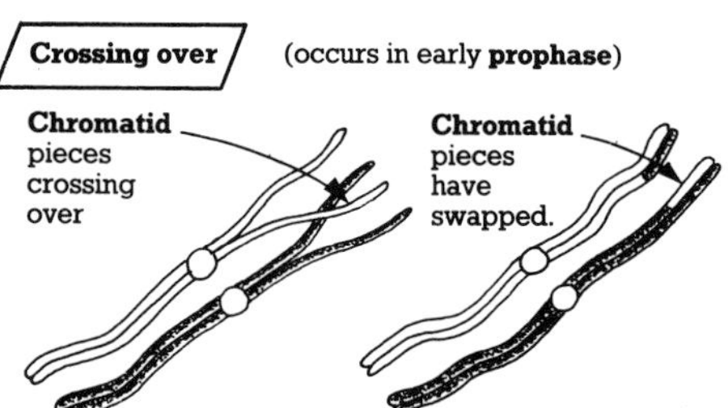

Chromatids of each **tetrad** cross over each other at places called **chiasmata** (sing. **chiasma**). Two chromatid pieces (one from each pair) break off and swap over. Causes mixing of **genes*** (helping to ensure new living things are never identical to parents, i.e. always a new variety of types).

First meiotic division

These pictures show an animal cell, but only four **chromosomes*** are shown.

Prophase (early stage)

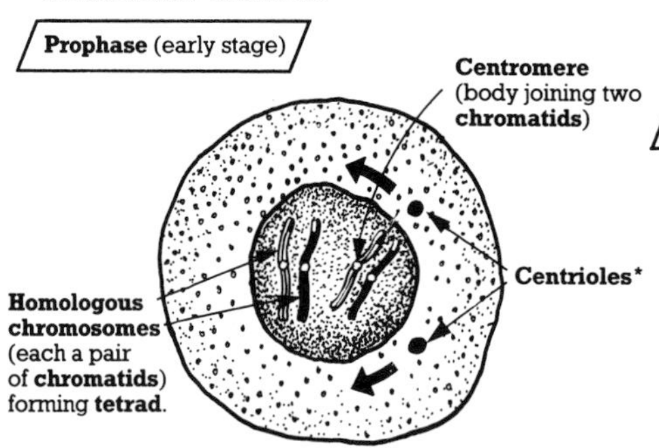

Threads of **chromatin*** in **nucleus*** coil up to form **chromosomes***. Paired chromosomes (**homologous chromosomes**) line up side by side, forming pairs called **bivalents**. Each chromosome duplicates, becoming a pair of **chromatids** (each group of four chromatids now called a **tetrad**). **Centrioles*** move to opposite poles of cell.

Prophase (later stage)

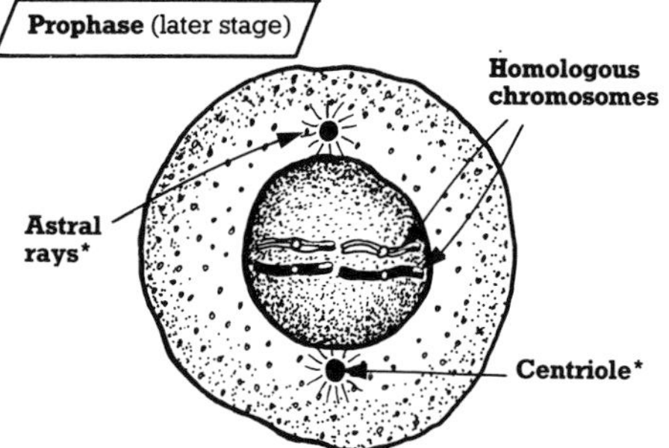

Homologous chromosomes (each a pair of **chromatids**) move together to equator of cell.

Metaphase

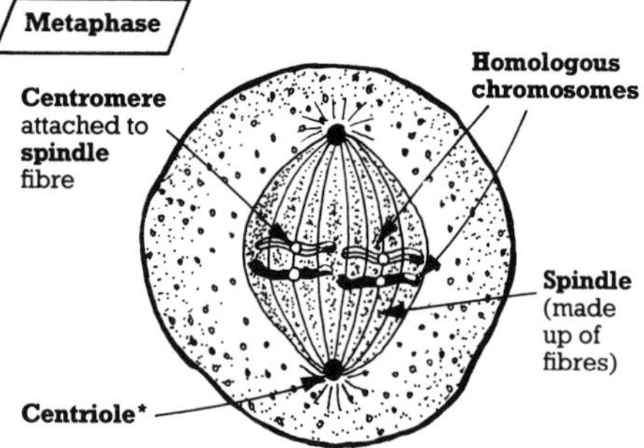

Nuclear membrane* disappears, two **centrioles*** form a **spindle** (see **metaphase** of **mitosis**, page 241). **Chromosomes*** (pairs of **chromatids**) become attached to spindle by **centromeres**.

* **Astral rays**, 241; **Centrioles**, 240; **Chromatin**, 238 (**Nucleus**); **Chromosomes**, 324; **Cytoplasm**, 238; **Gametes**, 321; **Genes**, 325; **Mitosis**, 240; **Nuclear membrane**, 238 (**Nucleus**); **Sexual reproduction**, 320; **Spores**, 320 (**Sporulation**).

Anaphase

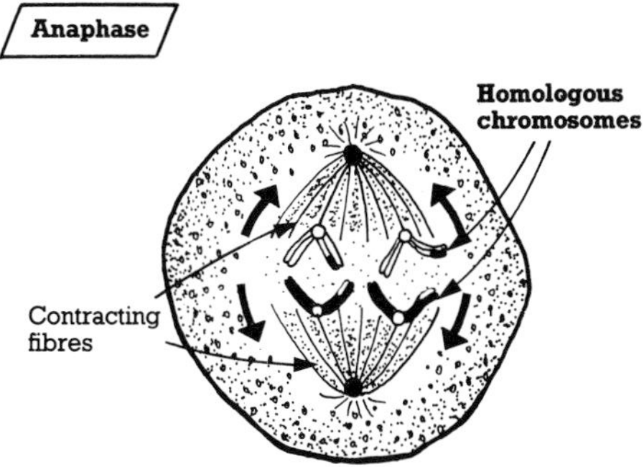

Homologous chromosomes (each still a pair of **chromatids**) separate (see **Law of segregation**, page 326), dragged apart by fibres of **spindle**.

Telophase

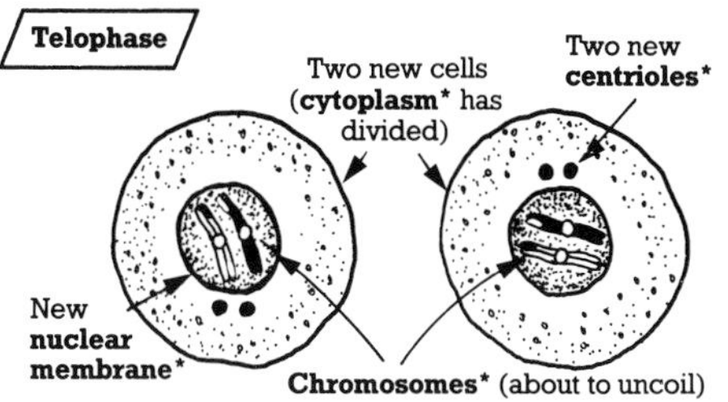

Spindle disappears, **centrioles*** duplicate. Happens in conjunction with **cytokinesis** (division of **cytoplasm***). Two new cells formed, each with half the original number of **chromosomes*** (each two **chromatids**). **Interphase*** (intervening period) usually follows, in which case **nuclear membranes*** form and chromosomes uncoil again to form threadlike mass (**chromatin***).

Second meiotic division

The **second meiotic division** happens in the cells produced by the **first meiotic division**. It occurs in exactly the same way, and with the same phases, as **mitosis*** (when the **nucleus*** divides as part of cell division for growth and repair) and is followed in the same way by the division of the **cytoplasm***. The only difference is that each dividing nucleus now has only the **haploid number** of **chromosomes*** (see **meiosis**) so the resulting new sex cells (**gametes***) will also be haploid. The second division differs according to whether male or female gametes are to be produced, and the final maturing of the gametes after the second division is different in animals and plants (see text right).

Gamete production (male)

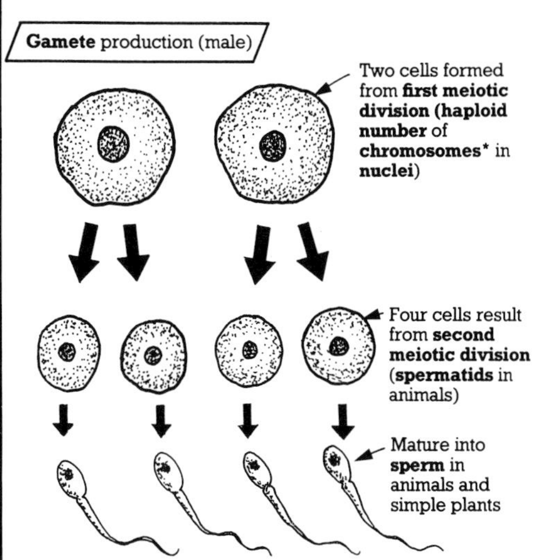

Two cells formed from **first meiotic division** divide again (see **second meiotic division**). In animals, resulting four cells called **spermatids** and mature into male **gametes*** (sex cells), or **sperm**. In simple plants, four cells either develop into sperm or into type of **spore*** involved in **alternation of generations***. In flowering plants, **nuclei*** of four cells each divide again (**mitosis***). Resulting cells (**pollen*** grains) each have two nuclei (one later divides again to form two **male nuclei***).

Gamete production (female)

Two cells formed from **first meiotic division** (**haploid number** of **chromosomes*** in **nuclei**)

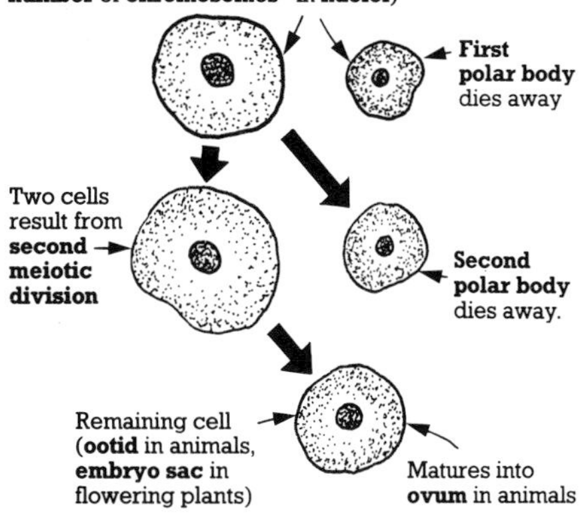

One of two cells formed by **first meiotic division** dies away (called **first polar body**). Other divides again (see **second meiotic division**). Of two resulting cells, one (**second polar body**) dies away. In animals, other one called **ootid** and matures into female **gamete*** (sex cell), or **ovum**. In flowering plants, other one called **embryo sac** and its **nucleus*** divides three more times (by **mitosis***). Of eight new nuclei, six have cells form around them, two stay naked. One of six cells is female gamete, or **egg cell** (see **ovule**, page 258). Formation of egg cell in simple plants is very similar.

* **Alternation of generations**, 321; **Centrioles**, 240; **Chromatin**, 238 (**Nucleus**); **Chromosomes**, 324; **Cytoplasm**, 238; **Gametes**, 321; **Interphase**, 241; **Male nuclei**, 321 (**Gametes**); **Mitosis**, 240; **Nuclear membrane**, 238 (**Nucleus**); **Pollen**, 258; **Spores**, 320 (**Sporulation**).

Genetics and heredity

Genetics is a branch of biology. It is the study of **inheritance** – the passing of characteristics from one generation to the next. The bodies which are instrumental in this process are called **chromosomes**. Each chromosome is made up of **genes** – the "coded" instructions for the appearance and constituents of an organism. For more about inheritance, see page 326.

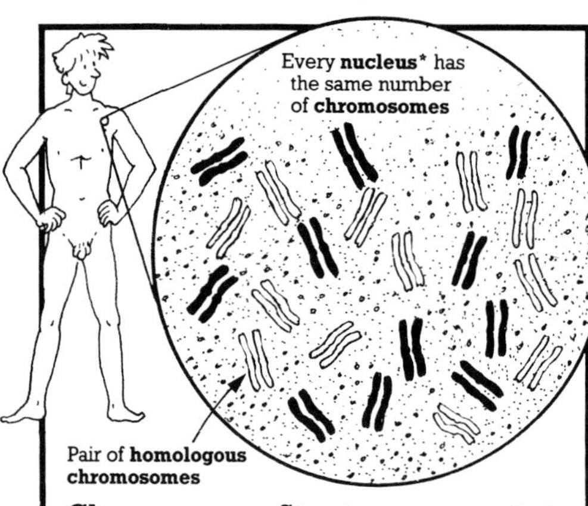

• **Chromosomes**. Structures present at all times in the **nuclei*** of all cells, though they only become independently visible (as thread-like bodies of differing shapes and sizes) when a cell is dividing (and has been stained with a dye). Each one is made of a single molecule of **DNA** (see **nucleic acids**), plus proteins called **histones**. The DNA molecule is a chain of many connected **genes**.

Every **species*** has its own number of chromosomes per cell, called the **diploid number** (humans have 46). These are arranged in pairs called **homologous chromosomes**.

• **Nucleic acids**. Two different acids, called **DNA** (**deoxyribonucleic acid**) and **RNA** (**ribonucleic acid**). Both are found in the **nuclei*** of all cells, hence their name (RNA is also found in the **cytoplasm*** – see **ribosomes**, page 239). Each molecule of a nucleic acid is very large, and is composed of many individual units called **nucleotides**. A DNA molecule consists of two chains of nucleotides twisted around each other, forming a shape called a **double helix** (rather like a twisted ladder). An RNA molecule consists of one chain of nucleotides (and looks like a ladder cut in half lengthwise).

Nucleic acid structure

DNA ▶

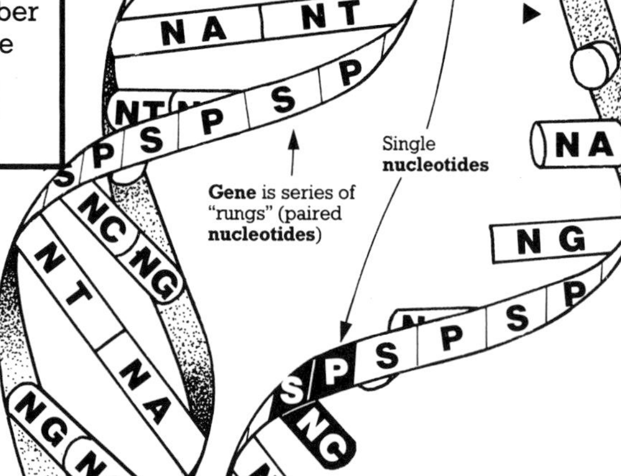

N = **nitrogen base** (linked nitrogen, carbon, hydrogen and oxygen atoms). 5 types:
A = **adenine** **T** = **thymine** (always paired in **DNA**)
G = **guanine** **C** = **cytosine** (always paired in **DNA**)
U = **uracil** (only found in **RNA**, replaces thymine of **DNA**)

S = sugar (linked carbon, hydrogen and oxygen atoms). **Deoxyribose** in **DNA**, **ribose** in **RNA**.
P = **phosphate group***.

* **Cytoplasm**, **Nucleus**, 238; **Phosphate group**, 333 (**ADP**); **Species**, 338.

- **Genes**. Sets of "coded" instructions which make up the **DNA** molecule of a **chromosome** (in humans, each DNA molecule is thought to contain about 1000 genes). Each gene is a connected series of about 250 "rungs" on the DNA "ladder". Since the order of the "rungs" varies, each gene has a different "code", relating to one specific characteristic (**trait**) of the organism, e.g. its **blood group*** or the composition of a **hormone***. With the exception of the **sex chromosomes**, the genes carried on paired **homologous chromosomes** (see **chromosomes**) are also paired, and run down the chromosomes in the same order (one member of each pair on each chromosome). These paired genes control the same characteristic and may give identical instructions. However, their instructions may also be different, in which case the instructions from one gene (the **dominant** gene) will "mask out" those from the other (the **recessive** gene), unless **incomplete dominance** or **codominance** is shown. Two such non-identical genes are called **alleles** or **allelomorphs**.

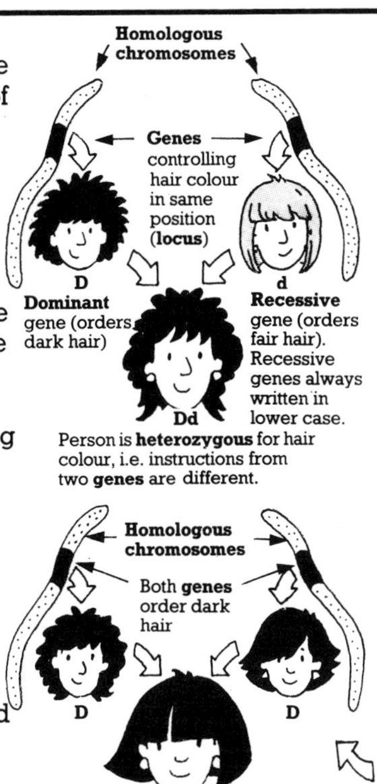

The two examples have different **genotypes** for hair colour, i.e. different sets of instructions (**DD** and **Dd**), but are the same **phenotype**, i.e. the resulting characteristic is the same (dark hair).

- **Codominance**. A special situation where a pair of **genes** controlling the same characteristic give different instructions, neither is **dominant** (see **genes**), but both are represented in the result. The human **blood group*** AB, for example, results from equal dominance between a gene for group A and one for group B.

- **Incomplete dominance** or **blending**. A situation where a pair of **genes** which control the same characteristic give different instructions, but neither is **dominant** (see **genes**) or obvious in the result. For example, a lack of dominance between a gene for red colour and one for white results in the intermediate roan colour of some cows.

- **Sex chromosomes**. One pair of **homologous chromosomes** (see **chromosomes**) in all cells (all the others are called **autosomes**). There are two different kinds of sex chromosomes, called the **X** and **Y chromosomes**. A male has one X and one Y. The Y chromosome carries the genetic factor (not a **gene** as such) determining maleness, thus all individuals with two X chromosomes are female.

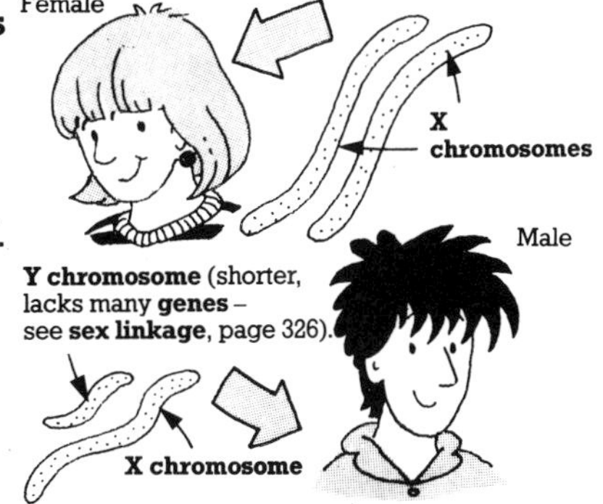

* **Blood group**, 287; **Hormones**, 334.

Inheriting genes

Every new organism inherits its **chromosomes*** (and **genes***) from its parents. In **sexual reproduction***, the **sperm*** and **ovum*** (sex cells) which come together to form this new individual have only half the normal number of chromosomes (the **haploid number** – see pages 322-323). This ensures that the **zygote*** (first new cell) formed from the two sex cells will have the normal number (see **chromosomes**, page 324). Two laws (**Mendel's laws**) point out genetic factors which are always true when cells divide to produce sex cells.

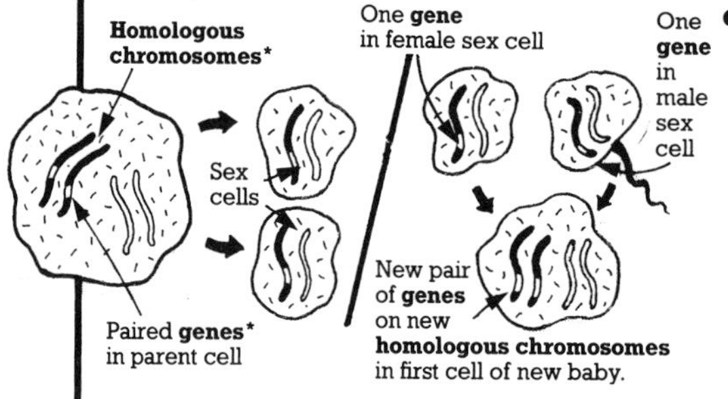

- **Law of segregation (Mendel's first law). Homologous chromosomes*** always separate when the **nucleus*** of a cell divides to produce **gametes*** (sex cells -see pages 322-323), hence so too do the paired **genes*** which control the same characteristic. The offspring thus always have paired genes (one member of each pair coming from each parent).

- **Law of independent assortment (Mendel's second law)**. Each member of a pair of **genes*** can join with either of the two members of another pair when a cell divides to form **gametes*** (sex cells). Hence all the different mixes are possible in a new individual.

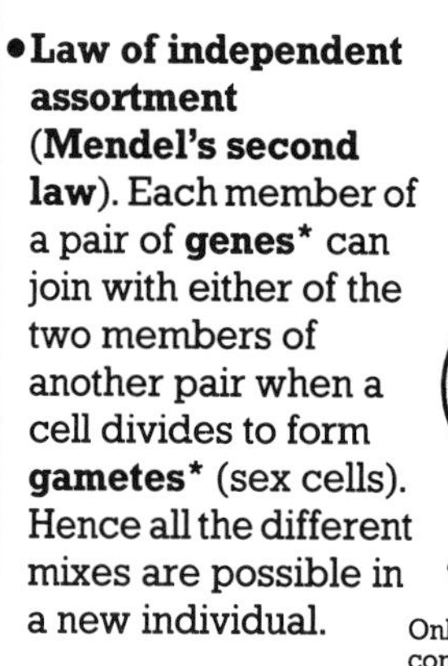

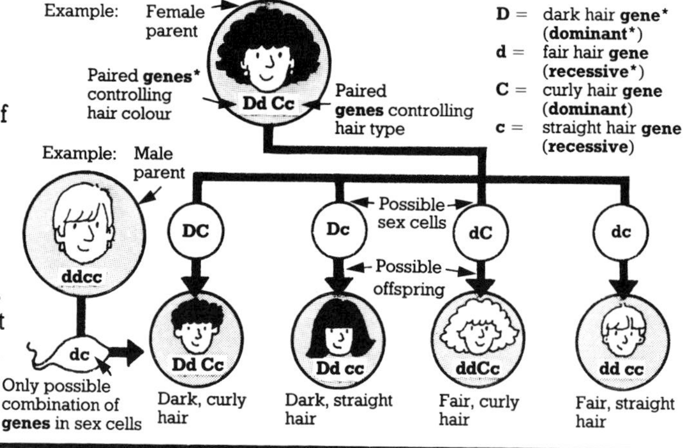

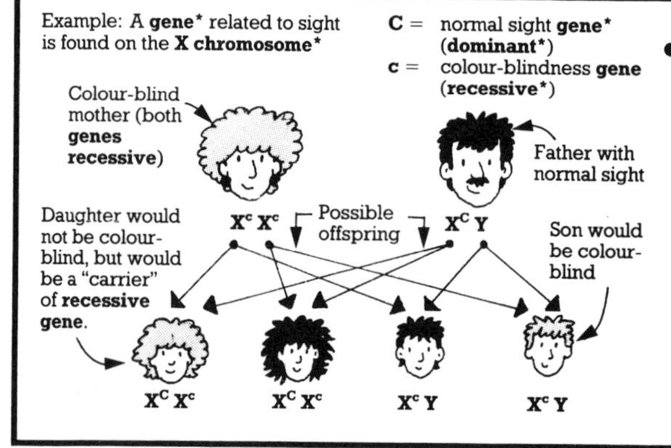

- **Sex linkage**. The two **sex (X) chromosomes*** in a female contain many paired **genes*** (like all **chromosomes***), but the **Y chromosome*** in a male lacks partners for most of the genes on its mate (the **X**). Thus any **recessive*** genes on the X will show up more often in males (see left). The unpaired genes on the X are **sex-linked genes**.

* **Homologous chromosomes**, 324 (**Chromosomes**); **Nucleus**, 238; **Ovum**, 321 (**Gametes**); **Recessive**, 325 (**Genes**); **Sexual reproduction**, 320; **Sperm**, 321 (**Gametes**); **X and Y chromosomes**, 325 (**Sex chromosomes**); **Zygote**, 321.

Fluid movement

The movement of substances around the body, especially their movement in and out of cells, is essential to the life of an organism. Food matter must be able to pass into the cells, and waste and harmful material must be able to move out. Most solids and liquids travel round the body in **solutions**, i.e. they (**solutes**) are dissolved in a fluid (the **solvent** – normally water).

- **Diffusion**. The movement of molecules of a substance from an area where they are in higher concentration to one where their concentration is lower. This is a two-way process (where the concentration of a **solute** is low, that of the **solvent** will be high, so its molecules will move the other way) and it ceases when the molecules are evenly distributed. Many substances, e.g. oxygen and carbon dioxide, diffuse into and out of cells.

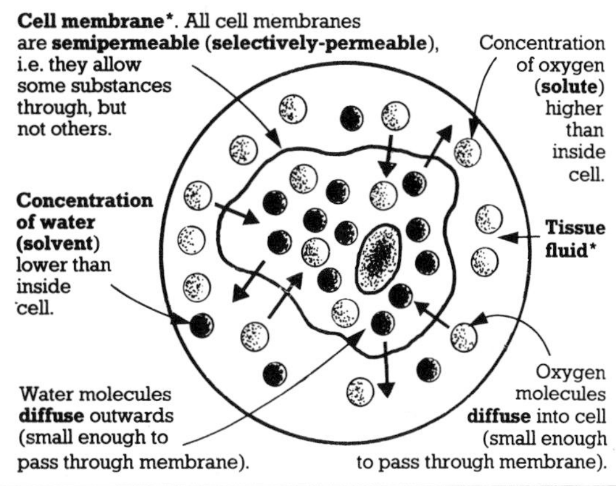

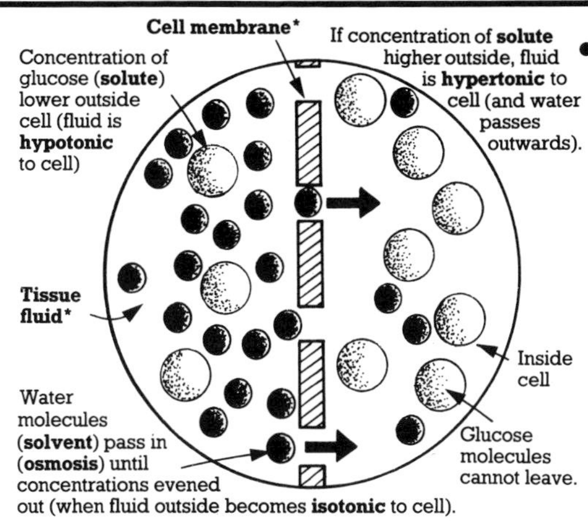

- **Osmosis**. The movement of molecules of a **solvent** through a **semipermeable** membrane (see above) in order to lower the concentration of a **solute** on the other side of the membrane, and even out the concentrations either side. This is a one-way type of **diffusion**, occurring when the molecules of the solute cannot pass the other way. **Osmotic pressure** is the pressure which builds up in an enclosed space, e.g. a cell, when a **solvent** enters by osmosis.

- **Active transport**. A process which occurs when substances need to travel in the opposite direction to that in which they would travel by **diffusion** (i.e. from low to high concentration, e.g. when cells take in large amounts of glucose for breakdown). It is not yet fully understood, but it is thought that special "carrier" molecules outside a cell "pick up" particles, "carry" them through the **cell membrane***, release them and return to the outside for more. Energy is needed for this action (since it opposes the natural tendency). This is supplied in the form of **ATP***.

- **Pinocytosis**. The taking in of a fluid droplet by inward-folding and separation of a section of **cell membrane*** (forming a **vacuole***). Most cells can do this.

* **ATP**, 333; **Cell membrane**, 238; **Tissue fluid**, 292; **Vacuoles**, 238.

Food and how it is used

Food is vital to all organisms, providing all the materials needed to be broken down for energy, to regulate cellular activities and to build and repair tissues (see pages 330-333). Of the various food substances, **carbohydrates**, **proteins** and **fats** are called **nutrients**, and **minerals**, **vitamins** (not needed by plants) and water are **accessory foods**. Plants build their own nutrients (by **photosynthesis***), and take in minerals and water; animals take in all the substances they need and break them down by digestion (see pages 336-337).

•**Carbohydrates**. A group of substances made up of carbon, hydrogen and oxygen, which exist in varying degrees of complexity (see "terms used", page 337 and also page 204). In animals, complex carbohydrates are taken in and broken down by digestion (see page 336) into the simple carbohydrate **glucose**. The breakdown of glucose (**internal respiration***) provides almost all the energy for life's activities. Plants build up glucose from other substances (by **photosynthesis***).

•**Proteins**. A group of substances made up of simpler units called **amino acids**. These contain carbon, hydrogen, oxygen, nitrogen and, in some cases, sulphur. Most protein molecules consist of hundreds, maybe thousands, of amino acids, joined together by links called **peptide links** into one or more chains called **polypeptides***. The many different types of protein each have a different arrangement of amino acids. They include the **structural proteins** (the basic components of new cells) and **catalytic proteins** (**enzymes***), which play a vital role in controlling cell processes.

Plants build up amino acids from the substances they take in (by **photosynthesis***), and then build proteins from them. Animals take in proteins and break them down into single amino acid molecules by digestion (see page 336). These are then transported in the blood to all the body cells and reassembled into the different proteins needed (see **ribosomes**, page 239 and also page 205).

•**Fats**. A group of substances made up of carbon, hydrogen and a small amount of oxygen (see also **lipids**, page 205). Plants build fats from the substances they take in and their seeds hold most as a store of food. This can be converted to extra **glucose** (see **carbohydrates**) to provide energy for the growing plant. Digestion of fats in animals produces **fatty acids** and **glycerol** (see page 336). If these need to be broken down for energy (as well as glucose), this occurs in the liver. This results in some products which the liver can convert to glucose, but others it cannot. These are instead converted elsewhere to a substance which forms a later stage of glucose breakdown. Fatty acids and glycerol not needed for energy are immediately recombined to form fat particles and stored in various body areas, e.g. under the skin (see **subcutaneous layer**, page 310).

* **Enzymes**, 331; **Internal respiration**, 332; **Photosynthesis**, 254; **Polypeptides**, 337.

• **Vitamins**. A group of substances vital to animals, though only needed in tiny amounts. The most important function of many vitamins is to act as **co-enzymes***, i.e. to help **enzymes*** catalyse chemical reactions. See page 337 for a list of vitamins and their functions.

• **Minerals**. Natural inorganic substances, e.g. phosphorus and calcium. They form a vital part of plant and animal tissue, e.g. in bones and teeth. Many are found in **enzymes*** and **vitamins**. They include **trace elements**, e.g. copper and iodine, present in tiny amounts.

• **Roughage** or **fibre**. Bulky, fibrous food, e.g. bran. Much of it is made up of **cellulose**, a **carbohydrate** found in plant **cell walls***. Unlike most carbohydrates, this cannot be digested by most animals, including humans, because they lack the necessary **digestive enzyme***, called **cellulase**. (Some animals, e.g. snails, do have this enzyme, and others, like cows, who must digest cellulose, do so in another way – see **rumen**, page 271.) The fact that roughage is bulky means that food can be gripped by intestinal muscles, and so moved on through the system.

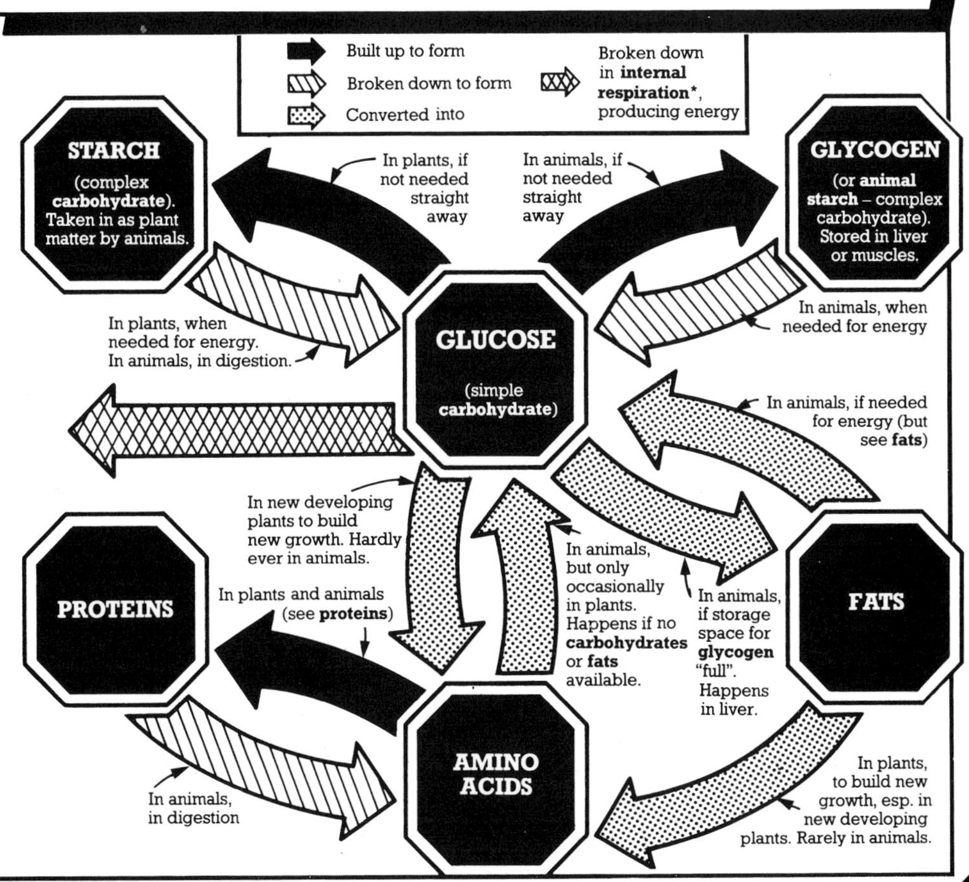

* **Cell wall**, 238; **Co-enzymes**, 331 (**Enzymes**); **Digestive enzymes**, 336; **Internal respiration**, 332.

Metabolism

Metabolism is a collective term for all the complex, closely-coordinated chemical reactions occurring inside an organism. These can be split into two opposing sets of reactions, called **catabolism** and **anabolism**. The rates of the reactions vary in response to variations in the organism's internal and external environments, and they play a major role in keeping internal conditions stable (see **homeostasis**, page 333).

• **Catabolism**. A collective term for all the reactions which break down substances in the body (**decomposition reactions**). One example is digestion in animals, which breaks down complex substances into simpler ones (see chart, page 336-337). Another is the further breakdown of these simple substances in the cells (**internal respiration***). Catabolism always liberates energy (in digestion, most is lost as heat, but in internal respiration, it is used for the body's activities). This is despite the fact that, as with all chemical reactions, catabolism itself requires energy. This energy (needed) is taken from the much greater amount of energy produced during the reactions. The rest of this is released, hence the overall result is always an energy "profit".

• **Anabolism**. A collective term for all the reactions which build up substances in the body (**synthesis reactions**). One example is the linking together of amino acids to form proteins (see page 328). Anabolism always needs energy to be taken in, since the small amount produced during the reactions is never enough (i.e. the overall result of anabolism is an energy "loss"). The extra energy is taken from the **catabolism** "profit".

• **Metabolic rate**. The overall rate at which metabolic reactions occur in an individual. In human beings, it varies widely from person to person, and in the same individual under different conditions. It increases under stress, when the body temperature rises and during exercise, hence the true and accurate measurement of a person's metabolic rate is a measurement taken when the subject is resting, has a normal body temperature, and has not recently exercised. This is called the **basal metabolic rate (BMR)** and is expressed in **kilojoules** per square metre of body surface per hour (see measuring method and calculations opposite).

People with high BMR can eat large amounts without putting on weight, because their **catabolism** of food matter (in the cells) happens so fast that not much fat is stored. This fast rate of reactions also often results in "excess" energy (i.e. energy not needed for **anabolism**), so they may appear to have a lot of "nervous energy". People with low BMR put on weight easily and often appear to have little energy.

The metabolic rate is influenced by a number of **hormones***, especially **STH**, **thyroxin**, **adrenalin** and **noradrenalin**. For more about these, see pages 334-335.

* **Hormones**, 334; **Internal respiration**, 332.

● **Kilojoule**. A unit of energy, specifically used in biology when referring to the amount of heat energy produced by the **catabolism** of food, and hence when measuring a person's **basal metabolic rate** (see **metabolic rate**). The calculations involved in measuring BMR combine certain known facts about the number of kilojoules produced by the breakdown of different substances with a measurement of oxygen consumption obtained under controlled conditions (see below).

To work out a person's **basal metabolic rate** (BMR = kJ/m^2/hr)

Facts known (worked out using a piece of apparatus called a calorimeter):

1. If 1 litre of oxygen is used to break down carbohydrates, c.21.21 kJ are produced (i.e. enough heat energy to heat c.5050g water by 1°C).
2. With fats, the result from 1 litre oxygen is c.19.74 kJ.
3. With proteins – c.19.32 kJ.

First calculation:
Heat energy generated when food (in general) is broken down using 1 litre of oxygen = the average of the three figures above, i.e. 20.09 kJ (provided subject measured has taken in equal amounts of the three foodstuffs).

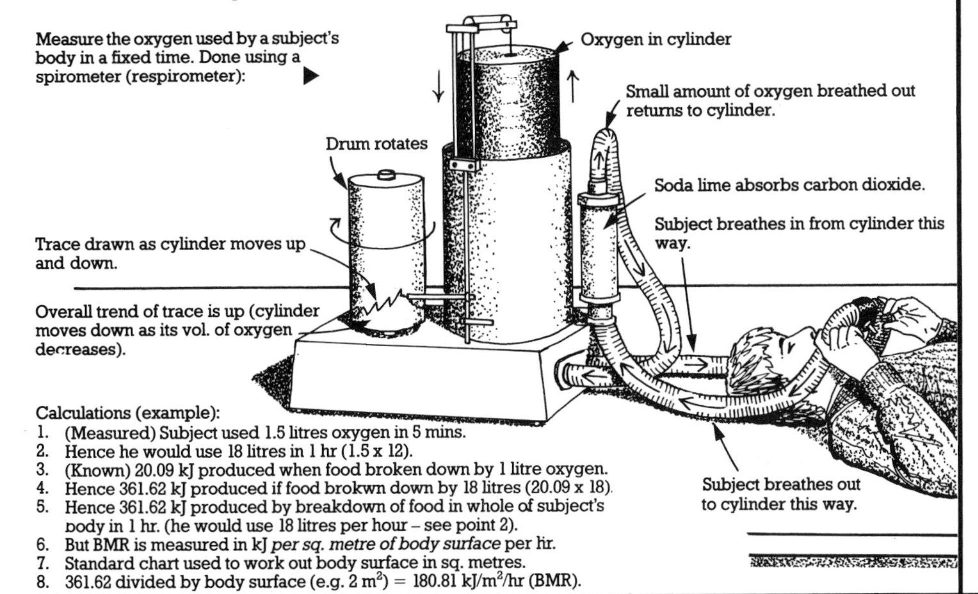

Calculations (example):

1. (Measured) Subject used 1.5 litres oxygen in 5 mins.
2. Hence he would use 18 litres in 1 hr (1.5 x 12).
3. (Known) 20.09 kJ produced when food broken down by 1 litre oxygen.
4. Hence 361.62 kJ produced if food brokwn down by 18 litres (20.09 x 18).
5. Hence 361.62 kJ produced by breakdown of food in whole of subject's body in 1 hr. (he would use 18 litres per hour – see point 2).
6. But BMR is measured in kJ *per sq. metre of body surface* per hr.
7. Standard chart used to work out body surface in sq. metres.
8. 361.62 divided by body surface (e.g. 2 m^2) = 180.81 kJ/m^2/hr (BMR).

● **Enzymes**. Special proteins (**catalytic proteins**) found in all living things and vital to the chemical reactions of life. They act as **catalysts***, i.e. they speed up reactions without themselves being changed. Many enzymes are aided by other substances, called **co-enzymes**, whose molecules are able to "carry" the products of one reaction (catalysed by an enzyme) on to the next reaction.

There are many different types of enzyme, e.g. **digestive enzymes**, which control the breakdown of complex food material into simple soluble substances (see text and chart, pages 336-337), and **respiratory enzymes**, which control the further breakdown of these simple substances in the cells to liberate energy (i.e. **internal respiration***).

* **Catalyst**, 161; **Internal respiration**, 332.

Energy for life and homeostasis

A living thing needs energy for its activities. This energy comes from a series of chemical reactions inside its cells, known as **internal respiration**, **tissue respiration** or **cellular respiration**. The cells contain various simple food substances, which are the results of digestive breakdown in animals (see pages 336-337) and **photosynthesis*** in plants. These substances all contain stored energy, which is released when internal respiration breaks them down. In almost all cases, glucose is the substance broken down (see **carbohydrates** and diagram, pages 328-329 and also equation, page 209). There are two kinds of internal respiration – **anaerobic** and **aerobic respiration**.

• **Anaerobic respiration**. A type of **internal respiration** which does not need free oxygen (oxygen taken into the body). It takes place in the cells of all organisms and releases a small amount of energy. In most organisms, it consists of a chain of chemical reactions called **glycolysis**, which break down glucose into **pyruvic acid**. In normal circumstances this is then immediately followed by **aerobic respiration**, which breaks down this poisonous acid in the presence of oxygen. This breakdown releases the bulk of the energy. In abnormal conditions, however, it may not be possible for the aerobic stage to follow immediately, in which case a further stage of anaerobic respiration occurs (see **oxygen debt**).

In some microscopic organisms, e.g. yeast and some bacteria, anaerobic respiration always runs through all its stages, providing enough energy for their needs without requiring oxygen.

• **Aerobic respiration**. A type of **internal respiration** which can only take place in the presence of free oxygen (oxygen taken into the body). It is the way in which most living things obtain the bulk of their energy and follows a stage of **anaerobic respiration**. Oxygen (brought by the blood) is taken into each cell and reacts in the **mitochondria*** with the **pyruvic acid** produced in anaerobic respiration. Carbon dioxide and water are the final products of the reactions, and chemical energy is released, which is then "stored" as **ATP**.

Aerobic respiration is an example of **oxidation** – the breakdown of a substance in the presence of oxygen.

• **Oxygen debt**. A situation which occurs when extreme physical exercise is undertaken by an organism which shows **aerobic respiration**. Under these circumstances, the oxygen in the organism's cells is used up faster than it can be taken in. This means that there is not enough to break down the poisonous **pyruvic acid** produced in the first, **anaerobic**, stage of respiration. Instead, the acid undergoes further anaerobic reactions to convert it to **lactic acid** (much less harmful). This begins to build up, and the organism is said to have acquired an oxygen debt. This is "paid off" later by taking in oxygen faster than usual to break down the lactic acid.

* **Mitochondria**, 239; **Photosynthesis**, 254.

• **ADP** (**adenosine diphosphate**) and **ATP** (**adenosine triphosphate**). Two substances which consist of a chemical grouping called **adenosine**, combined with two and three **phosphate groups** respectively. A phosphate group consists of linked phosphorus, oxygen and hydrogen atoms. It can combine with other substances (either by itself or linked with other phosphate groups in a chain). When **aerobic respiration** occurs, the chemical energy released is involved in reactions which result in a conversion of ADP molecules into ATP molecules (by attachment of a third phosphate group in each case). The energy taken in to effect these reactions can be regarded as being "stored" in the form of ATP. This is a substance which can be easily stored in all cells (it is found in especially large quantities in cells which require a lot of energy, e.g. muscle cells). When the energy is needed, reactions occur which convert ATP back into ADP. These reactions result in an overall release of energy – the "stored" energy. In this way, power is supplied for the cell's activities.

Homeostasis

Homeostasis is the maintenance, by an organism, of a stable **internal environment**, i.e. a constant temperature, stable composition, level and pressure of body fluids, constant **metabolic rate***, etc. This is vital if the organism is to function properly. It requires the detection of any deviation from the norm (caused by new internal or external factors) and the means to correct such deviations, and is practised most efficiently in birds and **mammals***, e.g. humans. Their detection of deviations is achieved by the **feedback** of information to controlling organs. The blood glucose level, for example, is constantly being detected by the pancreas (i.e. information is "fed back"). The correction of deviations is achieved by **negative feedback**, i.e. feedback which "tells" of deviations and results in a change of action. If the glucose level gets too high, for example, the pancreas reacts by producing more **insulin*** to reduce it (see also **antagonistic hormones**, page 334). Most homeostatic actions are, like this example, controlled by hormones, many of which are in turn controlled by the **hypothalamus*** in the brain. An example of the importance of the hypothalamus in homeostasis is the control of body temperature. All birds and mammals, e.g. humans, are **homiothermic** (warm-blooded), i.e. they can keep a constant temperature (about 37°C in humans) regardless of external conditions (the opposite is **poikilothermic**, or cold-blooded). A "thermostat" area of the hypothalamus, called the **preoptic area**, detects any changes in body temperature and sends impulses either to the **heat-losing centre** or to the **heat-promoting centre** (both also found in the hypothalamus). These areas then send out nervous impulses to cause various heat-losing or heat-promoting actions in the body.

* **Hypothalamus**, 303; **Insulin**, 334; **Mammals**, 341; **Metabolic rate**, 330.

Hormones

Hormones are special chemical "messengers" which control various activities inside an organism. These pages deal with the hormones produced by humans and their related groups. Plants also produce hormones (**phytohormones**), though these are not yet fully understood (see **abscission layer**, page 249, and **photoperiodism** and **growth hormones**, page 251). Human hormones are secreted by **endocrine glands***. Some act only on specific body parts (**target cells** or **target organs**), others cause a more general response. The principal controller of hormone production is the **hypothalamus*** (part of the brain). It controls the secretions of many glands, mainly through its control of the **pituitary gland***, which itself controls many other glands. The hypothalamus "tells" the pituitary to produce its hormones by sending **regulating factors** to its **anterior lobe** and nervous impulses to its **posterior lobe**. Hormone secretion is vital to **homeostasis***.

- **Regulating factors**. Special chemicals which control the production of a number of hormones, and hence many vital body functions. They are sent to the **anterior lobe** of the **pituitary gland*** by the **hypothalamus*** (part of the brain). There are two types – **releasing factors**, which make the gland secrete specific hormones, and **inhibiting factors**, which make it stop its secretion. For example, **FSHRF** (**FSH releasing factor**) and **LHRF** (**LH releasing factor**) cause the release of the hormones **FSH** and **LH** (see chart), and hence the onset of **puberty***. Many regulating factors are vital to **homeostasis***.

- **Antagonistic hormones**. Hormones that produce opposite effects. **Glucagon** and **insulin** (see chart) are examples. When the blood glucose level drops too far, the pancreas produces glucagon to raise it again. A high glucose level causes the pancreas to produce insulin to lower the level (see also **homeostasis**, page 333).

HORMONES
ACTH (**adrenocorticotropic hormone**) or **adrenocorticotropin**
TSH (**thyroid-stimulating hormone**) or **thyrotropin**
STH (**somatotropic hormone**) or **somatotropin** or **HGH** (**human growth hormone**)
FSH (**follicle-stimulating hormone**)
LH (**luteinizing hormone**). Also called **luteotropin** in women and **ICSH** (**interstitial cell stimulating hormone**) in men.
Lactogenic hormone or **PR** (**prolactin**)
Oxytocin
ADH (**anti-diuretic hormone**) or **vasopressin**
Thyroxin
TCT (**thyrocalcitonin**) or **calcitonin**
PTH (**parathyroid hormone**) or **parathyrin** or **parathormone**
Adrenalin or **adrenin** or **epinephrin** **Noradrenalin** or **norepinephrin**
Aldosterone
Cortisone **Hydrocortisone** or **cortisol**
Oestrogen (female **sex hormone**) **Progesterone** (female **sex hormone**)
Androgens (male **sex hormones**), esp. **testosterone**
Gastrin
CCK (**cholecystokinin**)
Secretin / **PZ** (**pancreozymin**)
Enterocrinin
Insulin
Glucagon

* **Endocrine glands**, 297; **Homeostasis**, 333; **Hypothalamus**, 303; **Pituitary gland**, 297; **Puberty**, 318.

WHERE PRODUCED	EFFECTS
Pituitary gland (p.297) (**anterior lobe**)	Stimulates production of hormones in **cortex** of **adrenal glands** (p.297).
Pituitary gland (p.297) (**anterior lobe**)	Stimulates production of **thyroxin** by **thyroid gland** (p.297).
Pituitary gland (p.297) (**anterior lobe**)	Stimulates growth by increasing rate at which amino acids are built up to make proteins in cells.
Pituitary gland (p.297) (**anterior lobe**)	In women, works with **LH** to stimulate development of **ova** in **ovarian follicles** (p.317) and secretion of **oestrogen** by follicles in early stages of **menstrual cycle** (p.318). In men, causes formation of **sperm** (p.321).
Pituitary gland (p.297) (**anterior lobe**)	Stimulates **ovulation** (p.318), formation of **corpus luteum** (p.318) and its secretion of **oestrogen** and **progesterone**. Works with oestrogen and progesterone to stimulate thickening of lining of **uterus** (p.317). In man, causes production of **androgens**.
Pituitary gland (p.297) (**anterior lobe**)	Works with **LH** to cause secretion of hormones by **corpus luteum** (p.318). Causes milk production after giving birth.
Hypothalamus (p. 303). Builds up in **pituitary gland** (**posterior lobe**)	Stimulates contraction of muscles of **uterus** (p.317) during labour and ejection of milk after giving birth.
Hypothalamus (p. 303). Builds up in **pituitary gland** (**posterior lobe**)	Increases amount of water re-absorbed into blood from **uriniferous tubules** (p.301) in kidneys.
Thyroid gland (p.297)	Increases rate of food breakdown, hence increasing energy and raising body temperature. Works with **STH** in the young to control rate of growth and development. Contains iodine.
Thyroid gland (p.297)	Decreases level of calcium and phosphorus in blood by reducing their release from bones (where they are stored).
Parathyroid glands (p.297)	Increases level of calcium in blood by increasing its release from bone (see above). Decreases phosphorus level.
Adrenal glands (p.297) (**medulla**). Also at nerve endings. Secreted at times of excitement or danger.	Stimulate liver to release more glucose into blood, to be broken down for energy. Stimulate increase in heart rate, faster breathing and blood vessel constriction.
Adrenal glands (p.297) (**cortex**)	Increases amount of sodium and water in blood by causing re-absorption of more from **uriniferous tubules** (p.301) in kidneys.
Adrenal glands (p.297) (**cortex**)	Stimulate increase in rate of food breakdown for energy, and thus increase resistance to stress. Lessen inflammation.
Mostly in **ovarian follicles** (p.317) and **corpus luteum** (p.318) in **ovaries** (female sex organs, p.317). Also in **placenta** (p.319) during pregnancy.	Oestrogen activates development of **secondary sex characters** at **puberty** (p.318), e.g. breast growth. Both prepare **mammary** (milk) **glands** for milk production and work with **LH** to cause thickening of lining of **uterus** (p.317). Progesterone dominates towards end of **menstrual cycle** (p.318) and during pregnancy, when it maintains uterus lining and mammary gland readiness.
Mostly in **interstitial cells** in **testes** (male sex organs, p.316).	Activate development and maintenance of **secondary sex characters** at **puberty** (p.318), e.g. beard growth.
	Stimulates production of **gastric juice** (p.108).
Cells in small intestine	Stimulates opening of **sphincter of Oddi**, contraction of **gall bladder** and release of **bile** (all p.297) into **duodenum** (p.295).
Cells in small intestine	Stimulate pancreas to produce **pancreatic juice** (p.336) and secrete it into **duodenum** (p.295)
Cells in small intestine	Stimulates production of **intestinal juice** (p.336)
Pancreas, when blood glucose level too high.	Stimulates liver to convert more glucose to glycogen for storage (p.329). Also speeds up transport of glucose to cells.
Pancreas, when blood glucose level too low.	Stimulates faster conversion of glycogen to glucose in liver (p.329), and conversion of fats and proteins to glucose.

Digestive juices and enzymes

All the **digestive juices*** of the human body (secreted into the intestines by **digestive glands***) contain **enzymes*** which control the breakdown of food into simple soluble substances. These are called **digestive enzymes** and can be divided into three groups. **Amylases** (or **diastases**) promote the breakdown of **carbohydrates*** (the final result being **monosaccharides** – see terms used, right) **Proteinases** (or **peptidases**) promote the breakdown of **proteins** into **amino acids*** by attacking the **peptide links** (see **proteins**, page 328). **Lipases** promote the breakdown of **fats** into **glycerol** and **fatty acids** (see **fats**, page 328). The chart below lists the different digestive juices of the body, together with their enzymes and the action of these enzymes.

Digestive juice: **Saliva**

Produced by: **Salivary glands*** in mouth

Digestive enzyme: **Salivary amylase** (or **ptyalin**)

Actions: Starts breakdown of **carbohydrates* starch** and **glycogen** (**polysaccharides** – see p.329)

Products: Some **dextrin** (shorter **polysaccharide**). See note 1.

Digestive juice: **Gastric juice**

Produced by: **Gastric glands*** in stomch lining. Secreted into stomach (see **gastrin**, p.334).

Digestive enzymes (and one other constituent):
1. **Pepsin** (**proteinase**). See note 2.
2. **Rennin** (**proteinase**). Found only in the young.
3. **Hydrochloric acid**
4. **Gastric lipase**. Found mainly in the young.

Actions:
1. Starts breakdown of **proteins*** (**polypeptides**).
2. Works (with calcium) to curdle milk, i.e. to act on its **protein** (**casein**). See note 3.
3. Activates **pepsin** (see note 2), curdles milk in adults (see note 3) and kills bacteria.
4. Starts breakdown of **fat*** molecules in milk.

Products:
1. Shorter **polypeptides**

2,3. **Curds**, i.e. milk solids

4. Intermediate compounds

Digestive juice: **Bile**

Produced by: Liver. Stored in **gall bladder***, secreted into small intestine (see **CCK**, p.334).

Constituents: **Bile salts** and **bile acids**

Actions: Break up **fats*** (and intermediate compounds) into smaller particles, a process called **emulsification**.

Digestive juice: **Pancreatic juice**

Produced by: Pancreas. Secreted into small intestine (see **secretin / PZ**, p.334).

Digestive enzymes:
1. **Trypsin** (**proteinase**). See note 2.
2. **Chymotrypsin** (**proteinase**). See note 2.
3. **Carboxypeptidase** (**proteinase**). See note 2.
4. **Pancreatic amylase** (or **amylopsin**)
5. **Pancreatic lipase**

Actions:

1, 2, 3. Continue breakdown of **proteins*** (long and shorter **polypeptides**).

4. Continues breakdown of **carbohydrates***.
5. Breaks down **fat*** particles.

Products:

1,2,3. **Dipeptides** and some **amino acids***.

4. **Maltose** (**disaccharide**)
5. **Glycerol** and **fatty acids** (see fats, p.328).

Digestive juice: **Intestinal juice** (or **succus entericus**)

Produced by: **Intestinal glands*** in small intestine lining. Final secretion into small intestine (see **enterocrinin**, p.334).

Digestive enzymes:
1. **Maltase** (**amylase**)
2. **Sucrase** (or **invertase** or **saccharase**) (**amylase**)
3. **Lactase** (**amylase**)
4. **Enterokinase**. See note 2.

Actions:
1. Breaks down **maltose** (**disaccharide**)
2. Breaks down **sucrose** (**disaccharide**)
3. Breaks down **lactose** (**disaccharide**)
4. Completes breakdown of **proteins*** (**dipeptides**)

Products:
1. **Glucose** (or **dextrose**) (**monosaccharide**)
2. **Glucose** and **fructose** (**monosaccharides**)
3. **Glucose** and **galactose** (**monosaccharides**)
4. **Amino acids***

Notes

1. Not much **dextrin** is produced at this stage, since food is not in the mouth long enough. Most carbohydrates pass through unchanged.

2. **Proteinases** are first secreted in inactive forms, to prevent them digesting the digestive tube (made of **protein***, like most of the body). Once in the tube (beyond a protective layer of **mucous membrane***), these are converted into active forms. **Hydrochloric acid** changes **pepsinogen** (inactive) into **pepsin**, **enterokinase** changes **trypsinogen** into **trypsin**, and trypsin then changes **chymotrypsinogen** and **procarboxypeptidase** into **chymotrypsin** and **carboxypeptidase**.

3. The action of **rennin** and **hydrochloric acid** in curdling milk is vital, since liquid milk would pass through the system too fast to be digested.

* **Amino acids**, 328 (**Proteins**); **Carbohydrates**, 328; **Digestive juices**, 296 (**Digestive glands**); **Enzymes**, 331; **Fats**, 328; **Gall bladder**, 297; **Gastric glands**, **Intestinal glands**, 296 (**Digestive glands**); **Mucous membrane**, 295; **Salivary glands**, 296.

Terms used

Polysaccharides. The most complex **carbohydrates***. Each is a chain of **monosaccharide** molecules. Most carbohydrates taken into the body are polysaccharides, e.g. **starch** (the main polysaccharide in edible plants) and **glycogen** (the main one in animal matter). For more about starch and glycogen, see page 329.

Disaccharides. Compounds of 2 **monosaccharide** molecules, either forming intermediate stages in the breakdown of **polysaccharides** or (in the case of **sucrose** and **lactose**) taken into the body as such. (Sucrose is found in sugar beet and sugar cane, lactose occurs in milk.)

Monosaccharides. The simplest **carbohydrates***. Almost all result from **polysaccharide** breakdown, though **fructose** is taken into the body as such (e.g. in fruit juices), as well as resulting from **sucrose** breakdown. **Glucose** is the final result of all action on carbohydrates (fructose and **galactose** are converted to glucose in the liver).

Polypeptides. The complex form taken by all **proteins** entering the body. Each is a chain of hundreds (or thousands) of **amino acid*** molecules (see **proteins**, page 328).

Dipeptides. Chains of 2 **amino acid*** molecules, forming intermediate stages in the breakdown of **polypeptides**.

Vitamins and their uses

Vitamin A (retinol)

Sources: Liver, kidneys, fish-liver oils, eggs, dairy products, margarine, **pigment*** (**carotene**) in green and yellow fruit and vegetables, esp. tomatoes, carrots (carotene converted to vitamin A in intestines).

Uses: Maintains general health of **epithelial*** cells (lining cells), aids growth, esp. bones and teeth. Essential for vision in dim light – involved in formation of light-sensitive **pigment*** (**rhodopsin**), found in **rods** of **retina***. Aids in resistance against infection.

Vitamin B complex

Group of at least 10 vitamins, usually occurring together. Include: **Thiamine** (or **aneurin**) (**B1**), **Riboflavin** (**B2**), **Niacin** (or **nicotinic acid** or **nicotinamide**) (**B3**), **Pantothenic acid** (**B5**), **Pyridoxine** (**B6**), **Cyanocobalamin** (or **cobalamin**) (**B12**), **Folic acid** (**Bc** or **M**), **Biotin** (sometimes called **vitamin H**), **Lecithin**.

Sources: All found in yeast, liver. All except B12 found in wholewheat cereals and bread, wheatgerm, green vegetables, e.g. beans (B12 not found in any vegetable products). B2 and B12 found especially in dairy products. Most also found in eggs, nuts, fish, lean meat, kidneys, potatoes. B6, folic acid and biotin also made by bacteria in intestines.

Uses: Most needed for growth and maintenance of healthy tissues, e.g. muscles (B1, B6) nerves (B1, B3, B6, B12), skin (B2, B3, B5, B6, B12), hair (B2, B5) and a number aid continuous function of body organs (B5, B6, lecithin). Most (B1, B2, B3, B5, B6, B12) are essential **co-enzymes***, aiding in breakdown of foods for energy (**internal respiration***). Many (esp. B2, B6, B12) also co-enzymes aiding build up of substances (**proteins***) for growth, regulatory or defence purposes. B12 and folic acid vital to formation of blood cells, B5 and B6 to manufacture of nerve chemicals (**neurotransmitters***).

Vitamin C (ascorbic acid)

Sources: Green vegetables, potatoes, tomatoes, citrus fruit, e.g. oranges, grapefruit, lemons.

Uses: Needed for growth and maintenance of healthy tissues, esp. skin, blood vessels, bones, gums, teeth. Essential **co-enzyme*** in many metabolic reactions, esp. **protein*** breakdown and build-up of **amino acids*** into new proteins (esp. **collagen** – see **connective tissue**, page 280). Aids in resistance against infection and healing of wounds.

Vitamin D (calciferol)

Sources: Liver, fish-liver oils, oily fish, dairy products, egg yolk, margarine, special substance (provitamin D3) in skin cells (converted to vitamin D when exposed to sunlight).

Uses: Essential for absorption of calcium and phosphorus, and their deposition in bones and teeth. May work with **PTH*** (**hormone**).

Vitamin E (tocopherol)

Sources: Meat, egg yolk, leafy green vegetables, nuts, dairy products, margarine, cereals, wholemeal bread, wheatgerm, seeds, seed and vegetable oils.

Uses: Not yet fully understood. Believed to be involved in formation of **DNA***, **RNA*** and red blood cells, and promotion of fertility and food breakdown in muscle cells.

Vitamin K (phylloquinone or menaquinone)

Sources: Liver, fruit, nuts, cereals, tomatoes, green vegetables, esp. cabbage, cauliflower, spinach. Also made by bacteria in intestines.

Uses: Essential to formation of **prothrombin*** in liver (needed to cause clotting of blood).

* **Amino acids**, 328 (**Proteins**); **Carbohydrates**, 328; **Co-enzymes**, 331 (**Enzymes**); **DNA**, 324 (**Nucleic acids**); **Epithelium**, 310 (**Epidermis**); **Internal respiration**, 332; **Neurotransmitters**, 305 (**Synapse**); **Pigments**, 255; **Prothrombin**, 287 (**Clotting**); **PTH**, 334; **Retina**, 313; **RNA**, 324 (**Nucleic acids**).

The classification of living things

Classification, or **taxonomy**, is the grouping together of living things according to characteristics which they share. The main, formal type of classification (**classical taxonomy**) bases its groups primarily on structural characteristics (but see also page 342). The resulting classification charts first list the largest groups (**Kingdoms**), and then go on to list the smaller and smaller divisions within these groups.

The first groups after the Kingdoms are called **Sub-kingdoms** and the next are **Phyla** (sing. **Phylum**) in the case of animals and **Divisions** in the case of plants (though some plant classification charts have no Sub-kingdoms). After these come **Classes**, **Orders**, **Families**, **Genera** (sing. **Genus**) and finally **Species** – the smallest groupings. Some Divisions or Phyla, especially those with only a few members, may not have all these groups (e.g. the next group after a Phylum may be an Order, Family, Genus or even a Species) and there are also further "mid-way" groups in some cases, e.g. **Sub-classes** or **Sub-phyla**. The charts on the next four pages only classify as far as Classes in most cases, with some Sub-classes and also the special **Infraclasses** of **mammals** (see page 341).

It should be noted that there are areas which are still under dispute in both plant and animal classification. The classification of plants in particular varies so much that one, two or even three groups are regarded by many taxonomists as completely separate Kingdoms, and not plants at all. The notes accompanying both the plant charts here and the animal chart on pages 340-341 cover some of the major differences. In addition, two of the most commonly-used plant classification charts are shown on these two pages, instead of just one, as given for the animal Kingdom.

The Plant Kingdom (Kingdom Plantae)

Chart 1

SUB-KINGDOM: **Thallophyta**. No roots, stems or leaves, no **embryo*** as such.

DIVISION: Schizophyta or **Schizomycophyta**. **Bacteria** (sing. **bacterium**). Single-celled organisms found everywhere in large numbers. Some are **pathogenic**, that is, they cause disease, but others are useful, decomposing dead organisms for example.

DIVISION: Myxomycophyta or **Myxomycota**. Slime moulds. Very simple organisms with no **cell walls*** and no **chlorophyll***. Live on decaying plants or animals. Reproduce with **spores***.

DIVISION: Eumycophyta or **Eumycota**. True **fungi** (sing. **fungus**). May be single-celled or made of intertwined threads called **hyphae** (sing. **hypha**) which form a mesh (**mycelium**) over dead material on which fungus feeds. Have **cell walls*** but no **chlorophyll***. Useful in causing decay, also in some industrial processes (e.g. brewing). Some are important antibiotics, e.g. Penicillium. Reproduce with **spores***. E.g. mushrooms.

All the remaining Divisions in the Sub-kingdom are types of **algae** (sing. **alga**) – simple plants found in salt or freshwater or damp places. All possess **chlorophyll*** (but see note 2) and the larger ones (seaweeds) each have a ribbon-like plant body called a **thallus**.

DIVISION: Cyanophyta. Blue-green algae. Primitive single-celled or many-celled algae with **cell walls***. Have bluish-green **pigment*** called **phycocyanin**. Found even in hot springs and arctic water.

DIVISION: Euglenophyta. Single-celled algae with no **cell walls***. Have **flagella***. Commonly found in freshwater.

DIVISION: Chrysophyta. Golden algae. Single-celled, with **cell walls***. Have **flagella***. Highly diverse group – salt or freshwater, also damp places.

DIVISION: Pyrrophyta. Fire algae. Single-celled, no **cell walls***. Have **flagella***.

DIVISION: Bacillariophyta. Diatoms. Single-celled, with silica "shells". Water-living (salt or freshwater). May occur singly, though often **colonial*** (living together in a mass).

DIVISION: Xanthophyta. Yellow-green algae. Mostly single-celled, with **cell walls*** and **pigment*** called **xanthophyll**. Found in salt or freshwater, also damp places.

DIVISION: Rhodophyta. Red algae (seaweeds). Many-celled with **cell walls*** and red and blue **pigments***. Mainly found in salt water. E.g. Dulse.

DIVISION: Phaeophyta. Brown algae (seaweeds). Many-celled, all with **cell walls***. Includes all common seaweeds, brown to olive-green in colour. Each has special disc-like attachment called a **holdfast**, which fixes it to a surface. E.g. kelps.

DIVISION: Chlorophyta. Green algae (seaweeds). Largest group of algae – includes many single-celled and many-celled algae. All have **cell walls***. Mostly freshwater, though some found in salt water and some in damp places, e.g. tree trunks and soil. Exist in enormous quantities (single-celled ones often **colonial** – see **Bacilliarophyta**).

SUB-KINGDOM: **Embryophyta**. All have **cell walls***, **chlorophyll***, roots, stems and leaves. Also a distinct protective layer of cells around new developing plant (**embryo***).

DIVISION: Bryophyta. Have some kind of roots, stems and leaves but no **vascular tissue***. Most have short stem-like structure called a **seta**, which bears tightly-packed leaves (in mosses) or flat leaf-like expansions (in liverworts). Have thread-like roots called **rhizoids** which cling to a surface instead of going into ground. Mainly land plants with wide distribution in damp areas. 3 Classes:

* **Cell wall**, 238; **Chlorophyll**, 255; **Embryo**, 321; **Flagella**, 268; **Pigment**, 255; **Spores**, 320 (**Sporulation**); **Vascular tissue**, 242.

Classes:

Hepaticae. Liverworts.

Musci. Mosses.

Anthocerotae. Hornworts.

DIVISION: Tracheophyta. Have roots, stems and leaves and **vascular tissue***.

Sub-Division: Pteridophyta. No flowers or seeds. 4 Classes:

Classes:

Psilotales. Primitive, loosely related to ferns.

Lycopodiales. Clubmosses. Creeping **evergreens***, related to ferns. Date back to prehistoric times.

Equisetales. Horsetails. Related to ferns, but can live with less moisture and shade. Giant forms abundant in prehistoric times.

Filicales. Ferns. Live in moist, shady areas. Have **fronds** – **bipinnate*** structures (stalks and leaves combined) which bear **spores***.

Sub-Division: Spermatophyta. Have seeds. 2 Classes:

Class:

Gymnospermae. Seeds not enclosed in a fruit, no flowers.

Sub-classes:

Cycadales. Cycads. Primitive, palm-like.

Coniferales. Conifers, e.g. firs. **Evergreens***, most with sharp pointed leaves (**needles**) and all with reproductive bodies called **cones**. **Ovules*** develop on external **scales** of female cones (no flowers), **pollen*** on scales of male cones.

Ginkgoales. Only one living member – Gingko (maidenhair tree).

Gnetales. Only 3 genera, e.g. Welwitschia (desert plant with large leathery leaves).

Class:

Angiospermae. Seeds enclosed in a fruit. Have flowers.

Sub-classes:

Dicotyledonae. Have 2 **cotyledons***, e.g. buttercup, rose.

Monocotyledonae. Have 1 **cotyledon***, e.g. grasses, lily.

Chart 2 (for descriptions, see chart 1)

Thallophytes is an informal term only.

DIVISION: Schizophyta or **Schizomycophyta**.

DIVISION: Myxomycophyta or **Myxomycota**.

DIVISION: Eumycophyta or **Eumycota**.

DIVISION: Cyanophyta.

DIVISION: Euglenophyta.

DIVISION: Chrysophyta.

DIVISION: Pyrrophyta.

DIVISION: Bacillariophyta.

DIVISION: Xanthophyta.

DIVISION: Rhodophyta.

DIVISION: Phaeophyta.

DIVISION: Chlorophyta.

Embryophytes is an informal term only.

DIVISION: Bryophyta.

Classes:

Hepaticae
Musci
Anthocerotae

Tracheophytes is an informal term only.

Pteridophytes is an informal term only.

DIVISION: Psilophyta. Was Class **Psilotales**.

DIVISION: Lycophyta. Was Class **Lycopodiales**.

DIVISION: Sphenophyta. Was Class **Equisetales**.

DIVISION: Pterophyta. Was Class **Filicales**.

Spermatophytes is an informal term only.

Gymnosperms is an informal term only.

DIVISION: Cycadophyta. Was Sub-class **Cycadales**.

DIVISION: Coniferophyta. Was Sub-class **Coniferales**.

DIVISION: Ginkgophyta. Was Sub-class **Ginkgoales**.

DIVISION: Gnetophyta. Was Sub-class **Gnetales**.

Angiosperms is an informal term only.

DIVISION: Anthophyta. Was Class **Angiospermae**.

Class:

Dicotyledonae. Was Sub-class **Dicotyledonae**.

Class:

Monocotyledonae. Was Sub-class **Monocotyledonae**.

Notes

1. The bacteria and blue-green algae (Divisions **Schizophyta** and **Cyanophyta**) do not have **nuclei*** and are thus not true plants or animals. For this reason, some classifications put them in a separate Kingdom (before plants and animals) called the Kingdom **Monera** or **Prokaryota (prokaryotic** means "without a nucleus", the opposite is **eukaryotic).**

2. Some of the single-celled algae (especially those of the Divisions **Euglenophyta**, **Chrysophyta** and **Pyrrophyta**) have both animal and plant characteristics (e.g. they can "eat" food as well as making it by **photosynthesis***, some have **flagella*** and some have no **cell walls***). For this reason, some classifications put them in a separate Kingdom called the Kingdom **Protista** (after **Monera** (see note 1) and before plants and animals). The Kingdom may also be expanded to include the **Protozoa** (see page 340).

3. The slime moulds and fungi (Divisions **Myxomycophyta** and **Eumycophyta**) have doubtful affinity with the plants (they lack **chlorophyll***) but no closer links with animals. For this reason, some classifications put them in a separate Kingdom called the Kingdom **Fungi** (after **Monera** and **Protista** (see notes 1 and 2) and before plants and animals).

* **Bipinnate**, 250; **Cell wall**, 238; **Chlorophyll**, 255; **Cotyledon**, 261; **Evergreen**, 236; **Flagella**, 268; **Nucleus**, 238; **Ovules**, 258; **Photosynthesis**, 254; **Pollen**, 258; **Spores**, 320 (**Sporulation**); **Vascular tissue**, 242.

The Animal Kingdom (Kingdom Animalia)

See introduction, page 338. As with the plant classification chart, this chart lists its members in order of complexity, beginning with the most primitive. The main characteristics which appear as animals become more complex are mentioned in this chart in the first instance they occur, but are assumed from then on. These are a true gut, a circulatory system, a nervous system, a true body cavity, some type of **segmentation***, some type of skeleton and the existence of lungs (see also pages 264-265). All other characteristics mentioned refer specifically to the group being defined.

In the classification of animals, there are a number of relatively primitive animals (especially particular kinds of worms) which belong together in small groups. These are normally only included in the most comprehensive classification charts (as minor **Phyla**), and are not listed here.

SUB-KINGDOM: Protozoa

PHYLUM: Protozoa. The only Phylum, with the same name as the Sub-kingdom. Single-celled animals. Mostly aquatic, though many are **parasitic***. E.g. Amoeba, Paramecium.

Classes: Mastigophora (or **Flagellata**), **Sarcodina, Ciliophora** (or **Ciliata**), **Sporozoa, Microspora.**

SUB-KINGDOM: Parazoa

PHYLUM: Porifera. The only Phylum. Sponges. Porous, non-mobile living mass, consisting of millions of single-celled organisms (see **colonial**, page 342).

Classes: Calcarea, Demospongiae, Sclerospongiae, Hexactinellida (or **Triaxonida** or **Hyalospongiae**).

SUB-KINGDOM: Metazoa. All the rest of the Animal Kingdom, i.e. all the many-celled (**multicellular**) animals.

PHYLUM: Coelenterata (coelenterates) or **Cnidaria.** Aquatic animals with **tentacles***. Only one body opening (for substances in and out). Move by muscular action. E.g. sea anemones, jellyfish, Hydra.

Classes: Hydrozoa, Scyphozoa, Anthozoa (or **Actinozoa**).

PHYLUM: Ctenophora. Marine, jelly-like animals, very similar to **coelenterates** but moving by means of tiny "hairs" (**cilia***).

Classes: Tentaculata, Nuda.

PHYLUM: Platyhelminthes. Flatworms. Have a mouth and a primitive excretory system. E.g. **parasitic*** flukes and tapeworms.

Classes: Turbellaria, Cestoidea (or **Cestoda** or **Eucestoda**), **Cestodaria, Monogenoidea** (or **Monogenea**), **Digenoidea** (or **Digenea**), **Aspidogastrea** (or **Aspidobothrea** or **Aspidocotylea**).

PHYLUM: Nemertea or **Rhynchocoela.** Ribbon worms. Marine worms with a true gut (i.e. running from mouth to **anus***), a primitive circulatory system and a sucking organ (**proboscis**) with a hooked end.

Classes: Anopla, Enopla.

PHYLUM: Aschelminthes. Worm-like aquatic animals. Mostly **parasitic***. E.g. roundworms, hookworms, threadworms.

Classes: Nematoda, Rotifera (or **Rotatoria**), **Gastrotricha, Kinorhyncha** (or **Echinodera**), **Priapulida, Nematomorpha** (or **Gordiacea**).

PHYLUM: Annelida (annelids) or **Annulata.** Most advanced worms. Have tubular, segmented bodies with a body cavity, circulatory and nervous system. Have fine bristles (**chaetae***) to grip soil or sand.

Classes: Aclitellata (marine worms), **Clitellata** (freshwater worms, earthworms and leeches).

PHYLUM: Mollusca (molluscs). Soft-bodied animals with a limey shell, a head and a muscular "foot" for creeping or digging. Mostly aquatic.

Classes: 3 minor ones – **Scaphopoda, Monoplacophora, Amphineura.** 3 more important:

Gastropoda (gastropods). Univalves, i.e. with a shell made of one piece only. E.g. snails, limpets.

Lamellibranchiata (lamellibranchs) or **Bivalvia (bivalves)** or **Pelecypoda.** Have a shell in two parts hinged together. E.g. oysters, clams, mussels.

Cephalopoda (cephalopods) or **Siphonopoda.** Molluscs with **tentacles*** and well-developed eyes. E.g. octopuses, squid, cuttlefish.

PHYLUM: Arthropoda (arthropods). Animals with many jointed limbs and a hard outer skeleton.

Sub-Phylum: Chelicerata (chelicerates). Common characteristics include pincer-like mouthparts (**chelicerae**, sing. **chelicera**).

Classes: 2 minor ones – **Merostomata** (king crabs), **Pycnogonida** (sea spiders). 1 more important:

Arachnida (arachnids). Animals with 8 legs. E.g. spiders, mites, scorpions.

Sub-Phylum: Crustacea (crustaceans). 1 Class with same name:

Class: Crustacea (crustaceans). Mostly aquatic animals, with **gills*** on their legs and 2 pairs of **antennae***. E.g. crabs, woodlice, lobsters.

Sub-Phylum: Uniramia. 1 pair of **antennae***, mostly land-living.

Classes: 3 minor ones – **Onychophora** (velvet worms), **Symphyla** (tiny, centipede-like), **Pauropoda** (tiny, millipede-like). 3 more important:

Chilopoda. Centipedes. Each body segment has 1 pair of legs. **Carnivorous***.

Diplopoda. Millipedes. Each body segment has 2 pairs of legs. **Herbivorous***.

Insecta (insects) or **Hexapoda.** Animals with 6 legs and usually 2 pairs of wings. E.g. beetles, moths, ants.

* **Antennae**, 274; **Anus**, 295 (**Large intestine**); **Carnivores**, 234; **Chaetae**, 268 (**Parapodia**); **Cilia**, 268; **Gills**, 273; **Herbivores**, 234; **Parasites**, 342; **Segmentation**, 264; **Tentacles**, 275.

PHYLUM: **Echinodermata (echinoderms)**. Marine animals, all with a limey skeleton just below the skin. Usually have a 5-rayed arrangement and a spiny skin.

Classes: **Asteroidea** (starfish), **Ophiuroidea** (brittle stars), **Echinoidea** (sea urchins), **Holothuroidea** (sea cucumbers), **Crinoidea** (sea lilies).

PHYLUM: **Chordata (chordates)**. At some time in their life, all have a **notochord** – a stiff "rod" of cells running lengthwise between the spinal cord and the gut.

Sub-Phyla: 2 minor ones – **Urochordata** (or **Tunicata**) (sea squirts), **Cephalochordata** (or **Acrania**) (lancets). 1 more important:

Craniata (craniates) or **Vertebrata (vertebrates)**. **Notochord** (see **Chordata**) replaced by spine (see note 7). Have well-developed brain.

Classes: 2 minor ones, both Classes of jawless fish – **Myxini** (hagfish), **Cephalaspidomorphi** (lampreys). 6 more important:

Elasmobranchiomorphi (elasmobranchs) or **Chondrichthyes**. Fishes with a skeleton made of **cartilage***. Have fins and breathe with **gills***. E.g. sharks, rays.

Osteichthyes. Fishes with a skeleton made of bone. Have fins and scales, and breathe with **gills***. E.g. sturgeon, cod, herring.

Amphibia (amphibians) or **Batrachia (batrachians)**. Animals which can live on land but must be near water. Most have lungs and lay eggs in water. E.g. frogs, newts, toads.

Reptilia (reptiles). Animals with dry, scaly skin. Live on land, and lay shelled eggs on land. E.g. snakes, lizards, crocodiles, turtles.

Aves. Birds. All have feathers and lay shelled eggs. E.g. penguins, robins, ostriches, hawks.

Mammalia (mammals). All females produce milk. Nearly all have hair or fur. Has 2 Sub-classes:

Sub-classes: **Prototheria**. Lay shelled eggs. Has just 1 Order – **Monotremata (monotremes)** – spiny anteaters and duck-billed platypuses.

Theria. Do not lay eggs. Has 2 special **Infraclasses** before the Orders:

Infraclasses: **Metatheria**, **Marsupalia (marsupials)** or **Didelphia**. Young develop in the **uterus*** for only a short time, completing their development attached to the openings of the mother's milk glands and normally held within a pouch of skin called a **marsupium**. E.g. kangaroos, opossums.

Eutheria or **Placentalia (placental mammals)**. Young develop in the **uterus*** until birth, attached by a highly-developed **placenta***. E.g. cows, mice, whales, humans.

Notes

1. In some charts, the Class **Sarcodina** of the Phylum **Protozoa** has 2 Sub-classes – **Rhizopoda** and **Actinopoda**. In others, these two are omitted as Sub-classes, and their members grouped together under the Class Sarcodina. In this case, the Class has the alternative name of Rhizopoda.

2. Some charts show another Sub-kingdom, called **Mesozoa**, between the **Parazoa** and the **Metazoa**. It has only one Phylum, also called Mesozoa, and is made up of obscure **parasites***. Its rank as a Sub-kingdom, or even as a Phylum, is in much doubt. Many people regard its members as peculiar forms of flatworms (Phylum **Platyhelminthes**).

3. The Classes **Monogenoidea** and **Digenoidea** of the Phylum **Platyhelminthes** are grouped together in some charts as the Class **Trematoda (trematodes)**.

4. The Class **Onychophora** of the Phylum **Arthropoda** is made a separate Phylum in some charts, since its members (velvet worms) show characteristics of both the Arthropoda and the **Annelida**.

5. In some charts, the Phylum **Arthropoda** does not have any Sub-phyla, merely the same 10 Classes. In others still, there are no Sub-phyla and only 7 Classes. The Classes **Pauropoda**, **Symphyla**, **Chilopoda** and **Diplopoda** are grouped together in one Class called **Myriapoda** ("many feet"). In most cases, however, the term **myriapods** is only an informal one.

6. The Sub-phyla **Urochordata** and **Cephalochordata** of the Phylum **Chordata** are sometimes known collectively as the **Protochordata (protochordates)**, though this term is purely an informal one. It is also sometimes extended to include the minor Phylum **Hemichordata** (acorn worms), since the members of this show certain features characteristic of the Chordata (the name implies a "half-way" stage).

7. The term **craniate** means "with a skull" – this is true of all members of the Sub-phylum **Craniata**. The alternative Sub-phylum name of **vertebrates** means "animals with backbones". This is not strictly true, since the most primitive Class – **Myxini** (hagfish) – do not have backbones.

8. The **invertebrates** are all the animals without backbones, i.e. everything on the chart up to the Sub-phylum **Craniata** – the **vertebrates**. (But see above, note 7.)

9. The two Classes **Myxini** (hagfish) and **Cephalaspidomorphi** (lampreys), the only jawless members of the Sub-phylum **Craniata**, are sometimes known collectively as the **Agnatha** and the remaining, jawed, Classes are called the **Gnathostomata**. However, these terms are only informal ones.

10. The Classes **Myxini** (hagfish), **Cephalaspidomorphi** (lampreys), **Elasmobranchiomorphi (cartilaginous fish)** and **Osteichthyes (bony fish)** of the Sub-phylum **Craniata** are sometimes known collectively as **Pisces** (fish), though this term is only an informal one.

11. The Classes of the Sub-phylum **Craniata** are sometimes put into two informal groups – the **Amniota** (amniotes – **Reptilia**, **Aves**, **Mammalia**) and **Anamniota** (all the other Classes). The amniotes are those whose **embryos*** (developing young) have an **amnion**, **chorion** and **allantois** (see pages 276 and 319).

* **Cartilage**, 281; **Embryo**, 321; **Gills**, 273; **Parasites**, 342; **Placenta**, 319; **Uterus**, 317.

Informal group terms

Listed here are the main terms used to group living things together according to their general life styles (i.e. their ecological similarities – see also page 237). These are informal terms, as opposed to the formal terms of the classification charts (pages 338-341), which are based on structural similarities.

Plants

- **Xerophytes**. Plants which can survive long periods of time without water, e.g. cacti.
- **Hydrophytes**. Plants which grow in water or very wet soil, e.g. reeds.
- **Mesophytes**. Plants which grow under average conditions of moisture.
- **Halophytes**. Plants which can withstand very salty conditions, e.g. sea pinks.
- **Lithophytes**. Plants which grow on rock, e.g some mosses.
- **Epiphytes**. Plants which grow on other plants, but only to use them for support, not to feed off them, e.g. some mosses.
- **Saprophytes**. Plants which live on decaying plants or animals, feed off them, but are not the agents of their decay, e.g. some fungi.

Animals

- **Predators**. Animals which kill and eat other animals (their **prey**), e.g. lions. Bird predators, e.g. hawks, are called **raptors**.
- **Detritus feeders**. Animals which feed on debris from decayed plant or animal matter, e.g. worms.
- **Scavengers**. Large **detritus feeders**, e.g. hyenas, which feed only on dead flesh (animal matter).
- **Territorial**. Holding and defending a **territory** (an area of land or water) either singly or in groups, e.g. many fish, birds and mammals. This is usually linked with attracting a mate and breeding.
- **Abyssal**. Living at great depths in a lake or the sea, e.g. oarfish and gulper eels.
- **Demersal**. Living at the bottom of a lake or the sea, e.g. angler fish and prawns.
- **Sedentary**. In the case of birds, this term describes those which do not **migrate***; with other animals, it is synonymous with **sessile**.
- **Nocturnal**. Active at night and sleeping during the day, e.g. owls and bats.

Plants and animals

- **Insectivores**. Specialized organisms which eat only insects, e.g. pitcher plants (which trap and digest insects) and hedgehogs.
- **Parasites**. Plants or animals which live in or on other living plants or animals (the **hosts**), and feed off them, e.g. mistletoe and fleas. Not all are harmful to the host.
- **Symbionts** or **symbiotes**. A pair of living things which associate closely with each other and derive mutual benefit from such close existence (**symbiosis**). **Lichens**, normally found on bare rock, are an example. Each is in fact two plants (a fungus and an alga). The alga produces food (by **photosynthesis***) for the fungus (which otherwise could not live on bare rock). The fungus uses its fine threads to hold the moisture the alga needs.
- **Commensals**. A pair of living things which associate closely with each other, may derive some benefit from such close existence (**commensalism**), e.g. shared food, but are not strictly **symbionts**. One type of worm, for instance, is very often found in the same shell as a hermit crab. One of the commonest examples of commensalism is the existence of house mice wherever there are humans.
- **Social** or **colonial**. Living together in groups. The two terms are synonymous in the case of plants, and refer to those which grow in clusters. In the case of animals, there is a difference of numbers between the terms. Lions, for example, are social, but their groups (**prides**) are not large enough to be called colonies. With true colonial animals, there is also a great difference in the level of interdependence between colony members. In a gannet colony, for example, this is relatively low (they only live close together because there is safety in numbers). In an ant colony, by contrast, different groups (**castes**) have very different jobs, e.g. gathering food or guarding the colony, so each member relies heavily on others. The highest level of colonial interdependence is shown by the tiny, physically inseparable, single-celled organisms which form one living mass, e.g. a sponge.
- **Sessile**. In the case of animals, this term refers to those which are not free to move around, i.e. they are permanently fixed to the ground or other solid object, e.g. sea anemones. With plants, it describes those without stalks, e.g. algae.
- **Pelagic**. Living in the main body of a lake or the sea, as opposed to at the bottom or at great depths. Pelagic creatures range from tiny **plankton** through medium-sized fishes and sharks to very large whales. The medium-sized and large ones are all animals, and are called **nekton**.
- **Plankton**. Microscopic aquatic animals and plants, vast numbers of which drift in lakes and seas, normally near the surface (plant plankton is **phytoplankton**, and animal plankton is **zooplankton**). Plankton is the food of many fishes and whales and is thus vital to the ecological balance (**food chains***) of the sea.
- **Littoral**. Living at the bottom of a lake or the sea near the shore, e.g. crabs and seaweed.
- **Benthos**. All **abyssal**, **demersal** and **littoral** plants and animals, i.e. all those which live in, on or near the bottom of lakes or seas.

* **Food chains**, 234 (**Food web**); **Migration**, 237; **Photosynthesis**, 254.

DICTIONARY OF SCIENCE

Glossary
and Index

Glossary

- **Abrasive**. A material which wears away the surface of another material.

- **Adhesive**. A substance which sticks to one or more other substances (see adhesion, page 23).

- **Alloy**. A mixture of two or more metals, or a metal and a non-metal. It has its own properties (which are metallic), independent of those of its constituents. For example, **brass*** is an alloy of copper and zinc, and **steel*** is an alloy of iron and carbon (different mixes give the steel different properties).

- **Amalgam**. An **alloy** of mercury with other metals. It is usually soft and may even be liquid.

- **Antacid**. A substance which counteracts excess stomach acidity by **neutralizing*** the acid. Examples are aluminium hydroxide and magnesium hydroxide.

- **Bleach**. A substance used to remove colour from a material or solution. Most strong **oxidizing*** and **reducing agents*** are good bleaches. The most common household bleach is a solution of sodium hypochlorite (also a highly effective **germicide**).

$$\underset{\text{Sodium hypochlorite}}{NaOCl} + \underset{}{\text{Coloured material}} \rightarrow \underset{\text{Sodium chloride}}{NaCl} + \text{Decolourized } \textbf{oxidized}^{*} \text{ material}$$

- **Calibration**. The "setting up" of a measuring instrument so that it gives the correct reading. The instrument is normally adjusted during manufacture so that it reads the correct value when it is measuring a known standard quantity, e.g. a balance would be adjusted to read exactly 1 kg when a standard 1 kg mass was on it.

- **Calorimetry**. The measurement of heat change during a chemical reaction or event involving heat transfer. For example, measuring the temperature rise of a known mass of a substance when it is heated electrically is used to find **specific heat capacity*** and the temperature rise of a mass of water can be used to calculate the energy produced by a fuel when it is burnt (see **bomb calorimeter** diagram, page 146).

- **Coefficient**. A **constant** for a substance, used to calculate quantities related to the substance by multiplying it by other quantities. For example, the force pushing two materials together multiplied by the **coefficient of friction*** for the surfaces gives the **frictional force***.

- **Constant**. A numerical quantity that does not vary. For example, in the equation $E = mc^2$ (see also page 84), the quantity c (the speed of light in a vacuum) is a constant. E and m are **variables** because they can change.

- **Coolant**. A fluid used for cooling in industry or in the home (see also **refrigerant**). The fluid usually extracts heat from one source and transfers it to another. In a **nuclear power station***, the coolant transfers the heat from the nuclear reaction to the steam generator, where the heat is used to produce steam. This turns turbines and generates electricity.

- **Cosine** (of an angle). the ratio of the length of the side adjacent to the angle to the length of the hypotenuse (the longest side) in a right-angled triangle. It depends on the angle.

- **Dehydrating agent**. A substance used to absorb moisture from another substance, removing water molecules if present, but also, importantly, hydrogen and oxygen atoms from the molecules of the substance. This leaves a different substance plus water (see also **drying agent**). Concentrated sulphuric acid is an example:

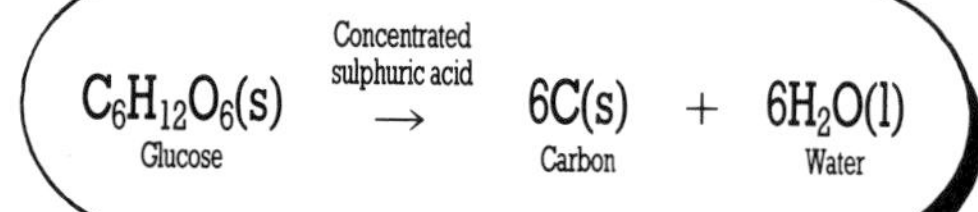

Concentrated sulphuric acid can also be used as a drying agent if it does not react with the substance added to it. For example, it is used to dry samples of chlorine gas, i.e. remove surrounding molecules of water vapour (see page 216).

- **Drying agent**. A substance used to absorb moisture from another substance, but which only removes water molecules from in and around the substance, not separate hydrogen and oxygen atoms from its molecules. The substance itself is not changed (see also **desiccation**, page 221 and **dehydrating agent**). **Phosphorus pentoxide** (**P_2O_5**) is an example:

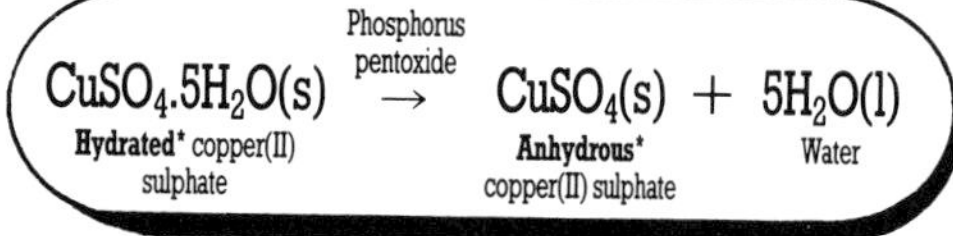

- **Ductile**. Describes a substance which can be stretched. It is normally used of metals which can be drawn out into thin wire, e.g. copper. Different substances show varying degrees of **ductility** (see page 165). See also **yield point**, page 23.

- **Fumigation**. The killing of pests such as insects by poisonous gas, e.g. sulphur dioxide, or smoke.

- **Fungicide**. A substance used to destroy harmful fungi, e.g. moulds and mildews growing on crops.

- **Germicide**. A substance used to destroy bacteria, especially those carrying disease (germs).

- **Graduations**. Equally-spaced marks used for measurement, e.g. those on a **measuring cylinder***.

- **Inert**. Describes an unreactive substance, i.e. one which does not easily take part in chemical reactions. Examples are the **noble** (or **inert**) **gases***.

- **Inversely proportional**. When applied to two quantities, this means that one is directly related to the **reciprocal** of the other.

- **Latex**. A milky fluid produced by plants, particularly that produced by the rubber tree, from which raw natural rubber is extracted (and which also forms the basis of some **adhesives**). Also certain similar **synthetic polymers***.

* **Anhydrous**, 155 (**Anhydrate**); **Brass**, 112, 175 (**Zinc**); **Coefficient of friction**, **Frictional force**, 7; **Hydrated**, 155 (**Hydrate**); **Measuring cylinder**, 224; **Neutralization**, 151; **Noble gases**, 189; **Oxidation**, **Oxidizing agent**, **Reducing agent**, 148; **Specific heat capacity**, 30; **Steel**, 174; **Synthetic polymers**, 200.

- **Malleable**. Describes a substance which can be moulded into different shapes. It is normally used of substances which can be hammered out into thin sheets, in particular many metals and **alloys** of metals. Different substances show varying degrees of **malleability** (see page 165).

- **Mean**. A synonym for average, i.e. the sum of a series of values divided by the number of values in the series.

- **Medium** (pl. **media**). Any substance through which a physical effect is transmitted, e.g. glass is a medium when light travels through it.

- **Meniscus**. The concave or convex surface of a liquid, e.g. water or mercury. It is caused by the relative attraction of the molecules to each other and to those of the container (see also **adhesion** and **cohesion**, page 23 and **parallax error**, page 102).

- **Ore**. A naturally-occurring mineral from which an element (usually a metal) is extracted, e.g. bauxite, which yields aluminium.

- **Organic solvent**. An organic liquid in which substances will dissolve.

- **Photocell** or **photoelectric cell**. A device used for the detection and measurement of light.

- **Proportional**. When applied to two quantities, this means that they are directly related, i.e. if X is proportional to Y, then X will change by a fixed amount for each fixed change in Y.

- **Rate**. The amount by which one quantity changes with respect to another, e.g. **acceleration*** is the rate of change of distance with time. Note that the second quantity is not necessarily time in all cases. If a graph of Y against X is plotted, the rate of change of Y with respect to X at a point is the gradient at that point.

- **Raw material**. A material obtained from natural sources for use in industry, e.g. iron **ore**, coke and limestone are the raw materials used to produce iron (see picture, page 174).

- **Reciprocal**. The value obtained from a number when one is divided by it, i.e. the reciprocal of any number x is $1/x$. For example, the reciprocal of 10 is 0.1.

- **Refrigerant**. A type of **coolant** used in refrigerators. It must be a liquid which **evaporates*** at low temperatures. The substances commonly used nowadays are the **chlorofluorocarbons**, or **freons**, although ammonia was widely used in the past.

- **Resins**. Substances used as **adhesives**. They are often insoluble in water. **Natural resins** are organic compounds secreted by certain plants and insects. **Synthetic resins** are plastic materials produced by **polymerization***.

- **Sine** (of an angle). The ratio of the length of the side opposite to the angle to the length of the hypotenuse (the longest side) in a right-angled triangle. It depends on the angle.

- **Spectrum** (pl. **spectra**). A particular distribution of wavelengths and frequencies, e.g. the wavelengths in the **visible light spectrum*** range from 4×10^{-7} to 7.5×10^{-7} m.

- **Superheated steam**. Steam above a temperature of 100°C. It is obtained by heating water under pressure.

- **System**. A set of connected parts which have an effect on each other and form a whole unit, e.g. a **digestive system*** or the substances involved in a reaction at **chemical equilibrium***.

- **Tangent** (of an angle). The ratio of the length of the side opposite to the angle to the length of the side adjacent to it in a right-angled triangle. It depends on the angle.

- **Tarnish**. To lose or partially lose shine due to the formation of a dull surface layer, e.g. silver sulphide on silver or lithium oxide on lithium. Tarnishing is a type of **corrosion***.

- **Variable**. A numerical quantity which can take any value. For example, in the equation $E = mc^2$ (see also page 84), E and m are variables since they can take any value (although the value of E depends on the value of m). The quantity c (the speed of light in a vacuum) is a **constant**.

- **Volatile**. Describes a liquid that **evaporates*** easily, e.g. petrol, or a solid that **sublimes*** easily, e.g. iodine.

- **Volume**. A measurement of the space occupied by a body. See page 101 for calculations of volume. The **SI unit*** of volume is the cubic metre (m^3).

- **Vulcanization**. The process of heating raw natural rubber (extracted from **latex**) with sulphur. Vulcanized rubber is harder, tougher and less temperature-sensitive than raw rubber (the more sulphur used, the greater the difference). This is because the sulphur atoms form cross-links between the chains of rubber molecules (see picture, page 201).

* **Acceleration**, 11; **Chemical equilibrium**, 163; **Corrosion**, 209; **Digestive system**, 294; **Evaporation**, 121; **Polymerization**, 200; **SI units**, 96; **Sublimation**, 121; **Visible light spectrum**, 55.

Index

The page numbers listed in the index are of three different types. Those printed in bold type (e.g. **261**) indicate in each case where the main definition(s) of a word (or words) can be found. Those in lighter type (e.g. 195) refer to supplementary entries. Page numbers printed in italics (e.g. *78*) indicate pages where a word (or words) can be found as a small print label to a picture.

If a page number is followed by a word in brackets, it means that the indexed word can be found inside the text of the definition indicated. If it is followed by (I), the indexed word can be found in the introductory text on the page given.

Bracketed singulars, plurals, symbols and formulae are given where relevant after indexed words. Synonyms are indicated by the word "see", or by an oblique stroke (/) if the synonyms fall together alphabetically.

A

B

C

D

E

G

H

I

J

K

L

N

O

P

Q

R

S

U

V

W

X

Y

Z

First published in 1988 by Usborne Publishing Ltd, Usborne House 83-85 Saffron Hill, London EC1N 8RT

Printed in Great Britain